Second Edition

Signal and Image Processing for Remote Sensing

Second Edition

Signal and Image Processing for Remote Sensing

Edited by
C.H. Chen

CRC Press
Taylor & Francis Group
Boca Raton London New York

CRC Press is an imprint of the
Taylor & Francis Group, an **informa** business

CRC Press
Taylor & Francis Group
6000 Broken Sound Parkway NW, Suite 300
Boca Raton, FL 33487-2742

First issued in paperback 2019

ISBN-13: 978-1-4398-5596-6 (hbk)
ISBN-13: 978-0-367-86614-3 (pbk)

Visit the Taylor & Francis Web site at
http://www.taylorandfrancis.com

and the CRC Press Web site at
http://www.crcpress.com

Contents

Part I Signal Processing for Remote Sensing

Part II   Image Processing for Remote Sensing

Preface

Since the publication of the first edition there has been considerable progress on signal processing and image processing in remote sensing. There is much temptation to publish a book titled with "advances" to document such progress in topical areas. Such a book may be useful for a short duration of 1 or 2 years, after that another book on "advances" is needed. We believe that despite the extensive use of the Internet and websites nowadays, printed books continue to play a major role in knowledge distribution. A technical book should have a useful shelf-life of at least 5 years. With this in mind the second edition of the book is self-reliant; that is, the readers do not have to read the first edition to understand the current edition. The book includes only four chapters reprinted from the first edition and the remaining 23 chapters are new or nearly entirely rewritten from the first edition. Thanks to the reviewers of the first edition, this book is not like a special issue of a journal that normally covers a narrow topic. Journal special issues tend to be highly condensed with focus on presenting new and original results for specialist readers. This book publication is aimed at a broad audience such that each chapter provides some tutorial materials in addition to the specific research results. However, it is impossible to cover all topics in a book of this size. It has been suggested to include fewer chapters with fewer authors on selected topics, who are leading experts in the field. Such an "ideal" edited book is hard to produce to my knowledge as leading experts are very busy with many writing commitments. In fact, this book is written by leading experts who tend to work in a team so that the team members are included in the list of authors. Another possibility is to write a handbook or encyclopedia that briefly goes over every topic in the field. Such books are not feasible as the field of remote sensing has not reached full maturity. Almost anything in a handbook can probably be found in Google or other search engines. It is the content of the book that counts the most and we believe that we have delivered a second edition book that takes into consideration all such factors while providing the most up-to-date activities on signal and image processing in remote sensing. We have recognized the roles of new and improved remote sensors that bring along new and challenging problems for signal and image processing. For industrial and academic researchers and graduate students alike, the chapters help to connect the "dots" in a topic area and the numerous references provided are always helpful. So much for the "justification" for the publication of this book.

While most data from satellites are in image forms and thus image processing has been used more often, signal processing can contribute significantly in extracting information from the remotely sensed data. In contrast to other books in this field that deal almost exclusively with the image processing for remote sensing, this book provides a good balance between the roles of signal and image processing in remote sensing. The book covers mainly methodologies of signal and image processing in remote sensing. New and even unconventional mathematical methods are considered. Each chapter presents a unique technique in dealing with a specific remote sensing problem. Furthermore, the term "remote sensing" is not limited to the problems with data from satellite sensors. Other sensors that acquire data remotely are also considered.

The book consists of 27 chapters divided into two parts. Part I comprises 11 chapters. The first chapter is a revision of the first edition with no change in figures. The normalized Hilbert–Huang transform makes the best use of time–frequency-domain information especially because of the inherently nonlinear process involved in the remote sensing data such as the infrared image. The empirical mode decomposition presented is of fundamental importance to signals in remote sensing problems. Chapter 2 continues with the presentation of empirical mode decomposition with applications to sea-level data and earthquake accelerogram data. Another remote sensing signal is the hydroacoustic signal to monitor the oceans for underwater nuclear tests. Chapter 3 deals specifically with the use of kernel-based algorithms for classification of hydroacoustic signals and provides a rigorous presentation on the support vector machine (SVM). As SVM is also used in some other chapters, the presentation is helpful to readers to understand such a powerful technique. Chapter 4 by Dr. Enders A. Robinson brings us back to the long-standing problem of constructing subsurface images for seismic exploration by Huygens construction and the Doppler effect in remote detection. A new topic area in remote sensing is the compressive remote sensing. For the large amount of remote sensing data, compressive sensing allows us to reduce the data size as well as to recover certain features from the undersampled data under certain constraints. The chapter presents results of compressive sensing applied to radar remote sensing and optical remote sensing. Despite the improvement of sensors, contextual factors that impact the target classification can be the unmeasured ambient variables such as soil moisture, or simply the background noise itself. These problems can be viewed as a special case of concept drift, which present a challenge to achieving robust sensing performance. Chapter 6 treats the background data as a source of supplementary contextual information and conditions the decision rules on this information. Target detection with ground-penetrating radars and hyperspectral imaging shows significant performance improvement with such context-dependent classification.

Recently, new areas in signal processing are nonnegative matrix factorization (NMF) and nonnegative tensor factorization (NTF), which are examined in Chapter 7 for extracting certain elementary features from image data with a given sparseness, for sea ice classification. As these new techniques were originally developed for signals, the chapter is contained in Part I. Chapters 8 and 9 are reprinted from the first edition to cover the use of time-series analysis and prediction-error filter in remote sensing problems. Following the same arrangement as in the first edition, two new chapters on the application of neural networks in the hyperspectral microwave sounding, and millimeter-wave retrieval of global precipitation are well presented in Chapters 10 and 11, respectively.

Part II comprises 16 chapters. Chapter 12 is concerned with the satellite SAR images simulation, image focusing, and a range-Doppler-based algorithm for target recognition. Although Chapter 13 is a reprint chapter, I believe it provides the most authoritative treatment of the polarimetric SAR techniques for remote sensing of an ocean surface. Chapter 14 on inverse synthetic aperture radar (ISAR) presents another way to look at the problem of forming a synthetic aperture without assuming that the target is static. ISAR imaging can provide acceptable solutions when SAR imaging fails. Four types of image inversion mapping are explored. An ISAR processor is then used to form well-focused images of the moving target.

Rather than grouping the chapters according to SAR image processing, hyperspectral image processing, and image classification, we place more emphasis on new approaches. Chapter 15 is about active learning methods in the classification of remote sensing images. Active learning allows us to have interaction between the user (human expert) and the automatic recognition system rather than expanding the number of training samples in a fully automated system. The chapter adopts a batch mode of active learning by selecting the samples of a batch on the basis of the uncertainty or least redundancy only. Chapter 16 presents the use of the marked-point processes to model the distribution of complex geometrical objects in a scene for detection of round and elliptical objects representing craters in planetary images. Along the same spirit, Chapter 17 presents among several mathematical approaches the polarimetric SAR amplitude joint pdf estimation via copulas along with joint pdf (probability density function) estimation and classification experiments. Chapter 18 provides a major update of the random forest classification of remote sensing data from the first edition. In Chapter 19, a new technique for hyperspectral image (HSI) target detection and classification based on sparse representation is introduced. Using the sparsity model, a test sample is represented by a sparse linear combination of training samples from the given dictionary. The sparse representation of a test spectral sample is recovered by solving a sparsity-constrained optimization problem via greedy pursuit algorithms. To improve the classification performance, the contextual information of HSI pixels is incorporated in the decision algorithm.

There is a constant need to improve spatial resolution in remote sensing due to the limitation of hyperspectral sensors, and numerous techniques have been explored. Chapter 20 presents a supervised super-resolution approach for HSI, which integrates the concepts of sub-pixel mapping into a supervised classification framework based on the use of support vector machines. The problem of signal subspace identification of hyperspectral data represents a critical first step in many hyperspectral processing algorithms such as target detection, change detection, classification, and unmixing. The identification of this subspace enables a correct dimensional reduction. Chapter 21 introduces the HySime method that estimates the signal and the noise correlation matrices and then selects the subset of eigenvalues that best represents the signal subspace in the maximum mean-squared error sense. It has long been recognized that incorporating the spatial relationships as contextual information in a Bayesian classification framework results in significant improvement in land cover classification accuracy. Chapter 22 presents new, intuitive, and efficient relationship models for modeling pairwise directional spatial relationships and the ternary between relationship using fuzzy mathematical morphology techniques. Both Chapters 23 and 24 deal with the image fusion problem in remote sensing, as in the first edition. Chapter 24 is completely rewritten from the first edition. Chapter 25 is the only chapter that covers specifically wavelet restoration of multi/hyperspectral images based on Baye's principles, which make use of prior statistical information. It also presents the fusion framework using wavelet transforms. Chapter 26 is a new but concise chapter by Professor Omatu and his team, on land cover estimation or classification by using neural network. The authors' effort is particularly appreciated right after the major earthquake disaster in Japan. Finally, a much-needed chapter for a book of this nature (Chapter 27) is a critical review of the new development of pansharpening which is a branch of image fusion that is receiving increasing attention in remote sensing. Here pan refers to panchromatic images. Progress in pansharpening methods have been substantially motivated by the advances in spaceborne instruments.

This preface provides a highly incomplete description of each chapter that should speak well for itself. However, it does give us a highlight of numerous topical areas of signal and image processing in remote sensing. I am most grateful to all the authors. Without their diligent efforts, the book would not have been possible. It has been my great honor to work with people whom I strongly believe are among the leaders in this field. This second edition is not intended to replace the first edition entirely and readers are encouraged to read both editions of the book for a more complete picture of signal and image processing in remote sensing.

C. H. Chen
North Dartmouth, Massachusetts

Editor

Chi Hau Chen received his PhD in electrical engineering from Purdue University in 1965, a MSEE from the University of Tennessee, Knoxville in 1962, and a BSEE from the National Taiwan University in 1959.

He is currently the chancellor professor emeritus of electrical and computer engineering at the University of Massachusetts Dartmouth, where he has taught since 1968. His research areas encompass statistical pattern recognition and signal/image processing with applications to remote sensing, geophysical, underwater acoustics and nondestructive testing problems; as well as computer vision for video surveillance; time-series analysis; and neural networks.

Dr. Chen has published 29 books in his areas of research. He is the editor of *Digital Waveform Processing and Recognition* (CRC Press, 1982), *Signal Processing Handbook* (Marcel Dekker, 1988) and the first edition of *Signal and Image Processing for Remote Sensing* (CRC Press, 2006). He is the chief editor of the *Handbook of Pattern Recognition and Computer Vision*, vols. 1–4 (World Scientific Publishing, 1993, 1999, 2005, 2010, respectively). He is the editor of *Fuzzy Logic and Neural Network Handbook* (McGraw-Hill, 1966). In the remote sensing field, he is the editor of *Information Processing for Remote Sensing* and *Frontiers of Remote Sensing Information Processing* (World Scientific Publishing, 1999 and 2003, respectively).

He served as associate editor of the *IEEE Transactions on Acoustics, Speech and Signal Processing* for 4 years, associate editor of the *IEEE Transactions on Geoscience and Remote Sensing* for 15 years, and since 2008 he has been a board member of *Pattern Recognition*.

Dr. Chen has been a Fellow of the Institute of Electrical and Electronic Engineers (IEEE) since 1988, a Life Fellow of the IEEE since 2003, a Fellow of the International Association of Pattern Recognition (IAPR) since 1996, and is a member of Academia NDT International.

Contributors

J. van Aardt
Department of Biosystems
Katholieke Universiteit Leuven
Leuven, Belgium

Bruno Aiazzi
Institute of Applied Physics
"Nello Carrara" of the National
    Research Council
Florence, Italy

H. Gökhan Akçay
Department of Computer Engineering
Bilkent University
Ankara, Turkey

Selim Aksoy
Department of Computer Engineering
Bilkent University
Ankara, Turkey

Luciano Alparone
Department of Electronics and
    Telecommunications
University of Florence
Florence, Italy

Alessio Bacci
Department of Information Engineering
University of Pisa
Pisa, Italy

Stefano Baronti
Institute of Applied Physics
"Nello Carrara" of the National
    Research Council
Florence, Italy

Jon Atli Benediktsson
Department of Electrical and Computer
    Engineering
University of Iceland
Reykjavik, Iceland

Fabrizio Berizzi
Department of Information Engineering
University of Pisa
and
Radar and Surveillance Systems (RaSS)
    National Laboratory
National Inter-University Consortium for
    Telecommunications
Pisa, Italy

José M. Bioucas-Dias
Instituto Superior, de Eugenharia
    de Lisbon
Lisbon, Portugal

William J. Blackwell
MIT Lincoln Laboratory
Lexington, Massachusetts

Lorenzo Bruzzone
Department of Information Engineering
    and Computer Science
University of Trento
Trento, Italy

Jocelyn Chanussot
Gronoble Institute of Technology
Saint Martin d'Heres, France

Kun-Shan Chen
National Central University
Chung-Li, Taoyuan, Taiwan

Yi Chen
Department of Electrical and
    Computer Engineering
The Johns Hopkins University
Baltimore, Maryland

R. Gökberk Cinbiş
Department of Computer Engineering
Bilkent University
Ankara, Turkey

Leslie M. Collins
Department of Electrical and Computer
    Engineering
Duke University
Durham, North Carolina

P. Coppin
Department of Biosystems
Katholieke Universiteit Leuven
Leuven, Belgium

Begüm Demir
Department of Information Engineering
    and Computer Science
University of Trento
Trento, Italy

John F. Doherty
Department of Electrical
    Engineering
Penn State University
University Park, Pennsylvania

Arno Duijster
Department of Physics
University of Antwer
Wilrijk, Belgium

Boris Escalante-Ramírez
Facultad de Ingenieria
Universidad National
    Autónoma de México
Mexico City, Mexico

Toru Fujinaka
Hiroshima University
Hiroshima, Japan

Andrea Garzelli
Department of Information
    Engineering
University of Siena
Siena, Italy

Elisa Giusti
Department of Information
    Engineering
University of Pisa
and

Radar and Surveillance Systems (RaSS)
    National Laboratory
National Inter-University Consortium for
    Telecommunications
Pisa, Italy

Norden E. Huang
Research Center for Adaptive Data
    Analysis
National Central University
Chungli, Taiwan

M. Yousuff Hussaini
Department of Mathematics
Florida State University
Tallahassee, Florida

Christian Igel
Department of Computer Science
University of Copenhagen
Copenhagen, Denmark

I. Jonckheere
Department of Biosystems
Katholieke Universiteit Leuven
Leuven, Belgium

P. Jönsson
Department of Biosystems
Katholieke Universiteit Leuven
Leuven, Belgium

Christian Jutten
Gronoble Institute of Technology
Saint Martin d'Heres, France

Juha Karvonen
Finnish Institute of Marine Research
Helsinki, Finland

Dayalan Kasilingam
Department of Electrical and
    Computer Engineering
University of Massachusetts Dartmouth
North Dartmouth, Massachusetts

A. Shaharyar Khwaja
Department of Electrical and Computer
    Engineering
Ryerson University
Toronto, Canada

Vladimir A. Krylov
INRIA
Sophia Antipolis, France

Jong-Sen Lee
Naval Research Laboratory (Retired)
Washington, DC

S. Lhermitte
Department of Biosystems
Katholieke Universiteit Leuven
Leuven, Belgium

Steven R. Long
NASA Goddard Flight Center
Wallops Flight Facility
Wallops Island, Virginia

Alejandra A. López-Caloca
Centro de Investigación en Geografía y
    Geomática
Mexico City, Mexico

Jianwei Ma
Department of Mathematics
Florida State University
Tallahassee, Florida

Marco Martorella
Department of Information
    Engineering
University of Pisa
and
Radar and Surveillance Systems (RaSS)
    National Laboratory
National Inter-University Consortium for
    Telecommunications
Pisa, Italy

Enzo Dalle Mese
Department of Information Engineering
University of Pisa
and
Radar and Surveillance Systems (RaSS)
    National Laboratory
National Inter-University Consortium for
    Telecommunications
Pisa, Italy

Kenneth D. Morton, Jr.
Department of Electrical and Computer
    Engineering
Duke University
Durham, North Carolina

Gabriele Moser
Dipartimento di Ingegneria Biofisica ed
    Elettronica
University of Genoa
Genoa, Italy

José M. P. Nascimento
Instituto Superior, de Eugenharia de Lisbon
Lisbon, Portugal

Nasser M. Nasrabdi
U.S. Army Research Laboratory
Adelphi, Maryland

Sigeru Omatu
Osaka Institute of Technology
Osaka, Japan

Claudio Persello
Department of Information Engineering
    and Computer Science
University of Trento
Trento, Italy

Mark Prior
Comprehensive Nuclear Test Ban Treaty
    Organization Preparatory
    Commission
Vienna, Austria

Christopher R. Ratto
Department of Electrical and Computer
    Engineering
Duke University
Durham, North Carolina

Enders A. Robinson
Columbia University (Retired)
Newburyport, Massachusetts

Arnab Roy
Department of Electrical Engineering
Penn State University
University Park, Pennsylvania

Paul Scheunders
Department of Physics
University of Antwerp
Wilrijk, Belgium

Dale L. Schuler
Naval Research Laboratory (Retired)
Broomfield, Colorado

Massimo Selva
Institute of Applied Physics
"Nello Carrara" of the National Research
    Council
Florence, Italy

Sebastiano Bruno Serpico
Dipartimento di Ingegneria Biofisica ed
    Elettronica
University of Genoa
Genoa, Italy

Anne H. S. Solberg
Department of Informatics
University of Oslo
Oslo, Norway

David H. Staelin
Massachusetts Institute of
    Technology
Cambridge, Massachusetts

Chinnawat "Pop" Surussavadee
Andaman Environment and Natural
    Disaster Research Center
Prince of Songkla University
Phuket, Thailand

Johannes R. Sveinsson
Department of Electrical and Computer
    Engineering
University of Iceland
Reykjavik, Iceland

Peter A. Torrione
Department of Electrical and Computer
    Engineering
Duke University
Durham, North Carolina

Trac D. Tran
Department of Electrical and Computer
    Engineering
The Johns Hopkins University
Baltimore, Maryland

Giulia Troglio
Dipartimento di Ingegneria Biofisica ed
    Elettronica
University of Genoa
Genoa, Italy

Yuta Tsuchida
Graduate School of Engineering
Osaka Prefecture University
Osaka, Japan

Matthias Tuma
Institut für Neuroinformatik
Ruhr-Universität Bochum
Bochum, Germany

Yu-Chang Tzeng
National Central University
Chung-Li, Taiwan

Jan Verbesselt
Department of Biosystems
Katholieke Universiteit Leuven
Leuven, Belgium

Alberto Villa
Faculty of Engineering
University of Iceland
Reykjavik, Iceland

Björn Waske
Institute of Geodesy and
    Geoinformation
University of Bonn
Bonn, Germany

Michifumi Yoshioka
Graduate School of Engineering
Osaka Prefecture University
Osaka, Japan

Sang-Ho Yun
Jet Propulsion Laboratory
Pasadena, California

Howard Zebker
Department of Geophysics
Stanford University
Stanford, California

Josiane Zerubia
INRIA
Sophia Antipolis, France

Yifan Zhang
Department of Physics
University of Antwerp
Wilrijk, Belgium

Part I

Signal Processing for Remote Sensing

1 On the Normalized Hilbert Transform and Its Applications to Remote Sensing

Steven R. Long and Norden E. Huang

CONTENTS

1.1 INTRODUCTION

The development of this new approach was motivated by the need to describe nonlinear distorted waves in detail, along with the variations of these signals that occur naturally in nonstationary processes (e.g., ocean waves). As has often been noted, natural physical processes are mostly nonlinear and nonstationary. Yet, historically there have been very few options in the available analysis methods to examine such data from nonlinear and nonstationary processes. The available methods have usually been either for linear, but nonstationary, or nonlinear, but stationary, and statistically deterministic processes. There is a need to examine data from nonlinear, nonstationary, and stochastic processes in the natural world, and this requires special treatment. The past approach of imposing a linear structure (by assumptions) on the nonlinear system is just not adequate. Other than periodicity, the detailed dynamics in the processes from the data also need to be determined. This is needed because one of the typical characteristics of nonlinear processes is its intrawave frequency modulation (FM), which indicates the instantaneous frequency (IF) changes within one oscillation cycle.

In the past, when the analysis was dependent on linear Fourier analysis, there was no means of depicting the frequency changes within one wavelength (the intrawave frequency variation) except by resorting to the concept of harmonics. The term "bound harmonics" was often used in this connection. Thus, the distortions of any nonlinear waveform have often been referred to as "harmonic distortions." The concept of harmonic distortion is a mathematical artifact resulting from imposing

3

a linear structure (through assumptions) on a nonlinear system. The harmonic distortions may thus have mathematical meaning, but there is no physical meaning associated with them, as discussed by Huang et al. [1–3]. For example, in the case of water waves, such harmonic components do not have any of the real physical characteristics of a water wave as it occurs in nature. The physically meaningful way to describe such data should be in terms of its IF, which will reveal the intrawave FMs occurring naturally.

It is reasonable to suggest that any such complicated data should consist of numerous superimposed modes. Therefore, to define only one IF value for any given time is just not meaningful (see Ref. [4] for comments on the Wigner–Ville distribution). To fully consider the effects of multicomponent data, a decomposition method should be used that can separate the naturally combined components completely and nearly orthogonally. In the case of nonlinear data, the orthogonality condition would need to be relaxed, as discussed by Huang et al. [1]. Initially, Huang et al. [1] proposed the empirical mode decomposition (EMD) approach to produce the intrinsic mode functions (IMF), which are both monocomponent and symmetric. This was an important step toward making the application truly practical. With the EMD satisfactorily determined, an important roadblock to truly nonlinear and nonstationary analysis was finally removed. However, the difficulties resulting from the limitations stated by the Bedrosian [5] and Nuttall [6] theorems must also be addressed in connection with this approach. Both limitations have firm theoretical foundations and must be considered; IMFs satisfy only the necessary condition, but not the sufficient condition. To improve the performance of the processing as proposed by Huang et al. [1], the normalized empirical mode decomposition (NEMD) method was developed as a further improvement on the earlier processing methods, as well as refinements by Huang et al. [7] and Wu and Huang [8].

1.2   REVIEW OF PROCESSING ADVANCES

1.2.1   NORMALIZED EMPIRICAL MODE DECOMPOSITION

The NEMD method was developed to satisfy the specific limitations set by the Bedrosian theorem, while also providing a sharper measure of local error when the quadrature differs from the Hilbert transform (HT) result.

From an example data set of a natural process, all the local maxima of the data are first determined. These local maxima are then connected with a cubic spline curve, which gives the local amplitude of the data, $A(t)$, as shown together in Figure 1.1. The envelope obtained through spline fitting is used to normalize the data by

$$y(t) = \frac{a(t)\cos\theta(t)}{A(t)} = \left(\frac{a(t)}{A(t)}\right)\cos\theta(t). \tag{1.1}$$

Here, $A(t)$ represents the cubic spline fit of all the maxima of the example data, and thus $a(t)/A(t)$ should normalize $y(t)$, forcing all maxima to be normalized to unity, as shown in Figure 1.2. As is apparent from Figure 1.2, a small number of the normalized data points can still have an amplitude in excess of unity. This is because the cubic spline is through the maxima only, so that at locations where the amplitudes are changing rapidly, the line representing the envelope spline can pass under some of the data points. These occasional misses are unavoidable, yet the normalization scheme has effectively separated the amplitude from the carrier oscillation. The IF can then be computed from this normalized carrier function $y(t)$ just obtained. Owing to the nearly uniform amplitude, the limitations set by the Bedrosian theorem are effectively satisfied. The IF computed in this way from the normalized data from Figure 1.2 is shown in Figure 1.3, together with the original example data. With the Bedrosian theorem limitations addressed, what of the limitations set by the Nuttall theorem?

If the HT can be considered to be the quadrature, then the absolute value of the HT performed on the perfectly normalized example data should be unity. Then any deviation from the absolute

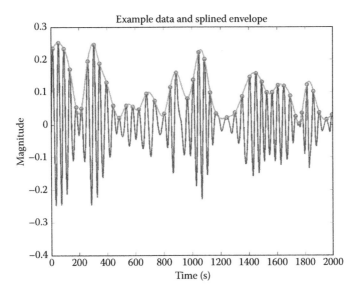

FIGURE 1.1 The best possible cubic spline fit to the local maxima of the example data. The spline fit forms an envelope as an important first step in the process. Note also how the frequency can change within a wavelength, and that the oscillations can occur in groups.

value of the HT from unity would be an indication of a difference between the quadrature and the HT results. An error index can thus be simply defined as

$$E(t) = \left[\text{abs}\left(\text{Hilbert transform}(y(t))\right) - 1 \right]^2. \tag{1.2}$$

This error index would be not only an energy measure as given in the Nuttall theorem but also a function of time as shown in Figure 1.4. Therefore, it gives a local measure of the error resulting

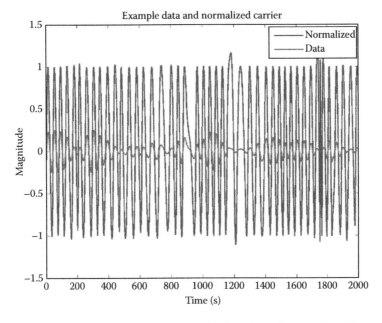

FIGURE 1.2 Normalized example data of Figure 1.1 with the cubic spline envelope. The occasional value beyond unity is due to the spline fit slightly missing the maxima at those locations.

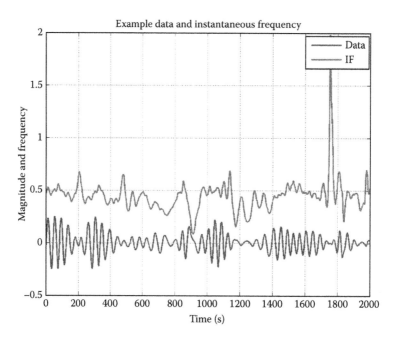

FIGURE 1.3   The instantaneous frequency determined from the normalized carrier function is shown with the example data. Data is about zero, and the instantaneous frequency varies about the horizontal 0.5 value.

from the IF computation. This local measure of error is both logically and practically superior to the integrated error bound established by the Nuttall theorem. If the quadrature and the HT results are identical, then it follows that the error should be zero. Based on experience with various natural data sets, the majority of the errors encountered here result from two sources. The first source is due to an imperfect normalization occurring at locations close to rapidly changing amplitudes, where the

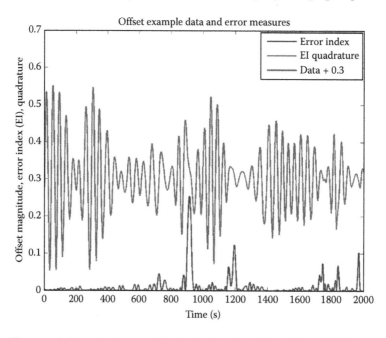

FIGURE 1.4   The error index as it changes with the data location in time. The original example data offset by 0.3 vertically for clarity is also shown. The quadrature result is not visible on this scale.

envelope spline-fitting is unable to turn sharply or quickly enough to cover all the data points. This type of error is even more pronounced when the amplitude is also locally small, thus amplifying any errors. The error index from this condition can be extremely large. The second source is due to nonlinear waveform distortions, which will cause corresponding variations of the phase function $\theta(t)$. As discussed by Huang et al. [1], when the phase function is not an elementary function, the differentiation of the phase determined by the HT is not identical to that determined by the quadrature. The error index from this condition is usually small (see Ref. [7]).

Overall, the NEMD method gives a more consistent, stable IF. The occasionally large error index values offer an indication where the method failed simply because the spline misses and cuts through the data momentarily. All such locations occur at the minimum amplitude with a resulting negligible energy density. Further improvements are discussed by Wang et al. [9].

1.2.2 Amplitude and Frequency Representations

In the initial methods [1,2,7], the main result of Hilbert spectral analysis (HSA) always emphasized the FM. In the original methods, the data were first decomposed into IMFs, as defined in the initial work. Then, through the HT, the IF and amplitude of each IMF were computed to form the Hilbert spectrum. This continues to be the method, especially when the data are normalized. The information on the amplitude or envelope variation is not examined. In the NEMD and HSA approach, it is justifiable not to pay too much attention to the amplitude variations. This is because if there is mode mixing, the amplitude variation from such mixed mode IMFs does not reveal any true underlying physical processes. However, there are cases when the envelope variation does contain critical information. An example of this is when there is no mode mixing in any given IMF, when a beating signal representing the sum of two coexisting sinusoidal ones is encountered. In an earlier paper, Huang et al. [1] attempted to extract individual components out of the sum of two linear trigonometric functions such as

$$x(t) = \cos at + \cos bt. \tag{1.3}$$

Two seemingly separate components were recovered after over 3000 sifting steps. Yet, the IMFs thus obtained were not purely trigonometric functions anymore, and there were obvious aliases in the resulting IMF components as well as the residue. The approach proposed then was unnecessary and unsatisfactory. The problem, in fact, has a much simpler solution: treating the envelope as an amplitude modulation (AM), and then processing just the envelope data. The function $x(t)$ as given in Equation 1.3 can then be rewritten as

$$x(t) = \cos at + \cos bt = 2\cos\left(\frac{a+b}{2}t\right)\cos\left(\frac{a-b}{2}t\right). \tag{1.4}$$

There is no difference between the sum of the individual components and the modulating envelope form; they are trigonometric identities. If both the frequency of the carrier wave, $(a+b)/2$, and the frequency of the envelope, $(a-b)/2$, can be obtained, then all the information in the signal can be extracted. This indicates the reason to look for a new approach to extracting additional information from the envelope. In this example, however, the envelope becomes a rectified cosine wave. The frequency would be easier to determine from the simple period counting than from the Hilbert spectral result. For a more general case when the amplitudes of the two sinusoidal functions are not equal, the modulation is not simple anymore. For even more complicated cases, when there are more than two coexisting sinusoid components with different amplitudes and frequencies, there is no general expression for the envelope and carrier. The final result could be represented as more than one frequency-modulated band in the Hilbert spectrum. It is then impossible to describe the individual components under this situation. In such cases, representing the signal as a carrier and envelope variation should still be meaningful, for the dual representations of frequency arise from the different definitions of frequency. The Hilbert-inspired view of amplitude and FMs still renders a correct representation of the signal, but this view is very different from that of Fourier analysis. In such cases, if one is sure of the stationarity and regularity of the signal, Fourier analysis could be used, which will give the more familiar results as suggested by Huang et al. [1]. The judgment for those cases is not on which one is correct, as both are correct; rather, it is on which one is more familiar and more revealing.

When more complicated data are present, such as in the case of radar returns, tsunami wave records, earthquake data, speech signals, and so on (representing a frequency "chirp"), the amplitude variation information can be found by processing the envelope data and treating the data as an approximate carrier. When the envelope of frequency chirp data, such as the example given in Figure 1.5, is decomposed through the NEMD process, IMF components are obtained as shown in Figure 1.6. Using these components (or IMFs), the Hilbert spectrum can be constructed as given in Figure 1.7, together with its FM counterpart. The physical meaning of the AM spectrum is not as clearly defined in this case. However, it does serve to illustrate the AM contribution to the variability of the local frequency. Improvements to the method are ongoing, as the works of Wu and Huang [8] and Tsui et al. [10] demonstrate.

1.2.3 INSTANTANEOUS FREQUENCY

It must be emphasized that IF is a very different concept from the frequency content of the data derived from Fourier-based methods, as discussed in great detail by Huang et al. [1]. The IF, as discussed here, is based on the instantaneous variation of the phase function from the HT of a data-adaptive decomposition, while the frequency content in the Fourier approach is an averaged frequency obtained from a convolution of the data with an *a priori* basis. Therefore, whenever the basis changes, the frequency content also changes. Similarly, when the decomposition changes, the IF also has to change. However, there are still persistent and common misconceptions on IF computed in this manner.

One of the most prevailing misconceptions about IF is that for any data with a discrete line spectrum, IF can be a continuous function. A variation of this misconception is that IF can give frequency

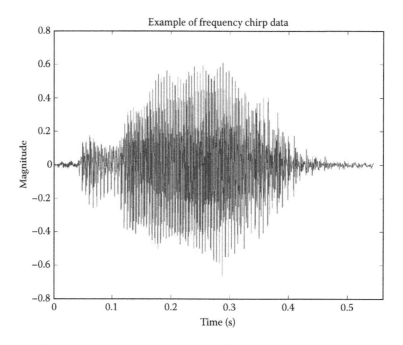

FIGURE 1.5   A typical example of complex natural data, illustrating the concept of frequency "chirps."

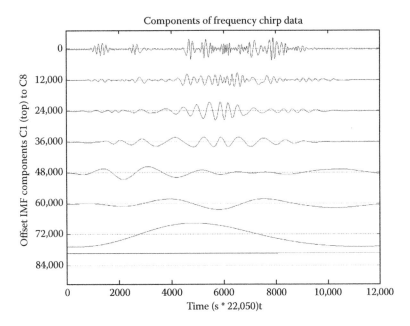

FIGURE 1.6 The eight IMF components obtained by processing the frequency chirp data of Figure 1.5, offset vertically form C1 (top) to C8 (bottom).

values that are not one of the discrete spectral lines. This dilemma can be resolved easily: In the nonlinear cases, when the IF approach treats the harmonic distortions as continuous intrawave FMs, the Fourier-based methods treat the frequency content as discrete harmonic spectral lines. In the case of two or more beating waves, the IF approach treats the data as AM and FM modulations, while the frequency content from the Fourier method will treat each constituting wave as a discrete

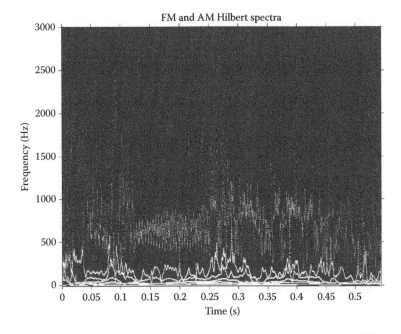

FIGURE 1.7 The AM and FM Hilbert spectral results from the frequency chirp data of Figure 1.5.

spectral line, *if* the process is stationary. Although they appear perplexingly different, they represent the same data.

Another misconception is on negative IF values. According to Gabor's [11] approach, the HT is implemented through two Fourier transforms: the first transforms the data into frequency space, while the second performs an inverse Fourier transform after discarding all the negative frequency parts [4]. Therefore, according to this argument, all the negative frequency content has been discarded, which then raises the question, how can there still be negative frequency values? This question arises due to a misunderstanding of the nature of negative IF from the HT. The direct cause of negative frequency in the HT is the consequence of multiple extrema between two zero-crossings. Then, there will be local loops not centered at the origin of the coordinate system as discussed by Huang et al. [1]. Negative frequency can also occur even if there are no multiple extrema. For example, this would happen when there are large amplitude fluctuations, which cause the Hilbert-transformed phase loop to miss the origin. Therefore, the negative frequency does not influence the frequency content in the process of the HT through Gabor's [11] approach. Both these causes are removed by the NEMD and normalized Hilbert transform (NHT) methods presented here.

The latest version of these methods (NEMD/NHT) consistently give more stable IF values. It satisfies the limitation set by the Bedrosian theorem, and offers a local measure of error sharper than the Nuttall theorem. Note here that in the initial spline of the amplitude done in the NEMD approach, the end effects again become important. The method used here is just to assign the end points as a maximum equal to the very last value. Other improvements using characteristic waves and linear predictions as discussed in Ref. [1] can also be employed. There could be some improvement, but the resulting fit will be very similar.

Ever since the introduction of the EMD and HSA methods by Huang et al. [1–3,7,12] and Ref. [8], the methods have attracted increasing attention. Some investigators, however, have expressed certain reservations. For example, Olhede and Walden [13] suggested that the idea of computing IF through the HT is good, but that the EMD approach is not rigorous. Therefore, they have introduced the wavelet projection as the method for decomposition and only adopt the IF computation from the HT. Flandrin et al. [14], however, suggest that the EMD is equivalent to a bank of dyadic filters, but refrain from using the HT. From the analysis presented here, it can be concluded that caution when using the HT is fully justified. The limitations imposed by Bedrosian and Nuttall certainly have solid theoretical foundations. The normalization procedure as shown here will help to remove any reservations about further applications of the improved HT methods in data analysis. The method, however, will offer relatively little help to the approach advanced by Olhede and Walden [13] because the wavelet decomposition definitely removes the nonlinear distortions from the waveform. The consequence of this, however, is that their approach should also be limited to nonstationary, but linear, processes. It only serves the limited purpose of improving the poor frequency resolution of the continuous wavelet analysis.

As clearly shown in Equation 1.1, to give a good representation of actual wave data or other data from natural processes by means of an analytical wave profile, the analytical profile will need to have IMFs, and also obey the limitations imposed by the Bedrosian and Nuttall theorems. In the past, such a thorough examination of the data has not been done. As reported by Huang et al. [2,12], most of the actual wave data recorded are not composed of single components. Consequently, the analytical representation of a given wave profile in the form of Equation 1.1 poses a challenging problem theoretically.

1.3 APPLICATION TO IMAGE ANALYSIS IN REMOTE SENSING

Just as much of the data from natural phenomena are either nonlinear or nonstationary, or both, it is also that way with the data that form images of natural processes. The methods of image processing are already well advanced, as can be seen in review works such as [12,13,15–17]. The NEMD/NHT methods can now be added to the available tools, for producing new and unique image products.

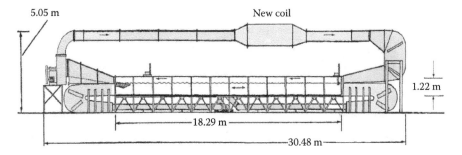

FIGURE 1.8 The NASA Air-Sea Interaction Research Facility's (NASIRF) main wave tank at Wallops Island, VA. The new coils shown were used to provide cooling and humidity control in the airflow over heated water.

Significant work has already been done in this new area by Nunes et al. [18], Linderhed [19–22], Nunes and Deléchelle [22], Bhuiyan et al. [24], and Wu et al. [25], among others. Because of the nonlinear and nonstationary quality of natural processes, the NEMD/NHT approach is especially well suited for image data, giving frequencies, inverse distances, or wave numbers as a function of time or distance, along with the amplitudes or energy values associated with these, as well as a sharp identification of imbedded structures. The various possibilities and products of this new analysis approach include, but are not limited to, joint and marginal distributions that can be viewed as isosurfaces, contour plots, and surfaces that contain information on frequency, inverse wavelength, amplitude, energy, and location in time, space, or both. Additionally, the concept of component images representing the intrinsic scales and structures imbedded in the data is now possible, along with a technique for obtaining frequency variations of structures within the images.

The laboratory used for producing the nonlinear waves for this example was the NASA Air-Sea Interaction Research Facility (NASIRF) located at the NASA Goddard Space Flight Center/Wallops Flight Facility, at Wallops Island, VA. The test section of the main wave tank is 18.3 m long and 0.9 m wide, filled to a depth of 0.76 m of water, leaving a height of 0.45 m over the water for airflow, if needed. The facility can produce wind and paddle-generated waves over a water current in either direction, and its capabilities, instruments, and software have been described in detail by Long [26,27], Long et al. [28,29], and Long and Klinke [30]. The basic description is shown as Figure 1.8, with an additional new feature indicated as *new coil* in Figure 1.8. This coil unit using both hot and cold water was installed to provide cold air of controlled temperature and humidity for experiments using cold air over heated water, in the FEDS4 experiments, the Flux Exchange Dynamics Study of 2004, a joint experiment involving the University of Washington/Applied Physics Laboratory, The University of Alberta, the Lamont-Doherty Earth Observatory of Columbia University, and NASA GSFC/Wallops Flight Facility. The cold airflow over heated water optimized conditions for the collection of infrared (IR) video images.

1.3.1 IR DIGITAL CAMERA AND SETUP

The camera used to acquire the laboratory image presented here as an example was provided by the University of Washington/Applied Physics Laboratory as part of FEDS4. The experimental setup is shown as Figure 1.9. For the example shown here, the resolution of the IR image was 640 × 512 pixels. The camera was mounted to look upwind at the water surface, so that its pixel image area covered a physical rectangle on the water surface on the order of 10 cm per side. The water within the wave tank was heated by four commercial spa heaters, while the air in the airflow was cooled and humidity controlled by NASIRF's new cooling and reheating coils. This produced a very thin layer of surface water that was cooled, so that whenever wave spilling and breaking occurred, it could be immediately seen by the IR camera.

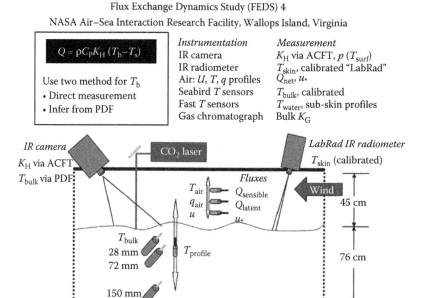

FIGURE 1.9 The experimental arrangement of FEDS4 (Flux Exchange Dynamics Study of 2004) used to capture IR images of surface wave processes. (Courtesy of A. Jessup and K. Phadnis of UW/APL.)

1.3.2 EXPERIMENTAL IR IMAGES OF SURFACE PROCESSES

With this imaging system in place, steps were taken to acquire interesting images of wave breaking and spilling due to wind and wave interactions. One such image is illustrated in Figure 1.10.

Using a horizontal line that slices through the central area of the image at the value of 275, Figure 1.11 illustrates the detail contained in the actual array of data values obtained from the IR

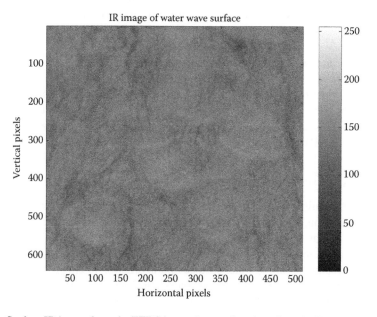

FIGURE 1.10 Surface IR image from the FEDS4 experiment. Grey bar gives the IR camera intensity levels, which can be converted to temperature. (Data courtesy of A. Jessup and K. Phadnis of UW/APL.)

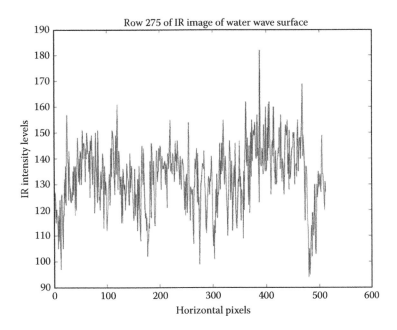

FIGURE 1.11 A horizontal slice of the raw IR image given in Figure 1.10, taken at row 275. Note the details contained in the IR image data, showing structures containing both short and longer length scales.

camera. This gives the IR camera intensity values stored in the pixels along the horizontal line. These can then be converted to the actual temperature when needed. A complex structure is evident here. Breaking wavefronts are seen in the crescent-shaped structures, where rolling, spilling breaking at the wave crest brings up the underlying warmer water. After processing, the resulting components produced from the horizontal row of Figure 1.11 are shown in Figure 1.12. As can be seen, the component with the longest scale, C9, contains the bulk of the intensity values. The shorter, riding scales are fluctuations about the levels shown in component C9. The sifting was done via the extrema approach discussed in the foundation articles, and produced a total of nine components.

Using this approach, the IR image was first divided into 640 horizontal rows of 512 values each. The rows were then processed to produce the components, each of the 640 rows producing a component set similar to that shown in Figure 1.12. From these basic results, component images can be assembled. This is done by taking the first component representing the shortest scale from each of the 640 component sets. These first components are then assembled together in order to produce an array that is also 640 rows by 512 columns and can also be visualized as an image. This is the first component image. This production of component images is then continued in a similar fashion with the remaining components representing progressively longer scales. To visualize the shortest component scales, component images 1 through 4 were added together, as shown in Figure 1.13. Throughout the image, streaks of short wavy structures can be seen to line up in the wind direction (along the vertical axis). Even though the image is formed in the IR camera by measuring heat at many different pixel locations over a rectangular area, the surface waves have an effect that can be thus remotely sensed in the image, either as streaks of warmer water exposed by breaking or more wave-like structures. If the longer-scale components are now combined using the fifth and sixth component images, a composite image is obtained as shown in Figure 1.14. Longer scales can be seen throughout the image area where breaking and mixing are occurring. Other wave-like structures of longer wavelengths are also visible. To produce a true wave number from images like these, one only has to convert using

$$\mathbf{k} = 2\pi/\lambda, \qquad (1.5)$$

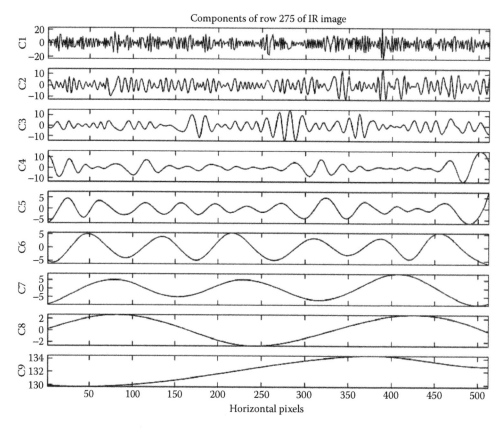

FIGURE 1.12 Components obtained by processing data from the slice shown in Figure 1.11. Note that component C9 carries the bulk of the intensity scale, while other components with shorter scales record the fluctuations about these base levels.

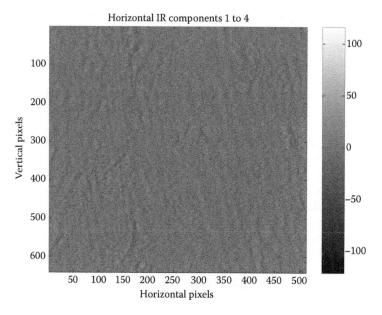

FIGURE 1.13 Composite of component images 1 to 4 from the horizontal rows representing the shortest scales. Grey scale can be converted to temperature.

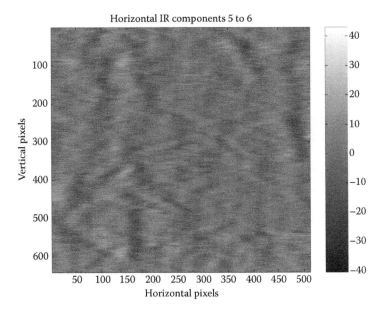

FIGURE 1.14 Composite of component images 5 to 6 from the horizontal rows representing the longer scales.

where **k** is the wave number (in 1/cm), and λ is the wavelength (in cm). This would require knowing the physical size of the image in cm or some other unit and its equivalent in pixels from the array analyzed.

Another approach to the raw image of Figure 1.10 is to separate the original image into columns instead of rows. This would make the analysis more sensitive to structures that were better aligned with that direction, and also with the direction of wind and waves. By repeating the steps leading to Figure 1.13, the shortest-scale component images in component images 3 through 5 can be combined to form Figure 1.15. Component images 1 and 2 developed from the vertical column analysis were not included here, after they were found to contain results of such a short scale uniformly spread throughout the image, and without structure. Indeed, they had the appearance of uniform noise. It is apparent that more structures at these scales can be seen by analyzing along the column direction. Figure 1.16 represents the longer scale in component image 6. By the sixth component image, the lamination process starts to fail somewhat in reassembling the image from the components. Further processing is needed to better match the results at these longer scales.

When the original data are a function of time, this new approach can produce the IF and amplitude as functions of time. Here, the original data are from an IR image, so that any slice through the image (horizontal or vertical) would be a set of camera values (ultimately temperature) representing the temperature variation over a physical length. Thus, instead of producing frequency (inverse time scale), the new approach here initially produces an inverse length scale. In the case of water surface waves, this is the familiar scale of the wave number, as given in Equation 1.5. To illustrate this, consider Figure 1.17, which shows the changes of scale along the selected horizontal row 400. The largest measures of IR energy can be seen to be at the smaller inverse length scales, which implies that it came from the longer scales of components 3 and 4. Figure 1.18 repeats this for the even longer length scales in components 5 and 6.

Returning to the column-wise processing at column 250 of Figures 1.15 and 1.16, further processing gives the contour plot of Figure 1.19, for components 3 through 5, and Figure 1.20 for components 4 through 6.

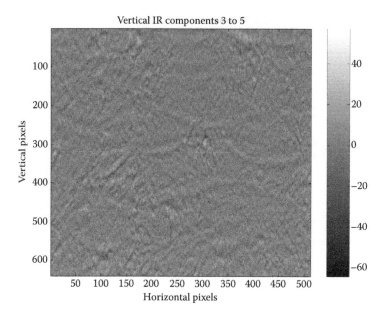

FIGURE 1.15   Component images 3 to 5 from the vertical columns here combined to produce a composite image representing the midrange scales.

1.3.3   Volume Computations and Isosurfaces

Many interesting phenomena take place in the flow of time, and thus it is interesting to note how changes occur with time in the images. To include time in the analysis, a sequence of images taken at uniform time steps can be used.

By starting with a single horizontal or vertical line from the image, a contour plot can be produced as was shown in Figures 1.17 through 1.20. Using a set of sequential images covering a known time

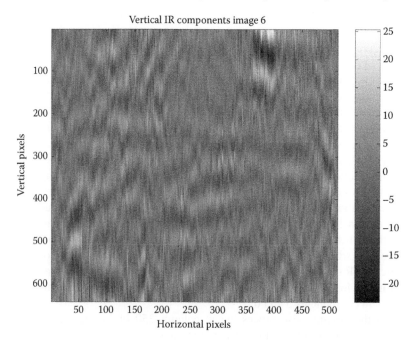

FIGURE 1.16   Component image 6 from the vertical columns representing the longer scale.

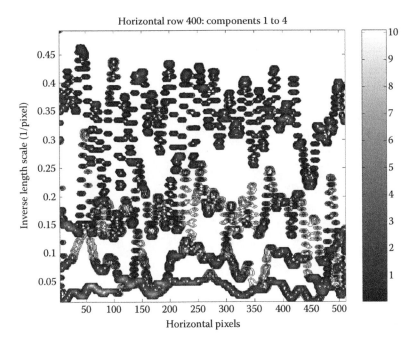

FIGURE 1.17   The results from the NEMD/NHT computation on horizontal row 400 for components 1 to 4, which resulted from Figure 1.13. Note the apparent influence of surface waves on the IR information. The most intense IR radiation can be seen at the smaller values of the inverse length scale, denoting the longer scales in components 3 and 4. A wavelike influence can be seen at all scales.

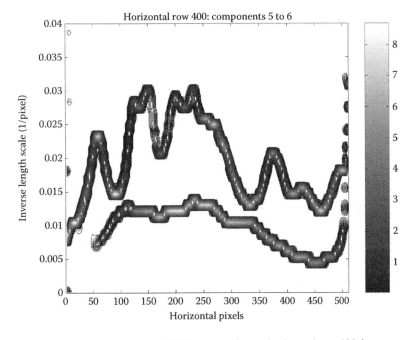

FIGURE 1.18   The results from the NEMD/NHT computation on horizontal row 400 for components 5 to 6, which resulted from Figure 1.14. Even at the longer scales, an apparent influence of surface waves on the IR information can still be seen.

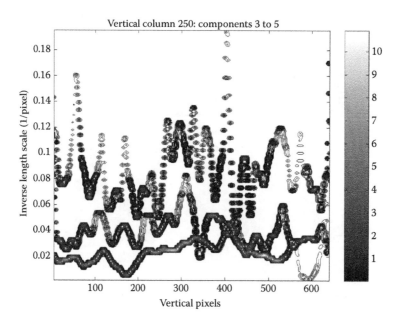

FIGURE 1.19 The contour plot developed from the vertical slice at column 250, using the components 3 to 5. The larger IR values can be seen at longer length scales (or shorter inverse length scales).

period and a pixel line of data from each (horizontal or vertical), a set of numerical arrays can be obtained from the NEMD/NHT analysis. Each array can be visualized by means of a contour plot, as already shown. The entire set of arrays can also be combined in sequence to form an array volume, or an array of dimension 3. Within the volume, each element of the array contains the amplitude or intensity of the data from the image sequence. The individual element location within the three-dimensional array specifies values associated with the stored data. One axis (call it x) of the volume can represent

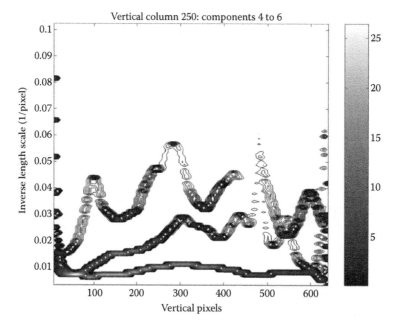

FIGURE 1.20 The contour plot developed from the vertical slice at column 250, using the components 4 through 6, as in Figure 1.19.

horizontal or vertical distance down the data line taken from the image. Another axis (call it y) can represent the resulting inverse length scale associated with the data. The additional axis (call it z) is produced by laminating the arrays together, and represents time, because each image was acquired in repetitive time steps. Thus, the position of the element in the volume gives location x along the horizontal or vertical slice, inverse length along the y-axis, and time along the z-axis.

Isosurface techniques would be needed to visualize this. This could be compared to peeling an onion, except that the different layers, or spatial contour values, are not bound in spherical shells. After a value of data intensity is specified, the isosurface visualization will make all array elements outside of the level of the value chosen transparent, while shading in the chosen value so that the elements inside that level (or behind it) cannot be seen. Some examples of this procedure can be seen in Ref. [27].

Another approach with the analysis of images is to reassemble lines from the image data using a different format. A sequence of sequential images taken in equal time steps is needed, and by using the same horizontal or vertical line from each image in the time sequence, laminate each line to its predecessor to build up an array that is the image length along the chosen line along one axis, and the number of images along the other axis, in corresponding units of time. Once complete, this two-dimensional array can be split into slices along the time axis. Each of these time slices, representing the variation in data values with time at a single pixel location, can then be processed with the new NEMD/NHT techniques. An example of this can also be seen in Ref. [27]. The NEMD/NHT techniques can thus reveal variations in frequency or time in the data at a specific location in the image sequence.

1.4   CONCLUSION

With the introduction of the normalization procedure, one of the major obstacles for NEMD/NHT analysis has been removed. Together with the establishment of the confidence limit [7] through the variation of stoppage criterion, and the statistically significant test of the information content for IMF [14,31], and the further development of the concept of IF [3], the new analysis approach has indeed approached maturity for applications empirically if not mathematically (for a recent overview of developments, see Ref. [32]). The new NEMD/NHT methods provide the best overall approach to determining the IF for nonlinear and nonstationary data. Thus, a new tool is available to aid in further understanding and gaining deeper insight into the wealth of data now possible by remote sensing and other means. Specifically, the application of the new method to data images was demonstrated. Further developments on this application can be seen in Ref. [23]

This new approach is covered by several U.S. patents held by NASA. Further information on obtaining the software and on the various U.S. patents can be found at the NASA-authorized commercial site http://www.fuentek.com/technologies/hht.htm. Information on upcoming meetings and recent developments may be obtained at http://rcada.ncu.edu.tw/intro.html.

ACKNOWLEDGMENTS

The authors wish to express their continuing gratitude and thanks to Dr. Eric Lindstrom of NASA Headquarters for his encouragement and support of the work, and their sincere thanks to Professor Chi Hau Chen for his patience and many helpful suggestions.

REFERENCES

1. Huang, N. E., Shen, Z., Long, S. R., Wu, M. C., Shih, S. H., Zheng, Q., Tung, C. C., and Liu, H. H. The empirical mode decomposition method and the Hilbert spectrum for non-stationary time series analysis, *Proc. Roy. Soc. London*, A454, 903–995, 1998.
2. Huang, N. E., Shen, Z., and Long, S. R. A new view of water waves—The Hilbert spectrum. *Ann. Rev. Fluid Mech.*, 31, 417–457, 1999.

3. Huang, N. E., Wu, Z., Long, S. R., Arnold, K. C., Chen, X., and Blank, K. On instantaneous frequency, *Advances in Adaptive Data Analysis*, 1(2), World Scientific, Singapore, 177–229, 2009.
4. Flandrin, P. *Time-Frequency/Time-Scale Analysis*, Academic Press, San Diego, 1999.
5. Bedrosian, E. On the quadrature approximation to the Hilbert transform of modulated signals, *Proc. IEEE*, 51, 868–869, 1963.
6. Nuttall, A. H. On the quadrature approximation to the Hilbert transform of modulated signals, *Proc. IEEE*, 54, 1458–1459, 1966.
7. Huang, N. E., Wu, M. L., Long, S. R., Shen, S. S. P., Qu, W. D., Gloersen, P., and Fan, K. L. A confidence limit for the empirical mode decomposition and the Hilbert spectral analysis, *Proc. of Roy. Soc. London*, A459, 2317–2345, 2003.
8. Wu, Z. and Huang, N. E. Ensemble empirical mode decomposition: A noise-assisted data analysis method, *Advances in Adaptive Data Analysis*, 1(1), World Scientific, Singapore, 1–41, 2009.
9. Wang, G., Chen, X.-Y., Qiao, F.-L., Wu, Z., and Huang, N. E. On intrinsic mode function, *Advances in Adaptive Data Analysis*, 2(3), World Scientific, Singapore, 277–293, 2010.
10. Tsui, P.-H., Chang, C.-C., and Huang, N. E. Noise-modulated empirical mode decomposition, *Advances in Adaptive Data Analysis*, 2(1), World Scientific, Singapore, 25–37, 2010.
11. Gabor, D. Theory of communication, *J. IEEE*, 93, 426–457, 1946.
12. Huang, N. E., Long, S. R., and Shen, Z. The mechanism for frequency downshift in nonlinear wave evolution. *Adv. Appl. Mech.*, 32, 59–111, 1996.
13. Olhede, S. and Walden, A. T. The Hilbert spectrum via wavelet projections, *Proc. R. Soc. Lond.*, A460, 955–975, 2004.
14. Flandrin, P., Rilling, G., and Gonçalves, P. Empirical mode decomposition as a filterbank. *IEEE Signal Proc. Lett.*, 11(2), 112–114, 2004.
15. Castleman, K. R. *Digital Image Processing*, Prentice Hall, New Jersey, 1996.
16. Russ, J. C. *The Image Processing Handbook, 5th Edition*, CRC Press, Florida, 2006.
17. Martesen, K. M. *Radiographic Image Analysis, 3rd Edition*, Saunders/Elsevier, Pennsylvania, 2010.
18. Nunes, J. C., Guyot, S., and Deléchelle, E. Texture analysis based on local analysis of the bidimensional empirical mode decomposition, *Machine Vision and Applications*, 16(3), Springer-Verlag, 177–188, 2005.
19. Linderhed, A. Compression by image empirical mode decomposition, *IEEE Int. Conf. Image Process.*, 1, 553–556, 2005.
20. Linderhed, A. Variable sampling of the empirical mode decomposition of two-dimensional signals, *Int. J. Wavelets Multi-resolution Inf. Process.*, 3, 435–452, 2005.
21. Linderhed, A. 2D empirical mode decompositions in the spirit of image compression, *Wavelet Independent Component Anal. Appl. IX, SPIE Proc.*, 4738, 1–8, 2002.
22. Nunes, J. C. and Deléchelle, É. Empirical mode decomposition: Applications on signal and image processing, *Advances in Adaptive Data Analysis*, 1(1), World Scientific, Singapore, 125–175, 2009.
23. Linderhed, A. Image empirical mode decomposition: A new tool for image processing, *Advances in Adaptive Data Analysis*, 1(2), World Scientific, Singapore, 265–307, 2009.
24. Bhuiyan, S. M. A., Attoh-Okine, N. O., Barner, K. E., Ayenu-Prah, A. Y., and Adhami, R. R. Bi-dimensional empirical mode decomposition using various interpolation techniques, *Advances in Adaptive Data Analysis*, 1(2), World Scientific, Singapore, 309–338, 2009.
25. Wu, Z., Huang, N. E., and Chen, X. The multi-dimensional ensemble empirical mode decomposition method, *Advances in Adaptive Data Analysis*, 1(3), World Scientific, Singapore, 339–372, 2009.
26. Long, S. R. NASA Wallops Flight Facility Air-Sea Interaction Research Facility, *NASA Reference Publication*, No. 1277, 29 pp., 1992.
27. Long, S. R. Applications of HHT in image analysis, *Hilbert-Huang Transform and Its Applications*, Interdisciplinary Mathematical Sciences, 5, World Scientific, Singapore, 289–305, 2005.
28. Long, S. R., Lai, R. J., Huang, N. E., and Spedding, G. R. Blocking and trapping of waves in an inhomogeneous flow, *Dynam. Atmos. Oceans*, 20, 79–106, 1993.
29. Long, S. R., Huang, N. E., Tung, C. C., Wu, M.-L. C., Lin, R.-Q., Mollo-Christensen, E., and Yuan, Y. The Hilbert techniques: An alternate approach for non-steady time series analysis, *IEEE GRSS*, 3, 6–11, 1995.
30. Long, S. R. and Klinke, J. A closer look at short waves generated by wave interactions with adverse currents, *Gas Transfer at Water Surfaces*, Geophysical Monograph 127, American Geophysical Union, 121–128, 2002.
31. Wu, Z. and Huang, N. E. A study of the characteristics of white noise using the empirical mode decomposition method, *Proc. Roy. Soc. London*, A460, 1597–1611, 2004.
32. Huang, N. E. Introduction to the Hilbert-Huang transform and its related mathematical problems, *Hilbert-Huang Transform and Its Applications*, Interdisciplinary Mathematical Sciences, 5, World Scientific, Singapore, 1–26, 2005.

2 Nyquist Pulse-Based Empirical Mode Decomposition and Its Application to Remote Sensing Problems

Arnab Roy and John F. Doherty

CONTENTS

2.1 INTRODUCTION

Empirical mode decomposition (EMD) is an adaptive signal-dependent decomposition with which any complicated signal can be decomposed into a series of constituents. Adding all the extracted constituents together reconstructs the original signal without information loss or distortion. Many methods exist that analyze signals simultaneously in the time and frequency domains, such as those based on wavelets, short-time Fourier transform (STFT), Wigner–Ville distribution (WVD), reduced interference distributions, and so on. These methods are based on the expansion of the signal into a set of basis functions that are defined by the method. The concept of EMD is to expand the signal into a set of functions defined by the signal itself. These decomposed constituents are called *intrinsic mode functions* (IMF). Signal adaptive decomposition by means of principal component analysis (PCA) [1] also expands the signal into a basis defined by the signal itself. PCA differs from EMD in that it is based on the signal statistics, while EMD is deterministic and is based on local properties. The EMD process allows time–frequency analysis of transient signals for which Fourier-based methods have been unsuccessful. Whenever we use the Fourier transform to represent frequencies we are limited by the uncertainty principle. For infinite signal length we can get exact information about the frequencies in the signal, but when we restrict ourselves to analyze a signal of finite length there is a bound on the precision of the frequencies that we can detect.

The instantaneous frequency represents the frequency of the signal at one time, without any information of the signal at other times. A problem with using instantaneous frequency, however, is that it provides a single value at each time. A multicomponent signal consists of many intrinsic frequencies and this is where the EMD is used, to decompose the signal into its IMFs, each with its own instantaneous frequency, so that multiple instantaneous frequencies of the signal components can be computed. Another advantage of EMD is that it results in an adaptive signal-dependent time-variant filtering procedure able to directly extract signal components which significantly overlap in time and frequency [2]. Moreover, the physical meaning of the intrinsic processes underlying the complex signal is often preserved in the decomposed signals. This is mainly due to the fact that the results are not influenced by predetermined bases and/or subband filtering processes.

EMD represents a totally different approach to signal analysis. EMD is an adaptive decomposition with which any complicated signal can be decomposed into a series of constituents. EMD is an analysis method that in many respects gives a better understanding of the physics behind the signals. Because of its ability to describe short-time changes in frequencies that cannot be resolved by Fourier spectral analysis, it can be used for nonlinear and nonstationary time series analysis. Each extracted signal admits well-defined instantaneous frequency. The original purpose for the EMD was to find a decomposition that made it possible to use the instantaneous frequency for time–frequency analysis of nonstationary signals. This chapter focuses on a modified EMD algorithm based on Nyquist pulse interpolation, its advantages, and remote-sensing applications.

2.2 EMPIRICAL MODE DECOMPOSITION

The elementary amplitude-modulated–frequency-modulated (AM–FM)-type signal components that are produced by the EMD procedure are called IMFs in the literature. An extracted component must satisfy two conditions to qualify as an IMF [3]:

1. The number of extrema and the number of zero crossings must be either equal to each other or differ at most by one.
2. The mean value of the envelopes defined by the local maxima and the local minima should be zero at any point, meaning that the functions should be symmetric with respect to the local zero mean.

Each of these IMFs is extracted by a process called *sifting*. The goal of sifting is to remove the higher-frequency components until only the low-frequency components remain. Given a signal $x(t)$, the sifting procedure divides it into a high-frequency detail, $d(t)$, and the low-frequency residual (or trend), $m(t)$, so that $x(t) = m(t) + d(t)$. This detail becomes the first IMF and the sifting process is repeated on the residual, $m(t) = x(t) - d(t)$. After K iterations of the sifting procedure the input signal can be represented as follows:

$$x(t) = \sum_{k=1}^{K} y_k(t) + m_K(t) \tag{2.1}$$

where $y_k(t)$, $k = 1, \ldots, K$, represent the IMFs and $m_K(t)$ is the residual, or the mean trend, after K sifting iterations. The effective algorithm of EMD can be summarized as follows [2]:

1. Identify all extrema of $x(t)$.
2. Interpolate between minima (resp. maxima), resulting in the envelope $e_{\min}(t)$ (resp. $e_{\max}(t)$).

3. Compute the mean $m(t) = (e_{\min}(t) + e_{\max}(t))/2$.
4. Extract the detail $d(t) = x(t) - m(t)$.
5. If $d(t)$ satisfies all IMF conditions, then set $y_1(t) = d(t)$, the first IMF; else, repeat the above steps with $d(t)$.
6. Evaluate the residual $m_1(t) = x(t) - y_1(t)$.
7. Iterate on the residual $m_1(t)$.

Steps 1 through 4 may have to be repeated several times until the detail $d(t)$ satisfies the IMF conditions. Practical methods to determine if $d(t)$ satisfies the IMF conditions, also called stopping criteria, are discussed next. In the original work [3], the sifting procedure for a particular IMF stops when the normalized difference in the extracted signal between two consecutive iterations is smaller than a predetermined threshold ε. A new stopping criterion was suggested in Ref. [4] where the iterations stop when the envelope mean signal is close enough to zero ($|m(t)| < \varepsilon$, $\forall t$). The reason for this choice is that forcing the envelope mean to zero will guarantee the symmetry of the envelope and the correct relation between the number of zero crossings and number of extremes that define the IMF. A modified version of this stopping criterion with two thresholds was introduced in Ref. [2], along with a discussion of typical threshold values. Yet another stopping criterion was introduced in Ref. [5] where sifting is stopped when the number of extrema and zero crossings remains constant over some predetermined number of iterations. The latter is the most commonly used criterion.

2.3 NYQUIST PULSE INTERPOLATION

The choice of cubic spline interpolation in EMD has been popular due to its reasonable performance and availability of computationally efficient software routines. Here, a new algorithm using Nyquist pulse interpolation is introduced. In communications theory, Nyquist's condition for distortionless transmission of a bandlimited signal is that [6]

$$
\begin{aligned}
p(0) &= 1 \\
p(nT) &= 0, \quad n = \pm 1, \pm 2, \dots
\end{aligned}
\tag{2.2}
$$

where $p(t)$ is a signaling pulse and T is the time duration between successive symbols. This condition guarantees that a sequence of pulses sampled at the optimum, uniformly spaced sampling instants, $n = 0, \pm 1, \pm 2, \dots$ will have zero intersymbol interference (ISI). Nyquist showed that pulses satisfying a vestigial sideband criterion, namely, that the pulse spectrum has odd symmetry about the corresponding ideally bandlimited spectrum band edge, will have this property. There is an infinite number of such pulses corresponding to different vestigial sidebands (see Refs. [7,8] for Nyquist pulse examples). Perhaps the most widely employed Nyquist pulse is the raised cosine pulse (of which the sinc pulse is a special case). Raised cosine interpolation has several advantages:

1. A finite impulse response (FIR) filter realization of the raised cosine filter simplifies hardware implementation.
2. Use of fast Fourier transform (FFT) ensures computationally efficient implementation.
3. Frequency resolution of the EMD technique can be controlled via external parameter.

The time- and frequency-domain expressions of the raised cosine pulse are [9]

$$
h(t) = \text{sinc}\left(\frac{\pi t}{T}\right) \frac{\cos(\pi \beta t / T)}{1 - \left(4\beta^2 t^2 / T^2\right)}
\tag{2.3}
$$

$$H(f) = \begin{cases} T; & |f| \le \dfrac{1-\beta}{2T} \\[2ex] \dfrac{T}{2}\left[1 + \cos\left(\dfrac{\pi T}{\beta}\left(|f| - \dfrac{1-\beta}{2T}\right)\right)\right]; & \dfrac{1-\beta}{2T} < |f| \le \dfrac{1+\beta}{2T} \\[2ex] 0; & \text{otherwise} \end{cases} \qquad (2.4)$$

The roll-off factor, β, is a measure of the excess bandwidth of the filter, that is, the bandwidth occupied beyond the Nyquist bandwidth of $1/2T$. Its value varies between 0 and 1. As one increases the value of β, the pass-band in the frequency domain increases and there is a corresponding decrease in the time-domain ripple level. This shows that the excess bandwidth of the filter can be reduced, but only at the expense of an elongated impulse response and this facilitates control over the performance of the interpolation scheme by the user. Figure 2.1 shows the raised cosine pulse in the time and frequency domains for several roll-off factors.

The EMD algorithm uses cubic spline interpolation. This is replaced by the raised cosine interpolation here and the modified algorithm is called raised cosine empirical mode decomposition (RCEMD). To explain this procedure we refer back to the EMD algorithm in Section 2.2. Step 2 of the algorithm requires evaluating the envelope values at intermediate points between successive extrema. In the original method, the cubic spline constructed of piecewise third-order polynomials is used. In the modified method, this is achieved by convolving the zero-padded sequence containing the extrema with the raised cosine pulse. This can be implemented using an FIR filter in hardware. The simplest member of this family, the sinc pulse ($\beta = 0$), has the added advantage of ideal low-pass frequency characteristics, for an infinite length sequence. However, a consequence of the ideal frequency behavior of the sinc pulse is that a filter implementing the sinc pulse is extremely sensitive to nonuniformity in sampling, resulting in poor performance relative to the cubic spline.

The sinc filter is highly sensitive to nonuniform sampling points due to slow decay of the pulse. The rate of decay of the tails of the raised cosine pulse increases with the roll-off factor and this makes it less sensitive to sampling point errors [9]. However, increasing the roll-off factor also has the effect of increased filter bandwidth, thereby reducing its frequency resolution. So, to summarize, increasing the roll-off factor 0 through 1 has two adverse consequences: it decreases the sensitivity of the filter to sampling point errors, but simultaneously reduces its frequency resolution. In Section 2.5.1, we introduce a two-tone signal model to study the performance of the two interpolating techniques, where it has been previously shown [10] that the uniformity of the spacing

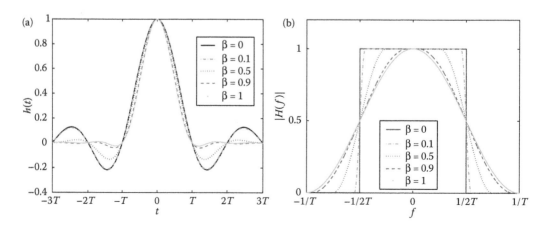

FIGURE 2.1   The roll-off factor dependence of raised cosine pulse in time (a) and frequency (b) domains.

between the sampling points (local maxima or minima) is determined by the ratio $(A_2/A_1)(f_2/f_1) \equiv \Gamma$, where $A_1, A_2, f_1,$ and f_2 represent the amplitudes and frequencies of the two tones. When $\Gamma \ll 1$, the sampling points are nearly equidistant and the choice of raised cosine pulse roll-off factor does not affect the frequency resolution of the algorithm. However, a pulse with a smaller roll-off factor (β close to 0) can resolve the signal components in fewer iterations and should therefore be preferred. When $\Gamma \ll 1$ is not satisfied, the spacings between consecutive local maxima and minima are no longer approximately uniform causing a raised cosine interpolator with small β to fail in signal separation and a larger roll-off factor is required.

FIR filter implementation is a major advantage of raised cosine interpolation. Filter construction requires two parameters: the roll-off factor, β, and the sampling period, T. The roll-off factor, β, is a predefined system parameter affecting frequency resolution of the algorithm. The sampling period, T, on the other hand, which is the mean duration between consecutive maxima (or minima), is a signal-dependent parameter that is estimated on a block-by-block basis as described in Section 2.4. The interpolated envelope is derived by convolving the zero-padded vector containing the maxima (or minima) with the filter tap values. Moreover, frequency-domain implementation reduces the computational complexity by replacing convolution by multiplication. So, an efficient implementation consists of transforming the zero-padded time-domain signals into frequency domain, filtering by the estimated interpolation filter coefficients and subsequent reconversion to time domain via inverse FFT.

2.4   RAISED COSINE EMPIRICAL MODE DECOMPOSITION

Signal interpolation using the raised cosine pulse requires approximately uniform spacing between control points or knots. A large β results in faster decay of the interpolant tails, thereby reducing its sensitivity to nonuniform sampling intervals. However, nonstationary signals result in larger variations in sampling intervals which reduces the effectiveness of a raised cosine pulse with large β as an interpolant. To solve this problem, the original signal is split into small, overlapping segments and interpolation using the raised cosine pulse is applied to each of them individually. By careful choice of the interpolation window length relative to the maximum rate of change of the signal instantaneous frequency, approximately uniform sampling intervals can be ensured.

We now enumerate the steps involved in the RCEMD algorithm. First, various parameters and variables are initialized.

1. Set the roll-off factor β, window shape w, and window size K (corresponding to the number of successive extrema to be included in the window). Further, N and M refer to the data length and the number of maxima or minima in the signal, respectively.
2. Initialize $\tilde{e}_{max}[n]$ and $\tilde{e}_{min}[n]$ to N-length zero-vectors.

Then the main loop of the algorithm is as follows:

1. Identify the extrema of $x[n]$.
2. For $i = 1: M - K$, do the following:
    a. Define index $q = u_i...u_{i+K}$, where u_j represents the position of the jth maxima of $x[n]$.
    b. Compute window coefficients $w[q]$.
    c. Compute the windowed upper envelope $\tilde{x}_{upper}[q] = w[q] \times x[u_i : u_i + K]$.
    d. Compute $T = [\sum_{k=1}^{K-1} x[u_{k+1}] - x[u_k]]/K - 1$.
    e. Compute raised cosine filter coefficients $h[q]$ using predefined β and estimated T.
    f. Compute $\tilde{e}_{max}[q] = \tilde{e}_{max}[q] + \tilde{x}_{upper}[q] * h[q]$.
    g. Compute $\tilde{e}_{min}[q]$ similarly, by first computing $\tilde{x}_{lower}[q]$. (Repeat steps (c–f) for $\tilde{e}_{min}[q]$.)

3. Calculate $m[n] = ((\tilde{e}_{min}[n] + \tilde{e}_{max}[n])/(2))$.
4. Extract the detail $d[n] = x[n] - m[n]$.
5. If $d[n]$ satisfies all IMF conditions, then set $y_1[n] = d[n]$, the first IMF; else repeat the above steps with $d[n]$.
6. Evaluate the residual $x_1[n] = x[n] - y_1[n]$.
7. Iterate on the residual $x_1[n]$.

The proposed algorithm differs from the original EMD algorithm in two crucial ways: first, a new interpolant is used, and second, signal filtering is performed at the local level. In our experiments, we have found that a window spanning five consecutive extrema ($K = 5$) produces good signal resolution for a wide variety of cases and has therefore been used consistently in all simulations. Signal resolution performance comparison with EMD, which is the focus of the following section, ignores boundary condition remediation operations such as signal mirrorizing. In general, the effect of signal boundary on the RCEMD algorithm was not found to be any worse than that for EMD, either in extent or in severity, for the examples considered here, but a detailed analysis of this phenomenon is beyond the scope of this work. Finally, $\beta = 1$ is used in all simulation results presented here to minimize the effect of nonuniform sampling.

2.5 SIGNAL DECOMPOSITION QUALITY OF RCEMD ALGORITHM

2.5.1 COMBINATION OF TONES

Frequency resolution of RCEMD and EMD algorithms for a combination of two tones is studied here. Mathematically, these signals are defined as

$$x(t) = s_1(t) + s_2(t) \tag{2.5}$$

where

$$\begin{aligned} s_1(t) &= A_1\sin(2\pi f_1 t + \phi_1) \\ s_2(t) &= A_2\sin(2\pi f_2 t + \phi_2) \end{aligned} \tag{2.6}$$

and the symbols have their usual meanings and $f_1 > f_2$. Two measures of similarity between the algorithm outputs (IMFs) and original components are defined as follows:

$$\Omega_1^k = \frac{\langle s_1(t)y_1(t)\rangle}{\langle s_1^2(t)\rangle} \tag{2.7}$$

$$\Omega_2^k = \frac{\langle s_2(t)y_2(t)\rangle}{\langle s_2^2(t)\rangle} \tag{2.8}$$

where $y_1(t)$ and $y_2(t)$ represent the IMFs generated after k iterations of the EMD or RCEMD algorithms and $\langle . \rangle$ denotes time-averaging. These quantities assume values between 0 and 1 with large values indicating better signal decomposition quality. Signal resolution quality of the two algorithms for bicomponent signals considered in the present and following examples is based on Ω_2^k. This choice is based on the observation that due to the presence of only two signal components and two extracted IMFs, a strong match between one signal component and a particular IMF implies strong match between the other component and the remaining IMF, due to which Ω_1^k, Ω_2^k or a combined metric (such as their mean) are equivalent measures of signal decomposition quality.

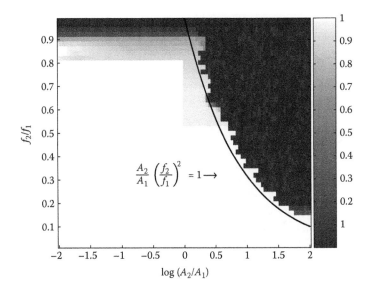

FIGURE 2.2 Final value of the performance metric after 100 iterations (Ω_2^{100}) of the RCEMD algorithm plotted for a range of amplitude and frequency ratios. A curve representing the theoretical limit for successful signal separation by EMD is also shown. (Adapted from G. Rilling and P. Flandrin, *IEEE Trans. Signal Process.* 56(1), 85–95, 2008.)

Here, signal separation performance of the EMD and RCEMD algorithms for combinations of tonal signals is examined based on Ω_2^k. Monte-Carlo simulations are performed by generating synthetic signals according to Equations 2.5 and 2.6, where $A_1 = 1$, $\phi_1 = 0$, and ϕ_2 varies uniformly over $[0,2\pi)$. The frequency resolution quality of the two techniques is studied here for a wide range of amplitude and frequency ratios, A_2/A_1 and f_2/f_1. Figure 2.2 shows the results of this experiment in a format similar to that used in Ref. [10] for EMD. In short, the intensity values in Figure 2.2 represent Ω_2^{100}, the signal separation quality after 100 iterations of the RCEMD algorithm, with a lighter shade representing a larger value (better signal separation quality). It is clear from Figure 2.2 that the transition between the regions of good and poor signal separation quality generally lie to the right of the curve representing the equation $(A_2/A_1)(f_2/f_1)^2 = 1$, which was considered as the theoretical limit for EMD performance [10]. Figure 2.3 shows results of direct comparison of signal separation qualities of the EMD and RCEMD algorithms for a range of signal parameters. Clearly, RCEMD performs better for the more difficult signal separation problem, namely, when the lower tone is stronger and the tones differ in frequency by a small amount.

2.5.2 Two Frequency-Modulated Components

Previously, a combination of pure tones was considered to study the RCEMD algorithm. Here, RCEMD performance is evaluated for a combination of FM signals. The following model allows signal separation quality comparison between EMD and RCEMD for nonstationary components. This is similar to the nonstationary signal model used in the EMD performance evaluation of Refs. [10,11].

The FM signals can be represented as

$$s_p(t) = A_p\cos\left(\omega_c t + k_f \int_{-\infty}^{t} m_p(\tau)\,d\tau + \theta_p\right) \tag{2.9}$$

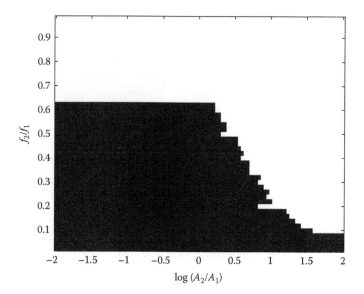

FIGURE 2.3 Differences between the final values of the performance metric after 100 iterations of the two techniques, $[\omega_2^{100}]_{\text{RCEMD}} - [\omega_2^{100}]_{\text{EMD}}$. The differences are quantized to three values: 1, signifying that performance of RCEMD is better ($[\omega_2^{100}]_{\text{RCEMD}} - [\omega_2^{100}]_{\text{EMD}} \geq 0.05$); −1, signifying that performance of EMD is better ($[\omega_2^{100}]_{\text{EMD}} - [\omega_2^{100}]_{\text{RCEMD}} \geq 0.05$); and 0, signifying that the performance of both techniques is about the same ($| [\omega_2^{100}]_{\text{RCEMD}} - [\omega_2^{100}]_{\text{EMD}} | < 0.05$). The regions representing the values 1 and 0 are in white and black, respectively. The value −1 does not appear in this graph.

and

$$m_p(t) = A_{m_p}\cos(\omega_p t + \phi_p) + B_{m_p} \tag{2.10}$$

where $p = 1,2$. Then, the instantaneous frequencies of the signals are given by

$$\omega_{I_p}(t) = \omega_c + k_f m_p(t), \quad p = 1,2 \tag{2.11}$$

In this signal model, each signal component has four adjustable parameters affecting the instantaneous frequency: A_{m_p}, B_{m_p}, ω_p, and ϕ_p. The relation between A_{m_p} and B_{m_p} and their effect on the instantaneous frequency are shown in Figure 2.4. Moreover, ω_p and ϕ_p control the starting phase and the rate of change of the instantaneous frequencies of the two signals, respectively. So, while ω_p determines the starting phase of the instantaneous frequencies in Figure 2.4, their oscillation frequencies depend on ω_p. From the figure it is clear that the parameters A_{m_p} and B_{m_p} determine the relative frequency separation between the components and, consequently, the level of difficulty for signal separation. In our experiments, we fix the parameters of signal $s_2(t)$ and vary those of $s_1(t)$ to achieve different frequency compositions of the component signals. Moreover, ω_p and ϕ_p are identical for the two components to prevent crossing of instantaneous frequencies.

Signal separation quality of the two algorithms is evaluated based on the metric Ω_2^{100} that was introduced earlier. Here, we set the signal parameters such that the the instantaneous frequencies of the two components bear a constant ratio at all times. To achieve this, the parameters A_{m_2}, B_{m_2}, ω_2, and ϕ_2 corresponding to $s_2(t)$ are assigned values first. Then, the instantaneous frequency of $s_1(t)$ is related to that for $s_2(t)$ as $f_{I_1} = \eta f_{I_2}$, where $\eta > 1$ is a constant, that then modulates the carrier signal. Twenty trials of the experiment are performed with different uniformly distributed values of A_{m_2}

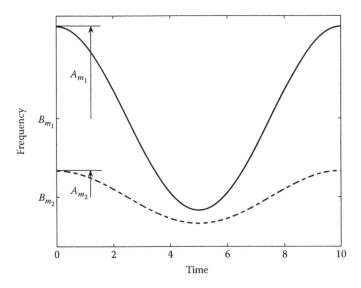

FIGURE 2.4 Instantaneous frequencies of the two frequency-modulated signal components. B_{m_1} and B_{m_2} represent the frequency offsets from the carrier frequency and A_{m_1} and A_{m_2} indicate the frequency spread around the offsets.

and B_{m_2}. The averaged results of this experiment are presented in Figure 2.5. Similar to the previous experiment, this figure shows the performance difference between EMD and RCEMD after 100 iterations of the algorithm. The advantage of the RCEMD technique for large instantaneous frequency ratios (fl_2/fl_1) and large amplitude ratios (A_2/A_1) results from the reduced sensitivity of the raised cosine filter (with $\beta = 1$) to irregular sampling, which is a significant problem for large frequency and amplitude ratios.

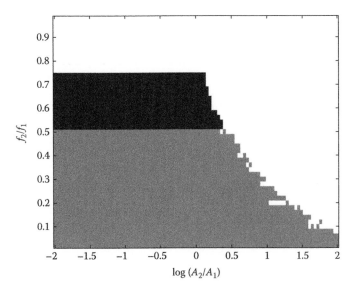

FIGURE 2.5 Similar to Figure 2.3, except that the component signals are frequency modulated in this case. Here, the regions representing the values 1, 0, and –1 are represented by white, gray, and black colors, respectively.

2.5.3 BICOMPONENT TRIGONOMETRIC FUNCTION

Next, we consider an example that has been previously examined in the EMD literature. This involves identifying the components at frequencies f and $3f$ in the signal $\cos^3(2\pi ft)$. Here, the performance of the two algorithms in identifying and isolating the two frequency components of the signal $\cos^3(2\pi ft)$ at frequencies f and $3f$ is compared. The RCEMD and the EMD algorithms are applied to the signal and their outputs are shown in Figure 2.6. The failure of the EMD algorithm to separate these signal components is unexpected considering that here $A_2/A_1(f_2/f_1)^2 = 1/3$. Our experiments indicate that the reason lies in the particular configuration of the starting phases of the components of this signal, which results in nonseparation by EMD.

2.5.4 MULTICOMPONENT SIGNAL

The validity of the RCEMD algorithm when signal has more than two components is demonstrated here. In this example all signal components have overlapping, time-varying instantaneous frequencies that are difficult to separate using traditional filtering techniques. Moreover, two of the signal components also have time-varying amplitudes, meaning that the amplitudes A_p in Equation 2.9 are now time-varying for two of the three signal components (here, $p = 1,2,3$). Successful signal decomposition using EMD for a similar signal has been previously demonstrated [10]. Here, we test the ability of the RCEMD algorithm to decompose the signals. The frequency- and time-domain signals are shown in Figures 2.7 and 2.8, respectively. Although, the signals have been correctly separated into their respective IMFs, some signal mixing is evident in regions of small instantaneous frequencies for this difficult signal separation problem. This is because there are fewer cycles of the signal over these intervals that results in an elongated RCEMD window, the length of which depends on interextrema spacing. The windowed signal no longer has constant instantaneous frequency over the extended interval, resulting in nonoptimum filtering. This effect, which is seen when the signal components have small instantaneous frequencies, is not observed for EMD.

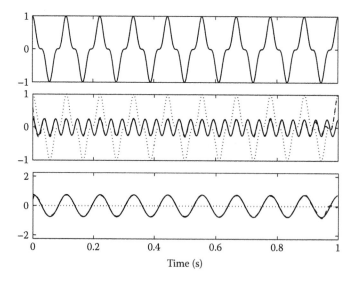

FIGURE 2.6 Signal decomposition quality comparison between RCEMD and EMD algorithms for $\cos^3(2\pi ft)$. The first panel shows the original signal. The other two panels show the signal components at frequencies $3f$ and f, along with the decomposition results from RCEMD (dashed) and EMD (dotted). The dashed line corresponding to RCEMD coincides with the corresponding signal components (in solid) and is therefore not clearly visible in the panels.

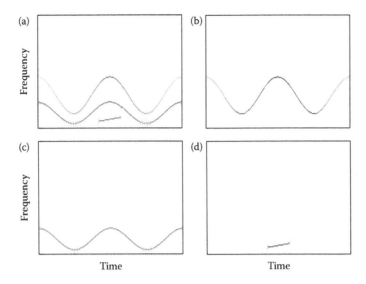

FIGURE 2.7 Frequency-domain signals for the multicomponent signal example. Panels (a), (b), (c), and (d) correspond to the combined signal and the three extracted components using the RCEMD algorithm, respectively.

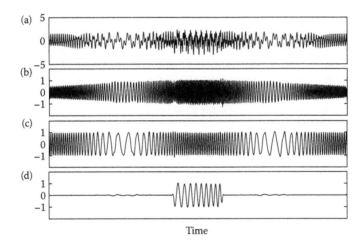

FIGURE 2.8 Time-domain signals for the multicomponent signal example. Panels (a), (b), (c), and (d) correspond to the combined signal and the three extracted components using the RCEMD algorithm, respectively.

2.6 LOW SAMPLING RATE PERFORMANCE OF RCEMD

Successful signal analysis by EMD requires a high degree of oversampling. This requirement arises from the need to precisely identify the local extrema of the signal to recover the signal envelope using natural cubic spline interpolation. However, raised cosine interpolation allows high fidelity reconstruction of the signal envelope even when the local extrema cannot be precisely identified due to low sampling rates. The ability of RCEMD to decompose signals at low sampling rates is demonstrated here using synthetic signals.

For this we consider a combination of FM signals of the form described in Section 2.5.2 with $f_{I_1} = 1.67 f_{I_2}$. This means that the instantaneous frequency of one of the signal components is always larger than the other by a factor of 1.67. We then compare the performance of the two algorithms at

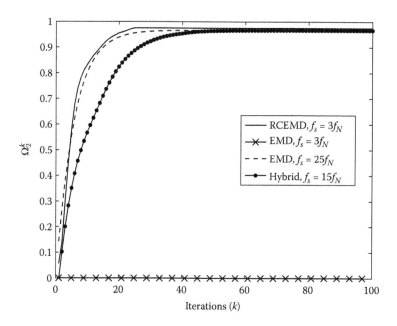

FIGURE 2.9 Signal analysis performance of the RCEMD, EMD, and hybrid techniques for a combination of frequency-modulated signals at different sampling rates.

different sampling rates. Here, a third algorithm, called hybrid algorithm, introduced in Ref. [12] to improve the low sampling rate performance of EMD, is also considered. Figure 2.9 shows the minimum sampling requirements for the three techniques for successful signal analysis. While the sampling rate requirement of the hybrid technique is lower than that of EMD, it is still much larger than RCEMD sampling requirement for successful signal separation.

2.7 REMOTE SENSING APPLICATION

This section describes some applications of the RCEMD algorithm to remote sensing problems. Two examples are considered: tidal component extraction and earthquake signal analysis.

2.7.1 TIDAL COMPONENT EXTRACTION

In this section, we validate the new RCEMD algorithm by applying it to real-world data. In this example, we apply the signal decomposition algorithm to sea-level measurements and expect to see components corresponding to diurnal and semidiurnal tides. Successful isolation of signal components with known physical interpretation is sought in this exercise. Sea-level data obtained from the Intergovernmental Oceanographic Commission database at Ref. [13] was used in this study. We used tide gauge data from Honolulu, Hawaii, spanning ~30 days for signal decomposition. Some signal preprocessing steps were carried out to prepare the data for the subsequent step. First, the 1-min sampled data was downsampled by a factor of 15 to reduce the data length. No useful information is lost in the process because the tidal phenomena occur at much longer timescales. A carefully designed noise filter that has a flat response at tidal frequencies is next applied to the data. The presence of noise causes mode mixing in EMD and related algorithms, and should be minimized before decomposition. This requirement is related to the peak-finding step in the decomposition procedure and is common to both EMD and RCEMD. Finally, the RCEMD algorithm is applied to the filtered data and the results are shown in Figure 2.10. Clean separation into two components—one with an ~12-h period and the other with an ~24-h period is observed with smooth amplitude

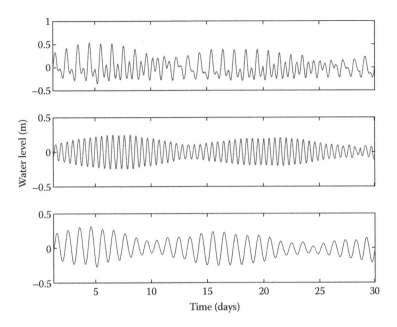

FIGURE 2.10   Application of RCEMD technique to sea-level data. The RCEMD algorithm is applied to the subsampled and noise-removed time series data shown in the top panel and the generated components are shown in the subsequent panels. The first extracted component corresponds to a roughly 12-h period signal and the other to a superimposed variation of period that is twice as long. The reason for the diurnal inequity is due to inclination of the lunar orbit with respect to the earth's equator.

variations in each case corresponding to the shifting configurations of the sun and the moon, the two major planetary bodies affecting sea levels. The diurnal variations arise due to the moon's declination effect (change in angle relative to the equator) and the diurnal variations are themselves amplitude modulated due to the roughly monthly cycle of movement of the moon between the two hemispheres of the earth.

2.7.2   EARTHQUAKE SIGNAL ANALYSIS

The RCEMD algorithm is applied to seismogram data to highlight its unique signal analysis capabilities. We will study the signal recorded for a magnitude 7.1 earthquake that struck the South Island of New Zealand on September 3, 2010 at 16:35 UTC. Publicly available accelerogram data recorded at Wainuiomata, New Zealand from Ref. [14], with sampling rate of 20 sample/s was used here. The time series data is shown in Figure 2.11. Time–frequency representation of this signal in the form of reassigned spectrogram is shown in Figure 2.12. In Figure 2.13, we present the reassigned spectrograms computed individually for the first six IMFs generated using RCEMD algorithm. Signal decomposition by RCEMD and EMD has similar result such that successive IMFs have monotonically decreasing frequency content. This is evident from the time–frequency representations of the IMFs in Figure 2.13 where the time variation of the frequency components is visible. Such a capability allows researchers to study seismological signals with greater detail, which is not possible by studying a time–frequency representation of the composite signal due to the presence of cross-terms from multiple signal components.

2.8   CONCLUSIONS

In this chapter, the use of the Nyquist pulse for EMD was explored. The family of Nyquist pulses, due to zero-ISI property, is ideal for signal interpolation, and can therefore replace cubic spline

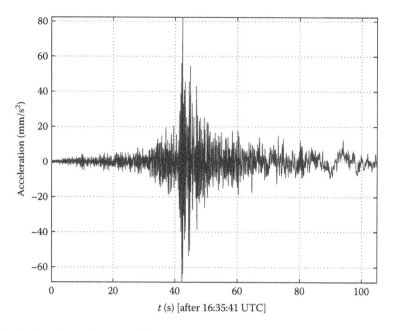

FIGURE 2.11   Earthquake accelerogram data.

interpolation in EMD. Such a modification based on the raised cosine pulse, which is the most popular Nyquist pulse, is introduced here and is called RCEMD. This technique has some powerful advantages over the EMD algorithm, namely, better frequency resolution and lower sampling requirements. These advantages were demonstrated using synthetic signals. Finally, application of RCEMD algorithm to remote sensing problems was considered via two examples, namely, tide component analysis and earthquake signal analysis. The unique capabilities provided by this signal decomposition procedure were highlighted for these real-world applications.

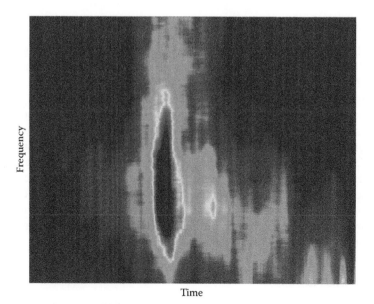

FIGURE 2.12   (**See color insert.**) Time–frequency representation of earthquake accelerogram data.

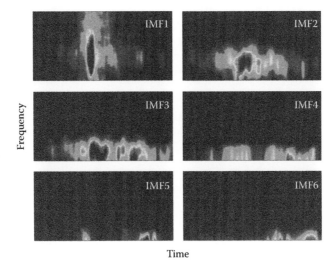

Time

FIGURE 2.13 **(See color insert.)** Time–frequency representations of first six IMFs produced by RCEMD operating on earthquake accelerogram data.

REFERENCES

1. I. T. Jolliffe, *Principal Component Analysis*, Springer Series in Statistics. New York, NY: Springer, 2002.
2. G. Rilling, P. Flandrin, and P. Gonçalvés, On empirical mode decomposition and its algorithms, In *Proc. IEEE-EURASIP Work. Nonlinear Signal Image Process. (NSIP)*, Grado, Italy, June 8–11, 2003.
3. N. E. Huang, Z. Shen, S. R. Long, M. C. Wu, H. H. Shih, Q. Zheng, N.-C. Yen, C. C. Tung, and H. H. Liu, The empirical mode decomposition and the Hilbert spectrum for nonlinear and non-stationary time series analysis, *Proc. R. Soc. Lond. A*, 454(1971), 903–995, 1998.
4. A. Linderhed, Adaptive image compression with wavelet packets and empirical mode decomposition, PhD dissertation, Linkoping, Sweden: Linkoping University, 2004.
5. N. E. Huang, M.-L. C. Wu, S. R. Long, S. S. Shen, W. Qu, P. Gloersen, and K. L. Fan, A confidence limit for the empirical mode decomposition and Hilbert spectral analysis, *Proc. R. Soc. Lond. A: Math. Phys. Eng. Sci.* 459(2037), pp. 2317–2345, 2003. Available online at: http://rspa.royalsocietypublishing.org/content/459/2037/2317.abstract.
6. H. Nyquist, Certain topics in telegraph transmission theory, *Trans. Am Institute Elect. Eng.*, 47(2), 617–644, 1928.
7. N. Beaulieu, C. Tan, and M. Damen, A "better than" Nyquist pulse, *Commun. Lett. IEEE*, 5(9), 367–368, 2001.
8. A. Assalini and A. Tonello, Improved Nyquist pulses, *Commun. Lett. IEEE*, 8(2), 87–89, 2004.
9. J. G. Proakis, *Digital Communications*, 4th ed. New York, NY: McGraw-Hill, 2000.
10. G. Rilling and P. Flandrin, One or two frequencies? The empirical mode decomposition answers, *IEEE Trans. Signal Process*, 56(1), 85–95, 2008.
11. Y. Kopsinis and S. McLaughlin, Investigation and performance enhancement of the empirical mode decomposition method based on a heuristic search optimization approach, *IEEE Trans. Signal Process*, 56(1), 1–13, 2008.
12. Z. Xu, B. Huang, and F. Zhang, Improvement of empirical mode decomposition under low sampling rate, *Signal Process.*, 89(11), 2296–2303, 2009.
13. IOC Sea Level Monitoring Facility, 2011. Available online at: http://www.ioc-sealevelmonitoring.org.
14. GeoNet M 7.1, Darfield (Canterbury), September 4, 2010, 2011. Available online at: http://www.geonet.org.nz/earthquake/historic-earthquakes/top-nz/quake-13.html.

3 Hydroacoustic Signal Classification Using Support Vector Machines*

Matthias Tuma, Christian Igel, and Mark Prior

CONTENTS

3.1 INTRODUCTION

While the fundamental scientific principles underlying many of the Earth-monitoring systems in operation today have been well known for decades, sensor hardware and sensor deployment are

---

* The views expressed herein are solely those of the authors and do not necessarily reflect the views of the CTBTO Preparatory Commission.

steadily evolving further. Simultaneously, advances in data transmission and storage leave their own and distinct marks on the field. Today, vast amounts of raw data are routinely being collected and transferred. One key challenge on the receiving end is to reliably extract knowledge that is both meaningful and yet sufficiently condensed, in order to facilitate human (or other) interpretation of the information flow. This especially holds for systems performing real-time monitoring. In general, we obtain information about real-world processes in the form of nonlinear, noisy, multidimensional signals. Both the underlying phenomenon as well as all elements of the propagation path are generally too complex to be fully captured by a physical model. Therefore, best practice data processing and analysis rules have to be inferred from observations and validated empirically. Here, *machine learning* comes into play, which is a branch of computer science and applied statistics covering software that improves its performance on a given task based on sample data or experience. This chapter considers *supervised learning* for classification, which refers to automatically assigning signals to predefined categories.

In the supervised learning scenario, the learning machine is provided with sample input–output pairs during a preparational training phase. The learner's task then is to infer from the training data a function that relates any admissible input (not only those seen during training) to a corresponding output. Such a function is called the learner's hypothesis. We demand that all examples used for training as well as later in operations be generated independently of each other by the same probability distribution. Relying on this assumption, the learner should choose a hypothesis which with high confidence will perform accurately on unseen test data. One central challenge in any supervised learning task is to settle a trade-off between on the one hand well fitting the hypothesis to the training examples, and on the other hand hedging against *overfitting*, which occurs when overly adapting to misleading peculiarities of the training data. Section 3.2 formally introduces the supervised learning task, regularized risk minimization, and support vector machines (SVMs) for classification. We introduce SVMs step by step from linear classification to the full kernelized case. This chapter emphasizes aspects deemed relevant to practitioners and at the same time strives to include a reasonably systematic overview over regularized risk minimization, SVM classification, and SVM model selection. The application of these concepts is illustrated through a classification task in acoustic remote sensing. We learn to discriminate signals stemming from the verification network for the Comprehensive Nuclear-Test-Ban Treaty (CTBT). The CTBT's permanently installed International Monitoring System (IMS) consists of several hundred geophysical sensors and relies on four different monitoring technologies. In particular, we are concerned with distinguishing explosive-like and nonexplosive signals recorded by the IMS underwater sensor network. Section 3.3 provides background on the IMS as well as preprocessing routines relevant to the application. We then approach the actual problem of CTBT hydroacoustic signal classification by drawing on the generic concepts established earlier. Section 3.4 concludes this chapter.

3.2 SUPERVISED LEARNING AND SUPPORT VECTOR MACHINE CLASSIFICATION

We next formalize the supervised learning and classification task. In the standard setting, we have obtained training data S from the same data-generating process to which we intend to apply the trained learning machine. The data S consists of N exemplary input–output pairs $(x_i, y_i) \in X \times Y$, $1 \leq i \leq N$. The input and output domains X and Y can in general be any (nonempty) set. Given S, the output of a learning machine is a prediction function or hypothesis $h: X \to Y$ from the input set X to the output set Y. For each possible input $x \in X$ we present to h, it will yield its prediction $h(x) \in Y$ of what output value should most likely be associated with x. It is not sufficient to learn the training data by heart. We rather want h to perform as well as possible for the entirety of input values in X, and especially for those that occur often. When we know the true label y of a sample x, we can compare it to the learner's prediction $h(x)$. Feeding both to a *loss function L*

$$L : \mathcal{Y} \times \mathcal{Y} \to \mathbb{R}_0^+,$$
$$\text{with the property: } L(h(x), y) = 0 \quad \text{for } h(x) = y, \tag{3.1}$$

assigns a "cost" to predicting $h(x)$ when the true label is y. Formally, the goal of supervised learning may now be expressed as finding a function h that minimizes the overall cost, or *risk* $\mathcal{R}_p$, when evaluated over the entire probability distribution $p(x, y)$ underlying our data-generating process:

$$h = \underset{\hat{h} \in \mathcal{H}}{\arg\min} \, \mathcal{R}_p(\hat{h}) = \underset{\hat{h} \in \mathcal{H}}{\arg\min} \int_{\mathcal{X} \times \mathcal{Y}} L(\hat{h}(x), y) \, \mathrm{d}p(x, y), \tag{3.2}$$

where $\mathcal{H}$ would ideally be the space of all (measurable) functions mapping from $\mathcal{X}$ to $\mathcal{Y}$. A hypothesis minimizing Equation 3.2 is called a *Bayes optimal* solution, the according risk the *Bayes risk* of p, and both depend on p and the loss function employed. Note that the Bayes risk is not necessarily zero. This is easy to see for finite $\mathcal{X}$. If an input pattern $x \in \mathcal{X}$ can belong to two different classes—$p(x, y_1) > 0$ and $p(x, y_2) > 0$ for different $y_1, y_2 \in \mathcal{Y}$—the best possible hypothesis still can map x only to one class and will inevitably make mistakes. In practice, nonzero Bayes risk frequently occurs in the case of noisy input signals, signals describing the underlying process incompletely, or uncertain labels.

Clearly, if we knew the underlying distribution p, we would have complete knowledge about the data-generating process. But p usually is unknown. Thus, for all practical purposes the overall risk $\mathcal{R}_p$ can neither be computed nor optimized directly, even if the integrals in Equation 3.2 were tractable. One step toward optimizing Equation 3.2 is to replace $\mathcal{R}_p$ by the equivalent quantity restricted to the training data. This defines the *empirical risk* $\mathcal{R}_S$ on S,

$$\mathcal{R}_S(\hat{h}) = \frac{1}{N} \sum_{i=1}^{N} L(\hat{h}(x_i), y_i), \tag{3.3}$$

which is simply the average loss on the training data. A minimizer of $\mathcal{R}_S$ is a hypothesis as consistent with the training examples as possible. Minimum empirical risk can, for example, be achieved by a function that reproduces the labels of all training examples and merely returns one single, arbitrary output for all other possible inputs. Evidently, this would be a poor prediction function, and a better objective than to just minimize Equation 3.3 over all functions is needed.

3.2.1 REGULARIZED RISK MINIMIZATION

A hypothesis h should not solely reflect peculiarities of the given training data, but work well for examples previously unencountered. What kind of quantity can assist us—without knowing more about the data-generating distribution than S—in automatically deciding which hypothesis may have *overfitted* to the training data and which may *generalize* well to unseen data? It is not only intuitive to look for a simple hypothesis still yielding a reasonably low empirical risk, but a range of according theorems in statistical learning theory (e.g., [34,39]) formalize and justify this concept. These theorems typically provide bounds on the risk in the following form. If, of all functions in a certain function or hypothesis space $\mathcal{H}$, h is the minimizer of $\mathcal{R}_S$, then with probability of at least $1 - \delta$ it holds that $\mathcal{R}_p(h) \leq \mathcal{R}_S(h) + B(N, \delta, \mathcal{H})$. The function B bounds the extent to which the true risk might exceed the empirical risk. An increase in both the number of training examples N and the uncertainty δ leads to a decrease in B. Most importantly, B increases with the complexity of $\mathcal{H}$. We argue that any function $B(N, \delta, \mathcal{H})$ which permits inequalities like the above provides a measure of the complexity of a function space $\mathcal{H}$—and therefore simplicity of a hypothesis class can be (nonuniquely) defined in precise ways. Such theorems confirm that if we permit the minimizer of $\mathcal{R}_S$ to stem from a highly complex class, this expressive power might be exploited for overfitting on the training data rather than be helpful in producing better hypotheses. We thus want to enforce the preference that if the learning machine draws from a more complex hypothesis class, it should have a very good justification in terms of—sufficiently—high associated decrease in training error. The hypothesis spaces considered in this chapter can be endowed with a norm $\| \|_{\mathcal{H}}$ serving as a measure of complexity of a hypothesis. Then we can express the aforementioned trade-off within the *regularized risk minimization* paradigm, in which h is found by minimizing the regularized risk $\mathcal{P}_S$,

$$\mathcal{P}_S(\hat{h}) = \|\hat{h}\|_{\mathcal{H}^2} + C\sum_{i=1}^{N} L(\hat{h}(x_i), y_i). \tag{3.4}$$

Here, $C \in \mathbb{R}^+$ is the so-called *regularization parameter* and balances the preference for low training error (right summand) against keeping the hypothesis simple (left summand), where complexity is assumed to correlate to the norm in $\mathcal{H}$, cf. Refs. [34,39].

3.2.2 SUPPORT VECTOR MACHINES

In general, a multitude of supervised learning algorithms exist, which may or may not fall into the framework of regularized risk minimization presented in the previous section. In this chapter, we focus on SVMs [7, 29], which are most commonly used for classification, but also applicable to regression and density estimation tasks. SVMs can be seen as composed of building blocks from originally different areas of research and, for example, are linked to functional analysis and convex optimization. Next, we introduce SVMs step by step as composition of concepts, namely straightforward linear classification seeking for large separating margins; allowing for margin violations in a regularized risk minimization framework; and nonlinear classification via kernel functions, which replace the scalar product in the original input space by a scalar product in another, unrealized feature space.

3.2.3 LINEAR CLASSIFICATION

From now on we assume, for clarity, that all inputs are represented by an m-dimensional, real-valued feature vector $x \in \mathbb{R}^m = \mathcal{X}$. Further, we restrict our considerations to two-class or binary classification and set $\mathcal{Y} = \{-1, 1\}$. A two-class hypothesis function $h: \mathcal{X} \rightarrow \{-1, 1\}$ with a linear decision boundary can be realized through an affine linear *decision function* $f: \mathcal{X} \rightarrow \mathbb{R}$, $f(x) = \langle x, w \rangle + b$, by taking the sign of f:

$$h(x) = \mathrm{sgn}(f(x)) = \mathrm{sgn}\left(\langle x, w \rangle + b\right). \tag{3.5}$$

Here, the weight vector $w \in \mathcal{X}$ lies in the same space as the input data, $b \in \mathbb{R}$ is a real-valued offset term, and sgn() is the standard signum function except for an argument of zero, in which case it returns $+1$. The decision boundary is the subspace of all points x for which $\langle x, w \rangle + b = 0$. For two-dimensional input, this subspace is a line, and a hyperplane of dimension $m - 1$ in general. The vector w is perpendicular to the decision boundary. With an offset or bias term of zero, $b = 0$, the decision surface passes through the origin. For $b \neq 0$, the decision surface is shifted from 0 by a distance of $\frac{b}{\|w\|}$.

The quantity $y_i(\langle w, x_i \rangle + b) = y_i f(x_i)$ is positive if the ith pattern (x_i, y_i) is classified correctly by the hypothesis sgn($f(x)$). We call this quantity the *functional margin* of (x_i, y_i) with respect to the linear decision boundary induced by (w, b). The *geometric margin* of (x_i, y_i) with respect to the linear decision boundary induced by (w, b) is given by $y_i f(x_i)/\|w\|$. The absolute value of the geometric margin is the distance of x_i from the decision boundary in the input space. A collection S of N two-class data points is called linearly separable if there exists a linear classifier (w, b) separating both classes without error. This implies $y_i(\langle w, x_i \rangle + b)/\|w\| \geq \rho > 0$ for all $1 \leq i \leq N$. The largest ρ for which this holds true defines the geometric margin of the linear classifier (w, b) with respect to S. Its value is determined by the data point having the smallest margin. In the following, we will not explicitly distinguish between functional and geometric margin when the meaning is clear from the context.

3.2.4 LINEAR SUPPORT VECTOR MACHINES

We first introduce large-margin SVM classification for linearly separable data. Figure 3.1 shows a separable two-class problem in two dimensions, and in the upper left a "barely separating"

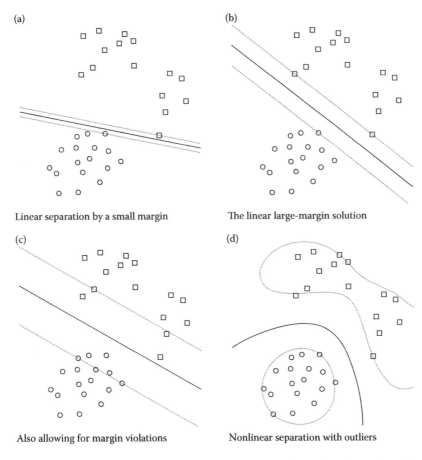

(a) Linear separation by a small margin

(b) The linear large-margin solution

(c) Also allowing for margin violations

(d) Nonlinear separation with outliers

FIGURE 3.1 Example of linearly separable two-class data, and different hypotheses for a discrimination boundary between them. (a) A hypothesis (solid line) that might have been proposed by an algorithm not maximizing the margin (distance to dotted lines) between samples and the discrimination boundary. (b) A maximum-margin solution found by the linear hard-margin SVM of Equation 3.6. We can expect the solution in (a) to be less reliable when classifying data from the same generating distribution. (c) The hypothesis produced by a soft-margin SVM using $C > 0$ in Equation 3.7. Note that its different slope and position could not have been reached in Equation 3.6. (d) A curved decision surface obtained by solving the canonical, nonlinear SVM optimization problem (Equation 3.10), using $C > 0$ and a radial basis function kernel (cf. Section 3.2.5) with $\gamma > 0$. While (a) clearly is not a large-margin classifier, all three hypotheses in (b) through (d) are valid solutions to the full SVM optimization problem. Solution (b) can be seen as a special case of (c) with a strong preference against margin violations expressed through a large regularization parameter. Both linear solutions are further special cases of the nonlinear one, using the scalar product in the original feature space as kernel function. In other words, (b) through (d) have all been obtained by solving the SVM optimization problem (Equation 3.10) for different choices of the regularization parameter C and kernel function k.

hyperplane. Intuitively, we see that this hypothesis is quite vulnerable, as already little noise on samples close to the boundary would lead to their misclassification. Therefore, *hard margin* SVMs for separable training data S yield a hypothesis for which the smallest margin of a training data point—the distance between the decision hyperplane and the closest data point in S—is maximal. It can be shown that solving the optimization problem

$$\text{minimize}_{w,b} \quad \frac{1}{2}\langle w, w \rangle$$
$$\text{subject to} \quad y_i\left(\langle w, x_i \rangle + b\right) \geq 1, \quad 1 \leq i \leq N,$$

(3.6)

leads to a decision function with a geometric margin of $\frac{1}{\|w\|}$, so that the objective of problem (3.6) exactly ensures a hypothesis with maximum margin [29]. The term "hard margin" refers to the fact that the constraints in Equation 3.6 strictly enforce a functional margin of at least one for each training pattern, which implies correct classification of all examples. Applied to the previous example in Figure 3.1, the large-margin objective (Equation 3.6) generates the hypothesis shown in the upper right. In practice, few datasets are linearly separable in the input space. But even for those, it may be beneficial to allow for misclassification of training patterns if in turn a more appealing hypothesis can be established with respect to the overall data. Obviously, if the Bayes risk is nonzero and the pool of training data sufficiently large, the Bayes optimal hypothesis makes errors on the training data. We thus want to allow for *margin violations*, that is, for training patterns (x_i, y_i) for which $y_i(\langle w, x_i \rangle + b) < 1$. This includes misclassified patterns, for which $y_i(\langle w, x_i \rangle + b) < 0$ indicates that they lie on the "wrong" side of the hyperplane. We, for these reasons, return to the concept of regularized risk minimization established in Section 3.2.1. To make usable an objective of the form of Equation 3.4, a loss function $L(h(x), y)$ has to be chosen. For classification we would ideally like to use the 0-1-loss, which returns 0 for correct classification, $h(x) = y$, and 1 otherwise. Incorporating the nonconvex 0-1-loss into Equation 3.6 would, however, complicate the optimization procedure by voiding convexity of the overall problem. For this reason, SVMs commonly rely on the *hinge loss* $L(f(x), y) = \max(0, 1 - yf(x))$ defined on the SVM decision function f as a convex *surrogate loss function*. The hinge loss also is the tightest convex upper bound on the 0-1-loss. Introducing the possibility for margin violations into Equation 3.6 and penalizing them by the hinge loss yields

$$\text{minimize}_{\xi,w,b} \quad \frac{1}{2}\langle w, w \rangle + C\sum_{i=1}^{N} \xi_i$$
$$\text{subject to} \quad y_i\left(\langle w, x_i \rangle + b\right) \geq 1 - \xi_i, \quad 1 \leq i \leq N$$
$$\xi_i \geq 0, \quad 1 \leq i \leq N.$$

(3.7)

This is the linear *soft margin* SVM optimization problem. Each variable $\xi_i \geq 0$ measures the margin violation of pattern (x_i, y_i). Their sum accounts for all violations of the separability paradigm by exactly the sum of the corresponding hinge losses, which in turn is penalized within the objective function. The lower left of Figure 3.1 shows the result of solving Equation 3.7 for $C > 0$. While the difference on the toy dataset is not drastic, it exemplifies the fact that Equation 3.7 allows for solutions not reachable through Equation 3.6. The hard-margin solution can however still be obtained by letting C tend to infinity. We are now only an additional step short of obtaining the canonical nonlinear SVM formulation as we will also employ in the experimental Section 3.3.

3.2.5 KERNELIZED SUPPORT VECTOR MACHINES

Solving Equation 3.7 will yield a regularized large-margin hypothesis, however always using a linear decision function. This can be a disadvantage, for example, when imagining a two-dimensional classification problem in which samples of one class all lie within a circle around the origin, and samples of the second class surround them in a ring-like fashion, as illustrated on the left of Figure 3.2. SVMs incorporate nonlinear hypotheses by implicitly transforming the input data to an unrealized, possibly high- or infinite-dimensional dot product space via kernel functions. In this feature space, different from the input space, SVMs perform linear classification. This, in general, gives rise to nonlinear decision surfaces in the original input space. From another angle, one might imagine subjecting the input data to a nonlinear transformation and only then solving Equation 3.7 for the transformed input. In the right of Figure 3.2, the result of carrying out such a nonlinear transformation from the input space into a higher-dimensional space is shown. If the new feature space, however, is of high dimension, carrying out computations dimension-wise is time consuming, and the transformation might be as well. Instead, we use the fact that a solution w of Equation 3.7 admits a representation of the form $w = \sum_i^N \alpha_i x_i$ with $\alpha \in \mathbb{R}^N$ [9,31]. In other words, it is guaranteed that the solution is a linear expansion of the training examples.

Substituting into Equation 3.7 yields

$$\text{minimize}_{\xi,\alpha,b} \quad \frac{1}{2}\sum_{i,j=1}^{N}\alpha_i\alpha_j\langle x_i,x_j\rangle + C\sum_{i=1}^{N}\xi_i$$

$$\text{subject to} \quad y_i\left(\sum_{j=1}^{N}\alpha_j\langle x_j,x_i\rangle + b\right) \geq 1-\xi_i, \quad 1\leq i\leq N \tag{3.8}$$

$$\xi_i \geq 0, \quad 1\leq i\leq N.$$

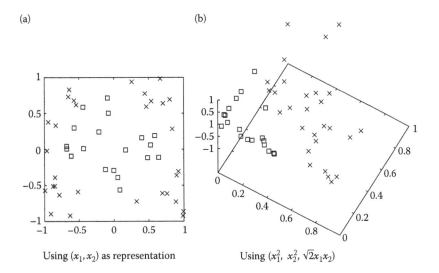

(a) (b)

Using (x_1, x_2) as representation   Using $(x_1^2, x_2^2, \sqrt{2}x_1x_2)$

FIGURE 3.2 Example of how an embedding into another feature space can turn a linearly nonseparable dataset into a linearly separable one. Squares and crosses indicate examples of two different classes. (a) Samples are represented by their raw input coordinates (x_1, x_2). (b) A nonlinear feature map $\phi: \mathbb{R}^2 \to \mathbb{R}^3$ is used to change the representation from (x_1, x_2) to $(x_1^2, x_2^2, \sqrt{2}x_1x_2)$.

Looking at Equation 3.8 we see that it is solely formulated in terms of scalar products between training examples. Where the examples mapped to another dot product space $\mathcal{F}$ by the map ϕ: $\mathbb{R}^m \to \mathcal{F}$ before solving Equation 3.8, each occurrence of $\langle x_i, x_j \rangle$ would simply have to be replaced by $\langle \phi(x_i), \phi(x_j) \rangle_F$. In order to allow for efficient computation of the dot product in $\mathcal{F}$, we can use a kernel function k: $\mathbb{R}^m \times \mathbb{R}^m \to \mathbb{R}$ with $k(x_i, x_j) = \langle \phi(x_i), \phi(x_j) \rangle_{\mathcal{F}}$. More importantly, one can proceed the other way around and specify a kernel k in order to solve Equation 3.8 in another dot product space. The theory of reproducing kernels [1,2,29] establishes the requirements k has to fulfill in order to be sure that substituting $k(x_i, x_j)$ for the scalar product will actually correspond to solving Equation 3.8 in some valid dot product space. The only condition on a symmetric function k is that k must be *positive definite*, in the sense that for every collection of points from $\mathcal{X}$, the matrix K of kernel entries between these points, $K_{ij} = k(x_i, x_j)$, must be positive definite. In detail, for a nonempty input set $\mathcal{X}$ and a function k: $\mathcal{X} \times \mathcal{X} \to \mathbb{R}$, the following holds:

$$\forall x, z \in \mathcal{X} : k(x,z) = k(z,x) \quad \text{and} \quad \forall n \in \mathbb{N}, x_1, \ldots, x_n \in \mathcal{X}, c_1, \ldots, c_n \in \mathbb{R} : \sum_{i,j=1}^{n} c_i c_j k(x_i, x_j) \geq 0$$

$$\Rightarrow \exists (\mathcal{F}, \phi : \mathcal{X} \to \mathcal{F}) \forall x, z \in \mathcal{X} : k(x,z) = \langle \phi(x), \phi(z) \rangle_{\mathcal{F}}. \tag{3.9}$$

Equation 3.9 states an equivalence between a kernel function k and a scalar product in some dot product space $\mathcal{F}$ as long as k is symmetric and positive definite. Subsequently, we list commonly used families of positive-definite kernels. Completing the kernelization of the SVM optimization problem, we state the final objective function which allows for both misclassified training examples and nonlinear decision surfaces using kernel functions:

$$\text{minimize}_{\xi, \alpha, b} \quad \frac{1}{2} \sum_{i,j=1}^{N} \alpha_i \alpha_j k(x_i, x_j) + C \sum_{i=1}^{N} \xi_i \tag{3.10}$$

$$\text{subject to} \quad y_i \left(\sum_{j=1}^{N} \alpha_j k(x_j, x_i) + b \right) \geq 1 - \xi_i, \quad 1 \leq i \leq N$$

$$\xi_i \geq 0, \quad 1 \leq i \leq N.$$

This constitutes the canonical SVM optimization problem [7]. The decision function f of an SVM is linear in the kernel-induced feature space. In the original input space, the SVM's final hypothesis h takes the form

$$h(x) = \text{sgn}(f(x)) = \text{sgn}\left(\sum_{i=1}^{N} \alpha_i k(x, x_i) + b \right). \tag{3.11}$$

The lower right of Figure 3.1 shows a classifier obtained from Equation 3.10 using $C > 0$ and a radial basis function kernel of $\gamma > 0$ (see below). The solution would differ considerably for different values, and even more so for other kernel function families. The final SVM hypothesis does not depend on correctly classified training examples with a distance to the decision hyperplane larger than the safety margin. Their coefficients α_i will hence be zero. All samples with nonzero coefficients α_i will lie on the margin or violate it (i.e., have $y_i f(x_i) = 1$ or $y_i f(x_i) < 1$, respectively). These are called *support vectors*, and only they contribute to the sum in Equation 3.11. This *sparsity property* of SVMs reduces the computational burden when evaluating the hypothesis on unseen examples.

3.2.5.1 Kernel Functions

Question arises as how to choose a proper kernel function. We first introduce two families of kernels commonly used for real-valued input vectors.

Polynomial kernel. For a nonnegative, real-valued offset parameter c and a positive integer exponent d, the function $k(x, z) = (\langle x, z \rangle + c)^d$ is a kernel on real-valued input vectors $x, z \in \mathbb{R}^m$. For $c = 0$ and $d = 1$, it reduces to the standard scalar product.

Gaussian kernel. A general Gaussian kernel on real-valued input vectors $x, z \in \mathbb{R}^m$ is given by $k(x,z) = e^{-(x-z)^T Q(x-z)}$, where Q is a positive-definite $m \times m$ matrix. The most common variant is the radial basis function (RBF) kernel using a positive scaling of the identity matrix I, $Q = \gamma I$, $\gamma \in \mathbb{R}^+$, which introduces γ as single free parameter. A theoretically appealing property of RBF kernels is that they fulfill a necessary condition for an SVM to be *universally consistent* [35]: an SVM using an RBF kernel will under mild conditions converge to the Bayes optimal hypothesis as the collection of training examples grows. Another variant is the automatic relevance detection (ARD) kernel, for which $Q_{ij} = \delta_{ij}\gamma_i$, with Kronecker delta δ. The ARD kernel owes its name to the fact that learning values for the m positive parameters $\gamma_i \in \mathbb{R}^+$ can provide insight into the relevance of individual features for classification. As a drawback, the ARD kernel introduces as many free parameters as there are input space dimensions. However, efficient parameter optimization has been demonstrated for both the ARD and the general Gaussian kernel ([14,15], also see Section 3.2.7).

Individual application domains, for example in biology or natural language processing, can require the use of highly specific and task-tailored kernel functions. Their design and analysis often constitute an active area of research of its own (e.g., [13,30]). Positive-definite kernels further exhibit convenient closure properties in the sense that several operations between kernels again yield a valid kernel [2]. For two kernels k_1 and k_2, their product $k_1 \cdot k_2$ and sum $k_1 + k_2$ is positive definite. Similarly, all following operations on a kernel k retain positive definiteness: scaling by a positive constant to ak, $a \in \mathbb{R}^+$; taking k as exponent e^k; or normalizing k to $\bar{k}(x,z) = k(x,z)/ \cdot$ $\bar{k}(x,z) = k(x,z)/\sqrt{k(x,x)k(z,z)}$. Further, closure under sum and product also imply closure under direct sum and direct product: let $k_a(x, z)$ be a kernel on $A \times A$ and $k_b(u, w)$ a kernel on $B \times B$. Then, kernels $k((x, u), (z, w))$ on $(A \times B) \times (A \times B)$ are both given by $k_a(x, z) \cdot k_b(u, w)$ and $k_a(x, z) + k_b(u, w)$. This, for example, is useful when working with combinations of features from different domains.

Convolution kernels. Not a standard choice of kernel function as such, we list Haussler's convolution kernel* [18] in preparation for an application in Section 3.3. Suppose the input space $\mathcal{X} = \mathbb{R}^m$ can be divided into a product of g subspaces. For simplicity, we assume that these subspaces are equivalent and write $\mathcal{X} = S^g$. This yields a partitioning of an input vector x into g subvectors $x^{(i)} \in S = \mathbb{R}^{\left(\frac{m}{g}\right)}$, $1 \leq i \leq g$, of equal length. Further, assume that we have g corresponding subkernels $k^{(i)}: S \times S \to \mathbb{R}$ defined on the subspace S. By the above composition rules we can construct a composite kernel on $\mathcal{X} \times \mathcal{X}$ by, for example, adding all subkernels, $k(x,z) = \sum_i k^{(i)}(x^{(i)}, z^{(i)})$, or multiplying them, $k(x, z) = \prod_i k^{(i)}(x^{(i)}, z^{(i)})$. The convolution or ANOVA kernel k_D of order D, $D \in \{1, \ldots, g\}$, generalizes from these two exemplary compositions by viewing both as sums over all possible monomials (multiples excluded) of degree D:

$$k_D(x,z) = \sum_{1 \leq j_1 < \cdots < j_D \leq g} \prod_{d=1}^{D} k^{(j_d)}(x^{(j_d)}, z^{(j_d)}). \tag{3.12}$$

Here, the sum runs over all possibilities to draw unique subsets of size D from $\{1, \ldots, g\}$. Clearly, the convolution kernel k_D is the direct sum kernel for $D = 1$ and the direct product kernel for $D = g$. In general, k_D yields the sum of all monomials (multiples excluded) of order D.

---

* The special form of convolution kernel as we consider here has also been studied in [28,39] as ANOVA kernels. We refer the reader to Haussler's overarching construction as the most systematic one.

3.2.6 Support Vector Machine Optimization

One of the canonical approaches to SVM optimization, that is, to solving problem 3.10, is to derive the corresponding *dual program* via Lagrange multipliers [7,39]. The resulting constrained quadratic optimization problem would be solvable using off-the-shelf methods, but highly efficient tailored methods have been derived. Decomposition methods for SVMs [21,25] are iterative algorithms that operate on a fixed subset of (usually two) variables per iteration. It is generally accepted that such solvers have a runtime complexity between quadratic and cubic in the number of training examples [21]. When working with large datasets, fast solvers specialized on linear kernels can be employed, at the cost of purely linear decision functions (e.g., [10,33]). Alternatives that also support nonlinear kernels are *online* SVM solvers (e.g., [3]), which usually aim for an approximate solution.

3.2.6.1 Support Vector Machine Execution

In general, the hypothesis put forth by an SVM is *sparse*: it only depends on the fraction of training examples that are support vectors. SVM execution can thus be faster than, for example, naive implementations of nearest-neighbor classification. It has, however, been shown that the number of support vectors itself grows linearly with the number of training examples if the Bayes risk is nonzero [36]. A number of different approaches have been proposed for approximating the final solution when classification speed is important (e.g., [24,32]).

3.2.7 Model Selection

Maybe the most important aspect in SVM usage is that of model selection—the process of choosing the regularization parameter C and a kernel function family as well as values for the kernel parameters. A multitude of methods for hyperparameter selection exist. Most of them optimize an approximation of, bound on, or heuristic substitute for the generalization error, that is, for Equation 3.2 using the 0-1-loss. We present the standard method, grid search on the cross-validation error, together with a gradient-based maximum-likelihood approach better suited for kernels with more than a few parameters.

3.2.7.1 Direct Search

One common estimator of the generalization error is the n-fold cross-validation error. By partitioning the available training data S into n different parts or *folds*, one obtains n possibilities for training a classifier on $n - 1$ parts of S. The average of the n validation errors on each remaining single evaluation fold is the n-fold cross-validation error (CV-n). It can be shown that the CV-n is a slightly biased estimator of the generalization error, and choices of $n = 5$ or $n = 10$ have proven reasonable in practice [17]. As the CV-n is not differentiable, the hyperparameter space is generally probed at multiple locations using some direct (zeroth-order) search heuristic.

Grid search. The most common SVM model selection procedure, grid search, defines a multidimensional grid of points in the hyperparameter search space, where the grid points may, for example, be spaced evenly on a linear or logarithmic scale. For each point on the grid, n SVMs are trained (each on a different union of $n - 1$ folds) using the corresponding hyperparameter vector. The parameter combination yielding lowest CV-n over the entire grid is in turn chosen to train the final classifier, now using all data in S. Due to the curse of dimensionality, CV-n gets prohibitive for more than a few free SVM hyperparameters. Sometimes variations such as nested grid search are employed, where the grid resolution iteratively increases while focusing around the previously best point. Beyond grid search, more elaborate direct search techniques, such as evolution strategies, have successfully been applied to SVM model selection [11,19].

3.2.7.2 Gradient-Based Model Selection

A lot of research has been devoted to developing and evaluating differentiable estimates of, bounds on, or heuristic substitutes for the generalization error (see e.g., [5]). Even if such an objective is not convex (i.e., does not prevent gradient-based optimizers from getting stuck in suboptimal local optima) gradient descent is still appealing. First, the directional information provided by the gradient guides the search. Second, the derivative of the objective might be faster to evaluate at a given point in the hyperparameter space than training n SVMs for CV-n. As a consequence, gradient-based approaches can have significant advantages over direct search, especially when the parameter space has more than a few dimensions.

Maximum-likelihood model selection. We present one recent gradient-based model selection algorithm which has been shown to outperform several established methods on a large benchmark set [15]. The classification error in general is a nondifferentiable quantity with respect to the parameters of a deterministic hypothesis. In contrast, assume a probabilistic classifier approximating the probability $P(y|x)$ of observing class $y \in \mathcal{Y}$ given input $x \in \mathcal{X}$ by some model $\hat{P}(y|x)$, where $\hat{P}$ depends smoothly on its parameters and the hyperparameters of the learning algorithm. A typical approach for learning these parameters is maximizing the *logarithmic likelihood function* $\mathcal{L} = \sum_{(x_i,y_i) \in S} \log \hat{P}(y_i|x_i)$ with respect to the adaptive parameters, which can be done using gradient ascent.

Learning a model of $P(y|x)$ is more general and therefore usually more difficult task than "just" learning a hypothesis for classification. While a perfect model gives the Bayes optimal hypothesis by $h(x) = \arg\max_{y \in \mathcal{Y}} P(y|x)$, a bad model leads to bad classification results. In the SVM framework, one therefore searches for a proper hypothesis directly without estimating $P(y|x)$. This at the same time prevents us from using the maximum-likelihood approach for model selection as described above. Therefore, Glasmachers and Igel [15] use a probabilistic interpretation of the output of an already-trained SVM solely for the purpose of model selection. They follow an approach by Platt [26], who proposed to estimate class membership probabilities $P(y = +1 \mid f(x))$ from SVM decision functions f by fitting a simple sigmoid $\sigma_{r,s}(f(x)) = 1/(1 + \exp(s \cdot f(x) + r))$ around f, where $s \in \mathbb{R}^-$ and $r \in \mathbb{R}$ are the scaling and offset parameter, respectively. This fitting can be done by gradient-based optimization of a cross-validation estimate of the likelihood. For an SVM decision function f, a sigmoid $\sigma_{r,s}$ around f, and validation data S', the likelihood is

$$\mathcal{L}(S', \sigma_{r,s}, f) = \sum_{\substack{(x',y') \in S' \\ y'=+1}} \log \sigma_{r,s}(f(x')) + \sum_{\substack{(x',y') \in S' \\ y'=-1}} \log(1 - \sigma_{r,s}(f(x'))). \tag{3.13}$$

In Ref. [15], this quantity is optimized with respect to the kernel parameters and the regularization parameter C of the SVM using gradient ascent. This requires the kernel function to depend smoothly on its parameters. The derivative of the SVM with respect to its hyperparameters and the kernel can be computed using the procedure proposed in Ref. [22]. It has to be stressed that the probabilistic interpretation of the SVM output is a heuristic, because the SVM (on purpose) does not aim at learning proper probabilities. However, the probabilistic interpretation is solely used to guide model selection. In practice, this maximum-likelihood approach to SVM model selection achieves state-of-the-art results and especially performs well when optimizing flexible kernels on small datasets [15].

The model selection algorithms described in this chapter are all available as part of the open-source machine learning library Shark [20].

3.2.8 SUMMARY

To conclude our introduction to SVMs for classification, we emphasize the following properties of these powerful learning machines:

- *Convex optimization.* When all of an SVM's hyperparameters are fixed, SVM training corresponds to solving a convex quadratic optimization problem, which is free from sub-optimal local extrema. In other words, SVMs always return some best solution possible given a fixed set of hyperparameters.
- *Consistency.* Under relatively mild conditions, standard SVMs are known to be universally consistent. As the number of training examples increases, the SVM solutions converge to the Bayes optimal hypothesis, which is the best classifier possible given the data-generating distribution [35].
- *Kernel trick.* As long as a kernel function can be defined between them, the input patterns may be arbitrary elements. For example, text documents, graphs, or trees can directly serve as input to an SVM if a valid kernel function between them is defined. Kernel-based algorithms in general can be seen as elegantly separating the general part of a learning machine from the problem-specific part. The kernel function (as well as the regularization parameter) allows for incorporation of domain-specific prior knowledge into the learning process, which is necessary to achieve well-generalizing hypotheses.
- *Model selection problem.* SVMs—like most other learning machines—are not parameter-free in the sense that they do not intrinsically determine all entities their behavior is influenced by. The regularization parameter and a kernel function with additional free parameters must be specified externally. However, there cannot be a "universal" learning machine that excels across all possible problems (e.g., [4]). That is, one has to incorporate prior knowledge to tailor a learning machine to some given task. For SVMs this includes choosing the SVM hyperparameters such as a proper family of kernel functions.
- *Multiclass classification.* There is no canonical extension of the binary SVM formulation (Equation 3.10) to multiclass problems with $|\mathcal{Y}| > 2$. Many application studies in the multiclass case rely on training and combining a number of binary machines. At the same time, several multiclass SVM formulations exist that solve the multicategory classification task in one joint optimization problem [8,23,40]. All these have different properties and require solvers distinct from those used for purely binary problems.
- *Training and execution time.* Training time in general is between quadratic and cubic in the number of training examples and additionally influenced by input dimension and SVM hyperparameters [21]. Evaluating an SVM on an unseen example benefits from the sparseness property and is linear in the number of support vectors times the number of operations per kernel function evaluation. If the Bayes risk is nonzero, the number of support vectors scales linearly with the number of training examples [36]. Several extensions or modifications have been proposed to reduce training and execution times. The latter can be achieved, for instance, by approximating the SVM solution after training [27,37].

In summary, consistency and convexity constitute convenient theoretical advantages that provide certain guarantees on the solution obtained by an SVM. In order for a practitioner to obtain truly meaningful results, however, some familiarity with the model selection problem and standard techniques to approach it are essential. While SVM learning for nonstandard domains—such as trees, graphs, or text—is well supported conceptually, this often gives rise to highly expertized fields of research in itself.

3.3   HYDROACOUSTIC SIGNAL CLASSIFICATION

We consider an application task of hydroacoustic signal classification using remote sensor data from the CTBT verification network.

3.3.1   NUCLEAR-TEST-BAN VERIFICATION

International arms control treaties must be verifiable with high confidence. A verification regime is a set of technological and administrative measures that discourages attempts toward treaty violation from the start by making actual violations detectable with high probability. The CTBT is an international agreement banning all nuclear explosions. As of 2011, it awaits formal ratification by nine further states before it will enter into force. Since 1996, the *Preparatory Commission for the Comprehensive Nuclear-Test-Ban Treaty Organization* (CTBTO) is tasked with building up a global verification system for operation under the treaty. Nuclear weapons have been detonated underground, underwater, and in different altitudes of the atmosphere. While the latter two forms are banished under the Partial Nuclear Test Ban Treaty of 1963, the CTBT would also forbid, and hence require verification against, underground tests.

3.3.1.1   International Monitoring System

At the heart of the CTBT's verification regime lies the *International Monitoring System* (IMS), a network of 321 geophysical monitoring stations positioned around the globe. Both the sensor technologies they employ and their locations have been defined by the treaty and its annexes. Any additional on-site inspection would be limited to an area of 1000 square kilometers through the treaty, which poses implicit constraints on the localization performance of the IMS. The IMS relies on four different monitoring technologies to achieve its goal [16,38]. In full operation, around 15 gigabyte of incoming data are expected each day, mostly transmitted in real time through a global VSAT satellite network. The largest subnetwork of 170 seismic stations, including arrays for enhanced detection capability, measures seismic energy traveling through Earth. Further, 60 infrasound stations record low-frequency pressure variations in the atmosphere, and 11 hydroacoustic stations measure energy transmitted through the world's oceans. These three sensor types constitute the so-called waveform technologies. While the analysis of waveform data can indicate that a detected event might not be of natural origin, the fourth sensor network, radionuclide measurements, can provide valuable evidence for a certain event being a nuclear explosion rather than an earthquake, chemical explosion, or other source not violating the treaty.

3.3.1.2   Hydroacoustic Monitoring

The hydroacoustic network's main purpose is to monitor the oceans for underwater nuclear tests. The main signature of a test would be water pressure waves generated by the underwater explosion. Other sources of underwater sound are natural events such as iceberg calving, suboceanic earthquakes, underwater volcanoes, and marine mammals. Man-made events include intentional or accidental chemical explosions (e.g., in military exercises or dynamite fishing), seismic air-gun surveys, and marine vessels. As a secondary use case, the hydroacoustic network can also contribute to processing of signals originating from the continents. Conceptually, global hydroacoustic monitoring relies on a natural phenomenon of underwater sound propagation. The deep sound channel (DSC) or sound fixing and ranging (SOFAR) channel is a certain depth region in the ocean around a minimum in the vertical sound speed profile. While local channel depths vary from close to the ocean's surface to around 1200 m below, the DSC as a whole exhibits waveguide properties for underwater sound on a global scale. Due to this phenomenon, the relatively small number of IMS hydroacoustic stations are sufficient for global ocean coverage. The IMS hydroacoustic network consists of 11 stations. Six of these use hydrophones placed underwater at the local deep sound channel axis, and five use seismometers residing near the coastline of steep-sloped oceanic islands. Figure 3.3 shows

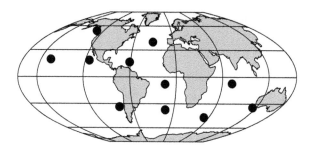

FIGURE 3.3   Locations of hydroacoustic stations defined by the treaty.

the locations of all hydroacoustic stations as defined by the treaty. Every on- and offshore station consists of several individual sensors, facing different sides of the island which hosts the communication infrastructure.

We classify signals recorded by IMS in-ocean hydrophones, learning to label them as either having explosive or nonexplosive signature. Information about sensor or source location is not taken into account, and also no assumption is made that the same signal might have been recorded by multiple sensors. As a consequence, we solve a classical classification task as if all signals had been obtained independently and identically distributed through a single sensor.

3.3.2   PREPROCESSING

We want to evaluate the baseline potential of SVM hydroacoustic signal classification while relying on as few as possible modifications of the existing processing pipeline. Hence, for representation, we use a set of precalculated features provided by the CTBTO's Provisional Technical Secretariat. The raw sensor data were sampled at 250 Hz and from the outset filtered into eight partially overlapping frequency bands between 1 and 100 Hz. Table 3.1 shows the filter band lower and upper frequency limits. In each band, detection and feature extraction are carried out independently, but according to the same algorithm. Within the continuous data stream, a detection algorithm for each band monitors if the ratio of a short-term average (10 s window) to long-term average (150 s window) exceeds a station-specific threshold. If detections in different frequency bands occur close enough in time, they are grouped into a signal, which is hence defined by one or more contemporaneous detections across frequency bands. For each signal a fixed set of 16 identically calculated features is extracted from every band with an associated detection. The union of all extracted features then serves as representation of the event which triggered the detections. Each event is thus represented by $\ell \cdot 16$ real numbers with $\ell \in \{1, \ldots, 8\}$. As a consequence, the feature sets of any two given signals can differ and need not even overlap. The 16 features extracted in every frequency band are listed in Table 3.2. They can be grouped into (i) time related, (ii) energy related, (iii) statistical moments, and (iv) cepstral features. The power cepstrum is a good indicator of periodicity in the signal's power spectrum, that is, the presence of harmonics. For this reason, cepstral features may be good indicators of the bubble pulses possibly accompanying underwater explosions. They are also in general considered to well separate the propagation Green's function from the source

TABLE 3.1

List of the Eight Frequency Bands Used for Feature Extraction

	Frequency Filter Bands (Hz)							
Lower limit	1	2	3	6	8	16	32	64
Upper limit	2	80	6	12	16	32	64	100

TABLE 3.2

Features Extracted from Each Band with a Detection

Temporal	Energy	Statistical	Cepstral Peak (2×)
Peak time	Peak level	Time spread	Position
Mean arrival	Total energy	Skewness	Level
Total duration	Average noise	Kurtosis	Variance
Zero crossing rate			

function. The cepstral features listed in Table 3.2 were calculated in two variants, once from a low-pass-filtered spectrum and once from a detrended spectrum.

3.3.3 CLASSIFYING INCOMPLETE DATA

The classification task described above falls into the group of missing data problems (e.g., [12]). For such it is common to consider a missingness matrix $R \in \{0, 1\}^{mN}$ of the same size as the matrix formed by all feature vectors $x_i \in S$. Each of its Boolean entries $R_{ij} \in \{0, 1\}$ indicates whether the corresponding feature x_{ij} is present in the sample x_i or not. The missingness matrix R of given training data S is then often interpreted as one specific instance or draw from an underlying *missingness distribution p_R*, just as S is assumed to be drawn from a data-generating distribution p. This presumes that the missing values actually existed and would have been observable, but were for some reason not obtained. Such missing data problems are often grouped according to three possible relations between these two generating distributions. If p_R is completely independent of p, the data is coined *missing completely at random* (MCAR). A situation where p_R depends on p, but is conditionally independent of all missing values, is termed as *missing at random* (MAR). For all other relations between p_R and p, one speaks of data *missing not at random* (MNAR). In addition, there are cases where a feature's absence is not simply due to nonobservation, but the entity to be observed was undefined or did not exist for some reason. This is referred to as *structural absence*. CTBT hydroacoustic data constitute a mixture of features being MNAR and structurally absent.

The most convenient way to deal with samples holding missing values is to eliminate them from the training data, or to discard all features which exhibit missing values. Since the missingness ratios in the hydroacoustic data are high across both features and samples, and incomplete test cases with possibly unencountered missingness pattern have to be classifiable, neither is a viable option. In general, deletion can significantly bias both classifier and results, and hence a wide range of more systematic approaches have been suggested, see, for example, Ref. [12] for a review. Among these, imputation techniques are the most common, where missing values are filled in according to some heuristic. Traditional imputation techniques include constant value imputation (e.g., imputing zero or the mean feature value), and guessing a good imputation value, possibly through regression or by using the value of another example that is similar in some sense. More elaborate techniques include, for example, multiple imputation, a Monte Carlo technique for which an imputation model first has to be specified. Then, multiple complete datasets are generated by sampling from the imputation model. Analysis is carried out on each of the imputed datasets and then combined into one final result. In maximum-likelihood-based approaches, an underlying model for the data-generating process is assumed and its parameters are estimated from the available data. Besides imputation, learning algorithms can also be modified to directly deal with incomplete input. For example, an elegant SVM variant by Chechik et al. [6] avoids imputation by altering the margin interpretation for samples with values that are structurally absent.

In practice, it is often tedious to identify among all possible approaches one that performs well on a given application problem. In addition, many of the above methods have been formulated for MAR and MCAR settings rather than MNAR or structural absence which are relevant for CTBT hydroacoustic data. We here use a straightforward approach and impute a single value of zero for all missing values. For many features in Table 3.2, this would constitute a physically or statistically plausible continuation for the limit case of a zero-threshold detector. For example, peak and total energy as well as skewness of a nonpresent signal might be well represented by zero. For others, such as the average noise level, zero cannot be seen as a logical continuation and our choice is far from ideal. In addition to zero-imputation, we use the "flag" approach which was found to be a well-performing baseline method in Ref. [6]. For each of the eight frequency bands, a Boolean variable is added to the imputed feature set, indicating whether or not that band has been detected and its features extracted. We further extend this "flag" approach by using kernel functions which have a bipartite structure, with one subkernel operating on the Boolean missingness representation and another on the real-valued, imputed features.

3.3.3.1 Heterogeneous Kernel Functions

Let S_r denote the zero-imputed hydroacoustic training data and x_r one of its samples, a 128-dimensional real-valued vector. Then we write x_b for the corresponding missingness vector of eight Booleans indicating whether the features of each band are present or not. Thus, each sample $x = (x_b, x_r)$ is represented by a vector in a 136-dimensional joint feature space $\mathcal{X}$, which is the Cartesian product $\mathcal{X}_b \times \mathcal{X}_r$ of the space of the according Boolean and real-valued feature vectors. In order to allow the machine to incorporate information held by the Boolean and real values differently, we employ kernel functions with a *bipartite structure*, where one subkernel k_b operates on the Boolean features $x_b \in \mathcal{X}_b$ and another subkernel k_r on their real counterparts $x_r \in \mathcal{X}_r$. On $\mathcal{X}_r$ we employ a standard radial basis function (RBF) kernel $k(x_r, z_r) = e^{-\gamma \|x_r - z_r\|^2}$. For $\mathcal{X}_b$, a polynomial kernel $k(x_b, z_b) = (\langle x_b, z_b \rangle + c)^d$ is used. When combining the subkernels k_b and k_r into one joint kernel k through a function f,

$$k : (\mathcal{X}_b \times \mathcal{X}_r) \times (\mathcal{X}_b \times \mathcal{X}_r) \rightarrow \mathbb{R}, \quad k(x,z) = f(k_b(x_b, z_b), k_r(x_r, z_r)), \tag{3.14}$$

f must be of such a form that the overall kernel k remains positive definite. Recalling the kernel composition rules of Section 3.2.5, we consider two intuitive possibilities. First, a direct product kernel

$$k_p(x,z) = \left(\langle x_b, z_b \rangle + c \right)^d \cdot e^{-\gamma \|x_r - z_r\|^2}, \tag{3.15}$$

and a weighted direct sum kernel

$$k_s(x,z) = \left(\langle x_b, z_b \rangle + c \right)^d + w e^{-\gamma \|x_r - z_r\|^2} \tag{3.16}$$

with weighting factor $w \in \mathbb{R}^+$. The structure of k_p and k_s stresses the similarity of features across all bands on the one hand and inherent differences between the Boolean and real-valued features on the other.

Independent of the combination of k_b and k_r, we might desire an overall kernel family that is more suitable for our hydroacoustic application task. Especially given that underwater signal propagation is frequency dependent, the overall kernel could better account for the fact that the input vector concatenates information from eight different frequency bands. Mirroring the steps above, we view $\mathcal{X}$ as the Cartesian product of the band-wise feature spaces: $\prod_{i=1}^{8} (\mathcal{X}_b^{(i)} \times \mathcal{X}_r^{(i)})$. For each band's subspace $\mathcal{X}^{(i)}$, a *band-wise bipartite direct product kernel* $k^{(i)}$ can in analogy to Equation 3.15 be defined as

$$k^{(i)}(x^{(i)}, z^{(i)}) = k_b^{(i)} \cdot k_r^{(i)} = \left(\langle x_b^{(i)}, z_b^{(i)} \rangle + c \right)^d \cdot e^{-\gamma^{(i)} \|x_r^{(i)} - z_r^{(i)}\|^2}. \tag{3.17}$$

Here, the polynomial kernel parameters c and d are chosen identical across all subkernels $k^{(i)}$. The RBF bandwidth parameters $\gamma^{(i)}$ are, however, allowed to vary from band to band since we expect different feature distributions across the frequency bands. Note that in each subkernel defined in Equation 3.17, the Boolean subkernel operates on two single Boolean values only. The corresponding scalar product can hence yield only one (if both samples have a detection in band i) or zero (if only one or none of them do). In the special case of $c = 0$, this has the effect of switching on or off the contribution of the overall subkernel: if two samples both have detections for band i, $k^{(i)}$ returns the RBF kernel evaluation between their real-valued features in band i. If only one or none hold detections in band i, $k^{(i)}$ returns zero. For this reason, and because it would again introduce more hyperparameters, we consider the direct product kernel (Equation 3.15) on the subspaces $\mathcal{X}^{(i)}$ rather than the weighted direct sum kernel (Equation 3.16). For combining all subkernels $k^{(i)}$ into one overall kernel, both addition and multiplication are feasible options. As seen in Section 3.2.5, convolution kernels allow for multiplication as well as addition of subkernels and also cover an intermediate range by varying the integer degree D as single free parameter. The convolution kernel k_D of degree D on the eight subkernels $k^{(i)}$ is

$$k_D(x, z) = \sum_{1 \leq j_1 < \cdots < j_D \leq 8} \prod_{d=1}^{D} k^{(j_d)}(x^{(j_d)}, z^{(j_d)}). \tag{3.18}$$

With Equations 3.15, 3.16, and 3.18, we have three candidate families of kernels for the imputed, Boolean-augmented data S on $\mathcal{X}$. All three kernels have as free parameters the real-valued polynomial offset $c \in \mathbb{R}_0^+$ and integer degree $d \in \mathbb{N}^+$. For Equations 3.15 and 3.16, the RBF kernels introduce the bandwidth $\gamma \in \mathbb{R}^+$ as a single parameter, while Equation 3.18 holds one RBF parameter $\gamma^{(i)} \in \mathbb{R}^+$ for each frequency band. An SVM using kernel (Equation 3.18) has 11 free kernel parameters plus the SVM regularization parameter C. In the following, we describe our experimental setup, including the model selection procedure to determine these SVM hyperparameters.

3.3.4 EXPERIMENTS

In total, 778 expert-labeled samples were available for classifier training, validation, and testing. Of these, less than 5% had values for all 128 features, while 91% of samples held values for the most common frequency band between 6 and 12 Hz. For all experiments described below we obtained the test error as an average over five different splits into 80% training and 20% test data. Within each of these 80% of training data, we used another "inner" fivefold cross-validation procedure for SVM model selection. The best hyperparameters were used to re-train an SVM on the entire 80% before obtaining the test error on the remaining 20% of test data. For the direct sum and direct product kernel, we conducted simple grid search on the fivefold cross-validation error CV-5 to find the best values for c, d, (w), γ, and C. For the convolution kernel, grid search is far from feasible and we employed the maximum-likelihood-based approach described in Section 3.2.7.2. We optimized the real-valued parameters only and repeated this for different combinations of integer values for the kernel parameters d and D.

3.3.5 RESULTS

We compared SVMs using the three candidate kernels (Equations 3.15, 3.16, and 3.18) on the zero-imputed and Boolean-augmented dataset S to two baseline methods which operated on the zero-imputed, but not Boolean-augmented dataset S_r only. Table 3.3 shows the results obtained. In the first column, *LDA* refers to a baseline linear discriminant analysis [17]. The SVM baseline classifier

TABLE 3.3

Average Classification Test Errors

	Classifier				
	LDA	*svm*	*svm-s*	*svm-p*	*svm-c*
Error (%)	5.2	4.9	4.9	4.8	4.3

svm used one single RBF kernel and was optimized by grid search as well. The three entries *svm-s*, *svm-p*, and *svm-c* all operated on the zero-imputed and Boolean-augmented training data S, and correspond to the candidate SVMs with a direct sum kernel (Equation 3.16), direct product kernel (Equation 3.15), and convolution kernel of degree one (Equation 3.18), respectively. In summary, all SVMs performed better than the linear approach, and the two bipartite kernels from Equations 3.15 and 3.16 were on par with the baseline RBF kernel not having access to the Boolean-encoded missingness pattern. Additionally, passing the Boolean indicators to the baseline *svm* did not influence its performance. The convolution kernel of degree one, which corresponds to summing up all bandwise subkernels, performed best among all approaches. With increasing degree however, error rates tended to increase as well. At the highest value of $D = 8$, which corresponds to multiplying all bandwise subkernels, the test error with 5.9% was higher than that of LDA.

It should be noted that SVMs allow to control the trade-off between sensitivity and specificity, or false-positive and false-negative rates, by penalizing positive and negative misclassification differently through two different values for the regularization parameter C. Table 3.3 should hence be seen as ranking the different approaches at some generic operation point rather than providing actual error rates for the practical application. It might be desirable to operate such a classifier at high sensitivity at the cost of more false alarms having to be rejected during human analyst review.

3.4   SUMMARY

We introduced SVMs as one specific approach to solving supervised classification tasks. Motivated by the concept of regularized risk minimization, we iteratively refined the SVM optimization problem from linear large-margin classification to the canonical kernelized case. Emphasizing several properties of SVMs relevant to practitioners, we in particular discussed the model selection problem and two approaches to solve it: simple grid search and one gradient-based method for hyperparameter selection. In the second part of the chapter, we described an exemplary application task of classifying hydroacoustic signals recorded by the sensor network for verification of the CTBT. We combined information from different frequency bands via task-specific kernel functions also incorporating information about a sample's missingness pattern. This custom classifier, in combination with parameter optimization through a maximum-likelihood approach to model selection, showed improved performance over baseline linear methods as well as SVMs using standard kernels.

REFERENCES

1. Aronszajn, N., Theory of reproducing kernels. *Transactions of the American Mathematical Society* 68(3), 337–404, 1950.
2. Berg, C., Christensen, J., Ressel, P., *Harmonic Analysis on Semigroups: Theory of Positive Definite and Related Functions*. Berlin, Heidelberg: Springer, 1984.
3. Bordes, A., Ertekin, S., Weston, J., Bottou, L., Fast kernel classifiers with online and active learning. *Journal of Machine Learning Research* 6, 1579–1619, 2005.
4. Bousquet, O., Boucheron, S., Lugosi, G., Introduction to statistical learning theory. In: Bousquet, O., von Luxburg, U., Rätsch, G. (eds.) *Advanced Lectures in Machine Learning, LNCS*, vol. 3176, pp. 169–207. Berlin, Heidelberg: Springer, 2004.

5. Chapelle, O., Vapnik, V., Bousquet, O., Mukherjee, S., Choosing multiple parameters for support vector machines. *Machine Learning* 46(1), 131–159, 2002.
6. Chechik, G., Heitz, G., Elidan, G., Abbeel, P., Koller, D., Max-margin classification of data with absent features. *Journal of Machine Learning Research* 9, 1–21, 2008.
7. Cortes, C., Vapnik, V., Support-vector networks. *Machine Learning* 20(3), 273–297, 1995.
8. Crammer, K., Singer, Y., On the algorithmic implementation of multiclass kernel-based vector machines. *Journal of Machine Learning Research* 2, 265–292, 2002.
9. De Vito, E., Rosasco, L., Caponnetto, A., Piana, M., Verri, A., Some properties of regularized kernel methods. *Journal of Machine Learning Research* 5, 1363–1390, 2004.
10. Fan, R.E., Chang, K.W., Hsieh, C.J., Wang, X.R., Lin, C.J., LIBLINEAR: A library for large linear classification. *Journal of Machine Learning Research* 9, 1871–1874, 2008.
11. Friedrichs, F., Igel, C., Evolutionary tuning of multiple SVM parameters. *Neurocomputing* 64(C), 107–117, 2005.
12. Garcia-Laencina, P., Sancho-Gomez, J., Figueiras-Vidal, A., Pattern classification with missing data: A review. *Neural Computing and Applications* 19(2), 263–282, 2010.
13. Gärtner, T., *Kernels for Structured Data*. Toh Tuck Link, Singapore; Hackensack, NJ; London: World Scientific Publishing, 2009.
14. Glasmachers, T., Igel, C., Gradient-based adaptation of general Gaussian kernels. *Neural Computation* 17(10), 2099–2105, 2005.
15. Glasmachers, T., Igel, C., Maximum likelihood model selection for 1-norm soft margin SVMs with multiple parameters. *IEEE Transactions on Pattern Analysis and Machine Intelligence* 32, 1522–1528, 2010.
16. Hafemeister, D., Progress in CTBT monitoring since its 1999 senate defeat. *Science and Global Security* 15, 151–183, 2007.
17. Hastie, T., Tibshirani, R., Friedman, J.H., *The Elements of Statistical Learning: Data Mining, Inference, and Prediction*. New York: Springer, 2003.
18. Haussler, D., Convolution kernels on discrete structures. Technical Report, UCS-CRL-99–10, University of California at Santa Cruz., 1999.
19. Igel, C., Evolutionary kernel learning. In: Sammut, C., Webb, G.I. (eds.) *Encyclopedia of Machine Learning*. Berlin: Springer, 2010.
20. Igel, C., Glasmachers, T., Heidrich-Meisner, V., Shark. *Journal of Machine Learning Research* 9, 993–996, 2008.
21. Joachims, T., Making large-scale SVM learning practical. In: Schölkopf, B., Burges, C., Smola, A. (eds.) *Advances in Kernel Methods—Support Vector Learning*, pp. 169–184. Cambridge, MA: MIT Press, 1998.
22. Keerthi, S.S., Sindhwani, V., Chapelle, O., An efficient method for gradient-based adaptation of hyperparameters in SVM models. In: Schölkopf, B., Platt, J.C., Hoffman, T. (eds.) *Advances in Neural Information Processing Systems 19 (NIPS)*, pp. 673–680. Cambridge, MA: MIT Press, 2006.
23. Lee, Y., Lin, Y., Wahba, G., Multicategory support vector machines: Theory and application to the classification of microarray data and satellite radiance data. *Journal of the American Statistical Association* 99(465), 67–82, 2004.
24. Maji, S., Berg, A.C., Malik, J., Classification using intersection kernel support vector machines is efficient. *IEEE Conference on Computer Vision and Pattern Recognition*, pp. 1–8, 2008.
25. Platt, J., Fast training of support vector machines using sequential minimal optimization. In: Schölkopf, B., Burges, C., Smola, A. (eds.) *Advances in Kernel Methods—Support Vector Learning*, pp. 185–208. Cambridge, MA: MIT Press, 1998.
26. Platt, J., Probabilistic outputs for support vector machines and comparisons to regularized likelihood methods. In: Smola, A.J., Bartlett, P. L., Schölkopf, B., Schuurmans, D. (eds.) *Advances in Large Margin Classifiers*, pp. 61–74. Cambridge, MA: MIT Press, 2000.
27. Romdhani, S., Torr, P., Schölkopf, B., Blake, A., Efficient face detection by a cascaded support vector machine expansion. *Proceedings of the Royal Society of London. Series A: Mathematical, Physical and Engineering Sciences* 460(2051), 3283–3297, 2004.
28. Saunders, C., Gammerman, A., Vovk, V., Ridge regression learning algorithm in dual variables. In: *Proceedings of the Fifteenth International Conference on Machine Learning*, pp. 515–521. San Francisco, CA: Morgan Kaufmann, 1998.
29. Schölkopf, B., Smola, A.J., *Learning with Kernels*. Cambridge, MA: MIT Press, 2002.
30. Schölkopf, B., Tsuda, K., Vert, J., *Kernel Methods in Computational Biology*. Cambridge, MA: MIT Press, 2004.

31. Schölkopf, B., Herbrich, R., Smola, A., A generalized representer theorem. In: Helmbold, D., Williamson, B. (eds.) *Computational Learning Theory, LNCS*, vol. 2111, pp. 416–426. Berlin, Heidelberg: Springer, 2001.

32. Schölkopf, B., Mika, S., Burges, C., Knirsch, P., Müller, K.R., Rätsch, G., Smola, A., Input space versus feature space in kernel-based methods. *IEEE Transactions on Neural Networks* 10, 1000–1017, 1999.

33. Shalev-Shwartz, S., Singer, Y., Srebro, N., Cotter, A., Pegasos: Primal estimated sub-gradient solver for SVM. *Mathematical Programming* 127, 3–30, 2011.

34. Shawe-Taylor, J., Cristianini, N., *Kernel Methods for Pattern Analysis*. Cambridge, UK: Cambridge University Press, 2004.

35. Steinwart, I., Support vector machines are universally consistent. *Journal of Complexity* 18, 768–791, 2002.

36. Steinwart, I., Sparseness of support vector machines—Some asymptotically sharp bounds. In: Thrun, S., Saul, L., Schölkopf, B. (eds.) *Advances in Neural Information Processing Systems* 16. pp. 1069–1076. Cambridge, MA: MIT Press, 2004.

37. Suttorp, T., Igel, C., Resilient simplification of kernel classifiers. In: Marques de Sá , J., Alexandre, L.A., Duch, W., Mandic, D. (eds.) *Proceedings of the 17th International Conference on Artificial Neural Networks (ICANN). LNCS*, vol. 4668, pp. 139–148. Berlin, Heidelberg: Springer, 2007.

38. Thunborg, A. (ed.), *Science for Security. Verifying the Comprehensive Nuclear-Test-Ban Treaty.* Preparatory Commission for the CTBTO, Vienna, Austria, 2009.

39. Vapnik, V., *Statistical Learning Theory*. New York: Wiley, 1998.

40. Weston, J., Watkins, C., Support vector machines for multi-class pattern recognition. In: *Proceedings of the 7th European Symposium in Artificial Neural Networks (ESANN)*, pp. 219–224, 1999.

4 Huygens Construction and the Doppler Effect in Remote Detection

Enders A. Robinson

CONTENTS

4.1 INTRODUCTION

The construction of excellent images of the subsurface is the aim of every geophysicist engaged in seismic exploration. Geophysicists and astronomers make use of tools of remote detection. Planets outside the solar system are such faint light sources that only about 10 have been directly imaged by telescopes. However, hundreds have been detected by the use of the relativistic Doppler effect. Essential to every geophysicist is Huygens construction, which not only gives a geometrical explanation of how waves travel but also gives a means of imaging the data by a process known as seismic migration. Just as the Huygens construction can be used to explain seismic migration, it can also be used to explain the relativistic Doppler effect. A symmetry is a physical or mathematical feature of a system that is preserved under some change. The fact that both seismic migration and the relativistic Doppler effect can each be expressed in terms of the Huygens construction shows that both systems exhibit the same symmetry. In the case of seismic imaging, the symmetry is that observer and observed can be interchanged and still produces the same final image. In the case of the relativistic Doppler effect, the symmetry is that the behavior of two objects depends not upon their absolute motion, but only upon their relative motion. Such symmetry is called relativistic symmetry.

4.2 REMOTE DETECTION

Remote detection is the acquisition of information about a target by the use of devices not in direct contact with the target. Remote detection makes it possible to collect data about dangerous or inaccessible regions. The target could be an aircraft, an internal bodily organ, an animal, an oil reservoir, or a galaxy. There are two kinds of remote detection. In passive remote detection, the observer waits for signals emitted by the target itself. In active remote detection, the observer initiates signals and collects the reflected or backscattered signals from the target. The most common signals used in remote detection are in the form of traveling waves. For animals, the two prime detectors are the eyes and ears. The eyes make use of electromagnetic waves, more specifically light waves. The ears make use of mechanical waves, more specifically sound waves.

Radar represents a method of active remote detection that uses electromagnetic waves as signals. A radar system has a transmitter that emits microwaves. When these waves come into contact with the target, the waves are scattered in all directions. The signal is thus partly reflected back to the observer. The receiver is usually, but not always, in the same location as the transmitter. Although the signal returned is weak, it can be amplified through use of electronics. Radar can detect objects at ranges where other emissions, such as sound or visible light, would be too weak. The distance to an object can be determined by transmitting a short pulse, and measuring the time it takes for its reflection to return. The distance is one-half the product of the round trip time (because the signal has to travel to the target and then back to the receiver) and the speed of the signal. The signal speed is the speed of light (which is 299,792,458 m/s in vacuum). The existing system for measuring distance, combined with a memory capacity to see where the target last was, is enough to measure speed. However, if the transmitter's output is coherent (sinusoidal), there is another effect that can be used to make almost instantaneous speed measurements known as the Doppler effect. Doppler radar is based on frequency modulation. If the target is moving, the reflected waves undergo a change of wavelength (and thus frequency). In other words, the returned signals are shifted away from the base frequency via the Doppler effect, thereby enabling the calculation of the speed of the object relative to the radar. The Doppler effect is only able to determine the relative speed of the target along the line of sight from the radar equipment to the target. Any component of target velocity perpendicular to the line of sight cannot be determined by using the Doppler effect alone, but it can be determined by tracking the target's azimuth over time.

Underwater acoustics (hydroacoustics) is the study of underwater sound. Sonar is a method of remote detection that uses sound propagation as the signal. There are two kinds of sonar: active and passive. Active sonar may be characterized as a method of echolocation of targets. The frequencies used in sonar systems vary from very low (infrasonic) to extremely high (ultrasonic). Sonar operations are affected by variations in sound speed, particularly in the vertical plane. Ocean temperature varies with depth. Between 30 and 100 m there is often a marked change, called the thermocline, dividing the warmer surface water from the cold, still water that make up the depth of the ocean. This phenomenon can disturb sonar, because a sound ray originating on one side of the thermocline is bent, or refracted, in its passage through the thermocline.

Biosonar (biological sonar) is the term for the echolocation used by animals such as dolphins, shrews, most bats, and most whales. Two bird groups also employ this system for navigating through caves. Biosonar, which uses sounds made by animals, works like active sonar. Ranging is done by measuring the time delay between the animal's own sound emission and the echoes that return from the environment.

A redshift is a Doppler shift in the spectrum of an astronomical object toward longer wavelengths, or toward the red end of the spectrum. A redshift occurs whenever a light source moves away from the observer. A blueshift is a shift in the spectrum toward shorter wavelengths. A blueshift occurs whenever a light source moves toward the observer. Spectroscopic astrophysics uses these Doppler shifts to determine the movement of distant astronomical objects. The Doppler shift is now being used to locate planets of stars other than the sun.

As its name implies, geophysics makes use of physical principles in measuring Earth and planets. In delineating the subsurface of Earth, geophysicists are generally restricted to measurements made close to the surface of the ground or in boreholes. As a result, geophysicists must make use of methods of remote detection. The signals that can penetrate the inaccessible regions of Earth are seismic waves. Reflection seismology represents active remote detection: man-made sources emit seismic signals, and detectors (placed in regard to the sources) record the resulting reflected seismic signals.

Earthquake seismology represents passive remote detection: an earthquake emits signals, and detectors (permanently in place) record the resulting seismic signals.

Galileo [1] wrote: "Philosophy is written in this grand book, the universe ... It is written in the language of mathematics, and its characters are *triangles, circles, and other geometric figures* without which it is humanly impossible to understand a single word of it; without these one is wandering in a dark labyrinth." In this chapter, we use triangles and circles to their fullest extent.

4.3 HUYGENS PRINCIPLE

Wave motion is basic to remote detection, and in particular, to geophysics. We want to examine how Christiaan Huygens used circles to reveal how waves are propagated. Huygens [2] states that every point of a wavefront may be considered the source of spherical wavelets that spread out in all directions with a speed equal to the speed of propagation of the waves. The primary wavefront at some later time is the envelope of these wavelets. This fundamental mechanism for wave propagation is known as Huygens principle or Huygens construction (see Figure 4.1). In this chapter, we limit ourselves to two dimensions, in which case the spherical wavelets of Huygens reduce to circles.

Huygens used his construction to demonstrate the mechanism for the refraction of light (see Figure 4.2).

In order to get to the essence of the construction, we block out various lines on Figure 4.2 to obtain the abbreviated diagram shown in Figure 4.3. In this figure, the circle represents the spherical wavelet of Huygens. The horizontal line OR is called the equatorial line. The trailing point of the Huygens circle is labeled S and the leading point is labeled R. The center of the circle is labeled T. Note that T is the midpoint of S and R. Let t be the length of the line segment OT. Let s be the length of the line segment OS. Let r be the length of the line segment OR. It is seen that t is the arithmetic mean of s and r, that is, $t = (s + r)/2$.

The point of tangency of the sloping line with the circle is labeled P. Let p be the length of the tangent OP. An important theorem in plane geometry states: If a secant and a tangent are drawn to a circle from the same point, then the tangent is the mean proportional between the whole secant

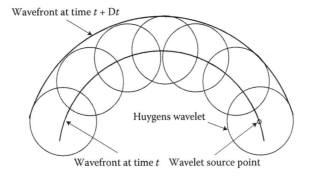

Wavefront at time t + Dt

Huygens wavelet

Wavefront at time t Wavelet source point

FIGURE 4.1   Huygens principle or Huygens construction.

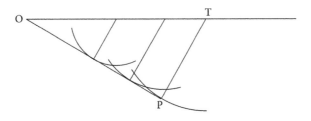

FIGURE 4.2   Huygens diagram for the refraction of light.

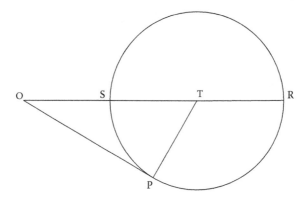

FIGURE 4.3   Huygens circle and triangle.

and its exterior segment. If we apply this theorem to Figure 4.3, then line OP is the tangent line and line OR is the secant line. The tangent is $p = OP$. The whole secant is $r = OR$, and the exterior segment is $s = OS$. The term *mean proportional* and the term *geometric mean* both designate the same thing. It follows that p is the geometric mean of s and r, that is, $p = \sqrt{sr}$.

4.4   SEISMIC EXPLORATION

Seismic waves travel through Earth and are reflected and refracted at geologic interfaces. Reflection seismology is a method of remote detection that can determine the location of an underground interface without digging or drilling. We shall deal with plane interfaces tilted to the horizontal. The interface may be regarded as a mirror that reflects seismic waves (see Figure 4.4). At the shot point on the surface of Earth, a small charge of dynamite is detonated at time s, which is called the sending time. The resulting seismic pulse travels down to an interface where it is reflected. The reflected pulse returns to the shot point at some later time, which is called the receiving time r. The difference of these two times is called the two-way travel time. The one-way time is one-half of the two-way time. If we multiply the one-way time by the speed of the seismic wave, we obtain the distance x from the source point to the reflection point.

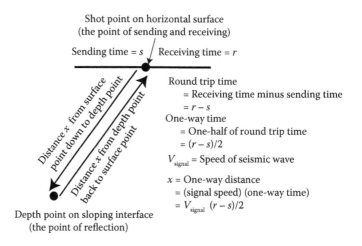

FIGURE 4.4   Round-trip path from surface to reflector.

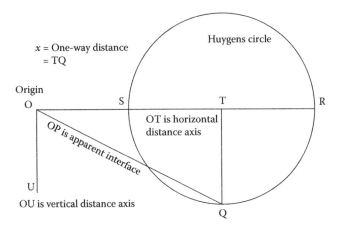

FIGURE 4.5 Classical seismic case.

The natural place to start is by examining the geometry of the problem (see Figure 4.5). The horizontal line OR depicts the surface of the ground. The interface outcrops at point O and then slopes downward to the right. The explosion is set off at shot point T. Now we must draw in the path of the seismic pulse from T to the reflection point on the dipping interface. Construct a circle with center T and radius equal to the distance x from the source point to the reflection point. The circle is the Huygens circle. The reflection point can be anywhere on the Huygens circle of radius x. We will consider two cases: the classical seismic case and the migrated seismic case.

In the classical seismic case (as shown in Figure 4.5), it is assumed that the reflection point Q lies directly under the shot point T. As a result, the apparent interface is given by the line OQ. This classical case is not correct, but it can serve as an approximation for interfaces that slope very gently. Under the assumption of the classical seismic case, a prospector could rightfully claim that the apparent length OQ of the interface is greater than the surface length OT.

The situation shown in Figure 4.5 is corrected in the migrated seismic case (as shown in Figure 4.6). Further discussion on seismic migration can be found in books on seismic processing [3]. The essential fact is that the actual interface acts like a mirror. It follows that the raypath must be at right angles to the interface. Such a raypath may be determined as follows. Construct the line that passes through the origin and that is tangent to the Huygens circle. Let P be the point of tangency. The seismic raypath is the line TP, which is perpendicular to the interface. The actual (or proper) interface is OP. Under the assumption of the migrated seismic case, a prospector would know that the length

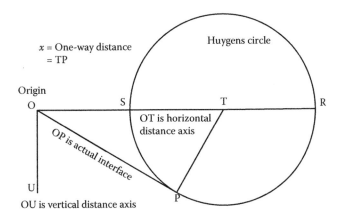

FIGURE 4.6 Migrated seismic case with observer on surface OR.

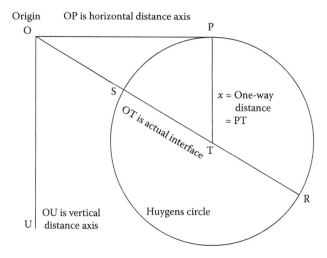

FIGURE 4.7   Migrated seismic case with observer on sloping interface OR.

OP of the interface is less than the surface length OT. For this reason the surface length OT is called the *dilated length of the interface* OP. The Huygens method described here is basic to the imaging of seismic data.

Symmetry is an important idea in seismic imaging (see Figure 4.7). In order to demonstrate the symmetry, we will now place the observer on the sloping interface and let us determine the location of the surface interface. The essential requirement is that the surface interface acts like a mirror. It follows that the raypath must be at right angles to the surface interface. Such a raypath may be determined as follows. Construct the line that passes through the origin and that is tangent to the Huygens circle. Let P be the point of tangency. The seismic raypath is the line TP, which is perpendicular to the surface interface OP.

4.5   CLASSICAL ONE-WAY DOPPLER SHIFT: CASE 1 AND CASE 2

The seismic case discussed above makes use of two spatial dimensions, namely horizontal distance along the surface of Earth and vertical distance into Earth. We will now turn our attention to the Doppler effect. The Doppler effect is exhibited by all types of wave motion. We will again deal with two dimensions, but now, one dimension is that of time and the other dimension is that of distance, namely the straight-line distance between two objects. Again, we will consider two cases: the classical Doppler case and the relativistic Doppler case.

The effect was discovered by Christian Johann Doppler [4]. He theorized that since the pitch of sound from a moving source varies for a stationary observer, the color of the light from a star should change, according to the star's velocity relative to Earth. Doppler's critics tested his mathematical results by an experiment in Holland. For some days, they had a locomotive pull a flat car back and forth at different speeds. On the flat car were trumpeters sounding various notes. On the ground, musicians with a good perception of absolute frequency recorded the note as the train approached and as it receded. When the train approached, these musicians found that the frequency of the note received was greater than the frequency of the note sent. When the train went away, they found that the frequency of the note received was less than the frequency of the note sent. Doppler's equations were confirmed.

Typically, the Doppler effect is expressed in terms of the frequencies of the waves involved. Often in geophysics, and especially in seismology, we deal with signals expressed as a function of time. The period is equal to the reciprocal of the frequency. For example, if the frequency is 100 Hz, the period is $1/100 = 0.01$ s. We will present the Doppler effect in terms of periods instead of frequencies.

The advantage of this approach is that we can illustrate our reasoning by means of diagrams involving triangles and circles.

In order to illustrate the classical Doppler effect, we assume that the two objects are submarines, labeled A and B. The water is the medium, which we assume is not moving, so the medium has zero velocity. Submarine A is resting on the bottom, and submarine B is moving at a constant velocity V. We use the convention that V is a positive number if B is moving away from A, and that V is a negative number if B is moving toward A. For the purposes of exposition we always assume that V is a positive number so that B is moving away from A. However, the results we obtain also hold when V is a negative number, in which case B is moving toward A.

The signal speed is designated by the V_{signal}, which we always take as a positive number. We want to express both time and distance in the same unit, which we take to be 1 s. The conversion factor between time and distance is the signal speed. The resulting unit of distance becomes 1 sonic second. For example, if the speed of sound in salt water is 1550 m/s, then 1 sonic second is 1550 m. An object 3100 m away would be at a distance of 2 sonic seconds. We also want to normalize the velocity V by the signal speed. If the velocity of the submarine is 10 m/s then the normalized speed of the submarine is

$$v = \frac{V}{V_{signal}} = \frac{V}{1550} = \frac{10}{1550} = 0.00645$$

For the Doppler effect to hold, signal speed must be greater that the speed of any object, so the normalized velocity v (in magnitude) is necessarily less than 1. In other words, v will lie between -1 and $+1$. The normalized signal speed is unity.

We will assume that each submarine has the ability to emit a monochromatic wave (a sinusoidal wave characterized by a single fixed frequency). We will consider two cases. In case 1, the fixed submarine A sends out a sonar signal of period s. This source signal travels to the moving submarine B where it is received as a signal of period r. In case 2, the moving submarine B sends out a sonar signal of period s. This source signal travels to the fixed submarine A where it is received as a signal of period r. In each case, the normalized velocity of the moving submarine B is v.

Let us treat case 1 where the source A has zero velocity, the receiver B has velocity v, and the signal is sent from A to B (see Figure 4.8). In the time interval $r = OR$, the submarine B travels a distance of $x = RQ$. Recall that we express both time and distance in the same unit. In the time interval SR, the sonic signal travels a distance of RQ. Because the normalized signal speed is unity, the distance RQ is equal to the time interval SR, that is, $x = RQ = SR = r - s$. It is for this reason that we can draw the Huygens circle of radius x shown in the Figure 4.8.

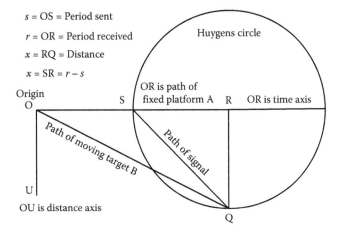

$s = OS = $ Period sent
$r = OR = $ Period received
$x = RQ = $ Distance
$x = SR = r - s$
Origin
O
Huygens circle
OR is path of fixed platform A
R   OR is time axis
S
Path of moving target B
Path of signal
U
OU is distance axis
Q

FIGURE 4.8 The classical one-way shift. Case 1: Fixed source A and moving receiver B.

The relationships between the velocity, the period sent s, the period received r, the frequency sent $f_s = 1/s$, and the frequency received $f_r = 1/r$ are

$$v = \frac{x}{r} = \frac{r - s}{r} = \frac{(1/s) - (1/r)}{1/s} = \frac{f_s - f_r}{f_s}$$

Thus, $f_s v = f_s - f_r$, so

$$f_r = (1 - v)f_s$$

We see that, if v is positive (B moving away from stationary A), then the received frequency is less than the source frequency. If v is negative (B approaching stationary A), then the received frequency is greater than the source frequency.

The Doppler factor is defined as the quotient of the received period divided by the sending period. Thus, the Doppler factor for case 1 is

$$k_1 = \frac{r}{s} = \frac{f_s}{f_r} = \frac{1}{1 - v}$$

If we define the frequency increment as $\Delta f = f_r - f_s$ then we have

$$v = \frac{V}{V_{signal}} = -\frac{\Delta f}{f_s}$$

We see that in the case of a redshift (i.e., the frequency increment is negative), the velocity is positive. Conversely, in the case of a blueshift (i.e., the frequency increment is positive), the velocity is negative.

Let us treat case 2 where the source B has velocity v, the receiver A has zero velocity, and the signal is sent from B to A (see Figure 4.9). In the time interval $s = OS$, the submarine B travels a distance of $x = QS$. Recall that we express both time and distance in the same unit. In the time interval SR, the sonic signal travels a distance of QS. Because the normalized sonic speed is unity, the distance QS is equal to the time interval SR, that is, $x = QS = SR = r - s$. It is for this reason that we can draw the Huygens circle of radius x shown in the Figure 4.9.

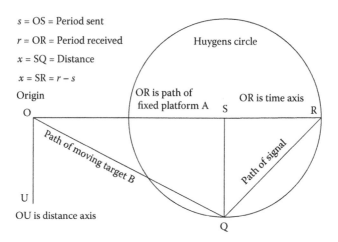

FIGURE 4.9   The classical one-way shift. Case 2: Moving source B and fixed receiver A.

The relationships between the velocity, the period sent, the period received, the frequency sent, and the frequency received are

$$v = \frac{x}{s} = \frac{r-s}{s} = \frac{(1/s)-(1/r)}{1/r} = \frac{f_s - f_r}{f_r}$$

This equation gives $f_r v = f_s - f_r$, so

$$f_r = \frac{1}{1+v} f_s$$

We see that, if v is positive (B moving away from stationary A), then the received frequency is less than the source frequency. If v is negative (B approaching stationary A), then the received frequency is greater than the source frequency.

The Doppler factor for case 2 is

$$k_2 = \frac{r}{s} = \frac{f_s}{f_r} = 1 + v$$

If we define the incremental frequency as $\Delta f = f_r - f_s$, then we have

$$v = \frac{V}{V_{signal}} = -\frac{\Delta f}{f_r}$$

We see that in the case of a redshift (i.e., the frequency increment is negative), the velocity is positive. Conversely, in the case of a blueshift (i.e., the frequency increment is positive), the velocity is negative.

Let us summarize. The classical Doppler effect or Doppler shift relates the frequency f_r of the wave received to the frequency f_s of the wave sent. The general relationship is

$$f_r = \frac{V_{signal} - V_{receiver}}{V_{signal} + V_{source}} f_s$$

where V_{signal} is the velocity of signal in the medium and V_{source} is the velocity of the source relative to the medium. This velocity is positive if the source is moving away from the receiver.

$V_{receiver}$ is the velocity of the receiver relative to the medium. This velocity is positive if the receiver is moving away from the source.

We considered two special cases. In case 1, we let the velocity of the source be zero, and we let $V_{receiver} = V$, so the relationship simplified to

$$f_r = \frac{V_{signal} - V}{V_{signal}} f_s = (1-v) f_s$$

In case 2, we let the velocity of the receiver be zero, and we let $V_{source} = V$, so the relationship simplified to

$$f_r = \frac{V_{signal}}{V_{signal} + V} f_s = \frac{1}{1+v} f_s$$

4.6 CLASSICAL TWO-WAY DOPPLER SHIFT

We now want to discuss the question of remote detection. The stationary submarine is called the fixed platform, and the moving submarine is called the moving target. An outgoing signal of period s is sent from the fixed platform to the moving target. The signal incident on the target is reflected. The reflected signal has the same period as the incident signal. Let t denote this period. The reflected signal leaves the moving target and travels back to the fixed platform where it is received. The period of the received signal is r.

We can join the results of case 1 and case 2 to obtain the two-way case required for active remote detection. Figure 4.10 depicts the joining of the two cases. The sending period s of case 1 is the same as the sending period s of the two-way case. The receiving period r of case 1 becomes the period t of the incident wave of the two-way case. The sending period s of case 2 becomes the period t of the reflected wave of the two-way case. The receiving period r of case 2 is the same as the receiving period r of the two-way case. The Huygens circle can be drawn as before.

An observer on the fixed platform records the period s of the signal transmitted and also records the period r of the signal received. From the two quantities s and r, the observer on the fixed platform wants to compute the velocity v of the moving target. In the two-way case, the period $s = \mathrm{OS}$ of the signal transmitted by the fixed platform is now called the *sending period* and the period $r = \mathrm{OR}$ of the signal reflected back to the fixed platform is now called the *receiving period*. Thus, the sending frequency at the fixed platform is $f_s = 1/s$, and the receiving frequency at the fixed platform is $f_r = 1/r$.

The sonic ray emitted at time $s = \mathrm{OS}$ travels to the moving target and its reflection travels back to the fixed platform arriving at time $r = \mathrm{OR}$. The raypath of the signal is SQ on the outward journey and QR on the return journey. The elapsed time ST for the outward journey is the same as the elapsed time TR for the return journey. In other words, ST = SR, so point T is the midpoint of point S and point R. The time interval $t = \mathrm{OT}$ is called the midpoint period. The midpoint period t is the period of the signal incident on the moving target and it is also the period of the signal reflected by the moving target. In other words, the process of reflection does not change the period of the wave.

The distance $x = \mathrm{TQ}$ is the separation distance of the two objects at time $t = \mathrm{OT}$. The constant velocity of the moving target is thus $v = x/t$. Because

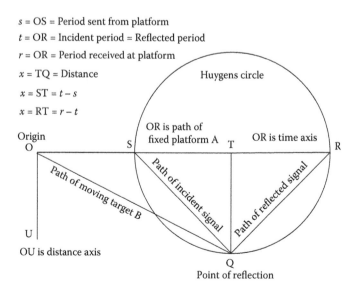

FIGURE 4.10 The classical two-way Doppler shift.

$$x = \frac{r - s}{2}, \quad t = \frac{r + s}{2}$$

it follows that the velocity is

$$v = \frac{x}{t} = \frac{r - s}{r + s} = \frac{(1/s) - (1/r)}{(1/s) + (1/r)} = \frac{f_s - f_r}{f_s + f_r}$$

This equation shows how the velocity of the moving target is computed from the two measurements s and r made on the fixed platform.

The Doppler factor k is defined as the ratio of the period received at one location divided by the period sent from another location. With reference to Figure 4.10, the period sent from the fixed platform is $s = OS$ and the period received at the moving target is $t = OT$. Thus, the one-way Doppler factor from fixed object to moving object is $k_1 = t/s$. Again with reference to Figure 4.10, the period sent from the moving target is $t = OT$ and the period received at the fixed platform is $r = OR$. Thus, the one-way Doppler factor from moving object to fixed object is $k_2 = r/t$. We can write

$$k_1 = \frac{t}{s} = \frac{r - s}{2s} = \frac{(1/s) - (1/r)}{2(1/r)} = \frac{1}{2}\left(\frac{f_s - f_r}{f_r}\right),$$

$$k_2 = \frac{r}{t} = \frac{2r}{r - s} = \frac{2(1/s)}{(1/s) - (1/r)} = 2\left(\frac{f_s}{f_s - f_r}\right)$$

For example, suppose the sending period were $s = 4$ and the receiving period were $r = 16$. The midpoint period would be $t = (4 + 16)/2 = 10$. Thus,

$$k_1 = \frac{t}{s} = \frac{10}{4} = 2.5, \quad k_2 = \frac{r}{t} = \frac{16}{10} = 1.6$$

The signal in each case covers the same distance x. In the case of k_1, the signal goes from the fixed platform to the moving target. In the case of k_2, the signal goes from the moving target to the fixed platform. Thus, an important feature to the classical two-way Doppler case is that the two one-way Doppler factors are different.

In the case of active remote detection, the observer on the fixed platform sends out a signal and receives its reflection. Thus, the observer can measure both the period s of the transmitted signal and the period r of the received reflected signal. From these two observations he can compute the velocity of the moving target.

4.7 RELATIVISTIC DOPPLER SHIFT

In explaining the Doppler effect for sound waves, we must take into account the velocities of the source and the receiver relative to the medium. In the case of electromagnetic waves traveling through vacuum, there is no medium. A symmetry is a physical or mathematical feature of a system that is preserved under some change. The type of symmetry known as relativistic symmetry requires that the behavior of two objects remain the same regardless of their absolute motion. Their behavior would only depend upon their relative motion. Relativistic symmetry is amenable to mathematical formulation and so it can be used to simplify many problems in physics. In the case of relativistic problems, the signal speed is the speed of light in vacuum. The relativistic Doppler effect must be used when the relative velocity between two bodies is close to the speed of light.

In the above presentation of the classical two-way case, we have seen that the Doppler factor k_1 (called the outward Doppler factor) from fixed object to moving object is not the same as the Doppler factor k_2 (called the inward Doppler factor) from moving object to fixed object. In order to have relativistic symmetry, both Doppler factors must be the same. We will now explore how we can obtain this objective. The requirement is that the outward Doppler factor must be the same as the inward Doppler factor. One fact that we do know is that the product of the two Doppler factors is

$$k_1 k_2 = \frac{t}{s} \frac{r}{t} = \frac{r}{s}$$

If we require

$$k = k_1 = k_2$$

then we have

$$k_1 k_2 = k^2 = \frac{r}{s}$$

As a result, the required symmetric Doppler factor is

$$k = \sqrt{\frac{r}{s}}$$

At the reflector (the moving target), the period of the incident wave is the same as the period of the reflected wave. Call this unknown period p. Then, we can write

$$k = \frac{p}{s}, \quad k = \frac{r}{p}$$

If we set these two equations to each other we obtain

$$\frac{p}{s} = \frac{r}{p}$$

which we can solve to obtain

$$p = \sqrt{s \, r}$$

Thus, the required period p is equal to the geometric mean of the sending period and the receiving period. The period p is called the proper period.

Figure 4.11 depicts the relativistic Doppler case with the Huygens circle. We see that the required period is given by the tangent line OP. Here, distance TP must be expressed in terms of light-seconds. In the case of the classical Doppler effect (Figure 4.10), the line OR is the time axis for both the fixed object and the moving object. Such a case is in keeping with the Newton point of view, and that is the reason this Doppler effect is called classical. In the case of the relativistic Doppler effect (Figure 4.11), the line OR is the time axis for the fixed object and line OP is the time axis for the moving object. Such a case is in keeping with the Einstein point of view, and that is the reason this

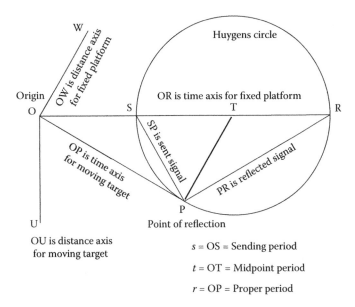

FIGURE 4.11 The relativistic Doppler shift with the Huygens circle. This is a two-dimensional diagram with two sets of oblique time–distance coordinates, namely the set (OR, OW) for the fixed platform and the set (OP, OU) for the moving target.

Doppler effect is called relativistic. From the Einstein point of view, only the relative motion of the two objects matters. The two oblique distance axes can also be drawn. The line OW is the distance axis for fixed platform. Line OW is perpendicular to line OP. The line OU is the distance axis for the moving target. Line OU is perpendicular to line OR. In the relativistic case, the fixed platform has its own set of oblique axes. Similarly, the moving target has its own set of oblique axes. The two sets are different.

If only relative motion matters, why do we speak of a fixed platform and a moving target? The answer is that we use these two designations as a matter of convenience. The fixed platform is the object upon which the observer resides, and the moving target is the remote object observed.

4.8 RELATIVISTIC SYMMETRY

Symmetry represents a helpful scheme that is often used in physics. In the theory of the relativistic Doppler effect, the symmetry involves motion. Relativistic symmetry requires that the behavior of two moving bodies does not depend on their individual velocities but only on their relative velocity. In classical (or Newton) time, the timelines of the two objects are always parallel. It follows that the outgoing Doppler factor k_1 differs from the returning Doppler factor k_2. In other words, there is a lack of symmetry in Newtonian mechanics. In relativistic (or Einstein) time, the timelines of the two objects in relative motion are not parallel. Such a situation is required in order that the outgoing Doppler factor k is the same as the returning Doppler factor k. In other words, there is a symmetry in Einstein mechanics. The requirement of symmetry in the relativistic Doppler factor means that the geometric mean must be used as the proper period.

In Figure 4.12, person A and person B are at the same spatial point at time zero. They are moving apart. It is a relative decision as to which person is moving and which person is standing still. Let us take the point of view of A. In other words, A is the observer. He assumes that he is stationary and B is moving. When A is at time 5, he takes the line that is perpendicular to B's timeline as the distance. This distance is the shortest distance from time 5 on A's timeline to B's timeline. This distance line meets B's timeline at time 4. According to A, his clock has ticked five times in 5 s, so A's

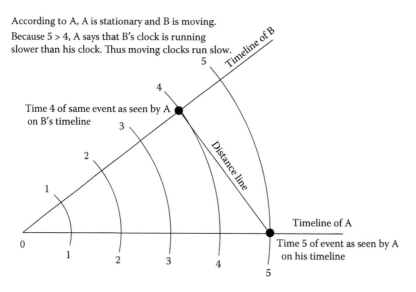

FIGURE 4.12 The observer A is stationary and the target B is moving.

clock has the rate of 5/5, which is 1 tick/s. According to A, B's clock has ticked four times in 5 s, so B's clock has the rate of 4/5, which is 0.8 tick/s. Because 0.8 < 1, A perceives that the clock of B has a slower rate than the clock of A. In other words, the moving clock is running at a slower rate than the stationary clock.

Let us take the point of view of B. In other words, B is the observer (see Figure 4.13). He assumes that he is stationary and A is moving. When B is at time 5, he takes the line that is perpendicular to A's timeline as the distance. This distance is the shortest distance from time 5 on B's timeline to A's timeline. This distance line meets A's timeline at time 4. According to B, his clock has ticked five times in 5 s, so B's clock has the rate of 5/5, which is 1 tick/s. According to B, A's clock has ticked four times in 5 s, so A's clock has the rate of 4/5, which is 0.8 tick/s. Because 0.8 < 1, B perceives that the clock of A is has a slower rate that the clock of B. In other words, the moving clock is running at a slower rate than the stationary clock.

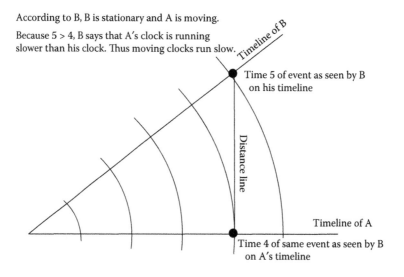

FIGURE 4.13 The observer B is stationary and the target A is moving.

Moving clocks are perceived to run slow. There is a symmetric disagreement about clock rates. If you go back and forth from the point of view of observer A to the point of view of observer B, you see the symmetry. The symmetry involves motion. Relativistic symmetry requires that the relativistic Doppler effect from A to B be the same as that from B to A. This same type of symmetry requires that geophysicists must migrate seismic data in order to get a correct image of the subsurface.

4.9 CONCLUSION

Seismic migration and the relativistic Doppler effect can each be expressed in terms of the Huygens construction. This duality shows that both systems exhibit the same symmetry, namely relativistic symmetry. In the case of seismic imaging, the symmetry is that the observer can be on either interface but still obtain the same overall image of the geologic section. In the case of the relativistic Doppler effect, the symmetry is that the behavior of two objects depends only upon their relative motion.

Let us summarize. In classical physics, time and space are absolute. In other words, the same timeline serves both the observer (on the fixed platform) and the observed (on the moving target). The fixed platform sends a signal of period s and receives the returning reflection of period r. The observer on the fixed platform naturally assumes that the period of the reflection at the moving target is the midpoint period t, namely the arithmetic mean of the two numbers s and r.

In modern physics, time and space are not absolute. In other words, one timeline serves the observer (on the fixed platform) and another (nonparallel) timeline serves the observed (on the moving target). The fixed platform sends a signal of period s and receives the returning reflection of period r. The observer on the fixed platform must assume that the period of the reflection at the moving target is the proper period p, namely the geometric mean of the two numbers s and r. We say that the midpoint time t is the dilated value of the proper time p. This peculiarity is called Einstein's dilation of time.

REFERENCES

1. Galilei, G., *Opere Il Saggiatore*, Rome, 1623.
2. Huygens, C., *Traité de la Lumière*, Paris, 1690.
3. Robinson, E. A. and S. Treitel, *Geophysical Signal Analysis*, Society of Exploration Geophysicists, Tulsa, OK, 2000.
4. Doppler, C., *Concerning the Colored Light of Double Stars*, Vienna, 1842.

5 Compressed Remote Sensing

Jianwei Ma, A. Shaharyar Khwaja, and M. Yousuff Hussaini

CONTENTS

5.1 INTRODUCTION

Remote sensing consists of data acquisition by means of imaging sensors and has many important applications, for example, imaging of Earth's surface and ocean floors, assessment of crop conditions, deep-space exploration, and so on. Sensor systems, which may be airborne or satellite borne, are classified into two categories [1]. The first one is passive sensors, such as aerial photography, infrared imaging, passive microwave radiometers, and so on, that make use of natural radiation emitted or reflected by the surface of the object being observed. The second category is that of active sensors, which have a transmitter and operate in the microwave range of the electromagnetic spectrum. These active sensors are represented by imaging and nonimaging radars, altimeters, scatterometers, and so on.

Remote sensing involves two basic stages: gathering data by the sensor and transmitting the digital data back to a processing center. Based on the conventional imaging principle, we can see that this process generates a large amount of data, for example, millions of pixels have to be stored momentarily when we take a picture using a megapixel camera and more pixels are often needed for higher resolution. The data that are stored digitally also have to satisfy the Nyquist criterion. Thus, the pixels require a huge storage space in memory or hard disk. In order to reduce the storage requirements, often an immediate data compression takes place inside the camera by an embedded microprocessor performing a discrete cosine transform for the JPEG 1992 format or a discrete wavelet transform for the JPEG 2000 format. For instance, a 6.1-megapixel digital camera senses 6.1×10^6 samples to construct an image, but in fact, a compressed image with average size smaller than 1 MB is saved in memory. The data compression process involves discarding a lot of small transform coefficients and reconstructing the remaining significant coefficients to obtain compressed pictures. The procedure is extremely wasteful of battery power for massive data acquisitions, particularly for large-scale aerospace remote sensing. Moreover, in order to transmit the data collected by satellites back to Earth, we have to achieve huge compression ratios, which introduce inevitable distortions and mosaic artifacts. These problems and constraints force us to think whether we can directly record fewer pixels without the use of an additional compression step to improve storage, power consumption, and transmission, without degrading spatial resolution and qualities.

Compressed sensing (CS) [2–5] provides a solution to the above-mentioned problems. It is a new concept allowing recovery of signals that have been sampled below the traditional Nyquist sampling rate. CS states that a compressible unknown signal can be recovered/reconstructed from incomplete sets of linear measurements by a specifically designed nonlinear recovery algorithm. It has been explained in detail by a number of references: Candès and Tao [2] show that measurement of a coded signal corrupted by errors can be recovered using linear programming (LP) under certain constraints on the coding matrix. Candès et al. [3] explain that a signal can be recovered from a random number of its known Fourier coefficients. Donoho [4] states that if an object has a sparse representation in a certain basis, a lesser number of nonadaptive measurements contains enough information to reconstruct the object via Basis pursuit. Candès and Wakin [5] propose a Nonuniform Sampler and Random Modulation Preintegration architecture for acquisition equipment based on CS as well as suggests applications for data compression, inverse problems, channel coding, and so on. Baraniuk [6] provides a tutorial review of CS and suggests its application for single-pixel imaging.

CS has been applied to optical and radar remote-sensing applications for instrument design as well as to signal and image processing. We have renamed it as compressed remote sensing (CRS). We find it necessary to mention that in the following literature review of CRS applications, we shall skip many useful articles on CS imaging that are not applied to remote sensing, for example, the Rice single-pixel camera [7], optical architectures for compressive imaging [8,9], and magnetic resonance imaging [10] to name a few. In Ref. [11], the author applies the single-pixel CS mechanism to remote sensing, which uses a single photodiode but carries out measurements multiple times. Multiple sensor-based single-time parallel Fourier domain imaging was also considered in Ref. [11]. Mahalanobis and Muise [12] use the CS architecture for application in surveillance systems. Bobin et al. [13] show that the use of CS allows improved resolution for astronomical data compression. Some extensions of CRS involving deblurring and data-fusion flying sensors have been presented in Refs. [14,15]. Similarly, CS has recently been applied to radar in a number of references. Herman and Strohmer [16] present high-resolution radar based on CS in the case of a small number of targets. Chen and Vaidyanathan [17] extend CS for multiple-input multiple-output (MIMO) radar. The CS concept has also been extended to Synthetic Aperture and Inverse Synthetic Aperture imaging radars: Huang et al. [18] use CS for through-wall imaging and Rilling et al. [19] present results for the case of Synthetic Aperture Radar (SAR) raw data compression and subsequent recovery using wavelet transforms. Wu et al. [20] and Lin et al. [42] propose estimation of velocities of moving targets using two-channel along-track interferometric SAR after cancelling the clutter signal from the received raw data. Zhang et al. [21] and Patel et al. [23] use CS to focus inverse SAR (ISAR) images of airplanes and achieve higher resolution, whereas Ender [22] use CS coupled with traditional image formation to focus the image of a LEO satellite. Stojanovic and Karl [24] present SAR imaging of moving targets in a high signal-to-clutter ratio (SCR) case and give examples using simulated data. In Refs. [23,24], the authors also show that, in the case of data loss, CS-based image formation performs much better than traditional image formation and results in higher resolution. Potter et al. [25] carry out a survey of the use of sparse reconstruction algorithms and randomized measurement strategies of CS in radar processing. Khwaja and Ma [26] describe two possible applications of CS for SAR: (1) estimation of moving target velocities in the case of high and low SCRs making use of clutter cancellation and (2) onboard compression of processed SAR images.

CRS essentially trades the online measurement and compression costs for offline computational cost. It offers many advantages such as the possibility of hardware simplification, reduction of data acquisition times and cost, achieving higher resolution and signal compression for optical remote sensing. CRS includes two stages: onboard encoding imaging and offline decoding recovery. The encoding imaging can be implemented by single or multiple optical sensors, which not only saves the cost of having additional sensors but also reduces the size and weight of onboard imaging instruments. CRS can save power consumption as the use of an immediate compression step is avoided. The encoding captures the compressed form of a scene directly, which is important in saving the computational cost when the data have to be transmitted back to Earth from satellites within a limited bandwidth. The data storage requirement is also reduced without degrading the spatial resolution and quality of pictures. Another potential advantage of CS imaging is that it can work much more easily in low light and outside the visible light spectrum due to the use of only one photon detector, so that it can be used for night vision and infrared imaging [7]. It can lead to new instruments with less storage space, less power consumption, and smaller size than currently used charge-coupled device cameras, which will match effective needs particularly for sensors sent very far away from Earth. For imaging radars such as SAR and ISAR, CRS can again help in reducing data storage space required as well as simplification of hardware. It can also lead to some specific applications such as moving target parameter estimation.

The application of CS depends on two main issues: the measurement process and a sparsity promoting basis/transform or a scenario where inherent sparsity present in a scene can be exploited. In this chapter, we present a few applications of CS. These applications concern compression and recovery of optical images as well as SAR and ISAR data. The use of a more general form of measurement matrix as well as the curvelet transform [27–29] is demonstrated along with recovery using the iterative curvelet transform. Similarly, for SAR and ISAR data, CS is described for a sparse scene as well as for a general multiple moving target parameter estimation problem in the cases of both high and low SCRs.

The remainder of this chapter is organized as follows. Section 5.2 describes the basics of CS, including the mathematical model and decoding methods. Section 5.3 relates to the CS applied to radar remote sensing, where we introduce SAR moving target parameter estimation, ultrawideband (UWB) through-wall imaging, ISAR imaging, and onboard compression of SAR data. In Section 5.4, we introduce the CS applied to optical remote sensing, including single-pixel remote sensing, CS deblurring and data-fusion-based CS imaging. Conclusions are drawn in Section 5.5.

5.2 BASICS OF COMPRESSED SENSING

5.2.1 Mathematical Model

CS can be described mathematically as a fundamental problem of recovering a finite signal x of size $N \times 1$ from a small set of measurements y. Let ϕ be a $K \times N$ CS measurement matrix. Here, $K \ll N$, that is, the rows of the measurement matrix are much fewer than the columns of the matrix. In practice, ϕ can denote an optical imaging lens. The encoding can be described as

$$y = \phi x + \varepsilon \tag{5.1}$$

where ε denotes possible measurement errors or noise. K denotes the number of measurements, and N denotes the dimension of the signal x. It seems hopeless to solve the ill-posed underdetermined linear systems of equations since the number of equations is much smaller than the number of unknown variables. Such problems are common in a variety of cases, for example, the number of sensors is limited or the measurements are extremely expensive and time consuming.

However, very often, x is sparse in a certain basis or compressible by a sparse transform ψ; thus, we have

$$y = \phi\psi\upsilon + \varepsilon, \quad x = \psi\upsilon \tag{5.2}$$

If the measurement matrix ϕ is noiselike incoherent in the ψ domain, that is, $\phi\psi$ satisfies a sufficient condition called the restricted isometry property [2–4], CS theory states that the sparse coefficients υ can be accurately recovered by solving a sparsity-promoting l_1-*norm* minimization or basis pursuit [2]

$$\min_{\upsilon} \left\| y - \phi\psi\upsilon \right\|_2^2 + \lambda \left\| \upsilon \right\|_{l_1} \tag{5.3}$$

The first term in Equation 5.3 is a penalty that represents closeness of the solution to the observed scenes. The second term is a regularization term that represents *a priori* sparse information of the original scenes. λ is a regularization parameter that can be tuned, and the best choice for λ depends on the variance of the noise and problem size parameters.

CS in fact extends traditional pixel sensing to linear projection sensing, that is, inner products between the scene and a set of test functions are measured. This can be explained by considering two simple cases: (1) if ϕ is an identity matrix or its sensing waveforms are spike-like Dirac functions, the method degenerates to traditional space-domain measurements obeying the Nyquist criterion. (2) If the sensing waveforms are sinusoids, y can be easily seen to be a vector of Fourier coefficients.

The choice of ϕ and ψ is critical for applications of CS to any problem. In general, we can select the measurement matrix ϕ by observing the following rules [5,11]:

1. *Universality:* ϕ should be noiselike incoherent/uncorrelated with a variety of sparse transform bases ψ for natural scenes. That is to say, ϕ and ψ are as different as possible; thus, sampling waveforms have a dense representation in ψ.
2. *Optimal performance:* For a "good" ϕ, the minimal number of necessary measurements to obtain a superresolution result should be close to the theoretical bound of $O(S \log N)$ for an S-sparse N-dimension signal. S-sparse means that an object has at most S nonzero entries in a transform domain. The greater the incoherence between ϕ and ψ, the smaller the number of measurements needed. Incoherence means that all the measurements make the same contribution to the recovery of the observed signal.
3. *Fast computation:* Recovery algorithms involve repeated applications of $\phi\psi$. Therefore, a rapidly computable ϕ is desirable for both encoding imaging and decoding recovery.
4. *Physically realizable:* ϕ should be conveniently implemented by hardware such as an optical or analog system.

Frequently used measurement matrices ϕ are random matrices (e.g., Gaussian random matrix, jittered sampling matrix [30], sparse binary sampling [31]), random partial orthogonal matrices, and random-like transforms (e.g., Noiselet transform [32]), because they are incoherent for most sparse transforms. However, to design optimal and deterministic measurement matrices satisfying all of the aforementioned properties is still an open research area. The commonly used sparse transforms ψ include wavelets, curvelets, contourlets, or a combination of these transforms depending upon the features present in an observed scene.

5.2.2 Decoding Method

Different decoding/recovery algorithms can be used such as LP [2], orthogonal matching pursuit [33], iterative shrinkage/thresholding (IST) [34–37], and so on. The choice of the recovery algorithm may depend on various factors, such as the size of the basis, computational complexity, and correlation between the basis elements. In this chapter, we mainly apply iterative curvelet thresholding (ICT), that is, using curvelets in IST for recovery of CS data [11,14,37].

The curvelet transform [27–29] is a new geometric multiscale transform that can be interpreted as an anisotropic geometric wavelet transform allowing an optimal sparse representation of objects with smooth-curve singularities. Unlike wavelets, the system of curvelets is indexed by three parameters: j, l, k for scale, orientation, and location, respectively.

1. A scale, $2^{-j}, j \in N_0$
2. An equispaced sequence of rotation angles, $\theta_{j,l} = 2\pi l 2^{-j/2}, 0 \leq l \leq 2^{-j/2} - 1$
3. A position, $x_k^{(j,l)} = R_{\theta_{j,l}}^{-1}(k_1 2^{-j}, k_2 2^{-j/2})^{\mathrm{T}}, (k_1, k_2) \in Z^2$

where $R_{\theta_{j,l}}$ denotes the rotation matrix with angle $\theta_{j,l}$. The curvelets are defined by

$$\psi_{j,l,k}(x) := \psi_j\left(R_{\theta_{j,l}}\left(x - x_k^{(j,l)}\right)\right) \tag{5.4}$$

with $x = (x_1, x_2) \in R^2$. Let $\mu = (j, l, k)$ be the collection of the triple index. The curvelet coefficients are given by $c_\mu(f) := \langle f, \psi_\mu \rangle$. The forward and inverse curvelet transforms have the same computational cost of $O(N^2 \log N)$ for an $N \times N$ image [28]. The motivation to use the curvelet transform for CS is that most natural scenes consisting of line-singularity edges are sparse in the curvelet domain.

The recovery is carried out by defining a thresholding transform function as

$$S_\tau(f, \psi) = \sum_\mu \tau\left(c_\mu(f)\right)\psi_\mu \tag{5.5}$$

where τ can be a hard thresholding function given as

$$\tau_\sigma(x) = \begin{cases} x, |x| \geq \sigma \\ 0, |x| < \sigma \end{cases} \tag{5.6}$$

In the context of CS recovery, the thresholding function can be written as follows [8]:

$$x_{p+1} = S_\tau\left(x_p + \phi^{\mathrm{T}}(y - \phi x_p), \psi\right) \tag{5.7}$$

where ψ in this case is the curvelet transform and p denotes the iteration number. The iterative process is stopped if the difference of the recovered x between two successive iterations is very small, that is, $\|x_{p+1} - x_p\| < \zeta$. This decoding in CRS is implemented by ground digital computers that undertake most of computational complexity.

5.3 COMPRESSED SENSING APPLIED TO RADAR REMOTE SENSING

In the following, we describe CS applications to radar remote sensing. In these applications, the observed scene is inherently sparse, that is, it contains a few strong scatterers. First, we describe the image acquisition geometry and the definition of the basis in the CS scenario. This is followed by a discussion of a few applications such as moving target parameter estimation, through-wall imaging, focusing of ISAR data, as well as onboard compression.

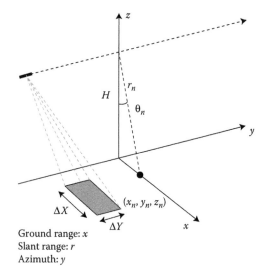

FIGURE 5.1   SAR acquisition geometry.

A general acquisition geometry of SAR is shown in Figure 5.1. The sensor moves along the azimuth/cross-range direction y with a velocity V, emitting pulses at carrier frequency f_c in the range time direction denoted by t. Consider a moving point n at initial slant-range position r_n and azimuth position y_n having a unit reflectivity. The acquired demodulated raw data for such a point can be written as

$$s_n(t, y) = p(t - 2d_{\text{movn}}(y)/c)\exp(-j2\pi f_c t) \tag{5.8}$$

where

$$d_{\text{movn}}(y) = \sqrt{(x_n - v_{xn}y/V)^2 + H^2 + (y_n - y - v_{yn}y/V)^2} \tag{5.9}$$

is the distance between the sensor position and the moving point. v_{xn} represents the ground range velocity of the point at ground position x_n while v_{yn} represents the azimuth velocity of the point at azimuth position y_n. $p(t)$ is the transmitted pulse. For a static point, the same equation can be used by substituting $v_{xn} = 0$ and $v_{yn} = 0$. The acquired raw data are processed by an image formation process that consists of a matched filtering and interpolation process [38,39].

If $p(t)$ is a chirp signal of length T_p and K is the chirp rate, then we have

$$p(t) = \text{rect}(t/T_p)\exp(j2\pi f_c t + j\pi Kt^2) \tag{5.10}$$

where

$$\text{rect}(t/T_p) = \begin{cases} 1, |t| \leq T_p/2 \\ 0, |t| > T_p/2 \end{cases}$$

The raw data are described by

$$s_n(t, y) = \exp\left(-j4\pi f_c d_{\text{movn}}(y)/c - j\pi K\left(t - 2d_{\text{movn}}(y)/c\right)^2\right) \tag{5.11}$$

The effect of v_{xn} in Equation 5.11 is to cause a shift of a point in the azimuth direction while v_{yn} causes a smearing effect in the azimuth direction when processed by normal image formation considering stationary points [40,41]. In case of multiple moving targets, the total raw data are given as a superposition of raw data from every target:

$$s(t, y) = \sum_n s_n(t, y) \tag{5.12}$$

Note that, in the case of ISAR, the sensor remains stationary; however, the basic geometric notations and processing procedure remain the same.

CS can be applied to process the raw data by defining the following elements [26].

5.3.1 MEASUREMENT MATRIX ϕ

The measurement matrix ϕ is normally chosen as a random matrix as it satisfies the restricted isometry property (RIP) property. However, hardware for such a measurement matrix is not easy to design at analog-to-digital converter level and this matrix is more suitable for compression purposes when data have already been acquired. In the case of imaging radar, ϕ has been used as a downsampling operator in the across-track direction in Ref. [26], along-track direction in Ref. [20] as well as both across- and along-track directions in Ref. [18]. These types of measurement matrices can represent an analog-to-digital converter operating at a lower sampling rate than the traditional rate as well as a system acquiring fewer samples in the azimuth direction. Using CS along with such measurement matrices offer the advantage of raw data processing/parameter extraction with simpler acquisition equipment and lesser data compared to the traditional case.

5.3.2 DEFINITION OF BASIS ψ

It is assumed that the observed scene is sparse; therefore, the basis consists of raw data for all the possible motion parameters, including static points in the observed scene. Consider the scene to consist of $n_r \times n_y$ pixels, where the array size depends on the sampling time in the range and azimuth directions. The ground range and azimuth velocities can be arranged in two one-dimensional vectors V_x and V_y, respectively, as follows [26]:

$$V_x = \left[0 \ldots v_{N_{vx}} \right] \tag{5.13}$$

$$V_y = \left[0 \ldots v_{N_{vy}} \right] \tag{5.14}$$

where N_{vx} and N_{vy} give the total number of velocity data points in the ground range and azimuthal directions. Thus, these vectors consist of possible velocities over which the reconstruction process can be carried out. Raw data for a single point and for one element v_x of V_x and one element v_y of V_y can be rewritten as

$$s_n(t, y, v_x, v_y) = \exp\left(-j4\pi f_c d_{\text{movn}}(y)/c - j\pi K\left(t - 2d_{\text{movn}}(y)/c \right)^2 \right) \tag{5.15}$$

where $d_{\text{movn}}(y)$ is given by Equation 5.9. These raw data can be stacked columnwise for ease of implementation. Therefore, $s_n(t, y, v_x, v_y)$, which is composed of $N_r \times N_y$ pixels, is stacked as an $N_r N_y \times 1$ vector. For each velocity combination, the basis $\psi(v_x, v_y)$ consisting of $N_r N_y \times n_r n_y$ pixels is given as follows [26]:

$$\psi(v_x, v_y) = \left[s_1(t, y, v_x, v_y) \ldots s_{n_r n_y}(t, y, v_x, v_y) \right] \tag{5.16}$$

The final basis consists of generating $\psi(v_x,v_y)$ for all the possible velocity combinations and can be stored as a matrix of size $N_r N_y \times N_{v_x} N_{v_y} n_r n_y$. This is given as [26]

$$\psi = [\psi(v_x = 0, v_y = 0) \dots \psi(v_x = v_{N_{vx}}, v_y = v_{N_{vy}})] \tag{5.17}$$

The reconstruction process can be carried out by searching over this basis. From an implementation point of view, it is necessary that the processing system should have enough memory to contain all the basis elements. This memory requirement increases with the number of pixels that is governed by the size of a scene, the number of velocity combinations, as well as the raw data size that depends upon the system parameters and configuration.

5.3.3 Definition of Reconstructed Estimate υ

The reconstructed estimate υ will give the amplitudes of estimates for each point in the scene and for every possible velocity combination given in the basis [26], that is,

$$\upsilon = [\upsilon(v_x = 0, v_y = 0) \dots \upsilon(v_x = v_{N_{vx}}, v_y = v_{N_{vy}})] \tag{5.18}$$

The estimate will have a nonzero value at a location that corresponds to an active basis element, whereas a zero value will show an inactive basis element. Its total size is $N_{v_x} N_{v_y} n_r n_y \times 1$ pixels and can be rearranged into $N_{v_x} N_{v_y}$ matrices of size $n_r n_y$ in order to show the estimated reflectivities at different velocity combinations, for example, if a basis is defined as in Equation 5.17, the first matrix of size $n_r n_y$ formed from υ, that is, $\upsilon(v_x = 0, v_y = 0)$ will indicate reconstructed stationary points in the observed scene. Each combination of motion parameter is given by a different matrix; therefore, it is possible to estimate motion parameters and reflectivities of more than one point present within the same pixel but having different motion parameters.

Note that in case of a different pulse or acquisition scenario, the method of basis definition remains the same, only Equation 5.17 will change to take into account different pulse or acquisition types, such as ISAR. The main constraint is that the observed scene should be sparse, described as follows for four possible scenarios.

5.3.3.1 Moving Target Parameter Estimation

A scene consisting of strong moving targets embedded in clutter can be considered as sparse. Such an environment is called a high SCR environment and is an ideal candidate for CS. In the case of several moving targets, each target will undergo a different amount of azimuth shift and defocusing due to different velocities. In order to counteract these effects and get focused images, these velocities should be taken into account during processing and for this purpose, the velocities should be estimated first. One traditional method is to carry out processing using different matched filters taking into account different velocities until a point is clearly focused. This process is computationally intensive and has to be repeated for each target in the case of multiple moving targets.

There are two main issues associated with scenes consisting of moving targets: estimation of velocities and focusing of images using the velocity estimates. CS provides a convenient way of estimation of motion parameters as well as reflectivities of moving targets using lesser amount of data. It also helps in focusing by giving original positions of moving targets. Moreover, CS offers the possibility of estimating multiple moving target parameters.

Application of CS to moving target parameter estimation and focusing consists of definition of the relevant undersampling matrix and basis according to Equation 5.17, followed by reconstruction. However, in case of low SCR, the assumption of the observed scene being sparse is no longer valid and application of CS will not lead to correct results as demonstrated by Figure 5.2, where the

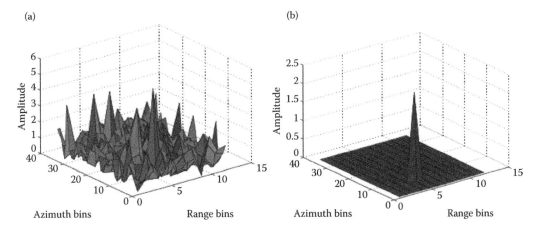

FIGURE 5.2 (a) A scene with low SCR. (b) CS-based reconstruction. (Adapted from A. S. Khwaja and J. Ma, *IEEE Trans. Instrum. Meas.*, 60(8), 2848–2860, 2011.)

observed scene consists of four moving points in a low SCR environment. Raw data are simulated corresponding to this scene with undersampling in the range direction and reconstructed using CS. It can be observed from the reconstructed results that only a single moving point is recovered. Therefore, it is necessary to apply a preprocessing step for clutter cancellation that enhances the sparsity of the scene enabling the application of CS [22,26].

Clutter cancellation can be applied in a straightforward way to physically separated multichannel along-track interferometric SAR so as to recover moving target parameter estimation as shown in Refs. [20,42]. Ref. [20] also provides one further application of CS, that is, to recover moving targets when the raw data suffer from aliasing due to undersampling in azimuth direction. However, depending upon the type of clutter cancellation, if part of the signal from moving targets is also suppressed, it is necessary to modify the basis ψ [22]. Thus, if the clutter cancellation operator is defined as F, the modified basis should be defined as $F\psi$. For a single-channel SAR system, F can be defined as a low-pass filter if f_{PRF}, that is, the sampling frequency in the azimuth direction, is considerably greater than the bandwidth B_y generated due to the antenna beamwidth in the azimuth direction. All the data from clutter are contained within B_y, while the data obtained from moving targets are shifted in azimuth frequency and there will be some part that will be outside $-B_y/2$ or $B_y/2$. Therefore, the low-pass filter should attenuate the frequencies ranging from $-B_y/2$ to $B_y/2$ to cancel the clutter. The frequencies ranging from $-f_{PRF}/2$ to $-B_y/2$ and $B_y/2$ to $f_{PRF}/2$ will contain part of the signals from the moving targets. This is shown in Figure 5.3.

In situations where f_{PRF} is not much greater than B_y, this type of filter cannot be applied as most of the signal of moving targets will be filtered out along with that of stationary clutter. However, one particular case where such a clutter cancellation filter is practically applicable is that of polarimetric data where HV and VH channels can be combined to get data sampled at twice the f_{PRF}, that is, the combined data will consist of frequencies ranging from $-f_{PRF}$ to f_{PRF} [43]. In this case, F can be formed as a low-pass filter attenuating lower frequencies from $-f_{PRF}/2$ (or $-B_y/2$) to $f_{PRF}/2$ (or $B_y/2$), which will filter out the signal from stationary clutter. The filtered data will contain azimuth frequencies ranging from $-f_{PRF}$ to $-f_{PRF}/2$ (or $-B_y/2$) and $f_{PRF}/2$ (or $B_y/2$) to f_{PRF} and thus will contain part of the signals from the moving targets. As an example, a scene consisting of four moving targets in a low SCR environment is shown in Figure 5.4. Raw data are generated from such a scene with undersampling in range direction [26] and clutter cancellation is applied before CS-based reconstruction. The reconstructed results demonstrate the presence of all four points present in the scene that is an improvement compared to the situation shown in Figure 5.2, where only one point was identified. However, a limitation of the type of clutter cancellation filter is that the data from

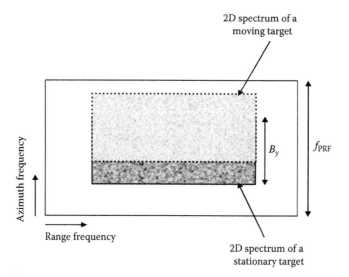

FIGURE 5.3 Bandwidth of stationary and moving targets.

targets moving only in the azimuth direction will be cancelled along with clutter as data from such type of targets do not undergo any shift in azimuth frequency.

5.3.3.2 UWB Through-Wall Imaging

CS can be applied to UWB through-wall imaging to reduce data volume as well as acquisition time. This has been shown in Ref. [18] where the observed scenes are assumed to comprise of a few point like scatterers. As mentioned in the preceding section, the basis can be modified to take into account different configurations, for example, for the case of through-wall imaging, the transmitted signal is a stepped-frequency signal, resulting in the following raw data for the nth point:

$$s_n(f, y) = \exp(-j2\pi f d_n(y)/c) \tag{5.19}$$

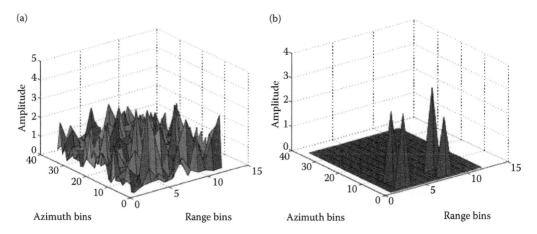

FIGURE 5.4 (a) A scene with low SCR. (b) CS-based reconstruction after clutter cancellation. (Adapted from A. S. Khwaja and J. Ma, *IEEE Trans. Instrum. Meas.*, 60(8), 2848–2860, 2011.)

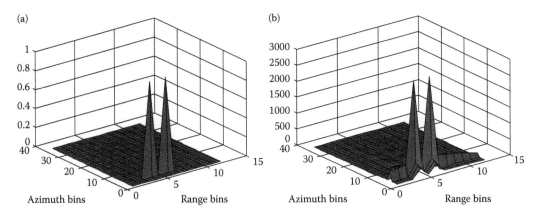

FIGURE 5.5 (**See color insert.**) (a) Result of CS-based reconstruction. (b) Result of traditional reconstruction.

where f represents the frequency of the transmitted signal and $d_n(y)$ represents the sensor-target distance.

Based on the raw data, the corresponding basis can be written as follows for a total number of N points:

$$\psi = \left[s_1(f,y) \ldots s_N(f,y) \right] \tag{5.20}$$

CS-based reconstruction for such a scene also offers the advantage of obtaining images with better resolution as well as lower side lobes compared to traditional image formation algorithms as shown in Figure 5.5. Here, the observed scene consists of two points; CS-based reconstruction shows two sharp peaks, whereas, traditional reconstruction shows peaks having noticeable side lobes that can hamper identification of weaker objects.

5.3.3.3 ISAR Imaging

ISAR is especially suited for CS application as it contains a few point scatterers in a weak stationary background. CS offers the advantage of giving better resolution compared to traditional methods using a lesser number of measurements in the cross-range direction as shown in Figure 5.6. Here,

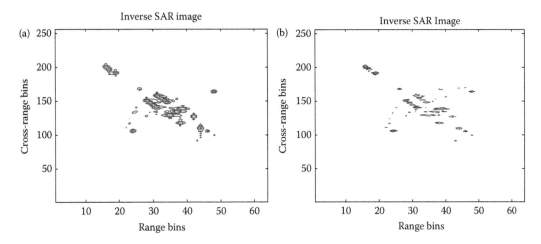

FIGURE 5.6 (**See color insert.**) (a) Traditional reconstruction (30% samples). (b) CS-based reconstruction (30% samples).

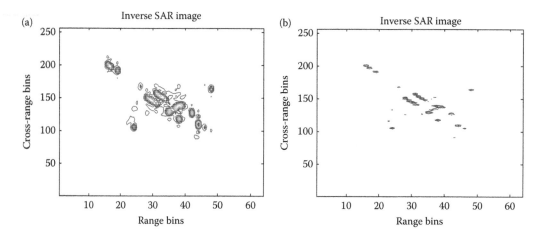

FIGURE 5.7 (**See color insert.**) (a) Traditional reconstruction (12% samples). (b) CS-based reconstruction (12% samples).

traditional image reconstruction with Fourier transforms in the cross-range directions using 30% of the cross-range data from Ref. [40] gives poorer resolution compared to CS-based image formation from the same data. This also has important implications in radar design as described in Ref. [22]. CS has been applied to ISAR data in Refs. [21–23] mainly to demonstrate the advantage of CS-based reconstruction over Fourier transform-based reconstruction. In the ISAR imaging scenario, assuming the data to be already focused in the range direction, raw data in the cross-range direction are defined as follows for the nth point:

$$s_n(y) = \exp(-j2\pi f_n y/V) \tag{5.21}$$

This results in the following construction for the basis, considering a total number of N points:

$$\psi = \left[s_1(y) \dots s_N(y)\right] \tag{5.22}$$

CS is useful for image formation in the case of a maneuvring target. Such a target, when viewed over a shorter cross-range duration, may be considered as static. However, shorter cross-range observation time means lesser cross-range data that leads to poorer resolution when using Fourier-based reconstruction, whereas CS can give a higher-resolution image using the same amount of data. Zhang et al. [21] also shows that the number of scatterers that are recovered successfully using CS decreases with the number of samples: For an S-sparse N-dimensional signal, the number of required measurements for successful reconstruction is equal to $O(S \log N)$. Using lesser number of measurements will recover lesser number of scatterers as shown in Figure 5.7, where 12% of cross-range data were used for image formation. Compared to traditional reconstruction, the cross-range resolution of the image obtained using CS-based reconstruction is better but the overall form of the image is less defined. This phenomenon will become worse in case of low signal-to-noise ratio (SNR). Thus, whereas CS offers the advantage of achieving higher resolution using a lesser number of samples, it may also lead to partially reconstructed images.

5.3.3.4 Onboard Compression of SAR Image

CS provides a possibility for fast onboard compression because it only needs a random downsampling or a multiplication with a random matrix. The ICT for offline recovery has been shown to improve reconstructed image quality in case of SAR images, which are not highly compressible due

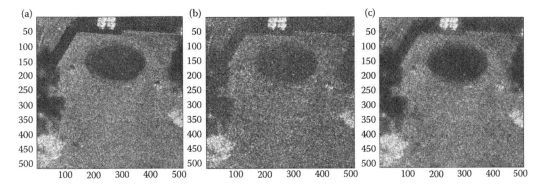

FIGURE 5.8 CS decoding of a MiniSAR image. (a) Original image. (b) Reconstructed by wavelets. (c) Reconstructed by curvelets. (Adapted from A. S. Khwaja and J. Ma, *IEEE Trans. Instrum. Meas.*, 60(8), 2848–2860, 2011.)

to the presence of multiplicative speckle noise. This is seen in Figure 5.8c, where it can be remarked that using the curvelet transform results in better image quality compared to the result using the wavelet transform shown in Figure 5.8b [26]. Here, we consider 50% of the measurements using a random matrix for an image formed from data acquired by Sandia Research Laboratories' MiniSAR sensor. The size of the considered image is 512×512 pixels and the size of the compressed image is 256×512 pixels. The algorithm in Equation 5.7 is used for decoding. Thus, CS offers the feature of computationally simple online compression of SAR images or data that can be used for reducing data storage size as well as transmission bandwidth.

5.4 COMPRESSED SENSING APPLIED TO OPTICAL REMOTE SENSING

In this section, we describe applications of CS for optical remote sensing. The observed scene in this case is sparse in wavelet or curvelet basis. We briefly describe a single-pixel remote sensing system, CS deblurring, and a data-fusion-based flying sensor.

5.4.1 SINGLE-PIXEL REMOTE SENSING

Single-pixel remote sensing is based on the single-pixel camera presented in Refs. [6,7,11] using 0/1 random binary matrices for the observation matrix ϕ. This camera is basically an optical computer comprising two lenses, a single photon detector, an analog-to-digital converter, and a digital micromirror device (DMD). For a mathematical interpretation, we first reshape the 2-D image into a 1-D signal $x = \{x_1, \ldots, x_i, \ldots, x_N\}$ and assemble the 2-D ϕ matrix using the random vector $\phi_m = \{\phi_{m,1}, \ldots, \phi_{m,i}, \ldots, \phi_{m,N}\}$ ($m = 1, \ldots, K$) in each row. The random vector ϕ_m is steered by orienting the bacterium-sized mirrors in DMD. The reflecting light field is focused on the single photodiode by lens in order to get one measurement $y_m = \sum_{i=1}^{N} \phi_{m,i} x_i$. This process is repeated K times using different ϕ_m to obtain all measurements $y = \phi x$. Such a repetitive operation can be accomplished in a momentary time by tuning the directions of mirrors in DMD. The camera is single-pixel but multitime (SPMT) that carries out the sequential imaging, which is slower than the parallel imaging of a classical charged coupled device camera. However, the SPMT does not need an additional compression step that is required by classical imaging.

As an alternative, one can apply a 2-D Noiselet transform [13,32] or jittered sampling [15,30] as the measurement matrix ϕ. Figure 5.9 shows a case of SPMT imaging of Moon's surface. Figure 5.9a is the original unknown scene. Figure 5.9b and c is obtained by the same decoding method of ICT but with different measurement matrices. In this case, the Noiselet measurement matrix used in Figure 5.9c leads to better results than random matrix used in Figure 5.9b as seen by the SNR.

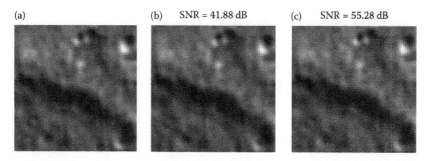

FIGURE 5.9 SPMT imaging for Moon's surface. (a) Original unknown scene. (b) Random measurement encoding and curvelet decoding. (c) Noiselet measurement encoding and curvelet decoding. (Adapted from J. Ma, *IEEE Geosci. Remote Sens. Lett.*, 6(2), 676–680, 2009.)

5.4.2 CS Deblurring

Deblurring is an inverse problem encountered in a wide variety of signal and image processing fields, including physical, optical, medical, and astronomical applications. For instance, images observed by satellites or telescopes are usually blurred due to the effects of the limitations of aperture, focal length, motion, or atmospheric turbulence. Here, we further consider a so-called CS deblurring problem [14]: Can we still obtain clear images from highly incomplete measurements when blurring disturbances occur?

The problem of CS deblurring can be symbolically stated as [14]

$$y = \phi H f + \varepsilon \tag{5.23}$$

Compared to Equation 5.1, we have a low-pass linear blurring operator H in Equation 5.23. The goal of deblurring is to retrieve f from y. The nonlinear deblurring processing is a mathematically ill-posed problem, which can be interpreted as the inversion of a low-pass filtering, a backward–forward diffusion, or a minimization of entropy. One serious problem is the amplification of noise by H^{-1}. In Ref. [14], a decoding algorithm based on the Poisson singular integral (PSI) and ICT is proposed to correct the deblurring problem with surprisingly incomplete measurements.

We give an illustration of primary results from Ref. [14]. Consider atmospheric remote-sensing data with detailed features. Figure 5.10a is the original cloud image, and Figure 5.10b is a blurred image. Figure 5.10c is the zero-filling reconstruction, that is, using the PSI operator without ICT in the CS deblurring framework. Amplified noise can be seen obviously in this result. Figure 5.10d is the result recovered by our method (i.e., using both PSI and ICT), in which detailed features have been recovered well from the 30% incomplete measurements.

5.4.3 Data-Fusion CS Imaging

CS can flexibly take advantage of multiple observations. In this section, we consider a flying sensor problem with highly overlapping observations [13,15]. Let X be a full scene to be observed. The raster scan delivers a sequence of L highly overlapping images $\{x_i\}$, $i = 1,\ldots,L$, each of which is a small area Ω_i of X. Let P_{Ω_i} be the projector from $X \rightarrow x_i$ (i.e., a small patch x_i from the full sky X); then, we have

$$x_i = P_{\Omega_i}(X) \tag{5.24}$$

The multiple observations are

$$y_i = \phi_i P_{\Omega_i}(X) + \varepsilon_i \tag{5.25}$$

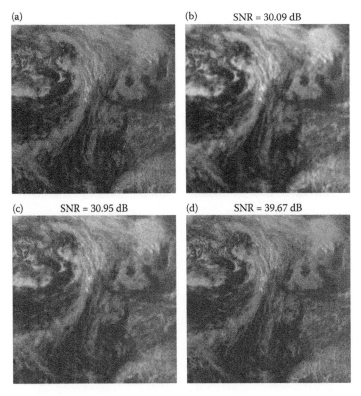

FIGURE 5.10 CS deblurring for atmospheric remote sensing by multipixel but single-time (MPST). (a) Original cloud image. (b) Blurred image (SNR = 30.09 dB). (c) Deblurring by zero-filling reconstruction, that is, using PSI but without ICT (SNR = 30.95 dB). (d) Deblurring using both PSI and ICT (SNR = 39.67 dB). (Adapted from J. Ma and F. X. Le Dimet, *IEEE Trans. Geosci. Remote Sens.*, 47(3), 792–802, 2009.)

where y_i and ϕ_i denote the observation and corresponding CS measurement matrices for the ith patch and ε_i represents possible measurement noise. Assuming that the scene X is sparse in a transform ψ, the CS decoding procedure is as follows [13]:

$$\min_X \left\{ \left\| \psi X \right\|_{l_1} + \lambda \sum_{i=1}^{L} \left\| y_i - \phi_i P_{\Omega_i}(X) \right\|_{l_2} \right\} \tag{5.26}$$

In the experiments, we use the Noiselet transform with jittered downsampling and the Fourier transform with pseudo-random downsampling for the CS measurement matrix ϕ [15]. Recovering X directly from L observations is better than recovering x_i independently and then merging them to obtain X [13,15]. Figure 5.11 shows a comparison of results by decoding of Moon's surface from one-column-shift overlapping observations using different CS measurement matrices. Figure 5.11a and b is obtained by sparse binary sampling [31]. Figure 5.11c and d is obtained by the Noiselet transform with 2D hexagonal jittered sampling. Figure 5.11e and f is obtained by the Noiselet transform with 1D jitter-based circular sampling. Figure 5.11g and h is obtained by the Fourier transform with density-varied downsampling [15].

5.5 CONCLUSION

In this chapter, we introduced the basic knowledge of CS in remote sensing. We renamed it as CRS(compressed remote sensing, or computational remote sensing), and reviewed some recent

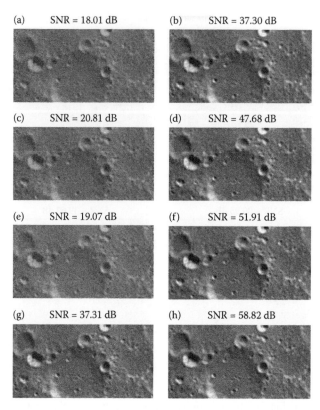

FIGURE 5.11   Decoding of Moon's surface from one-column-shift overlapping observations using different CS measurement matrices. Left column: 5% measurements. Right column: 10% measurements. (a, b) Sparse binary sampling. (c, d) Noiselet transform with 2D hexagonal jittered sampling. (e, f) Noiselet transform with 1D jitter-based circular sampling. (g, h) Fourier transform with density-varied downsampling. (Adapted from J. Ma and M. Y. Hussaini, *IEEE Trans. Instrum. Meas.*, 60(9), 3128–3139, 2011.)

applications involving SAR moving target parameter estimation, UWB through-wall imaging, ISAR imaging, single-pixel optical imaging, CS deblurring, and data-fusion CS imaging. Primary results show the CS theory to be promising for reducing online measurement cost in remote sensing and can also lead to applications for parameter extraction using reduced amount of data in case of sparse scenes. It can produce new concepts for building next-generation mechanisms and instrumentation for imaging and processing. Some application areas that can further benefit from advances in CRS are improvements in compression of data acquired by imaging radars, design of suitable sampling matrices, 3D radar imaging [44], and differential SAR tomography [45] as well as exploitation of joint sparsity in polarimetric and interferometric SAR data.

REFERENCES

1. G. Franceschetti and R. Lanari, *Synthetic Aperture Radar Processing*, Boca Raton, FL: CRC Press, 1999.
2. E. Candès and T. Tao, Decoding by linear programming, *IEEE Trans. Inform. Theory*, 51(12), 4203–4215, 2005.
3. E. Candès, J. Romberg, and T. Tao, Robust uncertainty principles: Exact signal reconstruction from highly incomplete frequency information, *IEEE Trans. Inform. Theory*, 52(2), 489–590, 2006.
4. D. Donoho, Compressed sensing, *IEEE Trans. Inform. Theory*, 52(4), 1289–1306, 2006.
5. E. Candès and M. Wakin, An introduction to compressive sampling, *IEEE Signal Process. Mag.*, 25(2), 21–30, 2008.
6. R. Baraniuk, Compressed sensing, *IEEE Signal Process. Mag.*, 24(4), 14–20, 2007.

7. M. Duarte, M. Davenport, D. Takhar, J. Laska, T. Sun, K. Kelly, and R. Baraniuk, Single-pixel imaging via compressive sampling, *IEEE Signal Process. Mag.*, 25(2), 83–91, 2008.
8. M. Neifeld and J. Ke, Optical architectures for compressive imaging, *Appl. Opt.*, 46(22), 5293–5303, 2007.
9. R. Marcia, R. Willett, and Z. Harmany, Compressive optical Imaging: Architectures and algorithms. In: *Optical and Digital Image Processing: Fundamentals and Applications* (eds G. Cristobal, P. Schelkens, and H. Thienpont), Wiley-VCH Verlag GmbH & Co. KGaA, Weinheim, Germany, pp. 1–22, 2010.
10. M. Lustig, D. Donoho, and J. Pauly, Sparse MRI: The application of compressed sensing for rapid MR imaging, *Magn. Reson. Med.*, 58(6), 1182–1195, 2007.
11. J. Ma, Single-pixel remote sensing, *IEEE Geosci. Remote Sens. Lett.*, 6(2), 676–680, 2009.
12. A. Mahalanobis and R. Muise, Object specific image reconstruction using a compressive sensing architecture for application in surveillance systems, *IEEE Trans. Aerospace Electron. Syst.*, 45(3), 1167–1180, 2009.
13. J. Bobin, J. Starck, and R. Ottensmer, Compressed sensing in astronomy, *IEEE J. Sel. Topics Signal Process.*, 2(5), 718–726, 2008.
14. J. Ma and F. X. Le Dimet, Deblurring from highly incomplete measurements for remote sensing, *IEEE Trans. Geosci. Remote Sens.*, 47(3), 792–802, 2009.
15. J. Ma and M. Y. Hussaini, Extensions of compressed imaging: Flying sensor, coded mask, and fast decoding, *IEEE Trans. Instrum. Meas.*, 60(9), 3128–3139, 2011.
16. M. Herman and T. Strohmer, High-resolution radar via compressed sensing, *IEEE Trans. Signal Process.*, 57(6), 2275–2284, 2009.
17. C. Chen and P. Vaidyanathan, Compressed sensing in MIMO radar, *Asilomar Conference on Signal, Systems, and Computers*, Pacific Grove, CA, November 2008.
18. Q. Huang, L. Qu, B. Wu, and G. Fang, UWB through-wall imaging based on compressive sensing, *IEEE Trans. Geosci. Remote Sens.*, 48(3), 1408–1415, 2010.
19. G. Rilling, M. Davies, and B. Mulgrew, Compressed sensing based compression of SAR raw data, *Proceedings of SPARS'09 (Signal Processing with Adaptive Sparse Structured Representations)*, Saint-Malo, France, April 2009.
20. Q. Wu, M. Xing, C. Qiu, B. Liu, Z. Bao, and T. Yeo, Motion parameter estimation in the SAR system with low PRF sampling, *IEEE Geosci. Remote Sens. Lett.*, 7(3), 450–454, 2010.
21. L. Zhang, M. Xing, C. Qiu, J. Li, and Z. Bao, Achieving higher resolution ISAR imaging with limited pulses via compressed sensing, *IEEE Geosci. Remote Sens. Lett.*, 6(3), 567–571, 2009.
22. J. Ender, On compressive sensing applied to radar, *Signal Process.*, 90(5), 1402–1414, 2010.
23. V. Patel, G. Easley, D. Healy, and R. Chellappa, Compressed synthetic aperture radar, *IEEE J. Selected Topics Signal Process.*, 4(2), 244–254, 2010.
24. I. Stojanovic and W. Karl, Imaging of moving targets with multi-static SAR using an overcomplete dictionary, *IEEE J. Selected Topics Signal Process.*, 4(1), 164–176, 2010.
25. L. Potter, E. Ertin, J. Parker et al., Sparsity and compressed sensing in radar imaging, *Proc. IEEE*, 98(6), 1006–1020, 2010.
26. A. S. Khwaja and J. Ma, Applications of compressed sensing for SAR moving-target velocity estimation and image compression, *IEEE Trans. Instrum. Meas.*, 60(8), 2848–2860, 2011.
27. E. Candès and D. Donoho, New tight frames of curvelets and optimal representations of objects with piecewise C^2 singularities, *Commun. Pure Appl. Math.*, 57(2), 219–266, 2004.
28. E. Candès, L. Demanet, D. Donoho, and L. Ying, Fast discrete curvelet transforms, *Multiscale Model. Simul.*, 5(3), 861–899, 2006.
29. J. Ma and G. Plonka, The curvelet transform, *IEEE Signal Process. Mag.*, 27(2), 118–133, 2010.
30. G. Tang, R. Shahidi, J. Ma, and F. Herrmann, Application of randomized sampling schemes to curvelet based sparsity-promoting seismic data recovery, Preprint, UBC, Vancouver, Canada.
31. R. Berinde and P. Indyk, Sparse recovery using sparse random matrices, preprint, *MIT CSAIL TR-2008-001*, 2008.
32. R. Coifman, F. Geshwind, and Y. Meyer, Noiselets, *Appl. Comput. Harmon. Anal.*, 10(1), 27–44, 2001.
33. J. Tropp and A. Gilbert, Signal recovery from random measurements via orthogonal matching pursuit, *IEEE Trans. Inform. Theory*, 53(12), 4655–4666, 2007.
34. I. Daubechies, M. Defrise, and C. De Mol, An iterative thresholding algorithm for linear inverse problems with a sparsity constraint, *Comm. Pure Appl. Math.*, 57(11), 1413–1457, 2004.
35. M. Elad and M. Zibulevsky, Iterative shrinkage algorithms and their acceleration for l1–l2 signal and image processing applications, *IEEE Signal Process. Mag.*, 27(3), 78–88, 2010.

36. M. Fornasier and H. Rauhut, Iterative thresholding algorithms, *Appl. Comput. Harmon. Anal.*, 25(2), 187–208, 2008.
37. J. Ma, Improved iterative curvelet thresholding for compressed sensing, *IEEE Trans. Instrum. Meas.*, 60(1), 126–136, 2011.
38. M. Soumekh, *Synthetic Aperture Radar Signal Processing*, NY: John Wiley and Sons, Inc., 1999.
39. I. Cumming and F. Wong, *Digital Processing of Synthetic Aperture Radar Data*, Norwood, MA: Artech House, 2005.
40. V. Chen and H. Ling, *Time-Frequency Transforms for Radar Imaging and Signal Analysis*, Norwood, MA: Artech House, 2002.
41. K. Raney, Synthetic aperture imaging radar and moving targets, *IEEE Trans. Aerosp. Electron. Syst.*, AES-7(3), 499–505, 1971.
42. Y. Lin, B. Zhang, W. Hong, and W. Hu, Along-track interferometric SAR imaging based on distributed compressed sensing, *IET Electron. Lett.*, 46(12), 858–860, 2010.
43. C. Liu and C. Gierull, A new application for PolSAR imagery in the field of moving target indicator/ship detection, *IEEE Trans. Geosci. Remote Sens.*, 45(11), 3426–3436, 2007.
44. C. Austin, E. Ertin, and R. Moses, Sparse signal methods for 3-D radar imaging, *IEEE J. Selected Topics Signal Process.*, 5(3), 408–423, 2011.
45. X. Zhu and R. Bamler, Tomographic SAR inversion by l_1-norm regularization—The compressive sensing approach, *IEEE Trans. Geosci. Remote Sens.*, 48(10), 3839–3846, 2010.

6 Context-Dependent Classification

An Approach for Achieving Robust Remote Sensing Performance in Changing Conditions

Christopher R. Ratto, Kenneth D. Morton, Jr.,
Leslie M. Collins, and Peter A. Torrione

CONTENTS

6.1 INTRODUCTION: THE PROBLEM OF CONCEPT DRIFT IN REMOTE SENSING

The goal of many remote sensing problems is to robustly and accurately identify objects of interest that are embedded within a background scene. Despite the high fidelity of signals and images recorded by modern sensors, responses from objects of interest are often dependent on interactions between the object of interest and the background. Furthermore, unmeasured ambient variables (e.g., amount of sunlight, soil moisture) can affect responses from both the objects of interest as well as the underlying background. These problems can be viewed as a special case of *concept drift* [1] that present a challenge to achieving robust sensing performance.

For several target detection problems, the negative effects of concept drift have been partially mitigated by developing features and classification algorithms that are invariant with respect to changing background and background/object interactions. Various methods for data normalization and whitening (e.g., [2–4]) and robust feature extraction (e.g., [5–9]) can be viewed as versions of this approach to achieving robustness. However, it may be difficult for data normalization approaches to adequately mitigate the nonlinear effects that ambient variables can have on sensor responses, and generating features that are invariant under these same effects are also challenging. Furthermore, the search for a single set of invariant features ignores the possibility that one set of features might be optimal under one given set of ambient conditions (e.g., "sunny" or "dry"), while another set of features may provide better performance under different conditions (e.g., "dark" or "wet").

An alternative approach to mitigating the effects of ambient factors is to treat the background data as a source of supplementary contextual information and subsequently condition the decision rules on this information. In this chapter, this approach is referred to as *context-dependent classification*. By explicitly incorporating models of ambient contextual factors, this approach treats context as a nuisance parameter and integrates a decision rule over its posterior uncertainty.

In this chapter, a generic framework for context-dependent classification is presented, and then to illustrate the general efficacy of the approach, the framework is applied to two distinct target detection problems using hyperspectral imagery (HSI) and ground-penetrating radar (GPR) sensing modalities. First, in order to more clearly motivate the problem, examples of the contextual factors that affect target detection with HSI and GPR are reviewed in the following sections.

6.1.1 CONTEXTUAL FACTORS IN HSI REMOTE SENSING

HSI sensors collect measurements of upwelling spectral radiance from a large number of contiguous spectral bands. Many different types of HSI sensors exist with technical specifications that vary according to the application. For example, the NASA Airborne Visible/Infrared Imaging Spectrometer (AVIRIS) sensor collects images from wavelengths between 400 and 2500 nm in 224 contiguous spectral bands, and has been used in various geological, agricultural, and urban mapping applications [10]. Meanwhile, the University of Hawaii's Airborne Hyperspectral Imager (AHI) is designed for subsurface sensing problems such as landmine detection and coastal imaging, and is therefore constrained to the long-wave infrared spectrum of 8–12 μm [11].

The contextual factors that are relevant to remote sensing in HSI are related to the illumination and atmospheric conditions under which data are collected as they impact the radiance of objects of interest in addition to the background. These factors include atmospheric transmittance, the presence of occlusions such as cloud cover or pollution, and ambient solar radiance [12]. Furthermore, the background in wide-area HSI can be highly spatially variable; the spatial distributions of ground cover by vegetation, bodies of water, and geological formations may be very important to consider in anomaly detection and image segmentation problems [13].

Figure 6.1 illustrates examples of HSI data that exhibit the effects of several potential contextual factors. The top row consists of three band-averaged image chips centered on buried antitank landmines at three different locations, observed in the morning, afternoon, and evening. The image chips show that the geometry of the target/hole spectral response varies with respect to location. In addition, as seen in the top-left image, surrounding materials may yield high-magnitude spectra that may interfere with that of the target. The three bottom plots show the full spectra, across 70 hyperspectral bands, of the center pixel of each image chip to illustrate the effects of solar radiance. In the morning and evening samples, the spectra have relatively low magnitude. But

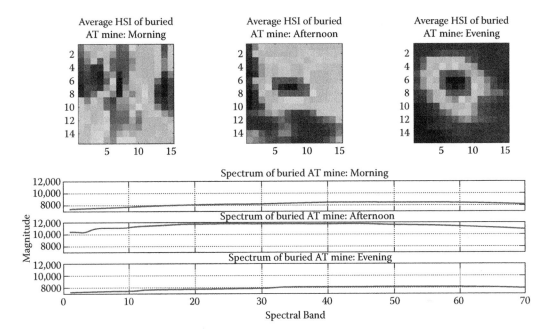

FIGURE 6.1 **(See color insert.)** Top: Band-averaged HSI chips for buried antitank mines at three different locations at morning, afternoon, and evening times. Bottom: Spectrum of center pixel for each of the HSI chips shown in the top figure.

in the afternoon, presumably after the area has been heated by solar radiation, the spectrum magnitude has increased substantially.

6.1.2 Contextual Factors in GPR Remote Sensing

GPR sensors operate by measuring the reflections of transmitted electromagnetic pulses at subsurface dielectric interfaces. GPR has been used in a wide variety of subsurface sensing applications, including landmine detection, forensics, oil and gas exploration, and archeology [14]. The advent of ultra-wideband GPR, over the past two decades, has made it an attractive choice for high-resolution subsurface imaging. However, a variety of environmental factors affect the phenomenology of GPR, and much research has been performed in identifying and mitigating these effects.

Two environmental factors, whose effects on GPR have been studied extensively, are surface texture and soil dielectric properties. These factors contribute significantly to signal degradation, which can potentially cause missed detections, and also contribute to signal clutter, which can potentially result in false alarms. Scattering from the ground surface is a major source of clutter in GPR signals collected under rough soil conditions; additionally, removing the ground reflection under these conditions becomes difficult (e.g., [15–17]) and random scattering from the surface can alter target signatures through distortion and time delay (see [18,19]). The dielectric properties of the soil are also important contextual factors to consider, since the dielectric contrast between a subsurface target and the background determines the magnitude of the target's response, and dispersion effects determine the response shape. While soil properties are partially determined by grain size, density, and composition, moisture also has a significant impact on soil permittivity and conductivity. Signatures for several targets have been found to vary significantly, in both magnitude and shape, with respect to soil composition and moisture in both field and laboratory scenarios [20].

Figure 6.2 illustrates examples of GPR *B-scans* (images of received signal as a function of time and antenna position) of low-metal, antitank landmines emplaced in dirt, gravel, asphalt, and

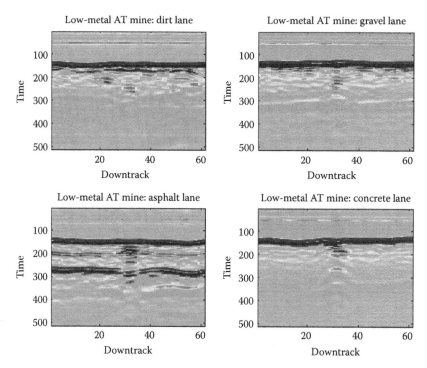

FIGURE 6.2   Sample GPR B-scans of buried antitank landmines in various types of road construction.

concrete. Although landmines have a distinctive hyperbolic signature that is seen in each image, the effects of several contextual factors are also evident. In the top two images, the rougher surfaces that characterize dirt and gravel roads contribute to the random scattering effects seen after the ground reflection, occurring between time samples 150 and 200. Furthermore, the B-scan collected over wet dirt illustrates a landmine signature with lower amplitude than the others, most likely due to the soil having higher conductivity and lower dielectric contrast with the target. In the bottom-left image, secondary reflections at interfaces between the asphalt and road bed, and the road bed and soil can be seen at time indices 200 and 300. In the concrete lane, the paved-over hole yields a nonhyperbolic signature just below the ground bounce that does not occur in the other B-scans.

6.2 GENERIC FRAMEWORK FOR CONTEXT-DEPENDENT CLASSIFICATION

Context-dependent classification involves determining a hidden set of latent variables, that is, the *contexts*, which help determine the appropriate classifier for distinguishing the classes of interest. In this chapter, only binary classifiers are considered to distinguish between the classes H_1 and H_0, but the framework is easily extended to multiclass problems. The underlying contexts may be character-ized by extracting physics-based *contextual features* ($\mathbf{f}_C$) from the sensor data. Consider the syn-thetic data example shown in Figure 6.3; the left plot illustrates four contexts corresponding to distinct bivariate Gaussian distributions in $\mathbf{f}_C$. Under our proposed model, each observation corre-sponds to both a vector of contextual features as well as a separate vector of *target features* ($\mathbf{f}_T$), which are drawn conditioned on both the context and the underlying object class. As can be seen in the rightmost plot in Figure 6.3, it is very difficult to accurately separate hypotheses H_1 and H_0 using target features in aggregate. However, within each particular context, separating H_1 from H_0 using each observation's target features is a simple linear discrimination task, as shown in the four center plots. Thus, given the underlying context of each observation, and the fact that the distributions of $\mathbf{f}_T$ are conditionally independent, the optimal approach to classification would be to model each

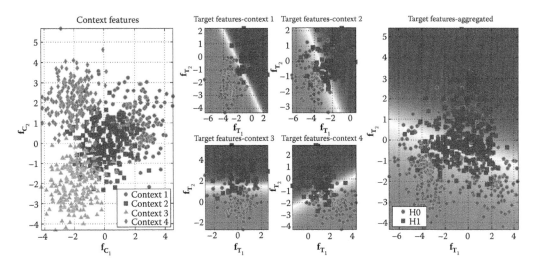

FIGURE 6.3 (**See color insert.**) A synthetic data example of a context-dependent classification problem. Left: Two-dimensional (2D) context features $\mathbf{f_C}$, colored by context. Center: The context-specific linear classification problems in target features $\mathbf{f_T}$. Right: Linear classification boundaries obtained without incorporating contextual information.

target detection problem independently. This could be accomplished by first obtaining contextual ground truth, and then applying the appropriate classifier.

Unfortunately, in many realistic scenarios the underlying context is unknown *a priori*. In such cases, it may be difficult to achieve classification performance on the aggregate target features. The rightmost plot of Figure 6.3 illustrates the decision regions learned by a linear classifier on the aggregate target features. As shown in the plot, this approach is suboptimal as it fails to exploit the underlying contextual information.

When the underlying context is unknown *a priori*, context-dependent learning can be used to probabilistically infer the underlying context, and then the decisions of an ensemble of classifiers for context-specific classification problem can be weighted appropriately. Figure 6.4 illustrates the receiver operating characteristic (ROC) curves, which show probability of detection (PD) versus probability of false alarm (PF), for each classification approach in the synthetic data example. The performance of the context-specific classifiers, attainable only through *a priori* knowledge of the underlying contexts, is illustrated by the black line. The performance of the single classifier on the aggregate target features is shown by the gray line. Context-dependent learning, whose performance is shown by the light gray line, can yield substantial improvements in performance to the single-classifier approach by inferring the underlying contexts probabilistically.

Figure 6.5 illustrates a flowchart describing the generic context-dependent learning framework used in this chapter for target detection applications. This framework can be used to statistically model both the *context identification* problem in $\mathbf{f_C}$ and the *target classification* problem in $\mathbf{f_T}$. In the context identification step, context features are modeled by a C-ary classifier that yields *context posteriors*, $p(c_i|\mathbf{f_C})$, for each context c_i, $i = 1, 2, \ldots, C$. In the target classification step, binary classifiers for discriminating targets from nontargets yield *within-context target posteriors* $p(H_l|\mathbf{f_T}, c_i)$. Finally, a posterior confidence for each observation can be calculated using

$$p(H_1 \mid \mathbf{x}) = \sum_{i=1}^{C} p(H_1 \mid \mathbf{f_T}, c_i) p(c_i \mid \mathbf{f_C}). \tag{6.1}$$

The remainder of this chapter highlights applications of the context-dependent learning framework to target detection problems with GPR and HSI sensors. An example of context-dependent

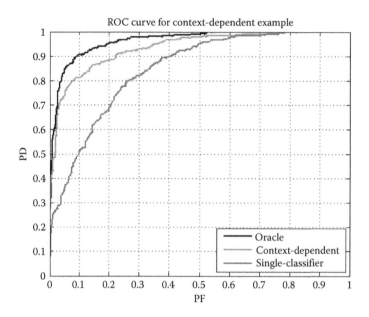

FIGURE 6.4   ROC curves corresponding to the example context-dependent problem shown in Figure 6.3.

learning using a supervised model (i.e., one that requires contextually labeled training data) for context identification will be presented first, followed by a technique using an unsupervised context model (i.e., one that does not utilize context labels in training). Supervised context models are useful when relevant contextual ground truth is available with training data. However, unsupervised learning can also be utilized when contextual truth is unavailable, and may lead to further performance improvements.

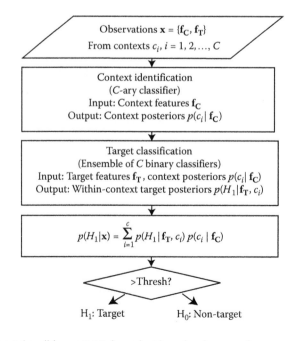

FIGURE 6.5   Flowchart describing context-dependent learning framework.

6.3 CONTEXT-DEPENDENT LEARNING WITH A SUPERVISED CONTEXT MODEL

If contextual ground truth is available to accompany the training data, a fully supervised approach to context-dependent learning can be used. Contextual ground truth could be provided by qualitative observations of environmental and operating conditions (e.g., "dirt," "wet," or "daytime"), or through quantitative measurements of these factors (e.g., volumetric soil water content, soil permittivity, or solar radiance). In the following section, examples of supervised learning techniques for context identification and target classification are presented and results of applications to landmine detection with GPR and HSI are discussed.

6.3.1 SUPERVISED CONTEXT IDENTIFICATION: C-ARY GAUSSIAN HYPOTHESIS TEST

A simple technique for C-ary supervised classification in a multidimensional feature space is to determine maximum-likelihood estimates of the parameters of each class's probability density function, and use Bayes' theorem to calculate the context posterior. As an example of such a technique, consider the Gaussian hypothesis test with a uniform prior, which yields context posteriors given by

$$p(c_i \mid \mathbf{f_C}) = \frac{\mathcal{N}(\mathbf{f_C} \mid \boldsymbol{\mu}_i, \boldsymbol{\Gamma}_i^{-1})}{\sum_{j=1}^{C} \mathcal{N}(\mathbf{f_C} \mid \boldsymbol{\mu}_j, \boldsymbol{\Gamma}_j^{-1})}, \tag{6.2}$$

where $\mathcal{N}(\mathbf{f_C} \mid \boldsymbol{\mu}_i, \boldsymbol{\Gamma}_i^{-1})$ denotes the Gaussian probability density function, with mean $\boldsymbol{\mu}_i$ and precision matrix (inverse covariance matrix) $\boldsymbol{\Gamma}_i$ evaluated at $\mathbf{f_C}$.

6.3.2 TARGET CLASSIFICATION: VARIATIONAL RELEVANCE VECTOR MACHINES

The relevance vector machine (RVM) [21,22] is a sparseness-promoting solution to the logistic discriminant classifier given by

$$y_n = \sigma[\mathbf{w}^T \phi(\mathbf{f_T})], \tag{6.3}$$

where σ denotes the logistic sigmoid (logit) function and $\phi(\mathbf{f_T})$ denotes a D-dimensional kernel transformation of the features $\mathbf{f_T}$. The likelihood of the binary class labels $\mathbf{T} = [\mathbf{t}_1, \mathbf{t}_2, \dots, \mathbf{t}_N]^T$ is given by

$$p(\mathbf{T} \mid \mathbf{F_T}, \mathbf{w}) = \prod_{n=1}^{N} \sigma(y_n)^{t_n} [1 - \sigma(y_n)]^{1-t_n}, \tag{6.4}$$

where $\mathbf{F_T}$ is the matrix of all observations' target features. The goal of RVM training is posterior inference of the discriminant weights $\mathbf{w}$. The weights are assumed to have a Gaussian prior given by

$$p(\mathbf{w} \mid \alpha) = \prod_{m=0}^{D} \mathcal{N}(w_m \mid 0, \alpha_m^{-1}), \tag{6.5}$$

where α_m is assumed to have a sparseness-promoting Gamma hyperprior,

$$p(\alpha) = \prod_{m=1}^{D} \Gamma(\alpha_m \mid a, b). \tag{6.6}$$

Posterior inference of the model parameters can be performed using Bayes' theorem, which requires integration over the priors:

$$p(\mathbf{T} \mid \mathbf{F_T}) = \iint p(\mathbf{T} \mid \mathbf{F_T}, \mathbf{w}) p(\mathbf{w} \mid \boldsymbol{\alpha}) p(\boldsymbol{\alpha}) \, d\mathbf{w} \, d\boldsymbol{\alpha}. \tag{6.7}$$

In this case, the functional forms of the likelihood and priors make this integral intractable. There are several approaches to approximating this integral, including type-II maximum likelihood [22] or variational Bayes (VB) inference [21]. In the examples shown in this chapter, the VB approach is used in which a lower bound to the log-evidence is maximized:

$$\ln p(\mathbf{T} \mid \mathbf{F_T}) = \ln \iint p(\mathbf{T} \mid \mathbf{F_T}, \mathbf{w}) p(\mathbf{w} \mid \boldsymbol{\alpha}) p(\boldsymbol{\alpha}) \, d\mathbf{w} \, d\boldsymbol{\alpha} \geq \mathcal{L}$$

$$= \iint q_{\mathbf{w}}(\mathbf{w}) q_{\alpha}(\boldsymbol{\alpha}) \ln \left[\frac{f(\mathbf{T}, \mathbf{F_T}, \mathbf{w}, \boldsymbol{\xi}) p(\mathbf{w} \mid \boldsymbol{\alpha}) p(\boldsymbol{\alpha})}{q_{\mathbf{w}}(\mathbf{w}) q_{\alpha}(\boldsymbol{\alpha})} \right] d\mathbf{w} \, d\boldsymbol{\alpha}, \tag{6.8}$$

where $q_{\mathbf{w}}(\mathbf{w})$ is the variational density of the weights, $q_{\alpha}(\boldsymbol{\alpha})$ is the variational density of the hyperparameters, and $f(\mathbf{T}, \mathbf{F_T}, \mathbf{w}, \boldsymbol{\xi})$ is a lower bound to Equation 6.4 given by

$$p(\mathbf{T} \mid \mathbf{F_T}, \mathbf{w}) \geq f(\mathbf{T}, \mathbf{F_T}, \mathbf{w}, \boldsymbol{\xi}) = \prod_{n=1}^{N} \sigma(\xi_n) \exp\left(\frac{\beta_n - \xi_n}{2} - \lambda(\xi_n)(\beta_n^2 - \xi_n^2) \right),$$

where

$$\tag{6.9}$$

$$\beta_n = (2t_n - 1) y_n,$$
$$\lambda(\xi_n) = (1/4\xi_n) \tanh(\xi_n/2),$$

where ξ are variational parameters. Free-form maximization of the right-hand side of Equation 6.8 yields the following update equations for the variational parameters and posterior estimates for the discriminant weights:

$$\xi_n^2 = \phi(\mathbf{f_{T}}_n)^T E[\mathbf{w}\mathbf{w}^T] \phi(\mathbf{f_{T}}_n), \tag{6.10}$$

$$q_{\mathbf{w}}(\mathbf{w}) = \mathcal{N}(\mathbf{w} \mid \mathbf{m}, \mathbf{S}),$$

where

$$\mathbf{S} = \left(\mathbf{A} + 2 \sum_{n=1}^{N} \frac{1}{4\xi_n} \tanh\left(\frac{\xi_n}{2}\right) \phi(\mathbf{f_{T}}_n) \phi(\mathbf{f_{T}}_n)^T \right), \tag{6.11}$$

$$\mathbf{m} = \frac{1}{2} \mathbf{S} \left(\sum_{n=1}^{N} (2t_n - 1) \phi(\mathbf{f_{T}}_n) \right)$$

$$q_{\alpha}(\boldsymbol{\alpha}) = \prod_{m=0}^{D} \Gamma(\alpha_m \mid \tilde{a}, \tilde{b}_m),$$

where

$$\tag{6.12}$$

$$\tilde{a} = a + \frac{1}{2},$$

$$\tilde{b}_m = b + \frac{1}{2} E[w_m^2].$$

During reestimation, the sparseness-promoting priors force many of the hyperparameters to infinity, thus forcing the corresponding weight density to be infinitely peaked at zero. Therefore, only a sparse subset of the vectors in $\phi(\mathbf{f}_T)$ (i.e., the *relevance vectors*) is used in the discriminant function. Note that using a direct kernel, that is, $\phi(\mathbf{f}_T) = [\mathbf{1}, \mathbf{f}_T]$, will force the weighting of individual features to zero, thus performing *de facto* feature selection [23]. In context-dependent learning, this property of the RVM can lead to the classifier utilizing different features depending on the underlying context.

Although the above equations describe the training of a single RVM, in context-dependent learning, an ensemble of direct kernel RVMs is employed. In learning scenarios in which a supervised context model was used, the target features $\mathbf{f}_T$ used in training were partitioned according to their known context labels. Then, individual RVMs were trained on target features from each context.

6.4 CONTEXT-DEPENDENT LEARNING WITH AN UNSUPERVISED CONTEXT MODEL

In some scenarios, it may be difficult and/or expensive to obtain contextual ground truth when collecting training data. Furthermore, the contextual ground truth that is available may not be relevant to the ultimate classification problem. In these situations, a supervised context model may neither be feasible nor desirable for the intended application. Instead, unsupervised learning techniques can be used to learn a context model without the use of discrete contextual labels. In the following section, an example of an unsupervised context model based on Gaussian mixtures and the Dirichlet process (DP), which requires no *a priori* information in the number of contexts present in the data, is presented. As an unsupervised context model *softly* partitions the training data into contexts, a technique for variational inference of a set of RVMs is also presented for context-based target classification.

6.4.1 Unsupervised Context Identification: DP Gaussian Mixture Model

Gaussian mixture models (GMMs) are a common technique for modeling multimodal data through unsupervised learning. A GMM models the overall density as a weighted sum of Gaussian distributions with mixture proportions π_i, $i = 1, 2, \ldots, C$

$$p(\mathbf{f}_C \mid \boldsymbol{\pi}, \boldsymbol{\mu}, \boldsymbol{\Gamma}) = \sum_{i=1}^{C} \pi_i \mathcal{N}(\mathbf{f}_C \mid \boldsymbol{\mu}_i, \boldsymbol{\Gamma}_i^{-1}). \tag{6.13}$$

A possible unsupervised extension of the context-dependent learning framework discussed previously would be to learn a GMM on the context features, and allow each component to describe a unique context. However, in situations where the number of contexts is unknown, it would be desirable for an unsupervised learning procedure to automatically determine the number of contexts. This is possible by expressing uncertainty in the number of GMM components through a DP prior [24].

The DP can be described as a distribution on probability distributions. Consider a continuous distribution G from a DP with base distribution G_0 and scaling parameter α, and a continuous random variable θ drawn from G:

$$G \sim DP(G_0, \alpha), \tag{6.14}$$

$$\theta_n \sim G, \quad n = 1, 2, \ldots, N \tag{6.15}$$

Next, let each $\theta_n = \{\boldsymbol{\mu}_n, \boldsymbol{\Gamma}_n^{-1}\}$ be the parameters of the distribution of the nth observation in a data set, and let θ_i^* be the unique values of θ_n. The distribution G can be modeled by the *stick-breaking* representation of the DP:

$$v_i \sim \text{Beta}(1, \alpha), \quad i = 1, 2, \ldots, \infty \tag{6.16}$$

$$\theta_i^* \sim G_0, \quad i = 1, 2, \ldots, \infty \tag{6.17}$$

$$\pi_i(\mathbf{v}) = v_i \prod_{j=1}^{i-1} (1 - v_j), \quad i = 1, 2, \ldots, \infty \tag{6.18}$$

$$G = \sum_{i=1}^{\infty} \pi_i(\mathbf{v}) \delta_{\theta_i^*} \tag{6.19}$$

In Equation 6.18, the mixture proportion π_i represents the ith fraction of a stick that has been broken into an infinite number of pieces, with the relative size of each piece drawn from the Beta distribution given by Equation 6.16. The distribution G then consists of a countably infinite set of atoms at each θ_i^*, but the vast majority of the atoms have negligible proportions. When used as a nonparametric prior on the distribution of observations $\mathbf{f}_C$, a *DP Gaussian Mixture Model* (DPGMM) can be formulated as a model in which the number of mixture components (determined by the number of unique θ_i^*) is random and increases with the number of observed clusters. Furthermore, G_0 is Normal-Wishart and therefore conjugate to the Gaussian-likelihood function for each mixture component. The data-generating process can be described by the following:

1. Draw $v_i | \alpha \sim \text{Beta}(1, \alpha)$, $i = 1, 2, \ldots$
2. Draw $\theta_i^* | G_0 \sim \mathcal{N}(\mu_i^* | \mathbf{m}_0, (\beta_0 \Gamma_i^*)^{-1}) \mathcal{W}(\Gamma_i^* | \mathbf{W}_0, v_0)$, $\quad i = 1, 2, \ldots$
3. For $n = 1, 2, \ldots, N$
   a. Draw indicator $c_n | \mathbf{v} \sim \text{Multi}\{\pi(\mathbf{v})\}$.
   b. Draw data $\mathbf{f}_{C_n} | c_n \sim \mathcal{N}(\mathbf{f}_{C_n} | \theta_{c_n}^* = \mu_{c_n}^*, \Gamma_{c_n}^{-1*})$.

Furthermore, the DPGMM-likelihood function is given by

$$p(\mathbf{f}_C | \mathbf{v}, \theta^*) = \sum_{i=1}^{\infty} \pi_i(\mathbf{v}) \mathcal{N}(\mathbf{f}_C | \mu_i^*, \Gamma_i^{-1*}). \tag{6.20}$$

The context posterior is given by the mixture of Student-t distributions,

$$p(c_i | \mathbf{f}_C) = \frac{t_{\omega_i}(\mathbf{f}_C | \rho_i, \Lambda_i)}{\sum_{j=1}^{T} t_{\omega_j}(\mathbf{f}_C | \rho_j, \Lambda_j)}, \tag{6.21}$$

where a uniform prior is assumed and the parameters ρ and Λ are functions of the prior and variational parameters [25].

The variational inference approach was used for training the model with a truncation level of $T = 20$ mixture components (see [24,25,41]). In both examples shown later in this chapter, the DPGMM determined that there are less than 20 mixture components (contexts) of nonnegligible proportions.

6.4.2 Target Classification: Context-Sensitive RVM Training

In context-dependent learning with unsupervised models, the training data cannot be partitioned into distinct contexts on which to train each of the C RVMs. Therefore, the approach to RVM learning must be modified to consider the relative importance of each observation. Recall the approximation to the log-likelihood of a single RVM given by Equation 6.9. For an ensemble of C contextually dependent RVMs, the approximate log-likelihood is given by

$$p(\mathbf{T},\mathbf{Z} \mid \mathbf{F}_{\mathbf{T}},\mathbf{w},\boldsymbol{\pi}) \geq F(\mathbf{T},\mathbf{Z},\mathbf{F}_{\mathbf{T}},\mathbf{w},\boldsymbol{\xi},\boldsymbol{\pi}) = \prod_{n=1}^{N}\prod_{i=1}^{C}\left[\sigma(\xi_{ni})\exp\left(\frac{\beta_{ni} - \xi_{ni}}{2} - \lambda(\xi_{ni})(\beta_{ni}^2 - \xi_{ni}^2)\right)\right]^{Z_{ni}},$$

where

$$\beta_{ni} = (2t_{ni} - 1)y_{ni}, \tag{6.22}$$
$$\lambda(\xi_{ni}) = (1/4\xi_{ni})\tanh(\xi_{ni}/2).$$

In Equation 6.22, the ensemble is governed by the proportions and binary latent variables $\mathbf{Z}$ for which the expected value is the context posterior, that is, $E_q[z_{ni}] = p(c_i|\mathbf{f}_{\mathbf{C}n})$. To investigate the effect of these mixture parameters on the variational update equations, consider rewriting the lower bound in Equation 6.8 as

$$\mathcal{L} = \iint q_{\mathbf{w}}(\mathbf{w})q_{\alpha}(\boldsymbol{\alpha})\ln[F(\mathbf{T},\mathbf{Z},\mathbf{F}_{\mathbf{T}},\mathbf{w},\boldsymbol{\xi},\boldsymbol{\pi})p(\mathbf{w}\mid\boldsymbol{\alpha})p(\boldsymbol{\alpha})]\mathrm{d}\mathbf{w}\,\mathrm{d}\boldsymbol{\alpha}$$
$$+ H[q_{\mathbf{w}}(\mathbf{w})] + H[q_{\alpha}(\boldsymbol{\alpha})]. \tag{6.23}$$

In Equation 6.23, $H[q_{\mathbf{w}}(\mathbf{w})]$ and $H[q_{\alpha}(\boldsymbol{\alpha})]$ denote the entropy of the variational posteriors. Since the lower bound is to be maximized with respect to $q_{\mathbf{w}}(\mathbf{w})$, $q_{\alpha}(\boldsymbol{\alpha})$ is assumed to be known and all terms not dependent on the weights may be considered constants. This simplification yields

$$\mathcal{L} = \int q_{\mathbf{w}}(\mathbf{w})\langle\ln[F(\mathbf{T},\mathbf{Z},\mathbf{F}_{\mathbf{T}},\mathbf{w},\boldsymbol{\xi},\boldsymbol{\pi})p(\mathbf{w}|\boldsymbol{\alpha})p(\boldsymbol{\alpha})]\rangle\,\mathrm{d}\mathbf{w} + H[q_{\mathbf{w}}(\mathbf{w})] + \mathrm{const.}$$
$$= \int q_{\mathbf{w}}(\mathbf{w})\ln q_{\mathbf{w}}^*(\mathbf{w})\mathrm{d}\mathbf{w} - \int q_{\mathbf{w}}(\mathbf{w})\ln q_{\mathbf{w}}(\mathbf{w})\mathrm{d}\mathbf{w} + \mathrm{const.}$$
$$= -KLD[q_{\mathbf{w}}(\mathbf{w})\|q_{\mathbf{w}}^*(\mathbf{w})] + \mathrm{const.}, \tag{6.24}$$

where

$$\ln q_{\mathbf{w}}^*(\mathbf{w}) = \langle\ln[F(\mathbf{T},\mathbf{Z},\mathbf{F}_{\mathbf{T}},\mathbf{w},\boldsymbol{\xi},\boldsymbol{\pi})p(\mathbf{w}|\boldsymbol{\alpha})p(\boldsymbol{\alpha})]\rangle,$$

and <•> denotes ensemble average. Therefore, the variational lower bound may be written as a negative Kullback–Leibler divergence between two densities, and is maximized when the two densities are equal. Therefore, the variational log-posterior may be written as

$$\ln q_{\mathbf{w}}(\mathbf{w}) = \sum_{i=1}^{C}\sum_{n=1}^{N}E_q[Z_{ni}]\left\{\ln\sigma(\xi_{ni}) + \frac{1}{2}(\beta_{ni} - \xi_{ni})\right.$$
$$\left. - \lambda(\xi_{ni})\left(\beta_{ni}^2 - \xi_{ni}^2\right)\right\} + \sum_{m=0}^{D}\ln\sqrt{\frac{\alpha_m}{2\pi}} - \frac{\alpha_m}{2}w_m^2$$
$$= \sum_{i=1}^{C}\sum_{n=1}^{N}p(c_i \mid \mathbf{f}_{\mathbf{C}n})\left\{\frac{2t_{ni} - 1}{2}\mathbf{w}_i^T\phi(\mathbf{f}_{\mathbf{T}n})\right.$$
$$\left. - \lambda(\xi_{ni})\mathbf{w}_i^T\phi(\mathbf{f}_{\mathbf{T}n})\phi(\mathbf{f}_{\mathbf{T}n})^T\mathbf{w}_i\right\} - \sum_{m=0}^{D}\frac{\alpha_m}{2}w_m^2 + \mathrm{const.}$$
$$= \sum_{i=1}^{C}\left\{\mathbf{w}_i^T\left[\sum_{n=1}^{N}p(c_i \mid \mathbf{f}_{\mathbf{C}n})\frac{2t_{ni} - 1}{2}\phi(\mathbf{f}_{\mathbf{T}n})\right]\right.$$
$$\left. - \mathbf{w}_i^T\left[\sum_{n=1}^{N}p(c_i \mid \mathbf{f}_{\mathbf{C}n})\phi(\mathbf{f}_{\mathbf{T}n})\phi(\mathbf{f}_{\mathbf{T}n})^T\lambda(\xi_{ni})\right]\mathbf{w}_i - \frac{1}{2}\mathbf{w}_i^T\mathbf{A}\mathbf{w}_i + \mathrm{const.}\right\}. \tag{6.25}$$

By inspection, it can be seen that $q_{\mathbf{w}}(\mathbf{w})$ is a mixture of C independent Gaussian distributions, each with parameters $\mathbf{m}_i$ and $\mathbf{S}_i$:

$$q_{\mathbf{w}}(\mathbf{w}_i \mid c_i) = \mathcal{N}(\mathbf{w}_i \mid \mathbf{m}_i, \mathbf{S}_i),$$

where

$$\mathbf{S}_i = \left[\mathbf{A} + 2\sum_{n=1}^{N} p(c_i \mid \mathbf{f}_{\mathbf{C}n})\lambda(\xi_{ni})\phi(\mathbf{f}_{\mathbf{T}n})\phi(\mathbf{f}_{\mathbf{T}n})^T \right]^{-1}$$

$$\mathbf{m}_i = \frac{1}{2}\mathbf{S}_i\left[\sum_{n=1}^{N} p(c_i \mid \mathbf{f}_{\mathbf{C}n})(2t_n - 1)\phi(\mathbf{f}_{\mathbf{T}n}) \right]. \tag{6.26}$$

Therefore, by updating the posteriors using Equation 6.26, terms corresponding to observations with a high probability of being from context i are weighted more heavily in learning the RVM for classifying targets in context i, and terms corresponding to observations with low probability of being from context i are less influential. This stands in contrast to the original RVM posterior updates in Equation 6.11, where all points are weighted equally in learning.

6.5   FEATURES FOR CONTEXT IDENTIFICATION AND TARGET DISCRIMINATION

To maximize the benefit from the proposed context-dependent framework, it is important to make use of contextual features that are motivated by aspects of sensor phenomenology that contribute to shifting environmental conditions. In the following section, physics-based features for context identification and target classification for HSI and GPR sensors are discussed.

6.5.1   HSI FEATURES

6.5.1.1   Contextual Features

In the HSI literature, various techniques have been proposed for unmixing hyperspectral data into its basis spectra by identifying endmember spectra, which are motivated by the pure materials present in the scene (e.g., [26,27]). Linear mixing assumptions suggest that a D-dimensional pixel from a hyperspectral image ($\mathbf{x}$) can be modeled as a linear combination of the spectral signatures of M endmembers ($\mathbf{e}$), with additive noise (ϵ):

$$\mathbf{x}_i = \sum_{k=1}^{M} p_{ik}\mathbf{e}_k + \epsilon_i, \tag{6.27}$$

$$\sum_{k=1}^{M} p_{ik} = 1, \tag{6.28}$$

$$p_{ik} \geq 0, \quad k = 1, 2, \ldots, M. \tag{6.29}$$

In Equation 6.27, p_{ik} indicates the weighting proportion of endmember $\mathbf{e}_k$ in pixel $\mathbf{x}_i$. The proportions are constrained to be greater than 0 and also must sum to unity. A visual interpretation of endmembers is possible by considering the vertices of the M-simplex that encloses all hyperspectral pixels when plotted in a D-dimensional space. For example, consider the synthetic data problem with endmembers at points $(0, -2)$, $(\sqrt{2}, 0)$, and $(-\sqrt{2}, 0)$. Figure 6.6 illustrates an example of 1000 "pixels" drawn from random proportions of these endmembers, and shows the 3-simplex (triangle) that encloses them.

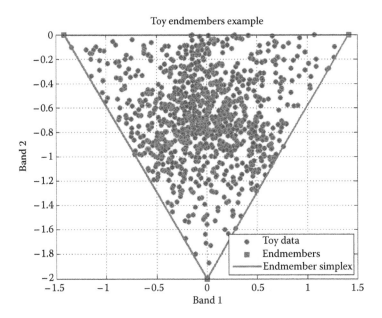

FIGURE 6.6 Synthetic data example illustrating endmembers and the enclosing simplex.

The iterative constrained endmembers (ICEs) algorithm is one of many endmember identification techniques available in the literature [26]. The ICE algorithm solves for a set of endmembers that satisfies a trade-off between minimizing the residual sum-of-squares (RSS) and sum-of-squared distances (SSD) given by

$$\text{RSS} = \sum_{i=1}^{N} \left(\mathbf{x}_i - \sum_{k=1}^{M} p_{ik}\mathbf{e}_k \right)^T \left(\mathbf{x}_i - \sum_{k=1}^{M} p_{ik}\mathbf{e}_k \right), \tag{6.30}$$

$$\text{SSD} = \sum_{k=1}^{M-1} \sum_{l=k+1}^{M} (\mathbf{e}_k - \mathbf{e}_l)^T (\mathbf{e}_k - \mathbf{e}_l), \tag{6.31}$$

The trade-off between minimizing RSS and SSD is presented in an objective function with a parameter (μ) for determining preference toward solutions that minimize either RSS or SSD (which is proportional to V, the sum of endmember variances):

$$\text{RSS}_{\text{reg}} = (1 - \mu)\frac{\text{RSS}}{N} + \mu V. \tag{6.32}$$

In the HSI classification examples discussed in this chapter, classification was performed on 15×15 image "chips" surrounding anomalies identified by an RX detector [2]. For each chip, the pixels outside of the 5×5 center region were averaged to form a background spectral vector. ICE was run on these background spectra for all image chips to identify four endmembers, and the proportions of each. As the endmember proportions exist in a manifold of dimensionality $M - 1$, they were projected to three dimensions via principal components analysis (PCA) to facilitate contextual modeling.

6.5.1.2 Target Classification Features

Since each of the image chips were extracted from RX anomalies, the spectral signatures of targets were constrained to the center region of pixels of each chip. Feature vectors for target classification were extracted by averaging the spectral vectors for the 5×5 center region of pixels into a 70-dimensional feature vector and normalizing to zero mean and unit variance.

6.5.2 GPR FEATURES

6.5.2.1 Contextual Features

In general, before feature-based processing is performed on GPR data, a real-time prescreening algorithm (e.g., [4]) is used to identify subsurface anomalies in the data. These anomalous locations are then saved as "prescreener alarms," and features are extracted from the surrounding data. The physics-based features used in training a contextual model were originally proposed in [28] and consist of 10 measurements of ground reflection arrival time, 10 estimates of the reflection coefficient using a simple transmission model, and 10 linear prediction errors from GPR images. The ground reflection arrival times ($\mathbf{t}_{GB}$) approximate the variations in antenna-to-ground distance over a given area [15], the reflection coefficients ($\mathbf{\Gamma}$) are indicative of the soil/air dielectric contrast [29], and the linear prediction errors ($\mathbf{P}_{LP}$) are indicative of the stochastic variations in subsurface scattering [30]. The three sets of contextual features were concatenated to yield a 30-dimensional (30D) feature vector ($\mathbf{f}_C = [\mathbf{t}_{GB}, \mathbf{\Gamma}, \mathbf{P}_{LP}]^T$) for characterizing the context of a given alarm.

Measurements of $\mathbf{t}_{GB}$ were made at 10 locations over a spatial distance of 1 m behind a given prescreener alarm, so as to not potentially confuse the target signature with the ground reflection (see Figure 6.7). For the purposes of this experiment, the ground reflection was assumed to be the global maximum of the received time-domain signal $a_{ij}(t)$ at a given location $i = 1, 2, \ldots, 10$, and channel $j = 1, 2, \ldots, 51$ (for this 51-channel radar system). This signal is known in the GPR literature as an *A-scan*. Measurements were made at each location, and averaged over all 51 channels. Therefore, the ground reflection feature t_{GB} for each location was calculated by

$$t_{\max_{ij}} = \arg\max a_{ij}(t), \quad i = 1, 2, \ldots, 10, \quad j = 1, 2, \ldots, 51 \tag{6.33}$$

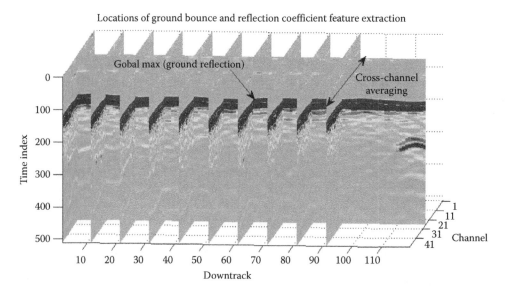

FIGURE 6.7 Locations of ground reflection time and reflection coefficient measurements relative to potential target position. Measurement locations are denoted by cross-track slices.

$$t_{GB_i} = \frac{1}{51} \sum_{j=1}^{51} t_{max_{ij}} \tag{6.34}$$

Measurements of Γ were calculated based on the radar range equation [31], which assumes plane-wave propagation through free-space and farfield scattering:

$$\Gamma_i = \frac{1}{51} \sum_{j=1}^{51} d_{ij}^2 \sqrt{P_{R_{ij}} A}, \quad i = 1, 2, \ldots, 10 \tag{6.35}$$

In Equation 6.35, A is a constant that includes the transmitted power, transmit and receive antenna gain, and radar cross-sectional terms. The idealized antenna-to-ground distance (d) is calculated in Equation 6.36 by dividing the ground reflection time by the spatial sampling rate of the radar system (512 samples/m in air). The reflection power (P_R) from the ground is calculated in Equation 6.37 by integrating the squared A-scan multiplied by a Gaussian window given by Equation 6.38. To isolate the ground reflection, the window is centered between the A-scan global maximum and minimum, with width determined by the temporal separation of the global maximum and minimum.

$$d_{ij} = \frac{t_{max_{ij}}}{512 \text{ samples/m}}, \quad i = 1, 2, \ldots, 10, \ j = 1, 2, \ldots 51 \tag{6.36}$$

$$P_{R_{ij}} = \sum_{t=1}^{512} w_{ij}(t) a_{ij}^2(t), \quad i = 1, 2, \ldots, 10, \ j = 1, 2, \ldots, 51 \tag{6.37}$$

$$w_{ij}(t) = \frac{1}{\sqrt{4\pi(t_{min_{ij}} - t_{max_{ij}})}} \exp\left\{ \frac{\left[t - (t_{min_{ij}} - t_{max_{ij}}) \right]^2}{4(t_{min_{ij}} - t_{max_{ij}})} \right\} \tag{6.38}$$

Contextual features were also extracted by applying linear predictors (i.e., autoregressive models) to time slices extracted from spatio-temporal GPR images (B-scans). For a particular alarm, the 100 previous A-scans (spanning a distance of 5 m) were concatenated to form an image, and their global maxima were aligned to the top of the image, with excess data discarded. Linear predictors of order $M = 10$ were then applied to 10 of the top 200 rows $a_n(T)$, where T is a time index of the aligned image:

$$a_n(T) = -\sum_{k=1}^{M} w_k a_{n-k}(T) + v(n), \quad n = 1, 2, \ldots, 100, \ T = 1, 21, \ldots, 181 \tag{6.39}$$

In the linear predictor given by Equation 6.39, w_k are the prediction weights and $v(n)$ is a white noise process. The weights that satisfy the minimum mean-square error criterion can be determined via the well-known Yule–Walker method [32], and the prediction error is given by

$$P_{LP_n} = \text{Var}[\mathbf{a}_n(T)] - \mathbf{w}_T^T \mathbf{p}_T + \mathbf{w}_T^T \mathbf{R}_T \mathbf{w}_T, \quad n = 1, 2, \ldots, 100, \ T = 1, 21, \ldots, 181, \tag{6.40}$$

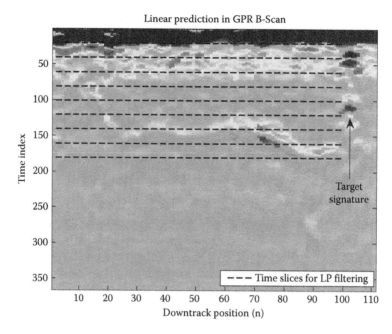

FIGURE 6.8   Aligned GPR B-scan at the prescreener alarm location. Time-slices (rows) extracted for feature extraction via linear prediction are denoted by dashed black lines.

where $\mathbf{R}_T$ and $\mathbf{p}_T$ are the correlation matrix and forward-prediction cross-correlation vector, respectively, of $\mathbf{a}_n(T)$. An example B-scan of GPR data extracted prior to an alarm is shown in Figure 6.8, with the rows extracted for linear prediction marked.

After all contextual features were calculated, they were normalized to zero mean and unit variance. Normalizing the features ensures equal scaling of all dimensions, reduces the eigenvalue spread, and prevents features with large scale from dominating smaller-scale features in classifier training. To facilitate classifier training and analysis, the dimensionality of the 30D feature space $\mathbf{f}_C = [\mathbf{t}_{GB},\ \Gamma,\ \mathbf{P}_{LP}]^T$ was reduced via PCA. Context identification was performed using the three-dimensional (3D) PCA projection of $\mathbf{f}_C$.

6.5.2.2   Target Classification Features

A wide variety of feature extraction and classification algorithms have been developed over the past decade for classifying GPR responses as landmine signatures or clutter/background (e.g., [4,6,7,9]). However, it has been shown that their relative performance varies with respect to the operating environment [33], suggesting that algorithm fusion may benefit from a context-dependent approach. For the GPR examples in this chapter, the target discrimination features $\mathbf{f}_T$ consisted of the prescreener confidence [4], a 40-dimensional edge histogram and scalar confidence value extracted by the edge histogram descriptor (EHD) algorithm [9], a 20D projection of the A-scan energy density spectrum and scalar confidence value extracted by the subspace spectral correlation feature (SPSCF) algorithm [7], an 8D observation map and scalar confidence value extracted with the hidden Markov model (HMM) algorithm [6]. Because each feature set was calculated on a different scale, the target discrimination features were normalized to zero mean and unit variance prior to classification.

6.6   EXPERIMENTAL RESULTS

6.6.1   Context-Dependent Target Detection with HSI

The HSI target detection experiment was conducted on data collected by the AHI sensor as part of the U.S. Army's Wide Area Airborne Mine Detection (WAAMD) system. Eight HSI images were

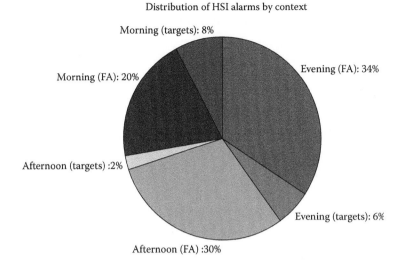

FIGURE 6.9   Pie chart illustrating distribution of image chips by context in HSI data set.

collected over an arid site with buried antitank landmines, empty holes, and vegetation present. Images were collected at various times throughout the day, loosely labeled as "morning," "afternoon," and "evening." Contextual and target classification features were extracted from the 4591 image chips corresponding to anomalies identified by the RX algorithm. For scoring purposes, detected holes were counted as targets due to the similarity in the spectral signatures of disturbed earth. The percentages of chips recorded in each of the three temporal contexts are illustrated in the pie chart in Figure 6.9.

Figure 6.10 summarizes the results of supervised context identification. The left plot illustrates a scatterplot of the 3D PCA of endmember proportions, labeled according to the time of day when the

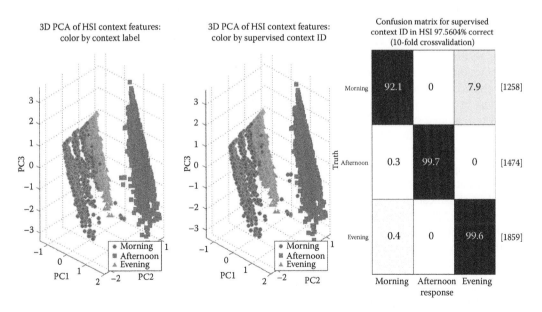

FIGURE 6.10   Left: Scatterplot of 3D PCA of HSI context features, colored according to temporal context label. Center: Scatterplot of 3D PCA of HSI context features, colored according to MAP context obtained through supervised learning. Right: Confusion matrix for MAP supervised context identification.

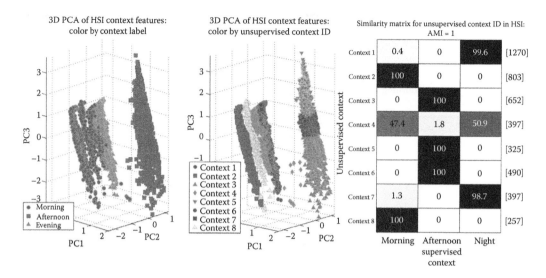

FIGURE 6.11 Left: Scatterplot of 3D PCA of HSI context features, colored according to temporal context label. Center: Scatterplot of 3D PCA of HSI context features, colored according to MAP context obtained through unsupervised learning. Right: Similarity matrix for MAP unsupervised context identification.

chips were recorded. Separation between the morning, afternoon, and evening contexts is clearly visible, suggesting that supervised context identification can exploit this information. The center plot illustrates the same data, but with points colored according to the contexts assigned through supervised context modeling. The corresponding confusion matrix is shown at right, with 98% of chips' temporal context identified correctly.

The results of unsupervised context identification are summarized in Figure 6.11. The left plot illustrates the contextual features colored by temporal label, and the center plot illustrates the clusters found by the unsupervised DPGMM. The similarity matrix at the right side shows that all unsupervised contexts, with the exception of context 4, correspond directly to a single temporal label. Each temporal context was therefore subdivided into three subcontexts by the DPGMM, and the adjusted mutual information (AMI) (see [34]) between the temporal labels and the unsupervised clustering was equal to 1.

The performance of context-sensitive RVM training is summarized in Figure 6.12. Each stem plot shows the discriminant weight assigned to each of the 70 target-region spectral bands for each of the eight contexts identified by the DPGMM. Because the RVM incorporates sparseness-promoting priors in learning, most of the weights are zero. The individual plots show that the RVM converged to a unique subset of bands receiving nonzero weight in each context, and no band was universally relevant. An additional interesting observation here is that the two most populated contexts (1 and 2) utilized many bands, while the smaller contexts required only a few.

Finally, Figure 6.13 shows ROC curves illustrating the performance of a single RVM (no contextual information), context-dependent classification with a supervised context model, and context-dependent classification with an unsupervised context model. Both context-dependent techniques achieve higher PD at lower PFs than the single RVM, but the approach using the unsupervised context model achieved better performance. It is possible that by partitioning the temporal contexts into several subcontexts, the unsupervised context model was able to exploit contextual information that was relevant for classification but not known *a priori*.

6.6.2 CONTEXT-DEPENDENT TARGET DETECTION WITH GPR

The GPR data under consideration were collected with the NIITEK [35] vehicle-mounted radar in March 2009 on dirt, gravel, asphalt, and concrete lanes at a temperate Eastern U.S. test site. The

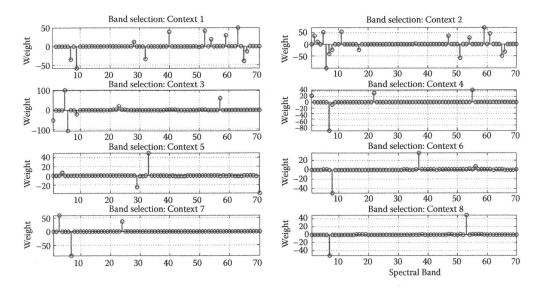

FIGURE 6.12 Stem plots illustrating the discriminant weight assigned to each feature by the RVM for each of the eight HSI contexts learned through unsupervised context modeling.

radar vehicle maintained constant speed and antenna height throughout the data collection. Data were collected on several test lanes at each site in which 10 types of antitank landmine targets and 155 mm artillery shells were buried alongside empty holes, which were considered clutter. The total data collection area was 12,383 m². An offline prescreening algorithm recorded a total of 2404 alarms for further processing. The percentages of alarms due to targets and false alarms in each of the four lane contexts are summarized by the pie chart in Figure 6.14.

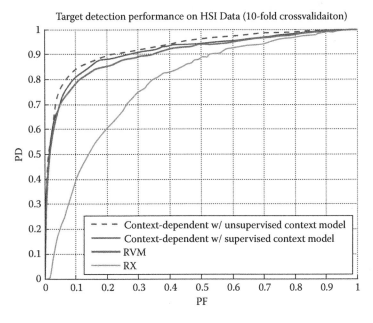

FIGURE 6.13 ROC curves illustrating performance of target detection in the HSI data set for context-dependent classification using an unsupervised context model (dashed), context-dependent classification using a supervised context model (black), a single RVM (gray), and the RX algorithm (light gray). Curves are plotted as PD versus PF.

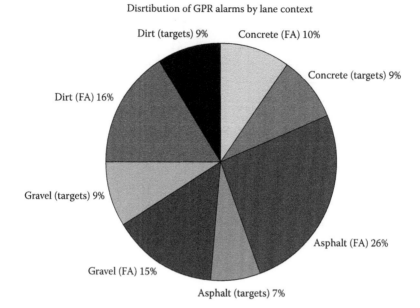

FIGURE 6.14 Pie chart illustrating distribution of prescreener alarms by context in the GPR data set.

Contextual and target discrimination features were extracted from each prescreener alarm using the techniques described in Section 6.5.2. To evaluate the context identification and alarm classification performance, 10-fold cross-validation was used to ensure that alarms recorded over the same object were not included in both training and testing of the models. The results of supervised context identification are summarized in Figure 6.15. The left plot within Figure 6.15 illustrates a scatterplot of the first three principal components of the contextual features, with points colored according to the labeled contexts of their corresponding alarms. This figure shows that alarms from the same

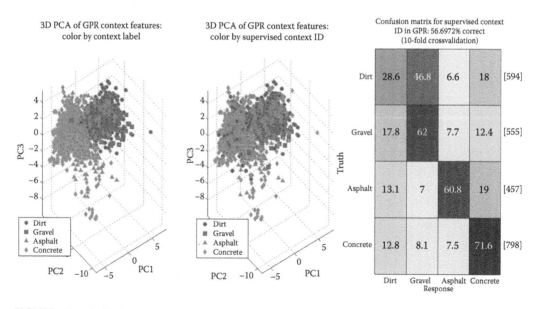

FIGURE 6.15 Left: Scatterplot of 3D PCA of GPR context features, colored according to lane context label. Center: Scatterplot of 3D PCA of GPR context features, colored according to MAP context obtained through supervised learning. Right: Confusion matrix for MAP supervised context identification.

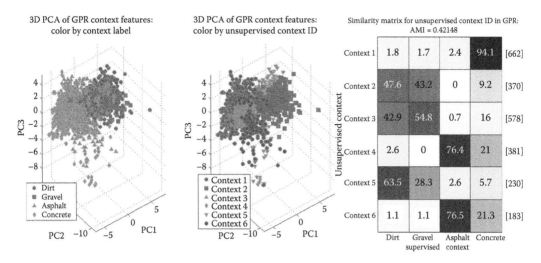

FIGURE 6.16 Left: Scatterplot of 3D PCA of GPR context features, colored according to lane context label. Center: Scatterplot of 3D PCA of GPR context features, colored according to MAP context obtained through unsupervised learning. Right: Similarity matrix for MAP unsupervised context identification.

lane tend to cluster in this contextual feature space, although an interesting observation is that dirt and gravel tend to overlap more while asphalt and concrete form more distinct clusters. The center plot is the same as the left plot, but with points colored according to the contexts assigned by the supervised context model. Results are summarized by the confusion matrix at right. A total of 57% of alarm's contexts were correctly identified, with dirt and gravel confused more often than asphalt and concrete.

The performance of unsupervised context identification is summarized in Figure 6.16. The left plot again illustrates the first three principal components of the contextual features, with points colored by lane context label. Training the DPGMM on the entire data set identified six clusters in the contextual feature space, and the center plot illustrates points colored according to the DPGMM clusters. The similarity matrix at the right side shows that there is substantial overlap between the DPGMM contexts and the lane labels. Although context 1 is almost entirely concrete data, the other five contexts are split among two lane types: contexts 2, 3, and 5 are split between dirt and gravel, and contexts 4 and 6 are split between asphalt and concrete. The AMI [34] between the unsupervised clustering and the context labels was 0.41026. This result suggests that although the cluster assignments were similar, the DPGMM result may exploit similarities between the lanes that are not evident through use of discrete labels.

The performance of context-specific classifier training with unsupervised contexts are summarized by the stem plots shown in Figure 6.17, which show the discriminant weight learned for each feature in each of the six contexts identified by the DPGMM. However, a unique subset of features was assigned nonzero weight in each context, and no single feature was universally relevant. These results invoke similarities to previous investigations which showed that different feature sets may perform best under specific environmental conditions [33].

Finally, classification performance was evaluated with 10-fold cross-validation. Figure 6.18 depicts the pseudo-ROC of PD versus false alarm rate (FAR), expressed in false alarms per square meter, of both types of context-dependent classification (supervised or unsupervised context modeling) and compares performance to a single RVM trained on the target discrimination features only, and therefore used no contextual information. Results illustrate that both context-dependent approaches outperformed the single RVM, but use of the unsupervised context model yielded the best improvement in FAR at high PD (>0.85).

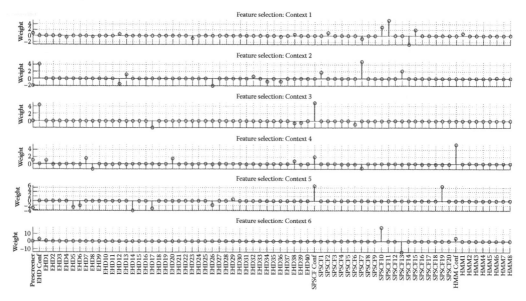

FIGURE 6.17 Stem plots illustrating the discriminant weight assigned to each feature by the RVM for each of the six GPR contexts learned through unsupervised context modeling.

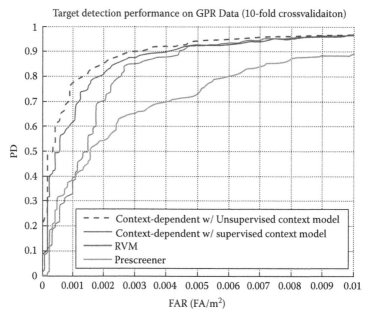

FIGURE 6.18 Pseudo-ROC curves illustrating performance of target detection in the GPR data set for context-dependent classification using an unsupervised context model (dashed), context-dependent classification using a supervised context model (black), a single RVM (gray), and the prescreening algorithm (light gray). Curves are plotted as PD versus FAR.

6.7 DISCUSSION

This chapter presented context-dependent classification as a potential technique for achieving robust target detection performance for remote sensing problems that exhibit concept drift across changing environmental conditions. Context-dependent classification methods differ from conventional approaches in that they first mitigate the effects of the underlying physical factors before applying a decision rule. This is achieved by applying a C-ary classifier to physics-based contextual features extracted from the background data to identify the context, then applying one of C classifiers to a separate set of target features for classifying targets from background in that context. Although the example problems that were considered utilized binary classifiers for target classification, the context-dependent framework could be easily extended to multiclass problems. Both supervised and unsupervised context models were considered; the supervised context model utilized a Gaussian hypothesis test, while the unsupervised model utilized a DPGMM. In both cases, RVMs were used as the target classifiers. In examples considering target detection with GPR and HSI sensors, context-dependent classification yielded improved performance over a conventional RVM-based classification technique. However, the context-dependent approach based on unsupervised context identification yielded the best improvement in performance. The advantages to unsupervised contextual modeling are twofold; an unsupervised learning technique can incorporate information beyond the scope of qualitative contextual labels, and the DP allows for automatic determination of the number of contexts present in the data.

Various other approaches to context-dependent classification have been considered in the remote sensing literature but were not described in this chapter. Many of these approaches use *discriminative* context models, which stand in contrast to the *generative* models used in this work. These include context-dependent classification based on jointly performing fuzzy clustering and linear discrimination [36] or by learning the intersections of various random sets [37]. The DP has also been applied to a discriminative contextual learning framework in the form of *multitask learning*, in which an optimal number of "tasks" based on different subsets of training data are learned [38,39].

The contextual modeling approach presented in this chapter could potentially be extended by incorporating spatial information. In applications to vehicular GPR, a context model based on an HMM has been proposed for consolidating closely located observations into the same context for classification purposes [40]. An extension of this method to HSI may be possible by considering 2D spatial variations in context, effectively making the context model a Markov random field [13]. Finally, applications of context-dependent classification to newly collected data in fielded systems will require the learning of new contexts as they are encountered while forgetting previously seen contexts have become irrelevant. Therefore, online context learning is an important direction of future research which will help incorporate context-dependent learning in fielded systems.

REFERENCES

1. A. Tsymbal, The problem of concept drift: Definitions and related work, Technical Report TCD-CS-2004-15, Department of Computer Science, Trinity College Dublin, Dublin, 2004.
2. I.S. Reed and X. Yu, Adaptive multiple-band CFAR detection of an optical pattern with unknown spectral distribution, *IEEE Transactions on Acoustics, Speech and Signal Processing*, 38(10), 1760–1770, 1990.
3. R. Mayer, F. Bucholtz, and D. Scribner, Object detection by using "whitening/dewhitening" to transform target signatures in multitemporal hyperspectral and multispectral imagery, *IEEE Transactions on Geoscience and Remote Sensing*, 41(5), 1136–1142, 2003.
4. P.A. Torrione, C.S. Throckmorton, and L.M. Collins, Performance of an adaptive feature-based processor for a wideband ground penetrating radar system, *IEEE Transactions on Aerospace and Electronic Systems*, 42(2), 644, 2006.
5. C. Lee and D.A. Landgrebe, Feature extraction based on decision boundaries, *IEEE Transactions on Pattern Analysis and Machine Intelligence*, 15(4), 388–400, 1993.
6. P.D. Gader and M.Y. Zhao, Landmine detection with ground penetrating radar using hidden Markov models, *IEEE Transactions on Geoscience and Remote Sensing*, 39(6), 1231–1244, 2001.

7. K.C. Ho, L. Carin, P.D. Gader, and J.N. Wilson, An investigation of using the spectral characteristics from ground penetrating radar for landmine/clutter discrimination, *IEEE Transactions on Geoscience and Remote Sensing*, 46(4), 1177–1191, 2008.

8. P. Torrione and L.M. Collins, Texture features for antitank landmine detection using ground penetrating radar, *IEEE Transactions on Geoscience and Remote Sensing*, 45(7), 2374–2382, 2007.

9. H. Frigui and P. Gader, Detection and discrimination of land mines in ground-penetrating radar based on edge histogram descriptors and a possibilistic K-nearest neighbor classifier, *IEEE Transactions on Fuzzy Systems*, 17(1), 185–199, 2009.

10. J.A. Benediktsson, J.R. Sveinsson, and K. Amason, Classification and feature extraction of AVIRIS data, *IEEE Transactions on Geoscience and Remote Sensing*, 33(5), 1194–1205, 1995.

11. P.G. Lucey et al., Three years of operation of AHI: The University of Hawaii's Airborne Hyperspectral Imager, *Proceedings of SPIE*, 4369, 112–120, 2001.

12. G. Healey and D. Slater, Invariant recognition in hyperspectral images, in *Proceedings of IEEE Computer Society Conference on Computer Vision and Pattern Recognition*, Collins, CO, USA, Vol. 1, p. 443, 1999.

13. Q. Jackson and D. Landgrebe, Adaptive Bayesian contextual classification based on Markov random fields, in *IEEE International Geoscience and Remote Sensing Symposium*, Toronto, Ontario, Canada, Vol. 3, pp. 1422–1424, 2002.

14. D.J. Daniels, *Ground Penetrating Radar*. London: Institution of Electrical Engineers, 2004.

15. K. Bradbury, P.A. Torrione, and L.M. Collins, Real-time Gaussian Markov random field-based ground tracking for ground penetrating radar data, *Proceedings of SPIE*, (7303), Orlando, FL, USA, 7303–7372, 2009.

16. P.A. Torrione and L.M. Collins, Ground response tracking for improved landmine detection in ground penetrating radar data, in *IEEE Geoscience and Remote Sensing Symposium*, Denver, CO, USA, pp. 153–156, 2006.

17. A. van der Merwe and I.J. Gupta, A novel signal processing technique for clutter reduction in GPR measurements of small, shallow land mines, *IEEE Transactions on Geoscience and Remote Sensing*, 38(6), 2627–2637, 2000.

18. R. Firoozabadi, E.L Miller, C.M Rappaport, and A.W Morgenthaler, Subsurface sensing of buried objects under a randomly rough surface using scattered electromagnetic field data, *IEEE Transactions on Geoscience and Remote Sensing*, 45(1), 104–117, 2007.

19. M. El-Shenawee and C.M. Rappaport, Quantifying the effects of different rough surface statistics for mine detection using the FDTD technique, in *Proceedings of the SPIE*, Orlando, FL, USA, 2000, 4038, 966–975.

20. T.W. Miller, J.M. Hendrickx, and B. Borchers, Radar detection of buried landmines in field soils, *Vadose Zone Journal*, 3(4), 1116–1127, 2004.

21. C.M Bishop and M.E Tipping, Variational relevance vector machines, in *Proceedings of the 16th Conference on Uncertainty in Artificial Intelligence*, Stanford, CA, USA, pp. 46–53, 2000.

22. M.E. Tipping, Sparse Bayesian learning and the relevance vector machine, *Journal of Machine Learning Research*, 1, 211–244, 2001.

23. Y. Li, C. Campbell, and M. Tipping, Bayesian automatic relevance determination algorithms for classifying gene expression data, *Bioinformatics*, 18(10), 1332–1339, 2002.

24. D.M. Blei and M.I. Jordan, Variational inference for Dirichlet process mixtures, *Bayesian Analysis*, 1(1), 121–144, 2006.

25. H. Attias, A variational Bayesian framework for graphical models, *Advances in Neural Information Processing Systems*, 12(1–2), 209–215, 2000.

26. M. Berman et al., ICE: A statistical approach to identifying endmembers in hyperspectral images, *IEEE Transactions on Geoscience and Remote Sensing*, 42(10), 2085–2095, 2004.

27. A. Zare and P. Gader, PCE: Piecewise Convex Endmember detection, *IEEE Transactions on Geoscience and Remote Sensing*, 48(6), 2620–2632, 2010.

28. C.R. Ratto, P.A. Torrione, and L.M. Collins, "Exploiting ground-penetrating radar phenomenology in a context-dependent framework for landmine detection and discrimination, *IEEE Transactions on Geoscience and Remote Sensing*, 49(5), 1–12, May 2011.

29. J.A. Huisman, S.S. Hubbard, J.D. Redman, and A.P. Annan, Measuring soil water content with ground penetrating radar: A review, *Vadose Zone Journal*, 2(4), 476–491, 2003.

30. C.R. Ratto, P.A. Torrione, and L.M. Collins, Estimation of soil permittivity through autoregressive modeling of time-domain ground-penetrating radar data, in *Proceedings of IEEE International Conference on Wireless Information Technology Systems*, Honolulu, HI, pp. 1–4, 2010.

31. M. Skolnik, *Radar Handbook*, 2nd edition. McGraw-Hill, Boston, MA, USA, 1990.

32. S. Haykin, *Adaptive Filter Theory*, Thomas Kailath, ed. Upper Saddle River, NJ: Prentice-Hall, 1996.
33. J.N. Wilson, P. Gader, W.H. Lee, H. Frigui, and K.C. Ho, A large-scale systematic evaluation of algorithms using ground-penetrating radar for landmine detection and discrimination, *IEEE Transactions on Geoscience and Remote Sensing*, 45(8), 2560–2572, 2007.
34. N.X. Vinh, J. Epps, and J. Bailey, Information theoretic measures for clusterings comparison: Is a correction for chance necessary?, in *Proceedings of the 26th Annual International Conference on Machine Learning*, Montreal, Quebec, Canada, pp. 1073–1080, 2009.
35. NIITEK. NIITEK web site. http://www.niitek.com.
36. H. Frigui, L. Zhang, and P.D. Gader, Context-dependent multisensor fusion and its application to land mine detection, *IEEE Transactions on Geoscience and Remote Sensing*, 48(6), 2528–2543, 2010.
37. J. Bolton and P. Gader, Random set framework for context-based classification with hyperspectral imagery, *IEEE Transactions on Geoscience and Remote Sensing*, 47(11), 3810–3821, 2009.
38. Y. Xue, X. Liao, L. Carin, and B. Krishnapuram, Multi-task learning for classification with Dirichlet process priors, *Journal of Machine Learning Research*, 8, 35–63, 2007.
39. K.D. Morton, P. Torrione, and L.M. Collins, Dirichlet process based context learning for mine detection in hyperspectral imagery, in *2nd IEEE Workshop on Hyperspectral Image and Signal Processing: Evolution in Remote Sensing (WHISPERS)*, Reykjavik, Iceland, pp. 1–4, 2010.
40. C. Ratto, P. Torrione, K. Morton, and L. Collins, Context-dependent landmine detection withground-penetrating radar using a hidden Markov context model, in *Proceedings of IEEE International Geoscience and Remote Sensing Symposium*, Honolulu, HI, USA, pp. 4192–4195, 2010.
41. T.S Jaakkola and M.I Jordan, Bayesian parameter estimation via variational methods, *Statistics and Computing*, 10(1), 25–37, 2000.

7 NMF and NTF for Sea Ice SAR Feature Extraction and Classification

Juha Karvonen

CONTENTS

7.1 INTRODUCTION

Sea ice synthetic aperture data (SAR) data are an important source of navigational information in the ice-covered seas, one of these areas being the northern parts of the Baltic Sea, especially Gulf of Finland with a very heavy winter sea traffic. To automate the interpretation of SAR images, several sea ice classification algorithms have been developed at the Finnish Meteorological Institute. To improve our classification schemes, more detailed information on sea ice types is required. One way to better utilize the information present in sea ice SAR data is to better utilize the textural information. We have studied the use of the independent component analysis [1] to extract basic textural primitives from SAR data to be used in classification [2]. To improve this feature extraction, we have also studied some novel methods to extract some elementary texture features from sea ice SAR data.

In this chapter, sparseness-constrained nonnegative matrix factorization (SC-NMF) and nonnegative tensor factorization (NTF) for sea ice SAR feature extraction and classification are studied. In this framework, the conventional nonnegative matrix factorization (NMF) can be interpreted as a special case of SC-NMF with no constraints. NMF and NTF can, among other things, be used to extract certain elementary features from image data. SC-NMF additionally adjusts the sparseness of these elementary features, and thus makes it possible to select sets of features or reconstruction weights with a given sparseness. NTF has two advantages compared to NMF: it does not lose the spatial redundancy, which occurs in the vectorization process performed for image data in the case of NMF, and it is also unique.

We also present some useful gradient-based features for clustering NMF and NTF basis vectors. These features alone can also be used for faster classification of SAR scenes than correlating with elementary features in multiple orientations.

7.2 NONNEGATIVE FACTORIZATION METHODS

7.2.1 NONNEGATIVE MATRIX FACTORIZATION

Nonnegative matrix factorization [3–5] is a method to find an approximate factorization for a nonnegative data matrix V such that $V \approx WH$, where W and H are nonnegative (matrix) factors. One useful feature of NMF is that it usually produces a sparse representation of the data. Given the data matrix V, the optimal choice of W and H is defined as the nonnegative matrices minimizing the reconstruction error between V and WH. A common error criterion is the Euclidean distance (L_2 norm), which leads to the minimization problem:

$$\min_{V,H} \left\| V - WH \right\|^2 ,$$
$$\text{s.t. } W, H \geq 0 \text{ (matrices nonnegative).} \tag{7.1}$$

A widely used class of NMF and NTF algorithms is the gradient descent algorithms with multiplicative update rules [4–6]. In our experiments here, we have used the Euclidean (L_2 norm) distance. The update rules for W and H then become

$$W \leftarrow W \otimes (VH^{\mathrm{T}}) \oslash (WHH^{\mathrm{T}}) \quad \text{and} \tag{7.2}$$

$$H \leftarrow H \otimes (W^{\mathrm{T}}V) \oslash (W^{\mathrm{T}}WH), \tag{7.3}$$

where $\otimes$ and $\oslash$ denote element-wise multiplication and division, respectively.

Also, faster algorithms based on alternating least squares (ALS) exist. The ALS approach holds all variables except for one fixed. There also exist algorithms for SC-NTF [7,8]. Also, other measures of difference between matrices, for example, the Kullback–Leibler divergence, can be used [5], but here we use the L_2 norm.

Sparseness can be defined based on the L_1 and L_2 norms such that for evenly distributed values it is zero and for one nonzero distribution component it is one (discrete case). Sparseness is thus defined as [9]

$$S(x) = \frac{\sqrt{n} - \left(\sum |x_i| \right) \Big/ \sqrt{\sum x_i^2}}{\sqrt{n} - 1} , \tag{7.4}$$

where n is the dimension of the vector x. Now the NMF with sparseness constraints becomes a minimization task to minimize the objective of Equation 10.1 with constraints:

$$S(w_i) = S_w$$
$$S(h_i) = S_h, \tag{7.5}$$

where w_i is the ith column of W and h_i is the ith row of H, and S_w and S_h are the desired sparseness values for W and H. The data matrix V is of the size $N \times T$ and W and H of the sizes $N \times M$ and $M \times T$, respectively. The values M, S_w, and S_h are defined by the user. The gradient descent algorithm to compute NMF with sparseness constraints is described in detail in Ref. [9]. The SC-NMF algorithm is a projected gradient-descent algorithm, which takes a step to the direction of the negative gradient and projects onto the constraint space, controlling the step size such that the objective function is reduced at each step. The projection operation enforces the desired sparseness. The algorithm can also be applied such that constraints are applied only either to W (containing the basis vectors) or to H, or no constraints at all (leading to the standard NMF).

7.2.2 Nonnegative Tensor Factorization

In NTF, a set of images forms a 3-D array (tensor) G, for example, a video clip, consisting of a set of video frames, in which case the third dimension is the time, and G is estimated as a sum of outer products of vectors, that is,

$$G \approx \sum_{i=1}^{k} w_i \otimes u_i \otimes v_i, \tag{7.6}$$

where G is the tensor consisting of the image data, k is the rank of the tensor, and w_i, u_i, and v_i are the components of the $K \times N_{\text{samples}}$, $K \times N$, and $K \times N$ matrices W, U, and V, respectively $\otimes$ denotes the outer product. Now, the minimization problem becomes

$$\min_{w_i, u_i, v_i} \frac{1}{2} \left\| G - \sum_{i=1}^{k} w_i \otimes u_i \otimes v_i \right\|_F^2, \tag{7.7}$$

$$\text{s.t. } w_i, u_i, v_i \geq 0.$$

In the general case, the dimension of G would be $M \geq 2$ and the number of the components would also be M, but for the image window data, $M = 3$.

The multiplicative NTF update rules for the elements of the three factors W, U, and V, are [6]

$$W_{sl} \leftarrow \frac{W_{sl} \sum_{i_U, i_V} G_{li_U i_V} U_{si_U} V_{si_V}}{\sum_{j=1}^{k} W_{jl} \left(U_j^T U_s \right) \left(V_j^T V_s \right)} \tag{7.8}$$

$$U_{sl} \leftarrow \frac{U_{sl} \sum_{i_W, i_V} G_{i_W l i_V} W_{si_W} V_{si_V}}{\sum_{j=1}^{k} U_{jl} \left(W_j^T W_s \right) \left(V_j^T V_s \right)} \tag{7.9}$$

$$V_{sl} \leftarrow \frac{V_{sl} \sum_{i_W, i_U} G_{i_W i_U l} W_{si_W} U_{si_u}}{\sum_{j=1}^{k} V_{jl} \left(W_j^T W_s \right) \left(U_j^T U_s \right)}. \tag{7.10}$$

K is the number of desired components, N_{samples} is the number of samples used in the training, that is, number of image windows, and the sample window size is $N \times N$.

7.3 SAR DATA PROCESSING

We used Envisat ASAR wide swath mode, Radarsat-1 and Radarsat-2 ScanSAR Wide (SCW) mode data acquired over the Baltic Sea ice in our tests. The data are C-band SAR data with a spatial resolution of about 100 m.

We have rectified the data to Mercator projection using WGS-84 as the reference ellipsoid and with the reference latitude (latitude of the correct scale) of $61°$ 40 min north to ensure compatibility with the ice charts of the Finnish Ice Service. Mercator projection is also used in Marine charts.

After rectification, a land masking is performed, and after land masking, a linear incidence angle correction based on the coefficients defined in Ref. [10] is performed. This correction removes the statistical dependency of the C-band backscattering from sea ice areas on the incidence angle. However, the incidence angle correction does not work for open water, because for open water the surface roughness affecting the backscattering changes depending on the wind conditions.

The SAR data used as NMF and NTF inputs were first filtered to reduce the effect of speckle, using a simple iterative median filtering with a window size 3×3 pixels and three iterations. Then, sample windows of the SAR data were randomly selected to be used in training the SC-NMF. We only included windows in which the standard deviation and the absolute value of the kurtosis exceed given threshold values, $T_1 = 5.0$ and $T_2 = 0.3$, respectively. This is done to avoid including areas of level ice only, because these areas do not contain much structural textural information. It can, however, carry some statistical textural information. The window radius was 13 pixels and the windows were round shaped. The window data were then vectorized, to produce vectors of the length 145, and fed into the SC-NMF algorithm. We conducted several tests, we run the NMF without any constrains and with selected different sparseness constraints for H. The sparseness of W was not constrained in these studies because the experimental results for SAR data in this case were not very promising. For comparison, we also performed the same computations for some visual image texture data, consisting of six Brodatz textures [11], that is, D17, D21, D34, D38, D77, and D93. These tests with the Brodatz showed that the studied methods actually capture the different elementary features of the texture data sets.

7.4 EXPERIMENTAL RESULTS

7.4.1 FEATURE EXTRACTION

We have studied the use of SC-NMF and NTF for feature extraction from both optical and SAR data and compared the results with each other. In the first phase, a representative set of randomly selected image data windows (typically at least a few thousand image windows) is sampled from a training data set. In the case of SC-NMF, the matrix V is formed of the vectorized image windows and the factors are defined by the SC-NMF algorithm. In the case of NTF, the tensor G is formed of the image windows, the third dimension being the number of the samples, and the factors are defined by the NTF algorithm. We have used round-shaped data windows; because of their symmetry, it is very simple, for example, to rotate such windows.

Optical texture data were studied just to test the method for texture feature extraction. The optical test data were extracted from a widely used optical texture image set, the collection of Brodatz textures [11]. The tests proved that the methods are useful for texture feature extraction for manmade textures. The SAR data were randomly selected areas of our typical operational Radarsat-1 ScanSAR Wide mode images over the Baltic Sea ice. The SAR images were preprocessed before feeding them to the algorithms. First, a median filtering was performed, and then only windows having an edge in the middle of the window were selected. The edges were located on the basis of the Canny edge detection algorithm [12]. We also required high-enough contrast within an acceptable data window. Finally, we rotated the image windows such that they have the main principal component of the edge pixels inside the window in the vertical direction. The idea behind this preprocessing is to reduce the complexity of the inputs due to rotation of similar features in the data set.

We varied the desired SC-NMF sparseness of H, containing the weights to reconstruct the original data windows (in V) from the produced set of basis vectors (in W), and noticed that for our SAR data, without any constraints and for small sparseness values, the basis vectors represent more complicated and rather sharp edge-like shapes and pieces of edges. As the sparseness of H is increased, smoother and rather straight edge-like vectors were generated. The mean sparseness of W decreases as the sparseness of H increases. The basis vectors produced with low sparseness values of H represent higher-frequency features and the basis vectors with higher sparseness values of H represent lower-frequency information (ramps or smooth edges). For the SAR data, the sparseness pairs ($S(H)$, $S(W)$) were (0.2, 0.352), (0.4, 0.221), (0.6, 0.110), and (0.8, 0.005), the latter values are means over the whole data sets. SC-NMF features (basis vectors) derived from SAR data with different sparseness values $S(H)$ are shown in Figure 7.1.

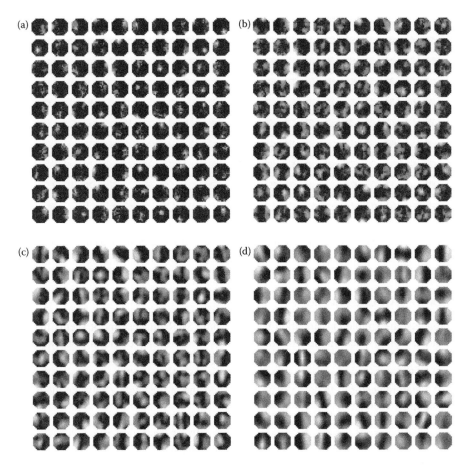

FIGURE 7.1 SC-NMF features extracted from sea ice SAR data, sparseness of H constrained to (a) 0.2, (b) 0.4, (c) 0.6, and (d) 0.8. The number of basis vectors $K = 100$.

In this study, we have only studied NTF with no sparseness constraints, and it seems that it is more difficult to visually compare the generated basis vectors to actual SAR data features. To produce more SAR-like features, we performed a clustering for the weights in the matrix W to the same number of cluster centers as the reduced dimension K. This clustering was performed using the K-means algorithm [13]. Then we reproduced a set of vectors based on these clustered weight vectors by multiplying the basis matrices produced by NTF. The basis matrices are formed as outer products between the components of U and V, and an NTF basis vector matrix B can be formed of these matrices by vectorizing them to be rows of B. B is then a $K \times M$ matrix. The matrix B_c containing the feature vectors, based on cluster reconstruction is

$$B_c = W_c B, \tag{7.11}$$

where W_c is the matrix containing the clustered weights. Probably, a more sophisticated clustering algorithm could yield even better results, but these results already show that reasonable features can be produced by this kind of clustering. A set of basis vectors for SAR data produced by NTF and the corresponding cluster reproduction, describing a set of assumed SAR elementary feature vectors, is shown in Figure 7.2.

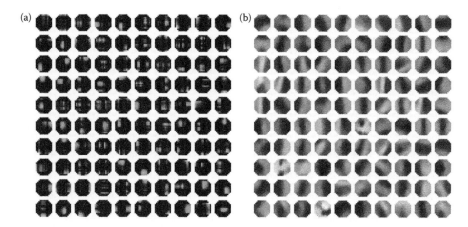

FIGURE 7.2   A set of SAR basis vectors produced by NTF (a) and a set of SAR elementary features produced by cluster reconstruction (b).

7.4.2   Sea Ice SAR Classification

We have made some simple classification tests for some SAR images, and it seems that different feature vectors, or filters, here in the sense that we compute the cosine of the angle between the basis vector and a data window, that is, the direction cosine, produce different response for different types of SAR areas, depending on the type of the edges present in the area. The measure of similarity, or response of a feature vector w_i to data x, was also the cosine of the angle γ between the basis vector w_i and the sampled vector x:

$$D(w_i, x) = \cos(\gamma) = \frac{w_i \cdot x}{|w_i||x|}. \tag{7.12}$$

This measure is one as the vectors have the same direction, and is zero as they are perpendicular to each other.

For classification, we first divided the SC-NMF basis vectors into four classes and then studied the class-wise responses around each SAR pixel to classify the SAR images. The classes for SC-NMF were based on four different desired sparseness values of H (0.2, 0.4, 0.6, and 0.8). In the case of NTF, the basis vectors were classified based on a set of features computed for each thresholded feature matrix. We also studied a feature classification based on vector sparseness computed using Equation 10.5. Finally, because in the training stage we tried to remove the rotation from the input data set, we produced an extended set of basis vectors, by rotating each produced basis vector in increments of 45°, thus producing eight differently oriented vectors for each of the basis vectors. The responses D at each pixel location for the four classes were computed, resulting in four features at each SAR pixel. The feature for each class is the highest response for all the feature vectors within each class. The values of the features can be visualized as false color images by placing features or their projections on the R, G, and B channels. We have used the four features corresponding to the four classes and the SAR pixel value with incidence angle correction as a set of features and computed visualized images which have their three principal components corresponding to the three highest absolute values of eigenvalues on the R, G, and B channels. According to our earlier studies [2,14], the uniform area SAR pixel intensity distributions for our SAR data are nearly Gaussians, and thus first- and second-order statistics can be used to completely describe the uniform areas. Actually, even the mean and the standard deviation are correlated, and here we have only included the pixel intensity value in our tests. In nonuniform (and non-Gaussian) areas, we get additional texture information from the class-wise NMF and NTF features. The classification results show that we get different responses at different kinds of edges and the ice type distributions over larger SAR segments can then be defined based on these responses. An example of the NMF classification is shown in Figure 7.3; the values $D(w_{opt}, x)$, corresponding to the best-matching basis vector w_{opt}, are also shown. It can be seen that the best matching is achieved in the relatively smooth areas where the image is best described by the low-frequency basis vector class. At the edges, the higher-frequency features describe the image data better, depending on the nature of the edges.

The filters of different SC-NMF classes match to slightly different ice edge types in the SAR images. The lowest-frequency filters match best to relatively smooth ice, producing high responses in those areas. The other three sets of classified filters or basis vectors give their best response in edges of medium sharpness (class 2), sharp edges (class 3), and very sharp edges (class 4). The local type of the ice can be classified based on the combined response. For example, an old and very deformed ice field has typically high values for all the three classes corresponding to edges, and a small feature, for example, a single ridge or lead, has high values only for the one or two classes representing the sharper edges, depending on the size of the feature. Similar behavior can also be found with the classified NTF feature vectors, depending on the classification method of the feature vectors. If the classification is different and not based on the sparseness, the interpretation of the result is also different.

For the NTF basis vectors we performed an automated clustering resulting into five classes. We used five gradient-based features for this purpose. The features are described further in detail in the subsection discussing the gradient features. This is just one example of clustering the basis vectors into classes, and for certain kinds of data sets more suitable selections can be used; also, performing the classification based on visual interpretation is a good alternative because then the expert interpretation is included. Here, we just try to show the steps of the proposed SAR classification scheme. The ordered basis and the corresponding classes are shown in different colors in Figure 7.4 and all the extended NTF basis vectors used in the classification tests are shown in Figure 7.5.

We also performed some tests for the classification based on the NTF basis vectors. The results show that we can locate different types of edges in SAR images using these basis vectors or filters. The class three (light green in the Figure 7.6c) basis vectors represent smooth changes and the best responses of level ice and open water are typically in this class. The classes one and two (green and light blue) represent sharper edges, and the classes four and five (black end red) represent even sharper edges. The classes four and five typically indicate highly deformed ice or an edge between a crack or level ice and deformed ice. The classification results for two SAR image cuts are shown in Figures 7.6 and 7.7. In the figures, we also show the scaled values of D for three classes (1, 3, and 5) as three-channel color images. It can be seen that in the areas with ice floes and cracks, the values of D are low for all the three features; in the level ice areas, feature three has the highest response, and in the deformed areas with high concentration both

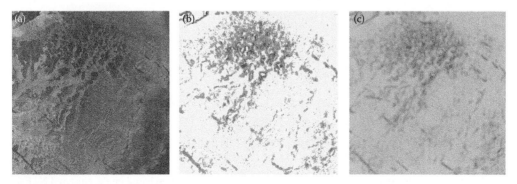

FIGURE 7.3 An example of SAR classification. The SAR image (a). The classification into the four classes (the darker values represent the classes corresponding to lower sparseness values $S(H)$ and the value is increased as a function of $S(H)$, (b). The mapping similarity $D(w_{opt}, x)$ (c).

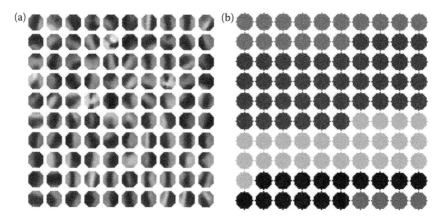

FIGURE 7.4 (**See color insert.**) Ordered NTF basis vectors (a) and the classes shown as color codes (b).

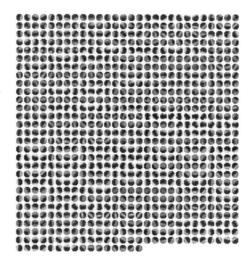

FIGURE 7.5 Extended set of the ordered NTF basis vectors. The vectors of the original basis have been rotated in steps of 45°.

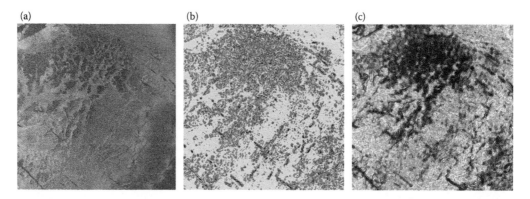

FIGURE 7.6 (**See color insert.**) A SAR image area (a) and its classification based on the clustered NTF features. (b) shows the best-matching class of the five classes using the same coloring as shown in Figure 7.4, and (c) image with the correlations D with the best-matching vectors of three from the five classes (1, 3, 5) on the R, G, and B channels, respectively.

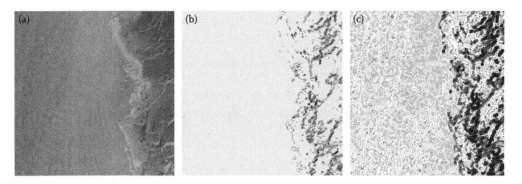

FIGURE 7.7 Another SAR image area with a large open water area (a) and its classification based on the clustered NTF features. (b) shows the best-matching class of the five classes using the same coloring as shown in Figure 7.4, and (c) image with the correlations with the best-matching vectors from three of the five classes (1, 3, 5) on the R, G, and B channels, respectively.

features one and five have higher response than feature three. In the zones of new ice with some deformation, the responses for all the three features are almost similar but higher than for the areas with floes and cracks.

7.4.3 Gradient Features and Their Stand-Alone Use

We have used the gradient features described in the following for clustering the basis vectors produced by SC-NMF and NTF. Additionally, we have also studied their performance for classifying SAR images instead of the basis vector matching. Because the matching (correlating) with the NMF/NTF elementary features requires much computing, we have tried to develop methods to classify the edges in the SAR data with features derived from the NMF/NTF elementary features. These features are based on the local gradient within a round-shaped image window around each SAR pixel.

SAR scene classifications based on features can be computed faster compared to correlating with preclassified NMF/NTF elementary features in multiple orientations and they seem to give reasonable classification results for our sea ice SAR data. The features we have been using are the relative amount of high gradients within a window (GF_1), the optimal threshold between the high- and low-gradient parts (GF_2), the high-gradient area shape (FG_3), and the distance between the high- and low-gradient values (FG_4). First, the absolute gradients over the whole image are computed, and then for each fixed-size (e.g., with a radius $R = 5$ pixels) circular window around a pixel, the features are computed. It is assumed that each pixel contains an edge and some background pixels. The pixels with high absolute gradient values are the edge pixels and the pixels with the low-gradient values are the background pixels. We also assume that there are pixels from one edge in each small window and that both the edge and background absolute gradient values have a Gaussian distribution. Based on the assumption, we apply the expectation-maximization (EM) algorithm [15] to find the parameters of the two Gaussian distributions for each window. Based on the EM algorithm we get the estimated value pairs representing the two Gaussian distributions, that is, means and standard deviations for both the distributions, (μ_1, σ_1) and (μ_2, σ_2), and also estimates for the mixing proportions p_1 and p_2. For the estimated distribution parameters and mixing proportions, we can estimate the optimal threshold (T) between them based on the Bayesian criteria. And after this operation we divide the pixels within the window into high- and low-gradient pixels by thresholding. After thresholding, we still apply a simple filtering to remove small separate high-gradient areas and only leaving the larger areas with a size than a given threshold (given number of pixels) to reduce the speckle effect in the results. The amount of high-gradient pixels is then N_h and the amount of low-gradient pixels is N_l, and we can get GF_1 and and GF_2 as

$$GF_1 = N_h/(N_h + N_l) = N_h/N \tag{7.13}$$

$$GF_2 = T. \tag{7.14}$$

GF_2 more or less describes the gradient magnitude. The shape factor is computed from the eigenvalues of the covariance matrix of the locations of the high-gradient pixels in the window as

$$GF_3 = \sqrt{\lambda_1/\lambda_2}, \tag{7.15}$$

where λ_1 is the larger of the two eigenvalues and λ_2 the smaller. Thus, for a spherical edge shape $GF_3 = 1$ and for a linear-directed structure, it is larger than one, depending on the edge shape. This feature is computed only for those locations where the high-gradient area is centered with respect to the window center. The fourth feature is the distance between the peaks (corresponding to means in the case of Gaussians) of the two distributions:

$$GF_4 = m_2 - m_1, \tag{7.16}$$

where $m_2 > m_1$. The fifth feature is the ratio of the number of filtered high-gradient pixels, that is, those belonging to high-gradient segments with its size larger than a given threshold, N_{hf}, and the segment size:

$$GF_5 = N_{hf}/N. \tag{7.17}$$

We have performed some classification tests over the Baltic Sea based on these gradient features. Based on the test classifications, we can see that these features can be used to distinguish multiple sea ice classes: slightly deformed ice, highly deformed ice, level ice, and open water. For the classes of open water and level ice, the features are often rather similar and they cannot completely be distinguished by these features. However, for example, the land fast ice zone (which typically is rather undeformed), clearly visible in the coastal zone of the classification image 10.8 as darker areas, can be located based on the type of the surrounding ice fields. In the more deformed ice fields, all the four features produce higher values than in open water or level ice, and also more detailed analysis of the ice type is possible based on this set of features, for example, large ridges produce high values for SG_2 and SG_3. An example of these features for a SAR mosaic can be seen in Figure 7.8.

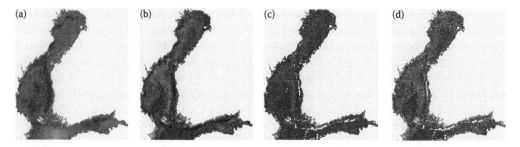

FIGURE 7.8 (**See color insert.**) A SAR image mosaic over the northern Baltic Sea with incidence angle correction applied, February 14, 2011 (a), an RGB image representing the features GF_1, GF_2, and GF_3 on R, G, and B channels, respectively (b), and the features GF_4 (c) and GF_5 (d). GF_4 has a large correlation with GF_2.

7.5 CONCLUSIONS

According to our experience it seems that the SC-NMF and NTF algorithms can produce useful elementary features to be used in classifying our sea ice SAR data over the Baltic Sea based on texture. The textured areas containing edges can be classified and different areas can be distinguished from each other. However, geophysical interpretation of the results is difficult and needs more studying, based on both visual inspection and comparisons to *in situ* measurements. The advantage of SC-NMF is that it directly produces reasonable classes of the feature vectors based on sparseness. In the case of NTF without any sparseness constraints, the feature vectors must first be classified to be used in a similar manner.

It also seems that SAR speckle is not affecting the classification very much. The results for speckle-filtered SAR data are rather similar to the results for the unfiltered SAR data. This is due to the fact that these features represent larger two-dimensional features or structures and the correlation is applied to a larger area than just one pixel. Naturally, the correlations for noisy data are reduced, but the same thing occurs for all the correlations and their magnitude order is preserved.

Because we have used only deformed ice areas as inputs for the SC-NMF and NTF algorithms, the ability of these basis vectors to distinguish between open water and level ice is not good. However, we have earlier shown that statistical textural information can be used for open water detection successfully [16].

The current classification algorithms are computationally rather heavy, but they can be made faster by reducing the sampling rate of the data windows. The number of feature vectors can be reduced by finding the feature vectors best describing the data, and also removing vectors, which are very much alike with other vectors in the set. Parallelizing of the algorithms is also straightforward.

We have not studied all the possible features for classification yet. For example, the ratios of responses between different classes could give additional information on the SAR content. The use of these features will be studied in the future.

We have also studied the application of the features we originally developed for clustering the basis vectors to SAR data directly, and the results of applying these features to sea ice SAR data have been promising. Using these features we can get useful information on the ice deformation and the ice structure.

In future, we are also interested to study the use of SC-NTF [7,8]. The advantage of this algorithm would be that the feature vectors corresponding to different sparseness values can then be considered as different classes, just like with SC-NMF, and no feature vector classification step is required. We have also made some preliminary studies of applying the projective NMF [17] to SAR data. Also, studies of feature extraction and classification in multiple resolutions using these methods are our future interests.

The presented methods could also be very useful in SAR scene interpretation of other targets than sea ice. For example, SAR images over forests would be a suitable application. Also, classification of data from imaging remote sensing instruments other than SAR are potential applications of the methods presented in this chapter.

REFERENCES

1. A. Hyvarinen, J. Karhunen, and E. Oja. *Independent Component Analysis*. John Wiley & Sons, New York, 2001. ISBN: 978-0-471-40540-5.
2. J. Karvonen and M. Simila. ICA-based classification of sea ice sar images. In *Proceedings of the 23rd European Association of Remote Sensing Laboratories (EARSeL) Annual Symposium*, pp. 211–217. EARSeL, 2003.
3. P. Paatero and U. Tapper. Positive matrix factorization: A non-negative factor model with optimal utilization of error estimates of data values. *Environmetrics*, 5:111–126, 1994.
4. D. D. Lee and H. S. Seung. Learning the parts of objects by non-negative matrix factorization. *Nature*, 401(6755):788–791, 1999.

5. D. D. Lee and H. S. Seung. Algorithms for non-negative matrix factorization. *Advances in Neural Information Processing 13 (Proceedings of NIPS 2000)*, 2001.

6. T. Hazan, S. Polak, and A. Shashua. Sparse image coding using a 3d non-negative tensor factorization. In *Proceedings of the Tenth IEEE International Conference on Computer Vision 2005, ICCV 2005*, vol. 1, pp. 50–57, 2005.

7. M. Heiler and C. Schnorr. Controlling sparseness in non-negative tensor factorization. *Lecture Notes in Computer Science*, 3951:56–67, 2006.

8. A. Cichocki, R. Zdunek, S. Choi, R. J. Plemmons, and S. Amari. Novel multi-layer non-negative tensor factorization with sparsity constraints. *Lecture Notes in Computer Science*, 4432:271–280, 2007.

9. P. O. Hoyer. Non-negative matrix factorization with sparseness constraints. *Journal of Machine Learning Research*, 5:1457–1469, 2004.

10. M. Simila, J. Karvonen, M. Makynen, T. Manninen, and M. Hallikainen. Incidence angle dependence of the statistical properties of the c-band hh-polarization backscattering signatures of the Baltic Sea ice. *IEEE Transactions on Geoscience and Remote Sensing*, 40(12):2593–2605, 2002.

11. P. Brodatz. *Textures—A Photographic Album for Artists and Designers.* Dover Publications, New York, 1966.

12. J. Canny. A computational approach to edge detection. *IEEE Transactions Pattern Analysis and Machine Intelligence*, 8(6):679–698, 1986.

13. Y. Linde, A. A. Buzo, and R. M. Gray. An algorithm for vector quantizer design. *IEEE Transactions on Communication*, 28(1):84–95, 1980.

14. J. Karvonen. *Compaction of C-Band Synthetic Aperture Radar Based Sea Ice Information for Navigation in the Baltic Sea.* Otamedia, 2006. Dr. Sci. (Tech.) Dissertation, Helsinki University of Technology, ISBN 951-22-8472-3.

15. N. M. Laird, A. P. Dempster, and D. B. Rubin. Maximum likelihood from incomplete data via the em algorithm. *Journal of the Royal Statistical Society Series*, 39(1):1–38, 1977.

16. J. Karvonen. A comparison of two c-band SAR ice/open water algorithms. In *Proceedings of the ESA SeaSAR 2010 Workshop, SP-679*. ESA, 2010.

17. Z. Yuan and E. Oja. Projective nonnegative matrix factorization for image compression and feature extraction. In *Proceedings of the 14th Scandinavian Conference on Image Analysis (SCIA 2005)*, pp. 333–342, 2002.

8 Relating Time Series of Meteorological and Remote Sensing Indices to Monitor Vegetation Moisture Dynamics

Jan Verbesselt, P. Jönsson, S. Lhermitte, I. Jonckheere,
J. van Aardt, and P. Coppin

CONTENTS

8.1 INTRODUCTION

The repeated occurrence of severe wildfires, which affect various fire-prone ecosystems of the world, has highlighted the need to develop effective tools for monitoring fire-related parameters. Vegetation water content (VWC), which influences the biomass burning processes, is an example of one such parameter [1–3]. The physical definitions of VWC vary from water volume per leaf or ground area (equivalent water thickness) to water mass per mass of vegetation [4]. Therefore, VWC could also be used to infer vegetation water stress and to assess drought conditions that linked with fire risk [5]. Decreases in VWC due to the seasonal decrease in available soil moisture can induce severe fires in most ecosystems. VWC is particularly important for determining the behavior of fires in savanna ecosystems because the herbaceous layer becomes especially flammable during the dry season when the VWC is low [6,7].

Typically, VWC in savanna ecosystems is measured using labor-intensive vegetation sampling. Several studies, however, indicated that VWC can be characterized temporally and spatially using meteorological or remote sensing data, which could contribute to the monitoring of fire risk [1,4]. The meteorological Keetch–Byram drought index (KBDI) was selected for this study. This index was developed to incorporate soil water content in the root zone of vegetation and is able to assess the seasonal trend of VWC [3,8]. The KBDI is a cumulative algorithm for the estimation of fire potential from meteorological information, including daily maximum temperature, daily total precipitation, and mean annual precipitation [9,10]. The KBDI has also been used for the assessment of VWC for vegetation types with shallow rooting systems, for example, the herbaceous layer of the savanna ecosystem [8,11].

The application of drought indices, however, presents specific operational challenges. These challenges are due to the lack of meteorological data for certain areas, as well as spatial interpolation techniques that are not always suitable for use in areas with complex terrain features. Satellite data provide sound alternatives to meteorological indices in this context. Remotely sensed data have significant potential for monitoring vegetation dynamics at regional to global scale, given the synoptic coverage and repeated temporal sampling of satellite observations (e.g., SPOT VEGETATION (SPOT VGT) or NOAA AVHRR) [12,13]. These data have the advantage of providing information on remote areas where ground measurements are impossible to obtain on a regular basis.

Most research in the scientific community using optical sensors (e.g., SPOT VGT) to study biomass burning has focused on two areas [4]: (1) the direct estimation of VWC and (2) the estimation of chlorophyll content or degree of drying as an alternative to the estimation of VWC. Chlorophyll-related indices are related to VWC based on the hypothesis that the chlorophyll content of leaves decreases proportionally to the VWC [4]. This assumption has been confirmed for selected species with shallow rooting systems (e.g., grasslands and understory forest vegetation) [14–16], but cannot be generalized to all ecosystems [4]. Therefore, only chlorophyll-related indices, such as the normalized difference vegetation index (NDVI), can be used in regions where the relationship among chlorophyll content, degree of curing, and water content has been established.

Accordingly, a remote sensing index that is directly coupled to the VWC is used to investigate the potential of hypertemporal satellite imagery to monitor the seasonal vegetation moisture dynamics. Several studies [4,16–18] have demonstrated that VWC can be estimated directly through the normalized difference of the near-infrared reflectance (NIR, 0.78–0.89 μm) ρ_{NIR}, influenced by the internal structure and the dry matter, and the short-wave infrared reflectance (SWIR, 1.58–1.75 μm) ρ_{SWIR}, influenced by plant tissue water content:

$$\text{NDWI} = \frac{\rho_{NIR} - \rho_{SWIR}}{\rho_{NIR} + \rho_{SWIR}} \tag{8.1}$$

The NDWI or normalized difference infrared index [19] is similar to the global vegetation moisture index [20].

The relationship between NDWI and KBDI time series, both related to VWC dynamics, is explored. Although the value of time-series data for monitoring vegetation moisture dynamics has been firmly established [21], only a few studies have taken serial correlation into account when correlating time series [6,22–25]. Serial correlation occurs when data collected through time contain values at time t, which are correlated with observations at time $t - 1$. This type of correlation in time series, when related to VWC dynamics, is mainly caused by the seasonal variation (dry–wet cycle) of vegetation [26]. Serial correlation can be used to forecast future values of the time series by modeling the dependence between observations but affects correlations between variables measured in time and violates the basic regression assumption of independence [22]. Correlation coefficients of serially correlated data cannot be used as indicators of goodness-of-fit of a model as the correlation coefficients are artificially inflated [22,27].

The study of the relationship between NDWI and KBDI is a nontrivial task due to the effect of serial correlation. Remedies for serial correlation include sampling or aggregating the data over longer time intervals, as well as further modeling, which can include techniques such as weighted regression [25,28]. However, it is difficult to account for serial correlation in time series related to VWC dynamics using extended regression techniques. The time series related to VWC dynamics often exhibit high non-Gaussian serial correlation and are more significantly affected by outliers and measurement errors [28]. Therefore, a sampling technique is proposed, which accounts for serial correlation in seasonal time series, to study the relationship between different time series. The serial correlation effect in time series is assumed to be minimal when extracting one metric per season (e.g., start of the dry season). The extracted seasonal metrics are then utilized to study the relationship between time series at a specific moment in time (e.g., start of the dry season).

The aim of this chapter is to address the effect of serial correlation when studying the relationship between remote sensing and meteorological time series related to VWC by comparing nonserially correlated seasonal metrics from time series. This chapter therefore has three defined objectives. First, an overview of time-series analysis techniques and concepts (e.g., stationarity, autocorrelation, autoregressive-integrated-moving-average (ARIMA), etc.) is presented and the relationship between time series is studied using cross-correlation and ordinary least square (OLS) regression analysis. Second, an algorithm for the extraction of seasonal metrics is optimized for satellite and meteorological time series. Finally, the temporal occurrence and values of the extracted nonserially correlated seasonal metrics are analyzed statistically to define the quantitative relationship between NDWI and KBDI time series. The influence of serial correlation is illustrated by comparing results from cross-correlation and OLS analysis with the results from the investigation of correlation between extracted metrics.

8.2   DATA

8.2.1   STUDY AREA

The Kruger National Park (KNP), located between 23°S and 26°S latitudes and 30°E and 32°E longitudes in the low-lying savanna of the northeastern part of South Africa, was selected for this study (Figure 8.1). Elevations range from 260 to 839 m above sea level, and mean annual rainfall varies between 350 mm in the north and 750 mm in the south. The rainy season within the annual climatic season can be confined to the summer months (i.e., November to April), and over a longer period can be defined by alternating wet and dry seasons [7]. The KNP is characterized by an arid savanna dominated by thorny, fine-leafed trees of the families Mimosaceae and Burseraceae. An exception is the northern part of the KNP where the Mopane, a broad-leafed tree belonging to the Ceasalpinaceae, almost completely dominates the tree layer.

8.2.2   CLIMATE DATA

Climate data from six weather stations in the KNP with similar vegetation types were used to estimate the daily KBDI (Figure 8.1). KBDI was derived from daily precipitation and maximum temperature data to estimate the net effect on the soil water balance [3]. Assumptions in the derivation of KBDI include a soil water capacity of ~20 cm and an exponential moisture loss from the soil reservoir. KBDI was initialized during periods of rainfall events (e.g., rainy season) that result in soils with maximized field capacity and KBDI values of zero [8]. The preprocessing of KBDI was done using the method developed by Janis et al. [10]. Missing daily maximum temperatures were replaced with interpolated values of daily maximum temperatures, based on a linear interpolation function [29]. Missing daily precipitation, on the other hand, was assumed to be zero. A series of

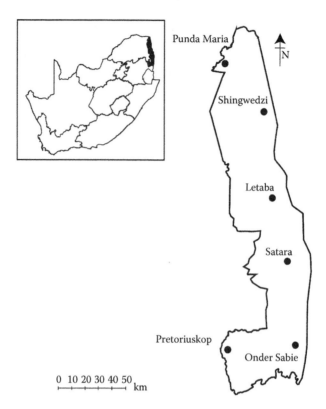

FIGURE 8.1 The Kruger National Park (KNP) study area with the weather stations used in the analysis (right). South Africa is shown with the borders of the provinces and the study area (top left).

error logs were automatically generated to indicate missing precipitation values and associated estimated daily KBDI values. This was done because zeroing missing precipitation may lead to an increased fire potential bias in KBDI. The total percentage of missing data gaps in rainfall and temperature series was maximally 5% during the study period for each of the six weather stations. The daily KBDI time series were transformed into 10-daily KBDI series, similar to the SPOT VGT S10 dekads (i.e., 10-day periods), by taking the maximum of each dekad. The negative of the KBDI time series (i.e., −KBDI) was analyzed in this chapter such that the temporal dynamics of KBDI and NDWI were related (Figure 8.2). The −KBDI and NDWI are used throughout this chapter. The Satara weather station, centrally positioned in the study area, was selected to represent the temporal vegetation dynamics. The other weather stations in the study area demonstrate similar temporal vegetation dynamics.

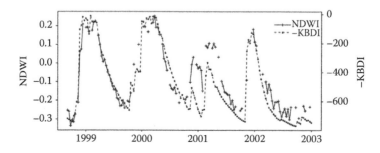

FIGURE 8.2 The temporal relationship between NDWI and −KBDI time series for the "Satara" weather station (Figure 8.1).

8.2.3   Remote Sensing Data

The data set used is composed of 10-daily SPOT VGT composites (S10 NDVI maximum value syntheses) acquired over the study area for the period April 1998 to December 2002. SPOT VGT can provide local-to-global coverage on a regular basis (e.g., daily for SPOT VGT). The syntheses result in surface reflectance in the blue (0.43–0.47 μm), red (0.61–0.68 μm), NIR (0.78–0.89 μm), and SWIR (1.58–1.75 μm) spectral regions. Images were atmospherically corrected using the simplified method for atmospheric correction (SMAC) [29]. The geometrically and radiometrically corrected S10 images have a spatial resolution of 1 km.

The S10 SPOT VGT time series were preprocessed to detect data that erroneously influence the subsequent fitting of functions to time series, necessary to define and extract metrics [6]. The image preprocessing procedures performed were:

- Data points with a satellite viewing zenith angle (VZA) above 50° were masked out as pixels located at the very edge of the image (VZA > 50.5°) swath are affected by resampling methods that yield erroneous spectral values.
- The aberrant SWIR detectors of the SPOT VGT sensor, flagged by the status mask of the SPOT VGT S10 synthesis, were also masked out.
- A data point was classified as cloud-free if the blue reflectance was <0.07 [30]. The developed threshold approach was applied to identify cloud-free pixels for the study area.

NDWI time series were derived by selecting savanna pixels based on the land cover map of South Africa [31] for a 3×3 pixel window centered at each of the meteorological stations to reduce the effect of potential spatial misregistration (Figure 8.1). Median values of the 9-pixel windows were then retained instead of single pixel values [32]. The median was preferred to average values as it is less affected by extreme values and therefore is less sensitive to potentially undetected data errors.

8.3   SERIAL CORRELATION AND TIME-SERIES ANALYSIS

Serial correlation affects correlations between variables measured in time, and violates the basic regression assumption of independence. Techniques that are used to recognize serial correlation, therefore, are discussed by applying them to the NDWI and −KBDI time series. Cross-correlation analysis is illustrated and used to study the relationship between time series of −KBDI and NDWI. Fundamental time-series analysis concepts (e.g., stationarity and seasonality) are introduced and a brief overview of the most frequently used method for time-series analysis to account for serial correlation, namely autoregression (AR), is presented.

8.3.1   Recognizing Serial Correlation

This chapter focuses on discrete time series, which contain observations made at discrete time intervals (e.g., 10 daily time steps of −KBDI and NDWI time series). Time series are defined as a set of observations, x_t, recorded at a specific time, t [26]. Time series of −KBDI and NDWI contain a seasonal variation which is illustrated in Figure 8.2 by a smooth increase or decrease of the series related to vegetation moisture dynamics. The gradual increase or decrease of the graph of a time series is generally indicative of the existence of a form of dependence or serial correlation among observations.

The presence of serial correlation systematically biases regression analysis when studying the relationship between two or more time series [25]. Consider the OLS regression line with a slope and an intercept:

$$Y(t) = a_0 + a_1 X(t) + e(t), \tag{8.2}$$

where t is time, a_0 and a_1 are OLS regression intercept and slope parameter, respectively, $Y(t)$ is the dependent variable, $X(t)$ is the independent variable, and $e(t)$ is the random error term. The standard error (SE) of each parameter is required for any regression model to define the confidence interval (CI) and derive the significance of parameters in the regression equation. The parameters a_0, a_1, and the CIs, estimated by minimizing the sum of the squared "residuals" are valid only if certain assumptions related to the regression and $e(t)$ are met [25]. These assumptions are detailed in statistical textbooks [33] but are not always met or explicitly considered in real-world applications. Figure 8.3 illustrates the biased CIs of the OLS regression model at a 95% confidence level. The SE term of the regression model is underestimated due to serially correlated residuals and explains the biased CI, where CI = mean ± 1.96 × SE.

The Gauss–Markov theorem states that the OLS parameter estimate is the best linear unbiased estimate (BLUE); that is, all other linear unbiased estimates will have a larger variance, if the error term, $e(t)$, is stationary and exhibits no serial correlation. The Gauss–Markov theorem consequently points to the error term and not to the time series themselves as the critical consideration [34]. The error term is defined as stationary when it does not present a trend and the variance remains constant over time [27]. It is possible that the residuals are serially correlated if one of the dependent or independent variables is also serially correlated, because the residuals constitute a linear combination of both types of variables. Both dependent and independent variables of the regression model are serially correlated (KBDI and NDWI), which explains the serial correlation observed in the residuals, as illustrated in Figure 8.3.

A sound practice used to verify serial correlation in time series is to perform multiple checks by both graphical and diagnostic techniques. The autocorrelation function (ACF) can be viewed as a graphical measure of serial correlation between variables or residuals. The sample ACF is defined when $x_1, \ldots, x_n$ are observations of a time series. The sample mean of $x_1, \ldots, x_n$ is [26]

$$\bar{x} = \frac{1}{n} \sum_{t=1}^{n} x_t. \tag{8.3}$$

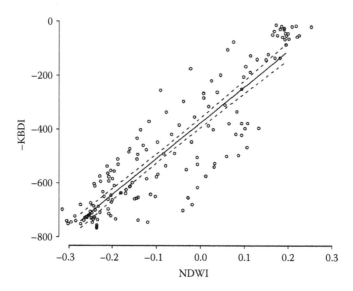

FIGURE 8.3 Result of the OLS regression fit between −KBDI and NDWI as dependent and independent variables, respectively, for the Satara weather station ($n = 157$). Confidence intervals (- - -) at a 95% confidence level are shown, but are "narrowed" due to serial correlation in the residuals.

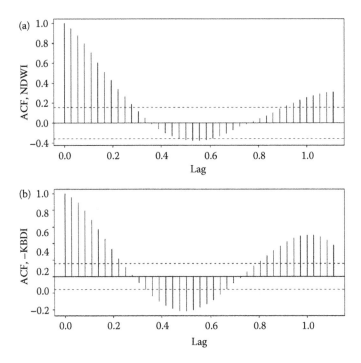

FIGURE 8.4   The autocorrelation function (ACF) for (a) −KBDI and (b) NDWI time series for the Satara weather station. The horizontal lines on the graph are the bounds $= \pm 1.96 / \sqrt{n}$ $(n = 157)$.

The sample autocovariance function with lag h and time t is

$$\hat{\gamma}(h) = n^{-1} \sum_{t=1}^{n-|h|} (x_{t+|h|} - \overline{x})(x_t - \overline{x}), \quad -n < h < n. \tag{8.4}$$

The sample ACF is

$$\hat{\rho} = \frac{\hat{\gamma}(h)}{\hat{\gamma}(0)}, \quad -n < h < n. \tag{8.5}$$

Figure 8.4 illustrates the ACF for time series of −KBDI and NDWI presented from the Kruger park data. The ACF clearly indicates a significant autocorrelation in the time series, as more than 5% of the sample autocorrelations fall outside the significance bounds $= \pm 1.96 / \sqrt{n}$ [26]. There are also formal tests available to detect autocorrelation such as the Ljung–Box test statistic and the Durbin–Watson statistic [25,26].

8.3.2   CROSS-CORRELATION ANALYSIS

The cross-correlation function (CCF) can be derived between two time series utilizing a technique similar to the ACF applied for one time series [27]. Cross-correlation is a measure of the degree of linear relationship existing between two data sets and can be used to study the connection between time series. The CCF, however, can be used only if the time series is stationary [27]. For example, when all variables are increasing in value over time, cross-correlation results will be spurious and subsequently cannot be used to study the relationship between time series.

Nonstationary time series can be transformed to stationary time series by implementing one of the following techniques:

- Differencing the time series by a period d can yield a series that satisfies the assumption of stationarity (e.g., $x_t - x_{t-1}$ for $d = 1$). The differenced series will contain one point less than the original series. Although a time series can be differenced more than once, one difference is usually sufficient.
- Lower-order polynomials can be fitted to the series when the data contain a trend or seasonality that needs to be subtracted from the original series. Seasonal time series can be represented as the sum of a specified trend, and seasonal and random terms. For example, for statistical interpretation results, it is important to recognize the presence of seasonal components and remove them to avoid confusion with long-term trends. Figure 8.5 illustrates the seasonal trend decomposition method using locally weighted regression for the NDWI time series [35].
- The logarithm or square root of the series may stabilize the variance in the case of a nonconstant variance.

Figure 8.6 illustrates the cross-correlation plot for stationary series of −KBDI and NDWI. −KBDI and NDWI time series became stationary after differencing with $d = 1$. The stationarity was confirmed using the "augmented Dickey–Fuller" test for stationarity [26,36] at a confidence level of 95% ($p < 0.01$; with stationarity as the alternative hypothesis). Note that ~95% confidence limits are shown for the autocorrelation plots of an independent series. These limits must be regarded with caution, since there exists an *a priori* expectation of serial correlation for time series [37].

Table 8.1 illustrates the coefficients of determination (i.e., multiple R^2) of the OLS regression analysis with serially correlated residuals −KBDI as dependent variable and NDWI as independent variable for all six weather stations in the study area. The Durbin–Watson statistic indicated that the residuals were serially correlated at a 95% confidence level ($p < 0.01$). These results will be

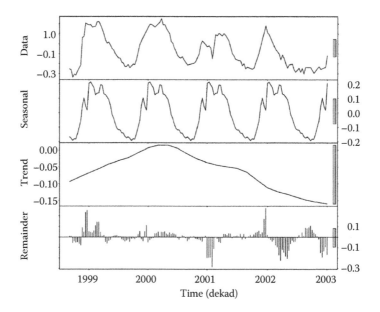

FIGURE 8.5 The results of the seasonal trend decomposition (STL) technique for NDWI time series of the Satara weather station. The original series can be reconstructed by summing the seasonal, trend, and remainder. In the y-axes, the NDWI values are indicated. The gray bars at the right-hand side of the plots illustrate the relative data range of the time series.

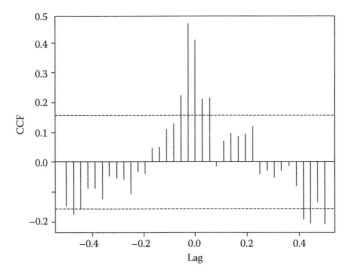

FIGURE 8.6   The cross-correlation plot between stationary −KBDI and NDWI time series of the Satara weather station, where CCF indicates results of the CCF. The horizontal lines on the graph are the bounds $(= \pm 1.96 / \sqrt{n})$ of the ~95% CI.

TABLE 8.1

Coefficients of Determination of the OLS Regression Model between −KBDI and NDWI ($n = 157$)

Station	R^2	Time Lag
Punda Maria	0.74	−1
Letaba	0.88	0
Onder Sabie	0.72	0
Pretoriuskop	0.31	0
Shingwedzi	0.72	−1
Satara	0.81	0

Note: The time expressed in dekads of maximum correlation of the cross-correlation between −KBDI and NDWI is also indicated.

---

compared with the method presented in Section 8.5. Table 8.1 also indicates the time lags at which correlation between time series was maximal, as derived from the cross-correlation plot. A negative lag indicates that −KBDI reacts prior to NDWI, for example, in the cases of Punda Maria and Shingwedzi weather stations, and subsequently can be used to predict NDWI. This is logical since weather conditions, for example, rainfall and temperature, change before vegetation reacts. NDWI, which is related to the amount of water in the vegetation, consequently lags behind the −KBDI. The major vegetation type in savanna vegetation is the herbaceous layer, which has a shallow rooting system. This explains why the vegetation in the study area quickly follows climatic changes and NDWI did not lag behind −KBDI for the other four weather stations.

8.3.3   Time-Series Analysis: Relating Time Series and Autoregression

A remedy for serial correlation, apart from applying variations in sampling strategy, is modeling of the time dependence in the error structure by AR. AR most often is used for purposes of forecasting and modeling of a time series [25]. The simplest AR model for Equation 8.2, where ρ is the result of the sample ACF at lag 1, is

$$e_t = \rho e_{t-1} + \varepsilon_t, \tag{8.6}$$

where ε_t is a series of serially independent numbers with mean zero and constant variance. The Gauss–Markov theorem cannot be applied and therefore OLS is not an efficient estimator of the model parameters if ρ is not zero [34].

Many different AR models are available in statistical software systems that incorporate time-series modules. One of the most frequently used models to account for serial correlation is the ARIMA model [26,37]. Briefly stated, ARIMA models can have an AR term of order p, a differencing (integrating) term (I) of order d, and a moving average (MA) term of order q. The notation for specific models takes the form of (p,d,q) [27]. The order of each term in the model is determined by examining the raw data and plots of the ACF of the data. For example, a second-order AR ($p = 2$) term in the model would be appropriate if a series has significant autocorrelation coefficients between x_t, and x_{t-1}, and x_{t-2}. ARIMA models that are fitted to time-series data using AR and MA parameters, p and q, have coefficients Φ and θ to describe the serial correlation. An underlying assumption of ARIMA models is that the series being modeled is stationary [26–27].

8.4   METHODOLOGY

The TIMESAT program is used to extract nonserially correlated metrics from remote sensing and meteorological time series [38,39]. These metrics are utilized to study the relationship between time series at specific moments in time. The relationship between time series, in turn, is evaluated using statistical analysis of extracted nonserially correlated seasonal metrics from time series (–KBDI and NDWI).

8.4.1   DATA SMOOTHING

It is often necessary to generate smooth time series from noisy satellite sensors or meteorological data to extract information on seasonality. The smoothing can be achieved by applying filters or by function fitting. Methods based on Fourier series [40–42] or least-square fits to sinusoidal functions [43–45] are known to work well in most instances. These methods, however, are not capable of capturing a sudden, steep increase or decrease of remote sensing or meteorological data values that often occur in arid and semiarid environments. Alternative smoothing and fitting methods have been developed to overcome these problems [38]. An adaptive Savitzky–Golay filtering method, implemented in the TIMESAT processing package developed by Jönsson and Eklundh [39], is used in this chapter. The filter is based on local polynomial fits. Suppose we have a time series (t_i, y_i), $i = 1, 2, \ldots, N$. For each point i, a quadratic polynomial

$$f(t) = c_1 + c_2 t + c_3 t^2 \tag{8.7}$$

is fit to all $2k + 1$ points for a window from $n = i - k$ to $m = i + k$ by solving the system of normal equations

$$\mathbf{A}^T \mathbf{A} c = \mathbf{A}^T b, \tag{8.8}$$

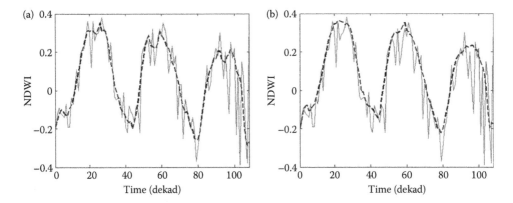

FIGURE 8.7   The Savitzky–Golay filtering of NDWI (——) is performed in two steps. First, the local polynomials are fitted using the weights from the preprocessing (a). Data points above the resulting smoothed function (– – –) from the first fit are attributed a greater importance. Second, the normal equations are solved with the weights of these data values increased by a factor 2 (b).

where

$$
A = \begin{pmatrix} w_n & w_n t_n & w_n t_n^2 \\ w_{n+1} & w_{n+1}t_{n+1} & w_{n+1}t_{n+1}^2 \\ & \vdots & \\ w_m & w_m t_m & w_m t_m^2 \end{pmatrix} \quad \text{and} \quad b = \begin{pmatrix} w_n y_n \\ w_{n+1}y_{n+1} \\ \vdots \\ w_m y_m \end{pmatrix}. \tag{8.9}
$$

The filtered value is set to the value of the polynomial at point i. Weights are designated as w in the above expression, with weights assigned to all of the data values in the window. Data values that were flagged in the preprocessing are assigned weight "zero" in this application and thus do not influence the result. The clean data values all have weights "one." Residual negatively biased noise (e.g., clouds) may occur for the remote sensing data and accordingly the fitting was performed in two steps [6]. The first fit was conducted using weights obtained from the preprocessing. Data points above the resulting smoothed function from the first fit are regarded more important, and in the second step the normal equations are solved using the weight of these data values, but increased by a factor of 2. This multistep procedure leads to a smoothed function that is adapted to the upper envelope of the data (Figure 8.7). Similarly, the ancillary metadata of the meteorological data from the preprocessing were also used in the iterative fitting to the upper envelope of the −KBDI time series [6].

The width of the fitting window determines the degree of smoothing, but it also affects the ability to follow a rapid change. It is sometimes necessary to locally tighten the window even when the global setting of the window performs well. A typical situation occurs in savanna ecosystems where vegetation, associated remote sensing, and meteorological indices respond rapidly to vegetation moisture dynamics. A small fitting window can be used to capture the corresponding sudden increase in data values. The data in the window are scanned and if a large increase or decrease is observed, the adaptive Savitzky–Golay method applied an automatic decrease in the window size. The filtering is then repeated using the new locally adapted size. Savitzky–Golay filtering with and without the adaptive procedure is illustrated in Figure 8.8. This figure shows that the adaptation of the window improves the fit at the rising edges and at narrow seasonal peaks.

8.4.2   Extracting Seasonal Metrics from Time-Series and Statistical Analysis

Four seasonal metrics were extracted for each of the rainy seasons. Figure 8.9 illustrates the different metrics per season for NDWI and KBDI time series. The beginning of a season, that is, 20% left of the

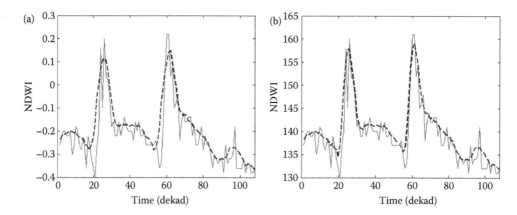

FIGURE 8.8 The filtering of NDWI (———) in (a) is done with a window that is too large to allow the filtered data (– – –) to follow sudden increases and decreases of underlying data values. The data in the window are scanned and if there is a large increase or decrease, an automatic decrease in the window size will result. The filtering is then repeated using the new locally adapted size (b). Note the improved fit at rising edges and narrow peaks.

rainy season, is defined from the final function fit as the point in time for which the index value has increased by 20% of the distance between the left minimum level and the maximum. The end of the season is defined in a similar way as the point 20% right of the rain season. The 80% left and right points are defined as the points for which the function fit has increased to 80% of the distance between the left and right minimum levels and the maximum, respectively. The current technique that is used to define metrics is also used by Verbesselt et al. [6] to define the beginning of the fire season.

The temporal occurrence and the value of each metric were extracted for further exploratory statistical analysis to study the relationship between time series. The SPOT VGT S10 time series consisted of four seasons (1998–2002) from which four occurrences and values per metric type were extracted. Twenty-four occurrence–value combinations per metric type were ultimately available for further analysis since six weather stations were used.

Serial correlation that occurs in remote sensing and climate-based time series invalidates inferences made by standard parametric tests, such as the Student's t-test or the Pearson correlation. All extracted occurrence–value combinations per metric type were tested for autocorrelation using the

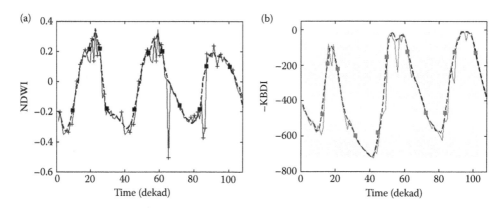

FIGURE 8.9 The final fit of the Savitzky–Golay function (– – –) to the NDWI (a) and −KBDI (b) series (———), with the four defined metrics, that is, 20% left and right, and 80% left and right (■), overlaid on the graph. Points with flagged data errors (+) were assigned weights of zero and did not influence the fit. A dekad is defined as a 10-day period.

Ljung–Box autocorrelation test [26]. Robust nonparametric techniques, such as the Wilcoxon's signed rank test were used in case of nonnormally distributed data. The normality of the data was verified using the Shapiro–Wilkinson normality test [36].

First, the distribution of the temporal occurrence of each metric was visualized and evaluated based on whether or not there was a significant difference between the temporal occurrence of the four metric types extracted from −KBDI and NDWI time series. Next, the strength and significance of the relationship between −KBDI and NDWI values of the four metric types were assessed with an OLS regression analysis.

8.5   RESULTS AND DISCUSSION

Figure 8.9 illustrates the optimized function fit and the defined metrics for the −KBDI and NDWI. Note that the Savitzky–Golay function could properly define the behavior of the different time series. The function was fitted to the upper envelope of the data by using the uncertainty information derived during the preprocessing step. The results of the statistical analysis based on the extracted metrics for −KBDI and NDWI are presented. The Ljung–Box statistic indicated that the extracted occurrences and values were not significantly autocorrelated at a 95% confidence level. All p-values were >0.1, failing to reject the null hypothesis of independence.

8.5.1   Temporal Analysis of the Seasonal Metrics

Figure 8.10 illustrates the temporal distribution of temporal occurrence of extracted metrics from time series of −KBDI and NDWI. The occurrences of extracted metrics were significantly nonnormally distributed at a 95% confidence level ($p > 0.1$), indicating that the Wilcoxon's signed rank test can be used. The Wilcoxon's signed rank test showed that −KBDI and NDWI occurrences of the 80% left and right, and 20% right were not significantly different from each other at a 95% confidence level ($p > 0.1$). This confirmed that −KBDI and NDWI were temporally related. It also corroborated the results of Burgan [11] and Ceccato et al. [4] who found that both −KBDI and NDWI were related to the seasonal vegetation moisture dynamics, as measured by VWC.

Figure 8.10, however, illustrates that the start of the rainy season (i.e., 20% left occurrence), derived from the −KBDI and NDWI time series, was different. The Wilcoxon's signed rank test confirmed that the −KBDI and NDWI differed significantly from each other at a 95% confidence level ($p < 0.01$). This phenomenon can be explained by the fact that vegetation in the study area starts growing before the rainy season starts, due to an early change in air temperature (N. Govender, Scientific Service Kruger National Park, South Africa, personal communication). This explained why the NDWI reacted before the change in climatic conditions as measured by the −KBDI, given that the NDWI is directly related to vegetation moisture dynamics [4].

8.5.2   Regression Analysis Based on Values of Extracted Seasonal Metrics

The assumptions of the OLS regression models between values of metrics extracted from −KBDI and NDWI time series were verified. The Wald test statistic showed nonlinearity to be not significant at a 95% confidence level ($p > 0.15$). The Shapiro–Wilkinson normality test confirmed that the residuals were normally distributed at a 95% confidence level ($p < 0.01$) [6,36]. Table 8.2 illustrates the results of the OLS regression analysis between values of metrics extracted from −KBDI and NDWI time series. The values extracted at the "20% right" position of the −KBDI and NDWI time series showed a significant relationship at a 95% confidence level ($p < 0.01$). The other extracted metrics did not exhibit significant relationships at a 95% confidence level ($p > 0.1$). A significant relationship between the −KBDI and NDWI time series was observed only at the moment when savanna vegetation was completely cured (i.e., 20% right-hand side). The savanna vegetation therefore reacted differently to changes in climate parameters such as rainfall and temperature, as measured by KBDI,

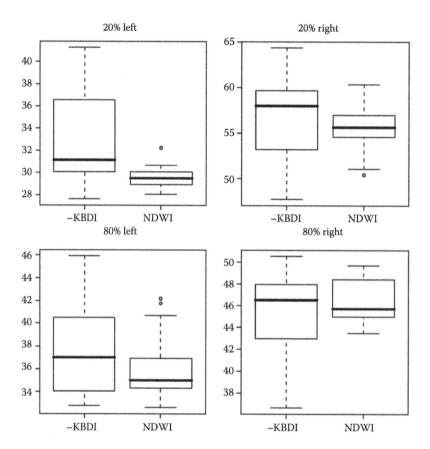

FIGURE 8.10 Box plots of the temporal occurrence of the four defined metrics, that is, 20% left and right, and 80% left and right, extracted from time series of −KBDI and NDWI. The dekads (10-day period) are shown on the y-axis and are indicative of the temporal occurrence of the metric. The upper and lower boundaries of the boxes indicate upper and lower quartiles. The median is indicated by the solid line (—) within each box. The whiskers connect the extremes of the data, which were defined as 1.5 times the interquartile range. Outliers are represented by (o).

depending on the phenological growing cycle. This phenomenon could be explained because a living plant uses defense mechanisms to protect itself from drying out, while a cured plant responds to climatic conditions [46]. These results consequently indicated that the relationship between extracted values of −KBDI and NDWI was influenced by seasonality. This is in corroboration with the results of Ji and Peters [23], who indicated that seasonality had a significant effect on the relationship

TABLE 8.2

Coefficients of Determination of the OLS Regression Models (NDWI ~ −KBDI) for the Four Extracted Seasonal Metric Values between −KBDI and NDWI Time Series ($n = 24$ per Metric)

NDWI ~ −KBDI	R^2	p-Values
20% Left	0.01	0.66
20% Right	0.49	<0.01
80% Left	0.00	0.97
80% Right	0.01	0.61

between vegetation as measured by a remote sensing index and drought index. These results further illustrated that the seasonal effect needs to be taken into account when regression techniques are used to quantify the relationship between time series related to vegetation moisture dynamics. The seasonal effect can also be accounted for by utilizing AR models with seasonal dummy variables, which take the effect of serial correlation and seasonality into account [23,26]. However, the proposed method to account for serial correlation by sampling at specific moments in time had an additional advantage; the influence of seasonality could be studied by extracting metrics at the specified moments, besides the fact that serial correlation was taken into account.

Furthermore, it was shown that serial correlation caused an overestimation of the correlation coefficient when results from Tables 8.1 and 8.2 were compared. All the coefficients of determination (R^2) of Table 8.1 were significant with an average value of 0.7, while in Table 8.2, only the correlation coefficient at the end of the rainy season (20% right-hand side) was significant ($R^2 = 0.49$). This confirmed the importance of accounting for serial correlation and seasonality in the residuals of a regression model, when studying the relationship between two time series.

8.5.3 TIME-SERIES ANALYSIS TECHNIQUES

Time-series analysis models most often are used for purposes of describing current conditions and forecasting [25]. The models use the serial correlation in time series as a tool to relate temporal observations. Future observations can be predicted by modeling the serial correlation structure of a time series [26]. Time-series analysis techniques (e.g., ARIMA) can be used to model a time series by using other, independent time series [27]. ARIMA subsequently can be used to study the relationship between −KBDI and NDWI. However, there are constraints that have to be considered before ARIMA models can be applied to a time series (e.g., drought index) by using other predictor time series (e.g., satellite index).

First, the goodness-of-fit of an ARIMA model will not be significant when changes in the satellite index precede or coincide with those in the drought index. The CCF can be used in this context to verify how time series are related to each other. Time lag results in Table 8.1 indicate that the −KBDI (drought index) precedes or coincides with the NDWI time series. This illustrates that ARIMA models cannot be used directly to predict the KBDI, with NDWI as the predictor variable. Consequently, other more advanced time-series analysis techniques are needed to model vegetation dynamics because they will precede or coincide with the dynamics monitored by remote sensing indices in most of the cases. Such more advanced time series analysis techniques, however, are not discussed since they are outside the scope of this chapter.

Second, availability of data is limited for time-series analysis, namely, from 1998 to 2002. This is an important constraint because two separate data sets are needed to parameterize and evaluate an ARIMA model. One set is needed for parameterization, while the other is used to forecast and validate the ARIMA model through comparison of the observed and expected values. Accordingly, it is necessary to interpolate missing satellite data that were masked out during preprocessing to ensure that adequate data are available for parameterization.

Third, the proposed sampling strategy investigated the time lag and correlation at a defined instant in time possible, as opposed to ARIMA or cross-correlation analysis, through which only the overall relationship between time series can be studied [25]. The applied sampling strategy is thus ideally suited to study the relationship between time series of climate and remote sensing data, characterized by seasonality and serial correlation. The sampling of seasonal metrics minimized the influence of serial correlation, thereby making the study of seasonality possible.

8.6 CONCLUSIONS

Serial correlation problems are not unknown in the field of statistical or general meteorology. However, the presence of serial correlation, found during analysis of a variable sampled sequentially at regular time intervals, seems to be disregarded by many agricultural meteorologists and remote sensing scientists. This is true despite abundant documentation available in the traditional meteorological and statistical literature. Therefore, an overview of the most important time-series analysis techniques and concepts was presented, namely, stationarity, autocorrelation, differencing, decomposition, AR, and ARIMA.

A method was proposed to study the relationship between a meteorological drought index (KBDI) and remote sensing index (NDWI), both related to vegetation moisture dynamics, by accounting for the serial correlation effect. The relationship between −KBDI and NDWI was studied by extracting nonserially correlated seasonal metrics, for example, 20% and 80% left- and right-hand side metrics of the rainy season, based on a Savitzky–Golay fit to the upper envelope of the time series. The serial correlation between the extracted metrics was shown to be minimal and seasonality was an important factor influencing the relationship between NDWI and −KBDI time series. Statistical analysis using the temporal occurrence of the extracted metrics revealed that NDWI and −KBDI time series are temporally connected, except at the beginning of the rainy season. The fact that the savanna vegetation starts regreening before the start of the rainy season explains this inability to detect the beginning of the rainy season. The values of the extracted seasonal metrics of NDWI and −KBDI were significantly related only at the end of the rainy season, namely, at the 20% right-hand side value of the fitted curve. The savanna vegetation at the end of the rainy season was cured and responded strongly to changes in climatic conditions monitored by the −KBDI, such as rain and temperature. The relationship between −KBDI and NDWI consequently changes during the season, which indicates that seasonality is an important factor that needs to be taken into account. Moreover, it was shown that correlation coefficients estimated by OLS regression analysis were overestimated due to the influence of serial correlation in the residuals. This confirmed the importance of taking the serial correlation of the residuals into account by sampling nonserially correlated seasonal metrics when studying the relationship between time series.

Consequently, the serial correlation effect was taken into account by the extraction of seasonal metrics from time series. The seasonal metrics in turn could be used to study the relationship between remote sensing and ground-based time series, such as meteorological or field measurements. A better understanding of the relationship between remote sensing and *in situ* observations at regular time intervals will contribute to the use of remotely sensed data for the development of an index that represents seasonal vegetation moisture dynamics.

ACKNOWLEDGMENTS

The SPOT VGT S10 data sets were generated by the Flemish Institute for Technological Development. The climate data were provided by the Weather Services of South Africa, whereas the National Land Cover Map (1995) was supplied by the Agricultural Research Centre of South Africa. We acknowledge the support of L. Eklundh, as well as the Craafoord and Bergvall foundations. We thank N. Govender from the Scientific Services of Kruger National Park for scientific input.

REFERENCES

1. Camia, A. et al., Meteorological fire danger indices and remote sensing, in *Remote Sensing of Large Wildfires in the European Mediterranean Basin*, E. Chuvieco, ed., Springer-Verlag, New York, 1999, p. 39.
2. Ceccato, P. et al., Estimation of live fuel moisture content, in *Wildland Fire Danger Estimation and Mapping: The Role of Remote Sensing Data*, E. Chuvieco, ed., World Scientific Publishing, New York, 2003, p. 63.
3. Dennison, P.E. et al., Modeling seasonal changes in live fuel moisture and equivalent water thickness using a cumulative water balance index, *Remote Sens. Environ.*, 88, 442, 2003.
4. Ceccato, P. et al., Detecting vegetation leaf water content using reflectance in the optical domain, *Remote Sens. Environ.*, 77, 22, 2001.

5. Jackson, T.J. et al., Vegetation water content mapping using Landsat data derived normalized difference water index for corn and soybeans, *Remote Sens. Environ.*, 92, 475, 2004.

6. Verbesselt, J. et al., Evaluating satellite and climate data derived indices as fire risk indicators in savanna ecosystems, *IEEE Trans. Geosci. Remote Sens.*, 44, 1622–1632, 2006.

7. Van Wilgen, B.W. et al., Response of savanna fire regimes to changing fire-management policies in a large African national park, *Conserv. Biol.*, 18, 1533, 2004.

8. Dimitrakopoulos, A.P. and Bemmerzouk, A.M., Predicting live herbaceous moisture content from a seasonal drought index, *Int. J. Biometeorol.*, 47, 73, 2003.

9. Keetch, J.J. and Byram, G.M., A drought index for forest fire control, USDA. Forest Service, Asheville NC, SE-38, 1988.

10. Janis, M.J., Johnson, M.B., and Forthun, G., Near-real time mapping of Keetch–Byram drought index in the south-eastern United States, *Int. J. Wildland Fire*, 11, 281, 2002.

11. Burgan, E.R., Correlation of plant moisture in Hawaii with the Keetch–Byram drought index, USDA. Forest Service Research Note, PSW-307, 1976.

12. Chuvieco, E. et al., Combining NDVI and surface temperature for the estimation of live fuel moisture content in forest fire danger rating, *Remote Sens. Environ.*, 92, 322, 2004.

13. Maki, M., Ishiahra, M., and Tamura, M., Estimation of leaf water status to monitor the risk of forest fires by using remotely sensed data, *Remote Sens. Environ.*, 90, 441, 2004.

14. Paltridge, G.W. and Barber, J., Monitoring grassland dryness and fire potential in Australia with NOAA AVHRR data, *Remote Sens. Environ.*, 25, 381, 1988.

15. Chladil, M.A. and Nunez, M., Assessing grassland moisture and biomass in Tasmania—The application of remote-sensing and empirical-models for a cloudy environment, *Int. J. Wildland Fire*, 5, 165, 1995.

16. Hardy, C.C. and Burgan, R.E., Evaluation of NDVI for monitoring live moisture in three vegetation types of the western US, *Photogramm. Eng. Remote Sens.*, 65, 603, 1999.

17. Chuvieco, E. et al., Estimation of fuel moisture content from multitemporal analysis of LAND-SAT thematic mapper reflectance data: Applications in fire danger assessment, *Int. J. Remote Sens.*, 23, 2145, 2002.

18. Fensholt, R. and Sandholt, I., Derivation of a shortwave infrared water stress index from MODIS near- and shortwave infrared data in a semiarid environment, *Remote Sens. Environ.*, 87, 111, 2003.

19. Hunt, E.R., Rock, B.N., and Nobel, P.S., Measurement of leaf relative water content by infrared reflectance, *Remote Sens. Environ.*, 22, 429, 1987.

20. Ceccato, P., Flasse, S., and Gregoire, J.M., Designing a spectral index to estimate vegetation water content from remote sensing data—Part 2: Validation and applications, *Remote Sens. Environ.*, 82, 198, 2002.

21. Myneni, R.B. et al., Increased plant growth in the northern high latitudes from 1981 to 1991, *Nature*, 386, 698, 1997.

22. Eklundh, L., Estimating relations between AVHRR NDVI and rainfall in East Africa at 10-day and monthly time scales, *Int. J. Remote Sens.*, 19, 563, 1998.

23. Ji, L. and Peters, A.J., Assessing vegetation response to drought in the northern great plains using vegetation and drought indices, *Remote Sens. Environ.*, 87, 85, 2003.

24. De Beurs, K.M. and Henebry, G.M., Land surface phenology, climatic variation, and institutional change: Analyzing agricultural land cover change in Kazakhstan, *Remote Sens. Environ.*, 89, 497, 2004.

25. Meek, D.W. et al., A note on recognizing autocorrelation and using autoregression, *Agric. Forest Meterol.*, 96, 9, 1999.

26. Brockwell, P.J. and Davis, R.A., *Introduction to Time Series and Forecasting*, 2nd edition, Springer-Verlag, New York, 2002, p. 31.

27. Ford, C.R. et al., Modeling canopy transpiration using time series analysis: A case study illustrating the effect of soil moisture deficit on *Pinus taeda*, *Agric. Forest Meterol.*, 130, 163, 2005.

28. Montanari, A., Deseasonalisation of hydrological time series through the normal quantile transform, *J. Hydrol.*, 313, 274, 2005.

29. Rahman, H. and Dedieu, G., SMAC—A simplified method for the atmospheric correction of satellite measurements in the solar spectrum, *Int. J. Remote Sens.*, 15, 123, 1994.

30. Stroppiana, D. et al., An algorithm for mapping burnt areas in Australia using SPOT-VEGETATION data, *IEEE Trans. Geosci. Remote Sens.*, 41, 907, 2003.

31. Thompson, M.A., Standard land-cover classification scheme for remote-sensing applications in South Africa, *South Afr. J. Sci.*, 92, 34, 1996.

32. Aguado, I. et al., Assessment of forest fire danger conditions in southern Spain from NOAA images and meteorological indices, *Int. J. Remote Sens.*, 24, 1653, 2003.

33. von Storch, H. and Zwiers, F.W., *Statistical Analysis in Climate Research*, Cambridge University Press, Cambridge, 1999, p. 483.
34. Thejll, P. and Schmith, T., Limitations on regression analysis due to serially correlated residuals: Application to climate reconstruction from proxies, *J. Geophys. Res. Atmos.*, 110, 2005.
35. Cleveland, R.B. et al., STL: A seasonal-trend decomposition procedure based on Loess, *J. Off. Stat.*, 6, 3, 1990.
36. R Development Core Team, *R: A Language and Environment for Statistical Computing*, R Foundation for Statistical Computing, Vienna, Austria, ISBN 3-900051-07-0, http://www.R-project.org, 2005.
37. Venables, W.N. and Ripley, B.D., *Modern Applied Statistics with S*, 4th edition, Springer-Verlag, New York, 2003, p. 493.
38. Jönsson, P. and Eklundh, L., Seasonality extraction by function fitting to time-series of satellite sensor data, *IEEE Trans. Geosci. Remote Sens.*, 40, 1824, 2002.
39. Jönsson, P. and Eklundh, L., TIMESAT—A program for analyzing time-series of satellite sensor data, *Comput. Geosci.*, 30, 833, 2004.
40. Menenti, M. et al., Mapping agroecological zones and time-lag in vegetation growth by means of Fourier-analysis of time-series of NDVI images, *Adv. Space Res.*, 13, 233, 1993.
41. Azzali, S. and Menenti, M., Mapping vegetation–soil–climate complexes in Southern Africa using temporal Fourier analysis of NOAA-AVHRR NDVI data, *Int. J. Remote Sens.*, 21, 973, 2000.
42. Olsson, L. and Eklundh, L., Fourier-series for analysis of temporal sequences of satellite sensor imagery, *Int. J. Remote Sens.*, 15, 3735, 1994.
43. Cihlar, J., Identification of contaminated pixels in AVHRR composite images for studies of land biosphere, *Remote Sens. Environ.*, 56, 149, 1996.
44. Sellers, P.J. et al., A global 1-degrees-by-1-degrees NDVI data set for climate studies. The generation of global fields of terrestrial biophysical parameters from the NDVI, *Int. J. Remote Sens.*, 15, 3519, 1994.
45. Roerink, G.J., Menenti, M., and Verhoef, W., Reconstructing cloudfree NDVI composites using Fourier analysis of time series, *Int. J. Remote Sens.*, 21, 1911, 2000.
46. Pyne, S.J., Andrews, P.L., and Laven, R.D., *Introduction to Wildland Fire*, 2nd edition, John Wiley & Sons, New York, 1996, p. 117.

9 Use of a Prediction-Error Filter in Merging High- and Low-Resolution Images

Sang-Ho Yun and Howard Zebker

CONTENTS

A prediction-error (PE) filter is an array of numbers designed to interpolate missing parts of data such that the interpolated parts have the same spectral content as the existing parts. The data can be a one-dimensional (1D) time series, a two-dimensional (2D) image, or a three-dimensional quantity such as subsurface material property. In this chapter, we discuss the application of a PE filter to recover missing parts of an image when a low-resolution image of the missing parts is available.

One of the research issues on PE filters is improving the quality of the image interpolation for nonstationary images, in which the spectral content varies with position. Digital elevation models (DEMs) are in general nonstationary. Thus, PE filters alone cannot guarantee the success of image recovery. However, the quality of the image recovery of a high-resolution image can be improved with independent data set such as a low-resolution image that has valid pixels for the missing regions of the high-resolution image. Using a DEM as an example image, we introduce a systematic method to use a PE filter incorporating the low-resolution image as an additional constraint, and show the improved quality of the image interpolation.

High-resolution DEMs are often limited in spatial coverage; they may possess systematic artifacts as well when compared to comprehensive low-resolution maps. We correct artifacts and interpolate regions of missing data in topographic synthetic aperture radar (TOPSAR) digital elevation models (DEMs) using a low-resolution shuttle radar topography mission (SRTM) DEM. Then, PE filters are to interpolate and fill missing data so that the interpolated regions have the same spectral content as the valid regions of the TOPSAR DEM. The SRTM DEM is used as an additional constraint in the interpolation. Using cross-validation methods, one can obtain the optimal weighting for the PE filter and the SRTM DEM constraints.

9.1　IMAGE DESCRIPTIONS

InSAR is a powerful tool for generating DEMs [1]. The TOPSAR and SRTM sensors are primary sources for the academic community for DEMs derived from single-pass interferometric data. Differences in system parameters such as altitude and swath width (Table 9.1) result in very different properties for derived DEMs. Specifically, TOPSAR DEMs have better resolution, while SRTM DEMs have better accuracy over larger areas. TOPSAR coverage is often not spatially complete.

9.1.1　TOPSAR DEM

TOPSAR DEMs are produced from cross-track interferometric data acquired with NASA's AIRSAR system mounted on a DC-8 aircraft. Although the TOPSAR DEMs have a higher resolution than other existing data, they sometimes suffer from artifacts and missing data due to roll of the aircraft, layover, and flight planning limitations. The DEMs derived from the SRTM have lower resolution, but fewer artifacts and missing data than TOPSAR DEMs. Thus, the former often provides information in the missing regions of the latter.

We illustrate joint use of these data sets using DEMs acquired over the Galápagos Islands. Figure 9.1 shows the TOPSAR DEM used in this study. The DEM covers Sierra Negra volcano on the island of Isabela. Recent InSAR observations reveal that the volcano has been deforming relatively rapidly [2,3]. InSAR analysis can require use of a DEM to produce a simulated interferogram required to isolate ground deformation. The effect of artifact elimination and interpolation for deformation studies is discussed later in this chapter.

The TOPSAR DEMs have a pixel spacing of about 10 m, sufficient for most geodetic applications. However, regions of missing data are often encountered (Figure 9.1), and significant residual artifacts are found (Figure 9.2). The regions of missing data are caused by layover of the steep volcanoes and flight planning limitations. Artifacts are large-scale and systematic, and most likely due to uncompensated roll of the DC-8 aircraft [4]. Attempts to compensate this motion include models of piecewise linear imaging geometry [5] and estimating imaging parameters that minimize the difference between the TOPSAR DEM and an independent reference DEM [6]. We use a nonparameterized direct approach by subtracting the difference between the TOPSAR and SRTM DEMs.

TABLE 9.1
TOPSAR Mission versus SRTM Mission

Mission	TOPSAR	SRTM
Platform	DC-8 aircraft	Space shuttle
Nominal	Altitude 9 km	233 km
Swath width	10 km	225 km
Baseline	2.583 m	60 m
DEM resolution	10 m	90 m
DEM coordinate system	None	Lat/Long

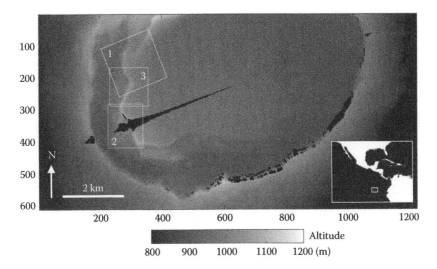

FIGURE 9.1　The original TOPSAR DEM of Sierra Negra volcano in Galápagos Islands (inset for location). The pixel spacing of the image is 10 m. The boxed areas are used for illustration later in this chapter. Note that there are a number of regions of missing data with various shapes and sizes. Artifacts are not identifiable due to the variation in topography. (From S.-H. Yun et al., *IEEE Transactions on Geoscience and Remote Sensing*, 43(7), 1682, 2005. With permission.)

9.1.2　SRTM DEM

The recent SRTM mission produced nearly worldwide topographic data at 90 m posting. SRTM topographic data are in fact produced at 30 m posting (1 arcsec); however, high-resolution data sets for areas outside of the United States are not available to the public at this time. Only DEMs at 90 m posting (3 arcsec) are available to download.

For many analyses, finer scale elevation data are required. For example, a typical pixel spacing in a spaceborne SAR image is 20 m. If the SRTM DEMs are used for topography removal in spaceborne interferometry, the pixel spacing of the final interferograms would be limited by the topography data to at best 90 m. Despite the lower resolution, the SRTM DEM is useful because it has fewer motion-induced artifacts than the TOPSAR DEM. It also has fewer data holes.

The merits and demerits of the two DEMs are in many ways complementary to each other. Thus, a proper data fusion method can overcome the shortcomings of each and produce a new DEM that combines the strength of the two data sets: a DEM that has a resolution of the TOPSAR DEM and large-scale reliability of the SRTM DEM. In this chapter, we present an interpolation method that uses both TOPSAR and SRTM DEMs as constraints.

9.2　IMAGE REGISTRATION

The original TOPSAR DEM, while in ground-range coordinates, is not georeferenced. Thus, we register the TOPSAR DEM to the SRTM DEM, which is already registered in a latitude–longitude coordinate system. The image registration is carried out between the DEM data sets using an affine transformation. Although the TOPSAR DEM is not georeferenced, it is already on the ground coordinate system. Thus, scaling and rotation are the two most important components. We have seen that the skewing component was negligible. Any higher-order transformation between the two DEMs would also be negligible. The affine transformation is as follows:

$$\begin{bmatrix} x_S \\ y_S \end{bmatrix} = \begin{bmatrix} a & b \\ c & d \end{bmatrix}\begin{bmatrix} x_T \\ y_T \end{bmatrix} + \begin{bmatrix} e \\ f \end{bmatrix},\tag{9.1}$$

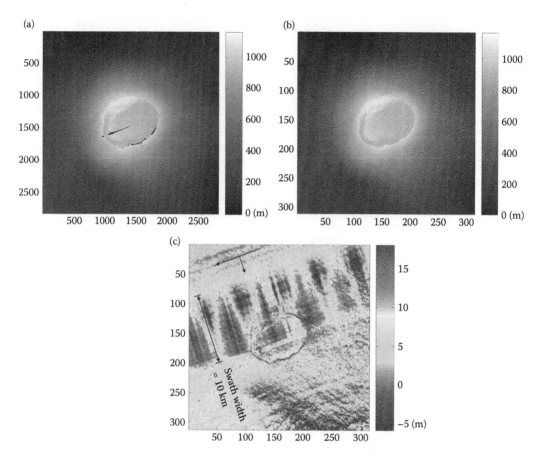

FIGURE 9.2   (a) TOPSAR DEM and (b) SRTM DEM. The tick labels are pixel numbers. Note the difference in pixel spacing between the two DEMs. (c) Artifacts obtained by subtracting the SRTM DEM from the TOPSAR DEM. The flight direction and the radar look direction of the aircraft associated with the swath with the artifact are indicated with long and short arrows, respectively. Note that the artifacts appear in one entire TOPSAR swath, while they are not as serious in other swaths.

where $[x_S/y_S]$ and $[x_T/y_T]$ are tie points in the SRTM and TOPSAR DEM coordinate systems, respectively. Since $[a\ b\ e]$ and $[c\ d\ f]$ are estimated separately, at least three tie points are required to uniquely determine them. We picked 10 tie points from each DEM based on topographic features and solved for the six unknowns in a least-square sense.

Given the six unknowns, we choose new georeferenced sample locations that are uniformly spaced; every ninth sample location corresponds to the sample location of SRTM DEMs. Those sample locations from $[x_S/y_S]$ and $[x_T/y_T]$ are calculated. Then, the nearest TOPSAR DEM value is selected and put into the corresponding new georeferenced sample location. The intermediate values are filled in from the TOPSAR map to produce the georeferenced 10-m data set.

It should be noted that it is not easy to determine the tie points in DEM data sets. Enhancing the contrast of the DEMs facilitated the process. In general, fine registration is important for correctly merging different data sets. The two DEMs in this study have different pixel spacings. It is difficult to pick tie points with higher precision than the pixel spacing of the coarser image. In our method, however, the SRTM DEM, the coarser image is treated as an averaged image of the TOPSAR DEM, the finer image. In our inversion, only the 9×9 averaged values of the TOPSAR DEM are compared with the pixel values of the SRTM DEM. Thus, the fine registration is less critical in this approach than in the case where a one-to-one match is required.

9.3 ARTIFACT ELIMINATION

Examination of the georeferenced TOPSAR DEM (Figure 9.2a) shows motion artifacts when compared to the SRTM DEM (Figure 9.2b). The artifacts are not clearly discernible in Figure 9.2a because their magnitude is small in comparison to the overall data values. The artifacts are identified by downsampling the registered TOPSAR DEM and subtracting the SRTM DEM. Large-scale anomalies that periodically fluctuate over an entire swath are visible in Figure 9.2c. The periodic pattern is most likely due to uncompensated roll of the DC-8 aircraft. The spaceborne data are less likely to exhibit similar artifacts, because the spacecraft is not greatly affected by the atmosphere. Note that the width of the anomalies corresponds to the width of a TOPSAR swath. As the SRTM swath is much larger than that of the TOPSAR system (Table 9.1), a larger area is covered under consistent conditions, reducing the number of parallel tracks required to form an SRTM DEM.

The maximum amplitude of the motion artifacts in our study area is about 20 m. This would result in substantial errors in many analyses if not properly corrected. For example, if this TOPSAR DEM is used for topography reduction in repeat-pass InSAR using ERS-2 data with a perpendicular baseline of about 400 m, the resulting deformation interferogram would contain one fringe (= 2.8 cm) of spurious signal.

To remove these artifacts from the TOPSAR DEM, we up-sample the difference image with bilinear interpolation by a factor of 9 so that its pixel spacing matches the TOPSAR DEM. The difference image is subtracted from the TOPSAR DEM. This process is described with a flow diagram in Figure 9.3. Note that the lower branch undergoes two low-pass filter operations when averaging and bilinear interpolation are implemented, whereas the upper branch preserves the high-frequency contents of the TOP-SAR DEM. In this way, we can eliminate the large-scale artifacts while retaining details in the TOPSAR DEM.

9.4 PREDICTION-ERROR FILTER

The next step in the DEM process is to fill in missing data. We use a PE filter operating on the TOPSAR DEM to fill these gaps. The basic idea of the PE filter constraint [7,8] is that missing data can be estimated so that the restored data yield minimum energy when the PE filter is applied. The PE filter is derived from training data, which are normally valid data surrounding the missing regions. The PE filter is selected so that the missing data and the valid data share the same spectral content. Hence, we assume that the spectral content of the missing data in the TOPSAR DEM is similar to that of the regions with valid data surrounding the missing regions.

9.4.1 DESIGNING THE FILTER

We generate a PE filter such that it rejects data with statistics found in the valid regions of the TOPSAR DEM. Given this PE filter, we solve for data in the missing regions such that the interpolated data are also nullified by the PE filter. This concept is illustrated in Figure 9.4.

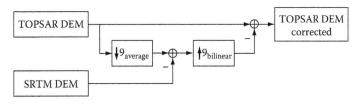

FIGURE 9.3 The flow diagram of the artifact elimination. (From S.-H. Yun et al., *IEEE Transactions on Geoscience and Remote Sensing*, 43(7), 1682, 2005. With permission.)

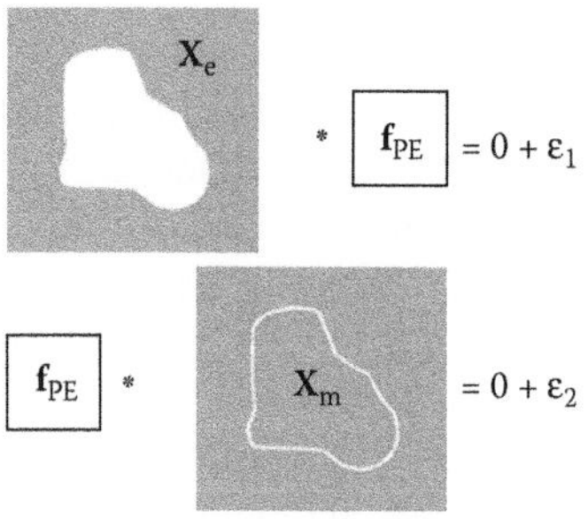

FIGURE 9.4 Concept of a PE filter. The PE filter is estimated by solving an inverse problem constrained with the remaining part, and the missing part is estimated by solving another inverse problem constrained with the filter. The ε_1 and ε_2 are white noise with small amplitude.

The PE filter, $\mathbf{f}_{PE}$, is found by minimizing the following objective function:

$$\left\| \mathbf{f}_{PE} * \mathbf{x}_e \right\|^2,$$ (9.2)

where $\mathbf{x}_e$ is the existing data from the TOPSAR DEM, and * represents convolution. This expression can be rewritten in a linear algebraic form using the following matrix operation:

$$\left\| \mathbf{F}_{PE} \, \mathbf{x}_e \right\|^2$$ (9.3)

or equivalently

$$\left\| \mathbf{X}_e \, \mathbf{f}_{PE} \right\|^2,$$ (9.4)

where $\mathbf{F}_{PE}$ and $\mathbf{X}_e$ are the matrix representations of $\mathbf{f}_{PE}$ and $\mathbf{x}_e$ for convolution operation. These matrix and vector expressions are used to indicate their linear relationship.

9.4.2 1D Example

The procedure of acquiring the PE filter can be explained with a 1D example. Suppose that a data set, $\mathbf{x} = [x_1, \ldots, x_n]$ (where $n \gg 3$) is given, and we want to compute a PE filter of length 3, $\mathbf{f}_{PE} = [1 \, f_1 \, f_2]$. Then we form a system of linear equations as follows:

$$\begin{bmatrix} x_3 & x_2 & x_1 \\ x_4 & x_3 & x_2 \\ \vdots & \vdots & \vdots \\ x_n & x_{n-1} & x_{n-2} \end{bmatrix} \begin{bmatrix} 1 \\ f_1 \\ f_2 \end{bmatrix} \approx 0.$$ (9.5)

The first element of the PE filter should be equal to 1 to avoid the trivial solution, $\mathbf{f}_{PE} = 0$. Note that Equation 9.5 is the convolution of the data and the PE filter. After simple algebra and with

$$\mathbf{d} \equiv \begin{bmatrix} x_3 \\ \vdots \\ x_n \end{bmatrix} \quad \text{and} \quad \mathbf{D} \equiv \begin{bmatrix} x_2 & x_1 \\ \vdots & \vdots \\ x_{n-1} & x_{n-2} \end{bmatrix}$$

we get

$$\mathbf{D}\begin{bmatrix} f_1 \\ f_2 \end{bmatrix} \approx -\mathbf{d} \tag{9.6}$$

and its normal equation becomes

$$\begin{bmatrix} f_1 \\ f_2 \end{bmatrix} = (\mathbf{D}^{\mathrm{T}}\mathbf{D})^{-1}\mathbf{D}^{\mathrm{T}}(-\mathbf{d}). \tag{9.7}$$

Note that Equation 9.7 minimizes Equation 9.2 in a least-square sense. This procedure can be extended to 2D problems, and more details are described in Refs. [7,8].

9.4.3 The Effect of the Filter

Figure 9.5 shows the characteristics of the PE filter in the spatial and Fourier domains. Figure 9.5a is the sample DEM chosen from Figure 9.1 (numbered Box 1) for demonstration. It contains various topographic features and has a wide range of spectral content (Figure 9.5d). Figure 9.5b is the 5×5 PE filter derived from Figure 9.5a by solving the inverse problem in Equation 9.3. Note that the first three elements in the first column of the filter coefficients are 0 0 1. This is the PE filter's unique constraint that ensures the filtered output to be white noise [7]. In the filtered output (Figure 9.5c), all the variations in the DEM were effectively suppressed. The size (order) of the PE filter is based on the complexity of the spectrum of the DEM. In general, as the spectrum becomes more complex, a larger size filter is required. After testing various sizes of the filter, we found a 5×5 size appropriate for the DEM used in our study. Figure 9.5d and e show the spectra of the DEM and the PE filter, respectively. These illustrate the inverse relationship of the PE filter to the corresponding DEM in

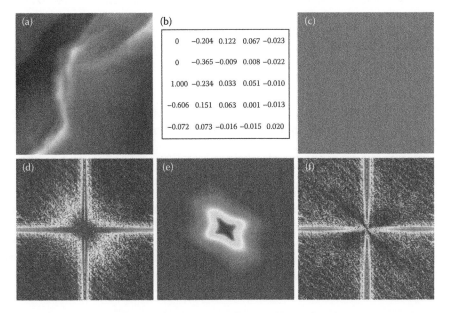

FIGURE 9.5 The effect of a PE filter. (a) Original DEM; (b) a 2D PE filter found from the DEM; (c) DEM filtered with the PE filter; and (d), (e), and (f) the spectra of (a), (b), and (c), respectively, plotted in dB. (a) and (c) are drawn with the same color scale. Note that in (c), the variation of the image (a) was effectively suppressed by the filter. The standard deviations of (a) and (c) are 27.6 and 2.5 m, respectively. (From S.-H. Yun et al., *IEEE Transactions on Geoscience and Remote Sensing*, 43(7), 1682, 2005. With permission.)

the Fourier domain, such that their product is minimized (Figure 9.5f). This PE filter constrains the interpolated data in the DEM to similar spectral content to the existing data.

All inverse problems in this study were derived using the conjugate gradient method, where forward and adjoint functional operators are used instead of the explicit inverse operators [7], saving computer memory space.

9.5 INTERPOLATION

9.5.1 PE FILTER CONSTRAINT

Once the PE filter is determined, we next estimate the missing parts of the image. As depicted in Figure 9.4, interpolation using the PE filter requires that the norm of the filtered output be minimized. This procedure can be formulated as an inverse computation minimizing the following objective function:

$$\left\| \mathbf{F}_{PE}\, \mathbf{x} \right\|^2, \tag{9.8}$$

where $\mathbf{F}_{PE}$ is the matrix representation of the PE filter convolution, and $\mathbf{x}$ represents the entire data set including the known and the missing regions. In the inversion process we update only the missing region, without changing the known region. This guarantees seamless interpolation across the boundaries between the known and missing regions.

9.5.2 SRTM DEM CONSTRAINT

As previously stated, 90-m posting SRTM DEMs were generated from 30-m posting data. This downsampling was done by calculating three "looks" in both the east and north directions. To use the SRTM DEM as a constraint to interpolate the TOPSAR DEM, we posit the following relationship between the two DEMs: each pixel value in a 90-m posting SRTM DEM can be considered equivalent to the averaged value of a 9×9 pixel window in a 10-m posting TOPSAR DEM centered at the corresponding pixel in the SRTM DEM.

The solution using the constraint of the SRTM DEM to find the missing data points in the TOPSAR DEM can be expressed as minimizing the following objective function:

$$\left\| \mathbf{y} - \mathbf{A}\mathbf{x}_m \right\|^2, \tag{9.9}$$

where $\mathbf{y}$ is an SRTM DEM expressed as a vector that covers the missing regions of the TOPSAR DEM, and $\mathbf{A}$ is an averaging operator generating nine looks, and $\mathbf{x}_m$ represents the missing regions of the TOPSAR DEM.

9.5.3 INVERSION WITH TWO CONSTRAINTS

By combining two constraints, one derived from the statistics of the PE filter and the other from the SRTM DEM, we can interpolate the missing data optimally with respect to both criteria. The PE filter guarantees that the interpolated data will have the same spectral properties as the known data. At the same time the SRTM constraint forces the interpolated data to have average height near the corresponding SRTM DEM. We formulate the inverse problem as a minimization of the following objective function:

$$\lambda^2 \left\| \mathbf{F}_{PE}\, \mathbf{x}_m \right\|^2 + \left\| \mathbf{y} - \mathbf{A}\mathbf{x}_m \right\|^2, \tag{9.10}$$

where λ set the relative effect of each criterion. Here, $\mathbf{x}_m$ has the dimensions of the TOPSAR DEM, while $\mathbf{y}$ has the dimensions of the SRTM DEM. If regions of missing data are localized in an image,

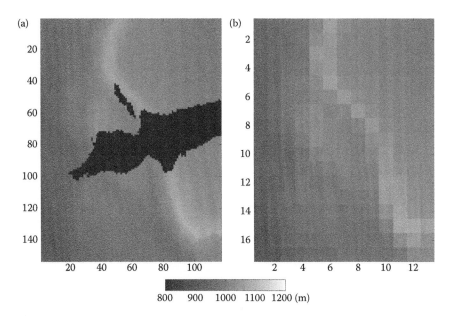

FIGURE 9.6 Example subimages of (a) TOPSAR DEM showing regions of missing data (black) and (b) SRTM DEM of the same area. These subimages are engaged in one implementation of the interpolation. The grayscale is altitude in meters. (From S.-H. Yun et al., *IEEE Transactions on Geoscience and Remote Sensing*, 43(7), 1682, 2005. With permission.)

the entire image does not have to be used for generating a PE filter. We implement interpolation in subimages to save time and computer memory space. An example of such a subimage is shown in Figure 9.6. The image is a part of Figure 9.1 (numbered Box 2). Figure 9.6a and b are examples of x_e in Equation 9.3 and y, respectively.

The multiplier λ determines the relative weight of the two terms in the objective function. As $\lambda \to \infty$, the solution satisfies the first constraint only, and if $\lambda = 0$, the solution satisfies the second constraint only.

9.5.4 OPTIMAL WEIGHTING

We used cross-validation sum of squares (CVSS) [9] to determine the optimal weights for the two terms in Equation 9.10. Consider a model $\mathbf{x}_m$ that minimizes the following quantity:

$$\lambda^2 \left\| \mathbf{F}_{PE}\ \mathbf{x}_m \right\|^2 + \left\| \mathbf{y}^{(k)} - \mathbf{A}^{(k)}\ \mathbf{x}_m \right\|^2 \quad (k = 1,\ldots,N), \tag{9.11}$$

where $\mathbf{y}^{(k)}$ and $\mathbf{A}^{(k)}$ are the $\mathbf{y}$ and the $\mathbf{A}$ in Equation 9.10 with the kth element and the kth row omitted, respectively, and N is the number of elements in $\mathbf{y}$ that fall into the missing region. Denote this model $\mathbf{x}_m^{(k)}(\lambda)$. Then we compute the CVSS defined as follows:

$$\text{CVSS}(\lambda) = \frac{1}{N} \sum_{k=1}^{N} (y_k - A_k \mathbf{x}_m^{(k)}(\lambda))^2, \tag{9.12}$$

where y_k is the omitted element from the vector $\mathbf{y}$ and A_k is the omitted row vector from the matrix $\mathbf{A}$ when the $\mathbf{x}_m^{(k)}(\lambda)$ was estimated. Thus, $A_k \mathbf{x}_m^{(k)}(\lambda)$ is the prediction based on the other $N-1$ observations. Finally, we minimize CVSS(λ) with respect to λ to obtain the optimal weight (Figure 9.7).

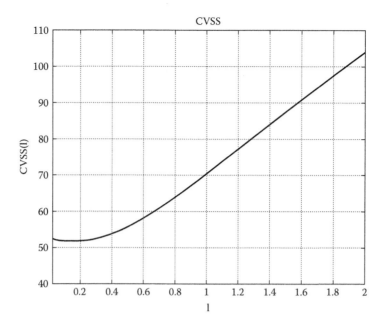

FIGURE 9.7 Cross-validation sum of squares. The minimum occurs when $\lambda = 0.16$. (From S.-H. Yun et al., *IEEE Transactions on Geoscience and Remote Sensing*, 43(7), 1682, 2005. With permission.)

In the case of the example shown in Figure 9.6, the minimum CVSS was obtained for $\lambda = 0.16$ (Figure 9.7). The effect of varying λ is shown in Figure 9.8. It is apparent (see Figure 9.8) that the optimal weight is the more "plausible" result than either of the end members, preserving aspects of both constraints.

In Figure 9.8a, the interpolation uses only the PE filter constraint. This interpolation does not recover the continuity of the ridge running across the DEM in north–south direction, which is observed in the SRTM DEM (Figure 9.6b). This follows from a PE filter obtained such that it eliminates the overall variations in the image. The variations include not only the ridge but also the accurate topography in the DEM.

The other end member, Figure 9.8c, shows the result for applying zero weight to the PE filter constraint. Since the averaging operator $\mathbf{A}$ in Equation 9.10 is applied independently for each 9×9 pixel group, it is equivalent to simply filling the regions of missing data with 9×9 identical values that are the same as the corresponding SRTM DEM (Figure 9.6b).

9.5.5 SIMULATION OF THE INTERPOLATION

The quality of cross-validation in this study is itself validated by simulating the interpolation process with known subimages that do not contain missing data. For example, if a known subimage is selected from Figure 9.1 (numbered Box 3), we can remove some data and apply our recovery algorithm. The subimage is similar in topographic features to the area shown in Figure 9.6. The process is illustrated in Figure 9.9. We introduce a hole as shown in Figure 9.9b and calculate the CVSS (Figure 9.9d) for each λ ranging from 0 to 2. Then we use the estimated λ, which minimizes the CVSS, for the interpolation process to obtain the image in Figure 9.9c. For each value of λ, we also calculate the root mean square (RMS) error between the known and the interpolated images. The RMS error is plotted against λ in Figure 9.9e. The CVSS is minimized for $\lambda = 0.062$, while the RMS error has a minimum at $\lambda = 0.065$. This agreement suggests that minimizing the CVSS is a useful method to balance the constraints. Note that the minimum RMS error in Figure 9.9e is about 5 m. This value is smaller than the relative vertical height accuracy of the SRTM DEM, which is about 10 m.

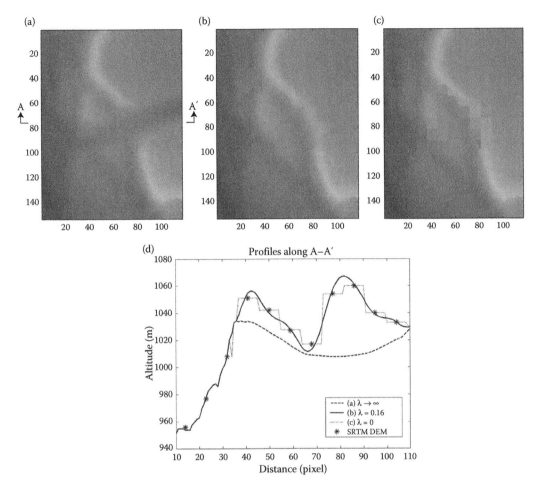

FIGURE 9.8   The results of interpolation applied to DEMs in Figure 9.6, with various weights. (a) $\lambda \to \infty$, (b) $\lambda = 0.16$, and (c) $\lambda = 0$. Profiles along A–A′ are shown in the plot (d). (From S.-H. Yun et al., *IEEE Transactions on Geoscience and Remote Sensing*, 43(7), 1682, 2005. With permission.)

9.6   INTERPOLATION RESULTS

The method presented in the previous section was applied to the entire image of Figure 9.1. The registered TOPSAR DEM contains missing data in regions of various sizes. Small subimages were extracted from the DEM. Each subimage is interpolated, and the results are reinserted into the large DEM. The locations and sizes of the subimages are indicated with white boxes in Figure 9.10a. Note the largest region of missing data in the middle of the caldera. This region is not only a simple large gap but also a gap between two swaths. The interpolation is an iterative process and fills up regions of missing data starting from the boundary. If valid data along the boundary (e.g., boundaries of a swath) contain edge effects, error tends to propagate through the interpolation process. In this case, expanding the region of missing data by a few pixels before interpolation produces better results. If there is a large region of missing data, the spectral content information of valid data can fade out as the interpolation proceeds toward the center of the gap. In this case, sequentially applying the interpolation to parts of the gap is one solution. Due to edge effects along the boundary of the large gap, the interpolation result does not produce topography that matches well with the surrounding terrain. Hence, we expand the gap by three pixels to eliminate edge effects. We divided the gap into multiple subimages, and each subimage was interpolated individually.

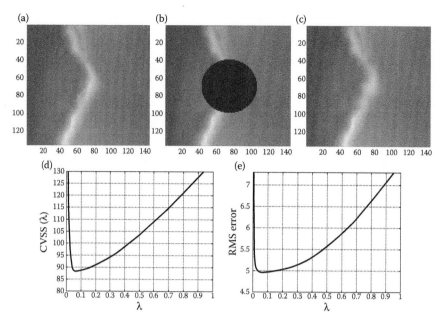

FIGURE 9.9 The quality of CVSS. (a) A sample image that does not have a hole, (b) a hole was made, (c) the interpolated image with an optimal weight, (d) CVSS as a function of λ. The CVSS has a minimum when $\lambda = 0.062$, and (e) RMS error between the true image (a) and the interpolated image (c). The minimum occurs when $\lambda = 0.065$. (From S.-H. Yun et al., *IEEE Transactions on Geoscience and Remote Sensing*, 43(7), 1682, 2005. With permission.)

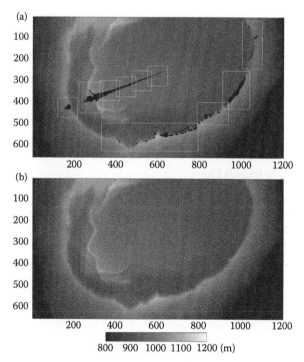

FIGURE 9.10 The original TOPSAR DEM (a) and the reconstructed DEM (b) after interpolation with a PE filter and SRTM DEM constraints. The grayscale is altitude in meters, and the spatial extent is about 12 km across the image. (From S.-H. Yun et al., *IEEE Transactions on Geoscience and Remote Sensing*, 43(7), 1682, 2005. With permission.)

9.7 EFFECT ON INSAR

Finally, we can investigate the effect of the artifact elimination and the interpolation on simulated interferograms. It is often easier to see differences in elevation in simulated interferograms than in conventional contour plots. In addition, simulated interferograms provide a measure of how sensitive the interferogram is to the topography. Figure 9.11 shows georeferenced simulated interferograms from three DEMs: the registered TOPSAR DEM, the TOPSAR DEM after the artifact elimination, and the TOPSAR DEM after the interpolation. In all interferograms, a C-band wavelength is used, and we assume a 452-m perpendicular baseline between two satellite positions. This perpendicular baseline is realistic [2]. The fringe lines in the interferograms are approximately height contour lines. The interval of the fringe lines is inversely proportional to the perpendicular baseline [10], and in this case one color cycle of the fringes represents about 20 m. Note that in Figure 9.11a the fringe lines are discontinuous across the long region of missing data inside the caldera. This is due to the artifacts in the original TOPSAR DEM. After eliminating these artifacts the discontinuity disappears (Figure 9.11b). Finally, the missing data regions are interpolated in a seamless manner (Figure 9.11c).

9.8 CONCLUSION

The aircraft roll artifacts in the TOPSAR DEM were eliminated by subtracting the difference between the TOPSAR and the SRTM DEMs. A 2D PE filter derived from the existing data and the SRTM DEM for the same region are then used as interpolation constraints. Solving the inverse problem constrained with both the PE filter and the SRTM DEM produces a high-quality interpolated map of elevation. Cross-validation works well to select optimal constraint weighting in the inversion. This objective criterion results in less biased interpolation and guarantees the best fit to the SRTM DEM. The quality of many other TOPSAR DEMs can be improved similarly.

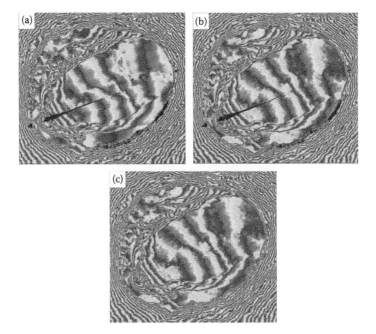

FIGURE 9.11 Simulated interferograms from (a) the original registered TOPSAR DEM, (b) the DEM after the artifact was removed, and (c) the DEM interpolated with a PE filter and the SRTM DEM. All the interferograms were simulated with the C-band wavelength (5.6 cm) and a perpendicular baseline of 452 m. Thus, one color cycle represents 20 m height difference. (From S.-H. Yun et al., *IEEE Transactions on Geoscience and Remote Sensing*, 43(7), 1682, 2005. With permission.)

REFERENCES

1. H.A. Zebker and R.M. Goldstein, Topographic mapping from interferometric synthetic aperture radar observations, *Journal of Geophysical Research*, 91(B5), 4993–4999, 1986.
2. F. Amelung, S. Jónsson, H. Zebker, and P. Segall, Widespread uplift and 'trapdoor' faulting on Galápagos volcanoes observed with radar interferometry, *Nature*, 407(6807), 993–996, 2000.
3. S. Yun, P. Segall, and H. Zebker, Constraints on magma chamber geometry at Sierra Negra volcano, Galápagos Islands, based on InSAR observations, *Journal of Volcanology and Geothermal Research*, 150, 232–243, 2006.
4. H.A. Zebker, S.N. Madsen, J. Martin, K.B. Wheeler, T. Miller, Y.L. Lou, G. Alberti, S. Vetrella, and A. Cucci, The TOPSAR interferometric radar topographic mapping instrument, *IEEE Transactions on Geoscience and Remote Sensing*, 30(5), 933–940, 1992.
5. S.N. Madsen, H.A. Zebker, and J. Martin, Topographic mapping using radar interferometry: Processing techniques, *IEEE Transactions on Geoscience and Remote Sensing*, 31(1), 246–256, 1993.
6. Y. Kobayashi, K. Sarabandi, L. Pierce, and M.C. Dobson, An evaluation of the JPL TOPSAR for extracting tree heights, *IEEE Transactions on Geoscience and Remote Sensing*, 38(6), 2446–2454, 2000.
7. J.F. Claerbout, *Earth Sounding Analysis: Processing versus Inversion*, Blackwell, 1992, http://sepwww.stanford.edu/sep/prof/index.html.
8. J.F. Claerbout and S. Fomel, *Image Estimation by Example: Geophysical Soundings Image Construction (Class Notes)*, 2002, http://sepwww.stanford.edu/sep/prof/index.html.
9. G. Wahba, *Spline Models for Observational Data*, Ser. No. 59, Applied Mathematics, Philadelphia, PA: SIAM, 1990.
10. H.A. Zebker, P.A. Rosen, and S. Hensley, Atmospheric effects in interferometric synthetic aperture radar surface deformation and topographic maps, *Journal of Geophysical Research*, 102(B4), 7547–7563, 1997.
11. S.-H. Yun, J. Ji, H. Zebker, and P. Segall, On merging high- and low-resolution DEMs from TOPSAR and SRTM using a prediction-error filter, *IEEE Transactions on Geoscience and Remote Sensing*, 43(7), 1682, 2005.

10 Hyperspectral Microwave Atmospheric Sounding Using Neural Networks

William J. Blackwell

CONTENTS

10.1 INTRODUCTION

Remote measurements of the Earth's atmospheric state using microwave and infrared wavelengths have been carried out for many years [1,2]. Physical considerations involving the use of these spectral regions include the relatively high cloud penetrating capability at microwave wavelengths and the relatively sharp weighting functions at infrared wavelengths, particularly in the shortwave region near 4 μm where Planck nonlinearity further increases temperature sensitivity. Infrared spectrometer technology has advanced markedly over the past 15 years or so to allow the simultaneous spectral sampling of thousands of bands spaced along narrow atmospheric absorption features [3]. The Atmospheric InfraRed Sounder (AIRS), launched in May 2002, measures 2378 channels from 3.7 to 15.4 μm [4] and the Infrared Atmospheric Sounding Interferometer, launched in 2006, measures 8461 channels from 3.6 to 15.5 μm [5]. These sensors, and similar sensors to be launched as part of the National polar-orbiting operational environmental satellite system Preparatory Project (NPP) and Metosat Third Generation systems, substantially improve atmospheric sounding through the use of hyperspectral measurements, which yield greater vertical resolution throughout the atmosphere [6].

A principal complication in the retrieval of geophysical parameters such as the global three-dimensional atmospheric temperature and moisture profile from satellite radiance observations is the nonlinear, non-Gaussian, and ill-posed physical and mathematical relationship between the radiance observed by a remote sensing instrument and the desired retrieved quantity. Great strides have recently been made to improve and better characterize the models that are used to capture these relationships, but these models are seldom invertible by direct means, usually due to the complex nature of the underlying physics of the relevant geophysical processes. Common inversion approaches involve iterated numerical optimization methodologies that minimize a cost function subject to constraints imposed by a set of regularization parameters constructed so that the optimization tends toward solutions that are more "statistically probable" and/or "physically realistic." These regularization parameters are often largely subjective, and the construction of effective retrieval algorithms therefore requires a substantial component of "black art" to balance the use of the information content in the measured upwelling atmospheric radiances with the plausibility of the retrieval.

A logistical drawback to iterated, model-based inversion techniques is the computational burden required to carry out the numerical optimizations. Modern thermal infrared sensors measure spectral radiances in tens of thousands of separate wavebands (sometimes termed "hyperspectral" or even "ultraspectral") for each observed pixel. The computational complexity of the optimization routines typically scales as the square (or cube) of the number of channels, and it is rare that all of the information available in the radiance spectrum is used. The vast presence of clouds further degrades performance, and therefore a separate preprocessing stage is often employed prior to (or in concert with) numerical inversion to correct the substantial radiance errors that can be introduced due to the high opacity of cloud formations in the infrared wavelengths.

An alternative approach to the numerical inversion approach described above is statistical regression (parameterized function approximation), where an ensemble of input/output pairs is used to empirically derive statistical relationships between the ensembles. In the case of linear regression, second-order statistical moments (covariances) are used to compute a linear fit that minimizes the sum-squared error between the fit and the data. A linear representation is seldom sufficient to fully characterize the complex statistical relationships endemic in atmospheric data, and nonlinear regression techniques must be used. An artificial neural network (NN) is a special class of nonlinear regression operators—the mathematical structure of an NN is chosen to afford several desirable properties, including scalability and differentiability. Patterned after the human nervous system, an artificial NN (hereafter, simply a neural net) consists of interconnected neurons, or nodes, that implement a simple, nonlinear function of the inputs. Usually, the inputs are linearly weighted (the weights modulate each input and the biases provide an offset) and passed through an activation function (often nonlinear). The power of NNs, both from the standpoint of their capabilities and the derivation of their free parameters, stems from the parallel structure of the computational elements. In this chapter, we primarily consider feedforward connections of layers of nodes with sigmoidal (softlimit) activation nodes. Many other variations can be used, but the feedforward variety is most common and the techniques described here are readily applied to other topologies.

The NN approach offers several substantial advantages over iterated, model-based inversion methodologies. Once the weights and biases are derived (during the training process), the network operates very quickly and can be easily implemented in software. This simplicity and speed greatly facilitates the development and maintenance, and therefore cost of complex geophysical retrieval systems that process high volumes of hyperspectral data. The trained NNs are continuous and differentiable, which simplifies error propagation and therefore performance sensitivity analyses. Finally, NNs can approximate functions with arbitrarily high degrees of nonlinearity with a sufficient number of nodes and layers. These advantages have spurred the recent use of NN estimation algorithms for geophysical parameter retrievals [15,27–30,49]. Methods based on NNs for data classification have also become commonplace, although we will focus on regression in this book. Many of the tips and techniques discussed, however, are directly applicable to both types of problems.

In this chapter, we explore the potential use of hyperspectral microwave observations and make three primary contributions. First, we propose a frequency multiplexing technique that can be used to realize hyperspectral microwave measurements with conventional receiver hardware. The approach is easily scalable and requires no new technology development. Second, we examine the relative merits of increased bandwidth versus increased channelization within an increased bandwidth and demonstrate that the optimal operating point depends on the available integration time. We also show that "multiplexed hyperspectral" operation always exceeds "sequential hyperspectral" operation (where a fixed intermediate frequency (IF) filter bank is swept across the measurement spectrum) for finite bandwidths and integration times. Third, we provide a set of comprehensive and global simulation analyses with state-of-the-art retrieval methods that illuminate many of the principal dimensions of the design and performance comparison trade space, including 60 GHz versus 118 GHz, hyperspectral microwave versus conventional microwave, and hyperspectral microwave versus hyperspectral infrared, and we also consider both geostationary and low-earth orbit configurations. Note that these areas are certainly not mutually exclusive, and an optimized sounding system will likely contain several of these components.

This chapter is organized as follows. The hyperspectral microwave concept is first introduced, and it is demonstrated that multiple receiver arrays can be used to multiplex a large set of channels onto a single spot on the ground. An overview of the data sets and physical models used in the simulation is provided. We next point out that opacity due to water vapor continuum absorption is a fundamental limitation of conventional millimeter-wave sounding and show how a hyperspectral millimeter-wave approach can be used to overcome this limitation. All retrievals presented in this chapter are carried out using NNs trained against cloud-resolving physical models, and we discuss the methodologies used to initialize, train, and evaluate the NN retrievals. We then define a variety of notional systems and compare the hyperspectral microwave approach with current and planned approaches that use microwave and hyperspectral infrared observations separately and in combination. Temperature, water vapor, and precipitation retrieval performance comparisons are then presented and the impact of correlated error sources on performance is examined. Finally, we summarize and provide suggestions for further research and development.

10.2 HYPERSPECTRAL MICROWAVE CONCEPT

A spate of recent technology advances driven in part by the gigabit wireless communications industry [7], the semiconductor industry [8], and the NASA Earth Science Technology Office [9] has significantly and profoundly changed the landscape of modern radiometry by enabling miniaturized, low-power, and low-noise radio-frequency receivers operating at frequencies up to 200 GHz. These advances enable the practical use of receiver arrays to multiplex multiple broad frequency bands into many spectral channels, and we explore the atmospheric sounding benefit of such systems in this chapter. We use the term "hyperspectral microwave" to refer generically to microwave sounding systems with ~100 spectral channels or more. In the infrared wavelength range, the term "hyperspectral" is used to denote the resolution of individual, narrow absorption features which are abundant throughout the infrared spectrum. In the microwave and millimeter wavelength range, however, there are substantially fewer spectral features and the spectral widths are typically broad, and an alternate definition is therefore appropriate.

We begin with an analysis of geostationary sounding systems, and we consider low-earth orbiting systems later in the chapter. Detailed studies of the geophysical products that could be derived from a geostationary microwave sensor and the radiometric requirements of such a sensor date back many years [10,11]. The persistent observations afforded by a geostationary platform would allow temporal sampling over most of the viewable earth hemisphere on time scales of ~15 min. The capability to sound in and around storms would significantly improve both regional and global numerical weather prediction models. Tropospheric information content of infrared observations, however, is compromised by clouds which attenuate the radiance to space from the atmosphere below the cloud level. The ability to model and forecast hurricanes would greatly improve with microwave measurements from geostationary orbit.

A simple example of a multiplexed hyperspectral microwave system is shown in Figure 10.1 with eight instantaneous fields of view (IFOV's), four near 118 GHz and four near 183 GHz. Each IFOV is sampled by a single feedhorn measuring two orthogonal polarizations (e.g., vertical and horizontal) that are each fed to a 10-channel spectrometer. The cross-hatching in the figure indicates slightly different frequency bands. The 183-GHz IFOV's each measure the same spectral channels. As the array is microscanned in 50-km (or finer) steps in two dimensions, each 50-km spot on the ground is

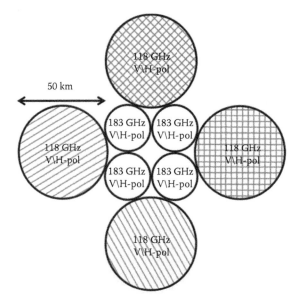

FIGURE 10.1 The instantaneous beam pattern (3-dB contours) on the earth for a notional multiplexed hyperspectral millimeter-wave array spectrometer.

eventually sampled by 80 channels near 118 GHz, and each 25-km spot on the ground is eventually sampled by 20 channels near 183 GHz for a total of 100 channels. Additional channels could be added by increasing the number of feeds and receiver banks, and hyperspectral microwave systems with hundreds of channels are therefore reasonable. Note that further channelization within a given receiver bandwidth quickly reaches a point of diminishing returns due to the increase in thermal noise as $1/\sqrt{B}$, where B is the bandwidth for a single channel. Detailed analyses by other investigators (see, e.g., [12]) demonstrate that a division into about 8–10 channels/receiver yields near optimal results.

We further illustrate the spectral multiplexing concept using temperature weighting functions. The temperature weighting function, given by the derivative of the transmittance function with respect to altitude, characterizes the degree to which each atmospheric layer contributes to the radiances viewed from space at the indicated frequencies. The weighting functions approach zero at high altitudes where the atmosphere becomes transparent, or at low altitudes where the overlying atmosphere is so thick as to be fully opaque. The temperature weighting functions at nadir incidence for a nominal eight-channel receiver operating in a 5-GHz bandwidth on the low-frequency side of the 118.75-GHz oxygen line are shown in Figure 10.2. The 1976 U.S. Standard Atmosphere over a nonreflective surface was used in the calculations. The channels are equally spaced in frequency, and the bandwidth of each channel is 560 MHz.

A hyperspectral microwave system can be constructed by using multiple receiver banks with a replicated, but frequency-shifted, version of the template channel set used to generate Figure 10.2. For example, eight receiver banks with eight channels each could be used to create the 64-channel system by progressively shifting the IF band of each receiver by 70 MHz.

We conclude by comparing the temperature weighting functions of the 64-channel multiplexed system to that of a current-generation microwave sounder, the Advanced Technology Microwave Sounder [13] scheduled to fly onboard the NPP in 2011. In the upper panel of Figure 10.3, the set of weighting functions for the 64-channel system is indicated by thin lines and the eight tropospheric Advanced Technology Microwave Sounder (ATMS) channels are indicated by heavy lines. The surface weights of all channels are shown in the lower panel of Figure 10.3. The frequencies for the eight ATMS channels show range from 50.3 GHz (most transparent) to 57.29 GHz.

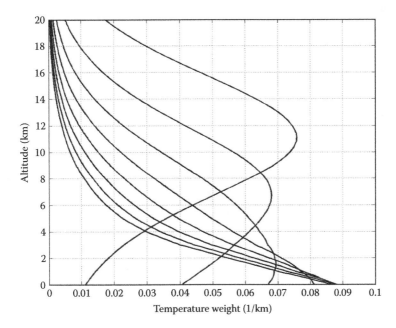

FIGURE 10.2 Template temperature weighting functions for channels near 118.75 GHz at nadir incidence. The 1976 U.S. Standard Atmosphere over a nonreflective surface was used in the calculations.

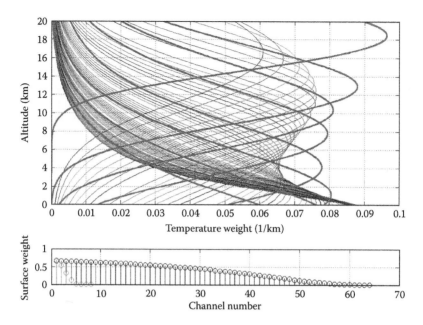

FIGURE 10.3 Weighting functions for 64 channels near 118.75 GHz and the eight ATMS tropospheric temperature sounding channels near 50–57 GHz at nadir incidence. The 1976 U.S. Standard Atmosphere over a nonreflective surface was assumed in the calculations.

The effective atmospheric vertical sampling density of the 64-channel multiplexed system is clearly superior to the conventional 8-channel system. Water-vapor-burden weighting functions are not shown here, as similar arguments apply. The temperature and water vapor profiling advantage resulting from fine vertical atmospheric sampling is discussed in Section 10.7.

The temperature weighting function for the ATMS 50.3-GHz channel is shown near the bottom of the top panel of Figure 10.3 (the temperature weight near the surface is ~0.06 km^{-1}), and this channel is significantly more transparent than the most transparent 118-GHz channel (the temperature weight near the surface is ~0.09 km^{-1}) due to water vapor continuum absorption, but the weighting function has a comparable shape. Water vapor absorption appears to be a significant handicap for mm-wave sounding [10]. Hyperspectral millimeter-wave systems, however, are not so adversely affected because of the very dense vertical spacing of the weighting functions in the atmospheric boundary layer. We address this important point in detail in Section 10.4.

10.3 OVERVIEW OF DATA SETS AND PHYSICAL MODELS

The performance comparisons we present in this chapter are all based on simulated observations derived using physical models and global ensembles of atmospheric states. The selection of the ensemble of atmospheric states is a critically important part of any simulation study, and we have taken great care to ensure that the profiles included in the analysis are sufficiently representative of the cloudy, moist atmospheres that challenge most atmospheric sounding systems.

10.3.1 NOAA88B ATMOSPHERIC PROFILE DATA SET

The NOAA88b radiosonde/rocketsonde data set contains 7547 profiles, globally distributed seasonally and geographically (see Figure 10.4). Atmospheric temperature, moisture, and ozone are given at 100 discrete levels from the surface to altitudes exceeding 50 km. Skin surface temperature is

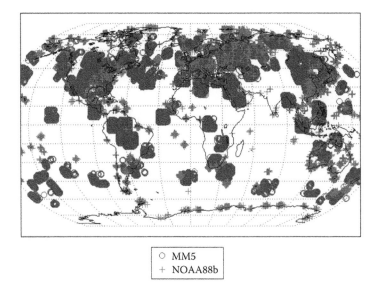

FIGURE 10.4 (See color insert.) Geographical locations of the pixels contained in the Mesoscale Model, version 5 (MM5) and NOAA88b data sets. Global coverage including a variety of land and ocean pixels is achieved by each set.

also recorded. NOAA88b water vapor measurements above ~10 km are of questionable quality and are not considered in this chapter.

Approximately 6500 profiles from the NOAA88b database were selected for inclusion in this study by eliminating cases for which the surface pressure was not equal to 1000 mbar. The mean and standard deviation of the temperature and water vapor profiles in the NOAA88b data set are shown in Figures 10.5 and 10.6, respectively.

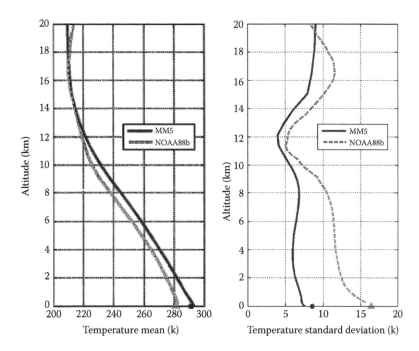

FIGURE 10.5 Mean and standard deviation of the temperature profiles in the MM5 and NOAA88b data sets.

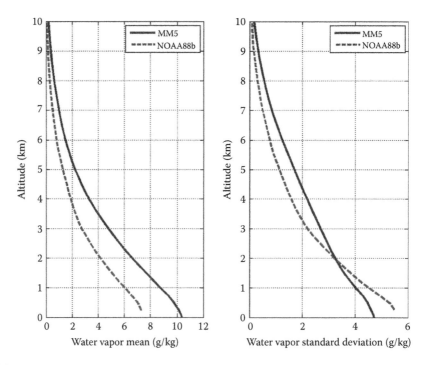

FIGURE 10.6   Mean and standard deviation of the water vapor profiles in the MM5 and NOAA88b data sets.

10.3.2   MIT MM5 Precipitation and Atmospheric Profile Data Set

The Massachusetts Institute of Technology (MIT) MM5 precipitation and atmospheric profile data set [14,15] is composed of meteorological parameters for 122 storms simulated using the fifth-generation National Center for Atmospheric Research (NCAR)/Penn State Mesoscale Model (MM5). These storms are globally distributed as illustrated in Figure 10.4 and span a year. Each storm has 190×190 picture elements (pixels) spaced on a rectangular 5-km grid with 42 pressure levels. About 46% of the 4.4 M pixels are precipitating with nonzero rain water or snow at 1000 mbar [14]. This percentage of precipitating pixels is high because only very cloudy regions were sampled. The validity of this ensemble of storms has been shown by the statistical agreement between its simulated brightness temperatures and those coincidentally observed by Advanced Microwave Sounding Unit (AMSU) instruments aboard NOAA-15, -16, and -17 satellites [14,15]. All radiative transfer calculations were performed at the native horizontal spatial resolution of the MM5 data and were then convolved with the appropriate geostationary millimeter-wave array spectrometer (GeoMAS) spatial response functions.

Approximately 50,000 nonprecipitating pixels selected from the 122 storm cases were used in the temperature and moisture profile retrieval study; approximately half of these pixels were cloudy (nonzero integrated cloud liquid water content). Precipitation was screened by requiring that the sum of vertically averaged water mixing ratios in the form of rain, graupel, and snow was <0.01 kg/kg. The pixels were spatially downsampled to ensure that the distance between any two pixels always exceeds 25 km. The mean and standard deviation of the temperature and water vapor profiles in the MM5 data set are shown in Figures 10.5 and 10.6, respectively. The histogram of integrated cloud liquid water (liquid water path) is shown in Figure 10.7.

The MM5 profiles are generally warmer and contain more water vapor and cloud liquid water than the NOAA88b profiles and are highly representative of the atmospheric thermodynamic state that would be encountered near precipitation. The NOAA88b profiles are characterized by a high level of variability due to the deliberate inclusion of more extreme cases.

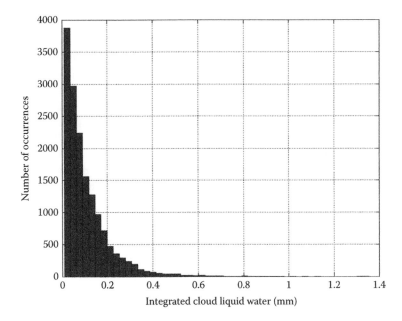

FIGURE 10.7   Histogram of the vertically integrated cloud liquid water content (i.e., liquid water path) in the MM5 storm data set. The NOAA88b data set does not include measurements of cloud liquid water. Rain becomes likely when integrated cloud liquid water content exceeds ~0.2 mm [16], and precipitating pixels have been removed from the data set.

10.3.3   MICROWAVE/MILLIMETER-WAVE NONSCATTERING RADIATIVE TRANSFER MODEL: **TBARRAY**

Simulated brightness temperature observations for atmospheric profiles in the NOAA88b data set were calculated using the TBARRAY software package of Rosenkranz [17]. TBARRAY is a line-by-line routine based on the Liebe Millimeter-Wave Propagation Model [18,19]. Scattering was not modeled because cloud liquid water content was not recorded in the NOAA88b data set. All radiative transfer calculations for the temperature and water vapor retrieval simulations were performed at a single angle at nadir incidence.

10.3.4   MICROWAVE/MILLIMETER-WAVE SCATTERING RADIATIVE TRANSFER MODEL: **TBSCAT**

Simulated brightness temperature observations for atmospheric profiles in the MIT MM5 data set were calculated using the TBSCAT software package of Rosenkranz [20]. TBSCAT is a multi-stream initial-value radiative transfer routine that includes both absorption and scattering. The collection of streams describe a system of coupled first-order differential equations, and TBSCAT approaches the solution as an initial-value problem starting from the top of the atmosphere. The solution uses the backward Euler method of finite differences. The absorption coefficients are calculated identically in TBARRAY and TBSCAT. The scattering calculations comprise the Mie coefficients, the inverse-exponential drop-size distribution (for precipitating cases only), the Liebe/Hufford permittivity model, and the Henyey–Greenstein phase function. The scattering coefficients are calculated using the Deirmendjian implementation in the Mie region and the Wiscombe implementation in the Rayleigh region. The clouds in the nonprecipitating MM5 data set were simulated using 10 streams and with both the cloud liquid water and cloud ice as scattering hydrometeors. The cloud liquid water was given a radius of 0.02 mm and the cloud ice was given a radius of 0.06 mm, as these values are consistent with recommendations from other investigators [21].

10.3.5   Ocean Surface Emissivity Model: FASTEM and FASTEM2

English and Hewison developed the FASTEM model [22], which parameterizes an "effective" ocean surface emissivity for frequencies between 10 and 220 GHz for earth incidence angles <60° and for oceanic surface wind speeds <20 m/s. FASTEM2, an updated version of FASTEM, uses an approach similar to that of Petty and Katsaros [23] to compute the surface emissivity. FASTEM and FASTEM2 both incorporate geometric optics, Bragg scattering, and foam effects. FASTEM was used for the simulations of the precipitating cases, and FASTEM2 (with the optical depth option set to zero) was used for the simulations of the nonprecipitating cases. FASTEM and FASTEM2 calculations used the MM5 oceanic wind speed at an altitude of 10 m, which ranged from 0.14 to 11 m/s. The oceanic surface wind speed is not recorded in the NOAA88b data set, and the FASTEM2 wind speed input for these cases was therefore randomized using a uniform distribution between 0.5 and 10 m/s.

10.3.6   Land Surface Emissivity Model

Land surface emissivity values were assigned randomly using a uniform distribution between 0.8 and 1.0. The same emissivity value was used for all frequencies. Recent work has shown that this simple model is fairly representative of most naturally occurring land emissivities [24], although improvements are planned in future work.

10.4   ATMOSPHERIC TRANSMITTANCE AT MILLIMETER WAVELENGTHS: IMPLICATIONS FOR GEOSTATIONARY SOUNDING

Atmospheric extinction increases as $\sim f^4$ due to Rayleigh scattering and as $\sim f^2$ due to absorption. Previous work has demonstrated that the relatively high sensitivity of the millimeter-wave bands to hydrometer scattering can be used to improve precipitation sensing [25–27]. However, the relatively high levels of atmospheric absorption present in millimeter-wave bands can hinder atmospheric sounding in the boundary layer near the surface [10]. We now explore this apparent handicap in the context of geostationary sounding.

The microwave/millimeter-wave absorption spectrum is shown in Figure 10.8 for a fixed amount of water vapor (15 mm) and no cloud liquid water. The characteristic decrease in transmittance with frequency is immediately apparent. The water vapor content in the 1976 U.S. Standard Atmosphere used to compute the absorption spectrum shown in Figure 10.8 is relatively low compared to a typical tropical atmosphere (~50 mm water vapor). Therefore, we now examine atmospheric transmittance with a focus on profiles from the MIT MM5 data set, which is characterized by high water content. Atmospheric transmittance at nadir incidence at 90 GHz is shown in Figure 10.9 as a function of the integrated water vapor content and the integrated cloud liquid water content, and the transmittance predictably decreases with increasing water content. To explore the frequency and water content dependence in tandem, we plot the difference between the transmittance at 50 GHz and at 90 GHz as a function of water content in Figure 10.10. For low water contents, the transmittance at 90 GHz exceeds than at 50 GHz, although the opposite is true for relatively high water contents.

The latter observation is a fundamental limitation of conventional (nonhyperspectral) millimeter-wave sounding. Hyperspectral millimeter-wave systems, however, are able to infer and correct for the increasing atmospheric absorption, even in atmospheres with high water content, because of the very dense vertical spacing of the weighting functions in the atmospheric boundary layer. This important distinction between conventional and hyperspectral sounding will be highlighted in Section 10.7.

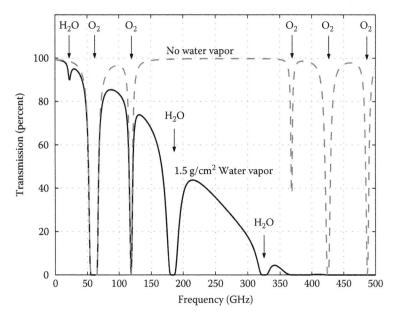

FIGURE 10.8 The microwave/millimeter-wave absorption spectrum. Two calculations for the percent transmission (nadir view) using the 1976 U.S. Standard Atmosphere are shown, one assuming no water vapor and one assuming 1.5 g/cm² (15 mm).

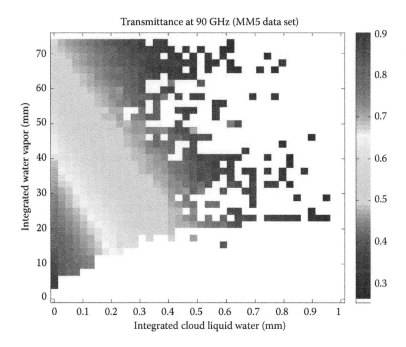

FIGURE 10.9 (**See colour insert.**) Atmospheric transmittance at nadir incidence at ~90 GHz as calculated from the MIT MM5 data set. Transmittance values are averaged over bins of integrated water vapor and integrated cloud liquid water content.

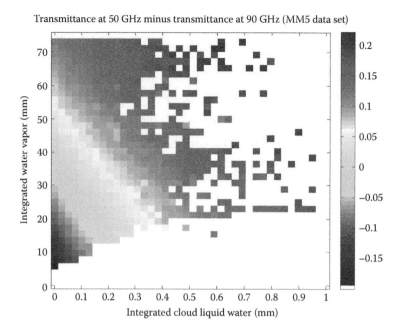

FIGURE 10.10 **(See colour insert.)** Atmospheric transmittance at nadir incidence at 50 GHz minus that at 90 GHz as calculated from the MIT MM5 data set. Transmittance values are averaged over bins of integrated water vapor and integrated cloud liquid water content.

10.5 HYPERSPECTRAL MICROWAVE PHYSICAL RETRIEVALS USING NEURAL NETWORKS

Recent work has demonstrated the utility of atmospheric profile retrievals based on feedforward multilayer perceptron NNs for both hyperspectral infrared and microwave observations [28,29]. The execution time of NN retrievals, once trained using physical models, for example, is typically several orders of magnitude faster than iterated retrievals while offering improved retrieval performance. NNs are used in this work to retrieve temperature and moisture profiles and precipitation rates. The methodology is now briefly reviewed, and the interested reader is referred to Ref. [30] for more details.

An NN is an interconnection of simple computational elements, or nodes, with activation functions that are usually nonlinear, monotonically increasing, and differentiable. NNs are able to deduce input–output relationships directly from the training ensemble without requiring underlying assumptions about the distribution of the data. Furthermore, an NN with only a single hidden layer of a sufficient number of nodes with nonlinear activation functions is capable of approximating any real-valued continuous scalar function to a given precision over a finite domain [31].

A multilayer feedforward NN consists of an input layer, an arbitrary number of hidden layers (usually one or two), and an output layer. The hidden layers typically contain sigmoidal activation functions of the form $z_j = \tanh a_j$, where $a_j = \sum_{i=1}^{d} w_{ji} x_i + b_j$. The output layer is typically linear. The weights w_{ji} and biases b_j for the jth neuron are chosen to minimize a cost function over a set of P training patterns. A common choice for the cost function is the sum-squared error, defined as

$$E(w) = \frac{1}{2} \sum_p \sum_k \left(t_k^{(p)} - y_k^{(p)} \right)^2$$

where $y_k^{(p)}$ and $t_k^{(p)}$ denote the network outputs and target responses, respectively, of each output node k given a pattern p, and w is a vector containing all the weights and biases of the network. The "training" process involves iteratively finding the weights and biases that minimize the cost function through some numerical optimization procedure. Second-order methods are commonly used to carry out the optimization.

10.5.1 PREPROCESSING WITH THE PROJECTED PRINCIPAL COMPONENTS TRANSFORM

Hyperspectral sounding systems typically measure atmospheric thermal emission in many spectral bands. The spectral information content is often correlated, and a linear preprocessing method such as the projected principal components (PPC) transform [28] can be effectively used to reduce the dimensionality and filter noise from the measured spectra, even in the presence of clouds [32].

Furthermore, the PPC transform can be used to optimally extract spectral radiance information that is correlated with a geophysical parameter, such as the temperature or water vapor profile. The r-rank linear operator that captures the most radiance information that is correlated to the profile is [28]

$$L_r = E_r E_r^T C_{TR} (C_{RR} + C_{\psi\psi})^{-1}$$

where

$$E_r = \begin{bmatrix} E_1 \mid E_2 \mid \cdots \mid E_r \end{bmatrix}$$

are the r most significant eigenvectors of $C_{TR}(C_{RR} + C_{\psi\psi})^{-1} C_{RT}$, C_{RR} is the spectral radiance covariance, $C_{\psi\psi}$ is the noise covariance, and C_{TR} is the cross-covariance of the atmospheric profile (temperature or water vapor) and the radiance. The hyperspectral millimeter-wave measurements processed in this work were transformed from ~100 channels to 25 PPCs for retrieval of both temperature and water vapor profiles. This factor of four reduction in the input dimensionality results in a significant improvement in both the network training time (typically a few hours on a 3-GHz Intel Xeon desktop workstation) and the generalization ability of the NNs.

10.5.2 NETWORK TOPOLOGIES

All the temperature and moisture retrievals in this work were implemented using NNs with a single hidden layer of 15 sigmoidal nodes and a linear output layer with ~10 nodes. Approximately five NNs were aggregated (to achieve ~50 total outputs) to estimate the entire profile. *Ad hoc* attempts were made to optimize the network topology, and this configuration resulted in the best performance. The networks estimate the atmospheric profile up to 20 km for temperature and up to 10 km for water vapor. These estimates and the corresponding "truth" were averaged over 1-km layers prior to the computation of error statistics.

10.5.3 NETWORK INITIALIZATION, TRAINING, AND PERFORMANCE EVALUATION

The profile data were randomly divided into three nonoverlapping, namely, the training set (80%), the validation set (10%), and the testing set (10%). The selection of profiles for the data sets was identical for all GeoMAS and synthetic thinned aperture radiometer (STAR) performance comparisons. The training set was used to derive the network weights and biases. The network training was stopped if the error on the validation set did not decrease after 10 consecutive training epochs or if 300 epochs were reached. Each NN was trained 10 separate times with random initializations, and

TABLE 10.1

GeoMAS Center Frequency Offsets from 118.75 GHz, Channel Bandwidths, and ΔT_{rms} for the Most Transparent Receiver Bank (One of Eight)

Ch.	Offset (MHz)	Bandwidth (MHz)	ΔT_{rms} (K)
1	−4720	560	0.19
2	−4160	560	0.19
3	−3600	560	0.19
4	−3040	560	0.19
5	−2480	560	0.19
6	−1920	560	0.19
7	−1360	560	0.19
8	−800	560	0.19
9	−29,750	1000	0.14

the validation set was used to select the best of the 10 networks. All the NN retrieval results presented in this chapter were derived using the testing set.

All networks were initialized using the Nguyen–Widrow method [33] and trained using the Levenberg–Marquardt optimization procedure [34,35]. The NETLAB NN software package [36] was used to train the networks. Random sensor noise (see the last column of Tables 10.1 and 10.2) was added to each simulated measurement at the beginning of each training epoch.

10.6 NOTIONAL SYSTEMS USED FOR COMPARISONS

We now assess the performance of the hyperspectral millimeter-wave concept in the geostationary context using two notional systems, an 88-channel system operating near 118 and 183 GHz and a 10-channel system operating near 60 and 183 GHz. The underlying physical and radiometric assumptions for both systems are identical, as discussed below, and identical NN retrieval algorithms are used. We make no claims of optimality for either of the two notional systems.

TABLE 10.2

STAR Center Frequencies, Channel Bandwidths, and ΔT_{rms}. The Individual STAR Channel Sensitivities are poorer than those for Comparable GeoMAS Channels Because the STAR System Described in [38] Must Sample Each Channel Separately, thus Decreasing the Available Integration Time

Ch.	Center Freq. (GHz)	Bandwidth (MHz)	ΔT_{rms} (K)
1	50.3	180	0.5
2	52.8	400	0.35
3	53.596	170	0.5
4	54.4	400	0.35
5	54.94	400	0.35
6	55.5	330	0.35
7	167	1000	0.71
8	183.31 ± 1	500	1.0
9	183.31 ± 3	1000	0.71
10	183.31 ± 7	2000	0.5

10.6.1   GEOSTATIONARY MILLIMETER-WAVE ARRAY SPECTROMETER

The nominal GeoMAS sensor configuration comprises a modest 88-channel hyperspectral millimeter-wave spectrometer (the Geostationary Millimeter-Wave Array Spectrometer, or GeoMAS) with 72 channels near the 118.75-GHz oxygen absorption line and 16 channels near the 183.31-GHz water vapor absorption line. This configuration could easily be realized using five dual-polarized antenna horns (sharing a common reflector) each feeding a simple nine-channel receiver bank. The spatial resolution of the water vapor channels could be doubled by arranging additional feeds in a 2×2 array configuration, thus raising the total number of feedhorns to eight (see Figure 10.1). Additional feeds and channels could be added to further improve the spatial resolution of both bands [e.g., additional Signal-to-Noise Ratio (SNR) margin could be used to sharpen the effective antenna beam], although we defer this analysis to a future paper.

   We assume an integration time of 22.5 ms, which would allow temperature profiles to be retrieved on a 200×200 grid spaced at 50 km ($10,000 \times 10,000$ km coverage area) in 15 min. The water vapor profiles would be retrieved on a 400×400 grid spaced at 25 km ($10,000 \times 10,000$ km coverage area) in 15 min.

   The GeoMAS channel properties for a single receiver bank (of eight) of the 118-GHz temperature band are summarized in Table 10.1. System temperatures of 650 K and 800 K were assumed for the 118- and 183-GHz systems, respectively. The eight 118-GHz receiver banks (each with eight channels split among a 5-GHz bandwidth near the 118.75-GHz O_2 line and a ninth channel at 89 GHz) are offset from one another in frequency by 70 MHz, and the band edge of the most opaque channel is ~50 MHz from the O_2 line center. A transparent channel at 89 GHz was added to each 118-GHz receiver bank to improve profiling performance at the surface and near the boundary layer, as a result of the discussion in Section 10.4. The two 183-GHz receiver banks (each with eight channels split among an 8-GHz bandwidth near the 183.31-GHz H_2O line) are offset from one another in frequency by 500 MHz, and the band edge of the most opaque channel is 10 MHz from the H_2O line center. The ΔT_{RMS} of each water vapor channel is 0.17 K. The frequency offset and overlap values were obtained using simple trial and error experiments. Further channel optimizations should improve GeoMAS retrieval performance but are beyond the scope of this chapter.

10.6.2   SYNTHETIC THINNED APERTURE RADIOMETER

A synthetic thinned aperture radiometer has recently been suggested for geostationary implementation [37,38] with ~900 receivers; 300 operating near 60 GHz and 600 operating near 183 GHz. We begin with identical fundamental assumptions used for the GeoMAS system: 40,000 temperature profiles (50-km grid spacing) and 160,000 water vapor profiles (25-km grid spacing) derived in 15 min. Synthetic and unsharpened filled aperture systems yield comparable image noise levels for comparable receivers and total integration times, provided that both systems survey the entire visible earth and have the same bandwidths and receiver noise temperatures [39].

   The STAR configuration follows closely from Ref. [38], although we have included broader bandwidths (therefore favorable to STAR performance) to facilitate comparisons with AMSU-A/B performance. Ten STAR channels are used: six identical to Advanced Microwave Sounding Unit-A (AMSU-A) temperature bands near 60 GHz, three identical to AMSU-B water vapor bands near 183 GHz, and a single band at 167 GHz. We assume system temperatures of 500 K for each 60-GHz receiver and 800 K for each 167/183-GHz receiver. Because each receiver can sample only a single channel at a time, the integration time available to a given channel is equal to the total integration time per pixel (22.5 ms for 60 GHz, 5.625 ms for 183 GHz) divided by the number of channels in the receiver (six near 60 GHz and four near 183 GHz). The STAR channel properties are summarized in Table 10.2.

10.7 TEMPERATURE AND MOISTURE SOUNDING PERFORMANCE COMPARISONS FOR NOTIONAL GEOSTATIONARY SYSTEMS

We now examine temperature and water vapor profile retrieval performance in nonprecipitating atmospheres for both the GeoMAS and STAR notional systems. Both ocean and land cases are included in equal proportion, and the NOAA88b and MM5 profile sets are analyzed separately. Performance is indicated by Root Mean Squared (RMS) errors in 1-km atmospheric layers at nadir incidence [off-nadir performance in Geostationary (GEO) differs negligibly].

10.7.1 TEMPERATURE

The temperature profile retrieval performance curves are shown in Figure 10.11. Also shown is the surface temperature RMS error. The GeoMAS performance is excellent both for the NOAA88b profile set (high variability) and for the MM5 profile set (high water content). GeoMAS performance exceeds STAR performance for both profile sets at all atmospheric levels, including the surface. The performance difference in the relatively moist MM5 profile set is substantial, with ~0.5 K difference in RMS error throughout the troposphere, including the critical atmospheric boundary layer. GeoMAS surface skin temperature retrievals are substantially superior to those of STAR, with a difference in RMS error of ~1 K. Note that the STAR performance closely resembles that of AMSU-A/B, as the primary tropospheric sounding channels are identical. The absence of relatively transparent channels at 23.8 and 31.4 GHz slightly degrades performance in the lower boundary layer, with the primary impact on water vapor retrieval over ocean.

The GeoMAS performance advantage as indicated in Figure 10.11 is due to two factors. First, the high density of weighting functions in the vertical dimension, as shown in Figure 10.3, enables GeoMAS measurements to capture profile information with high vertical resolution. Second, the

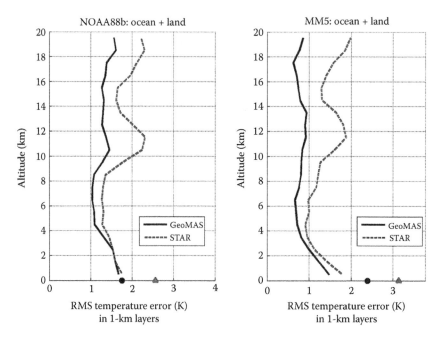

FIGURE 10.11 Comparison of GeoMAS (118/183 GHz) and STAR (60/183 GHz) temperature retrieval performance over land and ocean at nadir incidence. The left panel shows RMS error in 1-km layers using the NOAA88b data set (scattering was not modeled), and the right panel shows RMS error using the MIT MM5 data set (scattering was modeled). Surface temperature RMS error is indicated by a circle (GeoMAS) and a triangle (STAR).

large total bandwidth afforded by the GeoMAS system allows relatively high sensitivity to be achieved. We now explore separately the bandwidth benefit and the channelization benefit.

Figure 10.12 shows the RMS temperature profile averaged over all atmospheric layers from 0 to 20 km as a function of the number of receiver banks included in a hyperspectral microwave system. It is obvious that increased sensitivity due to increased bandwidth will always improve retrieval results, but we now examine two more interesting cases. First, a fixed amount of bandwidth is shared among all the receiver banks (the solid curves in Figure 10.12). Improvement in this case as receiver banks are added is only due to the advantage afforded by increased channelization, resulting in a more dense vertical spacing of the weighting functions. Note that the performance for this case is identical to that for a "sequential hyperspectral" system, whereby a range of channels (e.g., generated by sweeping the receiver local oscillator frequency) is sequentially measured. That is, receiver bank one is selected for one-eighth of the total integration time, then bank two for one-eighth, and so forth, until all eight receiver banks are utilized. The curves are shown for five different integration times, based on a scaling of the integration time used to generate Figure 10.11. The top curve corresponds to the same integration time used to generate Figure 10.11, and this integration time was increased by factors of 10, 100, 1000, and infinity (no noise) to generate the other curves. The second case (the dashed curves in Figure 10.12) assumes that a fixed amount of bandwidth is allocated for each receiver bank. We term this "multiplexed hyperspectral" operation, as all channels eventually view the scene for the duration of the total integration time. Both increased bandwidth and increased channelization positively impact performance in this case as receiver banks are added. Note that sequential hyperspectral operation is never superior to multiplexed hyperspectral operation, and the performance difference is most pronounced for short per-pixel integration times characteristic of geostationary sensing platforms.

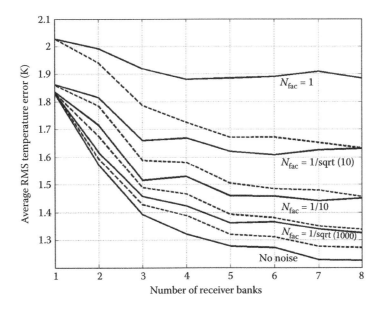

FIGURE 10.12 Incremental benefit of additional eight-channel receiver banks. All curves indicate the RMS temperature profile error averaged over all atmospheric layers from 0–20 km as a function of the number of receiver banks used. The solid curves were calculated using a fixed bandwidth that is shared among all receiver banks (sequential hyperspectral), and the dashed curves are calculated using a fixed bandwidth for each receiver bank (multiplexed hyperspectral). Each pair of dashed and solid curves was calculated using a different noise factor (N_{fac}) which is a simple scaling of the integration time. The dashed and solid curves labeled "no noise" are identical.

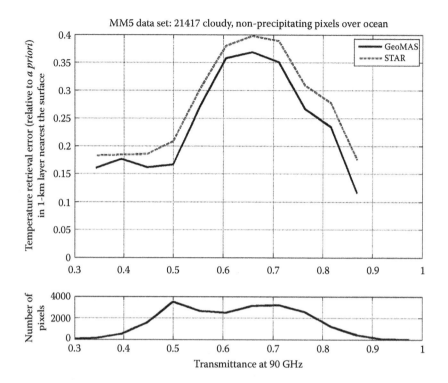

FIGURE 10.13 Comparison of GeoMAS (118/183 GHz) and STAR (60/183 GHz) temperature retrieval performance in the 1-km atmospheric layer nearest the surface as a function of transmittance at ~90 GHz. The errors are shown relative to the *a priori* standard deviation in the profile set.

These results are characterized by two interesting features. First, the marginal benefit of increased channelization is apparent in all cases. Second, the amount of channelization benefit is more pronounced as the integration time is increased.

We further examine the temperature profile retrieval performance by assessing the accuracy in the lower boundary layer (the 1-km layer nearest the surface) as a function of the atmospheric transmittance, shown in Figure 10.13. The errors shown have been normalized by the *a priori* standard deviation of the temperature over all the profiles included in each of 20 transmittance bins. That is, a value of 0.5 indicates that the RMS error is one half of the RMS error that would have resulted using the mean temperature value (over the bin) as the estimate. The bell-shape curve can be explained as follows. Relatively small errors toward the right of the figure result because the high atmospheric transmittance allows the surface temperature to be sensed with high accuracy, and the surface temperature tends to be correlated with the temperature in the lower boundary layer. As transmittance decreases, the surface is increasingly obscured and the error therefore increases. As transmittance decreases beyond 0.6 or so, the contribution from the lower boundary layer to the overall measured radiance for most channels is maximized, and the error therefore decreases. As transmittance approaches zero, the radiative contribution from the lowest 1-km layer decreases due to obscuration by opaque layers above.

10.7.2 Water Vapor

The water vapor profile retrieval performance curves are shown in Figure 10.14. The errors shown are the RMS profile retrieval error divided by the *a priori* standard deviation in the profile set. The

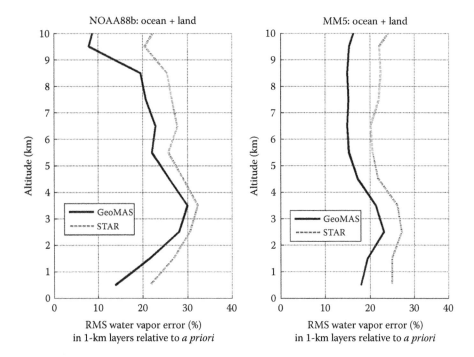

FIGURE 10.14 Comparison of GeoMAS (118/183 GHz) and STAR (60/183 GHz) water vapor retrieval performance over land and ocean at nadir incidence. The left panel shows RMS error (relative to the *a priori* variation in the validation set) in 1-km layers using the NOAA88b data set (scattering was not modeled), and the right panel shows RMS error using the MM5 storm data set (scattering was modeled).

GeoMAS performance is excellent both for the NOAA88b profile set (high variability) and for the MM5 profile set (high water content). GeoMAS performance exceeds STAR performance for both profile sets at all atmospheric levels.

10.8 INVESTIGATION OF ADDITIONAL ERROR SOURCES

The simulation analyses we have presented thus far have been performed under ideal circumstances to accentuate the relative merits of the various sensor systems. We now consider additional error sources that would be expected under realistic conditions. For example, simulation analyses often assume perfect sensor and atmospheric physics, as well as perfect ground truth, and these assumptions are invalid in practice. Of particular concern are correlated errors that could be especially detrimental to a hyperspectral microwave system comprising many channels with broad, overlapping weighting functions. None of the error sources considered here were presented to the NN retrievals during training.

10.8.1 TWO-DIMENSIONAL SIMULATION METHODOLOGY

We begin with an examination of two correlated error sources: (1) unknown array misalignment error and (2) spatial inhomogeneity error. Array misalignment could result due to mechanical and/or electrical imperfections in the antenna arrays used to produce the footprints shown in Figure 10.1. Subpixel spatial inhomogeneities introduced by atmospheric and surface features further degrade performance. We assess the impact of both of these error sources using the high-resolution gridded MM5 profile set.

10.8.2 ARRAY MISALIGNMENT ERROR

Static, unknown misalignment errors of ~10% of each temperature sounding footprint and 20% of each water vapor sounding footprint were introduced to the GeoMAS arrays during simulation of the brightness temperatures. This level of unknown misalignment error is expected to be much larger than would be encountered on an actual system. The STAR alignment was assumed to be perfect.

10.8.3 SPATIAL INHOMOGENEITY ERROR

Clouds, surface features, and water vapor are highly variable in the horizontal dimension, and sub-pixel inhomogeneities across 50-km temperature and 25-km water vapor footprints are likely. Gaussian antenna beam patterns for the GeoMAS and STAR systems were convolved with the high-resolution (5 km) brightness temperature fields calculated using the MM5 profile set to simulate the effects of spatial inhomogeneities.

10.8.4 SIMULATION ERROR

An uncorrelated Gaussian random error term with 0.2-K standard deviation and zero mean was added to all calculated brightness temperatures. This error accounts to first order for imperfections in the surface, transmittance, and radiative transfer models used to derive the retrievals, as well as any errors in the ground truth.

10.8.5 SIMULATION RESULTS

Temperature and water vapor retrieval results after inclusion of all the error sources above are shown in Figures 10.15 and 10.16, respectively. GeoMAS temperature profile retrieval RMS

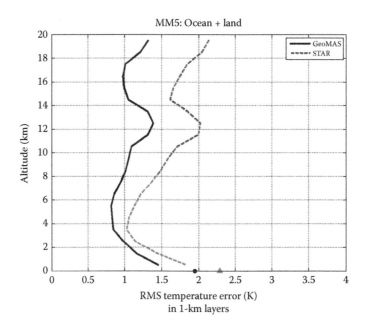

FIGURE 10.15 Comparison of GeoMAS (118/183 GHz) and STAR (60/183 GHz) two-dimensional temperature retrieval performance over land and ocean at nadir incidence using the MIT MM5 data set with detailed modeling of correlated error sources. Surface temperature RMS error is indicated by a circle (GeoMAS) and a triangle (STAR).

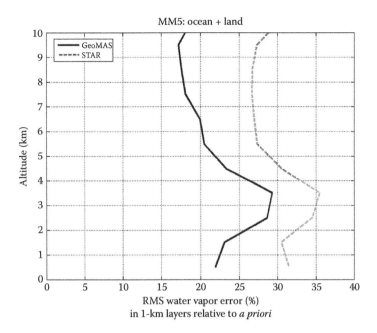

MM5: ocean + land

FIGURE 10.16 Comparison of GeoMAS (118/183 GHz) and STAR (60/183 GHz) two-dimensional water vapor retrieval performance (relative to the *a priori* variation in the validation set) over land and ocean at nadir incidence using the MIT MM5 data set with detailed modeling of correlated error sources.

performance exceeds 1.5 K in 1-km layers, surface temperature RMS performance exceeds 2 K, and water vapor profile retrieval performance exceeds 30% in 1-km layers. While performance does degrade relative to Figures 10.11 and 10.14, it is encouraging to note the high degree of robustness of the GeoMAS system to these error sources. Other sources of error are currently under study (e.g., IF filterbank stability) and associated analysis results will be reported in a future publication.

10.9 PRECIPITATION PERFORMANCE COMPARISON FOR NOTIONAL GEOSTATIONARY SYSTEMS

Airborne [40,41] and spaceborne [27,42–44] retrievals of precipitation from passive opaque millimeter-wave measurements have demonstrated the potential of a geostationary millimeter-wave sensor for precipitation mapping and tracking. We now compare rain-rate retrieval performance of the GeoMAS and STAR systems.

Brightness temperature simulations were carried out using TBSCAT and incorporated a fluffy-sphere ice scattering model with a wavelength-dependent density $F(\lambda)$ that was tuned to match total scattering cross-sections computed for spherical, hexagonal plate, and rosette hydrometeors using a discrete-dipole electromagnetic scattering model, DDSCAT [14]. Precipitation retrieval accuracies for GeoMAS and STAR were computed using NNs trained using 122 MM5 storms described in Section 10.3.2. The algorithms are the same as described in [39], employing three NNs. If the rain-rate estimates from the first network were over 8 mm/h, then the second NN was used to estimate the 15-min average precipitation rate; otherwise, the third network was used. Inputs used to train all three networks include the MM5 precipitation rates blurred to 25 km, a land/sea flag, and the current channel brightness temperatures and those observed 15 min earlier. The three layers of each network had 10, 5, and 1 neurons, respectively, where the first two layers used a hyperbolic tangent sigmoid function. The best of 100 tested networks was used for each network and task, where "best" means the minimum RMS retrieval errors over the full dynamic range.

TABLE 10.3

Rain-Rate Retrieval Performance (RMS Error in Millimeters per Hour) for the GeoMAS and STAR Systems at 25-km Spatial Resolution

Rain-Rate Range (mm/h)	GeoMAS (mm/h)	STAR (mm/h)
1–4	1.5	1.5
4–8	3.4	3.7
8–16	6.0	6.8
16–32	10.2	10.6
32–64	16.9	17.9

The results, shown in Table 10.3, indicate excellent performance of GeoMAS relative to STAR in light to moderately heavy rain.

10.10 TEMPERATURE AND MOISTURE RETRIEVAL PERFORMANCE SIMULATION COMPARISONS FOR HYPERSPECTRAL MICROWAVE AND AIRS/AMSUA/HSB

We now examine sounding performance for low-earth-orbit systems. Retrieval simulations were performed using the channel sets of the AIRS/AMSUA/Humidity Sounder for Brazil (HSB) sounding suite [4] currently flying on the NASA Aqua satellites and for two notional hyperspectral microwave systems. The notional Hyperspectral Microwave Array Spectrometer (HyMAS) systems include channels near 60 GHz (HyMAS$_{60}$) and near 118 GHz (HyMAS$_{118}$). Both HyMAS systems include 16 water vapor channels near 183 GHz. The HyMAS$_{118}$ and GeoMAS systems contain identical channels (see Section 10.6.1 and Table 10.1). The HyMAS$_{60}$ system contains 64 channels near oxygen lines in the 60-GHz region (see Table 10.4). Each receiver bank has offset in frequency by 240 MHz. No attempts were made to optimize the channelization of either of the HyMAS systems. As with the GeoMAS simulations, system temperatures at 60, 118, and 183 GHz were assumed to be 500, 650, and 800 K, respectively. We have assumed an integration time of 165 ms, which closely approximates that of the AMSU-A sensor. (The AMSU-A1 integration time is ~165 ms and the AMSU-A2 integration time is ~160 ms). Channel properties of the AIRS (2378 channels), AMSU-A (15 channels), and HSB (four channels) sensors were obtained from the Aqua Level 1B Version 5 channel list files. The AMSUA/HSB ΔT_{RMS} values have been improved to reflect the system temperatures mentioned above. Actual AMSU/HSB ΔT_{RMS} values are about twice those assumed here, where we have incorporated the recent improvements reported in Ref. [9].

TABLE 10.4

HyMAS$_{60}$ Center Frequencies, Channel Bandwidths, and ΔT_{rms} for a Single Receiver Bank

Ch.	Center (GHz)	Bandwidth (MHz)	ΔT_{rms} (K)
1	48.1	2220	0.03
2	50.0	1570	0.03
3	51.4	1140	0.04
4	52.3	840	0.04
5	53.0	620	0.05
6	53.6	470	0.06
7	54.0	360	0.06
8	54.4	280	0.07
9	89.0	1000	0.05

10.10.1 SIMULATION METHODOLOGY

The methods and assumptions used to carry out the retrieval performance simulations presented in this section are very similar to those used earlier in this chapter. Notable differences include the profile database and the simulation and retrieval methodology used for the combined microwave and infrared observations.

10.10.1.1 AIRS/AMSUA/HSB Level 2 Global Profile Database

Global atmospheric profile data derived from AIRS/AMSUA/HSB observations dating back to August 30, 2002 are available from the NASA Data and Information Services Center (http://disc.sci. gsfc.nasa.gov/AIRS/data-holdings). The "Version 5 Support Product'" (AIRX2SUP) was used as the ground truth in the performance simulation analyses presented in this section. This product includes temperature, moisture, and cloud liquid water profiles reported on 100 atmospheric levels with a 45-km horizontal spatial resolution at nadir. This product also includes the retrieved cloud-top pressures and cloud fractions for two cloud layers at each of the nine AIRS fields of view that comprise the AIRS/AMSUA/HSB field of regard. All of the available profile and cloud variables available in the AIRX2SUP product were used in the radiative transfer simulations. Furthermore, the AIRS Level 2 Version 5 quality flag "Pgood" was required to equal the surface pressure for the profile to be included in the data set. The resulting distribution of cloud fractions in the data set was approximately uniform from 0% to 90%, with a spike near zero with a relative frequency of ~10%.

AIRS Level 2 profiles and cloud products uniformly distributed from December 2004 to January 2006 were used to simulate a global training database of cloudy AIRS/AMSUA/HSB observations over ocean at nadir incidence. A separate validation database was constructed using profiles from seven AIRS focus days from September 6, 2002 to December 5, 2003. A 3×3 "golfball" with nine AIRS/HSB footprints and one AMSU-A footprint was simulated for each of the 80,000 profiles in the simulation training and validation data set (40,000 profiles were included in each set). The Stand-Alone Rapid Transmittance Algorithm [45] was used to simulate the AIRS observations and the TBARRAY algorithm [17] was used to simulate the AMSU-A, HSB, and HyMAS observations. This data set was also used to simulate an ensemble of 45-km $HyMAS_{60}$ and $HyMAS_{118}$ footprints. The retrieval methodology used for the HyMAS observations is identical to that presented earlier in the chapter for GeoMAS.

10.10.1.2 AIRS/AMSUA/HSB Cloud Clearing and Retrieval Methodology

The cloudy AIRS observations were first "cloud cleared" using the microwave observations together with the AIRS 3×3 field of regard to estimate the radiances that would have been observed by AIRS if the scene were cloud-free. The Stochastic Cloud Clearing (SCC) algorithm [46] was used in this work. The SCC algorithm also produces a degree-of-cloudiness estimate that was used to identify "mostly clear" scenes (approximately the clearest 30%) and "mostly cloudy" scenes (approximately the cloudiest 20%). The microwave and cloud-cleared infrared radiances were then used as inputs to the PPC/NN algorithm [28], and the temperature and moisture profiles were estimated. This methodology has been extensively validated with actual AIRS/AMSU observations [32,47] using a variety of performance metrics.

10.10.2 SIMULATED PERFORMANCE COMPARISONS OF HYMAS SYSTEMS WITH AIRS/AMSUA/HSB

The temperature and water vapor retrieval performance of the $HyMAS_{60}$, $HyMAS_{118}$, AIRS/AMSUA/HSB, and AMSUA/HSB-only systems are shown in Figures 10.17 and 10.18, respectively. We note that the AIRS/AMSUA/HSB results presented here agree well with those presented by other investigators [48,49]. The same set of 40,000 profiles (from the 7-day 2002–2003 validation set) was used to derive the performance of each of the four systems. It is interesting to note

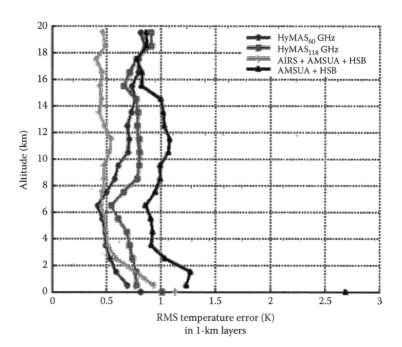

FIGURE 10.17 Comparison of HyMAS$_{60}$ (60/183 GHz), HyMAS$_{118}$ (118/183 GHz), AIRS/AMSUA/HSB, and AMSUA/HSB-only temperature retrieval performance over ocean at nadir incidence for seven focus days from 2002 to 2003. AMSUA/HSB noise has been reduced from on-orbit values for consistency with the T_{sys} values used throughout this chapter.

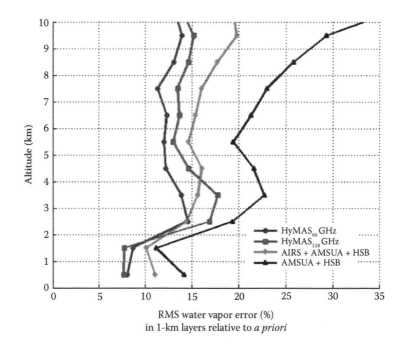

FIGURE 10.18 Comparison of HyMAS$_{60}$ (60/183 GHz), HyMAS$_{118}$ (118/183 GHz), AIRS/AMSUA/HSB, and AMSUA/HSB-only water vapor retrieval performance over ocean at nadir incidence for seven focus days from 2002 to 2003. AMSUA/HSB noise has been reduced from on-orbit values for consistency with the T_{sys} values used throughout this chapter.

that the performance of HyMAS$_{60}$ in the mid and lower troposphere is very similar to the AIRS/AMSUA/HSB system. HyMAS$_{118}$ water vapor sounding performance is also very competitive with AIRS/AMSUA/HSB. HyMAS improvements could be obtained by optimizing the channel sets (center frequencies and bandwidths) and/or by increasing the total number of temperature sounding receiver banks.

10.10.3 SIMULATED PERFORMANCE COMPARISONS OF HYMAS$_{60}$ WITH AIRS/AMSUA/HSB IN MOSTLY CLEAR AND MOSTLY CLOUDY CONDITIONS

We now examine performance in mostly clear scenes (favorable to infrared sounding) and in mostly cloudy scenes (favorable to microwave sounding). The temperature and water vapor retrieval performance of the HyMAS$_{60}$ and AIRS/AMSUA/HSB systems are shown in Figures 10.19 and 10.20, respectively. The same set of "mostly clear" or "mostly cloudy" profiles (from the 7-day 2002–2003 validation set) was used to derive the performance of each system. The mid-tropospheric temperature sounding performance of AIRS/AMSUA/HSB is slightly superior to HyMAS$_{60}$ in mostly clear scenes. HyMAS$_{60}$ shows a slight advantage in boundary layer temperature retrieval in mostly clear and mostly cloudy conditions and a pronounced advantage (0.4-K difference in RMS) for surface skin temperature retrieval in mostly cloudy conditions. HyMAS$_{60}$ water vapor sounding performance exceeds AIRS/AMSUA/HSB for both mostly clear and mostly cloudy scenes.

10.10.4 DISCUSSION OF OFF-NADIR PERFORMANCE

The results shown previously have been only for observations at nadir incidence. Additional worst-case analyses were performed at a sensor scan angle of 48°, corresponding to the edge-of-scan for AMSU-A. The scan-angle dependence of the observations is easily accommodated by the NN

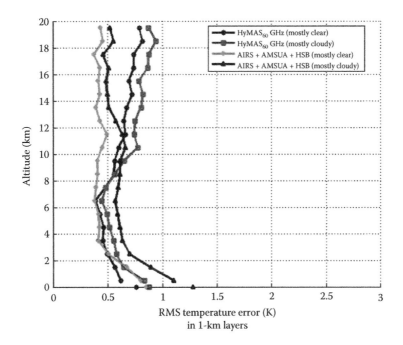

FIGURE 10.19 Comparison of HyMAS$_{60}$ (60/183 GHz) and AIRS/AMSUA/HSB temperature retrieval performance over ocean at nadir incidence for mostly clear and mostly cloudy conditions for seven focus days from 2002 to 2003. AMSUA/HSB noise has been reduced from on-orbit values for consistency with the T_{sys} values used throughout this chapter.

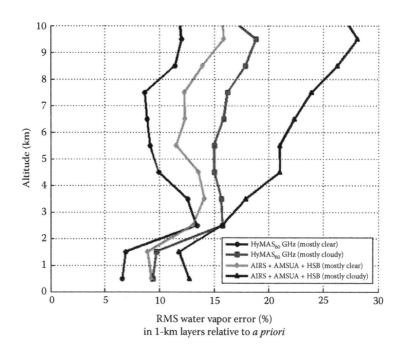

FIGURE 10.20 Comparison of $HyMAS_{60}$ (60/183 GHz) and AIRS/AMSUA/HSB water vapor retrieval performance over ocean at nadir incidence for mostly clear and mostly cloudy conditions for seven focus days from 2002 to 2003. AMSUA/HSB noise has been reduced from on-orbit values for consistency with the T_{sys} values used throughout this chapter.

retrieval [28]. The temperature profile RMS error degradation relative to the nadir-viewing configuration was no worse than 0.1 K throughout the troposphere for all cases shown in Figure 10.17 and the water vapor profile RMS error degradation relative to the nadir-viewing configuration was no worse than 1% (i.e., RMS_{nadir} in percent minus $RMS_{48°}$ in percent) throughout the troposphere for all cases shown in Figure 10.18. Skin surface temperature retrieval RMS error degradation for $HyMAS_{118}$ at 48° incidence degraded by 0.41 K relative to the nadir-viewing configuration. Skin surface temperature RMS error degradation for the AIRS + AMSUA + HSB and AMSUA + HSB cases was 0.16 and 0.14 K, respectively. The skin surface temperature RMS error degradation for $HyMAS_{60}$ was less than the estimated uncertainty of the retrieval simulation analyses, ~0.05 K.

10.11 SUMMARY, CONCLUSIONS, AND FUTURE DIRECTIONS

We have presented a new modality of atmospheric sounding using hyperspectral observations at microwave and millimeter-wave frequencies. The hyperspectral channel composition is achieved by multiplexing a number of independent spectrometers to build up a large number of densely spaced weighting functions. The performance of such a system, GeoMAS, was demonstrated using channels near 118 and 183 GHz, which are well suited for geostationary implementation due to the relatively small antenna reflector sizes that would be required for useful spatial resolution. Similar methodology could be adapted for low-earth-orbit sensors using frequencies near 60 GHz, and analysis presented in this chapter suggests that an 88-channel HyMAS system operating near 60, 89, and 183 GHz could rival the performance of state-of-the-art microwave + infrared sounding suites at a small fraction of the cost.

Hyperspectral microwave sounding and precipitation mapping performance could be further improved by spatial processing of Nyquist-sampled observations to sharpen the effective antenna beam. This sharpening amplifies sensor noise, but this increased noise could be offset by adding receiver banks. A promising area of current study is the investigation of the tradeoff between effective spatial resolution, receiver array complexity, and retrieval performance. The antenna reflector diameter required to meet a 50-km/25-km temperature/water-vapor resolution goal from geostationary orbit is ~2.7 m for a filled-aperture system. However, the results presented in this chapter together with other work on antenna beam sharpening [39] indicate that such requirements could be met with a GeoMAS system with an antenna reflector diameter <2 m if image sharpening is used during ground processing. Furthermore, the resolution versus noise trade can be dynamically optimized, meaning that the level of sharpening can be selected "on the fly" based on the atmospheric scene being viewed. Scenes with high spatial frequency content could be detected and sharpening could be used. The selection of the degree of sharpening could be made during ground processing and could be tailored by each user.

While the results presented in this chapter are encouraging and indicate great promise for future hyperspectral microwave sounding systems, additional analyses and concept demonstration are clearly necessary to study the advantages and challenges of this new sensing modality. Further studies of the error sensitivities of hyperspectral microwave systems are recommended, for example, and deployment of airborne prototype hyperspectral microwave sensors would provide insight into many of the noise and meteorological phenomena that are difficult to accurately model and simulate.

Another interesting area of further research is the potential synergy resulting from the combined use of hyperspectral microwave and hyperspectral infrared systems. For example, a temperature profile retrieval of high accuracy provided by a hyperspectral microwave system could be used to substantially improve the estimates of atmospheric carbon dioxide using an infrared sounder, as the relationship between temperature and carbon dioxide could be decoupled using microwave and infrared measurements.

ACKNOWLEDGMENTS

This work was sponsored by the National Oceanic and Atmospheric Administration under Air Force contract FA8721-05-C-0002. Opinions, interpretations, conclusions, and recommendations are those of the author and are not necessarily endorsed by the United States Government.

REFERENCES

1. D. H. Staelin, A. H. Barrett, J. W. Waters, F. T. Barath, E. J. Johnston, P. W. Rosenkranz, N. E. Gaut, and W. B. Lenoir, Microwave spectrometer on the Nimbus 5 satellite: Meteorological and geophysical data, *Science*, 182, 1339–1341, 1973.
2. W. L. Smith, Satellite techniques for observing the temperature structure of the atmosphere, *Bull. Am. Meteorol. Soc.*, 53(11), 1074–1082, 1972.
3. W. L. Smith, Atmospheric soundings from satellites—False expectation or the key to improved weather prediction? *Quarterly J. Royal Meteorol. Soc.*, 117(498), 267–297, 1991.
4. H. H. Aumann, M. T. Chahine, C. Gautier, M. D. Goldberg, E. Kalnay, L. M. McMillin, H. Revercomb et al., AIRS/AMSU/HSB on the Aqua mission: Design, science objectives, data products, and processing systems, *IEEE Trans. Geosci. Remote Sens.*, 41(2), 253–264, 2003.
5. G. Chalon, F. Cayla, and D. Diebel, IASI: An advanced sounder for operational meteorology, *Proceedings of the 52nd Congress of IAF*, October 2001, pp. 1–5.
6. W. L. Smith, H. Revercomb, G. Bingham, A. Larar, H. Huang, D. Zhou, J. Li, X. Liu, and S. Kireev, Evolution, current capabilities, and future advance in satellite nadir viewing ultra-spectral IR sounding of the lower atmosphere, *Atmos. Chem. Phys.*, 9, 5563–5574, 2009.
7. B. Razavi, A 60-GHz CMOS receiver front-end, *IEEE J. Solid-State Circuits*, 41(1), 17–22, 2006.
8. J. Powell, H. Kim, and C. Sodini, SiGe receiver front ends for millimeter-wave passive imaging, *IEEE Trans. Microwave Theory Tech.*, 56(11), 2416–2425, 2008.
9. P. Kangaslahti, D. Pukala, T. Gaier, W. Deal, X. Mei, and R. Lai, Low noise amplifier for 180 GHz frequency band, in *Microwave Symposium Digest, 2008 IEEE MTT-S International*, June 2008, pp. 451–454.

10. W. J. Blackwell and D. H. Staelin, Comparative performance analyses of passive microwave systems for tropospheric sounding of temperature and water vapor profiles, *GOES-8 and Beyond, SPIE*, 2812, 472–478, 1996.

11. D. Staelin, J. Kerekes, and F. J. Solman, Final report of the geosynchronous microwave sounder working group, MIT Lincoln Laboratory, Lexington, MA, August 22, 1997, prepared for NOAA/NESDIS GOES Program Office.

12. A. Lipton, Satellite sounding channel optimization in the microwave spectrum, *IEEE Trans. Geosci. Remote Sens.*, 41(4), 761–781, 2003.

13. C. Muth, P. Lee, J. Shiue, and W. Allan Webb, Advanced technology microwave sounder on NPOESS and NPP, *IEEE IGARSS Proc.*, 4, 2454–2458, 2004.

14. C. Surussavadee and D. H. Staelin, Comparison of AMSU millimeterwave satellite observations, MM5/TBSCAT predicted radiances, and electromagnetic models for hydrometeors, *IEEE Trans. Geosci. Remote Sens.*, 44(10), 2667–2678, 2006.

15. C. Surussavadee and D. H. Staelin, Millimeter-wave precipitation retrievals and observed-versus simulated radiance distributions: Sensitivity to assumptions, *J. Atmos. Sci.*, 64(11), 3808–3826, 2007.

16. F. J. Wentz, A well-calibrated ocean algorithm for SSM/I, *J. Geophys. Res.*, 102, 8703–8718, 1997.

17. P. W. Rosenkranz, Absorption of microwaves by atmospheric gases, in *Atmospheric Remote Sensing by Microwave Radiometry*, M. A. Janssen, Ed. New York: Wiley, 1993, ch. 2.

18. H. J. Liebe, MPM: An atmospheric millimeter-wave propagation model, *Int. J. Infrared Millimeter Waves*, 10(6), 631–650, 1989.

19. H. J. Liebe, P. W. Rosenkranz, and G. A. Hufford, Atmospheric 60-GHz oxygen spectrum: New laboratory measurements and line parameters, *J. Quant. Spectrosc. Ra.*, 48, 629–643, 1992.

20. P. W. Rosenkranz, Radiative transfer solution using initial values in a scattering and absorbing atmosphere with surface reflection, *IEEE Trans. Geosci. Remote Sens.*, 40(8), 1889–1892, 2002.

21. F. T. Ulaby, R. K. Moore, and A. K. Fung, *Microwave Remote Sensing Active and Passive: Volume I Microwave Remote Sensing Fundamentals and Radiometry*. Norwood, MA: Artech House Inc., 1981.

22. S. English and T. Hewison, A fast generic millimeter-wave emissivity model, *SPIE Proc.*, 3503, 288–300, 1998.

23. G. W. Petty and K. B. Katsaros, The response of the SSM/I to the marine environment. Part II: A parameterization of the effect of the sea surface slope distribution on emission and reflection, *J. Atmos. Oceanic Technol.*, 11, 617–628, 1994.

24. F. Karbou, C. Prigent, L. Eymard, and J. R. Pardo, Microwave land emissivity calculations using AMSU measurements, *IEEE Trans. Geosci. Remote Sens.*, 43, 948–959, 2005.

25. A. J. Gasiewski, Numerical sensitivity analysis of passive EHF and SMMW channels to tropospheric water vapor, clouds, and precipitation, *IEEE Trans. Geosci. Remote Sens.*, 30(5), 859–869, 1992.

26. C. Prigent, J. R. Pardo, and W. B. Rossow, Comparisons of the millimeter and submillimeter bands for atmospheric temperature and water vapor soundings for clear and cloudy skies, *J. Appl. Meteorol.*, 45, 1622–1633, 2006.

27. D. H. Staelin and F. W. Chen, Precipitation observations near 54 and 183 GHz using the NOAA-15 satellite, *IEEE Trans. Geosci. Remote Sens.*, 38(5), 2322–2332, 2000.

28. W. J. Blackwell, A neural-network technique for the retrieval of atmospheric temperature and moisture profiles from high spectral resolution sounding data, *IEEE Trans. Geosci. Remote Sens.*, 43(11), 2535–2546, 2005.

29. W. J. Blackwell, Neural network retrievals of atmospheric temperature and moisture profiles from high-resolution infrared and microwave sounding data, in *Signal and Image Processing for Remote Sensing*, C. C. Chen, Ed. Boca Raton, FL: Taylor & Francis, 2006, ch. 11.

30. W. J. Blackwell and F. W. Chen, *Neural Networks in Atmospheric Remote Sensing*, Boston, MA: Artech House, 2009.

31. K. M. Hornik, M. Stinchcombe, and H. White, Multilayer feedforward networks are universal approximators, *Neural Netw.*, 4(5), 359–366, 1989.

32. W. J. Blackwell, M. Pieper, and L. G. Jairam, Neural network estimation of atmospheric profiles using AIRS/IASI/AMSU in the presence of clouds, *SPIE Asia-Pacific Remote Sensing Symposium*, November 2008.

33. D. Nguyen and B. Widrow, Improving the learning speed of two-layer neural networks by choosing initial values of the adaptive weights, *IJCNN*, 3, 21–26, 1990.

34. D. Marquardt, An algorithm for least-squares estimation of nonlinear parameters, *J. Soc. Indus. Appl. Math.*, 11(2), 431–441, 1963.

35. M. T. Hagan and M. B. Menhaj, Training feedforward networks with the Marquardt algorithm, *IEEE Trans. Neural Netw.*, 5, 989–993, 1994.
36. I. T. Nabney, *NETLAB: Algorithms for Pattern Recognition*, London: Springer, 2002.
37. A. Tanner, W. Wilson, B. Lambrigsten, S. Dinardo, S. Brown, P. Kangaslahti, T. Gaier, C. Ruf, S. Gross, B. Lim, S. Musko, S. Rogacki, and J. Piepmeier, Initial results of the geostationary synthetic thinned array radiometer (GeoSTAR) demonstrator instrument, *IEEE Trans. Geosci. Remote Sens.*, 45(7), 1947–1957, 2007.
38. B. Lambrigtsen, S. Brown, T. Gaier, P. Kangaslahti, and A. Tanner, A baseline for the decadal-survey path mission, *IEEE IGARSS Proceedings*, vol. 3, July 2008, pp. 338–341.
39. D. H. Staelin and C. Surussavadee, Precipitation retrieval accuracies for geo-microwave sounders, *IEEE Trans. Geosci. Remote Sens.*, 45(10), 3150–3159, 2007.
40. W. J. Blackwell, J. W. Barrett, F. W. Chen, R. V. Leslie, P. W. Rosenkranz, M. J. Schwartz, and D. H. Staelin, NPOESS aircraft sounder testbed-microwave (NAST-M): Instrument description and initial flight results, *IEEE Trans. Geosci. Remote Sens.*, 39(11), 2444–2453, 2001.
41. R. V. Leslie and D. H. Staelin, NPOESS aircraft sounder testbed microwave: Observations of clouds and precipitation at 54, 118, 183, and 425 GHz, *IEEE Trans. Geosci. Remote Sens.*, 42(10), 2240–2247, 2004.
42. F. W. Chen and D. H. Staelin, AIRS/AMSU/HSB precipitation estimates, *IEEE Trans. Geosci. Remote Sens.*, 41(2), 410–417, 2003.
43. C. Surussavadee and D. H. Staelin, Global millimeter-wave precipitation retrievals trained with a cloud-resolving numerical weather prediction model, Part I: Retrieval design, *IEEE Trans. Geosci. Remote Sens.*, 46(1), 99–108, 2008.
44. C. Surussavadee and D. H. Staelin, Global millimeter-wave precipitation retrievals trained with a cloud-resolving numerical weather prediction model, Part II: Performance evaluation, *IEEE Trans. Geosci. Remote Sens.*, 46(1), 109–118, 2008.
45. L. Strow, S. Hannon, and S. Desouza-Machado, An overview of the AIRS radiative transfer model, *IEEE Trans. Geosci. Remote Sens.*, 41(2), 303–313, 2003.
46. C. Cho and D. H. Staelin, AIRS observations versus numerical weather predictions of cloud-cleared radiances, *J. Geophys. Res.*, 111, 1–10, 2006.
47. H. H. Aumann, E. Manning, C. Barnet, E. Maddy, and W. Blackwell, An anomaly correlation skill score for the evaluation of the performance of hyperspectral infrared sounders, *Atmospheric and Environmental Remote Sensing Data Processing and Utilization V: Readiness for GEOSS III. Proceedings of the SPIE*, 7456, 74 560T–74 560T–7, 2009.
48. J. Susskind, C. D. Barnet, and J. M. Blaisdell, Retrieval of atmospheric and surface parameters from AIRS/AMSU/HSB data in the presence of clouds, *IEEE Trans. Geosci. Remote Sens.*, 41(2), 390–409, 2003.
49. L. Shi, Retrieval of atmospheric temperature profiles from AMSUA measurement using a neural network approach, *J. Atmos. Oceanic Technol.*, 18, 340–347, 2001.

11 Satellite Passive Millimeter-Wave Retrieval of Global Precipitation

Chinnawat "Pop" Surussavadee and David H. Staelin

CONTENTS

11.1 INTRODUCTION

Accurate observation of global precipitation is important for water resource management, agriculture, hydrology, natural disaster monitoring and warning, weather forecasting, and climate study. Despite its importance, it is a difficult task since each common observational method has its own deficiencies. For example, global rain gauges are sparse, particularly over ocean and underpopulated areas, and can also be affected by wind. Remote sensors receive signals from hydrometeors aloft that may evaporate or change location before reaching the ground. Global radar coverage is sparse. Only satellites can provide global coverage and can repeat frequently. Satellite sensors observing at visible and infrared frequencies cannot penetrate clouds, which makes their precipitation estimates less accurate. Satellite passive millimeter-wave sensors can penetrate through clouds, but the development of accurate precipitation retrieval methods is handicapped by the lack of reliable global ground truth coincident with satellite observations and by the indirect physical relation between millimeter-wave observations and surface precipitation rates.

This chapter discusses the development of a global precipitation retrieval algorithm for the Advanced Microwave Sounding Unit (AMSU). The algorithm is called the AMSU MIT Precipitation (AMP) retrieval algorithm [1–8] and was developed by Surussavadee and Staelin. AMP is a neural network-based algorithm that retrieves global surface precipitation rates for both rain and snowfall (mm h^{-1}); water paths (mm) for rain, snow, graupel, cloud liquid water, cloud ice, and the sum of rain, snow, and graupel; and peak vertical wind (convective strength, m s^{-1}). AMP can be used with AMSU-A/AMSU-B aboard the U.S. National Oceanic and Atmospheric Administration (NOAA satellites), and AMSU-A/Microwave Humidity Sounder (MHS) aboard NOAA and Meteorological Operational Satellite Programme (MetOp) satellites. It can also be adapted to other passive millimeter-wave radiometers observing near oxygen and water vapor absorption bands, as discussed in Section 11.2. AMP was the first algorithm to successfully retrieve precipitation over snow-covered land and sea ice [5,6]. To compensate for the lack of reliable global ground truth, the development of AMP began with the development and validation of a global reference physical model [1,2], which is used for training neural networks (NNs). Principal component analysis (PCA) is used extensively during AMP preprocessing to filter out noise and surface signals that could cause false alarms [3–8].

This chapter is organized as follows. Section 11.2 describes the physical basis of passive millimeter-wave precipitation retrieval and Section 11.3 describes AMSU. Section 11.4 discusses the physical relationship between precipitation rate and millimeter-wave radiance spectra. Section 11.5 describes signal processing methods used in AMP and Section 11.6 deals with AMP in more detail, including its development and comparisons of its precipitation estimates with CloudSat radar echoes [9,10], globally distributed NOAA rain gauges [11], and Global Precipitation Climatology Project (GPCP) [12] estimates. Section 11.7 summarizes and concludes the chapter.

11.2 PHYSICAL BASIS OF PASSIVE MILLIMETER-WAVE PRECIPITATION RETRIEVAL

11.2.1 THERMAL RADIATION

The main source of the earth's energy is the sun. As solar radiation propagates through the atmosphere, part of its energy is absorbed or scattered by atmospheric constituents. The remainder reaches the earth's surface where some is absorbed and the rest is reflected back toward the atmosphere. When the atmosphere and terrestrial surface absorb electromagnetic energy their physical temperatures tend to increase. At the same time, both the atmosphere and surface radiate energy by means of thermal emission. The amount of thermal radiation is a function of wavelength and physical temperature. Planck's radiation law characterizes the thermal radiation of a blackbody, which is any material that absorbs all incident radiation without any reflection. It is hence a perfect absorber and also a perfect emitter. Such blackbody spectral radiation has intensity

$$I_{bb}(f,T) = \frac{2hf^3}{c^2}\left(\frac{1}{e^{hf/kT} - 1}\right)(\text{W m}^{-2}\,\text{sr}^{-1}\,\text{Hz}^{-1}), \qquad (11.1)$$

where h is Planck's constant (6.63×10^{-34} (J s)), f is frequency (Hz), k is Boltzmann's constant (1.38×10^{-23} (J K^{-1})), T is physical temperature (K), and c is the velocity of light ($\sim 3 \times 10^8$ (m s^{-1})).

In the Rayleigh–Jeans limit, where $hf/kT \ll 1$ for millimeter wavelengths observing the earth, the Taylor series expansion can be applied to Equation 11.1, which yields

$$I_{bb}(f,T) = \frac{2kT}{\lambda^2}(W\,m^{-2}\,sr^{-1}\,Hz^{-1}), \tag{11.2}$$

where λ is the wavelength (m). It is important to note that radiation intensity is directly proportional to physical temperature.

Real materials such as soil, water, and trees absorb only a fraction ε of the incident energy and the remainder gets reflected; such materials are called gray bodies. The same fraction ε of the Planck radiation (11.2) is therefore emitted. Hence, real materials emit less radiation than what a blackbody does at the same physical temperature. The radiation intensity emitted by a gray body at temperature T is

$$I(f,T) = \varepsilon\frac{2kT}{\lambda^2}(W\,m^{-2}\,sr^{-1}\,Hz^{-1}), \tag{11.3}$$

where ε is the emissivity of the gray body and $0 \le \varepsilon \le 1$.

For a homogeneous material with a uniform physical temperature of T (K), the brightness temperature based on Equation 11.2 is defined as

$$T_B = \frac{\lambda^2}{2k}I(f,T)(K). \tag{11.4}$$

Hence,

$$\varepsilon = \frac{T_B}{T}. \tag{11.5}$$

Since the fractions of radiant energy absorbed or reflected by a gray surface are ε and R, respectively, it follows that for a ray incident at any particular angle and polarization, the reflectivity $R = 1 - \varepsilon$.

11.2.2 Millimeter-Wave Interaction with Atmospheric Constituents

Electromagnetic waves propagating through the atmosphere are absorbed or scattered by two groups of atmospheric constituents, hydrometeors and atmospheric gases.

11.2.2.1 Millimeter-Wave Interaction with Hydrometeors

Hydrometeors are any condensed forms of atmospheric water vapor such as rain water, cloud liquid water, cloud ice, snow, and hail/graupel. When an electromagnetic wave with the power density S_i (W m^{-2}) is incident on a particle with cross-sectional area A (m^2), some incident power will be absorbed and the rest will be scattered by the particle. The absorption cross section Q_a (m^2) is defined as the ratio of absorbed power P_a to incident power density, S_i:

$$Q_a = \frac{P_a}{S_i}(m^2). \tag{11.6}$$

The scattering cross section Q_s is defined as

$$Q_s = \frac{P_s}{S_i}(m^2), \tag{11.7}$$

where P_s is scattered power.

The absorption and scattering efficiency factors are defined as $\xi_a = (Q_a/A)$ and $\xi_s = (Q_s/A)$, respectively. The extinction cross section, $Q_e = Q_a + Q_s$, and the extinction efficiency, $\xi_e = \xi_a + \xi_s$, characterize the total effect of absorption and scattering.

In a given volume, several hydrometeor species can coexist. If they are randomly located such that their scattered fields are incoherent, the scattering coefficient for the volume is the sum of the scattering cross sections contributed by all particles in the volume. For example, for a volume having single spherical hydrometeor species with the drop-size probability distribution $p(r)$, where the particles' radii range from r_1 to r_2 (m), the scattering coefficient for that volume is

$$\alpha_s = \int_{r_1}^{r_2} p(r)Q_s(r)\,dr\,(\mathrm{Np\,m^{-1}}), \qquad (11.8)$$

where Np signifies nepers, which correspond to one optical depth (1/e). Similarly, the absorption and extinction coefficients are

$$\alpha_a = \int_{r_1}^{r_2} p(r)Q_a(r)\,dr\,(\mathrm{Np\,m^{-1}}) \qquad (11.9)$$

and

$$\alpha_e = \int_{r_1}^{r_2} p(r)Q_e(r)\,dr\,(\mathrm{Np\,m^{-1}}), \qquad (11.10)$$

respectively.

The absorption and scattering depend on frequency, polarization, and hydrometeor mass, size, dielectric permittivity, shape, and orientation. When the wavelength of the incident wave is much longer than the particle, the Rayleigh approximation for scattering can be used, and otherwise the absorption and scattering cross sections of a dielectric sphere with radius r can be computed using Mie's theory [13] with an iterative method [14]. The absorption and scattering cross sections of arbitrary shapes can be approximated by the discrete-dipole approximation method [15].

Dielectric permittivities for water [16] and ice [17] are functions of temperature and frequency. For heterogeneous materials composed of ice and air, for example, snow and grauple, a mixing rule [18,19] can be used to compute the effective permittivity with the assumption that the scale of the inhomogeneity is much smaller than the wavelength. The effective permittivity is a function of the fractional volume of ice in the air matrix, called the ice factor, $F(\lambda)$.

11.2.2.2 Millimeter-Wave Interaction with Atmospheric Gases

At millimeter wavelengths, oxygen and water vapor are the only atmospheric gases that have significant absorption bands below the stratopause. The atmospheric optical depth (Np) at a zenith angle θ between the altitudes z_1 and z_2 is defined as

$$\tau_\theta(z_1, z_2) = \int_{z_1}^{z_2} \alpha_e(z)\sec\theta\,dz. \qquad (11.11)$$

Figure 11.1 shows the zenith opacity, $\tau_\theta = \int_0^\infty \alpha_e(z)\sec\theta\,dz$ for clear skies viewed from the ground at millimeter wavelengths observed by satellite-borne radiometers such as (1) AMSU-A/

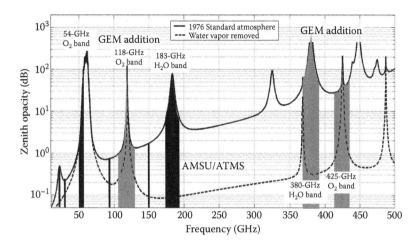

FIGURE 11.1   Zenith opacity in the clear sky condition viewed from the ground at millimeter wavelengths and spectral coverage of millimeter-wave radiometers.

AMSU-B aboard NOAA satellites, (2) AMSU/MHS aboard NOAA and MetOp satellites, (3) AMSU/ Humidity Sounder for Brazil (HSB) aboard the U.S. National Aeronautics and Space Administration (NASA) Aqua satellite, (4) the Special Sensor Microwave Imager/Sounder (SSMIS) aboard the U.S. Air Force Defense Meteorological Satellite Program (DMSP) satellites, (5) the U.S. National Polar-orbiting Operational Environmental Satellite System (NPOESS) Advanced Technology Microwave Sounder (ATMS), which is scheduled to fly on the NPOESS Preparatory Program (NPP) satellite and the subsequent Joint Polar Satellite System (JPSS) platforms, and (6) geostationary microwave (GEM) sounders proposed for geostationary satellites [20].

The solid line in the figure was computed using the 1976 standard atmosphere. Frequencies with high zenith opacity (>3 dB) are often called opaque frequencies, and those with low zenith opacity (<3 dB) are often called window frequencies. Absorption peaks in Figure 11.1 are near the oxygen absorption bands, for example, 54, 118, and 425 GHz, and water vapor resonances, for example, 22, 183, and 380 GHz. The dashed line corresponds to zero water vapor, thus showing the effects of water vapor absorption.

11.2.3   RADIATIVE TRANSFER EQUATION

The propagation of electromagnetic radiation through the atmosphere can be described mathematically by the radiative transfer equation. In a nonscattering atmosphere that only absorbs, the brightness temperature observed from space has four components, as illustrated in Figure 11.2 and given by Equation 11.12. For the case where a satellite observes from an altitude L above the surface, the cosmic background temperature is T_c (K), the atmospheric temperature profile is $T(z)$, the atmospheric absorption coefficient profile is $\alpha_a(z)$ (Np·m^{-1}), the earth's surface reflectivity is $R(\theta)$ and specular (the angle of incidence equals the angle of reflection), and the surface temperature is T_s (K), then the satellite-observed brightness temperature from zenith angle θ is

$$T_B = T_1 + T_2 + T_3 + T_4 \text{ (K)}, \tag{11.12}$$

where

$$T_1 = R(\theta)T_c e^{-2\tau(0,L)\sec\theta} \text{ (K)},$$

$$T_2 = R(\theta)\sec\theta \cdot e^{-\tau(0,L)\sec\theta} \int_0^L T(z)\alpha_a(z)e^{-\tau(0,z)\sec\theta} \,dz \text{ (K)},$$

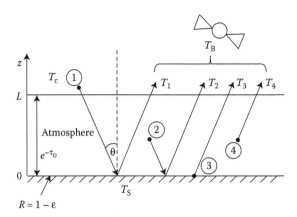

FIGURE 11.2 Geometry for contributions of satellite-observed brightness temperature.

$$T_3 = \left[1 - R(\theta)\right]T_s e^{-\tau(0,L)\sec\theta} = \varepsilon(\theta)T_s e^{-\tau(0,L)\sec\theta} \,(\text{K}),$$

$$T_4 = \sec\theta \int_0^L T(z)\alpha_a(z)e^{-\tau(z,L)\sec\theta}\,dz\,(\text{K}).$$

T_1 is the brightness temperature contributed from the cosmic background temperature, T_c, that propagates downward, reflects from the surface, and then propagates upward to the satellite. The factor $e^{-2\tau(0,L)\sec\theta}$ is due to the atmospheric attenuation both ways. T_2 is contributed by the atmospheric radiation that is emitted downward from all altitudes, $\int_0^L T(z)\,\alpha_a(z)e^{-\tau(0,z)\sec\theta}\,dz$, and then is reflected from the surface and propagates upward to the top of the atmosphere. T_3 is contributed by surface emission, which is equal to $\varepsilon(\theta)T_s$, which is attenuated as it propagates upward to the top of the atmosphere. T_4 is contributed by the atmospheric radiation at all altitudes that is emitted and propagates upward to the top of the atmosphere.

Measurements in the 50–60 GHz oxygen absorption band are used to derive temperature profiles in clear and cloudy atmospheres. Measurements near the 22-GHz water vapor line are used to obtain column water vapor and liquid water abundance, mostly over ocean. Nonresonant absorption by liquid hydrometeors is roughly proportional to $\sim f^2$ and can therefore be distinguished from resonant water vapor absorption. Measurements near the 183-GHz water vapor line are used to obtain humidity profiles. The effect of temperature and water vapor profiles on observed brightness temperatures can be illustrated by letting surface reflectivity $R = 0$ for simplicity. Equation 11.12 reduces to:

$$T_B = T_3 + T_4 = \varepsilon(\theta)\cdot T_s e^{-\tau(0,L)\sec\theta} + \sec\theta\int_0^L T(z)\,\alpha_a(z)e^{-\tau(z,L)\sec\theta\cdot e}\,dz$$

$$= \varepsilon(\theta)\cdot T_s e^{-\tau(0,L)\sec\theta} + \int_0^L T(z)W(z)\,dz\,(\text{K}), \tag{11.13}$$

where $W(z)$ is called the weighting function (WF) and is equal to $\sec\theta\int_0^L \alpha_a(z)e^{-\tau(z,L)\sec\theta}\,dz$. Equation 11.13 often remains a good approximation even when $R > 0$, although $W(z)$ then includes the reflected contributions, and the cosmic background contributions are presumed to be negligible.

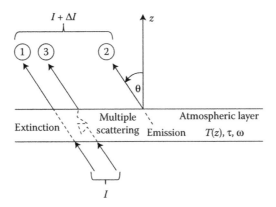

FIGURE 11.3 Radiation propagating through an atmospheric layer.

For general atmospheres, scattering has to be considered. Consider the radiation propagating through a layer of atmosphere, as shown in Figure 11.3. Its intensity, I, is changed by three processes, including (1) attenuation by extinction, (2) emission from the layer, and (3) multiple scattering, so that the radiative transfer equation can then be written as the sum of the three terms in order as

$$\frac{dI(z;\mu,\varphi)}{dz/\mu} = -\alpha_e I(z;\mu,\varphi) + \alpha_a I[T(z)]$$

$$+ \alpha_s \int_0^{2\pi} \int_{-1}^{1} I(z;\mu',\varphi') \frac{P(\mu,\varphi,\mu',\varphi')}{4\pi} d\mu' d\varphi', \qquad (11.14)$$

where μ equals $\cos\theta$, φ is the azimuth angle, $P(\mu, \varphi, \mu', \varphi')$ is the phase function describing the amount of intensity scattered from the direction (μ',φ') into the direction (μ, φ), $I[T(z)]$ is the radiation intensity from the atmospheric layer having physical temperature $T(z)$, and ω is the single-scattering albedo defined as $\omega = \alpha_s/\alpha_e$ or $1 - \omega = \alpha_a/\alpha_e$, where $\alpha_e = \alpha_a + \alpha_s$. Using the linear relationship between radiation intensity and brightness temperature from Equation 11.4, Equation 11.14 can be written as

$$-\mu \frac{dT_B(\tau;\mu,\varphi)}{d\tau} = -T_B(\tau;\mu,\varphi) + (1 - \omega)T(\tau)$$

$$+ \frac{\omega}{4\pi} \int_0^{2\pi} \int_{-1}^{1} T_B(\tau;\mu',\varphi')P(\mu,\varphi,\mu',\varphi')d\mu' d\varphi'. \qquad (11.15)$$

A two-stream variant of the radiative transfer model TBSCAT [21] is a component of the global reference physical model used in this chapter. TBSCAT assumes that the radiance is independent of the azimuthal coordinate and the phase function is scalar. It solves the radiative transfer Equation 11.15 by numerically integrating an ensemble of trial functions constructed such that the boundary conditions at the top of the atmosphere are satisfied. The boundary conditions at the surface are imposed after integration through the atmosphere.

11.3 DESCRIPTION OF AMSU

AMSU was first launched aboard NOAA-15 in May 1998. It is composed of two units, AMSU-A and AMSU-B, which have also flown on NOAA-16 and NOAA-17. MHS observes at the same

TABLE 11.1

AMSU Channel Frequencies and Weighting Function Peak Heights

Channel[a]	Center Frequencies (MHz)	WF (km)	Channel[a]	Center Frequencies[b] (MHz)	WF[c]
A1	23,800 ± 72.5	0	A11	57,290.344 ± 322.2 ± 48	24.5 km
A2	31,400 ± 50	0	A12	57,290.344 ± 322.2 ± 22	29.5 km
A3	50,300 ± 50	0	A13	57,290.344 ± 322.2 ± 10	34.5 km
A4	52,800 ± 105	0	A14	57,290.344 ± 322.2 ± 4.5	40.5 km
A5	53,596 ± 115	4	A15	89,000 ± 1000	0 km
A6	54,400 ± 105	8	B1 (H1)	89 ± 0.9 (89)	>70 mm
A7	54,940 ± 105	9.5	B2 (H2)	150 ± 0.9 (157)	41 mm
A8	55,500 ± 87.5	12.5	B3 (H3)	183.31 ± 1	0.75 mm
A9	57,290.344 ± 87.5	16.5	B4 (H4)	183.31 ± 3	3 mm
A10	57,290.344 ± 217	20.5	B5 (H5)	183.31 ± 7 (190.311)	9 mm

[a] A, B, and H stand for AMSU-A, AMSU-B, and MHS, respectively.

[b] MHS frequencies are in parenthesis.

[c] WF (mm) represents the water vapor burden at one optical depth based on the 1976 U.S. standard atmosphere at nadir over a nonreflective surface.

number of channels and approximately the same frequencies as AMSU-B and replaced it on NOAA-18, NOAA-19, and MetOp-A. Both the AMSU-A/AMSU-B and AMSU-A/MHS systems will be called AMSU in this chapter, unless stated otherwise.

AMSU-A observes with 50-km spatial resolution at nadir 12 channels near the 54-GHz oxygen absorption band, plus two channels near the 22.35-GHz water vapor resonance (23.8 and 31.4 GHz), and another near 89 GHz. AMSU-B observes five channels, mostly near the 183-GHz water vapor absorption band at 15-km spatial resolution at nadir. AMSU scans ~2200 km across the track with maximum scan angles of 48.33° and 48.95° from nadir for AMSU-A and AMSU-B (MHS), respectively. AMSU-A views 30 equally spaced angles and AMSU-B (MHS) views 90 angles. Table 11.1 shows AMSU channel characteristics, where abbreviations A, B, and H stand for AMSU-A, AMSU-B, and MHS, respectively. Frequencies in parenthesis are for MHS. The WF values listed in Table 11.1 correspond to the altitudes (km) where WFs peak for channels A1–A15 and to water vapor burden (mm) at approximately one optical depth for channels B1–B5 computed using the 1976 U.S. standard atmosphere at nadir over a nonreflective surface [6]. Figure 11.4 shows WFs for AMSU-A and AMSU-B.

Table 11.1 and Figure 11.4 suggest AMSU's temperature and humidity profile sounding capability. The temperature profile is sensed by AMSU-A channels observing thermal emission at different altitudes. A1–A4 and B1–B2 are window channels and are affected by the surface, whereas the rest are normally opaque channels. A9–A15 are mostly sensitive to altitudes above precipitation and hence are not useful for precipitation retrieval. AMSU-B channels are sensitive to middle and lower tropospheric humidity profiles and hydrometeors. Although B3–B5 are usually opaque, they can see through to the surface when the atmosphere is extremely dry, although for B3 this is usually restricted to rare dry polar winter events.

11.4 PHYSICS RELATING PRECIPITATION RATE TO MILLIMETER-WAVE RADIANCE SPECTRA

Precipitation rate is approximately proportional to the product of vertical wind and absolute humidity. Water vapor is raised rapidly by vertical wind toward lower temperatures and eventually

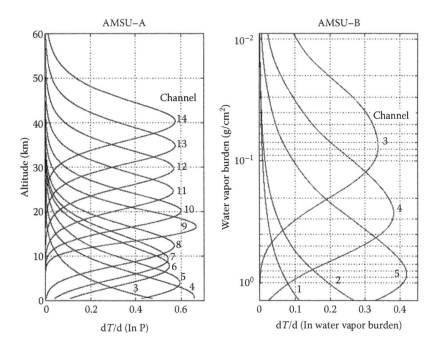

FIGURE 11.4   AMSU-A and AMSU-B weighting functions.

precipitates. AMSU's "altitude slicing" capability in both the 53- and 183-GHz bands, as discussed in Section 11.3, yields temperature and relative humidity profiles that can be combined to estimate absolute humidity profiles. Vertical wind can be inferred from hydrometeor size distributions, which depend on altitude, and cell-top altitudes. Hydrometeor size distributions can be sensed by scattering signatures in the 50–190-GHz band. Only large icy hydrometeors associated with heavy precipitation and raised by strong vertical wind scatter strongly in the 54-GHz band. Smaller icy hydrometeors associated with light precipitation have strong scattering signatures at 150–190 GHz. Scattering signatures from cloud ice particles and water droplets are very small. Cell-top altitudes can be estimated based on the strength of their scattering signature as a function of atmospheric transparency. For example, a strong scattering signature in the most opaque AMSU-B channel 3 implies a strong convective cell at high altitude. Cell tops at low altitude can only be sensed by window channels.

Precipitation rate can also be revealed by the absorption of water droplets, which produce a relatively warm signature over a cold reflective sea surface and a relatively cold signature over a hot land surface. These opposite signatures require separate algorithms for land and sea. Absorption signatures of water droplets smaller than a millimeter are usually less than a few tens of degrees. These small rain signatures are confounded by variations in frequency-dependent surface emissivity signatures over land. Over sea, the water-droplet absorption signatures are unambiguous and can be distinguished from water vapor by observations near the 22.35- and 183-GHz water vapor resonances and also in the adjacent spectral windows. For example, AMSU reveals the opacity of precipitation over sea by its emission in the 23.8, 31.4, and 50.2 GHz window channels.

Although precipitation can be indirectly sensed using the scattering and absorption signatures from hydrometeors aloft as described above, accurate surface precipitation estimation remains difficult since hydrometeors aloft may shift location or evaporate before reaching the surface. The need for evaporation corrections for active and passive millimeter-wave surface precipitation retrievals, particularly for undervegetated land such as desert, and associated correction methods are described in Section 11.6.6 [5–8].

11.5   SIGNAL PROCESSING METHODS

11.5.1   PRINCIPAL COMPONENT ANALYSIS

Brightness temperatures observed at different sensor channels are generally correlated. The multivariate brightness temperature dataset for each field of view (FOV) can be compressed to lower dimensionality with little loss of information using PCA, also known as the Karhunen–Loève or Hotelling transforms. PCA can also be used to attenuate noise or unwanted signals. PCA orthonormally transforms correlated variables into a new coordinate system spanning the subspace such that the new variables, called principal components (PCs), are uncorrelated.

Consider a zero-mean n-dimensional random vector $x = [x_1 \ x_2 \ \ldots \ x_n]^T$, where the random variable x_i represents the measurement for channel i. PCA transforms the random vector x to a reduced m-dimensional random vector $y = [y_1 \ y_2 \ldots y_m]^T = W^T x$ such that the mean-square reconstruction error

$$C(\cdot) = E[(x - \hat{x})^T (x - \hat{x})] \tag{11.16}$$

is minimized, where $E[\cdot]$ denotes the expectation, $\hat{x} = WW^T x$ is the reconstructed estimate of x, $W = [w_1 | w_2 | \ldots | w_m]$ is an $n \times m$ transform matrix, and w_i is a unit vector of size $n \times 1$. The first PC, $y_1 = w_1^T x$, has the largest variance, that is,

$$w_1 = \arg\max_{\|w\|=1} E\{[w^T x]^2\}. \tag{11.17}$$

The succeeding PCs have variances ordered from large to small, that is, the kth PC is $w_k^T x$, where

$$w_k = \arg\max_{\|w\|=1} E\left\{ \left[w^T \left(x - \sum_{i=1}^{k-1} w_i w_i^T x \right) \right]^2 \right\}, \tag{11.18}$$

w_i are eigenvectors of the covariance matrix, C_{xx}, of the original data x, where w_k is the eigenvector associated with the kth largest eigenvalue.

PCs for AMSU window channels were computed using estimated nadir brightness temperature spectra for 122 entire satellite orbits spanning a year and containing the 122 storms used in Refs. [1,2] and described later in Section 11.6.1. PCs that are sensitive to precipitation and are reasonably insensitive to the surface, residual viewing angle effects, or other unrelated variations were manually selected to be used in precipitation retrieval. This manual selection was based on many global AMSU images. Figures 11.5 and 11.6 show the AMSU land and sea PCs, respectively [3]. Only the land PC in Figure 11.5b met the selection criteria, whereas others were sensitive to the surface. Only the sea PC in Figure 11.6b met the selection criteria. The sea PC in Figure 11.6a was strongly affected by surface roughness, whereas that in Figure 11.6c was strongly angle dependent.

11.5.2   NEURAL NETWORKS

An NN is a mathematical model inspired by biological neural systems. Each NN is composed of interconnected artificial neurons, also called processing elements or nodes, working together. NNs can learn simple and complex nonlinear relationships between inputs and outputs from a training ensemble. They are useful for pattern recognition, classification, and estimation [22,23].

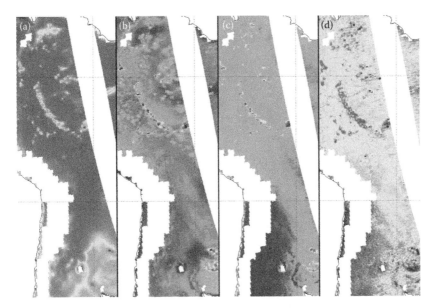

FIGURE 11.5   Land principal components for AMSU aboard NOAA-16 on July 26, 2002. (a) PC# 1, (b) PC# 2, (c) PC# 3, (d) PC# 5. Each image maps 55–75°W, 35°S–10°N. (From C. Surussavadee and D. H. Staelin, Global millimeter-wave precipitation retrievals trained with a cloud-resolving numerical weather prediction model, Part I: Retrieval design, *IEEE Trans. Geosci. Remote Sens.*, 46(1), 99–108, 2008. With permission. Copyright 2008 IEEE.)

Figure 11.7 shows a neuron, the fundamental unit of NNs. A neuron is characterized by weight w_i on each neural input x_i and a bias b which are summed by an adder, and an activation function f of that sum. The output, y, of the neuron is

$$y = f\left(\sum_{i=1}^{n} w_i x_i + b \right), \tag{11.19}$$

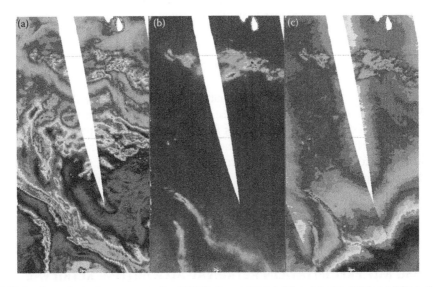

FIGURE 11.6   Sea principal components for AMSU aboard NOAA-16 on July 26, 2002. (a) PC# 1, (b) PC# 2, (c) PC# 6, Each image maps 50–90°E, 50°S–10°N. (From C. Surussavadee and D. H. Staelin, Global millimeter-wave precipitation retrievals trained with a cloud-resolving numerical weather prediction model, Part I: Retrieval design, *IEEE Trans. Geosci. Remote Sens.*, 46(1), 99–108, 2008. With permission. Copyright 2008 IEEE.)

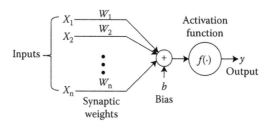

FIGURE 11.7 A neuron.

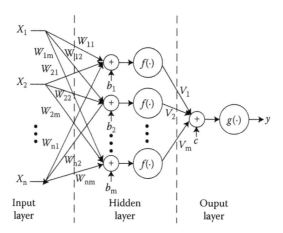

FIGURE 11.8 A multilayer feedforward neural network with one hidden layer and one output node.

where n is the total number of inputs. There are a variety of activation functions, for example, a linear function $f(x) = x$, a sigmoid function $f(x) = 1/(1 + e^{-x})$, a hyperbolic tangent function $f(x) = \tan h \, x = (e^x - e^{-x})/(e^x + e^{-x})$, and a hyperbolic tangent sigmoid function $f(x) = (2/(1 + e^{-2x})) - 1$.

A multilayer feedforward NN is composed of an input layer, hidden layer(s), and an output layer, as shown in Figure 11.8, with one hidden layer. The output of the NN in Figure 11.8 is

$$y = g\left(\sum_{j=1}^{m} v_j f\left(\sum_{i=1}^{n} w_{ij} x_i + b_j \right) + c \right), \tag{11.20}$$

where w_{ij} is the weight associated with the ith input to the jth node, and m is the number of neurons in the hidden layer. The AMP algorithms use three layers with 10, 5, and 1 neuron, respectively. The first two layers are hidden and the final layer is the output layer. The hyperbolic tangent sigmoid function was used for hidden layers and the linear function was used for the output layer.

NNs are trained to map inputs to their corresponding outputs using a training ensemble that should be broadly representative of all possible relationships. Weights and biases are adjusted using backpropagation to minimize an error metric, typically mean-square error. The training ensemble is generally divided into three independent sets, including (1) a training set used for adjusting weights and biases during training, (2) a validation set used to stop training to avoid overtraining, and (3) a testing set used for final accuracy evaluation.

11.6 AMSU MIT PRECIPITATION RETRIEVAL ALGORITHM: RETRIEVAL DESIGN AND PERFORMANCE EVALUATION

The AMP retrieval algorithm was developed by Surussavadee and Staelin through several steps and versions [1–8]. Figure 11.9 shows the AMP development strategy, which is composed of three main

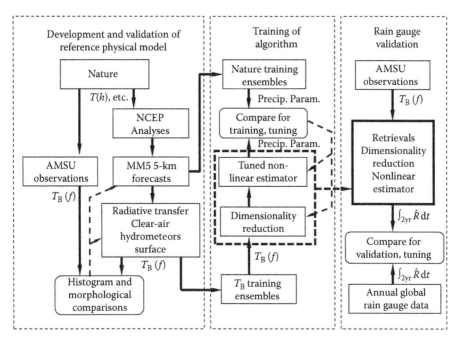

FIGURE 11.9 AMP development strategy, where $\hat{R}$ stands for precipitation estimate.

parts, including (1) the development and validation of a global reference physical model, (2) training of the algorithm, and (3) rain gauge validation. A global reference physical model is required for algorithm training since accurate global precipitation data coincident with satellite observations are not available. For example, radar data are indirect and geographically limited, and rain gauge data have insufficient time resolution. Model validation is required to ensure retrieval fidelity. A global reference physical model, called NCEP/MM5/TBSCAT/F(λ), was developed and validated by comparing model-predicted brightness temperatures (T_B) with those coincidentally observed by AMSU over 122 global representative storms spanning a year [1,2]. The model was then used as global ground-truth to train AMP to retrieve precipitation parameters [2–8]. PCA was used to filter out unwanted signals and noises, and perturbation analysis characterized by local brightness temperature features usually characterizing glaciated convective precipitation cells. NNs were used for estimators based on input brightness temperatures and associated preprocessing products. AMP-retrieved annual precipitation was validated against 787 global NOAA rain gauges, and corrections for AMP overestimates due to precipitation evaporation were derived [5–8].

11.6.1 Development and Validation of the Global Reference Physical Model, NCEP/MM5/TBSCAT/F(λ)

The global reference physical model, NCEP/MM5/TBSCAT/F(λ), is based on U.S. National Center for Environmental Prediction (NCEP) analyses [24], the fifth-generation National Center for Atmospheric Research/Penn State Mesoscale Model (MM5) [25], a radiative transfer model, TBSCAT [21], and electromagnetic scattering models for icy hydrometeors, F(λ) [1]. NCEP analyses at 1-degree resolution available every 6 h were used to initialize MM5. The MM5 configuration used in the study includes three cocentered nested domains at 45-, 15-, and 5-km resolution, respectively, from the outermost to the innermost domains. Kain–Fritsch 2 (KF2) [26] and Goddard [27] were used to treat unresolved and resolved precipitation, respectively. KF2 was not used for the innermost domain since the domain's resolution is sufficient to treat precipitation explicitly.

NCEP-initialized MM5 4–6 h forecasts of atmospheric parameters were fed to a two-stream variant of TBSCAT for computing brightness temperatures at AMSU frequencies. These forecasted brightness temperature fields were coincident with AMSU overpasses within 7 min. Atmospheric transmittance models [28,29] and complex permittivities for water [16] and ice [17] were used. Land emissivities were assumed to be uniformly distributed from 0.91 to 0.97, a typical range. Sea emissivities were computed using FASTEM [30], which incorporates geometric optics, Bragg scattering, and foam effects. Icy hydrometeors were modeled as homogeneous spheres having wavelength- and habit-dependent densities $F(\lambda)$ and total Mie scattering cross sections identical to those computed for equal-mass habits, that is, hexagonal plates (snow), 6-point bullet rosettes (graupel), and spheres (cloud-ice), using the discrete-dipole electromagnetic program DDSCAT6.1 [1,15].

To validate NCEP/MM5/TBSCAT/$F(\lambda)$, its predicted brightness temperatures were compared with those coincidently observed by AMSU aboard NOAA-15, NOAA-16, and NOAA-17 for 122 global representative storms spanning a year (July 2002–June 2003), as shown in Figure 11.10, where numbers 1–12 stand for January–December, and 14 indicates unglaciated cases, defined as those with microwave ice signatures too weak to be flagged as precipitation [31], but for which more than 0.25 mm of cloud liquid water is retrieved [32]. Each storm is a 2850-km square.

Brightness temperatures simulated using MM5 domain-2 outputs at 15-km resolution were convolved with a Gaussian function having a full-width at half-maximum (FWHM) of 50 km for comparison with AMSU-A. Figure 11.11 shows good agreement between AMSU-observed (left column) and NCEP/MM5/TBSCAT/$F(\lambda)$-predicted (right column) brightness temperatures at 183 ± 7 GHz for storm systems on December 31, 2002 at 2344 UTC (top row), and on January 2, 2003 at 1003 UTC (bottom row). Figure 11.12 shows good agreement between histograms of AMSU-observed and NCEP/MM5/TBSCAT/$F(\lambda)$-predicted brightness temperatures over 122 representative storm systems, where AMSU channels are arranged in order of increasing opacity, that is, 50.3, 89, 150, 183 ± 7, 183 ± 3, and 183 ± 1 GHz. [1,2] show strong sensitivity of NCEP/MM5/TBSCAT/$F(\lambda)$-predicted brightness temperatures to assumptions in the model, and show that the good agreement shown in Figure 11.12 can only be achieved when those assumptions are accurate.

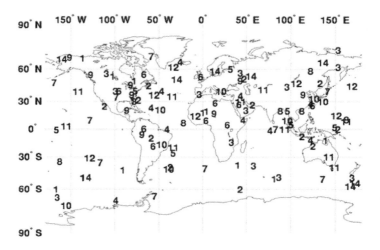

FIGURE 11.10 122 globally representative storms. The numbers 1–12 stand for January–December, and 14 indicates largely unglaciated cases. (From C. Surussavadee and D. H. Staelin, Comparison of AMSU millimeter-wave satellite observations, MM5/TBSCAT predicted radiances, and electromagnetic models for hydrometeors, *IEEE Trans. Geosci. Remote Sens.*, 44(10), 2667–2678, 2006. With permission. Copyright 2006 IEEE.)

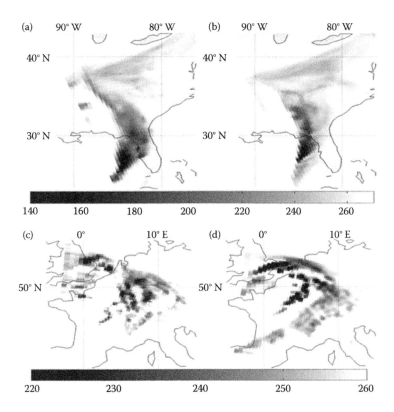

FIGURE 11.11 Comparisons of AMSU-observed (left column) and NCEP/MM5/TBSCAT/F(λ)-predicted (right column) brightness temperatures at 183 ± 7 GHz for two storm systems on December 31, 2002 at 2344 UTC (a, b), and on January 2, 2003 at 1003 UTC (c, d). (From C. Surussavadee and D. H. Staelin, *J. Atmos. Sci.*, 64(11), 3808–3826, 2007. With permission.)

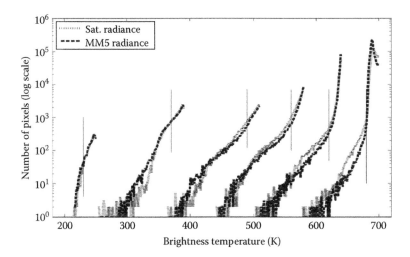

FIGURE 11.12 Histogram (pixels K^{-1}) comparisons of NCEP/MM5/TBSCAT/F(λ)-predicted and AMSU-observed T_B over 122 representative storm systems for channels near 50.3, 89, 150, 183 ± 7, 183 ± 3, and 183 ± 1 GHz, in order of increasing opacity from left to right. Only T_B values below 250 K are plotted. The absolute T_B values were shifted to the right by 0, 140, 260, 330, 390, and 450 K, respectively, for clarity. The vertical bar for each histogram represents 230 K. (From C. Surussavadee and D. H. Staelin, *J. Atmos. Sci.*, 64(11), 3808–3826, 2007. With permission.)

11.6.2 AMP Version 3

Since relationships between precipitation parameters and brightness temperatures are nonlinear and complex, as discussed in Sections 11.2 and 11.4, AMP employs NNs trained using NCEP/MM5/TBSCAT/F(λ). Due to lack of data storage capacity, 106 global storms, which were a random subset of the 122 storms described in Section 11.6.1, were used. NCEP/MM5/TBSCAT/F(λ)-predicted brightness temperatures at 5-km resolution were convolved with Gaussian functions having FWHM of 50 and 15 km to compute AMSU-A and AMSU-B brightness temperatures, respectively. NNs were trained to estimate precipitation parameters at 15-km resolution, which are the convolution of a Gaussian function having FWHM of 15 km and MM5 precipitation parameters at 5-km resolution. As discussed in Section 11.3, since AMSU-A channels 9–15 respond mostly to altitudes above most precipitation, AMP employs 13 AMSU channels, including AMSU-A channels 1–8 and AMSU-B channels 1–5.

The first version of AMP, AMP-1 [2], was developed and tested using MM5 to evaluate the sensitivity of brightness temperatures and retrievals to assumptions in NCEP/MM5/TBSCAT/F(λ). The second version of AMP, AMP-2 [3,4], works well with real AMSU observations globally except for (1) high-elevation surfaces ("too-high"), for example, Himalayan, Greenland, and Antarctic plateaus, (2) snow-covered land and sea ice ("icy surfaces"), and (3) too-cold and potentially dry atmospheres ("too-cold"). These three cases were omitted in AMP-2. Too-high surface elevations, h_{surf}, can be snow covered and are sensed more strongly, which could cause false alarms. The surface elevations are called "too-high" when 2 km $< h_{surf}$ for |lat| $< 60°$, or 1.5 km $< h_{surf}$ for $60° <$ |lat| $< 70°$, or 0.5 km $< h_{surf}$ for |lat| $> 70°$. The atmospheres are called "too-cold" when AMSU-A channel 5 (A5) is below 242 K. When the air is too-cold and potentially dry, even the most opaque water vapor channel, AMSU-B channel 3 (B3, 183 $\pm$ 1 GHz), could see through to the surface and cause false alarms. AMP-2 retrievals were evaluated using MM5 and were shown to generally agree with retrievals from AMSR-E [33,34] and AMSU employing NOAA's algorithm [35,36].

The third version of AMP, AMP-3 [5], improves AMP-2 and is the first algorithm that successfully retrieves precipitation over icy surfaces, for example, the North Pole. Precipitation retrieval over icy surfaces is difficult since signals from icy surfaces mimic those from precipitation. The keys to the success of AMP-3 include (1) use of multiple NNs trained for different surface types and climates, (2) selection of only those channels less sensitive to surfaces, and (3) use of PCA to filter out residual surface signals and other noise. AMP-3 successfully retrieves global precipitation except for (1) high elevation surfaces and (2) icy surfaces when the atmosphere is too cold.

Table 11.2 shows the AMP-3 algorithm, which is composed of six steps [5]. First, footprints with out-of-bounds brightness temperatures ($T_B < 50$ K or $T_B > 400$ K) are flagged as invalid and are omitted. Footprints with too-high surface elevations, h_{surf}, are omitted. Retrievals for too-cold footprints are set to zero.

Second, small biases for AMSU-A channels 5–8 (A5–A8) relative to NCEP/MM5/TBSCAT/F(λ) are corrected [1]. Surface is classified to land, sea, snow-covered land, and sea ice. Since brightness temperatures depend on the satellite zenith angle, to make the precipitation retrieval problem less complicated, AMP estimates brightness temperatures that AMSU would have seen at nadir using one NN per channel serving both land and sea [3,5,6]. NNs were trained using brightness temperatures simulated at nadir as the target and those at all satellite zenith angles for 106 MM5 storms as inputs. To estimate AMSU-A brightness temperature at nadir for a single pixel, inputs to NNs include the secant of the satellite zenith angle and brightness temperatures for AMSU-A channels 1–8 for the same pixel. Similarly, to estimate AMSU-B brightness temperature at nadir for a single pixel, the inputs to the NNs are the secant of the satellite zenith angle and brightness temperatures for AMSU-B channels 1–5 for the same pixel. Since there are no horizontal spatial offsets between inputs and target, atmospheric inhomogeneities introduce no errors. The root mean square (rms) discrepancies of the estimated nadir brightness temperatures evaluated using MM5 are small compared to most precipitation signatures [3]. The rms discrepancies are 0.33–0.56 K for AMSU-A channels 3–8 for zenith angles >50°. The worst rms discrepancy for all channels except for 89 GHz

TABLE 11.2
AMP-3 Algorithm

Number	Action
1.	Flag pixel (Omit) if: $T_B < 50$ K, or $T_B > 400$ K, or 2 km $< h_{surf}$; $|lat| < 60°$, or 1.5 km $< h_{surf}$; $60° < |lat| < 70°$, or 0.5 km $< h_{surf}$; $|lat| > 70°$. Retrieval = 0 if A5 < 242 K.
2.	Remove biases relative to MM5 simulations for A5–A8.
	Neural nets (NNs) correct T_B to nadir values (trained using 106 MM5 storms).
	Classify surface: land versus water using coordinates; ice/snow versus other (Grody algorithm).
3.	Bound scattering areas (convective cells) using (B5 < 0.667·(A5-248) + 258) if A5 ≥ 248, or (B4 < 247.5) if A5 < 248 K; then evaluate boundary-value T_B's.
	Compute ΔT_B relative to interpolated boundary values for A4-8.
	Compute scores for those PCs (see Step 4) that correlate well globally with rain but not with surface emissivity or humidity.
	Feed ΔT_B, secant θ_{zenith}, PCs and other inputs to NNs.

4.	Case	PCA input	PCA training	Other NN input	NN training
A	Land	A4-8	Land; 122 orbits	PC1, B3-4	106 MM5, land
B	Sea $|lat| < 45°$, A5 ≥ 248 K	A1-8, B1-5	Ice-free sea 122 orbits	PC2-5	106 MM5 ice-free sea
C	Other sea pixels	A4-8	A5 < 248 K, 122 orbits	PC1-2 B3-4	106 MM5 53.6 < 248 K
D	All sea	A4-8	Same as 4C	PC1-2 B3-4	106 MM5 sea

Number	Action
5.	Land: P (mm h^{-1}) = A, as given in Step 4 for Case A.
	Water: P (mm h^{-1}) = [$kB + (1 - k)C + D$]/2, where $k = 0$ for $|lat| > 50°$, $k = 1$ for $|lat| < 40°$, and $k = (50 - |lat|)/10$ for $40° < |lat| < 50°$, and B, C, and D are the Step 4 outputs.
6.	Omit pixels having $P > 3$ mm h^{-1} in Step 5 for surfaces classified in Step 2 as snow or ice, and any precipitation at pixels within ~30 km of such pixels.

Source: From C. Surussavadee and D. H. Staelin, *J. Appl. Meteorol. Climatol.*, 49(1), 124–135, 2010. With permission.
Note: P is surface precipitation rate, h_{surf} stands for surface elevation, and PC1 stands for scores of first PC.

is 1 K for zenith angles <50° and 1.57 K for zenith angles >50°. The residual nadir correction errors are attenuated further using PCA performed in the next step.

Third, brightness temperature perturbations, ΔT_B, for A4–A8 were detected and computed. These perturbations are relative to Laplacian interpolation of the immediately surrounding unperturbed brightness temperature field, where the perturbation borders are determined using the 183 ± 7 GHz channel, or in colder regions the 183 ± 3 GHz channel [31]. These ΔT_Bs are the cold signatures associated with scattering from large icy hydrometeors and are a key indicator of heavy convective precipitation. PCA was then applied to the channels listed in Table 11.2. Only PCs sensitive to precipitation, but nearly blind to surface, humidity, and angle effects, and to other noises were selected, as illustrated in Section 11.5.1.

Fourth, surface precipitation rates are estimated using four different NNs for different climates and surfaces, as listed in Table 11.2. Fifth, estimates in Step 4 were combined into a single estimate (P). Sixth, footprints with excessively high precipitation estimates ($P > 3$ mm h^{-1}) over snow and ice and nearby footprints are most likely false alarm and are omitted. Figure 11.13 shows examples of AMP-3 precipitation estimates, including (1) precipitation over the North Pole on July 15, 2003, where light pink shows snow-covered land and sea ice, (2) global annual precipitation accumulation (mm year^{-1}) for year 2006 computed using AMSU observations on NOAA-15 and NOAA-16, (3) a typhoon and precipitation in tropics on September 29, 2006, and (4) evolution of an Arctic storm system at 102-min intervals observed by NOAA-18 on June 22, 2008, where light pink shows snow-covered land and sea ice.

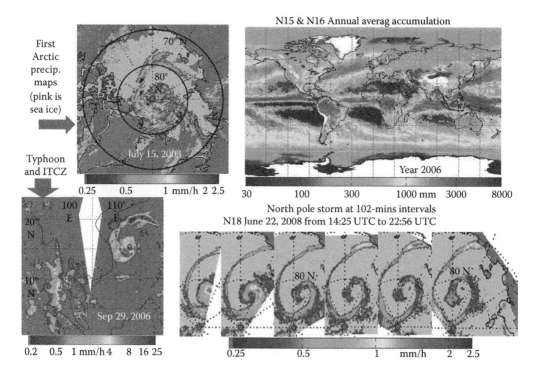

FIGURE 11.13 (**See color insert.**) Examples of AMP-3 precipitation estimates. Light pink shows snow-covered land and sea ice. (Adapted from C. Surussavadee and D. H. Staelin, *IEEE Geosci. Remote Sens. Lett.*, 7(3), 440–444, 2010. With permission. Copyright 2010 IEEE.)

11.6.3 COMPARISON OF AMP-3 RETRIEVALS WITH CLOUDSAT RADAR OBSERVATIONS

To determine whether AMP-3 precipitation retrievals over icy surface are due to surface effects or dense clouds, Figure 11.14 compares them with simultaneous observations from nadir-looking CloudSat radar echoes [9,10] for four storm systems, where time gaps between AMSU and CloudSat observations are ranged from 24 min to 3 h. CloudSat flies from green to red arrows with a highest altitude of ~12 km. Pink, dark pink, and black pixels indicate surface snow or ice, too-high surface elevations, and excessive estimates over icy surfaces, respectively. Gray is land or water. AMP-3 precipitation retrievals generally agree with CloudSat observations [5]. The differences are partly due to time differences.

11.6.4 CORRECTIONS FOR RADIO-FREQUENCY INTERFERENCE

Comparison of AMP-3 precipitation retrievals for NOAA-15 and NOAA-16 shows biases in AMSU-B brightness temperatures [6]. Biases are due to radio-frequency interference (RFI) from on-board down-link transponders. The AMSU data used in the study are from NOAA's Web site (http://www.class.ncdc.noaa.gov), where AMSU data prior to January 4, 2007 were obtained in December 2007, and the remainder was obtained in August 2008. RFI affecting NOAA-16 AMSU-B is only the left side of the swath and is negligible for AMSU-B channels 1–3. The RFI was found from the left–right asymmetry of globally monthly averaged brightness temperatures over nonprecipitating footprints.

RFI affecting NOAA-15 AMSU-B is more extensive for all AMSU-B channels. The RFI-corrected NOAA-16 AMSU-B brightness temperatures were used to calibrate NOAA-15 AMSU-B by comparing their daily observed average brightness temperatures within 40 min and 100 km for |lat| > 65° over full days. Figure 11.15 shows computed RFI affecting channels B4 (183 ± 3 GHz)

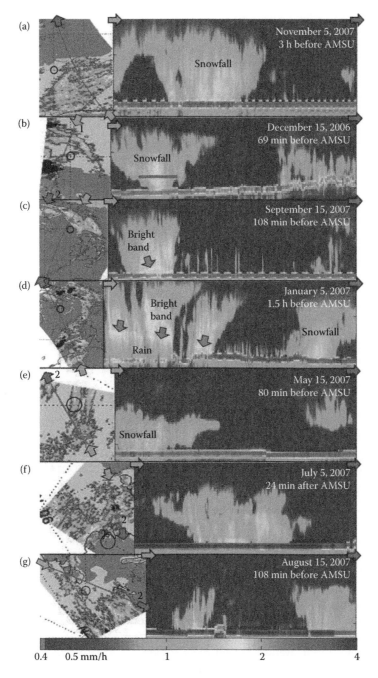

FIGURE 11.14 (See color insert.) Comparisons of AMP-3 surface-precipitation retrievals and CloudSat echo strengths. The CloudSat orbit is red and bounded by red and green arrows; its peak illustrated echo altitude is ~12 km. Pink, dark pink, and black pixels indicate surface snow or ice, excessive surface elevation, and untrustworthy retrievals, respectively. Gray is land or water. The 200-km-diameter black circles are centered at (a) 60°S/20°E over Antarctic ice shelf, (b) 50°N/110°W over Canadian snow, (c) 60°N/10°W near a North Atlantic front, (d) 60°N/70°W over Labrador coastline, and over Arctic ice pack at (e) 82°N/180°E, (f) 82°N/30°W, and (g) 82°N/90°E. (From C. Surussavadee and D. H. Staelin, Satellite retrievals of arctic and equatorial rain and snowfall rates using millimeter wavelengths, *IEEE Trans. Geosci. Remote Sens.*, 47(11), 3697–3707, 2009. With permission. Copyright 2009 IEEE.)

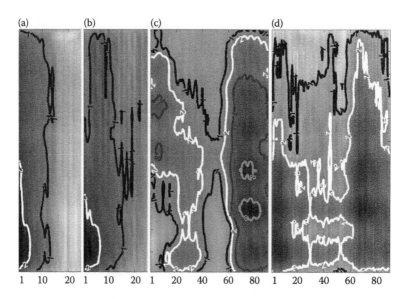

FIGURE 11.15 Average RFI (K) affecting AMSU-B as a function of scan position (bottom scale) from January 1, 2002 (top) to December 31, 2007 (bottom). (a) NOAA-16 channel B4, (b) NOAA-16 channel B5, (c) NOAA-15 channel B4, and (d) NOAA-15 channel B5. Contours are 1 (black), 2 (white), 4 (dark gray), and 8 K (light gray). RFI affected NOAA-16 channels B4 and B5 only on one edge of the scan. (From C. Surussavadee and D. H. Staelin, *J. Appl. Meteorol. Climatol.*, 49(1), 124–135, 2010. With permission.)

and B5 (183 ± 7 GHz) in units of Kelvin as a function of scan angle for January 1, 2002 (top) to December 31, 2007 (bottom) for NOAA-15 and NOAA-16.

11.6.5 COMPARISON OF AMP-3 RETRIEVALS WITH GLOBAL RAIN GAUGE OBSERVATIONS

The accuracy of RFI-corrected AMP-3 precipitation retrievals was estimated by comparing them with those observed by 787 globally distributed NOAA rain gauges [11] in nonhilly and not-too-high regions, where nonhilly was defined arbitrarily as those for which the surface elevation varied by <500 m within a box of $\pm 0.2°$ of longitude and latitude [6]. The definition for too-high surface elevation is described in Section 11.6.2. The RFI discussed in Section 11.6.4 was corrected prior to Step 1 of Table 11.2. AMP-3 annual precipitation (mm year^{-1}) for each gauge was computed from all passes of NOAA-15 and NOAA-16 in years 2006 and 2007, where estimates within $\pm 0.4°$ longitude and latitude of the gauge were averaged. To reduce AMSU's temporal sampling noise due to its observations ~4 times daily from NOAA-15 and NOAA-16, the comparison between AMP-3 and gauges was performed using 2-year annual averages for each site.

Table 11.3 shows 2-year average annual precipitation for AMP-3 and gauges, and AMP-3/gauge ratios for 12 surface classes defined using Advanced Very High Resolution Radiometer (AVHRR) infrared spectral images plus one additional class defined geographically as coastal (<55 km from coast line) [6,37]. The results show that (1) the well-populated areas of the globe tend to have ratios between 1.0 and 1.25, (2) AMP-3 overestimates are dependent on surface class, and (3) ratios are high for coastal sites and undervegetated land, including cultivated crops to grassland, shrubs over bare ground, and bare ground in an increasing order. Overestimates for the coastal sites are due to coastal meteorological effects and mixed pixels between land and sea requiring land/sea flag improvement. Column 5 (sat. rms) is the rms scatter σ of the individual rain gauges within each class relative to AMSU. Column 7 (region rms) equals $\sigma/\sqrt{N-1}$, where N is the number of gauges in that class (Column 6). The rms variations in Column 7 are generally

TABLE 11.3

AMP-3 versus Rain Gauge Annual Precipitation for Different Surface Classes

Surface Class	Sat. (mm year⁻¹)	Gauge (mm year⁻¹)	Ratio	Sat. rms (%)	No. Sites (N)	Region rms (%)
Tundra	429	490	0.88	16.0	8	6.1
Water	1379	1571	0.88	28.2	30	5.2
High-latitude deciduous forest	524	526	1.00	20.3	24	4.2
Broadleaf evergreen forest	2925	2831	1.03	4.85	2	4.8
Mixed coniferous forest	788	739	1.07	21.4	82	2.4
Coniferous evergreen forest	596	536	1.11	29.9	60	3.9
Broadleaf deciduous forest	1113	900	1.24	29.8	18	7.2
Wooded grassland	1238	988	1.25	37.3	66	4.6
Cultivated crops	949	693	1.37	43.5	102	4.3
Coastal (<55 km)	1439	928	1.55	57.8	260	3.6
Grassland	837	349	2.40	102.4	77	11.7
Shrubs, bare ground	771	252	3.06	119.4	22	26.1
Bare ground	559	62	9.08	379	29	71.6

Source: From C. Surussavadee and D. H. Staelin, *J. Appl. Meteorol. Climatol.*, 49(1), 124–135, 2010. With permission.

small relative to the differences between the means characterizing the different classes, so these surface-dependent biases appear to be statistically significant (>3σ) only for surfaces more arid than coniferous evergreen forests.

Figure 11.16 shows locations of 426 gauges that are nonhilly, not-too-high, noncoastal (≥55 km from coast line), and nondesert (gauges ≥300 mm year⁻¹). Pluses indicate 39 sites having AMP-3/gauge ratios >2 and squares indicate three sites having AMP-3/gauge ratios <0.5.

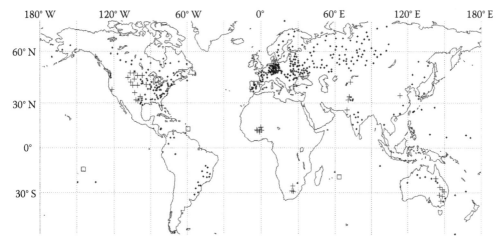

FIGURE 11.16   Locations of gauges that are nonhilly, noncoastal (≥55 km from coast line), and nondesert (gauges ≥300 mm year⁻¹). Pluses: sites with AMP-3/gauge ratios >2. Squares: sites with AMP-3/gauge ratios <0.5.

11.6.6   AMP-4 and AMP-5: AMP Versions with Correction Methods for Near-Surface Evaporation

The reason for AMP-3 overestimates over undervegetated land was investigated using MM5 simulation [6]. The black solid line in Figure 11.17 (left y-axis) shows high correlation between a virga metric (V metric) and the mean of simulated AMSU retrievals divided by MM5 truth over the 106 MM5 storms. The V metric is defined for a 15-km FOV as the highest MM5 layer-average hydrometeor density (kg m^{-3}) in air divided by that in the lowest MM5 atmospheric layer. Only FOVs where MM5 truth is >1 mm h^{-1} were plotted. The dashed line is for this ratio plus one standard deviation. The gray line is the probability density function of the V metric (right y-axis). The high correlation strongly suggests that AMP-3 overestimates are due to evaporation.

Two evaporation correction methods were developed [7,8] for passive and active microwave surface precipitation estimates using (1) annual precipitation totals measured by 509 global rain gauges located in nonhilly, not-too-high, and noncoastal (>55 km from coastline) sites, and (2) corresponding RFI-corrected AMP-3 precipitation estimates for NOAA-15, NOAA-16, and NOAA-18 for years 2006 and 2007. The rain gauges are a subset of the 787 gauges that satisfy these constraints.

The first method computes for each surface class a correction factor, which is the ratio between the means of rain gauge and AMP-3 annual accumulations (mm year^{-1}). The surface classes include the 11 AVHRR-derived classes plus a class for water. The correction factor for the water class was set to 1 since those sites are small islands ambiguously classed as water. Correction maps at 1-degree resolution were computed separately for years 2006 and 2007. Application of the corrections to RFI-corrected AMP-3 estimates resulted in the fourth version of AMP, AMP-4 [6–8].

The second method employs NNs trained to estimate AMP-3/gauge annual accumulation ratios based on 19 inputs that include the surface class (a single integer from 1 to 12), the base-10 logarithm of AMP-3 retrieved annual precipitation, and the base-10 logarithm of annual average relative humidity for 17 pressure levels ranging from 1000 to 300 mb. When rain gauge annual accumulations are <2000 mm year^{-1}, the ratios (RFI-corrected AMP-3)/(rain gauge) are highly negatively correlated

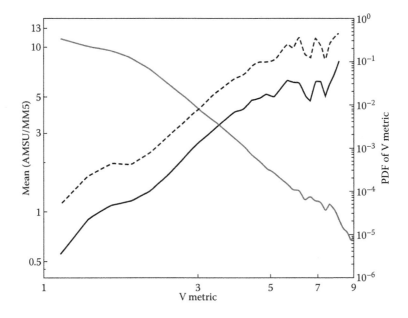

FIGURE 11.17   Dark solid line (left y-axis): (AMSU estimate)/(MM5 truth) ratio versus the MM5 virga metric (V metric). Dashed line: ratio plus one standard deviation. Gray line (right y-axis): probability density function of V. (From C. Surussavadee and D. H. Staelin, *J. Appl. Meteorol. Climatol.*, 49(1), 124–135, 2010. With permission.)

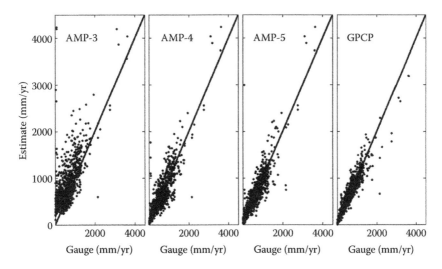

FIGURE 11.18 Scatter plots of AMP-3, AMP-4, AMP-5, and GPCP versus gauge annual precipitation accumulations. (From C. Surussavadee and D. H. Staelin, Correcting microwave precipitation retrievals for near-surface evaporation, *IEEE Int. Geosci. Remote Sens. Symp. Proc.*, pp. 1312–1315, July 2010. With permission. Copyright 2010 IEEE.)

with relative humidity profiles, particularly for altitudes below ~750 mb [7,8]. The correlation is low when annual accumulations are >2000 mm year^{-1}. Hence, the second method employs AMP-4 correction ratios when AMP-4 annual accumulations are >2000 mm year^{-1}. Application of 1-degree correction maps to RFI-corrected AMP-3 estimates results in the fifth version of AMP, AMP-5 [7,8].

In the development of these two methods, the 1-year gauge data were randomly separated into two independent halves for years 2006 and 2007. When one half was used for training, the other half was used for accuracy evaluation. Gauges used in the study were not corrected for wind loss since existing global wind-loss adjustments are outdated (e.g., many gauges have been replaced). Figure 11.18 compares AMP-3, AMP-4, AMP-5, and GPCP [12] estimates for annual accumulations (mm year^{-1}) with gauge measurements for the 509 gauges for years 2006 and 2007. Wind-loss adjustments [38] in GPCP data were removed for comparison. Table 11.4 shows rms errors, biases

TABLE 11.4
RMS Errors, Biases (Estimate–Gauge), and Correlation Coefficients for AMP-3, AMP-4, AMP-5, and GPCP for 509 Gauge Annual Observations

	Estimate	AMP-3	AMP-4	AMP-5	GPCP
rms error	All	433.59	223.39	209.46	151.62
	≤900 mm year^{-1}	439.61	191.07	173.29	112.63
	>900 mm year^{-1}	407.91	323.96	317.43	255.60
Bias	All	206.95	15.18	17.09	20.66
	≤900 mm year^{-1}	219.12	28.20	36.10	46.81
	>900 mm year^{-1}	156.92	− 38.31	− 61.00	− 86.80
Correlation coefficient	All	0.66	0.86	0.88	0.93
	≤900 mm year^{-1}	0.20	0.67	0.75	0.91
	>900 mm year^{-1}	0.74	0.81	0.83	0.83

Source: From C. Surussavadee and D. H. Staelin, Correcting microwave precipitation retrievals for near-surface evaporation, *IEEE Int. Geosci. Remote Sens. Symp. Proc.*, pp. 1312–1315, July 2010. With permission. © 2010 IEEE.

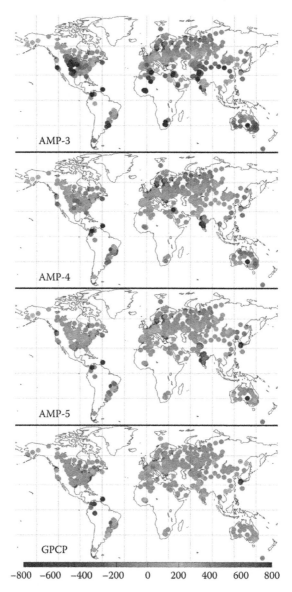

FIGURE 11.19 (**See color insert.**) Global maps of 2-year mean annual precipitation errors (estimate–gauge) for AMP-3, AMP-4, AMP-5, and (after removal of gauge wind-loss corrections) GPCP for 509 wind-loss uncorrected gauge observations. (From C. Surussavadee and D. H. Staelin, Correcting microwave precipitation retrievals for near-surface evaporation, *IEEE Int. Geosci. Remote Sens. Symp. Proc.*, pp. 1312–1315, July 2010. With permission. Copyright 2010 IEEE.)

(estimate–gauge), and correlation coefficients for AMP-3, AMP-4, AMP-5, and GPCP, as evaluated using rain gauges. Results show that AMP-4 and AMP-5 estimates are much more accurate and less scattered than AMP-3 estimates. By incorporating relative humidity profiles, AMP-5 performs better than AMP-4 for rms errors and correlation coefficients, and can be used over ocean. AMP-4 has lower biases than AMP-5 since it was corrected for biases for each surface class. AMP-5 retrieval accuracy approaches that of GPCP, which has the advantage that it incorporates rain gauge measurements from the Global Precipitation Climatology Centre (GPCC) gauge analyses that include many of the rain gauges used for our standard.

Figure 11.19 shows global maps of 2-year mean annual precipitation discrepancies (estimate–gauge) for AMP-3, AMP-4, AMP-5, and (after removal of gauge wind-loss corrections) GPCP for 509 gauges. AMP-4 and AMP-5 offer significant improvement over AMP-3. AMP-5 agrees with gauges better than AMP-4, particularly over central United States and south-east Australia. Site-specific problems such as the Australian dry lake can be individually fixed in the future.

11.7 SUMMARY AND CONCLUSIONS

This chapter discusses the development of the AMP retrieval algorithm. AMP is the first algorithm that successfully retrieved precipitation globally, including snow-covered land and sea ice. Two limitations of AMP retrievals include their inability to retrieve precipitation rates over (1) high-elevation lands such as the Himalayan, Greenland, and Antarctic plateaus, and (2) snow-covered land and sea ice when the atmosphere is so cold and dry that even the opaque channels see the surface.

AMP starts from the development and validation of a global reference model, NCEP/MM5/TBSCAT/F(λ), which is used for algorithm training. AMP employs linear combinations of NN estimates trained separately using NCEP/MM5/TBSCAT/F(λ) for different surface types and climates. NNs appear to successfully learn the complex and nonlinear relationships between AMSU brightness temperatures and precipitation parameters. PCA is shown to be a very useful preprocessing method for filtering out noise and unwanted signals in the observed brightness temperatures.

AMP precipitation detections over icy surfaces were shown to generally agree with observations from the CloudSat nadir-looking radar. Comparison of AMP-3 annual precipitation estimates and observations from 787 globally distributed rain gauges shows AMP-3 overestimates over arid regions. MM5 simulations suggest that the overestimates are most likely due to near-surface evaporation. Two methods for correcting millimeter-wave precipitation retrievals for near-surface evaporation were developed. AMP-4 employs surface class only. AMP-5 employs both surface class and climatological relative humidity profiles. Both methods significantly correct the AMP-3 overestimates. By incorporating relative humidity profiles, AMP-5 performs better than AMP-4 and provides corrections over ocean.

AMP can be used for AMSU-A/AMSU-B aboard NOAA satellites and AMSU-A/MHS aboard NOAA and MetOp satellites, and can be adapted to other passive millimeter-wave radiometers observing near oxygen absorption bands and water vapor resonances, for example, AMSU/HSB aboard NASA Aqua satellite, SSMIS aboard DMSP satellites, ATMS being launched aboard NPOESS and JPSS satellites, and GEM sounders. Similar evaporation corrections should also be applicable to other microwave remote sensing retrievals of precipitation that do not directly probe near-surface altitudes.

ACKNOWLEDGMENTS

The authors thank Pennsylvania State University and the University Corporation for Atmospheric Research for providing the MM5 model and technical support; the Alliance for Computational Earth Science, Massachusetts Institute of Technology, for assisting with computer resources; P. W. Rosenkranz for his forward radiance program, TBSCAT, and helpful discussions; B. T. Draine and P. J. Flatau for making DDSCAT6.1 available; the National Center for Atmospheric Research for assistance with NCEP global tropospheric analyses; L. von Bosau for assembling the NOAA rain gauge data; A. Graumann, C. Nichols, and L. Zhao of the NOAA/National Environmental Satellite, Data, and Information Service for help with AMSU data; and G. J. Huffman and D. T. Bolvin for help in understanding the GPCP data.

This work is supported in part by the National Aeronautics and Space Administration under Grant NAG5-13652 and Contract NAS5-31376, the National Oceanic and Atmospheric Administration under Contract DG133E-02-CN-0011, the National Aeronautics and Space Administration under Grant NNX07AE35G, and the Prince of Songkla University under Grant PSU 977/301.

REFERENCES

1. C. Surussavadee and D. H. Staelin, Comparison of AMSU millimeter-wave satellite observations, MM5/TBSCAT predicted radiances, and electromagnetic models for hydrometeors, *IEEE Trans. Geosci. Remote Sens.*, 44(10), 2667–2678, 2006.

2. C. Surussavadee and D. H. Staelin, Millimeter-wave precipitation retrievals and observed-versus-simulated radiance distributions: Sensitivity to assumptions, *J. Atmos. Sci.*, 64(11), 3808–3826, 2007.

3. C. Surussavadee and D. H. Staelin, Global millimeter-wave precipitation retrievals trained with a cloud-resolving numerical weather prediction model, Part I: Retrieval design, *IEEE Trans. Geosci. Remote Sens.*, 46(1), 99–108, 2008.

4. C. Surussavadee and D. H. Staelin, Global millimeter-wave precipitation retrievals trained with a cloud-resolving numerical weather prediction model, Part II: Performance evaluation, *IEEE Trans. Geosci. Remote Sens.*, 46(1), 109–118, 2008.

5. C. Surussavadee and D. H. Staelin, Satellite retrievals of arctic and equatorial rain and snowfall rates using millimeter wavelengths, *IEEE Trans. Geosci. Remote Sens.*, 47(11), 3697–3707, 2009.

6. C. Surussavadee and D. H. Staelin, Global precipitation retrievals using the NOAA AMSU millimeter-wave channels: Comparisons with rain gauges, *J. Appl. Meteorol. Climatol.*, 49(1), 124–135, 2010.

7. C. Surussavadee and D. H. Staelin, Correcting microwave precipitation retrievals for near-surface evaporation, *IEEE Int. Geosci. Remote Sens. Symp. Proc.*, Honolulu, Hawaii, pp. 1312–1315, July 2010.

8. C. Surussavadee and D. H. Staelin, Evaporation correction methods for microwave retrievals of surface precipitation rate, *IEEE Trans. Geosci. Remote Sens.*, in press, 2011.

9. G. G. Mace, R. Marchand, Q. Zhang, and G. Stephens, Global hydrometeor occurrence as observed by Cloud Sat: Initial observations from summer 2006, *Geophys. Res. Lett.*, 34(9), L09 808-1–L09 808-5, 2007.

10. 2B-GEOPROF, December 5, 2007. Available at http://www.cloudsat.cira.colostate.edu/.

11. NOAA, cited 2007: Monthly climatic data for the world. Available at http://www7.ncdc.noaa.gov/IPS/mcdw/mcdw.html.

12. G. J. Huffman, R. F. Adler, M. M. Morrissey, D. T. Bolvin, S. Curtis, R. Joyce, B. McGavock, and J. Susskind, Global precipitation at one degree daily resolution from multisatellite observations, *J. Hydrometeorol.*, 2(1), 36–50, 2001.

13. F. T. Ulaby, R. K. Moore, and A. K. Fung, *Microwave Remote Sensing: Active and Passive*. Addison-Wesley Pub. Co., Reading, MA, 1981.

14. D. Deirmendjian, *Electromagnetic Scattering on Spherical Polydispersions*, American Elsevier Publishing Co., New York, NY, 1969.

15. B. T. Draine and P. J. Flatau, *User Guide for the Discrete Dipole Approximation Code DDSCAT 6.1*, October 11, 2005. Available at http://arxiv.org/abs/astro-ph/0409262.

16. H. J. Liebe, G. A. Hufford, and T. Manabe, A model for the complex permittivity of water at frequencies below 1THz, *Int. J. Infra. Mill. Waves*, 12(7), 659–675, 1991.

17. G. Hufford, A model for the complex permittivity of ice at frequencies below 1 THz, *Int. J. Infra. Mill. Waves*, 12(7), 677–682, 1991.

18. A. H. Shivola, Self-consistency aspects of dielectric mixing theories, *IEEE Trans. Geosci. Remote Sens.*, 27(4), 403–415, 1989.

19. K. Karkainen, A. Sihvola, and K. Nikoskinen, Analysis of a three-dimensional dielectric mixture with finite difference method, *IEEE Trans. Geosci. Remote Sens.*, 39(5), 1013–1018, 2001.

20. D. H. Staelin and C. Surussavadee, Precipitation retrieval accuracies for geo-microwave sounders, *IEEE Trans. Geosci. Remote Sens.*, 45(10), 3150–3159, 2007.

21. P. W. Rosenkranz, Rapid radiative transfer model for AMSU/HSB channels, *IEEE Trans. Geosci. Remote Sens.*, 41(2), 362–368, 2003.

22. S. Haykin, *Neural Networks: A Comprehensive Foundation*, 2nd edition, Prentice-Hall, Upper Saddle River, NJ, 1999.

23. W. J. Blackwell and F. W. Chen, *Neural Networks in Atmospheric Remote Sensing*, Artech House, Boston, MA, 2009.

24. NCEP Global Tropospheric Analyses, 1×1 Daily September 15, 1999–Present. Available at http://dss.ucar.edu/datasets/ds083.2/.

25. J. Dudhia, D. Gill, K. Manning, W. Wang, and C. Bruyere, *PSU/NCAR Mesoscale Modeling System Tutorial Class Notes and Users' Guide* (*MM5 Modeling System Version 3*), January 2005. Available at http://www.mmm.ucar.edu/mm5/documents/tutorial-v3-notes.html.

26. J. S. Kain, The Kain–Fritsch convective parameterization: An update, *J. Appl. Meteorol.*, 43(1), 170–181, 2004.

27. W.-K. Tao and J. Simpson, Goddard cumulus ensemble model. Part I: Model description, *Terr. Atmos. Ocean. Sci.*, 4(1), 35–72, 1993.
28. H. J. Liebe, P. W. Rosenkranz, and G. A. Hufford, Atmospheric 60-GHz oxygen spectrum: New laboratory measurements and line parameters, *J. Quant. Spectrosc. Radiat. Transf.*, 48(5/6), 629–643, 1992.
29. P. W. Rosenkranz, Water vapor microwave continuum absorption: A comparison of measurements and models, *Radio Sci.*, 33(4), 919–928, 1998.
30. S. J. English and T. J. Hewison, Fast generic millimeter-wave emissivity model, *Proc. Int. Soc. Opt. Eng.*, 3503, 288–300, 1998.
31. F. W. Chen and D. H. Staelin, AIRS/AMSU/HSB precipitation estimates, *IEEE Trans. Geosci. Remote Sens.*, 41(2), 410–417, 2003.
32. P. W. Rosenkranz, Retrieval of temperature and moisture profiles from AMSU-A and AMSU-B measurements, *IEEE Trans. Geosci. Remote Sens.*, 39(11), 2429–2435, 2001.
33. R. Adler, T. Wilheit, Jr., C. Kummerow, and R. Ferraro, AE_Rain, *AMSR-E/Aqua L2B Global Swath Rain Rate/Type GSFC Profiling Algorithm*, May 17, 2007. http://nsidc.org/data/docs/daac/ae_rain_l2b. gd.html. [Online]. Available: http://nsdic.org/data/amsre/versions.html
34. R. Adler, T. Wilheit, Jr., C. Kummerow, and R. Ferraro, *AMSR-E/Aqua L2B Global Swath Rain Rate/Type GSFC Profiling Algorithm V001*. Boulder, CO: Nat. Snow and Ice Data Center, March–June 2004, updated daily, Digital media.
35. *MSPPS Orbital Data (MSPPS_ORB)*, May 2007. Available at http://www.class.noaa.gov.
36. R. R. Ferraro, F. Weng, N. C. Grody, L. Zhao, H. Meng, C. Kongoli, P. Pellegrino, S. Qiu, and C. Dean, NOAA operational hydrological products derived from the Advanced Microwave Sounding Unit, *IEEE Trans. Geosci. Remote Sens.*, 43(5), 1036–1049, 2005.
37. M. C. Hansen, R. S. DeFries, J. R. G. Townshend, and R. Sohlberg, Global land cover classification at 1 km spatial resolution using a classification tree approach, *Int. J. Remote Sens.*, 21, 1331–1364, 2000.
38. D. R. Legates, *A Climatology of Global Precipitation, Publications in Climatology*, Vol. 40, University of Delaware, 85 pp, 1987.
39. C. Surussavadee and D. H. Staelin, NPOESS precipitation retrievals using the ATMS passive microwave spectrometer, *IEEE Geosci. Remote Sens. Lett.*, 7(3), 440–444, 2010.

Part II

Image Processing for Remote Sensing

12 On SAR Image Processing
From Focusing to Target Recognition

Kun-Shan Chen and Yu-Chang Tzeng

CONTENTS

12.1 INTRODUCTION

Synthetic aperture radar (SAR) image understanding and interpretation are essential for remote sensing of Earth environment and target detection [1–12]. In the development of aided target recognition and identification system, the SAR image database with rich information content plays important roles. It is well recognized that these image base volumes are enormous and it is impractical to acquire them solely by *in situ* measurements. Hence, SAR simulation is one potential alternative to alleviating the problem. However, it is highly desirable to develop a full-blown SAR image simulation scheme with high verisimilitude including the sensor and target geolocation relative to the Earth, movement of the SAR sensor, SAR system parameters, radiometric and geometric characteristics of the target, and environment clutter. The simulation should at least include the computation of radar cross section (RCS) of targets, orbital parameters estimation, SAR echo signal generation, image focusing, and so on. Such a simulation scheme is also well suited for satellite SAR mission planning. Nevertheless, the simulation apparently is of high computational complexity and is heavy computational resource demanding. To release the heavy computational burden, a graphic processing unit (GPU)-based algorithm that explores and makes use of the graphic computation power is developed. As an application example of the simulated SAR images, the second part of this chapter deals with target recognition. Before doing so, feature enhancement and extraction are presented. It is well known that good feature enhancement is essential for target identification and recognition of

SAR images [13–18]. Novel algorithm proposed for spotlight mode SAR involves the formation of a cost function containing nonquadratic regularization constraints [19–20]. In stripmap mode, neither the radar antenna nor the target being mapped rotates, and the radar flight track is almost perpendicular to the look direction with a small squint angle. By reformulating the projection kernel and using it in an optimization equation form, an optimal estimate of the target's reflectivity field may be obtained [21]. As a result, the image fuzziness may be reduced and image fidelity was preserved. Thus, the target's features were adequately enhanced, and dominant scatterers could be well separated. Finally, target recognition from various commercial aircrafts was demonstrated by means of a simulated database and a neural classifier. The neural network is structured to allow its training by the Kalman filtering technique [22–26]. Performance evaluation was done from simulated images and various real images by Radarsat-2 and TerraSAR-X in stripmap mode.

12.2   SATELLITE SAR IMAGES SIMULATION

Figure 12.1 illustrates a functional block diagram of a satellite SAR image simulation and a follow-up target recognition. The simulation processing flow is basically adapted from Refs. [9,27,28]. Generally, the inputs include satellite and radar parameters setting, the target's computer aided design (CAD) model, and also the clutter model. The computation includes satellite orbit estimation, imaging geometry, target RCS, SAR echo, and raw signal generation. Also included is Doppler centroid and rate estimation, a critical step for image focusing. The most time-consuming part is echo and raw

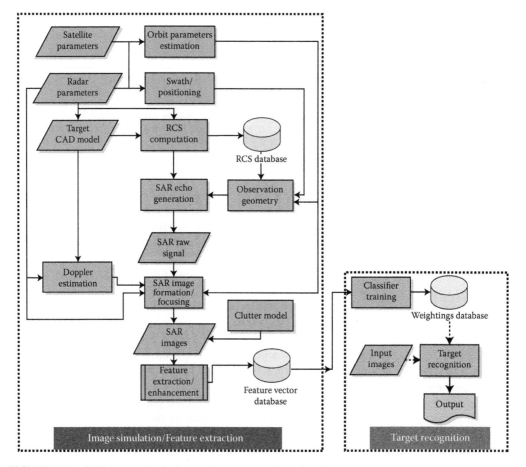

FIGURE 12.1   SAR image simulation and target recognition flowcharts.

signal generation, which is illustrated in Figure 12.1. Then a refined range-Doppler (RD) method is applied to perform the SAR image focusing. The outputs include sets of image for a desired target for a range of radar looking angles and orientation angles. Once the image database is built and made available, feature extraction and enhancement are performed, followed by a target recognition stage that includes a neural classifier training, neural weights storage for a later operation stage.

12.2.1 SAR Signal Model

A typical side-looking SAR observation geometry is shown in Figure 12.2 where the slant range between SAR and target, R, is a function of satellite moving time or slow time η or equivalently depends on the squint angle away from the zero-Doppler plane which corresponds to the slant range R_0. In SAR processing, it is essential to project R onto R_0.

An SAR transmits a linear frequency-modulated signal of the form

$$s_t(\tau) = w_r\left(\tau, T_p\right)\cos\left(2\pi f_c \tau + \pi \alpha_r \tau^2\right) \tag{12.1}$$

where α_r is the chirp rate, T_p is the pulse duration, f_c is the carrier frequency, τ is the ADC sampling time or fast time, and $w_r(\cdot)$ is a rectangular function

$$w_r\left(\tau, T_p\right) = \text{rect}\left(\frac{\tau}{T_p}\right) = \begin{cases} 1, & \left|\dfrac{\tau}{T_p}\right| \le 0.5 \\ 0, & \text{else.} \end{cases} \tag{12.2}$$

The received signal or raw data is a delayed version of the transmitted signal (Equation 12.1) [16],

$$\begin{aligned} s_r(\tau) &= A_0 s_t\left(\tau - \frac{2R}{c}\right) \\ &= A_0 w_r\left(\tau - \frac{2R}{c}, T_p\right) w_a(\eta)\cos\left\{2\pi f_0\left(\tau - \frac{2R}{c}\right) + \pi\alpha_r\left(\tau - \frac{2R}{c}\right)^2 + \varphi\right\}, \end{aligned} \tag{12.3}$$

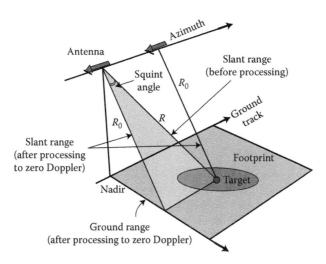

FIGURE 12.2 A typical stripmap SAR observation geometry. (From I. Cumming and F. Wong, *Digital Signal Processing of Synthetic Aperture Radar Data: Algorithms and Implementation*, Artech House, 2004. With permission.)

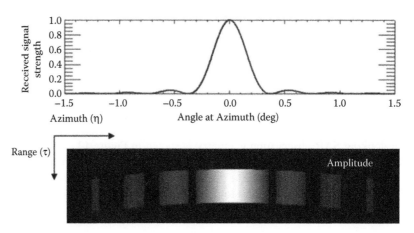

FIGURE 12.3 A typical antenna pattern along the azimuth direction and the received signal.

where R is the distance from the antenna to the target being observed, A_0 is the slant range backscatter coefficient of the target, φ is the phase term, and $w_a(\eta)$ is the antenna pattern and is a function of slow time. A commonly used pattern is of the form

$$w_a(\eta) \cong \mathrm{sinc}^2\left\{\frac{\sqrt{3}\theta(\eta)}{2\beta_{az}}\right\}, \tag{12.4}$$

where β_{az} is the azimuth beamwidth and $\theta(\eta)$ is the angle measured from boresight in the slant range plane. Figure 12.3 displays such an antenna pattern (upper) and received echo (lower) along the azimuth direction. The echo strength is varied according to antenna gain and is changed with flight time η. Note that the echo still comes with the carrier frequency that contains no target information and needs to be removed before further processing.

After demodulation, the received signal is given as [22]

$$s_0(\tau,\eta) = A_0 w_r\left(\tau - \frac{2R(\eta)}{c}, T_p\right)w_a(\eta)\exp\left\{-j\frac{4\pi f_c R(\eta)}{c} + j\pi\alpha_r\left(\tau - \frac{2R(\eta)}{c}\right)^2 + \varphi''\right\}, \tag{12.5}$$

where φ'' is the lumped sum of phase noise from the atmosphere, satellite altitude error, terrain, and so on, and is only of interest for interferometric SAR. Equation 12.5 serves as the fundamental signal model for the working process that follows.

12.2.2 RCS Computation

To facilitate the radar response from a target, we need the target's radar cross section under a radar observation scenario. The radar backscattering characteristics is taken into account. The coherent scattering process between a target and its background is neglected for the sake of simplicity. Also, fully polarimetric response is not considered, however. Radar Cross Section Analysis and Visualization System [29] is a powerful algorithm that utilizes the physical optics (PO), physical diffraction theory (PDT), and shooting and bouncing rays (SBRs) to compute the RCS of complex radar targets [30–34]. Single scattering and diffraction from a target are first computed by PO and PDT, followed by SBRs to account for multiple scattering and diffraction. The system outputs for a given three-dimensional (3D) CAD model of the target of interest. The CAD model contains

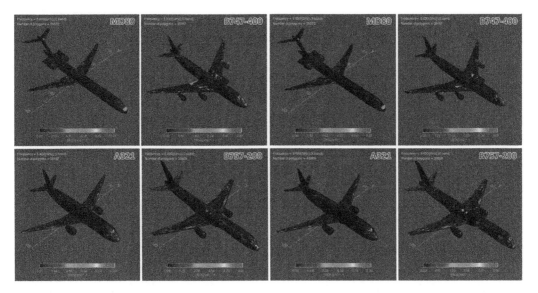

FIGURE 12.4 **(See color insert.)** Computed RCS of commercial aircrafts for C-band HH polarization (left) and X-band VV polarization (right).

numerous grids or polygons, each associated with computed RCS as a function of incident and aspect angles for a given set of radar parameters. The number of polygons is determined by the target's geometry complexity and its electromagnetic size. To realize the imaging scenario, each polygon must be properly oriented and positioned based on ECR coordinates. Figure 12.4 displays the computed RCSs of these aircraft for the cases of Radarsat-2 and TerraSAR-X.

12.2.3 CLUTTER MODEL

Although the coherent scattering process between a target and its background is neglected for the sake of simplicity, the clutter from background is incoherently integrated into the radar echo. Extensive studies on SAR speckle and its reduction have been documented [12,35–37]. Many studies [38–41] suggest that for surface-like clutter such as vegetation canopy [42,43] and airport runway, the radar signal statistics follows the Weibull distribution fairly well. For our aircraft targets of interest, we applied the Weibull distribution for the runway clutter model. Other models can be easily adopted in the simulation chain.

$$p(x) = \frac{\kappa}{\lambda}\left(\frac{x}{\lambda}\right)^{\kappa-1} \exp\left[-(x/\lambda)^\kappa\right], \tag{12.6}$$

where the shape parameter k and the scale parameter λ may be estimated from real SAR images over the runway with mean amplitude A [37]:

$$k = \frac{\pi}{\text{std}\left[\ln(A)\right]\sqrt{6}} \tag{12.7}$$

$$\lambda = \exp\left\{\left[<\ln(A)>\right] + \frac{0.5722}{\kappa}\right\} \tag{12.8}$$

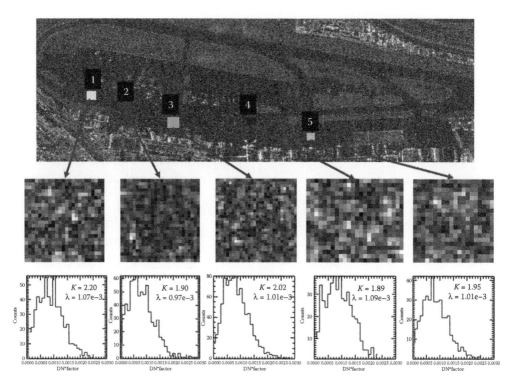

FIGURE 12.5 Samples of clutter and estimated model parameters on a TerraSAR-X image.

Figure 12.5 displays sample regions of clutter and estimated model parameters on a TerraSAR-X image. The model parameters are obtained by averaging several samples.

12.2.4 SATELLITE ORBIT DETERMINATION

In applying Equation 12.5, the range between a satellite in space and a target being imaged on the ground must be precisely determined for matched filtering. That is to say, we need to know the satellite and target position vectors in order to estimate the Doppler frequency and its rate. Before doing so, we have to define and then determine the reference systems for time and geodetic. Six fundamental parameters required to determine the orbital position include the semimajor axis a, eccentricity e, inclination angle i, right ascension of ascending node Ω, argument of perigee ω, and true anomaly v. Three of the parameters determine the orientation of the orbit or trajectory plane in space, while three locate the body in the orbital plane [44,45]. These six parameters are uniquely related to the position and velocity of the satellite at a given epoch.

To deduce the orbital elements of a satellite, at least six independent measurements are needed [44]. In this chapter, we adopted simplified perturbations models (SPM) [46–48] to estimate the orbital state vectors of satellite in the Earth Centered Inertial coordinate system. A C++ version program code by Henry [46] was used. Once the orbit parameters are determined, we can transform it to Earth Centered Rotation coordinate system, as described below. The transformation matrix is of the form [44]

$$\mathbf{U}^{ECI}_{ECR} = \Pi\Theta\mathbf{NP}, \tag{12.9}$$

where matrices Π, Θ, $\mathbf{N}$, and $\mathbf{P}$ represent, respectively, polar motion, Earth rotation, nutation, and precession matrices:

$$\Theta(t) = \mathbf{R}_z(\text{GAST}), \tag{12.10}$$

$$\mathbf{P}(T_1, T_2) = \mathbf{R}_z\left(-z(T,t)\right)\mathbf{R}_y\left(\vartheta(T,t)\right)\mathbf{R}_z\left(-\xi(T,t)\right), \tag{12.11}$$

where GAST is Greenwich apparent sidereal time. Referring to Figure 12.6, the nutation matrix is

$$\mathbf{N}(T) = \mathbf{R}_x(-\varepsilon - \Delta\varepsilon)\mathbf{R}_z(-\Delta\psi)\mathbf{R}_z(\varepsilon) \tag{12.12}$$

Note that from Ref. [44], in computing the derivative of the transformation, the precession, nutation, and polar motion matrices may be considered as constant.

$$\frac{d\mathbf{U}_{\text{ECR}}^{\text{ECI}}}{dt} \approx \Pi \frac{d\Theta}{dt}\mathbf{N}\mathbf{P}. \tag{12.13}$$

Then the state vectors in the transformation are

$$r_{\text{ECR}} = \mathbf{U}_{\text{ECR}}^{\text{ECI}}\mathbf{r}_{\text{ECI}}, \tag{12.14}$$

$$\mathbf{v}_{\text{ECR}} = \mathbf{U}_{\text{ECR}}^{\text{ECI}}\mathbf{v}_{\text{ECI}} + \frac{d\mathbf{U}_{\text{ECR}}^{\text{ECI}}}{dt}\mathbf{r}_{\text{ECI}}, \tag{12.15}$$

$$r_{\text{ECI}} = \left(\mathbf{U}_{\text{ECR}}^{\text{ECI}}\right)^{\text{T}}\mathbf{r}_{\text{ECR}}, \tag{12.16}$$

and

$$\mathbf{v}_{\text{ECI}} = \left(\mathbf{U}_{\text{ECR}}^{\text{ECI}}\right)^{\text{T}}\mathbf{v}_{\text{ECR}} + \frac{d\left(\mathbf{U}_{\text{ECR}}^{\text{ECI}}\right)^{\text{T}}}{dt}\mathbf{r}_{\text{ECR}}. \tag{12.17}$$

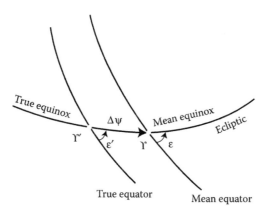

FIGURE 12.6 Illustration of nutation.

In summary, four key steps are needed:

1. Find the satellite position: calculated from two-line elements data and imaging time duration.
2. Find the radar beam pointing vector derived from the satellite attitude (pitch, roll, and yaw angle).
3. Locate the target center position derived from the satellite position and the line of sight.
4. Locate each target's polygon position derived from the target center position and the target aspect angle.

12.2.5 IMAGE FOCUSING

Several focusing algorithms have been proposed, including RD, omega-K, chirp-scaling, and so on [27,49,50] and their improved versions, with each bearing pros and cons. We adopted the refined RO algorithm with secondary range compression because of its fast computation while it maintains reasonably good spatial resolution and a small defocusing error. Figure 12.7 outlines the functional block diagram of processing steps in the refined RD algorithm.

The power balance method in conjunction with the average cross-correlation coefficient method was used to obtain a rough estimate of baseband Doppler frequency [51]. To obtain absolute Doppler centroid frequency, f_{DC}, algorithms such as multilook cross-correlation, multilook beat frequency, and wave number domain algorithm may be utilized to determine the ambiguity number [27]:

$$f_{DC} = f_{DC,base} + M_{amb}PRF, \tag{12.18}$$

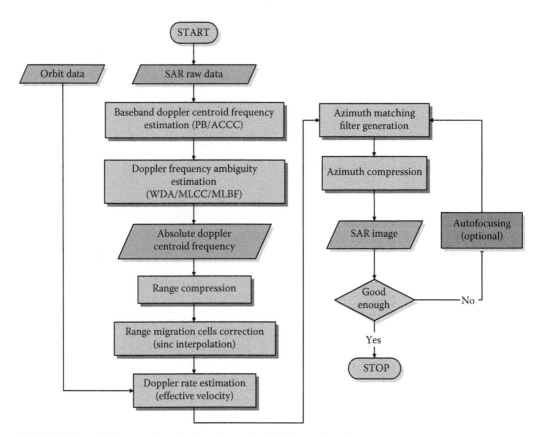

FIGURE 12.7 SAR image-focusing flowchart of an RD-based algorithm.

where $f_{DC,base}$ is the baseband part of PRF, M_{amb} is the ambiguity number, and PRF is the pulse repetition frequency.

After an effective satellite velocity, V_r, is estimated from orbit state vectors, the Doppler rate estimation can be computed via

$$f_R\left(R_0\right) = \frac{2V_r^2\left(R_0\right)\cos\left(\theta_{sq,c}\right)}{R_0}, \tag{12.19}$$

where $\theta_{sq,c}$ is the squint angle and R_0 is referred to as the zero-Doppler plane (see Figure 12.2). Note that in some cases, Equation 12.19 is just a rough estimate, and then autofocus via the phase gradient algorithm [27] may be further applied to refine the Doppler rate given above.

12.3 COMPUTATION COMPLEXITY AND VALIDATION

An SAR image is sensitive to the target's geometry, including orientation and aspects angles. For target recognition and identification, a more complete database for feature extraction is preferable to achieve better performance and reduce the false alarm rate. In SAR image simulation, suppose that n samples (incident angles and aspect angles) are desired; then, the computation complexity is $\theta(n^3)$. Table 12.1 lists CPU hours to complete one TerraSAR-X image simulation of MD80 aircraft using 25672 polygons representing RCS for just a pose (one incident angle and one aspect angle). It would take about 11 months to complete all poses for incident angles from $20°$ to $50°$ with $5°$ a step and aspect angles from $−180°$ to $\sim+180°$ with $1°$ a step (total 2520 poses). Apparently, finding methods to speed up the computation time is essential for practical use.

As for a GPU, work load is assumed to be highly parallel—many data to be processed by the same algorithm. Recently, applications of GPUs to SAR simulation are reported [52–53]. Based on that assumption, each processing unit is designed to handle many threads. Processing units work as a group to maximize throughput of all threads. Latency can be hidden by skipping stalled threads, if there are few compared to the number of eligible threads. Grouping means shared control logics and cache. The GPU-based SAR simulation is divided into a grid of blocks. Each block consists of a number of *threads*, which are executed in a *multiprocessor* (also named as *stream multiprocessor* or *block*). Compute unified device architecture (CUDA) developed by NVIDIA [54] is used for the proposed GPU-based implementation on a Linux operating system. The usage of the shared memory implemented in the proposed method is applied to the data-intensive computational tasks of the GPU-based computations [55]. The data with high dependency are assigned into the same block of the shared memory with a finer granularity of parallel implementation. To make use of this highly efficient memory architecture, we devised local variables and parameters by using these registers. The GPU-based experiments were performed on a low-cost 960 (240)-core NVIDIA Tesla personal supercomputer with one Intel Xeon 5504 quad-core CPU and four NVIDIA GTX295 (240 core) GPUs. Figure 12.8 shows the speed-up performance using the GPU algorithm against a CPU one, as a function of the number of samples

TABLE 12.1

CPU Hours to Complete a TerraSAR-X Simulated Image of MD80 for One Pose (with 25672 RCS Polygons)

OS	RAM (GB)	CPU (GHz)	Multithreading	CPU Time (Hours)
Win-XP 64 bits	1.97	2.4 (4 cores)	OFF	20.1
			ON	18.7
Win-XP 32 bits	3.24	2.8 (2 cores)	OFF	21.1

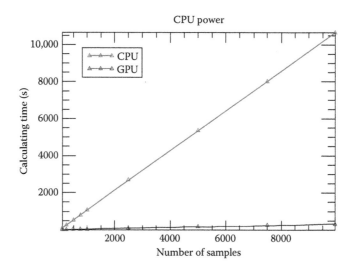

FIGURE 12.8 Speedup performance as a function of number of polygons using a GPU algorithm.

(polygons in target's RCS) for one pose. As the number of samples increases, CPU time grows very fast. The required computation time is absorbed by GPUs, as obviously seen from the chart. The average speed-up ratio to CPU is about 32. With a dual-GPU configuration, 65 times speed-up was boosted.

12.4 APPLICATION TO TARGET RECOGNITION

The proposed working flows and algorithms were validated by evaluating the image quality, including geometric and radiometric accuracy using simple point targets first, followed by simulating four types of commercial aircraft : A321, B757-200, B747-400, and MD80. The satellite SAR systems for simulation include ERS-1, ERS-2, Radarsat-2, TerraSAR-X, and ALOS PALSAR, but others can be easily realized. Table 12.2 lists the key image quality indices for a simulated image of TerraSAR-X and PLASAR satellite SAR systems. Quality indices include 3 dB azimuth beamwidth (m), peak-to-side lobe ratio (dB), integrated side lobe ratio (dB), and scene center accuracy in different coordinate systems. As can be seen from the table, the simulated images are all well within the nominal specifications for different satellite systems.

12.4.1 FEATURE ENHANCEMENT

The essential information for target classification, detection, recognition, and identification is through feature selection for the training phase. For our interest, depending on the spatial resolution offered by Radarsat-2 and TerraSAR-X in stripmap mode, we aim at target recognition. Before feature extraction, feature enhancement is performed first. To enhance the stripmap mode data using the nonquadratic regularization method [19,20], one has to modify the projection operator kernel accordingly. The received signal s_o in Equation 12.5 may be expressed as [19–21]:

$$s_o = Tf + s_n, \tag{12.20}$$

where s_n represents noise, and T is the projection operation kernel with dimension $MN \times MN$ which plays a key role in contrast enhancement if the signal in Equation 12.5 is of dimension $M \times N$. It has

TABLE 12.2
Comparison of Simulated Image Quality for a Point Target

Index		ALOS PALSAR		TerraSAR-X	
		Simulation	Nominal	Simulation	Nominal
3 dB Azimuth beamwidth (m)		4.26	5.10	3.49	4.51
PSLR (dB)	Slant range	−24.85	−20.5	−22.85	−13
	Azimuth	−25.42		−31.16	
ISLR (dB)	Slant range	−30.62	−15.0	−31.62	−17
	Azimuth	−25.62		−43.98	
Scene center	Long, Lat (°)	121.49927°E, 24.764622°N	121.49930°E, 24.764477°N	119.395090°E, 24.894470°N	119.39506°E, 24.894620°N
	ECR (x,y,z) (m)	−3027814.2, 4941079.3, 2655407.1	−3027808.4, 4941075.0, 2655421.7	−2817396.77, 5608995.60, 2830631.97	−2817397.45, 5608996.11, 2830630.29
	Long/Lat difference (°)	-3×10^{-5}, 1.45×10^{-4}		-3×10^{-6}, 1.5×10^{-4}	
	ECR difference (ground range, azimuth) (m)	0.98, −16.26		0.06, 16.97	

been shown that [3,4] nonquadratic regularization is practically effective in minimizing the clutter while emphasizing the target features via

$$\hat{\mathbf{f}} = \arg\min\left\{ \left\| \mathbf{s}_r - \mathbf{Tf} \right\|_2^2 + \gamma^2 \left\| \mathbf{f} \right\|_p^p \right\},$$ (12.21)

where $\| \; \|_p$ denotes ℓ_p-norm $(p \le 1)$, γ^2 is a scalar parameter, and $\left\{ \left\| \mathbf{s}_o - \mathbf{Tf} \right\|_2^2 + \gamma^2 \left\| \mathbf{f} \right\|_p^p \right\}$ is recognized as the cost or objective function.

To easily facilitate the numerical implementation, both $\mathbf{s}_o$ and $\mathbf{f}$ may be formed as long vectors, with $\mathbf{T}$ being a matrix. Then, from Equations 12.5 and 12.20, we may write the projection operation kernel for the stripmap mode as [21]

$$\mathbf{T} = \exp\left\{ -i\left[\omega_0 \frac{2\left(\mathbf{r} + \frac{(\mathbf{x} - \mathbf{x}_0)^2}{2\mathbf{r}} \right)}{c} - \alpha\left(\mathbf{t} - \frac{2\left(\mathbf{r} + \frac{(\mathbf{x} - \mathbf{x}_0)^2}{2\mathbf{r}} \right)}{c} \right)^2 \right] \right\},$$ (12.22)

where $\mathbf{t}$, $\mathbf{r}$, $\mathbf{x} - \mathbf{x}_0$ are matrices to be described below by first defining the following notations to facilitate the formation of these matrices:

N : the number of discrete sampling points along the slant range direction
M : the number of discrete sampling points along the azimuth direction
Δt : the sampling interval along the slant range
ℓ_r : the size of the footprint along the slant range
ℓ_x : the size of the footprint along the azimuth direction
$\mathbf{1}_l$: the column vector of dimension $l \times 1$ and all elements equal to one, in which $l = M$ or $l = MN$

With these notions, we can obtain explicit $\mathbf{t}$, $\mathbf{r}$, $\mathbf{x} - \mathbf{x}_0$ forms, after some mathematical derivations [21].

$$\mathbf{t} = [\mathbf{1}_M \otimes \mathbf{M}_1]_{MN \times MN},$$ (12.23)

$$\mathbf{v}_1 = \begin{bmatrix} 0 \\ 1 \\ \vdots \\ N-1 \end{bmatrix} \Delta t = \begin{bmatrix} 0 \\ \Delta t \\ \vdots \\ (N-1)\Delta t \end{bmatrix}_{N \times 1}, \tag{12.24a}$$

$$\mathbf{M}_1 = \mathbf{V}_1 \cdot \mathbf{1}_{MN}^T = \begin{bmatrix} 0 & \cdots & 0 \\ \Delta t & & \Delta t \\ \vdots & & \vdots \\ (N-1)\Delta t & \cdots & (N-1)\Delta t \end{bmatrix}_{N \times MN}, \tag{12.24b}$$

$$\mathbf{r} = [\mathbf{M}_2 \otimes \mathbf{V}_2^T]_{MN \times MN}, \tag{12.25}$$

$$\mathbf{V}_2 = \begin{bmatrix} \dfrac{-\ell_r}{2} + r_0 \\ \dfrac{-\ell_r}{2} + r_0 + \Delta\ell \\ \vdots \\ \dfrac{-\ell_r}{2} + r_0 + (N-1)\Delta\ell \end{bmatrix}_{N \times 1}, \quad \Delta\ell = \dfrac{\ell_r}{N}, \tag{12.26a}$$

$$\mathbf{M}_2 = \mathbf{1}_M^T \otimes \mathbf{1}_{MN} = \begin{bmatrix} \mathbf{1}_{MN} & \cdots & \mathbf{1}_{MN} \end{bmatrix}_{MN \times MN}, \tag{12.26b}$$

$$\mathbf{x} - \mathbf{x}_0 = [\mathbf{M}_3 \otimes \mathbf{M}_4]_{MN \times MN}, \tag{12.27}$$

$$\mathbf{V}_3 = \begin{bmatrix} 0 \\ 1 \\ \vdots \\ M-1 \end{bmatrix} \dfrac{\ell_x}{M} - \dfrac{\ell_x}{2} = \begin{bmatrix} \dfrac{-\ell_x}{2} \\ \dfrac{\ell_x}{M} - \dfrac{\ell_x}{2} \\ \vdots \\ (M-1)\dfrac{\ell_x}{M} - \dfrac{\ell_x}{2} \end{bmatrix}_{M \times 1}, \tag{12.28a}$$

$$\mathbf{V}_4 = \begin{bmatrix} \mathbf{V}_3[\delta - 1 : (M-1)]_{\delta \times 1} \\ \mathbf{0}_{(M-\delta) \times 1} \end{bmatrix}_{M \times 1}; \quad \delta = \left\lfloor \dfrac{M-1}{2} \right\rfloor + 1, \tag{12.28b}$$

$$\mathbf{V}_5 = \begin{bmatrix} \mathrm{Mirror}(\mathbf{V}_3[0 : \delta - 1])_{\delta \times 1} \\ \mathbf{0}_{(M-\delta) \times 1} \end{bmatrix}_{M \times 1}, \tag{12.28c}$$

$$\mathbf{M}_3 = \begin{bmatrix} \mathrm{Toep}(\mathbf{V}_5^T, \mathbf{V}_4^T) \end{bmatrix}_{M \times M}, \tag{12.28d}$$

$$\mathbf{M}_4 = \left[\mathbf{1}_{N \times 1} \cdot \mathbf{1}_{N \times 1}^T \right]_{N \times N}, \tag{12.28e}$$

In the above equations, each of the bold-faced letters denotes a matrix and $\otimes$ represents the Kronecker product. The operation $\mathbf{V}[m:n]$ takes element m to element n from vector $\mathbf{V}$, and Toep$(\cdot)$ converts the input into a Toeplitz matrix [56,57]. Note that in Equation 12.28c, by Mirror $[a:b, c:d]_{p \times q}$ we mean taking elements a to b along the rows and elements c to d along the columns, so that the resulting matrix is of size $p \times q$.

12.4.2 FEATURE VECTORS AND EXTRACTION

The feature vector contains two types: fractal geometry and scattering characteristics. In the fractal domain, the image is converted into a fractal image [58]. It has been explored that SAR signals may be treated as a spatial chaotic system because of the chaotic scattering phenomena [59–63]. Applications of fractal geometry to SAR analysis are studied in Refs. [64,65]. There are many techniques proposed to estimate the fractal dimension of an image. Among them, the wavelet approach proves both accurate and efficient. It stems from the fact that the fractal dimension of an N-dimensional random process can be characterized in terms of fractional Brownian motion (fBm) [58]. The power spectral density of the fBm is written as

$$P(f) \propto f^{-(2H+D)} \tag{12.29}$$

where $0 < H < 1$ is the persistence of the fBm and D is the topological dimension ($= 2$ in the image). The fractal dimension of this random process is given by $D = 3 - H$. As image texture, an SAR fractal image is extracted from SAR imagery data based on local fractal dimension. Therefore, wavelet transform can be applied to estimate the local fractal dimension of an SAR image. From Equation 12.29, the power spectrum of an image is therefore given by

$$P(u,v) = \upsilon \left(\sqrt{u^2 + v^2} \right)^{-2H-2}, \tag{12.30}$$

where υ is a constant. Based on the multiresolution analysis, the discrete detailed signal of an image I at a resolution level j can be written as [66–67]

$$\begin{aligned} D_j I &= \left\langle I(x,y), 2^{-j} \Psi_j \left(x - 2^{-j} n, y - 2^{-j} m \right) \right\rangle \\ &= \left(I(x,y) \otimes 2^{-j} \Psi_j (-x,-y) \right) \left(2^{-j} n, 2^{-j} m \right) \end{aligned} \tag{12.31}$$

where $\otimes$ denotes a convolution operator, $\Psi_j(x,y) = 2^{2j} \Psi(2^j x, 2^j y)$ and $\Psi(x,y)$ is a two-dimensional wavelet function. The discrete detailed signal, thus, can be obtained by filtering the signal with $2^{-j} \Psi_j(-x,-y)$ and sampling the output at a rate 2^{-j}. The power spectrum of the filtered image is given by [67]

$$P_j(u,v) = 2^{-2j} P(u,v) \left| \tilde{\Psi}_j(u,v) \right|^2, \tag{12.32}$$

where $\tilde{\Psi}_j(u,v) = \tilde{\Psi}(2^{-j} u, 2^{-j} v)$ and $\tilde{\Psi}(u,v)$ is the Fourier transform of $\Psi(u,v)$. After sampling, the power spectrum of the discrete detailed signal becomes

$$P_j^d (u,v) = 2^j \sum_k \sum_l P_j \left(u + 2^{-j}2k\pi, v + 2^{-j}2l\pi\right). \tag{12.33}$$

Let σ_j^2 be the energy of the discrete detailed signal:

$$\sigma_j^2 = \frac{2^{-j}}{(2\pi)^2} \iint P_j^d (u,v)\,du\,dv. \tag{12.34}$$

By inserting Equations 12.32 and 12.33 into Equation 12.34 and changing variables in this integral, Equation 12.34 may be expressed as $\sigma_j^2 = 2^{-2H-2}\sigma_{j-1}^2$.

Therefore, the fractal dimension of an SAR image can be obtained by computing the ratio of the energy of the detailed images:

$$D = \frac{1}{2}\log_2 \frac{\sigma_j^2}{\sigma_{j-1}^2} + 2 \tag{12.35}$$

A fractal image indeed represents information regarding spatial variation; hence, its dimension estimation can be realized by sliding a preset size window over the entire image. The selection of the window size is subject to reliability and homogeneity considerations with the center pixel of the window replaced by the local estimate of the fractal dimension. Once fractal image is generated, features of angle, target area, long axis, short axis, and axis ratio are extracted from the target of interest. As for scattering center, features of major direction [X], major direction [Y], minor direction [X], minor direction [Y], major variance, and minor variance are selected, in addition to radar look angle and aspect angle. Figure 12.9 displays such a feature vector from simulated MD80 and B757-200 aircrafts from Radarsat-2. Finally, the recognition is done by a dynamic learning neural classifier [22–26]. This classifier is structured using a polynomial basis function model. A digital Kalman filtering scheme [68] is applied to train the network. The necessary time to complete the

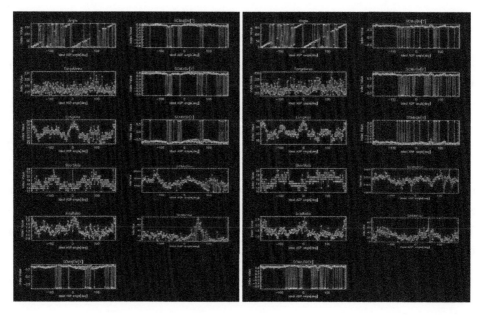

FIGURE 12.9 Selected features of simulated Radarsat-2 images of A321 (left) and B757-200 (right) for incidence angle of 35° and aspect angle from −180° ~ +180° (1° step).

training basically is not sensitive to the network size and is fast. Also, the network allows recursive training when new and updated training data sets are available without revoking training from scratch. The classifier is structured with 13 input nodes to feed the target features, 350 hidden nodes in each of the four hidden layers, and four output nodes representing four aircraft targets (A321, B757-200, B747-400, and MD80).

12.4.3 Performance Evaluation

12.4.3.1 Simulated SAR Images

Both simulated Radarsat-2 and TerraSAR-X images for four targets, as described above, are used to evaluate the classification and recognition performance. For this purpose, Radarsat-2 images with an incidence angle of 45° and TerraSAR-X with 30° of incidence were tested. Both are with a spatial resolution of 3 m in stripmap mode. The training data contain all 360 aspect angles (1° a step); among them 180 samples were randomly taken to test. Fast convergence of neural network learning is observed. The confusion matrix for Radarsat-2 and TerraSAR-X in classifying four targets is given in Tables 12.3 and 12.4, respectively. The overall accuracy and kappa coefficient are very satisfactory. Higher confusion between B757-200 and MD80 was observed. It is also noted that with Radarsat-2 and TerraSAR-X systems, comparable results were obtained as long as classification rate is concerned.

12.4.3.2 Real SAR Images

Data acquisition on May 8, 2008 from TerraSAR-X over an airfield was processed into four looks images with a spatial resolution of 3 m in stripmap mode. Feature enhancement by an operator kernel in Equation 12.22 was performed. Meanwhile, ground truth collections were conducted to identify the targets. Figure 12.10 displays such acquired images where the visually identified targets are marked and their associated image chips were later fed into neural classifier that is trained by the simulated image database. Among all the targets, three types of them are contained in simulated

TABLE 12.3

Confusion Matrix of Classifying Four Targets from Simulated Radarsat-2 Images

		A321	B747-400	B757-200	MD80	Producer Accuracy
Target	A321	165	2	6	7	0.917
	B747-400	0	175	3	2	0.972
	B757-200	4	1	164	11	0.911
	MD80	4	2	11	163	0.906
User accuracy		0.954	0.972	0.891	0.891	
		Overall accuracy: 0.926, kappa coefficient: 0.902				

TABLE 12.4

Confusion Matrix of Classifying Four Targets from Simulated TerraSAR-X Images

		A321	B747-400	B757-200	MD80	Producer Accuracy
Target	A321	166	0	5	9	0.922
	B747-400	0	179	0	1	0.994
	B757-200	2	0	168	10	0.933
	MD80	1	0	8	171	0.950
User accuracy		0.982	1.000	0.928	0.895	
		Overall accuracy: 0.950, kappa coefficient: 0.933				

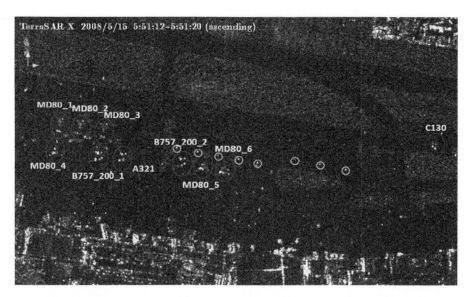

FIGURE 12.10 A TerraSAR-X image over an airfield acquired on May 15, 2008. The identified targets are marked from ground truth.

database: MD80, B757-200, and A321. From Table 12.5, these targets are well recognized, where the numeric value represents membership. Winner-takes-all approach was taken to determine the classified target and whether these targets are recognizable. More sophisticated schemes such as type-II fuzzy may be adopted in the future. It is realized that for certain poses, there exists confusion between MD80 and B757-200, as already demonstrated in the simulation test above. It was not able to recognize target C130, as expected.

As another example, a blind test was performed. Here, blind means the targets to be recognized are not known to the tester, nor is the image acquisition time. The ground truth was collected by the third party and was only provided after recognition operation was completed. This is very close to the real situation for recognition operation. A total of 12 targets (T1–T12) were chosen for the test, as indicated in Figure 12.11. Unlike in Figure 12.10, this image was acquired in descending mode but was unknown at the time of test. Table 12.6 gives the test results, where the recognized targets and truth are listed. It is readily indicated that all the MD80 targets were successfully recognized,

TABLE 12.5

Membership Matrix for Target Recognition on TerraSAR-X Image of Figure 12.10

	Output				
Input	A321 (%)	B747-400 (%)	B757-200 (%)	MD80 (%)	Recognizable
MD80_1	16.61	7.26	0	76.12	Yes
MD80_2	0	9.63	18.50	71.85	Yes
MD80_3	0	3.63	8.78	87.58	Yes
MD80_4	0	3.64	23.08	73.27	Yes
MD80_5	17.88	12.28	0	69.82	Yes
MD80_6	5.64	0	46.72	47.63	Yes
B757-200_1	12.05	11.84	76.09	0	Yes
B757-200_2	0	19.41	78.55	2.03	Yes
A321	63.28	15.09	0	21.61	Yes
C130	—	—	—	—	—

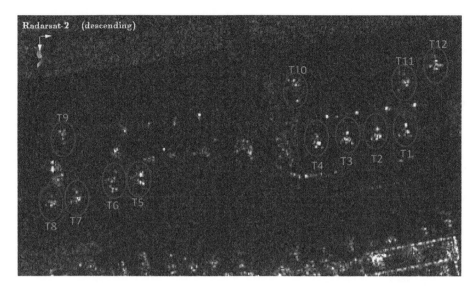

FIGURE 12.11 A Radarsat-2 image over an airfield. The identified targets are marked.

TABLE 12.6
Membership Matrix for Target Recognition on Raradsat-2 Image of Figure 12.11

	A321 (%)	B747-400 (%)	B757-200 (%)	MD80 (%)	Recognized	Truth
T1	19.25	0	22.88	57.86	MD80	MD80
T2	0.01	0	0	99.98	MD80	MD80
T3	14.99	0	11.59	73.40	MD80	MD80
T4	21.31	10.81	15.27	52.60	MD80	MD80
T5	0	3.09	95.10	1.79	B757-200	E190
T6	96.82	1.70	0	1.46	A321	B737-800
T7	0.13	0	0.17	99.69	MD80	MD80
T8	0.10	0.04	0	99.84	MD80	MD80
T9	63.28	15.09	0	21.61	A321	FOKKER-100
T10	6.48	2.26	0	91.24	MD80	MD80
T11	43.61	15.59	40.59	0	A321	DASH-8
T12						C130

while four types of target were wrongly recognized. This is mainly attributed to the lack of database. Again, the T12 target was not completely recognizable by the system for the same reason. Enhancement and updation of the target database are clearly essential.

12.5 CONCLUSIONS

In this chapter, we presented a full-blown satellite SAR image simulation scheme based on the GPU computation algorithm. The simulation steps included orbit state vector estimation, imaging scenario setting, target RCS computation, and clutter model, all specified by an SAR system specification. As an application example, target recognition has been successfully demonstrated using the simulated image database as training sets. Extended tests on both the simulated image and real images were conducted to validate the proposed algorithm. To this end, it is suggested that a more powerful target recognition scheme be explored for high-resolution SAR images. Extension to fully polarimetric SAR image simulation seems highly desirable as more such data are being available for much better target discrimination capability. In this aspect, further improvement on the computational efficiency by taking advantage of GPU power is critical.

REFERENCES

1. F. T. Ulaby, R. K. Moore, and A. K. Fung, *Microwave Remote Sensing: Active and Passive, Vol. 2, Radar Remote Sensing and Surface Scattering and Emission Theory*, Norwood, MA: Artech House, 1982.
2. A. W. Rihaczek and Steven J. Hershkowitz, *Radar Resolution and Complex-Image Analysis*, Norwood, MA: Artech House, 1996.
3. A. W. Rihaczek and Steven J. Hershkowitz, *Theory and Practice of Radar Target Identification*, Boston, MA: Artech House, 2000.
4. C. H. Chen, ed., *Information Processing for Remote Sensing*, Singapore: World Scientific Publishing Co., 2000.
5. J. C. Leachtenauer and R. G. Driggers, *Surveillance and Reconnaissance Systems: Modeling and Performance Prediction*, Boston, MA: Artech House, 2001.
6. C. Oliver and S. Quegan, *Understanding Synthetic Aperture Radar Images*, Raleigh, NC: SciTech Publishing, 2004.
7. J. R. Schott, *Remote Sensing: The Image Chain Approach*, Oxford, UK: Oxford University Press, 2007.
8. M. Skolnik, *Radar Handbook*, 3rd edition, New York, NY: McGraw-Hill, 2008.
9. J.-S. Lee and E. Pottier, *Polarimetric Radar Imaging: From Basics to Applications*, Boca Raton, FL: CRC Press, 2009.
10. G. Margarit, J.-J. Mallorqui, J. M. Rius, and J. Sanz-Marcos, On the usage of GRECOSAR, an orbital polarimetric SAR simulator of complex targets, to vessel classification studies, *IEEE Trans. Geosci. Remote Sens.*, 44, 3517–3526, 2006.
11. D. Lemoine Brunner, G. Bruzzone, and H. Greidanus, Building height retrieval from VHR SAR imagery based on an iterative simulation and matching technique, *IEEE Trans. Geosci. Remote Sens.*, 48, 1487–1504, 2010.
12. J. S. Lee, Digital image enhancement and noise filtering by use of local statistics, *IEEE Trans. Pattern Anal. Mach. Intell.*, 2, 165–168, 1980.
13. S. Musman, D. Kerr, and C. Bachmann, Automatic recognition of ISAR ship images, *IEEE Trans. Aerosp. Electron. Syst.*, 32, 1392–1404, 1996.
14. M. Migliaccio, G. Ferrara, A. Gambardella, F. Nunziata, and A. Sorrentino, A physically consistent speckle model for marine SLC SAR images, *IEEE J. Ocean. Eng.*, 32, 837, 2007.
15. G. Margarit, Mallorqui, J. J. Fortuny-Guasch, and C J. Lopez-Martinez, Phenomenological vessel scattering study based on simulated inverse SAR imagery, *IEEE Trans. Geosci. Remote Sens.*, 47, 1212–1223, 2009.
16. J. Wang and L. Sun, Study on ship target detection and recognition in SAR imagery, in *CISE '09 Proceedings of the 2009 First IEEE International Conference on Information Science and Engineering*, Nanjing, China, pp. 1456–1459, 2009.
17. B. Bhanu and G. Jones, Object recognition results using MSTAR synthetic aperture radar data, in *Proceedings of IEEE Workshop on Computer Vision Beyond the Visible Spectrum: Methods and Applications*, Hilton Head, South Carolina, pp. 55–62, 2000.
18. M. Nishimoto, X. Liao, and L. Carin., Target identification from multi-aspect high range-resolution radar signatures using a hidden Markov model, *IEICE Trans. Electron E Series C*, 87, 1706–1714, 2004.
19. M. Çetin and W. C. Karl, Feature-enhanced synthetic aperture radar image formation based on nonquadratic regularization, *IEEE Trans. Image Process.*, 10, 623–631, 2001.
20. M. Çetin, W. C. Karl, and A. S. Willsky, Feature-preserving regularization method for complex-valued inverse problems with application to coherent imaging, *Opt. Eng.*, 45, 017003-1 ~ 11, 2006.
21. C. Y. Chiang, K. S. Chen, C. T. Wang, and N. S. Chou, Feature enhancement of stripmap-mode SAR images based on an optimization scheme, *IEEE Geosci. Remote Sens. Lett.*, 6, 870–874, 2009.
22. Y. Tzeng, K. S. Chen, W. L. Kao, and A. K. Fung, A dynamic learning neural network for remote-sensing applications, *IEEE Trans. Geosci. Remote Sens.*, 32, 1096–1102, 1994.
23. K. S. Chen, Y. C. Tzeng, C. F. Chen, and W. L. Kao, Land-cover classification of multispectral imagery using a dynamic learning neural network, *Photogramm. Eng. Remote Sens.*, 61, 403–408 1995.
24. K. S. Chen, W. L. Kao, and Y. C. Tzeng, Retrieval of surface parameters using dynamic learning neural network, *Int. J. Remote Sens.*, 16, 801–809, 1995.

25. K. S. Chen, W. P. Huang, D. W. Tsay, and F. Amar, Classification of multifrequency polarimetric SAR image using a dynamic learning neural network, *IEEE Trans. Geosci. Remote Sens.*, 34, 814–820, 1996.
26. Y. C. Tzeng and K. S. Chen, A fuzzy neural network to SAR image classification, *IEEE Trans. Geosci. Remote Sens.*, 36, 301–307, 1998.
27. I. Cumming and F. Wong, *Digital Signal Processing of Synthetic Aperture Radar Data: Algorithms and Implementation*, Boston, MA: Artech House, 2004.
28. J. C. Curlander and R. N. McDonough, *Synthetic Aperture Radar: Systems and Signal Processing*, New York, NY: Wiley-Interscience, 1991.
29. Ching-Chian Chen, Ya Cheng, and Ming Ouhyoung, Radar cross section analysis and visualization system, in *Proceedings of Computer Graphics Workshop 1995*, pp. 12–16, Taipei, Taiwan, 1995.
30. S. W. Lee, H. Ling, and R. Chou, Ray-tube integration in shooting and bouncing ray method, *Microw. Opt. Technol. Lett.*, 1, 286–289, 1988.
31. H. Lee, R. C. Chou, and S. W. Lee, Shooting and bouncing rays: Calculating the RCS of an arbitrarily shaped cavity, *IEEE Trans. Antennas Propag.*, 37, 194–205, 1989.
32. R. Bhalla and H. Ling, 3D scattering center extraction using the shooting and bouncing ray technique, *IEEE Trans. Antennas Propag.*, 44, 1445–1453, 1996.
33. S. K. Jeng, Near-field scattering by physical theory of diffraction and shooting and bouncing rays, *IEEE Trans. Antennas Propag.*, 46, 551–558, 1998.
34. S. H. Chen and S. K. Jeng, An SBR/image approach for indoor radio propagation in a corridor, *Trans. IEICE Electron.*, E78-C(8), 1058–1062, 1995.
35. J. S. Lee, Refined filtering of image noise using local statistics, *Comput. Vis. Graph. Image Process.*, 15, 380–389, 1981.
36. J. S. Lee, M. R. Grunes, and R. Kwok, Classification of multi-look polarimetric SAR imagery based on complex Wishart distribution, *Int. J. Remote Sens.*, 15, 2299–2311, 1994.
37. J. S. Lee, P. Dewaele, P. Wambacq, A. Oosterlinck, and I. Jurkevich, Speckle filtering of synthetic aperture radar images: A review, *Remote Sens. Rev.*, 8, 313–340, 1994.
38. J. I. Marcum, A statistical theory of target detection by pulsed radar, *IEEE Trans. Inf. Theory*, IT-6, 59–267, 1960.
39. G. V. Trunk and S. F. George, Detection of targets in non-Gaussian clutter, *IEEE Trans. Aerosp. Electron Syst.*, AES-6, 620–628, 1970.
40. D. C. Schleher, Radar detection in Weibull clutter, *IEEE Trans. Aerosp. Electron. Syst.*, 12, 136–143, 1976.
41. Boothe, R. R., The Weibull distribution applied to the ground clutter backscattering coefficient, in Schleher, D. C., ed. *Automatic Detection and Radar-Data Processing*, pp. 435–450, Artech House, 1980.
42. A. K. Fung, *Microwave Scattering and Emission Models and Their Applications*, Artech House, 1994.
43. K. S. Chen and A. K. Fung, Frequency dependence of signal statistics from vegetation components, *IEEE Proc. Radar Sonar Navig.*, 142(6), 301–305, 1996.
44. O. Montenbruck and E. Gill, *Satellite Orbits: Models, Methods, and Applications*, Berlin, Germany: Springer-Verlag, 2000.
45. D. A. Vallado and W. D. McClain, *Fundamentals of Astrodynamics and Applications*, 3rd edition, Hawthorne, CA: Microcosm Press/Springer, 2007.
46. M. F. Henry, NORAD SGP4/SDP4 Implementations, Available at http://www.zeptomoby.com/satellites/.
47. D. Vallado, P. Crawford, R. Hujsak, and T. S. Kelso, Revisiting spacetrack report# 3, AIAA, vol. 6753, pp. 1–88, 2006.
48. N. Z. Miura, Comparison and design of simplified general perturbation models, California Polytechnic State University, San Luis Obispo 2009. Earth Orientation Centre. Available at http://hpiers.obspm.fr/eop-pc/.
49. C. Cafforio, C. Prati, and F. Rocca, SAR data focusing using seismic migration techniques, *IEEE Trans. Aerosp. Electron. Syst.*, 27, 194–207, 1991.
50. A. Moreira, J. Mittermayer, and R. Scheiber, Extended chirp scaling algorithm for air- and spaceborne SAR data processing in stripmap and scanSAR imaging modes, *IEEE Trans. Geosci. Remote Sens.*, 34, 1123–1136, 1996.
51. F.-K. Li, D. N. Held, J. C. Curlander, and C. Wu, Doppler parameter estimation for spaceborne synthetic-aperture radars, *IEEE Trans. Geosci. Remote Sens.*, GE-23, 47–56, 1985.
52. M. Lambers, A. Kolb, H. Nies, and M. Kalkuhl, GPU-based framework for interactive visualization of SAR data, in *Proceedings of IEEE Geoscience and Remote Sensing Symposium (IGARSS)*, Barcelona, Spain, pp. 4076–4079, July 2007.

53. T. Balz and U. Stilla, Hybrid GPU based single- and double-bounce SAR simulation, *IEEE Trans. Geosci. Remote Sens.*, 47, 3519–3529, 2009.

54. NVIDIA CUDA Compute unified device architecture programming guide 1.1. Internet draft, NVIDIA Corporation, 2008.

55. J. Sanders and E. Kandrot, *CUDA by Example: An Introduction to General-Purpose GPU Programming*, Reading, MA: Addison-Wesley, 2010.

56. G. Strang, *Introduction to Applied Mathematics*, Cambridge, MA: Wellesley-Cambridge Press, 1986.

57. D. Bini, *Toeplitz Matrices, Algorithms and Applications*, ECRIM News Online Edition, No. 22, July 1995.

58. A. B. Mandelbrolt and J. W. Van Ness, Fractional Brownian motion, fractional noises and applications, *IEEE Trans. Med. Imag.* 10, 422–437, 1968.

59. E. Ott, *Chaos in Dynamical Systems*, Cambridge, UK: Cambridge University Press, 1993.

60. R. C. Hilborn, *Chaos and Nonlinear Dynamics*, Oxford, UK: Oxford University Press, 1994.

61. H. Leung and T. Lo, A spatial temporal dynamical model for multipath scattering from the sea, *IEEE Trans. Geosci. Remote Sens.* 33, 441–448, 1995.

62. H. Leung, N. Dubash, and N. Xie, Detection of small objects in clutter using a GA-RBF neural network, *IEEE Trans. Aerosp. Electron. Syst.* 38, 98–117, 2002.

63. H. D. I. Abarbanal, *Analysis of Observed Chaotic Data*, Berlin, Germany: Spring-Verlag, 1996.

64. Y. C. Tzeng and K. S. Chen, Change detection in synthetic aperture radar images using a spatially chaotic model, *Opt. Eng.* 46, 086202, 2007.

65. N. S. Chou, Y. C. Tzeng, K. S. Chen, C. T. Wang, and K. C. Fan, On the application of a spatial chaotic model for detecting landcover changes in synthetic aperture radar images, *J. Appl. Remote Sens.*, 3, 033512, 1–16, 2009.

66. S. G. Mallat, A theory for multiresolution signal decomposition: The wavelet representation, *IEEE Trans. Pattern Anal. Mach. Intell.* 11, 674–693, 1989.

67. S. Mallat, *A Wavelet Tour of Signal Processing*, 2nd edition, New York, NY: Academic Press, 1999.

68. R. Brown and P. Hwang, *Introduction to Random Signal Analysis and Kalman Filtering*, New York, NY: Wiley, 1983.

13 Polarimetric SAR Techniques for Remote Sensing of the Ocean Surface

Dale L. Schuler, Jong-Sen Lee, and Dayalan Kasilingam

CONTENTS

13.1 INTRODUCTION

Selected methods that use synthetic aperture radar (SAR) image data to remotely sense ocean surfaces are described in this chapter. Fully polarimetric SAR (POLSAR) radars provide much more usable information than conventional single-polarization radars. Algorithms, presented here, to measure directional wave spectra, wave slopes, wave–current interactions, and current-driven surface features use this additional information.

Polarimetric techniques that measure directional wave slopes and spectra with data collected from a single aircraft, or satellite, collection pass are described here. Conventional single-polarization backscatter cross-section measurements require two orthogonal passes and a complex SAR modulation transfer function (MTF) to determine vector slopes and directional wave spectra.

The algorithm to measure wave spectra is described in Section 13.2. In the azimuth (flight) direction, wave-induced perturbations of the polarimetric orientation angle are used to sense the azimuth component of the wave slopes. In the orthogonal range direction, a technique involving an alpha parameter from the well-known Cloude–Pottier entropy/anisotropy/averaged alpha ($H/A/\bar{\alpha}$) polarimetric decomposition theorem is used to measure the range slope component. Both measurement types are highly sensitive to ocean wave slopes and are directional. Together, they form a means of using POLSAR image data to make complete directional measurements of ocean wave slopes and wave slope spectra.

NASA Jet Propulsion Laboratory (JPL) airborne SAR (AIRSAR) P-, L-, and C-band data obtained during flights over the coastal areas of California are used as wave-field examples. Wave parameters measured using the polarimetric methods are compared with those obtained using *in situ* NOAANational Data Buoy Center (NDBC) buoy products.

In a second topic (Section 13.3), polarization orientation angles are used to remotely sense ocean wave slope distribution changes caused by ocean wave–current interactions. The wave–current features studied include surface manifestations of ocean internal waves and wave interactions with current fronts.

A model [1], developed at the Naval Research Laboratory (NRL), is used to determine the parametric dependencies of the orientation angle on internal wave–current, wind-wave direction, and wind-wave speed. An empirical relation is cited to relate orientation angle perturbations to the underlying parametric dependencies [1].

A third topic (Section 13.4) deals with the detection and classification of biogenic slick fields. Various techniques, using the Cloude–Pottier decomposition and Wishart classifier, are used to classify the slicks. An application utilizing current-driven ocean features, marked by slick patterns, is used to map spiral eddies. Finally, a related technique, using the polarimetric orientation angle, is used to segment slick fields from ocean wave slopes.

13.2 MEASUREMENT OF DIRECTIONAL SLOPES AND WAVE SPECTRA

13.2.1 Single-Polarization versus Fully Polarimetric SAR Techniques

SAR systems conventionally use backscatter intensity-based algorithms [2] to measure physical ocean wave parameters. SAR instruments, operating at a single polarization, measure wave-induced backscatter cross-section, or sigma-0, modulations that can be developed into estimates of surface wave slopes or wave spectra. These measurements, however, require a parametrically complex MTF to relate the SAR backscatter measurements to the physical ocean wave properties [3].

Sections 13.2.3 through 13.2.6 outline a means of using fully POLSAR data with algorithms [4] to measure ocean wave slopes. In the Fourier-transform domain, this orthogonal slope information is used to estimate a complete directional ocean wave slope spectrum. A parametrically simple measurement of the slope is made by using POLSAR-based algorithms.

Modulations of the polarization orientation angle, θ, are largely caused by waves traveling in the azimuth direction. The modulations are, to a lesser extent, also affected by range traveling waves. A method, originally used in topographic measurements [5], has been applied to the ocean and used to measure wave slopes. The method measures vector components of ocean wave slopes and wave spectra. Slopes smaller than 1° are measurable for ocean surfaces using this method.

An eigenvector or eigenvalue decomposition average parameter $\bar{\alpha}$, described in Ref. [6], is used to measure wave slopes in the orthogonal range direction. Waves in the range direction cause modulation of the local incidence angle ϕ, which, in turn, changes the value of $\bar{\alpha}$. The alpha parameter is "roll-invariant." This means that it is not affected by slopes in the azimuth direction. Likewise, for ocean wave measurements, the orientation angle θ parameter is largely insensitive to slopes in the range direction. An algorithm employing both ($\bar{\alpha}$, θ) is, therefore, capable of measuring slopes in any direction. The ability to measure a physical parameter in two orthogonal directions within an individual resolution cell is rare. Microwave instruments, generally, must have a two-dimensional imaging or scanning capability to obtain information in two orthogonal directions.

Motion-induced nonlinear "velocity-bunching" effects still present difficulties for wave measurements in the azimuth direction using POLSAR data. These difficulties are dealt with by using the same proven algorithms [3,7] that reduce nonlinearities for single-polarization SAR measurements.

13.2.2 SINGLE-POLARIZATION SAR MEASUREMENTS OF OCEAN SURFACE PROPERTIES

SAR systems have previously been used for imaging ocean features such as surface waves, shallow-water bathymetry, internal waves, current boundaries, slicks, and ship wakes [8]. In all of these applications, the modulation of the SAR image intensity by the ocean feature makes the feature visible in the image [9]. When imaging ocean surface waves, the main modulation mechanisms have been identified as tilt modulation, hydrodynamic modulation, and velocity bunching [2]. Tilt modulation is due to changes in the local incidence angle caused by the surface wave slopes [10]. Tilt modulation is strongest for waves traveling in the range direction. Hydrodynamic modulation is due to the hydrodynamic interactions between the long-scale surface waves and the short-scale surface (Bragg) waves that contribute most of the backscatter at moderate incidence angles [11]. Velocity bunching is a modulation process that is unique to SAR imaging systems [12]. It is a result of the azimuth shifting of scatterers in the image plane, owing to the motion of the scattering surface. Velocity bunching is the highest for azimuth traveling waves.

In the past, considerable effort had gone into retrieving quantitative surface wave information from SAR images of ocean surface waves [13]. Data from satellite SAR missions, such as ERS and 2 and RADARSAT 1 and 2, had been used to estimate surface wave spectra from SAR image information. Generally, wave height and wave slope spectra are used as quantitative overall descriptors of the ocean surface wave properties [14]. Over the years, several different techniques have been developed for retrieving wave spectra from SAR image spectra [7,15,16]. Linear techniques, such as those having a linear MTF, are used to relate the wave spectrum to the image spectrum. Individual MTFs are derived for the three primary modulation mechanisms. A transformation based on the MTF is used to retrieve the wave spectrum from the SAR image spectrum. Since the technique is linear, it does not account for any nonlinear processes in the modulation mechanisms. It has been shown that SAR image modulation is nonlinear under certain ocean surface conditions. As the sea state increases, the degree of nonlinear behavior generally increases. Under these conditions, the linear methods do not provide accurate quantitative estimates of the wave spectra [15]. Thus, the linear transfer function method has limited utility and can be used as a qualitative indicator. More accurate estimates of wave spectra require the use of nonlinear inversion techniques [15].

Several nonlinear inversion techniques have been developed for retrieving wave spectra from SAR image spectra. Most of these techniques are based on a technique developed in Ref. [7]. The original method used an iterative technique to estimate the wave spectrum from the image spectrum. Initial estimates are obtained using a linear transfer function similar to the one used in Ref. [15]. These estimates are used as inputs in the forward SAR imaging model, and the revised image spectrum is used to iteratively correct the previous estimate of the wave spectra. The accuracy of this technique is dependent on the specific SAR imaging model. Improvements to this technique [17] have incorporated closed-form descriptions of the nonlinear transfer function, which relates the wave spectrum to the SAR image spectrum. However, this transfer function also has to be evaluated iteratively. Further improvements to this method have been suggested in Refs. [3,18]. In this method, a cross-spectrum is generated between different looks of the same ocean wave scene. The primary advantage of this method is that it resolves the 180° ambiguity [3,18] of the wave direction. This method also reduces the effects of speckle in the SAR spectrum. Methods that incorporate additional *a posteriori* information about the wave field, which improves the accuracy of these nonlinear methods, have also been developed in recent years [19].

In all of the slope-retrieval methods, the one nonlinear mechanism that may completely destroy wave structure is velocity bunching [3,7]. Velocity bunching is a result of moving scatterers on the ocean surface either bunching or dilating in the SAR image domain. The shifting of the scatterers in the azimuth direction may, in extreme conditions, result in the destruction of the wave structure in the SAR image.

SAR imaging simulations were performed at different range-to-velocity (*R/V*) ratios to study the effect of velocity bunching on the slope-retrieval algorithms. When the (*R/V*) ratio is artificially increased to large values, the effects of velocity bunching are expected to destroy the wave structure in the slope estimates. Simulations of the imaging process for a wide range of radar-viewing conditions indicate that the slope structure is preserved in the presence of moderate velocity-bunching modulation. It can be argued that for velocity bunching to affect the slope estimates, the (*R/V*) ratio has to be significantly larger than 100s. The two data sets discussed here are designated "Gualala River" and "San Francisco." The Gualala river data set has the longest waves and it also produces the best results. The *R/V* ratio for the AIRSAR missions was 59 s (Gualala) and 55 s (San Francisco). These values suggest that the effects of velocity bunching are present, but are not sufficiently strong to significantly affect the slope-retrieval process. However, for spaceborne SAR imaging applications, where the (*R/V*) ratio may be greater than 100 s, the effects of velocity bunching may limit the utility of all methods, especially in high sea states.

13.2.3 MEASUREMENT OF OCEAN WAVE SLOPES USING POLARIMETRIC SAR DATA

In this section, the techniques that were developed for the measurement of ocean surface slopes and wave spectra using the capabilities of fully polarimetric radars are discussed. Wave-induced perturbations of the polarization orientation angle are used to directly measure slopes for azimuth traveling waves. This technique is accurate for scattering from surface resolution cells where the sea return can be represented as a two-scale Bragg-scattering process.

13.2.3.1 Orientation Angle Measurement of Azimuth Slopes

It has been shown [5] that by measuring the orientation angle shift in the polarization signature, one can determine the effects of the azimuth surface tilts. In particular, the shift in the orientation angle is related to the azimuth surface tilt, the local incidence angle, and, to a lesser degree, the range tilt. This relationship is derived [20] and independently verified [6] as

$$\tan \theta = \frac{\tan \omega}{\sin \phi - \tan \gamma \cos \phi} \tag{13.1}$$

where θ, $\tan \omega$, $\tan \gamma$, and ϕ are the shifts in the orientation angle, the azimuth slope, the ground range slope, and the radar look angle, respectively. According to Equation 13.1, the azimuth tilts may be estimated from the shift in the orientation angle, if the look angle and range tilt are known.

The orthogonal range slope $\tan \gamma$ can be estimated using the value of the local incidence angle associated with the alpha parameter for each pixel. The azimuth slope $\tan \omega$ and the range slope $\tan \gamma$ provide complete slope information for each image pixel.

For the ocean surface at scales of the size of the AIRSAR resolution cell (6.6 m × 8.2 m), the averaged tilt angles are small and the denominator in Equation 13.1 may be approximated by $\sin \phi$

for a wide range of look angles, $\cos\phi$, and ground range slope, $\tan\gamma$, values. Under this approximation, the ocean azimuth slope, $\tan\omega$, is written as

$$\tan\omega \cong (\sin\phi)\cdot\tan\theta \qquad (13.2)$$

The above equation is important because it provides a direct link between POLSAR measurable parameters and physical slopes on the ocean surface. This estimation of ocean slopes relies only on (1) the knowledge of the radar look angle (generally known from the SAR viewing geometry) and (2) the measurement of the wave-perturbed orientation angle. In ocean areas where the average scattering mechanism is predominantly tilted-Bragg scatter, the orientation angle can be measured accurately for angular changes <1°, as demonstrated in Ref. [20].

POLSAR data can be represented by the scattering matrix for single-look complex data and by the Stokes matrix, the covariance matrix, or the coherency matrix for multilook data. An orientation angle shift causes rotation of all these matrices about the line of sight. Several methods have been developed to estimate the azimuth slope-induced orientation angles for terrain and ocean applications. The "polarization signature maximum" method and the "circular polarization" method have proven to be the two most effective methods. Complete details of these methods and the relation of the orientation angle to orthogonal slopes and radar parameters are given [21,22].

13.2.3.2 Orientation Angle Measurement Using the Circular-Pol Algorithm

Image processing was done with both the polarization signature maximum and the circular polarization algorithms. The results indicate that for ocean images a significant improvement in wave visibility is achieved when a circular polarization algorithm is chosen. In addition to this improvement, the circular polarization algorithm is computationally more efficient. Therefore, the circular polarization algorithm method was chosen to estimate orientation angles. The most sensitive circular polarization estimator [21], which involves RR (right-hand transmit, right-hand receive) and LL (left-hand transmit, left-hand receive) terms, is

$$\theta = \Big[\mathrm{Arg}\big(\langle S_{RR}S_{LL}^*\rangle\big) + \pi\Big]\Big/4 \qquad (13.3)$$

A linear-pol basis has a similar transmit-and-receive convention, but the terms (HH, VV, HV, VH) involve horizontal (H) and vertical (V) transmitted, or received, components. The known relations between a circular-pol basis and a linear-pol basis are

$$\begin{aligned}
S_{RR} &= (S_{HH} - S_{VV} + i2S_{HV})/2 \\
S_{LL} &= (S_{VV} - S_{HH} + i2S_{HV})/2
\end{aligned} \qquad (13.4)$$

Using the above equation, the Arg term of Equation 13.3 can be written as

$$\theta = \mathrm{Arg}\big(\langle S_{RR}S_{LL}^*\rangle\big) = \tan^{-1}\left(\frac{-4\,\mathrm{Re}\big(\langle (S_{HH} - S_{VV})S_{HV}^*\rangle\big)}{-\langle |S_{HH} - S_{VV}|^2\rangle + 4\langle |S_{HV}|^2\rangle}\right) \qquad (13.5)$$

The above equation gives the orientation angle, θ, in terms of three of the terms of the linear-pol coherency matrix. This algorithm has been proven to be successful in Ref. [21]. An example of the accuracy is cited from related earlier studies involving wave–current interactions [1]. In these studies, it has been shown that small wave slope asymmetries could be accurately detected as changes in the orientation angle. These small asymmetries had been predicted by theory [1] and their detection indicates the sensitivity of the circular-pol orientation angle measurement.

13.2.4 Ocean Wave Spectra Measured Using Orientation Angles

NASA/JPL/AIRSAR data were taken (1994) at L-band imaging a northern California coastal area near the town of Gualala (Mendocino County) and the Gualala River. This data set was used to determine if the azimuth component of an ocean wave spectrum could be measured using orientation angle modulation. The radar resolution cell had dimensions of 6.6 m (range direction) and 8.2 m (azimuth direction), and 3×3 boxcar averaging was done to the data inputted into the orientation angle algorithms.

Figure 13.1 is an L-band, VV-pol, pseudo color-coded image of a northern California coastal area and the selected measurement study site. A wave system with an estimated dominant wavelength of 157 m is propagating through the site with a wind-wave direction of 306° (estimates from wave spectra, Figure 13.4). The scattering geometry for a single average tilt radar resolution cell is shown in Figure 13.2. Modulations in the polarization orientation angle induced by azimuth traveling ocean waves in the study area are shown in Figure 13.3a and a histogram of the orientation angles is given in Figure 13.3b. An orientation angle spectrum versus wave number for azimuth direction waves

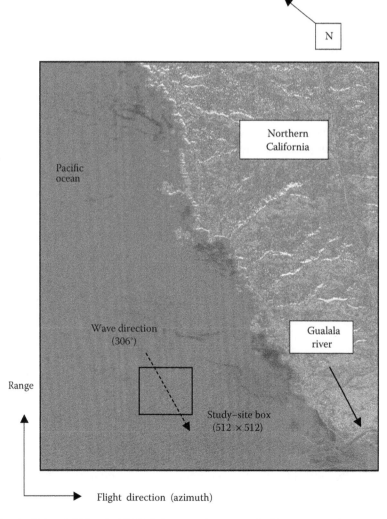

FIGURE 13.1 An L-band, VV-pol, AIRSAR image, of northern California coastal waters (Gualala River dataset), showing ocean waves propagating through a study-site box.

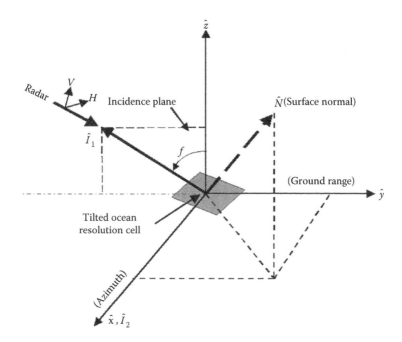

FIGURE 13.2 Geometry of scattering from a single, tilted, resolution cell. In the Gualala River dataset, the resolution cell has dimensions 6.6 m (range direction) and 8.2 m (azimuth direction).

propagating in the study area is given in Figure 13.4. The white rings correspond to ocean wavelengths of 50, 100, 150, and 200 m. The dominant 157 m wave is propagating at a heading of 306°. Figure 13.5a and b gives plots of spectral intensity versus wave number (a) for wave-induced orientation angle modulations and (b) for single-polarization (VV-pol)-intensity modulations. The plots are of wave spectra taken in the direction that maximizes the dominant wave peak. The orientation angle–dominant wave peak (Figure 13.5a) has a significantly higher signal and background ratio than the conventional intensity-based VV-pol-dominant wave peak (Figure 13.5b).

Finally, the orientation angles measured within the study sites were converted into azimuth direction slopes using an average incidence angle and Equation 13.2. From the estimates of these values, the ocean rms azimuth slopes were computed. These values are given in Table 13.1.

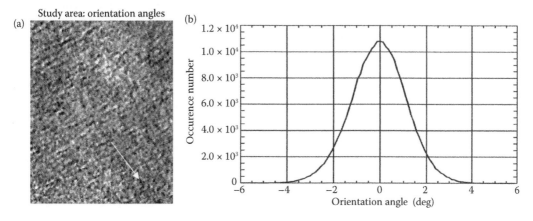

FIGURE 13.3 (a) Image of modulations in the orientation angle, θ, in the study site and (b) a histogram of the distribution of study-site θ values.

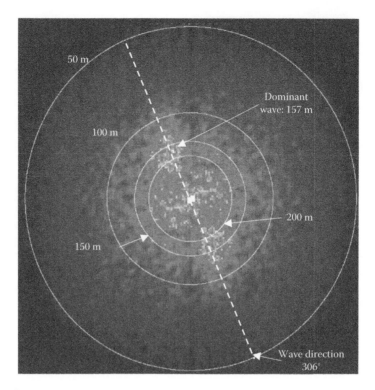

FIGURE 13.4 Orientation angle spectra versus wave number for azimuth direction waves propagating through the study site. The white rings correspond to 50, 100, 150, and 200 m. The dominant wave, of wavelength 157 m, is propagating at a heading of 306°.

13.2.5 TWO-SCALE OCEAN-SCATTERING MODEL: EFFECT ON THE ORIENTATION ANGLE MEASUREMENT

In Section 13.2.3.2, Equation 13.5 is given for the orientation angle. This equation gives the orientation angle θ as a function of three terms from the polarimetric coherency matrix T. Scattering has only been considered as occurring from a slightly rough, tilted surface equal to or greater than the size of the radar resolution cell (see Figure 13.2). The surface is planar and has a single tilt θ_s. This section examines the effects of having a distribution of azimuth tilts, $p(\varphi)$, within the resolution cell, rather than a single averaged tilt.

For single-look or multilook processed data, the coherency matrix is defined as

$$
T = \langle kk^{*\mathrm{T}} \rangle = \frac{1}{2}
\begin{bmatrix}
\langle |S_{HH} + S_{VV}|^2 \rangle & \langle (S_{HH} + S_{VV})(S_{HH} - S_{VV})^* \rangle & 2\langle (S_{HH} + S_{VV})S_{HV}^* \rangle \\
\langle (S_{HH} - S_{VV})(S_{HH} + S_{VV})^* \rangle & \langle |S_{HH} - S_{VV}|^2 \rangle & 2\langle (S_{HH} - S_{VV})S_{HV}^* \rangle \\
2\langle S_{HV}(S_{HH} + S_{VV})^* \rangle & 2\langle S_{HV}(S_{HH} - S_{VV})^* \rangle & 4\langle |S_{HV}|^2 \rangle
\end{bmatrix}
$$

$$
\text{with} \quad k = \frac{1}{\sqrt{2}}
\begin{bmatrix}
S_{HH} + S_{VV} \\
S_{HH} - S_{VV} \\
2S_{HV}
\end{bmatrix}
\tag{13.6}
$$

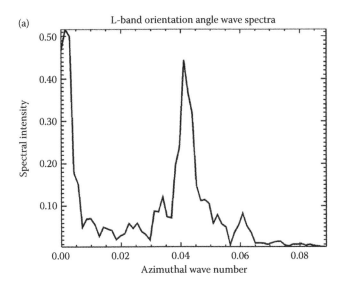

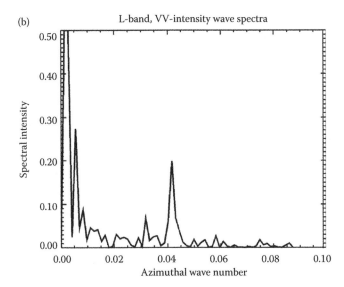

FIGURE 13.5 Plots of spectral intensity versus wave number (a) for wave-induced orientation angle modulations and (b) for VV-pol intensity modulations. The plots are taken in the propagation direction of the dominant wave (306°).

We now follow the approach of Cloude and Pottier [23]. The composite surface consists of flat, slightly rough "facets," which are tilted in the azimuth direction with a distribution of tilts,

$$p(\varphi) = \begin{cases} \frac{1}{2\beta} & |\varphi - s| \leq \beta \\ 0 & \text{otherwise} \end{cases} \tag{13.7}$$

where s is the average surface azimuth tilt of the entire radar resolution cell.

TABLE 13.1

Northern California: Gualala Coastal Results

| | *In Situ* Measurement Instrument | | | |
Parameter	Bodega Bay, CA 3-m Discus Buoy 46013	Point Arena, CA Wind Station	Orientation Angle Method	Alpha Angle Method
Dominant wave period (s)	10.0	N/A	10.03 from dominant wave number	10.2 from dominant wave number
Dominant wavelength (m)	156 from period, depth	N/A	157 from wave spectra	162 from wave spectra
Dominant wave direction (°)	320 Est. from wind direction	284 Est. from wind direction	306 from wave spectra	306 from wave spectra
rms slopes azimuth direction (°)	N/A	N/A	1.58	N/A
rms slopes range direction (°)	N/A	N/A	N/A	1.36
Estimate of wave height (m)	2.4 Significant wave height	N/A	2.16 Est. from rms slope, wave number	1.92 Est. from rms slope, wave number

Note: Date 7/15/94; data start time (UTC): 20:04:44 (BB, PA), 20:02:98 (AIRSAR); wind speed: 1.0 m/s (BB), 2.9 m/s (PA), mean = 1.95 m/s; wind direction: 320° (BB), 284° (PA), mean = 302°; Buoy: "Bodega Bay" (46013)= BB; location: 38.23 N 123.33 W; water depth: 122.5 m; wind station: "Point Arena" (PTAC–1) = PA; location: 38.96 N, 123.74 W; study-site location: 38°39.6′ N, 123°35.8′ W.

The effect on T of having both (1) a mean bias azimuthal tilt θ_s and (2) a distribution of azimuthal tilts β has been calculated in Ref. [22] as

$$T = \begin{bmatrix} A & B\sin c(2\beta)\cos 2\theta_s & -B\sin c(2\beta)\sin 2\theta_s \\ B^*\sin c(2\beta)\cos 2\theta_s & 2C(\sin^2 2\theta_s + \sin c(4\beta)\cos 4\theta_s) & C(1-2\sin c(4\beta))\sin 4\theta_s \\ -B^*\sin c(2\beta)\sin 2\theta_s & C(1-2\sin c(4\beta))\sin 4\theta_s & 2C(\cos^2 2\theta_s - \sin c(4\beta)\cos 4\theta_s) \end{bmatrix}$$

(13.8)

where $\sin c(x) = \sin (x)/x$ and

$$A = |S_{HH} + S_{VV}|^2, \quad B = (S_{HH} + S_{VV})(S_{HH}^* - S_{VV}^*), \quad C = 0.5|S_{HH} - S_{VV}|^2$$

Equation 13.8 reveals the changes due to the tilt distribution (β) and bias (θ_s) that occur in all terms, except in the term $A=|S_{HH} + S_{VV}|^2$, which is roll-invariant. In the corresponding expression for the orientation angle, all of the other terms, except the denominator term $\langle|S_{HH} - S_{VV}|^2\rangle$, are modified

$$\theta = \text{Arg}\left(\langle S_{RR}S_{LL}^*\rangle\right) = \tan^{-1}\left(\frac{-4\,\text{Re}\left(\langle(S_{HH} - S_{VV})S_{HV}^*\rangle\right)}{-\langle|S_{HH} - S_{VV}|^2\rangle + 4\langle|S_{HV}|^2\rangle}\right)$$

(13.9)

From Equations 13.8 and 13.9, it can be determined that the exact estimation of the orientation angle θ becomes more difficult as the distribution of ocean tilts β becomes stronger and wider because the ocean surface becomes progressively rougher.

13.2.6 ALPHA PARAMETER MEASUREMENT OF RANGE SLOPES

A second measurement technique is needed to remotely sense waves that have significant propagation direction components in the range direction. The technique must be more sensitive than current intensity-based techniques that depend on tilt and hydrodynamic modulations. Physically based POLSAR measurements of ocean slopes in the range direction are achieved using a technique involving the "alpha" parameter of the Cloude–Pottier polarimetric decomposition theorem [23].

13.2.6.1 Cloude–Pottier Decomposition Theorem and the Alpha Parameter

The Cloude–Pottier entropy, anisotropy, and the alpha polarization decomposition theorem [23] introduce a new parameterization of the eigenvectors of the 3×3 averaged coherency matrix $\langle |T| \rangle$ in the form

$$\langle |T| \rangle = [U_3] \cdot \begin{bmatrix} \lambda_1 & 0 & 0 \\ 0 & \lambda_2 & 0 \\ 0 & 0 & \lambda_3 \end{bmatrix} \cdot [U_3]^{*\mathrm{T}} \tag{13.10}$$

where

$$[U_3] = \mathrm{e}^{j\phi} \begin{bmatrix} \cos\alpha_1 & \cos\alpha_2 \mathrm{e}^{j\phi_2} & \cos\alpha_3 \mathrm{e}^{j\phi_3} \\ \sin\alpha_1 \cos\beta_1 \mathrm{e}^{j\delta_1} & \sin\alpha_2 \cos\beta_2 \mathrm{e}^{j\delta_2} & \sin\alpha_3 \cos\beta_3 \mathrm{e}^{j\delta_3} \\ \sin\alpha_1 \sin\beta_1 \mathrm{e}^{j\gamma_1} & \sin\alpha_2 \sin\beta_2 \mathrm{e}^{j\gamma_2} & \sin\alpha_3 \sin\beta_3 \mathrm{e}^{j\gamma_3} \end{bmatrix} \tag{13.11}$$

The average estimate of the alpha parameter

$$\bar{\alpha} = P_1\alpha_1 + P_2\alpha_2 + P_3\alpha_3 \tag{13.12}$$

where

$$P_i = \frac{\lambda_i}{\displaystyle\sum_{j=1}^{j=3} \lambda_j}. \tag{13.13}$$

The individual alphas are for the three eigenvectors and the Ps are the probabilities defined with respect to the eigenvalues. In this method, the average alpha is used and is, for simplicity, defined as $\bar{\alpha} \equiv \alpha$. For the ocean backscatter, the contributions to the average alpha are dominated, however, by the first eigenvalue or eigenvector term.

The alpha parameter, developed from the Cloude–Pottier polarimetric scattering decomposition theorem [23], has desirable directional measurement properties. It is (1) roll-invariant in the azimuth direction and (2) in the range direction, it is highly sensitive to wave-induced modulations of ϕ in the local incidence angle ϕ. Thus, the alpha parameter is well suited for measuring wave components traveling in the range direction, and discriminates against wave components traveling in the azimuth direction.

13.2.6.2 Alpha Parameter Sensitivity to Range Traveling Waves

The alpha angle sensitivity to range traveling waves may be estimated using the small perturbation scattering model as a basis. For flat, slightly rough scattering areas that can be characterized by Bragg scattering, the scattering matrix has the form

$$S = \begin{bmatrix} S_{HH} & 0 \\ 0 & S_{VV} \end{bmatrix} \tag{13.14}$$

Bragg-scattering coefficients S_{VV} and S_{HH} are given by

$$S_{HH} = \frac{\cos\phi_i - \sqrt{\varepsilon_r - \sin^2\phi_i}}{\cos\phi_i + \sqrt{\varepsilon_r - \sin^2\phi_i}} \quad \text{and} \quad S_{VV} = \frac{(\varepsilon_r - 1)(\sin^2\phi_i - \varepsilon_r(1 + \sin^2\phi_i))}{\left(\varepsilon_r\cos\phi_i + \sqrt{\varepsilon_r - \sin^2\phi_i}\right)^2} \tag{13.15}$$

The alpha angle is defined such that the eigenvectors of the coherency matrix T are parameterized by a vector k as

$$k = \begin{bmatrix} \cos\alpha \\ \sin\alpha\cos\beta e^{j\delta} \\ \sin\alpha\sin\beta e^{j\gamma} \end{bmatrix} \tag{13.16}$$

For Bragg scattering, one may assume that there is only one dominant eigenvector (depolarization is negligible) and the eigenvector is given by

$$k = \begin{bmatrix} S_{VV} + S_{HH} \\ S_{VV} - S_{HH} \\ 0 \end{bmatrix} \tag{13.17}$$

Since there is only one dominant eigenvector, for Bragg scattering, $\alpha = \alpha_1$. For a horizontal, slightly rough resolution cell, the orientation angle $\beta = 0$, and δ may be set to zero. With these constraints, comparing Equations 13.16 and 13.17 yields

$$\tan\alpha = \frac{S_{VV} - S_{HH}}{S_{VV} + S_{HH}} \tag{13.18}$$

For $\varepsilon \to \infty$,

$$S_{VV} = 1 + \sin^2\phi_i \quad \text{and} \quad S_{HH} = \cos^2\phi_i \tag{13.19}$$

which yields

$$\tan\alpha = \sin^2\phi_i \tag{13.20}$$

Figure 13.6 shows the alpha angle as a function of incidence angle for $\varepsilon \to \infty$ (upper) and for (lower) $\varepsilon = 80 - 70j$, which is a representative dielectric constant of sea water. The sensitivity to alpha values to incidence angle changes (this is effectively the polarimetric MTF) as the range slope estimation is dependent on the derivative of α with respect to ϕ_i.

For $\varepsilon \to \infty$,

$$\frac{\Delta\alpha}{\Delta\phi_i} = \frac{\sin 2\phi_i}{1 + \sin^4\phi_i} \tag{13.21}$$

Figure 13.7 shows this curve (upper) and the exact curve for $\varepsilon = 80 - 70j$ (lower). Note that for the typical AIRSAR range of incidence angles (20–60°), across the swath, the effective MTF is high (>0.5).

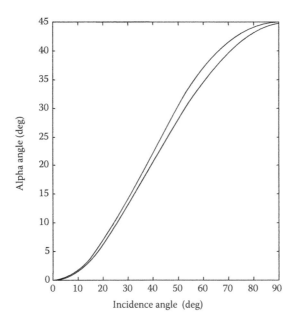

FIGURE 13.6 Small perturbation model dependence of alpha on the incidence angle. The lower curve is for a dielectric constant representative of sea water (80–70*j*) and the upper curve is for a perfectly conducting surface.

13.2.6.3 Alpha Parameter Measurement of Range Slopes and Wave Spectra

Model studies [6] result in an estimate of what the parametric relation α versus the incidence angle ϕ should be for an assumed Bragg-scatter model. The sensitivity (i.e., the slope of the curve of $\alpha(\phi)$) was large enough (Figure 13.6) to warrant investigation using real POLSAR ocean backscatter data.

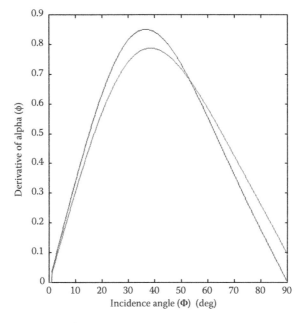

FIGURE 13.7 Derivative of alpha with respect to the incidence angle. The lower curve is for a sea water dielectric and the upper curve is for a perfectly conducting surface.

In Figure 13.8a, a curve of α versus the incidence angle φ is given for a strip of Gualala data in the range direction that has been averaged 10 pixels in the azimuth direction. This curve shows a high sensitivity for the slope of α(φ). Figure 13.8b gives a histogram of the frequency of occurrence of the alpha values.

The curve of Figure 13.8a was smoothed by utilizing a least-square fit of the α(φ) data to a third-order polynomial function. This closely fitting curve was used to transform the α values into corresponding incidence angle φ perturbations. Pottier [6] used a model-based approach and fitted a third-order polynomial to the α(φ) (lower curve) of Figure 13.6 instead of using the smoothed, actual, image α(φ) data. A distribution of φ values has been made and the rms range slope value has been determined. The rms range slope values for the data sets are given in Tables 13.1 and 13.2.

Finally, to measure an alpha wave spectrum, an image of the study area is formed with the mean of α(φ) removed line by line in the range direction. An FFT of the study area results in the wave

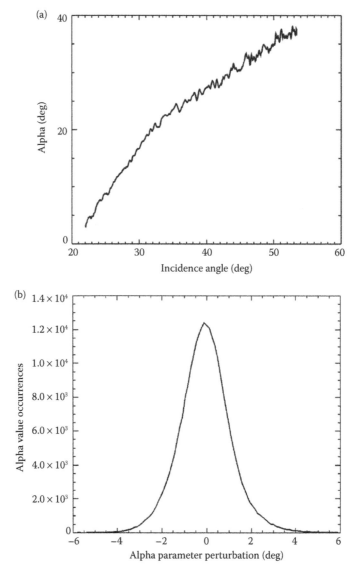

FIGURE 13.8 Empirical determination of the (a) sensitivity of the alpha parameter to the radar incidence angle (for Gualala River data) and (b) a histogram of the alpha values occurring within the study site.

spectrum that is shown in Figure 13.9. The spectrum of Figure 13.9 is an alpha spectrum in the range direction. It can be converted to a range direction wave slope spectrum by transforming the slope values obtained from the smoothed alpha, $\alpha(\phi)$, values.

13.2.7 Measured Wave Properties and Comparisons with Buoy Data

The ocean wave properties estimated from the L- and P-band SAR data sets and the algorithms are the (1) dominant wavelength, (2) dominant wave direction, (3) rms slopes (azimuth and range), and (4) average dominant wave height. The NOAA NDBC buoys provided data on the (1) dominant wave period, (2) wind speed and direction, (3) significant wave height, and (4) wave classification (swell and wind waves).

Both the Gualala and the San Francisco data sets involved waves classified as swell. Estimates of the average wave period can be determined either from buoy data or from the SAR-determined dominant wave number and water depth (see Equation 13.22).

The dominant wavelength and direction are obtained from the wave spectra (see Figures 13.4 and 13.9). The rms slopes in the azimuth direction are determined from the distribution of orientation angles converted to slope angles using Equation 13.2. The rms slopes in the range direction are determined by the distribution of alpha angles converted to slope angles using values of the smoothed curve fitted to the data of Figure 13.8a.

Finally, an estimate of the average wave height, H_d, of the dominant wave was made using the peak-to-trough rms slope in the propagation direction S_{rms} and the dominant wavelength λ_d. The estimated average dominant wave height was then determined from $\tan(S_{rms}) = H_d/(\lambda_d/2)$. This average dominant wave height estimate was compared with the (related) significant wave height provided by the NDBC buoy. The results of the measurement comparisons are given in Tables 13.1 and 13.2.

TABLE 13.2
Open Ocean: Pacific Swell Results

| | In Situ Measurement Instrument | | | |
Parameter	San Francisco, CA, 3 m Discus Buoy 46026	Half Moon Bay, CA, 3 m Discus Buoy 46012	Orientation Angle Method	Alpha Angle Method
Dominant wave period (s)	15.7	15.7	15.17 from dominant wave number	15.23 from dominant wave number
Dominant wavelength (m)	376 from period, depth	364 from period, depth	359 from wave spectra	362 from wave spectra
Dominant wave direction (°)	289 Est. from wind direction	280 Est. from wind direction	265 from wave spectra	265 from wave spectra
rms slopes azimuth direction (°)	N/A	N/A	0.92	N/A
rms slopes range direction (°)	N/A	N/A	N/A	0.86
Estimate of wave height (m)	3.10 significant wave height	2.80 significant wave height	2.88 Est. from rms slope, wave number	2.72 Est. from rms slope, wave number

Note: Date 7/17/88; data start time (UTC): 00:45:26 (Buoys SF, HMB), 00:52:28 (AIRSAR); wind speed: 8.1 m/s (SF), 5.0 m/s (HMB), mean = 6.55 m/s; wind direction: 289° (SF), 280° (HMB), mean = 284.5°; Buoys: "San Francisco" (46026) = SF; location: 37.75 N 122.82 W; water depth: 52.1 m; "Half Moon Bay" (46012) = HMB; location: 37.36 N 122.88 W; water depth: 87.8 m.

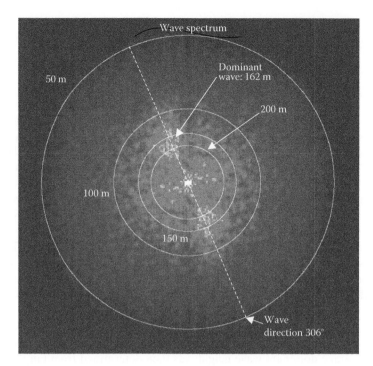

FIGURE 13.9 Spectrum of waves in the range direction using the alpha parameter from the Cloude–Pottier decomposition method. Wave direction is 306° and dominant wavelength is 162 m.

13.2.7.1 Coastal Wave Measurements: Gualala River Study Site

For the Gualala River data set, parameters were calculated to characterize ocean waves present in the study area. Table 13.1 gives a summary of the ocean parameters that were determined using the data set as well as wind conditions and air and sea temperatures at the nearby NDBC buoy ("Bodega Bay") and wind station ("Point Arena") sites. The most important measured SAR parameters were rms wave slopes (azimuth and range directions), rms wave height, dominant wave period, and dominant wavelength. These quantities were estimated using the full-polarization data and the NDBC buoy data.

The dominant wave at the Bodega Bay buoy during the measurement period is classified as a long wavelength swell. The contribution from wind wave systems or other swell components is small relative to the single dominant wave system. Using the surface gravity wave dispersion relation, one can calculate the dominant wavelength at this buoy location where the water depth is 122.5 m. The dispersion relation for surface water waves at finite depth is

$$\omega_{\mathrm{w}}^2 = gk_{\mathrm{w}} \tan \mathrm{h}(kH) \tag{13.22}$$

where ω_{w} is the wave frequency, k_{w} is the wave number ($2\pi/\lambda$), and H is the water depth. The calculated value for λ is given in Table 13.1.

A spectral profile similar to Figure 13.5a was developed for the alpha parameter technique and a dominant wave was measured having a wavelength of 156 m and a propagation direction of 306°. Estimates of the ocean parameters obtained using the orientation angle and alpha angle algorithms are summarized in Table 13.1.

13.2.7.2 Open-Ocean Measurements: San Francisco Study Site

AIRSAR P-band image data were obtained for an ocean swell traveling in the azimuth direction. The location of this image was to the west of San Francisco Bay. It is a valuable data set because

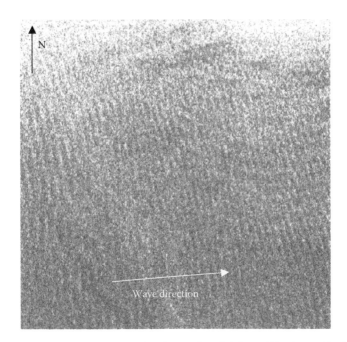

FIGURE 13.10   P-band span image of near-azimuth traveling (265°) swell in the Pacific Ocean off the coast of California near San Francisco.

its location is near two NDBC buoys ("San Francisco" and "Half-Moon Bay"). Figure 13.10 gives a span image of the ocean scene. The long-wavelength swell is clearly visible. The covariance matrix data was first Lee-filtered to reduce speckle noise [24] and was then corrected radiometrically.

A polarimetric signature was developed for a 512×512 segment of the image and some distortion was noted. Measuring the distribution of the phase between the HH-pol and VV-pol backscatter returns eliminated this distortion. For the ocean, this distribution should have a mean nearly equal to zero. The recalibration procedure set the mean to zero and the distortion in the polarimetric signature was corrected. Figure 13.11a gives a plot of the spectral intensity (cross-section modulation) versus the wave number in the direction of the dominant wave propagation. Figure 13.11b presents a spectrum of orientation angles versus the wave number. The major peak, caused by the visible swell, in both plots occurs at a wave number of 0.0175 m^{-1} or a wavelength of 359 m. Using Equation 13.22, the dominant wavelength was calculated at the San Francisco/Half Moon Bay buoy positions and depths. Estimates of the wave parameters developed from this data set using the orientation and alpha angle algorithms are presented in Table 13.2.

13.3   POLARIMETRIC MEASUREMENT OF OCEAN WAVE–CURRENT INTERACTIONS

13.3.1   INTRODUCTION

Studies have been carried out on the use of polarization orientation angles to remotely sense ocean wave slope distribution changes caused by wave–current interactions. The wave–current features studied here involve the surface manifestations of internal waves [1,25–32] and wave modifications at oceanic current fronts.

Studies have shown that POLSAR data may be used to measure bare surface roughness [33] and terrain topography [34,35]. Techniques have also been developed for measuring directional ocean wave spectra [36].

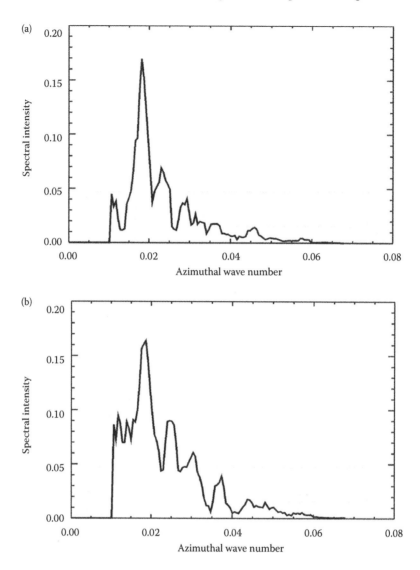

FIGURE 13.11 (a) Wave number spectrum of P-band intensity modulations and (b) a wave number spectrum of orientation angle modulations. The plots are taken in the propagation direction of the dominant wave (265°).

The POLSAR image data used in all of the studies are NASA JPL/AIRSAR, P-, L-, and C-band, quad-pol microwave backscatter data. AIRSAR images of internal waves were obtained from the 1992 Joint US/Russia Internal Wave Remote Sensing Experiment (JUSREX'92) conducted in the New York Bight [25,26].

AIRSAR data on current fronts were obtained during the NRL Gulf Stream Experiment (NRL-GS'90). The NRL experiment is described in Ref. [20]. Extensive sea-truth is available for both of these experiments. These studies were motivated by the observation that strong perturbations occur in the polarization orientation angle θ in the vicinity of internal waves and current fronts. The remote sensing of orientation angle changes associated with internal waves and current fronts are applications that have only recently been investigated [27–29]. Orientation angle changes should also occur for the related SAR application involving surface expressions of shallow-water bathymetry [37].

In the studies outlined here, polarization orientation angle changes are shown to be associated with wave–current interaction features. Orientation angle changes are not, however, produced by all types of ocean surface features. For example, orientation angle changes have been successfully used here to discriminate internal wave signatures from other ocean features, such as surfactant slicks, which produce no mean orientation angle changes.

13.3.2 ORIENTATION ANGLE CHANGES CAUSED BY WAVE–CURRENT INTERACTIONS

A study was undertaken to determine the effect that several important types of wave–current interactions have on the polarization orientation angle. The study involved both actual SAR data and an NRL theoretical model described in Ref. [38].

An example of a JPL/AIRSAR VV-polarization, L-band image of several strong, intersecting, internal wave packets is given in Figure 13.12. Packets of internal waves are generated from parent solitons as the soliton propagates into shallower water at, in this case, the continental shelf break. The white arrow in Figure 13.12 indicates the propagation direction of a wedge of internal waves (bounded by the dashed lines). The packet members within the area of this wedge were investigated.

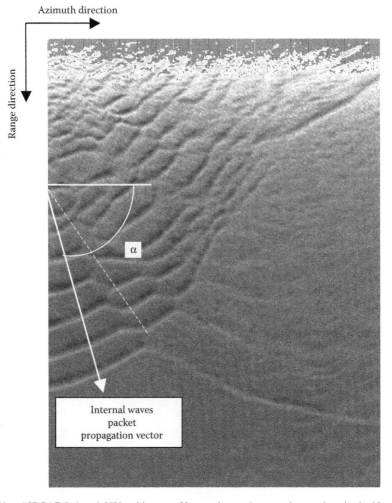

FIGURE 13.12 AIRSAR L-band, VV-pol image of internal wave intersecting packets in the New York Bight. The arrow indicates the propagation direction for the chosen study packet (within the dashed lines). The angle α relates to the SAR/packet coordinates. The image intensity has been normalized by the overall average.

Radar cross-section (σ_0) intensity perturbations for the type of internal waves encountered in the New York Bight have been calculated in Refs. [30,31,39] and others. Related perturbations also occur in the ocean wave height and slope spectra. For the solitons often found in the New York Bight area, these perturbations become significantly larger for ocean wavelengths longer than about 0.25 m and shorter than 10–20 m. Thus, the study is essentially concerned with slope changes to meter-length wave scales. The AIRSAR slant range resolution cell size for these data is 6.6 m, and the azimuth resolution cell size is 12.1 m. These resolutions are fine enough for the SAR backscatter to be affected by perturbed wave slopes (meter-length scales).

The changes in the orientation angle caused by these wave perturbations are seen in Figure 13.13. The magnitude of these perturbations covers a range $\theta = [-1 \text{ to } +1]$. The orientation angle perturbations have a large spatial extent (>100 m for the internal wave soliton width).

The hypothesis assumed was that wave–current interactions make the meter wavelength slope distributions asymmetric. A profile of orientation angle perturbations caused by the internal wave study packet is given in Figure 13.14a. The values are obtained along the propagation vector line of Figure 13.12. Figure 13.14b gives a comparison of the orientation angle profile (solid line) and a normalized VV-pol backscatter intensity profile (dotted-dash line) along the same interval. Note that the orientation angle positive peaks (white stripe areas, Figure 13.13) align with the negative troughs (black areas, Figure 13.12). In the direction orthogonal to the propagation vector, every point is averaged 5×5 pixels along the profile. The ratio of the maximum of θ caused by the soliton

−1.0°   0°   +1.0°

Orientation angle

FIGURE 13.13 The orientation angle image of the internal wave packets in the New York Bight. The area within the wedge (dashed lines) was studied intensively.

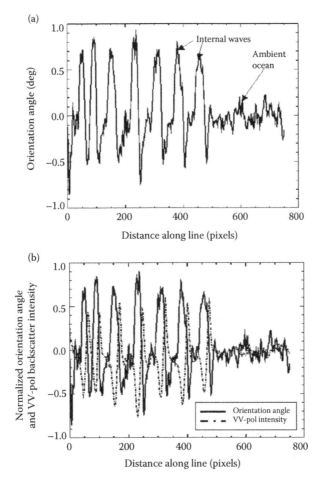

FIGURE 13.14 (a) The orientation angle value profile along the propagation vector for the internal wave study packet of Figure 13.12 and (b) a comparison of the orientation angle profile (solid line) and a normalized VV-pol backscatter intensity profile (dot–dash line). Note that the orientation angle positive peaks (white areas, Figure 13.13) align with the negative troughs (black areas, Figure 13.11).

to the average values of θ within the ambient ocean is quite large. The current-induced asymmetry creates a mean wave slope that is manifested as a mean orientation angle.

The relation between the tangent of the orientation angle θ, wave slopes in the radar azimuth and ground range directions ($\tan \omega$, $\tan \gamma$), and the radar look angle ϕ from Ref. [21] is given by Equation 13.1, and for a given look angle ϕ the average orientation angle tangent is

$$\langle \tan \theta \rangle = \int_0^\pi \int_0^\pi \tan \theta(\omega, \gamma) \cdot P(\omega, \gamma) \, d\gamma \, d\omega \tag{13.23}$$

where $P(\omega,\gamma)$ is the joint probability distribution function for the surface slopes in the azimuth and range directions. If the slopes are zero-meaned, but $P(\omega,\gamma)$ is skewed, then the mean orientation angle may not be zero even though the mean azimuth and range slopes are zero. It is evident from the above equation that both the azimuth and the range slopes have an effect on the mean orientation angle. The azimuth slope effect is generally larger because it is not reduced by the cos ϕ term, which

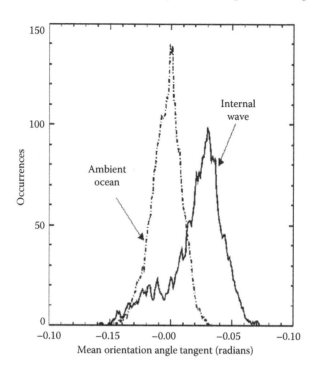

FIGURE 13.15 Distributions of orientation angles for the internal wave (solid line) and the ambient ocean (dot–dash–dot line).

only affects the range slope. If, for instance, the meter wavelength waves are produced by a broad wind-wave spectrum, then both ω and γ change locally. This yields a nonzero mean for the orientation angle. Figure 13.15 gives a histogram of orientation angle values (solid line) for a box inside the black area of the first packet member of the internal wave. A histogram for the ambient ocean orientation angle values for a similar-sized box near the internal wave is given by the dot–dash–dot line in Figure 13.15. Notice the significant difference in the mean value of these two distributions. The mean change in $\langle \tan(\theta) \rangle$ inferred from the bias for the perturbed area within the internal wave is 0.03 rad, corresponding to a θ value of 1.72°.

The mean water wave slope changes needed to cause such orientation angle changes are estimated from Equation 13.1. In the denominator of Equation 13.1, the value of $\tan(\gamma) \cos(\phi) \ll \sin(\phi)$ for the value $\phi (= 51°)$ at the packet member location. Using this approximation, the ensemble average of Equation 13.1 provides the mean azimuth slope value,

$$\langle \tan(\omega) \rangle \cong \sin(\phi) \langle \tan(\theta) \rangle \qquad (13.24)$$

From the data provided in Figure 13.15, $\langle \tan(\omega) \rangle = 0.0229$ rad or $\omega = 1.32°$. A slope value of this magnitude is in approximate agreement with slope changes predicted by Lyzenga and Bennett [32] for internal waves in the same area during an earlier experiment (SARSEX, 1988).

13.3.3 ORIENTATION ANGLE CHANGES AT OCEAN CURRENT FRONTS

An example of orientation angle changes induced by a second type of wave–current interaction, the convergent current front, is given in Figure 13.16a and b. This image was created using AIRSAR P-band polarimetric data.

The orientation angle response to this (NRL-GS'90) Gulf-Stream convergent-current front is the vertical white linear feature in Figure 13.16a and the sharp peak in Figure 13.16b. The perturbation

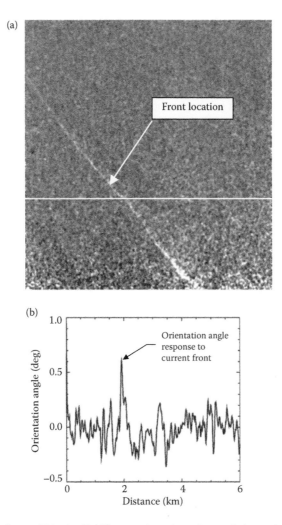

FIGURE 13.16 Current front within the Gulf Stream. An orientation angle image is given in (a) and orientation angle values are plotted in (b) (for values along the white line in (a)).

of the orientation angle at, and near, the front location is quite strong relative to angle fluctuations in the ambient ocean. The change in the orientation angle maximum is $\cong 0.68°$. Other fronts in the same area of the Gulf Stream have similar changes in the orientation angle.

13.3.4 MODELING SAR IMAGES OF WAVE–CURRENT INTERACTIONS

To investigate wave–current interaction features, a time-dependent ocean wave model has been developed that allows for general time-varying current, wind fields, and depth [20,38]. The model uses conservation of the wave action to compute the propagation of a statistical wind-wave system. The action density formalism that is used and an outline of the model are both described in Ref. [38]. The original model has been extended [1] to include calculations of polarization orientation angle changes due to wave–current interactions.

Model predictions have been made for the wind-wave field, radar return, and perturbation of the polarization orientation angle due to an internal wave. A model of the surface manifestation of an internal wave has also been developed. The algorithm used in the model has been modified from its original form to allow calculation of the polarization orientation angle and its variation throughout the extent of the soliton current field at the surface.

The values of both RCS ($\equiv \langle\sigma_0\rangle$) and $\langle\tan(\theta)\rangle$ are computed by the model. The dependence of $\langle\tan(\theta)\rangle$ on the perturbed ocean wavelength was calculated by the model. This wavelength dependence is shown in Figure 13.17. The waves resonantly perturb $\langle\tan(\theta)\rangle$ for wavelengths in the range of 0.25–10.0 m. This result is in good agreement with previous studies of sigma-0 resonant perturbations for the JUSREX'92 area [39].

Figure 13.18a and b shows the form of the soliton current speed dependence of $\langle\sigma_0\rangle$ and $\langle\tan(\theta)\rangle$. The potentially useful near-linear relation of $\langle\tan(\theta)\rangle_V$ with current U (Figure 13.18b) is important in applications where determination of current gradients is the goal. The near-linear nature of this relationship provides the possibility that, from the value of $\langle\tan(\theta)\rangle_v$, the current magnitude can be estimated. Examination of the model results has led to the following empirical model of the variation of $\langle\tan(\theta)\rangle$ as:

$$\langle\tan\theta\rangle = f(U, w, \theta_w) = (aU) \cdot (w^2 e^{-bw}) \cdot \sin(\alpha|\psi_w| + \beta\psi_w^2) \tag{13.25}$$

where U is the surface current maximum speed (in m/s), w is the wind speed (in m/s) at (standard) 19.5 m height, and ψ_w is the wind direction (in radians) relative to the soliton propagation direction. The constants are $a = 0.00347$, $b = 0.365$, $\alpha = 0.65714$, and $\beta = 0.10913$. The range of ψ_w is over $[-\pi, \pi]$.

Using Equation 13.25, the dashed curve in Figure 13.18 can be generated to show good agreement relative to the complete model. The solid lines in Figure 13.18 represent results from the complete model and the dashed lines are results from the empirical relation of Equation 13.25. This relation is much simpler than conventional estimates based on perturbation of the backscatter intensity.

The scaling for the relationship is a relatively simple function of the wind speed and the direction of the locally wind-driven sea. If the orientation angle and wind measurements are available, then Equation 13.25 allows the internal wave–current maximum U to be calculated.

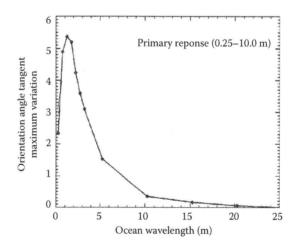

FIGURE 13.17 The internal wave orientation angle tangent maximum variation as a function of ocean wavelength as predicted by the model. The primary response is in the range of 0.25–10.0 m and is in good agreement with previous studies of sigma-0. (From Thompson, D.R., *J. Geophys. Res.*, 93(C10), 12371, 1988.)

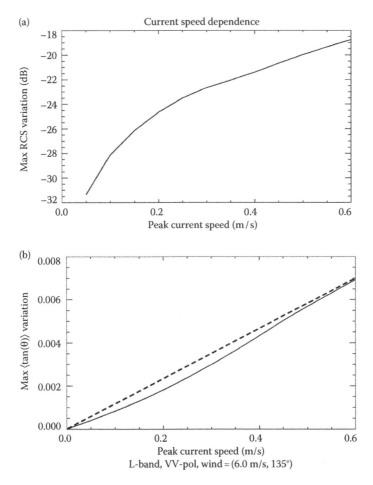

FIGURE 13.18 (a and b) Model development of the current speed dependence of the max RCS and $\langle\tan(\theta)\rangle$ variations. The dashed line in Figure 13.23b gives the values predicted by an empirical equation in Ref. [38]. (Adapted from Jansen, R.W. et al., Modeling of current features in Gulf Stream SAR imagery, Navel Research Laboratory Report NRL/MR/7234-93-7401, 1993.)

13.4 OCEAN SURFACE FEATURE MAPPING USING CURRENT-DRIVEN SLICK PATTERNS

13.4.1 INTRODUCTION

Biogenic and man-made slicks are widely dispersed throughout the oceans. Current-driven surface features, such as spiral eddies, can be made visible by associated patterns of slicks [40]. A combined algorithm using the Cloude–Pottier decomposition and the Wishart classifier [41] is utilized to produce accurate maps of slick patterns and to suppress the background wave field. This technique uses the classified slick patterns to detect spiral eddies. Satellite SAR instruments performing wave spectral measurements, or operating as wind scatterometers, regard the slicks as a measurement error term. The classification maps produced by the algorithm facilitate the flagging of slick-contaminated pixels within the image.

Aircraft L-band AIRSAR data (4/2003) taken in California coastal waters provided data on features that contained spiral eddies. The images also included biogenic slick patterns, internal wave packets, wind waves, and long wave swell.

The temporal and spatial development of spiral eddies is of considerable importance to oceanographers. Slick patterns are used as "markers" to detect the presence and extent of spiral eddies generated in coastal waters. In a SAR image, the slicks appear as black distributed patterns of lower return. The slick patterns are most prevalent during periods of low to moderate winds. The spatial distribution of the slicks is determined by local surface current gradients that are associated with the spiral eddies.

It has been determined that biogenic surfactant slicks may be identified and classified using SAR polarimetric decompositions. The purpose of the decomposition is to discriminate against other features such as background wave systems. The parameters entropy (H), anisotropy (A), and average alpha ($\bar{\alpha}$) of the Cloude–Pottier decomposition [23] were used in the classification. The results indicate that biogenic slick patterns, classified by the algorithm, can be used to detect the spiral eddies.

The decomposition parameters were also used to measure small-scale surface roughness as well as larger-scale rms slope distributions and wave spectra [4]. Examples of slope distributions are given in Figure 13.8b and that of wave spectra in Figure 13.9. Small-scale roughness variations that were detected by anisotropy changes are given in Figure 13.19. This figure shows variations in anisotropy at low wind speeds for a filament of colder, trapped water along the northern California coast. The air–sea stability has changed for the region containing the filament. The roughness changes are not seen in the conventional VV-pol image (Figure 13.19a), but are clearly visible in an

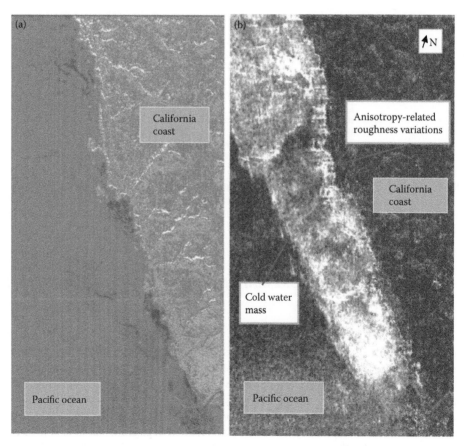

L-band, VV-pol image   Anisotropy- A image

FIGURE 13.19   (a) Variations in anisotropy at low wind speeds for a filament of colder, trapped water along the northern California coast. The roughness changes are not seen in the conventional VV-pol image, but are clearly visible in (b) an anisotropy image. The data are from coastal waters near the Mendocino County town of Gualala.

anisotropy image (Figure 13.19b). The data are from coastal waters near the Mendocino County town of Gualala.

Finally, the classification algorithm may also be used to create a flag for the presence of slicks. Polarimetric satellite SAR systems (e.g., RADARSAT-2, ALOS/PALSAR, SIR-C) attempting to measure wave spectra, or scatterometers measuring wind speed and direction can avoid using slick-contaminated data.

In April 2003, the NRL and the NASA JPL jointly carried out a series of AIRSAR flights over the Santa Monica Basin off the coast of California. Backscatter POLSAR image data at P-, L-, and C-bands were acquired. The purpose of the flights was to better understand the dynamical evolution of spiral eddies, which are generated in this area by interaction of currents with the Channel Islands. Sea-truth was gathered from a research vessel owned by the University of California at Los Angeles. The flights yielded significant data not only on the time history of spiral eddies but also on surface waves, natural surfactants, and internal wave signatures. The data were analyzed using a polarimetric technique, the Cloude–Pottier $\langle H/A/\alpha \rangle$ decomposition given in Ref. [23]. In Figure 13.20a, the anisotropy is again mapped for a study site east of Catalina Island, CA. For comparison, a VV-pol image is given in Figure 13.20b. The slick field is reasonably well mapped by anisotropy—but the image is noisy because of the difference in the two small second and third eigenvalues that are used to compute it.

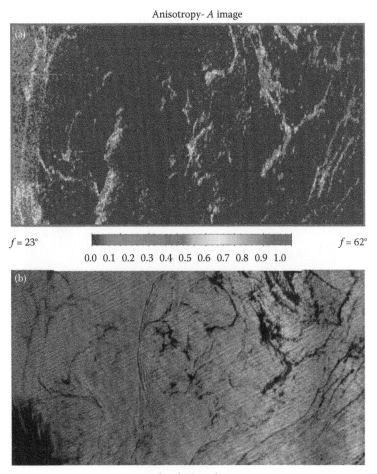

FIGURE 13.20 (a) Image of anisotropy values. The quantity, $1-A$, is proportional to small-scale surface roughness and (b) a conventional L-band, VV-pol image of the study area.

13.4.2 Classification Algorithm

The overall purpose of the field research effort outlined in Section 12.4.1 was to create a means of detecting ocean features such as spiral eddies using biogenic slicks as markers, while suppressing other effects such as wave fields and wind-gradient effects. A polarimetric classification algorithm [41–43] was tested as a candidate means to create such a feature map.

13.4.2.1 Unsupervised Classification of Ocean Surface Features

Van Zyl [44] and Freeman–Durden [45] developed unsupervised classification algorithms that separate the image into four classes: odd-bounce, even bounce, diffuse (volume), and an in-determinate class. For an L-band image, the ocean surface typically is dominated by the characteristics of the Bragg-scattering odd (single) bounce. City buildings and structures have the characteristics of even (double) scattering, and heavy forest vegetation has the characteristics of diffuse (volume) scattering. Consequently, this classification algorithm provides information on the terrain scatterer type. For a refined separation into more classes, Pottier [6] proposed an unsupervised classification algorithm based on their target decomposition theory. The medium's scattering mechanisms, characterized by entropy H, $\bar{\alpha}$ average alpha angle, and later anisotropy A, were used for classification. The entropy H is a measure of randomness of the scattering mechanisms, and the alpha angle characterizes the scattering mechanism. The unsupervised classification is achieved by projecting the pixels of an image onto the H–$\bar{\alpha}$ plane, which is segmented into scattering zones. The zones for the Gualala study-site data are shown in Figure 13.21. Details of this segmentation are given in Ref. [6]. In the alpha–entropy scattering zone map of the decomposition, backscatter returns from the ocean surface normally occur in the lower left zone of both alpha and entropy. Returns from slick-covered areas have higher entropy H and average alpha $\bar{\alpha}$ values, and occur in both the lowest zone and higher zones.

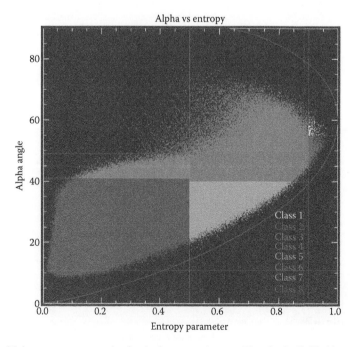

FIGURE 13.21 Alpha-entropy scatter plot for the image study area. The plot is divided into eight color-coded scattering classes for the Cloude–Pottier decomposition described in Ref. [6]. (Adapted from Pottier, E., in Proceedings of the 4th International Workshop Radar Polarimetry, IRESTE, Nantes, France, pp. 535–548, 1998.)

13.4.2.2　Classification Using Alpha–Entropy Values and the Wishart Classifier

Classification of the image was initiated by creating an alpha–entropy zone scatterplot to determine the $\bar{\alpha}$ angle and level of entropy H for scatterers in the slick study area. The image was then classified into eight distinct classes using the Wishart classifier [41]. The alpha–entropy decomposition method provides good image segmentation based on the scattering characteristics.

The algorithm used is a combination of the unsupervised decomposition classifier and the supervised Wishart classifier [41]. One uses the segmented image of the decomposition method to form training sets as input for the Wishart classifier. It has been noted that multilook data are required to obtain meaningful results in H and $\bar{\alpha}$, especially in the entropy H. In general, 4-look processed data are not sufficient. Normally, additional averaging (e.g., 5×5 boxcar filter), either of the covariance or of coherency matrices, has to be performed prior to the H and $\bar{\alpha}$ computation. This prefiltering is done on all the data. The filtered coherency matrix is then used to compute H and $\bar{\alpha}$. Initial classification is made using the eight zones. This initial classification map is then used to train the Wishart classification. The reclassified result shows improvement in retaining details. Further improvement is possible by using several iterations. The reclassified image is then used to update the cluster centers of the coherency matrices. For the present data, two iterations of this process were sufficient to produce good classifications of the complete biogenic fields. Figure 13.22 presents a completed classification map of the biogenic slick fields. Information is provided by the eight color-code classes in the image in Figure 13.22. The returns from within the largest slick (labeled as A) have classes that progressively increase in both average alpha and entropy as a path is made from clean water inward toward the center of the slick. Therefore, the scattering becomes less surface-like ($\bar{\alpha}$ increase) and also becomes more depolarized (H increase) as one approaches the center of the slick (Figure 13.22, Label A).

The algorithm outlined above may be applied to an image containing large-scale ocean features. An image (JPL/CM6744) of classified slick patterns for two-linked spiral eddies near Catalina Island, CA, is given in Figure 13.23b. An L-band, HH-pol image is presented in Figure 13.23a for comparison. The Pacific swell is suppressed in areas where there are no slicks. The waves do, however, appear in areas where there are slicks because the currents associated with the orbital motion of the waves alternately compress or expand the slick-field density. Note the dark slick patch to the left of label A in Figure 13.23a and b. This patch clearly has strongly suppressed the backscatter at

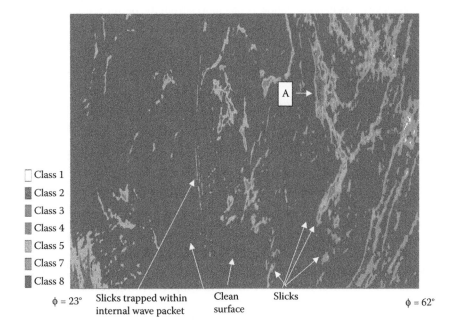

FIGURE 13.22　Classification of the slick-field image into $H/\bar{\alpha}$ scattering classes.

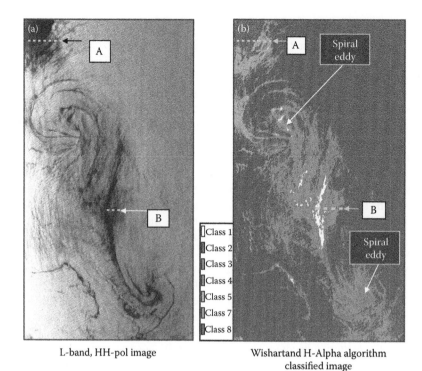

L-band, HH-pol image

Wishartand H-Alpha algorithm
classified image

Class 1
Class 2
Class 3
Class 4
Class 5
Class 7
Class 8

FIGURE 13.23 (a) L-band, HH-pol image of a second study image (CM6744) containing two strong spiral eddies marked by natural biogenic slicks and (b) classification of the slicks marking the spiral eddies. The image features were classified into eight classes using the H–$\overline{\alpha}$ values combined with the Wishart classifier.

HH-pol. The corresponding area of Figure 13.23b has been classified into three classes and colors (Class 7—salmon, Class 5—yellow, and Class 2—dark green), which indicate progressive increases in scattering complexity and depolarization as one moves from the perimeter of the slick toward its interior. A similar change in scattering occurs to the left of label B near the center of Figure 13.23a and b. In this case, as one moves from the perimeter into the slick toward the center, the classes and colors (Class 7—salmon, Class 4—light green, Class 1—white) also indicate progressive increases in scattering complexity and depolarization.

13.4.2.3 Comparative Mapping of Slicks Using Other Classification Algorithms

The question arises whether or not the algorithm using entropy–alpha values with the Wishart classifier is the best candidate for unsupervised detection and mapping of slick fields. Two candidate algorithms were suggested as possible competitive classification methods. These were (1) the Freeman–Durden decomposition [45] and (2) the ($H/A/\overline{\alpha}$)–Wishart segmentation algorithm [42,43], which introduce anisotropy to the parameter mix because of its sensitivity to ocean surface roughness. Programs were developed to investigate the slick classification capabilities of these candidate algorithms. The same amount of averaging (5 × 5) and speckle reduction was done for all of the algorithms. The results with the Freeman–Durden classification were poor at both L- and C-bands. Nearly all of the returns were surface, single-bounce scatter. This is expected because the Freeman–Durden decomposition was developed on the basis of scattering models of land features. This method could not discriminate between waves and slicks and did not improve on the results using conventional VV or HH polarization.

The ($H/A/\overline{\alpha}$)–Wishart segmentation algorithm was investigated to take advantage of the small-scale roughness sensitivity of the polarimetric anisotropy A. The anisotropy is shown (Figure

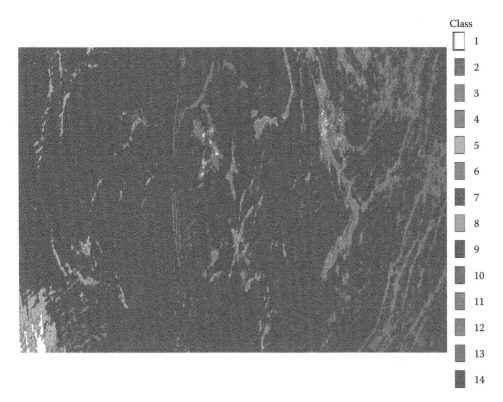

FIGURE 13.24 Classification of the slick-field image into H/A/$\overline{\alpha}$ 14 scattering classes. The Classes 1–7 correspond to anisotropy A values 0.5–1.0 and the Classes 8–14 correspond to anisotropy A values 0.0–0.49. The two lighter blue vertical features at the lower right of the image appear in all images involving anisotropy and are thought to be smooth slicks of lower concentration.

13.20b) to be very sensitive to slick patterns across the whole image. The ($H/A/\overline{\alpha}$)–Wishart segmentation method expands the number of classes from 8 to 16 by including the anisotropy A. The best way to introduce information about A in the classification procedure is to carry out two successive Wishart classifier algorithms. The first classification only involves $H/\overline{\alpha}$. Each class in the $H/\overline{\alpha}$ plane is then further divided into two classes according to whether the pixel's anisotropy values are >0.5 or <0.5. The Wishart classifier is then employed a second time. Details of this algorithm are given in Refs. [42,43]. The results of using the ($H/A/\overline{\alpha}$)–Wishart method and iterating it twice are given in Figure 13.24.

Classification of the slick-field image using the ($H/A/\overline{\alpha}$)–Wishart method resulted in 14 scattering classes. Two of the expected 16 classes were suppressed. The Classes 1–7 corresponded to anisotropy A values from 0.5 to 1.0 and the Classes 8–14 corresponded to anisotropy A values from 0.0 to 0.49. The new two lighter blue vertical features at the lower right of the image appeared in all images involving anisotropy and were thought to be a smooth slick of the lower surfactant material concentration. This algorithm was an improvement relative to the $H/\overline{\alpha}$–Wishart algorithm for slick mapping. All of the slick-covered areas were classified well and the unwanted wave-field intensity modulations were suppressed.

13.5 CONCLUSIONS

Methods that are capable of measuring ocean wave spectra and slope distributions in both the range and azimuth directions were described. The new measurements are sensitive and provide nearly direct measurements of ocean wave spectra and slopes without the need for a complex MTF. The orientation modulation spectrum has a higher dominant wave peak and background ratio than the intensity-based spectrum. The results determined for the dominant wave direction, wavelength, and wave height are comparable to the NDBC buoy measurements. The wave slope and wave spectra measurement methods that have been investigated may be developed further into fully operational algorithms. These algorithms may then be used by POLSAR instruments, such as ALOS/PALSAR and RADARSAT-2, to monitor sea-state conditions globally.

This work has also investigated the effect of internal waves and current fronts on the SAR polarization orientation angle. The results provide (1) a potential independent means for identifying these ocean features and (2) a method of estimating the mean value of the surface current and slope changes associated with an internal wave. Simulations of the NRL wave–current interaction model [38] have been used to identify and quantify the different variables such as current speed, wind speed, and wind direction, which determine changes in the SAR polarization orientation angle.

The polarimetric scattering properties of biogenic slicks have been found to be different from those of the clean surface wave field and the slicks may be separated from this background wave field. Damping of capillary waves, in the slick areas, lowers all of the eigenvalues of the decomposition and increases the average alpha angle, entropy, and the anisotropy.

The Cloude–Pottier polarimetric decomposition was also used as a new means of studying scattering properties of surfactant slicks perturbed by current-driven surface features. The features, for example, spiral eddies, were marked by filament patterns of slicks. These slick filaments were physically smoother. Backscatter from them was more complex (three eigenvalues nearly equal) and was more depolarized.

Anisotropy was found to be sensitive to small-scale ocean surface roughness, but was not a function of large-scale range or azimuth wave slopes. These unique properties provided an achievable separation of roughness scales on the ocean surface at low wind speeds. Changes in anisotropy due to surfactant slicks were found to be measurable across the entire radar swath.

Finally, POLSAR decomposition parameters alpha, entropy, and anisotropy were used as an effective means for classifying biogenic slicks. Algorithms, using these parameters, were developed for the mapping of both slick fields and ocean surface features. Selective mapping of biogenic slick fields may be achieved using either the entropy or the alpha parameters with the Wishart classifier or, by the entropy, anisotropy, or the alpha parameters with the Wishart classifier. The latter algorithm gives the best results overall.

Slick maps made using this algorithm are of use for satellite scatterometers and wave spectrometers in efforts aimed at flagging ocean surface areas that are contaminated by slick fields.

REFERENCES

1. Schuler, D.L., Jansen, R.W., Lee, J.S., and Kasilingam, D., Polarisation orientation angle measurements of ocean internal waves and current fronts using polarimetric SAR, *IEE Proc. Radar Sonar Navig.*, 150(3), 135–143, 2003.
2. Alpers, W., Ross, D.B., and Rufenach, C.L., The detectability of ocean surface waves by real and synthetic aperture radar, *J. Geophys. Res.*, 86(C7), 6481, 1981.
3. Engen, G. and Johnsen, H., SAR-ocean wave inversion using image cross-spectra, *IEEE Trans. Geosci. Remote Sens.*, 33, 1047, 1995.
4. Schuler, D.L., Kasilingam, D., Lee, J.S., and Pottier, E., Studies of ocean wave spectra and surface features using polarimetric SAR, in *Proceedings of the International Geoscience and Remote Sensing Symposium (IGARSS'03)*, Toulouse, France, IEEE, 2003.
5. Schuler, D.L., Lee, J.S., and De Grandi, G., Measurement of topography using polarimetric SAR Images, *IEEE Trans. Geosci. Remote Sens.*, 34, 1266, 1996.
6. Pottier, E., Unsupervised classification scheme and topography derivation of POLSAR data on the $\ll H/A/\alpha \gg$ polarimetric decomposition theorem, in *Proceedings of the 4th International Workshop Radar Polarimetry*, IRESTE, Nantes, France, pp. 535–548, 1998.

7. Hasselmann, K. and Hasselmann, S., The nonlinear mapping of an ocean wave spectrum into a synthetic aperture radar image spectrum and its inversion, *J. Geophys. Res.*, 96(10), 713, 1991.

8. Vesecky, J.F. and Stewart, R.H., The observation of ocean surface phenomena using imagery from SEASAT synthetic aperture radar—An assessment, *J. Geophys. Res.*, 87, 3397, 1982.

9. Beal, R.C., Gerling, T.W., Irvine, D.E., Monaldo, F.M., and Tilley, D.G., Spatial variations of ocean wave directional spectra from the SEASAT synthetic aperture radar, *J. Geophys. Res.*, 91, 2433, 1986.

10. Valenzuela, G.R., Theories for the interaction of electromagnetic and oceanic waves—A review, *Bound. Layer Meteorol.*, 13, 61, 1978.

11. Keller, W.C. and Wright, J.W., Microwave scattering and straining of wind-generated waves, *Radio Sci.*, 10, 1091, 1975.

12. Alpers, W. and Rufenach, C.L., The effect of orbital velocity motions on synthetic aperture radar imagery of ocean waves, *IEEE Trans. Antennas Propag.*, 27, 685, 1979.

13. Plant, W.J. and Zurk, L.M., Dominant wave directions and significant wave heights from SAR imagery of the ocean, *J. Geophys. Res.*, 102(C2), 3473, 1997.

14. Hasselmann, K., Raney, R.K., Plant, W.J., Alpers, W., Shuchman, R.A., Lyzenga, D.R., Rufenach, C.L., and Tucker, M.J., Theory of synthetic aperture radar ocean imaging: A MARSEN view, *J. Geophys. Res.*, 90, 4659, 1985.

15. Lyzenga, D.R., An analytic representation of the synthetic aperture radar image spectrum for ocean waves, *J. Geophys. Res.*, 93(13), 859, 1998.

16. Kasilingam, D. and Shi, J., Artificial neural network based-inversion technique for extracting ocean surface wave spectra from SAR images, in *Proceedings of the IGARSS'97*, Singapore, IEEE, pp. 1193–1195, 1997.

17. Hasselmann, S., Bruning, C., Hasselmann, K., and Heimbach, P., An improved algorithm for the retrieval of ocean wave spectra from synthetic aperture radar image spectra, *J. Geophys. Res.*, 101, 16615, 1996.

18. Lehner, S., Schulz-Stellenfleth, J., Schattler, B., Breit, H., and Horstmann, J., Wind and wave measurements using complex ERS-2 SAR wave mode data, *IEEE Trans. Geosci. Remote Sens.*, 38(5), 2246, 2000.

19. Dowd, M., Vachon, P.W., and Dobson, F.W., Ocean wave extraction from RADARSAT synthetic aperture radar inter-look image cross-spectra, *IEEE Trans. Geosci. Remote Sens.*, 39, 21–37, 2001.

20. Lee, J.S., Jansen, R., Schuler, D., Ainsworth, T., Marmorino, G., and Chubb, S., Polarimetric analysis and modeling of multi-frequency SAR signatures from Gulf Stream fronts, *IEEE J. Ocean. Eng.*, 23, 322, 1998.

21. Lee, J.S., Schuler, D.L., and Ainsworth, T.L., Polarimetric SAR data compensation for terrain azimuth slope variation, *IEEE Trans. Geosci. Remote Sens.*, 38, 2153–2163, 2000.

22. Lee, J.S., Schuler, D.L., Ainsworth, T.L., Krogager, E., Kasilingam, D., and Boerner, W.M., The estimation of radar polarization shifts induced by terrain slopes, *IEEE Trans. Geosci. Remote Sens.*, 40, 30–41, 2001.

23. Cloude, S.R. and Pottier, E., A review of target decomposition theorems in radar polarimetry, *IEEE Trans. Geosci. Remote Sens.*, 34(2), 498, 1996.

24. Lee, J.S., Grunes, M.R., and De Grandi, G., Polarimetric SAR speckle filtering and its implication for classification, *IEEE Trans. Geosci. Remote Sens.*, 37, 2363, 1999.

25. Gasparovic, R.F., Apel, J.R., and Kasischke, E., An overview of the SAR internal wave signature experiment, *J. Geophys. Res.*, 93, 12304, 1998.

26. Gasparovic, R.F., Chapman, R., Monaldo, F.M., Porter, D.L., and Sterner, R.F., Joint U.S./Russia internal wave remote sensing experiment: Interim results, Applied Physics Laboratory Report S1R-93U-011, Johns Hopkins University, 1993.

27. Schuler, D.L., Kasilingam, D., and Lee, J.S., Slope measurements of ocean internal waves and current fronts using polarimetric SAR, European Conference on Synthetic Aperture Radar (EUSAR'2002), Cologne, Germany, 2002 (The paper was presented at Session 7 of EUSAR'2002).

28. Schuler, D.L., Kasilingam, D., Lee, J.S., Jansen, R.W., and De Grandi, G., Polarimetric SAR measurements of slope distribution and coherence changes due to internal waves and current fronts, in *Proceedings of the International Geoscience and Remote Sensing Symposium (IGARSS'2002)*, Vol. 1, pp. 638–640, Toronto, Canada, 2002.

29. Schuler, D.L., Lee, J.S., Kasilingam, D., and De Grandi, G., Studies of ocean current fronts and internal waves using polarimetric SAR coherences, in *Proceedings of Progress in Electromagnetics Research Symposium (PIERS'2002)*, P. 3, Cambridge, MA, 2002.

30. Alpers, W., Theory of radar imaging of internal waves, *Nature*, 314, 245, 1985.

31. Brant, P., Alpers, W., and Backhaus, J.O., Study of the generation and propagation of internal waves in the Strait of Gibraltar using a numerical model and synthetic aperture radar images of the European ERS-1 satellite, *J. Geophys. Res.*, 101(14), 14237, 1996.

32. Lyzenga, D.R. and Bennett, J.R., Full-spectrum modeling of synthetic aperture radar internal wave signatures, *J. Geophys. Res.*, 93(C10), 12345, 1988.

33. Schuler D.L., Lee, J.S., Kasilingam, D., and Nesti, G., Surface roughness and slope measurements using polarimetric SAR data, *IEEE Trans. Geosci. Remote Sens.*, 40(3), 687, 2002.

34. Schuler, D.L., Ainsworth, T.L., Lee, J.S., and De Grandi, G., Topographic mapping using polarimetric SAR data, *Int. J. Remote Sens.*, 35(5), 1266, 1998.

35. Schuler, D.L, Lee, J.S., Ainsworth, T.L., and Grunes, M.R., Terrain topography measurement using multi-pass polarimetric synthetic aperture radar data, *Radio Sci.*, 35(3), 813, 2002.

36. Schuler, D.L. and Lee, J.S., A microwave technique to improve the measurement of directional ocean wave spectra, *Int. J. Remote Sens.*, 16, 199, 1995.

37. Alpers, W. and Hennings, I., A theory of the imaging mechanism of underwater bottom topography by real and synthetic aperture radar, *J. Geophys. Res.*, 89, 10529, 1984.

38. Jansen, R.W., Chubb, S.R., Fusina, R.A., and Valenzuela, G.R., Modeling of current features in Gulf Stream SAR imagery, Naval Research Laboratory Report NRL/MR/7234-93-7401, 1993.

39. Thompson, D.R., Calculation of radar backscatter modulations from internal waves, *J. Geophys. Res.*, 93(C10), 12371, 1988.

40. Schuler, D.L., Lee, J.S., and De Grandi, G., Spiral eddy detection using surfactant slick patterns and polarimetric SAR image decomposition techniques, in *Proceedings of the International Geoscience and Remote Sensing Symposium (IGARSS)*, Anchorage, Alaska, September, 2004.

41. Lee, J.S., Grunes, M.R., Ainsworth, T.L., Du, L.J., Schuler, D.L., and Cloude, S.R., Unsupervised classification using polarimetric decomposition and the complex Wishart classifier, *IEEE Trans. Geosci. Remote Sens.*, 37(5), 2249, 1999.

42. Pottier, E. and Lee, J.S., Unsupervised classification scheme of POLSAR images based on the complex Wishart distribution and the polarimetric decomposition theorem, in *Proceedings of the 3rd European Conference on Synthetic Aperture Radar (EUSAR'2000)*, Munich, Germany, 2000.

43. Ferro-Famil, L., Pottier, E., and Lee, J-S, Unsupervised classification of multifrequency and fully polarimetric SAR images based on the H/A/Alpha-Wishart classifier, *IEEE Trans. Geosci. Remote Sens.*, 39(11), 2332, 2001.

44. Van Zyl, J.J., Unsupervised classification of scattering mechanisms using radar polarimetry data, *IEEE Trans. Geosci. Remote Sens.*, 27, 36, 1989.

45. Freeman, A. and Durden, S.L., A three component scattering model for polarimetric SAR data, *IEEE Trans. Geosci. Remote Sens.*, 36, 963, 1998.

14 An ISAR Technique for Refocusing Moving Targets in SAR Images

Marco Martorella, Elisa Giusti, Fabrizio Berizzi, Alessio Bacci, and Enzo Dalle Mese

CONTENTS

14.1 INTRODUCTION

Synthetic Aperture Radar (SAR) has become a powerful tool for remotely observing the earth. Airborne SAR systems allow imaging medium-to-large areas with ad hoc missions whereas spaceborne SAR systems are mainly designed for continuous monitoring of vast areas at a global scale, although the monitoring of a selected area may occur only at a given prescheduled time (revisiting time). Anyway, recent spaceborne systems, such as Cosmo Skymed (CSK), have reduced the revisiting time by introducing the concept of SAR constellation and by making each satellite as reconfigurable as possible. Such short revisiting time allows using spaceborne systems, such as CSK, for homeland security and defense applications.

In such scenarios, man-made moving target imaging becomes very important. Nevertheless, many SAR processors, which are designed to form highly focussed images at very high resolution, are based on the assumption that the illuminated area is static during the synthetic aperture formation [1]. As a consequence of such an assumption, the existing techniques are unable to focus moving targets, leading to blurred and displaced images of any object that is not static during the synthetic aperture formation.

Inverse SAR (ISAR) proposes another way to look at the problem of forming a synthetic aperture to achieve high cross-range resolution of noncooperative targets [2,3]. ISAR techniques do not base their functioning on the assumption that the target is static during the synthetic aperture formation, but instead, they partly exploit the target's own motions to form the synthetic aperture. Although ISAR techniques do not make use of *a priori* information about the target's motion, some other constraints apply to the ISAR image formation. Such constraints may include the image size, the achievable cross-range resolution, the cross-range scaling problem, and the fact that the imaging system performance is not entirely predictable [4–6]. Nevertheless, ISAR imaging provides acceptable solutions when SAR imaging fails, as it will be proven in this work.

A functional block scheme is represented in Figure 14.1 that aims at describing a detection and moving target refocussing system [7]. With such a system, we aim at exploiting high-resolution Single Look

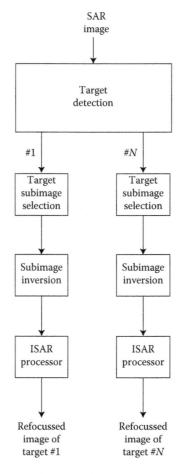

FIGURE 14.1 Block scheme of the detection and refocussing processor.

Complex (SLC) SAR images to detect moving targets, which typically appear defocussed. The same data are then used to form well-focussed images of the same targets by means of an ISAR technique.

In this work, we aim at providing technical details about the refocussing process. Specifically, four types of image inversion mapping are explored that are based on the Fourier transform, the Polar Format (PF), the Chirp Scaling (CS), and the ω–k (OK) algorithms. An ISAR processor is then used to form well-focused images of the moving target. In particular, the autofocus, the image formation, and the time window selection problems will be addressed and solutions will be given based on results presented in the literature. Results based on the use of CSK Spotlight SAR data will prove the effectiveness of the proposed moving target refocussing technique.

To make ISAR images effective for target classification and recognition, an image cross-range scaling is performed that is able to provide fully scaled target images. These images are displayed with respect to spatial coordinates, both along the range and cross-range axes. To obtain such a result, the target effective rotation vector must be estimated. A technique based on the scatterer's phase estimation proposed in Ref. [5] is applied to the refocussed images to display fully scaled ISAR images. The technique will be applied to images obtained by using different inversion techniques and results will be compared.

The remainder of this work is organized as follows. Section 14.2 recalls the main SAR image formation algorithms, whereas Section 14.3 provides details about the SAR to raw data inversion techniques. Section 14.4 deals with the ISAR processor used to form the refocussed images after the inversion process. Results are discussed in Section 14.5 and finally conclusions are drawn in Section 14.6.

14.2   SPOTLIGHT SAR ALGORITHMS

In this section we will recall some of the most commonly used SAR image formation algorithms and we will introduce the notation that will be used to define the inversion algorithms, which will be detailed in Section 14.3. Specifically the ω–k, the Polar Format (PF) and Chirp Scaling (CS) algorithms will be recalled. Let the geometry be represented in Figure 14.2, where the radar moves

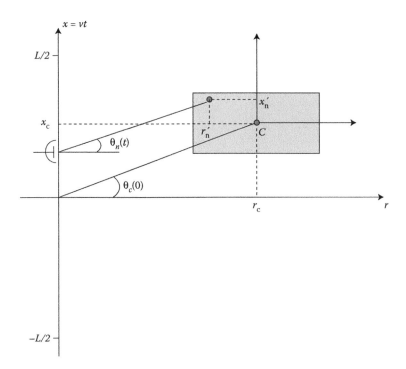

FIGURE 14.2   SAR spotlight geometry.

along the cross-range coordinate x with velocity v, r represents the slant-range coordinate, $\theta_c(t)$ and $\theta_n(t)$ are respectively the aspect angles of the scene centre and the nth scatterer.

14.2.1 ω–κ ALGORITHM

SAR image reconstruction algorithms typically make use of a plane-wave approximation. However, in some applications, the wavefront curvature must be taken into account. The ω–k algorithm, also known as the wavenumber algorithm, improves the resolution by accurately modeling the actual spherical wave. Depending on the imaging parameters, a very accurate interpolation must be performed to avoid artifacts in the resulting image [1].

The ω–k, in fact, shows aberrations at very high squint angles, as it depends on the interpolation accuracy and the target's location within the scene, as demonstrated in [8]. Such aberrations result in an inexact location of scatterers that are positioned close to the flight path. As these effects are evident only when squint angles are close to 90° (broadside looking antenna), the ω–k algorithm is largely used when dealing with airborne and spaceborne stripmap and spotlight SAR data.

The flow chart of the ω–k algorithm is shown in Figure 14.3, where $S_M(t_f, t)$ is the received signal at the output of the matched filter, t_f is the fast time variable, t is the slow time variable, and $k = 2\pi f/c$ is the wavenumber. Since the spatial frequency domain, which is defined by means of the variable pair (K_r, K_x), is irregularly sampled, a 2D interpolation is needed prior to applying a 2D inverse fast Fourier transform (IFFT).

The received signal can be written as follows, assuming that the target can be considered composed of a superposition of N_S independent and ideal scatterers:

$$s_R(t_f, t) = \sum_{n=1}^{N_S} \sigma_n p \left[t - \frac{2}{c}\sqrt{(r_n' + r_c)^2 + (x_n' + x_c - v \cdot t)^2} \right] \text{rect}\left(\frac{t}{T_{obs}}\right) \tag{14.1}$$

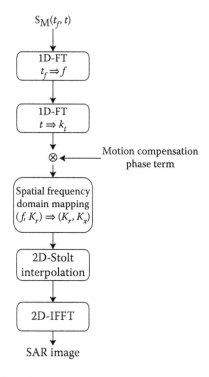

FIGURE 14.3 ω–k algorithm flow chart.

where (x_c, r_c) are the cross-range and slant-range coordinates of the scene center, (x'_n, r'_n) are the cross-range and slant-range coordinates of a generic scatterer in a reference system embedded on the scene center, $p(t)$ is the transmitted pulse, T_{obs} is the observation time, and σ_n is the complex reflectivity of the nth scatterer.

The ω–k algorithm can be summarized by means of the following steps:

- Fourier transform with respect to the fast time variable, t_f after matched filtering

$$S_{MF}(f,t) = |P(f)|^2 \sum_{n=1}^{N_2} \sigma_n e^{-j2k\sqrt{(r'_n+r_c)^2+(x'_n+x_c-v\cdot t)^2}} \text{rect}\left[\frac{t}{T_{obs}}\right]. \tag{14.2}$$

- Fourier transform with respect to the slow time variable, t, by applying the Stationary Phase Method [9–11]

$$S_{MF}(f,k_t) = |P(f)|^2 \sum_{n=1}^{N_S} \sigma_n e^{-j\sqrt{4k^2-k_t^2}\,(r'_n-r_c)-jk_t(x'_n-x_c)} I_n(f,k_t) \tag{14.3}$$

$$I_n(f,k_t) = \begin{cases} 1 & k_t \in \left[-2k\sin\left(\theta_n\left(-\frac{L}{2}\right)\right), 2k\sin\left(\theta_n\left(-\frac{L}{2}\right)\right)\right] \\ 0 & \text{otherwise} \end{cases}. \tag{14.4}$$

- Motion compensation of the radar platform. This step is performed by multiplying $S_{MF}(f,k_t)$ with the focussing function $S_c(f,k_t) = e^{-j\sqrt{4k^2-k_t^2}\,r_c - jk_t x_c}$.
- Coordinates transformation

$$\begin{cases} K_r = \sqrt{4k^2 - k_t^2} \\ K_x = k_t \end{cases}. \tag{14.5}$$

- Stolt interpolation in order to obtain a regularly sampled rectangular grid.
- 2D Fourier transform in order to obtain an estimate of the reflectivity function in a spatial grid.

14.2.2 POLAR FORMAT ALGORITHM

The PF algorithm is well known for processing stripmap and spotlight mode SAR data. This algorithm involves the assumption that the spherical wavefronts of the radar signal can be approximated by planar wavefronts around a central reference point in the imaged scene. The surface to which data are focussed is also assumed planar when compensating for platform motion. This assumption is quite good for small scenes. For large scenes, conversely, such approximations lead to errors in parts of the scene far away from the central reference point. Many algorithms have been proposed in the literature to compensate for the phase errors due to the wavefronts curvature, such as that proposed in [12]. The plane wave assumption, typically used when formulating the PF Algorithm, not only limits the scene size but also the squint angle.

However, in the following, only the traditional PF has been considered. Figure 14.4 shows the PFA block scheme.

Under the assumption that the scene dimension is smaller than the distance between the radar and the scene center, the PF algorithm can be used and can be summarized in the following steps:

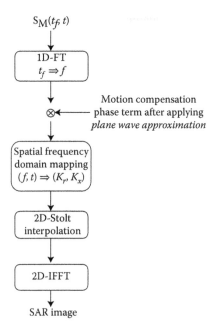

FIGURE 14.4 PFA algorithm flow chart.

- Fourier transform with respect to the fast time variable, t_f. By assuming the *straight-iso-range* approximation, Equation 14.2 can be rewritten as

$$S_{MF}(f,k_t) = |P(f)|^2 \sum_{n=1}^{N_S} \sigma_n e^{-j2k\left(\sqrt{r_c^2+(x_c-vt)^2} +\cos(\theta_c(t))r_n +\sin(\theta_c(t))x_n\right)} rect\left(\frac{t}{T_{obs}}\right)$$

- Radar platform motion compensation. The motion-compensated signal can be obtained by multiplying $S_{MF}(f, t)$ with the focussing function $S_c(f,t) = e^{j2k\sqrt{r_c^2+(x_c-vt)^2}}$

$$S_{MC}(f,t) = S_{MF}(f,t) \cdot S_C(f,t). \tag{14.6}$$

- Coordinates transformation. Under the *straight-iso-range* approximation the coordinates transformation is defined as follows, where $\theta_c(t)$ is the aspect angle with respect to the scene center

$$\begin{cases} K_r = 2k\cos(\theta_c(t)) \\ K_x = 2k\sin(\theta_c(t)). \end{cases} \tag{14.7}$$

- Stolt interpolation in order to obtain a regularly sampled rectangular grid.
- 2D Fourier transform in order to obtain the estimate of the reflectivity function.

14.2.3 CHIRP SCALING ALGORITHM

The CS algorithm has been proposed for high-quality SAR processing. This algorithm avoids any interpolation in the SAR processing chain and has been found suitable for the high-quality processing of several spaceborne SAR systems (e.g., SEASAT, ERS-1, RADARSAT). The CS algorithm basically consists of multiplying the SAR data in the Range Doppler (RD) domain by a quadratic phase function (chirp scaling) in order to equalize the range cell migration for a reference range, followed by an azimuth and range compression in the wavenumber domain. After transforming the signal back to the RD domain, a residual phase correction is carried out. Finally, azimuth IFFT is performed to generate a focussed image.

The chirp scaling multiplier or Range Perturbation Function (RPF) is exactly a linear Frequency Modulation (FM) (i.e., a quadratic function of range) when the following conditions are satisfied:

- The radar pulse is linear frequency modulated
- The azimuth FM rate parameter is range invariant

For high squint angle and large swath width the assumption that the azimuth FM rate parameter is range invariant is not satisfied, and therefore the range chirp may have a nonlinear component.

Furthermore, the azimuth frequency variable should be centered around the Doppler centroid frequency at each range cell. The classical implementation of the CS algorithm relies on the assumption that the Doppler centroid does not change from one range cell to the other. However, when the look angle variation is large (large swaths), the Doppler centroid can vary significantly from near to far range, and therefore it should be accounted for.

An Extended CS algorithm was developed that accounts for strong motions errors and variable Doppler centroids in range and also in azimuth that can be found in Refs. [13–15].

Nevertheless, only the classical CS algorithm is considered in this work. Figure 14.5 shows the CS block scheme.

Let the transmitted signal be a chirp signal, the received signal after demodulation can be written as follows:

$$s_R(t_f, t) = e^{-j2kr_n(t)} e^{j\pi K\left(t - \frac{2r_n(t)}{c}\right)^2} w_r\left(t - \frac{2r_n(t)}{c}\right) \tag{14.8}$$

where $r_n(t) = \sqrt{(r_n' + r_c)^2 + (x_n' + x_c - v \cdot t)^2}$ is the range history, within the observation time, of the nth scatterer, K is the frequency sweep rate, and $w_r(\cdot)$ is the received signal domain.

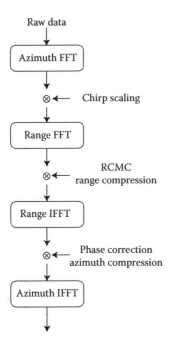

FIGURE 14.5 CS flow chart.

The CS algorithm can be summarized in the following steps:

- Azimuth Fourier transform and equalization of the range cell migration for each range cell. Since the range migration depends on the range, the first step of the algorithm consists of a range migration equalization. This operation aims at making the range migration look the same for all range cells, therefore eliminating the range migration dependence on range.
- Range Fourier transform, range compression, and Range Cell Migration Correction (RCMC).
- Range inverse Fourier transform.
- Removal of the residual phase term and azimuth compression.
- Azimuth inverse Fourier transform.

Details can be found in [15].

14.3 SAR INVERSION MAPPING

Moving targets are typically displaced and defocused in SAR images because of their relative motion with respect to the scene center. Since man-made targets of interests are usually noncooperative targets, the use of Non-Cooperative Target Imaging (NCTI) techniques, such as ISAR, becomes mandatory.

As one of the aims of this work is to refocus SAR images of moving targets, the first challenging problem to be tackled in this work is to transform the SAR image containing the defocussed target into raw data. This operation is mandatory as ISAR processors accept raw data at their input and not SAR images. Since a SAR image is obtained by processing very large amount of data, which may contain several targets, each one with its own motions, subimages containing separate targets have to be cut from the SAR image and treated separately.

The solution proposed in this chapter consists of inverting each SAR subimage in order to obtain an equivalent raw data, which contains only the target's echo with some residual background clutter/noise. Such raw data are then used as input of an ISAR processor which is devoted to producing well-focussed ISAR images. This concept is depicted in Figure 14.1.

In this section, the problem of obtaining raw data-like data from a SAR subimage is addressed. Specifically, four inversion techniques are proposed and analyzed, namely the Inverse ω–k (IOK), the Inverse PFA (IPFA), the Inverse CS (ICS), and the Inverse RD (IRD).

To obtain equivalent raw data from SAR images, a new spatial frequency domain must be defined for each SAR subimage. As the resolution of the whole SAR image and of that of the subimage are equal, the observation time and the bandwidth must be the same. On the other hand, the Pulse Repetition Frequency (PRF) and the frequency spacing in the equivalent raw data are different and can be calculated as follows:

$$\text{PRF} = \frac{N}{T_{\text{obs}}} \tag{14.9}$$

$$\Delta f = \frac{B}{M} \tag{14.10}$$

where N is the number of samples in the cross-range direction and M is the number of samples in the range direction of each subimage.

14.3.1 ω–k INVERSION

The ω–k algorithm makes use of an interpolation to obtain very-high-resolution SAR images. Unfortunately, such an operation on the data makes the ω–k algorithm not perfectly invertible, which means that some errors or artifacts are introduced in the data after attempting any type of

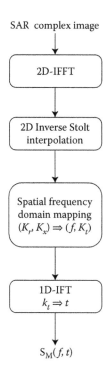

FIGURE 14.6 Inverse ω–k algorithm.

inversion. The Inverse ω–k (IOK) algorithm proposed in this study has been obtained by inverting each single step of the direct algorithm, as shown in Figure 14.6. As all steps in the direct algorithm are invertible except for the interpolator, we will expect that artifacts may be introduced by the interpolation inversion, which is implemented by means of an additional interpolator. Additional comments on the nonperfect inversion of the ω–k algorithm can be found in Section 14.5.

To get equivalent raw data from the SAR subimage, an interpolation step is required. This is necessary to remap the rectangular domain into the polar domain. This step involves a loss of information because there are regions of the circular sector that cannot be calculated by using an interpolator (data in the gray region shown in Figure 14.7). Although an extrapolator could be used to estimate the data in those regions, some artifacts may be introduced that could degrade the image quality.

14.3.2 POLAR FORMAT INVERSION

The PF algorithm [16] makes use of the assumption that radar pulses spherical wavefronts can be assumed planar around a reference point in the scene. This assumption is quite satisfied when the swath size is much smaller than the distance between the radar and the focusing point. The most relevant drawback of the PF algorithm is that it requires an interpolation step which is typically time consuming.

The IPF can be obtained by tracing back the steps used to form the image via the PF algorithm. Specifically, a 2D-IFFT is applied to the SAR subimage followed by a 2D Inverse Stolt Interpolation and a spatial frequency domain mapping.

The IPF algorithm flow chart is shown in Figure 14.8 where the steps involved in the PF algorithm are reversed.

14.3.3 CHIRP SCALING INVERSION

Differently from the PF, the CS algorithm does not need an interpolation step, therefore resulting in a more computationally efficient algorithm.

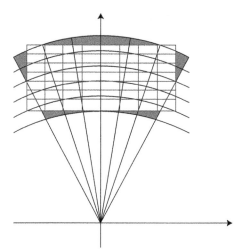

FIGURE 14.7 Polar and rectangular grid.

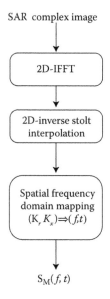

FIGURE 14.8 IPFA algorithm.

The CS algorithm addresses the problem of equalizing the range migration of all the point scatterers composing the target. Since all the scatterers follow the same trajectory, they can be compensated by a known phase term (details can be found in Ref. [17]). The ICS algorithm flow chart is shown in Figure 14.9, where the steps involved in the CS algorithm are reversed.

14.3.4 Fourier Inversion

When the total aspect angle variation is not too large and when the effective rotation vector is sufficiently constant during the observation time, the RD technique represents an accurate and computationally effective tool for SAR/ISAR image reconstruction. Under this constraint, the polar grid

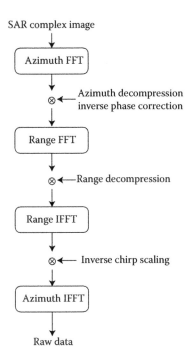

FIGURE 14.9 ICS algorithm.

in the spatial frequency domain can be assumed to be a nearly regularly sampled rectangular grid. Therefore, a two-dimensional FFT can be used to reconstruct the image. In this case the inversion algorithm, namely IRD, consists of a Fourier inversion, which is usually implemented via a two-dimensional IFFT.

14.4 ISAR PROCESSING

The ISAR processor used to refocus SAR subimages of moving targets is based on a RD approach. Specifically, a time window and an autofocus algorithm are applied to the data before using a Fourier transform to form the image. Autofocus, RD image formation and time windowing are briefly recalled below.

Figure 14.10 shows the ISAR geometry. The reference system T_ξ is embedded in the radar with the axis ξ_2 oriented along the line of sight (LOS). Without losing generality, it is assumed that the target moves along a trajectory that intersects the axis ξ_2 at the central instant $t = 0$ of the observation time T_{obs}. The target rotation due to the translation motion is denoted as the translational rotation vector $\Omega_{tr}(t)$. In practical conditions, external forces produce angular motions that are represented by the angular rotation vector $\Omega_a(t)$ applied to the center O of the target. The sum of these two rotation vectors yields the total angular rotation vector $\Omega_T(t)$. The projection of $\Omega_T(t)$ on the plane orthogonal to the LOS is called effective rotation vector $\Omega_{eff}(t)$, which is the rotation vector component that contributes to the target aspect angle variation. The imaging plane (x_1, x_2) is orthogonal to the effective rotation vector and is represented in Figure 14.10. The time-varying coordinate system T_x is chosen so as to have the x_2 axis oriented along the LOS, the x_3 axis along the effective rotation vector and the origin in O. With this choice, the x_1 and the x_2 axes become the cross-range and range coordinates of the imaging plane, respectively. It is worth noting that, in general, the imaging plane (x_1, x_2) is time-varying because the effective rotation vector varies with respect to time.

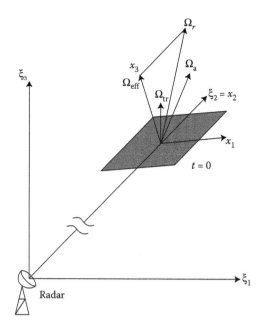

FIGURE 14.10 ISAR geometry.

14.4.1 Autofocus

The autofocus technique adopted here is the Image Contrast (IC)-Based Autofocus (ICBA) algorithm [18]. The ICBA is a parametric technique and it is based on IC maximization. Under the assumption that $|P(f)| \approx 1$ when $|f - f_0| < B$ and that the target can be modeled as the composition of N_s ideal and independent scatterers, after platform motion compensation and after applying a suitable ISAR algorithm, the signal received from a moving target can be approximated as follows as indicated in Equation 14.11. It should be noted that the signal in Equation 14.11 represents an approximation because the inversion process is not exact (due to the interpolation process).

$$S_M(f,t) = \sum_{n=1}^{N_s} \sigma_n e^{-j\frac{4\pi f}{c}R_n(t)} \text{rect}\left(\frac{t}{T_{obs}}\right) \text{rect}\left(\frac{f - f_0}{B}\right) \tag{14.11}$$

where B and f_0 are the bandwidth and the central frequency of the transmitted radar pulse, respectively and $R_n(t)$ is the residual distance between the nth scatterer on the target and the radar at the time instant t. The residual distance is approximately the difference between the actual distance and the distance between the radar and the SAR center scene. It is worth pointing out that such a distance accounts for the relative motion of the target and the stationary center scene, which must be compensated before forming an image of the moving target.

The *straight-iso-range* approximation can be applied when the target is much smaller than the radar–target distance. In practice, this means to be able to effectively approximate the residual distance as follows:

$$R_n(t) \approx R_0(t) + \mathbf{x}^{\mathrm{T}} \cdot \mathbf{i}_{LOS} \tag{14.12}$$

where $R_0(t)$ is the residual distance between an arbitrary point "O" on the target at the time t and the radar, $\mathbf{x}$ is the column vector that identify a scatterer on the target and $\mathbf{i}_{LOS}$ is the column unit vector that identify the radar LOS.

The autofocus technique aims at removing the term $R_0(t)$ due to the target's residual traslational motion. For a relatively short observation time interval T_{obs} and relatively smooth target motions, the radar–target residual distance can be expressed by means of a quadratic form, as follows:

$$R_0(t) = R_0 + v_R t + \frac{a_R}{2} t^2 \qquad (14.13)$$

where v_R and a_R are the radial velocity and acceleration of the target, respectively. The estimation of $R_0(t)$ resorts to the estimation of the target radial motion parameters. Let $\Theta = [v_R, a_R]$ be the vector containing the unknowns, the radial motion compensation problem can be recast as an optimization problem where the IC is maximized with respect to the unknown vector Θ, as defined in Equation 14.14

$$\hat{\Theta} = \underset{\Theta}{\mathrm{argmax}}\{IC(\Theta)\} \qquad (14.14)$$

where

$$IC(\Theta) = \frac{\sqrt{A\left\{[I(\Theta) - A[I(\Theta)]]^2\right\}}}{A[I(\Theta)]} \qquad (14.15)$$

and where $A[\cdot]$ indicates an average operation, $I(\Theta)$ is the ISAR image magnitude after compensating for the target translational motion by using Θ as focusing parameters. This can be expressed mathematically as follows:

$$I(\Theta) = \left| RD\left\{ S_M(f,t) \cdot e^{j\frac{4\pi f}{c} R_0(t)} \right\} \right| \qquad (14.16)$$

and RD$\{\cdot\}$ indicates the operation of image formation by means of the RD technique, as it will be clarified in Section 14.4.2.

14.4.2 IMAGE FORMATION

The image formation is based on a Fourier approach. The received signal after motion compensation can be written as follows:

$$S_{MC}(f,t) = \sum_{n=1}^{N_s} \sigma_n e^{-j\frac{4\pi f}{c} x^T \cdot i_{LoS}} \mathrm{rect}\left(\frac{t}{T_{obs}}\right) \mathrm{rect}\left(\frac{f - f_0}{B}\right) \qquad (14.17)$$

RD images are formed by applying a 2D-FT to the motion-compensated data [2,3]. The Fourier approach can be applied when the total aspect angle variation is not too large and when the effective rotation vector is sufficiently constant. Under these assumptions the polar grid in the spatial frequency domain can be approximated by a uniformly sampled rectangular grid. Let $I(\tau, v)$ be the complex-valued ISAR image, which is mathematically defined as follows:

$$I(\eta, v) = 2D - FT\{S_{MC}(f,t)\} \qquad (14.18)$$

where (η, v) are the range and Doppler coordinates, respectively [18].

14.4.3 TIME-WINDOWING

As already stated, the RD technique can be successfully applied when the effective rotation vector does not change significantly during the observation time. However, the target's own motion may induce a nonuniform target's rotation vector. In order to minimize target's rotation variation, the Coherent Processing Interval (CPI) is controlled via a Time-Windowing approach. The technique used here is the automatic time-windowing technique proposed in Ref. [6], where an automatic selection of the time window is proposed. Specifically, the time-window position across the data and its length are automatically chosen in order to obtain one or more images with the highest focus. The criterion used to define the highest-focused image is based on the IC.

Therefore, the optimal position and length of the time window are obtained by maximizing the IC with respect to the couple $(\bar{t}, \Delta t)$. Therefore, the following optimization problem can be formulated:

$$(\bar{t}_{opt}, \Delta t_{opt}) = \underset{(\bar{t}, \Delta t)}{\mathrm{argmax}}[\mathrm{IC}(\bar{t}, \Delta t)] \tag{14.19}$$

where $\mathrm{IC}(\bar{t}, \Delta t)$ is defined as in Equation 14.15.

14.4.4 CROSS-RANGE SCALING

ISAR generates two-dimensional high-resolution images of targets in time delay—Doppler domain. In order to determine the size of the target, it is preferable to have fully scaled image. The range scaling can be performed by using the well-known relationship $r = c\tau/2$, where r is the slant-range coordinate and τ is the time delay. On the other hand, cross-range scaling requires the estimation of the modulus of the target effective rotation vector. Recently, a novel algorithm has been proposed to solve this problem. This algorithm is based on the assumption of quasi-constant target rotation. When the target rotation vector can be assumed constant within the coherent integration time, the chirp rate produced by the scattering centers can be related to the modulus of the target effective rotation vector by means of an analytical expression. Therefore, each scattering center carries information about the modulus of the target rotation vector through its chirp rate. The signal components of the scattering centers are therefore processed by means of a polynomial Fourier transform to estimate the chirp rate. The use of such a method, namely local polynomial Fourier transform (LPFT), requires the solution of an optimization problem.

It has been largely proven in the literature that the use of the IC and the image entropy are good parameters to assess the quality of the reconstructed ISAR image. In the proposed algorithm, the IC will be used as a cost function to be maximized for estimating scattering centers chirp rate.

For the sake of clarity, the algorithm flow chart is in Figure 14.11 and can be summarized as follows:

- The ISAR image is segmented in order to extract N subimages $I_T^{(n)}(\eta, v)$ relative to N scattering centers. Their relative scattering center location C_n is also calculated.
- Each subimages is inversely Fourier transformed in the Doppler domain in order to obtain N signals $S_{MC}^{(n)}(\eta, t)$.
- The N signals $S_{MC}^{(n)}(\eta, t)$ are then used for the scattering chirp rate estimation by jointly using the 2-LPFT and the IC. Specifically, the chirp rate is estimated by maximizing the IC value of the image obtained by applying a second-order LPFT to the signal $S_{MC}^{(n)}(\eta, t)$.
- The chirp rate and the scattering location obtained in the previous steps are used for the estimation of the modulus of the effective rotation vector via a least-squares error (LSE) approach.

Further details can be found in Ref. [5].

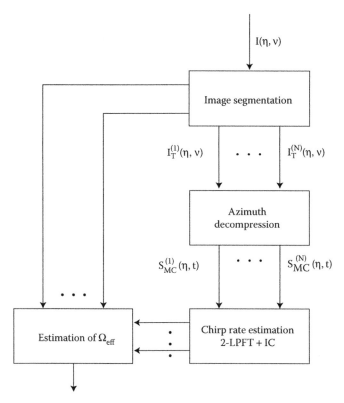

FIGURE 14.11 Cross-range scaling algorithm flow chart.

14.5 EXPERIMENTAL RESULTS

The results provided in this chapter are obtained by applying the proposed technique to CSK Spotlight data. It is worth pointing out that the CSK SAR images used here have been obtained by processing the raw data by means of the ω–k algorithm.

Since the Inverse ω–k requires a higher computational cost with respect to the Fourier Inversion algorithm, it becomes important to evaluate the benefits of applying the former to justify its use.

For this purpose, a subimage containing the target is back-transformed into the data domain by means of both the Inverse ω–k and the Fourier inversion algorithm. A comparison analysis between the two ISAR images, obtained by processing the two raw data like, is then carried out by considering the image quality and the IC value.

As previously stated, for long recorded data, the target rotation vector is far from being constant during the observation time. It is therefore necessary to select one or more shorter integration time intervals to obtain one or more focussed ISAR images of the target. In order to compare the results obtained by using the inverse Fourier algorithm with the results obtained by using the correct SAR inversion algorithm, it is necessary to compare the IC values of a large set of ISAR images. For these purposes, a time moving window of fixed length has been shifted along with the data, instead of using the time-windowing algorithm described in the previous section.

14.5.1 CSK PRODUCT DESCRIPTION

CSK is conceived as a dual-use end-to-end Earth Observation System. Dual use means that it can be used for both civilian and military purposes, such as risk management applications, NonCooperative Target Surveillance and recognition, cartography and planning applications, agriculture, forest, hydrology, geology, and so on.

The system consists of a constellation of four Low Earth Orbit mid-sized satellites, each equipped with a multimode high-resolution SAR operating at X-band. The system is completed by a Ground segment which provides the means and resources to manage and control the mission, to receive and process the data acquired by the satellites, and to disseminate and archive the generated products.

To ensure many combinations between image size and spatial resolution the system has been designed to operate in different modalities:

- A Spotlight mode providing low spatial resolution (<1 m for civilian consumers) and swath size of 10 km × 10 km.
- Two Stripmap modes: Himage mode and PingPong mode. Both modes provide spatial resolution of few meters and a swath size of one-tenth of a kilometer. The PingPong mode implements a strip acquisition by alternating a pair of polarization across burst.
- Two ScanSAR modes providing from medium-to-coarse spatial resolution over large swath.

For our purposes, Spotlight images have been processed since fine spatial resolutions allow a better understanding of man-made targets of moderate size, such as maritime targets and airplanes.

14.5.2 CSK SAR Images

The refocussing algorithm has been tested on maritime targets since in the available SAR images, some ships were clearly visible and no ground truth was needed to validate the results. Specifically two CSK SAR images have been processed, the first that covers the area of Messina and the latter that covers the area of Istanbul, which have been acquired respectively on April 23, 2008 and on April 14, 2008. Some specifications of both the SAR images are summarized in Table 14.1, where θ_{in} is the incidence angle. It is worth pointing out that both the CSK SAR image have been obtained by processing the raw data by means of the ω–k algorithm. CSK SAR images are shown in Figure 14.12.

Specifically some targets have been highlighted which are clearly visible in the SAR images. Each subimages will be back transformed into the data domain by using the Inverse ω–k algorithm but also by using the Fourier Inverse algorithm, and the results will then be compared.

14.5.3 Refocussing Results

Results obtained by applying ISAR image processing to SAR data are presented in this section. Specifically, the ICBA autofocussing technique is first applied to the entire subimage data to refocus moving targets with simple motions. The time-windowing technique is then applied to a case of a

TABLE 14.1
SAR Images Specifications

	SAR Image	
	Istanbul #1	Messina
PRF	3312.72 Hz	2866.97 Hz
B	199.29 MHz	198.41 MHz
T_{obs}	1.5158 s	1.5186 s
θ_{in}	57°	58°
Polarization	HH	HH
Look	right	Right
Direction	Desc	Asc

Note: HH, Horizontal horizontal.

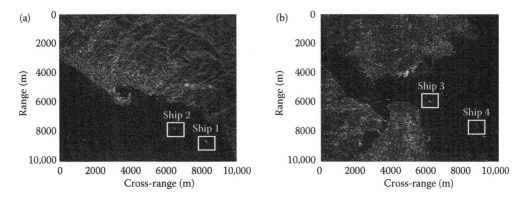

FIGURE 14.12   Quick-Look SAR image covering the area of Messina (a) and Istanbul (b).

moving target with complex motions, which cannot simply be refocussed by applying the autofocus technique to the entire data. Finally, cross-range scaling is applied to the refocussed ISAR images to obtain fully scaled ISAR images, which can directly lead to target size estimation.

14.5.3.1   ISAR Autofocusing Results

As stated previously, results will be shown in this section in terms of Image Contrast values and image quality from a visual point of view. Specifically, original SAR images and refocussed ISAR images will be shown for each selected target.

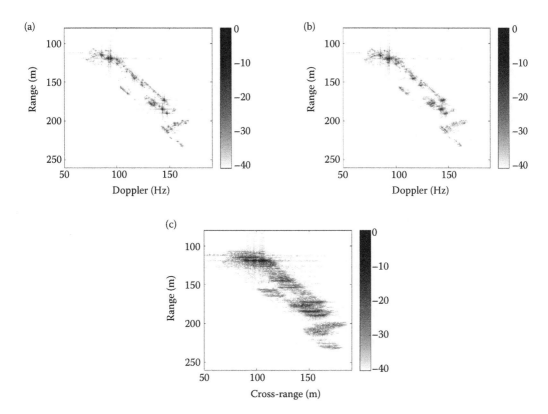

FIGURE 14.13   (**See color insert.**) ISAR image of ship1 obtained by using the whole observation time and the IRD (a) and the IOK (b), and SAR image of ship 1 (c).

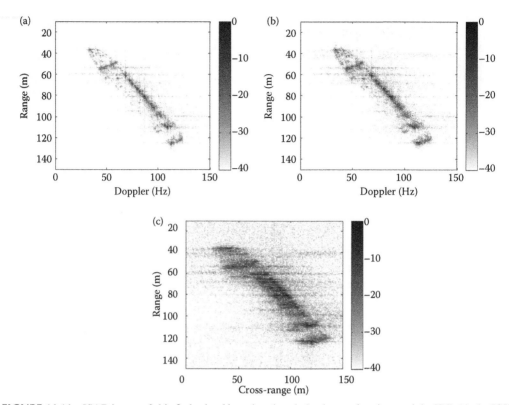

FIGURE 14.14   ISAR image of ship 2 obtained by using the whole observation time and the IRD (a), the IOK (b), and SAR image of ship 2 (c).

As a first case study, we will consider the problem of refocussing targets with simple motions, which usually do not require any time-windowing. These are typical cases where the target's motion is mainly due to its traslational movement, and therefore, the target's rotation vector can be assumed constant during the observation time. Three examples are shown in Figures 14.13 through 14.16, which prove the effectiveness of the proposed refocussing algorithm. Specifically, both the refocussed ISAR image obtained by inverting by means of the IRD and IOK are shown to demonstrate that both inversion algorithms are effective.

A more direct comparison is made between ISAR images obtained by using IRD and IOK SAR image inversion and the results are evaluated by means of the resulting IC. Table 14.2 contains the IC values of the SAR and the ISAR images obtained by using the whole observation time and both the IRD and IOK techniques for each target.

A larger set of ISAR images is compared to obtain some meaningful statistics that can be used to evaluate the performance of both inversion algorithms. To this purpose, a time moving window of fixed length is shifted along the data obtained after SAR image inversion. Specifically the time window length and the time shift used in this experiment are $T_{win} = 0.5$ s and $T_{step} = 0.1$ s, respectively. The IC values of the ISAR images obtained by processing the data selected by the moving window are shown in Figure 14.17 in terms of their difference.

To complete the result analysis, Table 14.2 contains the IC values of the SAR and the ISAR images obtained by using the whole observation time and both the IRD and the IOK techniques for each target.

By observing the previous results, the following remarks can be made:

- The proposed ISAR processing is able to produce well-focussed radar images of moving targets when it is applied to data obtained by using both the IRD and the IOK inversion

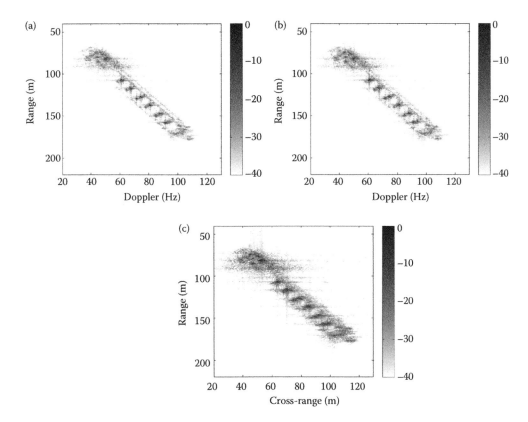

FIGURE 14.15   ISAR image of ship 3 obtained by using the whole observation time and the IRD (a), the IOK (b), and SAR image of ship 3 (c).

algorithms. Therefore, there is no evident reason to prefer one inversion algorithm to another one in terms of image focus (measurable from the image magnitude).

- Although the difference in terms of IC does not lead to very evident differences in the ISAR image magnitudes (from a visual inspection point of view), phase errors may be present that are introduced by the IRD, because it is an approximated way of inverting the SAR image, or the IOK, because of the application of an interpolation, which is not an invertible operation. It should be said that phase errors introduced by the inversion algorithm may destroy useful information for phase-related applications/postprocessing, such as interferometry [19], super-resolution [20,21], and so on.

14.5.3.2   Time-Windowing Results

As already stated, for long recorded data, the target rotation vector may not be constant during the observation time. Therefore, the RD technique used to form the ISAR images cannot produce well-focussed images. This issue can be overcome by using a time-windowing technique (subaperture formation). This technique allows controlling the CPI in order to keep the target rotation vector constant during the reduced CPI. Differently from the previous results, in the example below the necessity of using a time-windowing algorithm is rather evident.

The target is a small maneuvring target. As it can be noted by observing Figure 14.18a and b, the ISAR processing applied to the entire CPI is not able to produce well-focussed target images. Furthermore, it is quite clear, by observing Figure 14.18a and b, that the target experiences an evident yaw motion during the observation time.

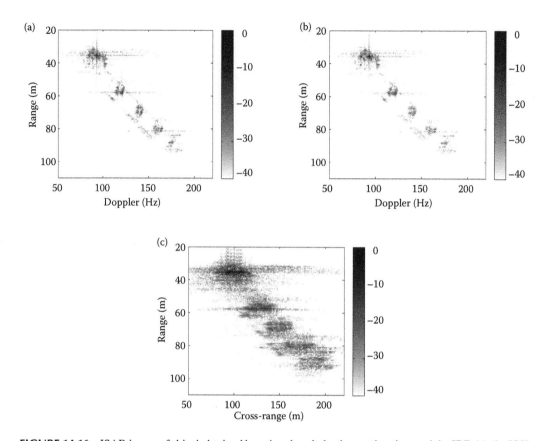

FIGURE 14.16 ISAR image of ship 4 obtained by using the whole observation time and the IRD (a), the IOK (b), and SAR image of ship 4 (c).

By applying the time-windowing technique, an increase in the IC value can be measured, which typically indicates a better image reconstruction. In Figure 14.19, several ISAR images of the target are shown that are obtained by using both the IRD and the IOK, each one obtained from a different time interval. Specifically these images have been obtained by using $T_{win} = 0.2$ s and $T_{step} = 0.1$ s. By observing Figure 14.19, it can be noted that the vessel orientation changes slightly from frame to frame.

The target ISAR images obtained by applying the automatic time-windowing algorithm described in Section 14.4.3 are shown in Figure 14.20. The IC values of both the target images are $IC_{IRD} = 8.41$ and $IC_{IOK} = 8.93$, respectively. As it can be noted, although good images are obtained by manually choosing the window length, better results and better Doppler resolutions are obtained by using the automatic time-windowing algorithm.

TABLE 14.2
IC Values

IC	SAR	ISAR–IRD	ISAR–IOK
Ship #1	6.2912	10.3545	9.8767
Ship #2	4.3425	6.5967	6.0036
Ship #3	6.9902	7.6421	7.0361
Ship #4	5.6503	7.6923	7.9231

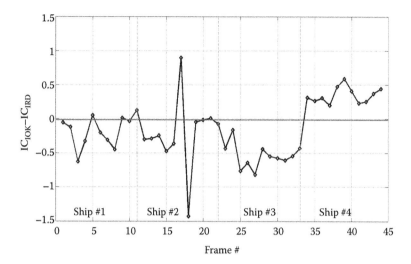

FIGURE 14.17 $IC_{IOK}-IC_{IRD}$ for each frame of each target.

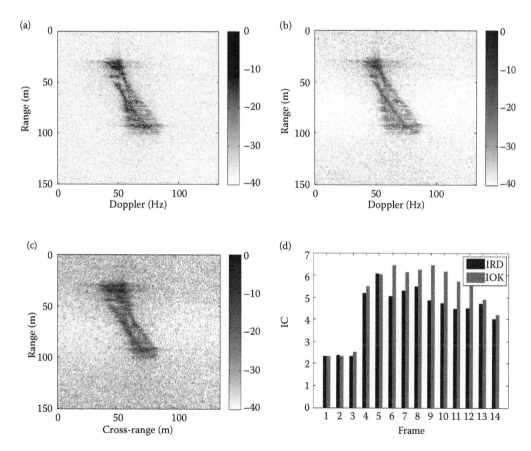

FIGURE 14.18 (**See color insert.**) ISAR image of ship10 obtained by using the IRD (a) the IOK (b), SAR image of the target (c), and IC values relative to each frame (d).

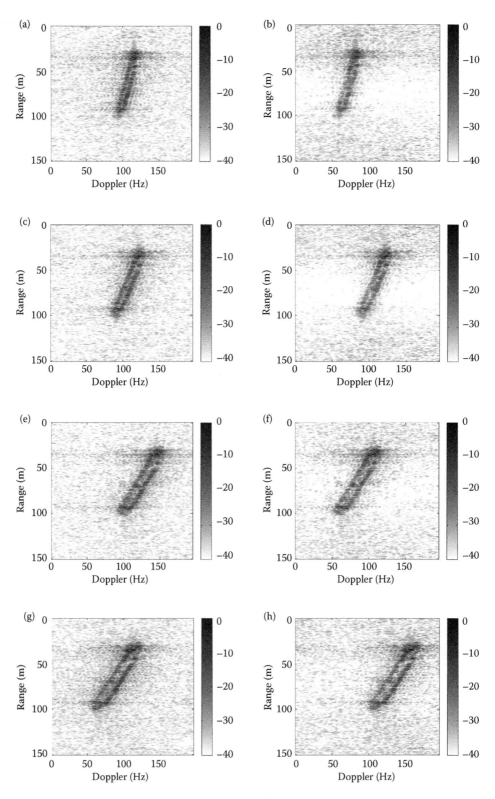

FIGURE 14.19 ISAR image sequences obtained by windowing the data (time subapertures)—IRD (a, c, e, g) and IOK (b, d, f, h).

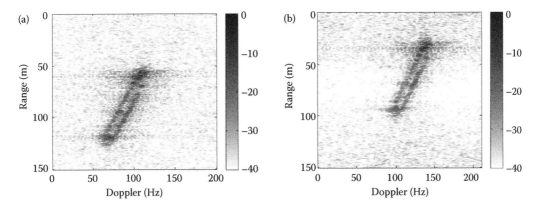

FIGURE 14.20 Most focussed ISAR image obtained by applying the IC-based Time-Windowing technique—IRD (a) and IOK (b).

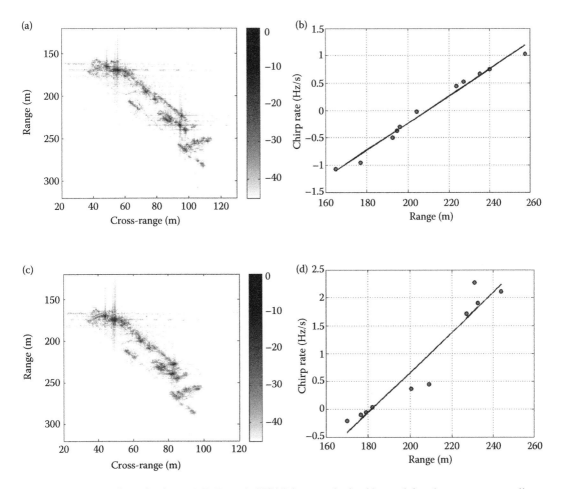

FIGURE 14.21 (**See color insert.**) Fully scaled ISAR images obtained by applying the cross-range scaling technique and chirp rate estimates—IRD (a, b) and IOK (c, d).

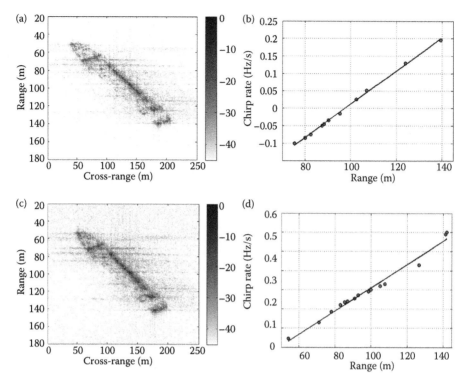

FIGURE 14.22   Fully scaled ISAR images obtained by applying the cross-range scaling technique and chirp rate estimates—IRD (a, b) and IOK (c, d).

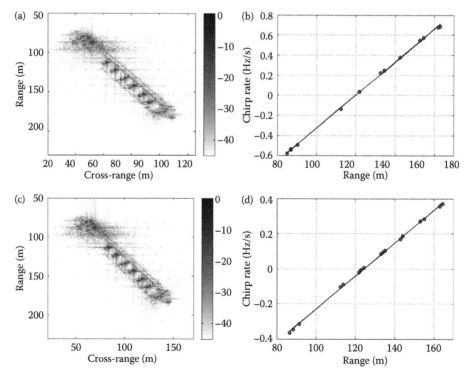

FIGURE 14.23   Fully scaled ISAR images obtained by applying the cross-range scaling technique and chirp rate estimates—IRD (a, b) and IOK (c, d).

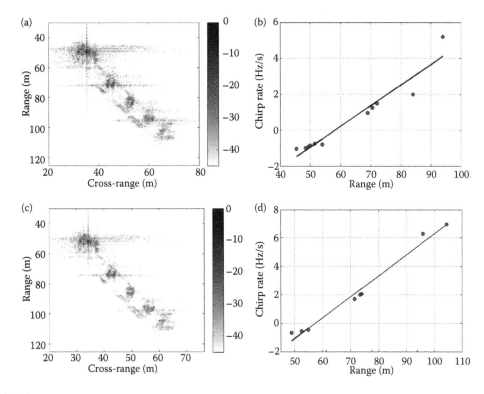

FIGURE 14.24 Fully scaled ISAR images obtained by applying the cross-range scaling technique and chirp rate estimates—IRD (a, b) and IOK (c, d).

14.5.3.3 Cross-Range Scaling Results

Figure 14.21a and c, Figure 14.22a and c, Figure 14.23a and c, Figure 14.24a and c shows the fully scaled ISAR images of the targets obtained by using both the IRD and the IOK algorithms. Figure 14.21b and d, Figure 14.22b and d, Figure 14.23b and d, Figure 14.24b and d shows the estimated chirp rates (dots) and the regression straight line, whose slope is proportional to the effective rotation vector modulus square value. As can be noted, for each target, the chirp rate estimates fit the LSE straight line very tightly. Table 14.3 shows the estimated length of the vessels.

It is worth noting that, although the actual length of the ships is unknown, the estimates obtained by applying the cross-range scaling algorithm described in Section 14.4.4 are all likely values. It must also be pointed out that targets are projected onto the ISAR image plane. Therefore, any estimated distance between two points on the target is actually the result of a projection, thus producing an underestimated value. It should also be remembered that the ISAR image plane is unknown, as

TABLE 14.3
Estimated Length of the Vessels

	IRD	IOK
Ship #1	131 m	124 m
Ship #2	178 m	164 m
Ship #3	139 m	125 m
Ship #4	78 m	67 m

it depends on the orientation of the effective rotation vector, and therefore such projection cannot be determined *a priori*.

It is important to note that the adopted cross-range scaling algorithm is based on phase information. The results showed that the phase information is nicely preserved after inverting the SAR image back to the data domain, both when using the IRD and the IOK. Although, ISAR interferometry as well as other applications strongly based on the phase information have not been tested after applying the proposed algorithm, the results shown relatively to the cross-range scaling algorithm are surely encouraging.

14.6 CONCLUSIONS

It has been demonstrated in this work that spaceborne ISAR imaging of noncooperative targets can be enabled starting from SLC SAR images. The idea of using SAR images as a starting point followed by inversion mapping and ISAR processing has been developed and demonstrated by using CSK Spotlight SAR data of maritime scenes. Results have shown that

- Moving targets on the sea surface appear defocussed and displaced when imaged by conventional SAR algorithms.
- Moving targets can be refocussed by inverting a SAR subimage back to the data domain and by running an ISAR processor.
- Moving targets ISAR images can be fully scaled by using a cross-range scaling technique.

Several inversion techniques have been proposed that can be applied to relevant SAR data. In the example provided, as the SAR images were obtained by means of an ω–k algorithm, a comparison between the ω–k inversion and a simple Fourier inversion has been made. The results have shown that for small scenes that contain relatively small targets, the benefits of using ad hoc inversion techniques rather than a simple Fourier inversion can be counteracted by errors introduced by the inversion process (interpolator). It should be noted that the phase estimation results necessary for the cross-range scaling technique confirm that both inversion mappings preserve the phase information adequately.

ACKNOWLEDGMENT

This work has been partially sponsored by the Italian Space Agency (ASI) under the CSK® AO, Project ID 2143.

REFERENCES

1. W. C. Carrara, R. S. Goodman, and R. M. Majewsky, *Spotlight Synthetic Aperture Radar: Signal Processing Algorithms*. Boston: Artech House, 1995.
2. D. A. Ausherman, A. Kozma, J. L. Walker, H. M. Jones, and E. C. Poggio, Developments in radar imaging, *IEEE Transactions on Aerospace and Electronic Systems*, 20, 363–400, 1984.
3. J. L. Walker, Range-Doppler imaging of rotating objects, *IEEE Transactions on Aerospace and Electronic Systems*, 16, 23–52, 1980.
4. V. Chen and W. Miceli, Simulation of ISAR imaging of moving targets, *IEE Proceedings: Radar, Sonar and Navigation*, 148(3), pp. 160–166, 2001.
5. M. Martorella, A novel approach for ISAR image cross-range scaling, *IEEE Transactions on Aerospace and Electronic Systems*, 44(1), 281–294, 2008.
6. M. Martorella and F. Berizzi, Time windowing for highly focused isar image reconstruction, *IEEE Transactions on Aerospace and Electronic Systems*, 41, 992–1007, 2005.
7. M. Martorella, F. Berizzi, D. Pastina, and P. Lombardo, Exploitation of Cosmo Skymed SAR images for maritime traffic surveillance, *IEEE Radar Conference* 2011, pp. 108–112, 2011.

8. J. A. C. Lee and D. C. Munson, Runway imaging from an approaching aircraft using synthetic aperture radar, in *Proceedings of International Conference on Image Processing*, pp. 915–918, Vol. 3, 1996.

9. A. Papoulis, *Signal Analysis*. New York: McGraw-Hill, 1984.

10. S. Silver, *Microwave Antenna Theory and Design*. New York: Dover, 1965.

11. A. Li and O. Loffeld, Two-dimensional SAR processing in the frequency domain, in *Proceedings of the IEEE International Geoscience and Remote Sensing Symposium*, pp. 1065–1068, 1991.

12. C. W. Chen, Modified polar format algorithm for processing spaceborne SAR data, in *Proceedings of the IEEE Radar Conference*, Philadelphia, pp. 44–49, 2004.

13. R. K. Raney, H. Runge, R. Bamler, I. G. Cumming, and F. H. Wong, Precision SAR processing using chirp scaling, *IEEE Transaction on Geoscience and Remote Sensing*, 32, 786, 1994.

14. A. Moreira and Y. Huang, Airborne SAR processing of highly squinted data using a chirp scaling approach with integrated motion compensation, *IEEE Transactions on Geoscience and Remote Sensing*, 32, 1029, 1994.

15. A. Moreira, J. Mittermayer, and R. Scheiber, Extended chirp scaling algorithm for air and spaceborne SAR data processing in stripmap and scansar imaging modes, *IEEE Transactions on Geoscience and Remote Sensing*, 34, 1123, 1996.

16. J. C. Curlander, *Synthetic Aperture Radar: Systems and Signal Processing*. Wiley, New York, 1991.

17. A. Khwaja, L. Ferro-Famil, and E. Pottier, SAR raw data generation using inverse SAR image formation algorithm, *IEEE Geoscience and Remote Sensing Symposium*, 4191–4194, 2006.

18. M. Martorella, F. Berizzi, and B. Haywood, A contrast maximization based technique for 2D ISAR autofocusing, *IEE Proceedings—Radar, Sonar and Navigation*, 152(4), pp. 253–262, 2005.

19. M. M. N. Battisti, Interferometric phase and target motion estimation for accurate 3D reflectivity in ISAR system, *IEEE Radar Conference 2010*, pp. 108–112, 2010.

20. M. Martorella, N. Acito, and F. Berizzi, Statistical clean technique for ISAR imaging, *IEEE Transactions on Geoscience and Remote Sensing*, 45(11), 3552–3560, 2007.

21. J. Tsao and B. Steinberg, Reduction of sidelobe and speckle artifacts in microwave imaging: The CLEAN technique, *IEEE Transactions on Antennas and Propagation*, 36, 543–556, 1988.

15 Active Learning Methods in Classification of Remote Sensing Images

Lorenzo Bruzzone, Claudio Persello, and Begüm Demir

CONTENTS

15.1 INTRODUCTION

Land cover maps are usually obtained from remote sensing (RS) images by using automatic supervised classification techniques, which require a set of labeled samples for training the classification algorithm. However, the accuracy of the thematic maps that can be obtained with these techniques strongly depends on the quality and quantity of the available training samples, whose collection is costly and time consuming. Accordingly, in real classification problems, the available training samples are often not enough for an adequate learning of the classifier. A possible approach to address this problem is to exploit unlabeled samples in the learning of the classification algorithm according to a semisupervised classification procedure. The semisupervised approach has been widely investigated in the recent years in the RS community [1–6] and has proved to be effective in several application domains. However, the convergence to a correct solution of semisupervised methods is not always guaranteed. Even if some strategies have been proposed for detecting the correct convergence of semisupervised algorithms [6], the use of these techniques in operational applications should be done carefully. A different approach to both enrich the information given as input to the supervised classifier and improve the statistic of the classes is to iteratively expand the original training set according to a process that requires an interaction between the user and the automatic recognition system. This approach is known in the machine learning community as active learning (AL) and, in the recent few years, it has also been studied in the RS community for the classification of different types of RS images. The AL process is conducted according to an iterative process. At each iteration, the most informative unlabeled samples are automatically chosen by the learning algorithm for a manual labeling from a human expert (supervisor) and the supervised algorithm is retrained with the additional labeled samples. In this way, the unnecessary and redundant labeling of samples that are not informative for the classifier is avoided, greatly reducing the labeling cost and time. Moreover, AL allows one to reduce the computational complexity of the training phase. In this chapter, we focus our attention on AL methods.

In RS classification problems, the collection of labeled samples can be derived according to: (1) *in situ* ground surveys, which are associated to high cost and require significant time; (2) interpretation of color composites by experts (image photointerpretation), which is relatively cheap and fast; or (3) hybrid solutions where both photointerpretation and ground surveys are used. The choice of the labeling strategy depends on the considered problem and image type. For example, we can reasonably assume that for the classification of optical very-high-resolution (VHR) images, the labeling of samples can be easily carried out by photointerpretation. The metric or submetric resolution of these images allows a human expert to identify and label the objects on the ground on the basis of the inspection of their geometric and spectral properties in real or false color compositions. In cases when medium (or low)-resolution multispectral images and hyperspectral data are considered, the land cover classes are characterized only on the basis of their spectral signatures (the geometric properties of the objects are not visible and cannot be used by the photointerpreter) and usually cannot be recognized with high reliability by a human expert. For example, hyperspectral data, thanks to a dense sampling of the spectral signature, allows one characterizing several different land cover classes (e.g., associated to different arboreal species) that cannot be recognized by a visual analysis of different false color compositions. Thus, in these cases, ground surveys are necessary for the labeling of samples.

On the basis of the aforementioned considerations, depending on both the type of classification problem and the type of data, the cost and time associated to the labeling process significantly changes. These different scenarios require the definition of different AL schemes: we expect that in cases where photointerpretation is possible, several iterations of the labeling step are feasible; whereas in cases where ground-truth surveys are necessary, only few iterations of the AL process can be done, because of both high cost and required time associated with *in situ* data collection.

Most of the studies in AL have focused on selecting the single most informative unlabeled sample to be included in the training set at each iteration, by assessing its uncertainty [7–13]. This can be inefficient, since the classifier has to be retrained for each new labeled sample added to the training set. This approach can be inappropriate for RS image classification tasks for the above-mentioned reasons. Thus, in this chapter we focus on batch-mode AL, where a batch of $h > 1$ unlabeled samples is queried at each iteration. The problem with such an approach is that by selecting the samples of the batch on the basis of the uncertainty only, some of the selected samples could be similar to each other, and thus do not provide additional information for the model updating with respect to other samples in the batch. The key issue of batch-mode AL is to select sets of samples with little redundancy, so that they can provide the highest possible information to the classifier. Thus, the query function adopted for selecting the batch of the most informative samples should take into account two main criteria: (1) uncertainty and (2) diversity of samples [14–17]. The uncertainty criterion is associated to the confidence of the supervised algorithm in correctly classifying the considered sample, while the diversity criterion aims at selecting a set of unlabeled samples that are as more diverse (distant one another) as possible, thus reducing the redundancy among the selected samples. The combination of the two criteria results in the selection of the potentially most informative set of samples at each iteration of the AL process.

The aim of this chapter is to review different AL techniques proposed in the literature for the classification of RS images. In particular, we focus our attention on AL techniques based on support vector machines (SVMs) by considering and comparing different query functions and strategies to assess the uncertainty, as well as different diversity criteria in the multiclass case. The investigated techniques are theoretically described and experimentally compared among them in the classification of VHR images and hyperspectral data. On the basis of this comparison, some guidelines are derived on the use of AL techniques for the classification of different types of RS images.

The rest of this chapter is organized as follows. Section 15.2 reviews the background on AL methods and their application to RS problems. Section 15.3 presents a selected set of the most recent and effective AL techniques. Section 15.4 presents the description of the two considered data sets and Section 15.5 analyzes the obtained experimental results. Finally, Section 15.6 presents the conclusion of this chapter.

15.2   BACKGROUND ON ACTIVE LEARNING

15.2.1   ACTIVE LEARNING PROCESS

A general AL process can be described considering the quintuple (G, Q, S, T, U) [7]. G is a supervised classifier, which is trained on the labeled training set T. Q is a query function used to select the most informative unlabeled samples from a pool U of unlabeled samples. S is a supervisor who can assign the true class label to any unlabeled sample of U. The AL process is an iterative process, where the supervisor S (i.e., a human expert) interacts with the system by iteratively labeling the most informative samples selected by the query function Q at each iteration. At the initial stage, an initial training set T of few labeled samples is required for the first training of the classifier G. After initialization, the query function Q is used to select a set of samples X from the pool U and the supervisor S assigns them the true class label. Then, these new labeled samples are included into T and the classifier G is retrained using the updated training set. The closed loop of querying and retraining continues for some predefined iterations or until a stop criterion is satisfied. Algorithm 1 gives a description of a general AL process.

ALGORITHM 1:   ACTIVE LEARNING PROCEDURE

1. Train the classifier G with the initial training set T
2. Classify the unlabeled samples of the pool U

Repeat
3. Query a set of samples (with query function Q) from the pool U
4. A label is assigned to the queried samples by the supervisor S
5. Add the new labeled samples to the training set T
6. Retrain the classifier
Until a stopping criterion is satisfied.

The query function Q is of fundamental importance in AL and constitutes the core of each AL technique. Several query functions have been proposed so far in the machine learning literature. A probabilistic approach to AL is presented in Ref. [8], which is based on the estimation of the posterior probability density function of the classes both for obtaining the classification rule and to estimate the uncertainty of unlabeled samples. In the two-class case, the query of the most uncertain samples is obtained by choosing the samples closest to 0.5 (half of them below and half above this probability value). The query function proposed in Ref. [18] is designed to minimize future errors, that is, the method selects the unlabeled pattern that, once labeled and added to the training data, is expected to result in the lowest error on test samples. This approach is applied to two regression models (i.e., weighted regression and mixture of Gaussians) where an optimal solution for minimizing future error rates can be obtained in closed form. Unfortunately, this solution is intractable to calculate the expected error rate for most classifiers without specific statistical models. A statistical learning approach is used in Ref. [19] for regression problems with multilayer perceptron. In Ref. [20], a method is proposed that selects the next example according to an optimal criterion (which minimizes the expected error rate on future test samples), but solves the problem by using a sampling estimation. The authors in Ref. [20] present two techniques for estimating future error rate. In the first technique, the future error rate is estimated by log-loss using the entropy of the posterior class distribution on the set of unlabeled samples. In the second technique, a 0–1 loss function using the posterior probability of the most probable class for a set of unlabeled samples is used. Instead of estimating the expected error over the full distribution, the error is measured over the samples in the pool U. Furthermore, the estimation of the error is obtained using the learner at the previous iteration. The query function causes the selection of the examples which maximize the sharpness of the learner's existing belief over the unlabeled examples. The method is implemented using naive Bayes.

Another popular paradigm is given by committee-based active learners. The "query by committee" approach [21–23] is a general AL algorithm that has theoretical proofed guarantees on the reduction in prediction error with the number of queries. A committee of classifiers using different hypothesis about parameters is trained to label a set of unknown examples. The algorithm selects the samples where the disagreement between the classifiers is maximal. In Ref. [24], two query methods are proposed that combine the idea of query by committee and that of boosting and bagging.

An interesting category of AL approaches, which have gained significant success in numerous real-world learning tasks, is based on the use of SVMs [9–15]. The SVM classifier [25,26] is particularly suited to AL due to its intrinsic high generalization capabilities and because its classification rule can be characterized by a small set of support vectors (SVs) that can be easily updated over successive learning iterations [13]. One of the most popular (and effective) query heuristic for active SVM learning is margin sampling (MS), which selects the data point closest to the current separating hyperplane. This method results in the selection of the unlabeled sample with the lowest confidence, that is, the maximal uncertainty on the true information class. The query strategy proposed in Ref. [11] is based on the splitting of the version space [11,14]: the points which split the current version space into two halves having equal volumes are selected at each step, as they are likely to be the actual SVs. Three heuristics for approximating the above criterion are described; the simplest among them selects the point closest to the hyperplane as in Ref. [9]. In Ref. [7], an approach is proposed that estimates the uncertainty level of each sample according to the output score of a classifier and selects only those samples whose outputs are within the uncertainty range. In Ref. [12], the authors present possible generalizations of the active SVM approach to multiclass problems.

It is important to observe that the above-mentioned methods consider only the uncertainty of samples, which is an optimal criterion only for the selection of one sample at each iteration. Selecting a batch of $h > 1$ samples exclusively on the basis of the uncertainty (e.g., the distance to the classification hyperplane) may result in the selection of similar (redundant) samples that do not provide additional information. However, in many problems it is necessary to speed up the learning process by selecting batches of more than one sample at each iteration. In order to address this shortcoming, in Ref. [14], an approach is presented especially designed to construct batches of samples by incorporating a diversity measure that considers the angles between the induced classification hyperplanes (more details on this approach are given in the next section). Another approach to consider the diversity in the query function is the use of clustering [15,16]. In Ref. [15], an AL heuristic is presented, which explores the clustering structure of samples and identifies uncertain samples avoiding redundancy (details of this approach are given in the next section). In Refs. [27,28], the authors present a framework for batch-mode AL that applies the Fisher information matrix to select a number of informative examples simultaneously.

Nevertheless, most of the above-mentioned approaches are designed for binary classification and thus are not suitable for most of the RS classification problems. In this chapter, we focus on multiclass SVM-based AL approaches that can select a batch of samples at each iteration for the classification of RS images. The next subsection provides a discussion and a review on the use of AL for the classification of RS images.

15.2.2 ACTIVE LEARNING FOR THE CLASSIFICATION OF RS DATA

AL has been applied mainly to text categorization and image-retrieval problems. However, the AL approach can be adopted for the interactive classification of RS images by taking into account the specific features of this domain. In RS problems, the supervisor S is a human expert that can derive the land cover type of the area on the ground associated with the selected patterns according to the three possible strategies identified in the introduction, that is, photointerpretation, ground survey, or hybrid strategies. Here, these different strategies are associated with significantly different costs and times, and the choice of the strategy (and thus the costs and times) depends on the considered classification problem. The image photointerpretation is relatively cheap but it strongly depends on the expert's ability to reliably identify the correct label of selected samples. The cost of ground surveys is normally much higher and depends on the considered area. According to these strategies, the AL approach can be run as (1) interactive expert-guided classification tool or (2) *in situ* ground surveys planning and supervised classification tool. In Ref. [29], the AL problem is formulated considering a spatially dependent label acquisition costs. We observe that the labeling cost mainly depends on the type of the RS data, which affects the aforementioned labeling strategy. For example, in the case of multispectral VHR images, often the labeling of samples can be carried out by photointerpretation, while in the case of medium/low-resolution multispectral images and hyperspectral data, expensive ground surveys are necessary. No particular restrictions are usually considered for the definition of the initial training set T and its size $|T|$, since we expect that the AL process can be started up with few samples for each class without affecting the convergence capability. The pool of unlabeled samples U can be associated with the whole considered image or to a portion of it (for reducing the computational time associated to the query function and/or for considering only the areas of the scene accessible for labeling). An important issue is related to the capability of the query function to select batches of $h > 1$ samples, which results to be of fundamental importance for the adoption of AL in real-world RS problems. It is worth noting here the importance of the choice of the h value in the design of the AL classification system, as it affects the number of iterations and thus both the performance and the cost of the classification system. In general, we expect that for the classification of VHR images (where photointerpretation is possible), several iterations of the labeling step may be carried out and small values for h can be adopted; whereas in cases where ground-truth surveys are necessary, only few iterations of the AL process are possible and large h values are necessary.

In the RS domain, AL was applied to the detection of subsurface targets, such as landmines and unexploded ordnance in Refs. [30,31]. In Ref. [31], an efficient AL procedure is developed, based on a mutual information measure. In this procedure, one initially performs excavation with the purpose of acquiring labels to improve the classifier, and once this AL phase is completed, the resulting classifier is applied to the remaining unlabeled signatures to quantify the probability that each item is an unexploded ordnance. Some preliminary works about the use of AL for RS classification problems can be found in Refs. [13,17,32,33]. The technique proposed in Ref. [13] is based on MS and selects the most uncertain sample for each binary SVM in a one-against-all (OAA) multiclass architecture (i.e., querying $h = n$ samples, where n is the number of classes). In Ref. [32], an AL technique is presented, which selects the unlabeled sample that maximizes the information gain between the *a posteriori* probability distribution estimated from the current training set and the training set obtained by including that sample into it. The information gain is measured by the Kullback–Leibler (KL) divergence. This KL-maximization technique can be implemented with any classifier that can estimate the posterior class probabilities. However, this technique can be used to select only one sample at each iteration. In Ref. [33], two batch-mode AL techniques for multiclass RS classification problems are proposed. The first technique is MS by closest SV (MS-cSV), which considers the smallest distance of the unlabeled samples to the n hyperplanes (associated to the n binary SVMs in a (OAA) multiclass architecture) as the uncertainty value. At each iteration, the most uncertain unlabeled samples, which do not share the closest SV, are added to the training set. This technique will be presented in the next section. The second technique, called entropy query-by bagging (EQB), is a classifier-independent approach based on the selection of unlabeled samples according to the maximum disagreement between a committee of classifiers. The committee is obtained by bagging: different training sets (associated with different EQB predictors) are drawn with replacement from the original training data. In Ref. [33], each training set is used to train the OAA SVM architecture to predict the different labels for each unlabeled sample. Finally, the entropy of the distribution of the different labels associated to each sample is calculated to evaluate the disagreement among the classifiers on the unlabeled samples. The samples with maximum entropy (i.e., those with maximum disagreement among the classifiers) are added to the current training set. In Ref. [17], different batch-mode AL techniques for the classification of RS images with SVM are investigated. The investigated techniques exploit different query functions, which are based on both the uncertainty and diversity criteria. Moreover, a query function that is based on a kernel-clustering technique for assessing the diversity of samples and a strategy for selecting the most informative representative sample from each cluster is proposed. A detailed description of the two methods presented in Ref. [17] (i.e., (1) multiclass-level uncertainty with angle-based diversity (MCLU-ABD) and (2) multiclass-level uncertainty with enhanced clustering-based diversity (MCLU-ECBD)) is given in the following section.

15.3 INVESTIGATED ACTIVE LEARNING METHODS

In this section, we present different query functions Q based on SVM for multiclass RS classification problems. SVM is a binary classifier, whose goal is to divide the d-dimensional feature space into two subspaces (one for each class) using a separating hyperplane. Let us assume that a training set T made up of N pairs $\left(\mathbf{x}_i, y_i\right)_{i=1}^{N}$ is available, where $\mathbf{x}_i$ are the training samples and $y_i \in \{+1; -1\}$ are the associated labels. The decision rule used to find the membership of an unknown sample is based on the sign of the discrimination function $f(\mathbf{x}) = \langle \mathbf{w} \cdot \mathbf{x} \rangle + b$ associated with the hyperplane. An important property of SVMs is related to the possibility to project the original data into a higher-dimensional feature space via a kernel operator $K(\cdot, \cdot)$, which satisfies the Mercer's conditions [34]. The training phase of the classifier can be formulated as a minimization problem by using the Lagrange optimization theory, which leads to the calculation of the values of Lagrange multipliers α_i associated with the original training patterns $\mathbf{x}_i \in \mathcal{X}$. After the training, the discrimination function is given by

$$f(\mathbf{x}) = \sum_{i \in \mathrm{SV}} y_i \alpha_i K(\mathbf{x}_i \cdot \mathbf{x}) + b \tag{15.1}$$

where SV is the set of support vectors, that is, the training samples associated with $\alpha_i > 0$. In order to address multiclass problems on the basis of binary classifiers, the general approach consists of defining an ensemble of binary classifiers and combining them according to some decision rules [26]. The design of the ensemble of binary classifiers involves the definition of a set of two-class problems, each modeled with two groups of classes. The selection of these subsets depends on the kind of approach adopted to combine the ensemble, for example, according to OAA or *one-against-one* strategies [26]. In this work, we adopt the OAA strategy, which involves a parallel architecture made up of n SVMs, one for each information class. Each SVM solves a two-class problem defined by one information class against all the others. We refer the reader to Ref. [26] for greater details on SVM in RS.

In this chapter, we present query functions, defined in the framework of SVM, which are based on (1) uncertainty and (2) the combination of uncertainty and diversity.

15.3.1 UNCERTAINTY-BASED QUERY FUNCTIONS

The uncertainty criterion aims at selecting the unlabeled samples that have maximum uncertainty about their correct label among all samples in the unlabeled sample pool U. Since the most uncertain samples have the lowest probability to be correctly classified by the current classification model, they are the most useful to be included in the training set. In this chapter, we investigate two possible techniques in the framework of multiclass SVM: (a) binary-level uncertainty (BLU) (which evaluates uncertainty at the level of binary SVM classifiers) and (b) multiclass-level uncertainty (MCLU) (which analyzes uncertainty within the considered OAA architecture).

15.3.1.1 Binary-Level Uncertainty

BLU technique separately selects a batch of the most uncertain unlabeled samples from each binary SVM on the basis of the MS query function. In the technique adopted in Ref. [13], at each iteration only the (single) unlabeled sample closest to the hyperplane of each binary SVM was added to the training set (i.e., $h = n$). In the presented BLU technique, at each iteration the most uncertain q ($q > 1$) samples are selected from each binary SVM (instead of a single sample). In greater detail, n binary SVMs are initially trained with the current training set and the functional distance $f_i(\mathbf{x})$, $i = 1, \ldots, n$ of each unlabeled sample $\mathbf{x} \in U$ to the hyperplane is obtained. Then, the set of q samples $\left\{\mathbf{x}_{1,i}^{BLU}, \mathbf{x}_{2,i}^{BLU}, \ldots, \mathbf{x}_{q,i}^{BLU}\right\}$, $i = 1, 2, \ldots, n$ closest to margin of the corresponding hyperplane are selected for each binary SVM, where $\mathbf{x}_{j,i}^{BLU}$, $j = 1, 2, \ldots, q$, represents the selected jth sample from the ith SVM. Totally, $\rho = qn$ samples are taken. Since some unlabeled samples can be selected by more than one binary SVM, the redundant samples are removed. Thus, the total number m of selected samples can actually be smaller than ρ (i.e., $m \le \rho$). The set of m most uncertain samples $\{\mathbf{x}_1^{BLU}, \mathbf{x}_2^{BLU}, \ldots, \mathbf{x}_m^{BLU}\}$ is forwarded to the diversity step. Figure 15.1 shows the architecture of the BLU technique.

15.3.1.2 Multiclass-Level Uncertainty

The MCLU technique selects the most uncertain samples according to a confidence value $c(\mathbf{x})$, $\mathbf{x} \in U$, which is defined on the basis of their functional distance $f_i(\mathbf{x})$, $i = 1, \ldots, n$ to the n decision boundaries of the binary SVM classifiers included in the OAA architecture [33,35]. In this technique,

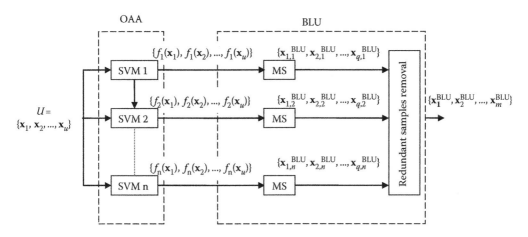

FIGURE 15.1 Multiclass architecture for the BLU technique.

the distance of each sample $\mathbf{x} \in U$ to each hyperplane is calculated and a set of n distance values $\{f_1(\mathbf{x}), f_2(\mathbf{x}), \ldots, f_n(\mathbf{x})\}$ is obtained. Then, the confidence value $c(\mathbf{x})$ can be calculated using different strategies. Here, we consider two strategies: (1) the minimum distance function $c_{\min}(\mathbf{x})$ strategy, which is obtained by taking the smallest distance to the hyperplanes (as absolute value), that is, Ref. [33]

$$c_{\min}(\mathbf{x}) = \min_{i=1,2,\ldots,n} \left\{ \text{abs}[f_i(\mathbf{x})] \right\} \tag{15.2}$$

and (2) the difference function $c_{\text{diff}}(\mathbf{x})$ strategy, which considers the difference between the first largest and the second largest distance values to the hyperplanes, that is, Ref. [35]

$$r_{1\max} = \arg\max_{i=1,2,\ldots,n} \left\{ f_i(\mathbf{x}) \right\}$$
$$r_{2\max} = \arg\max_{j=1,2,\ldots,n,\, j \neq r_{1\max}} \left\{ f_j(\mathbf{x}) \right\} \tag{15.3}$$
$$c_{\text{diff}}(\mathbf{x}) = f_{r_{1\max}}(\mathbf{x}) - f_{r_{2\max}}(\mathbf{x})$$

The $c_{\min}(\mathbf{x})$ function models a simple strategy that computes the confidence of a sample $\mathbf{x}$ taking into account the minimum distance to the hyperplanes evaluated on the basis of the most uncertain binary SVM classifier. Differently, the $c_{\text{diff}}(\mathbf{x})$ strategy assesses the uncertainty between the two most likely classes. If this value is high, the sample $\mathbf{x}$ is assigned to $r_{1\max}$ with high confidence. On the contrary, if $c_{\text{diff}}(\mathbf{x})$ is small, the decision for $r_{1\max}$ is not reliable and there is a possible conflict with the class $r_{2\max}$ (i.e., the sample $\mathbf{x}$ is very close to the boundary between class $r_{1\max}$ and $r_{2\max}$). Thus, this sample is considered uncertain and is selected by the query function for better modeling the decision function in the corresponding position of the feature space. Once the $c(\mathbf{x})$ value of each $\mathbf{x} \in U$ is obtained based on one of the two above-mentioned strategies, the m samples $\mathbf{x}_1^{\text{MCLU}}, \mathbf{x}_2^{\text{MCLU}}, \ldots, \mathbf{x}_m^{\text{MCLU}}$ with lower $c(\mathbf{x})$ are selected to be forwarded to the diversity step. Note that $\mathbf{x}_j^{\text{MCLU}}$ denotes the selected jth most uncertain sample based on the MCLU strategy. Figure 15.2 shows the architecture of the MCLU technique.

15.3.2 Query Functions Based on the Combination of Uncertainty and Diversity

In this section, we present three effective query functions recently proposed in the RS literature in the framework of SVM, which combine the uncertainty and diversity criteria: (1) MS-cSV [33], (2) MCLU-ABD criterion [17], and (3) MCLU-ECBD [17].

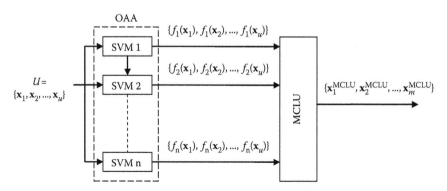

FIGURE 15.2 Architecture for the MCLU technique.

15.3.2.1 Margin Sampling with Closest Support Vector

MS-cSV technique is a modification of MS in order to select more than one sample at each iteration of the AL process by exploiting a diversity criterion [33]. This technique assesses the uncertainty as the smallest distance to the hyperplanes, that is, the MCLU method is used with the minimum distance function $c_{\min}(\mathbf{x})$ strategy in the uncertainty step to assign a confidence value $c(\mathbf{x})$ to each unlabeled sample $\mathbf{x} \in U$. Then, the distance of each SV to each unlabeled sample $\mathbf{x} \in U$ is calculated and the information on the closest SV of the each sample is stored for the selection of diverse samples. This is done by selecting the unlabeled samples that have the lowest $c(\mathbf{x})$ value and, at the same time, do not share the same closest SV (i.e., no sample is added which has the same closest SV with respect to the samples already included in the batch X). The MS-cSV algorithm technique is summarized in Algorithm 2.

ALGORITHM 2:   MS-cSV

Inputs:
m (the number of samples selected on the basis of their uncertainty)
h (batch size)
SVs (support vectors)

Output:
X (set of unlabeled samples to be included in the training set)
1. Compute the minimum distance function $c_{\min}(\mathbf{x})$ for each sample $\mathbf{x} \in U$.
2. Initialize X to the empty set.
3. Include in X the most uncertain sample (the one that has the lowest $c(\mathbf{x})$ value).
4. Estimate the closest support vector to each sample $\mathbf{x} \in U$.

Repeat
5. Select the unlabeled sample that has the next lowest $c(\mathbf{x})$ value as well as does not share the same closest SV with the samples included in X (i.e., if an uncertain unlabeled sample shares the same closest SV with the samples included in X, do not include this sample in X and query next most uncertain sample).
6. Include that unlabeled sample in X.

Until $|X| = h$
7. The supervisor S adds the label to the set of samples $\mathbf{x}_v^{\text{MS-cSV}} \in X$, $v = 1, 2, \ldots, h$, and these samples are added to the current training set.

15.3.2.2 Multiclass-Level Uncertainty with Angle-Based Diversity

MCLU-ABD technique assesses the uncertainty on the basis of MCLU and selects the m most uncertain samples. Then, in the diversity step, the most diverse $h < m$ samples are chosen based on angle-based diversity (ABD) [17]. Cosine angle distance is a similarity measure between two samples defined in the kernel space by Ref. [14],

$$\left| \cos\left(\angle(\mathbf{x}_i, \mathbf{x}_j) \right) \right| = \frac{\left| \phi(\mathbf{x}_i) \cdot \phi(\mathbf{x}_j) \right|}{\left\| \phi(\mathbf{x}_i) \right\| \left\| \phi(\mathbf{x}_j) \right\|} = \frac{K(\mathbf{x}_i, \mathbf{x}_j)}{\sqrt{K(\mathbf{x}_i, \mathbf{x}_i) K(\mathbf{x}_j, \mathbf{x}_j)}}$$

$$\angle(\mathbf{x}_i, \mathbf{x}_j) = \cos^{-1}\left(\frac{K(\mathbf{x}_i, \mathbf{x}_j)}{\sqrt{K(\mathbf{x}_i, \mathbf{x}_i) K(\mathbf{x}_j, \mathbf{x}_j)}} \right)$$

(15.4)

where $\phi(\cdot)$ is a nonlinear mapping function and $K(\cdot, \cdot)$ is the kernel function. The cosine angle distance in the kernel space can be constructed using only the kernel function without considering the direct knowledge of the mapping function $\phi(\cdot)$ (kernel trick). The angle between two samples is small (cosine of angle is high) if these samples are close to each other and vice versa. In the MCLU-ABD technique, the MCLU and ABD criteria are combined based on a weighting parameter λ [14]. On the basis of this combination, a new sample $\mathbf{x}_t$ is included in the selected batch X according to the results of the following optimization problem:

$$x_t = \arg\min_{i=1,...,m}\left\{\lambda\left|c(\mathbf{x}_i)\right| + (1-\lambda)\left[\max_{x_j\in X}\frac{K(\mathbf{x}_i,\mathbf{x}_j)}{\sqrt{K(\mathbf{x}_i,\mathbf{x}_i)K(\mathbf{x}_j,\mathbf{x}_j)}}\right]\right\} \quad (15.5)$$

where λ provides the trade-off between uncertainty and diversity. The cosine angle distance between each sample selected in the uncertainty step (m most uncertain samples) and the samples included in X is calculated and the maximum value is taken as the diversity value of the corresponding sample. Then, the sum of the uncertainty and diversity values weighted by λ is considered to define the combined value. The unlabeled sample $\mathbf{x}_t$ that minimizes such a value is included in X. This process is repeated until the number of samples of the set X ($|X|$) is equal to h. This technique guarantees that the selected samples in X are diverse regarding their angles in the kernel space. Since the initial size of X is zero, the first sample included in X is always the most uncertain sample. It is worth noting that this method is a generalized version of the method proposed in Ref. [14] to multiclass architectures.

ALGORITHM 3:   MCLU-ABD

Inputs:
λ (weighting parameter that tune the trade-off between uncertainty and diversity)
m (number of samples selected on the basis of their uncertainty)
h (batch size)

Output:
X (set of unlabeled samples to be included in the training set)
1. Compute $c(\mathbf{x})$ for each sample $\mathbf{x} \in U$.
2. Select the set of m unlabeled samples with lower $c(\mathbf{x})$ value (most uncertain) $\{\mathbf{x}_1^{MCLU}, \mathbf{x}_2^{MCLU}, \ldots, \mathbf{x}_m^{MCLU}\}$.
3. Initialize X to the empty set.
4. Include in X the most uncertain sample (the one that has the lowest $c(\mathbf{x})$ value).

Repeat
5. Compute the combination of uncertainty and diversity with the following equation formulated for the multiclass architecture:

$$x_t = \arg\min_{i=1,...,m}\left\{\lambda\left|c(\mathbf{x}_i)\right| + (1-\lambda)\left[\max_{x_j\in X}\frac{K(\mathbf{x}_i,\mathbf{x}_j)}{\sqrt{K(\mathbf{x}_i,\mathbf{x}_i)K(\mathbf{x}_j,\mathbf{x}_j)}}\right]\right\} \quad (15.6)$$

where we consider the m most uncertain samples selected at the step 2 and $c(\mathbf{x})$ is calculated as explained in the MCLU subsection (with $c_{min}(\mathbf{x})$ or $c_{diff}(\mathbf{x})$ strategy).
6. Include the unlabeled sample $\mathbf{x}_t$ in X.

Until $|X| = h$
7. The supervisor S adds the label to the set of samples $\{\mathbf{x}_1^{MCLU-ABD}, \mathbf{x}_2^{MCLU-ABD}, \ldots, \mathbf{x}_h^{MCLU-ABD}\} \in X$ and these samples are added to the current training set T.

15.3.2.3   Multiclass-Level Uncertainty with Enhanced Clustering-Based Diversity

In the MCLU-ECBD technique, MCLU is used with the $c_{diff}(\mathbf{x})$ strategy in the uncertainty step to select the m most uncertain samples and the ECBD method is used to select the most diverse $h < m$

samples on the basis of clustering [17]. Clustering techniques evaluate the distribution of the samples in a feature space and group the similar samples into the same clusters. The aim of using clustering in the diversity step is to consider and analyze the distribution of uncertain samples. Since the samples within the same cluster are correlated and provide similar information, a representative sample is selected for each cluster. The advantage of this approach is that cluster prototypes are implicitly sparse in the feature space, that is, distant from each other. The ECBD technique works in the kernel space by applying the kernel k-means clustering [36,37] to the m samples obtained in the uncertainty step to select the $h < m$ most diverse patterns. The kernel k-means clustering iteratively divides the m samples into $k = h$ clusters ($C_1, C_2, \ldots, C_h$) in the kernel space. At the first iteration, initial clusters C_1, $C_2, \ldots, C_h$ are constructed assigning initial cluster labels to each sample [36]. In the next iterations, a pseudo-center is chosen as the cluster center (the cluster centers in the kernel space $\phi(\mu_1), \phi(\mu_2), \ldots, \phi(\mu_h)$ cannot be expressed explicitly). Then the distance of each sample from all cluster centers in the kernel space is computed and each sample is assigned to the nearest cluster. The Euclidean distance between $\phi(\mathbf{x}_i)$ and $\phi(\mu_v)$, $v = 1, 2, \ldots, h$, is calculated as [36,37]:

$$D^2(\phi(\mathbf{x}_i),\phi(\mu_v)) = \left\| \phi(\mathbf{x}_i) - \phi(\mu_v) \right\|^2$$

$$= \left\| \phi(\mathbf{x}_i) - \frac{1}{|C_v|} \sum_{j=1}^{m} \delta(\phi(\mathbf{x}_j), C_v)\phi(\mathbf{x}_j) \right\|^2$$

$$= K(\mathbf{x}_i,\mathbf{x}_i) - \frac{2}{|C_v|} \sum_{j=1}^{m} \delta(\phi(\mathbf{x}_j), C_v)K(\mathbf{x}_i,\mathbf{x}_j)$$

$$+ \frac{1}{|C_v|^2} \sum_{j=1}^{m} \sum_{l=1}^{m} \delta(\phi(\mathbf{x}_j), C_v)\delta(\phi(\mathbf{x}_l), C_v)K(\mathbf{x}_j,\mathbf{x}_l) \tag{15.7}$$

where $\delta(\phi(\mathbf{x}_j), C_v)$ shows the indicator function. $\delta(\phi(\mathbf{x}_j), C_v) = 1$ only if $\mathbf{x}_j$ is assigned to C_v, otherwise $\delta(\phi(\mathbf{x}_j), C_v) = 0$. $|C_v|$ denotes the total number of samples in C_v and is calculated as $|C_v| = \sum_{j=1}^{m} \delta(\phi(\mathbf{x}_j), C_v)$. As mentioned before, $\phi(\cdot)$ is a nonlinear mapping function from the original feature space to a higher-dimensional space and $K(\cdot,\cdot)$ is the kernel function. The kernel k-means algorithm can be summarized as follows [36]:

1. The initial value of $\delta(\phi(\mathbf{x}_i), C_v)$, $i = 1, 2, \ldots, m$, $v = 1, 2, \ldots, h$, is assigned and h initial clusters $\{C_1, C_2, \ldots, C_h\}$ are obtained.
2. Then $\mathbf{x}_i$ is assigned to the closest cluster.

$$\delta(\phi(\mathbf{x}_i), C_v) = \begin{cases} 1 & \text{if } D^2(\phi(\mathbf{x}_i),\phi(\mu_v)) < D^2(\phi(\mathbf{x}_i),\phi(\mu_j)) \quad \forall j \neq v \\ 0 & \text{otherwise} \end{cases} \tag{15.8}$$

3. The sample that is closest to μ_v (the Euclidean distance is calculated in the kernel space by Equation 15.7) is selected as the pseudo-center η_v of C_v.

$$\eta_v = \arg\min_{\mathbf{x}_i \in C_v} D^2(\phi(\mathbf{x}_i),\phi(\mu_v)) \tag{15.9}$$

4. The algorithm is iterated until converge, which is achieved when samples do not change clusters anymore.

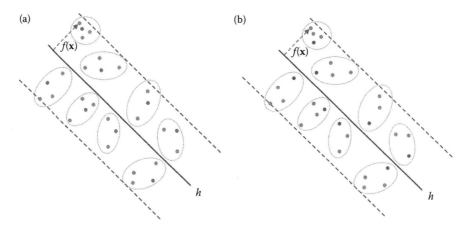

FIGURE 15.3 (**See color insert.**) Comparison between the samples selected by (a) the CBD technique presented in Ref. [15] and (b) the ECBD technique.

After $C_1, C_2, \ldots, C_h$ are obtained, the most informative (i.e., uncertain) sample is selected as the representative sample of each cluster. This sample is defined as

$$\mathbf{x}_v^{\text{MCLU-ECBD}} = \arg \min_{\phi(\mathbf{x}_i) \in C_v} \left\{ c_{\text{diff}}(\mathbf{x}_i^{\text{MCLU}}) \right\} \quad v = 1,2,\ldots,h \tag{15.10}$$

where $\mathbf{x}_v^{\text{MCLU-ECBD}}$ represents the vth sample selected using the MCLU-ECBD and is the most uncertain sample of the vth cluster (i.e., the sample that has minimum $c_{\text{diff}}(\mathbf{x})$ in the vth cluster). Totally, h samples are selected, one for each cluster, using Equation 15.10.

It is worth noting that the uncertainty and clustering-based diversity (CBD) were combined for binary SVM AL in Ref. [15] and this technique is generalized to multiclass problems in Ref. [17] by exploiting MCLU in the uncertainty step. CBD differs from ECBD for two reasons: (1) the standard k-means clustering [40] is applied to the original feature space and not in the kernel space where the SVM separating hyperplane operates, and (2) the medoid sample of each cluster is selected in the diversity step as the corresponding cluster representative sample. In order to better understand the difference in the selection of the representative sample of each cluster between the query function of the CBD presented in Ref. [15] (which selects the medoid sample as cluster representative) and the ECBD query function (which selects the most uncertain sample of each cluster), Figure 15.3 presents a qualitative example. Note that, for simplicity, the example is presented for a binary SVM in order to visualize the confidence value $c_{\text{diff}}(\mathbf{x})$ as the functional distance (MS is used instead of MCLU). The uncertain samples are first selected based on MS for both techniques, and then the diversity step is applied. The query function presented in Ref. [15] selects the medoid sample of each cluster (reported in blue in the figure), which, however, is not in agreement with the idea to choose the most uncertain sample in the cluster. On the contrary, the MCLU-ECBD considers the most uncertain sample of each cluster (reported in red in the figure), which in the binary example is the sample closest to the SVM hyperplane. This is a small difference with respect to the algorithmic implementation but a relevant difference from a theoretical viewpoint and for possible implications on the results.

The MCLU-ECBD algorithm can be summarized as follows:

ALGORITHM 4: MCLU-ECBD

Inputs:
m (the number of samples selected on the basis of their uncertainty)
h (batch size)

Output:

X (set of unlabeled samples to be included in the training set)

1. Compute $c(\mathbf{x})$ for each sample $\mathbf{x} \in U$.
2. Select the set of m unlabeled samples with lower $c(\mathbf{x})$ value (most uncertain) $\{\mathbf{x}_1^{\text{MCLU}}, \mathbf{x}_2^{\text{MCLU}}, ...,$
   $\mathbf{x}_m^{\text{MCLU}}\}$.
3. Apply the kernel k-means clustering (diversity criterion) to the selected m most uncertain samples with $k = h$.
4. Select the representative sample $\mathbf{x}_v^{\text{MCLU-ECBD}}$, $v = 1, 2, ..., h$ (i.e., the most uncertain sample) of each cluster according to Equation (15.10).
5. Initialize X to the empty set and include in X the set of samples $\mathbf{x}_v^{\text{MCLU-ECBD}} \in X$, $v = 1, 2, ..., h$.
6. The supervisor S adds the label to the set of samples $\mathbf{x}_v^{\text{MCLU-ECBD}} \in X$, $v = 1, 2, ..., h$, and these samples are added to the current training set.

15.4 DATA SET DESCRIPTION AND DESIGN OF EXPERIMENTS

15.4.1 DATA SET DESCRIPTION

In our experiments, we used one VHR and one hyperspectral data set. The first data set is a Quickbird multispectral image acquired on the city of Pavia (northern Italy) on June 23, 2002. This image includes the four pan-sharpened multispectral bands and the panchromatic channel with a spatial resolution of 0.7 m. The image size is 1024×1024 pixels. The reader is referred to Ref. [39] for greater details on this data set. The available labeled data (6784 samples) were collected by photointerpretation. These samples were randomly divided to derive a validation set V of 457 samples, a test set TS of 4502 samples, and a pool U of 1825 samples. According to Ref. [40], test pixels were collected on both homogeneous areas TS_1 and edge areas TS_2 of each class. Four percent of the samples of each class in U are randomly selected as initial training samples, and the rest are considered as unlabeled samples. Table 15.1 shows the land cover classes and the related number of samples used in the experiments.

The second data set is a hyperspectral image acquired on the Kennedy Space Center (KSC), Florida, on March 23, 1996. This image consists of 512×614 pixels and 224 bands with a spatial resolution of 18 m. The number of bands is initially reduced to 176 by removing water absorption and low signal-to-noise bands. The available labeled data (5121 samples) were collected using land cover maps derived from color infrared photography provided by KSC and Landsat Thematic Mapper imagery. The reader is referred to Ref. [41] for greater details on this data set. After the elimination of noisy samples, the labeled samples were randomly divided to derive a validation set V of 513 samples, a test set TS of 2556

TABLE 15.1
Number of Samples of Each Class in U, V, TS_1, and TS_2
for the Pavia Data Set

Class	U	V	TS_1	TS_2
Water	58	14	154	61
Tree areas	111	28	273	118
Grass areas	103	26	206	115
Roads	316	79	402	211
Shadow	230	57	355	311
Red buildings	734	184	1040	580
Gray buildings	191	48	250	177
White building	82	21	144	105
Total	1825	457	2824	1678

TABLE 15.2

Number of Samples of Each Class in *U*, *V*, and TS for the KSC Data Set

Class	*U*	*V*	TS
Scrub	305	76	380
Willow swamp	97	24	120
Cabbage palm hammock	102	26	128
Cabbage palm/oak hammock	101	25	125
Slash pine	65	16	80
Oak/broadleaf hammock	92	23	114
Hardwood swamp	42	11	52
Graminoid marsh	173	43	215
Spartina marsh	208	52	260
Cattail marsh	151	38	188
Salt marsh	168	42	209
Mud flats	185	46	231
Water	363	91	454
Total	2052	513	2556

samples, and a pool *U* of 2052 samples. Four percent of the samples of each class are randomly chosen from *U* as initial training samples and the rest are considered as unlabeled samples. The land cover classes and the related number of samples used in the experiments are shown in Table 15.2.

15.4.2 DESIGN OF EXPERIMENTS

In our experiments, without loss in generality, we adopt an SVM classifier with radial basis function (RBF) kernel. The values for the regularization parameter *C* and the spread γ of the RBF kernel parameters are chosen performing a grid-search model selection only at the first iteration of the AL process. Indeed, initial experiments revealed that, if a reasonable number of initial training samples is considered, performing the model selection at each iteration does not increase significantly the classification accuracies at the cost of a much higher computational burden. The MCLU step is implemented with different *m* values defined on the basis of the value of *h* (i.e., $m = 2h, 4h, 6h, 10h$), with $h = 5, 10, 40, 100$. In the BLU technique, the $q = h$ most uncertain samples are selected for each binary SVM. Thus, the total number of selected samples for all SVMs is $\rho = qn$. After removing repetitive patterns, $m \leq \rho$ samples are obtained. The value of λ used in the MCLU-ABD (for computing Equation 15.6) is varied as $\lambda = 0.3, 0.5, 0.6, 0.8$. The total cluster number *k* for kernel *k*-means clustering is fixed to *h*.

All experimental results are referred to the average accuracies obtained in 10 trials according to 10 initial randomly selected training sets. Results are provided as learning rate curves, which show the average classification accuracy versus the number of training samples used to train the SVM classifier. In all the experiments, the size of final training set $|T|$ is fixed to 472 for the Pavia data set and 483 for the KSC data set. The total number of iterations is given by the ratio between the number of samples to be added to the initial training set and the predefined value of *h*.

15.5 EXPERIMENTAL RESULTS

We carried out different kinds of experiments in order to: (1) compare the effectiveness of the ·query functions defined on the basis of the uncertainty criterion only (i.e., MCLU and BLU [17]), (2) assess the effectiveness of the query functions defined on the basis of the combination of the uncertainty and diversity criteria (i.e., MS-cSV [33], MCLU-ABD [17], and MCLU-ECBD [17]), and (3) analyze and compare the two aforementioned approaches.

15.5.1 Comparison among Query Functions Based on the Uncertainty Criterion

In the first set of trials, we analyze the effectiveness of the MCLU and BLU uncertainty techniques, and compare them with each other. Figure 15.4 reports the behavior of the overall accuracies versus the number of labeled samples obtained by the MCLU and BLU techniques. In the MCLU case, $m = 20$ samples are selected for both data sets. In the BLU case, ρ is fixed to 40 and 65 for the Pavia and the KSC data sets, respectively. Then, the most uncertain $h = 5$ samples (which have smallest confidence value), are selected without considering any diversity criteria. The confidence value is calculated with the $c_{\text{diff}}(\mathbf{x})$ strategy for MCLU, as preliminary tests pointed out that it is more effective than the $c_{\text{min}}(\mathbf{x})$ strategy. From Figure 15.4, one can observe that the MCLU technique is more effective than the BLU in the selection of the most uncertain samples on both data sets, that is, the average accuracies provided by the MCLU are in general significantly higher than those obtained by the BLU.

15.5.2 Comparison among Query Functions Based on the Combination of the Uncertainty and Diversity Criteria

In the second set of trials, we compare the effectiveness of the MS-cSV [33], MCLU-ECBD [17], and MCLU-ABD [17] techniques. Figure 15.5 shows the behavior of the average accuracies versus

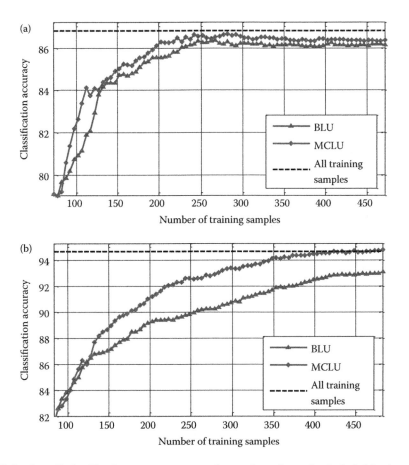

FIGURE 15.4 Overall classification accuracy versus the number of samples included in the training set obtained by the MCLU and BLU for (a) Pavia and (b) KSC data sets.

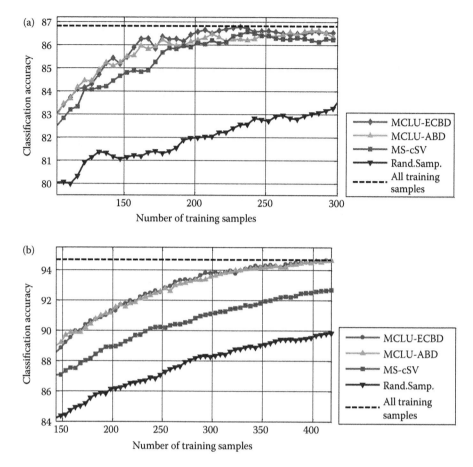

FIGURE 15.5 Overall classification accuracy obtained by the MCLU-ECBD, MCLU-ABD, and MS-cSV techniques for (a) Pavia and (b) KSC data sets. The learning curves are reported starting from 92 samples for Pavia and 142 samples for KSC data sets to better highlight the differences. The line "all training samples" reported in (a) and (b) shows the accuracy obtained using the full pool as the training set.

the number of training samples obtained in the case of $h = 5$ for both data sets. In Figure 15.5, we report the highest average accuracy (obtained with the best values of the parameters λ and m) for each technique. For the KSC data set, the highest accuracies for MCLU-ECBD are obtained with $m = 30$ (while $k = 5$), whereas the best results for MCLU-ABD are obtained with $\lambda = 0.6$ and $m = 20$. For the Pavia data set, the highest accuracies for MCLU-ECBD are obtained with $m = 20$ (while $k = 5$), whereas the best results for MCLU-ABD are obtained with $\lambda = 0.6$ and $m = 20$.

The results obtained on the Pavia data set (see Figure 15.5a) show that the proposed MCLU-ECBD technique leads to the highest accuracies in most of the iterations; furthermore, it achieves convergence in less iterations (and thus with a smaller number of labeled samples) than the other techniques. The MCLU-ECBD technique yields an accuracy of 86.77% with only 232 samples, while using the full pool as training set (1825 samples) we obtain an accuracy of 86.82%. The MCLU-ABD method provides in general slightly lower accuracy than the MCLU-ECBD technique and higher accuracies than the MS-cSV technique.

The results obtained on the KSC data set (Figure 15.5b) show that the MCLU-ECBD provides similar results compared with MCLU-ABD, and both of them significantly outperform the MS-cSV method. The MCLU-ECBD technique reaches an accuracy of 94.64% with only 413 samples, while using the full pool as training set (2052 samples) we obtain an accuracy of 94.68%.

TABLE 15.3

Examples of Computational Time (in Seconds) Taken from the MCLU-ECBD, MCLU-ABD, and MS-cSV Techniques

Data Set	Technique	h			
		5	10	40	100
Pavia	MCLU-ECBD	10	6	7	11
	MCLU-ABD	10	5	6	10
	MS-cSV	384	193	—	—
KSC	MCLU-ECBD	17	14	15	19
	MCLU-ABD	18	13	15	19
	MS-cSV	977	614	134	—

For a more detailed comparison, additional experiments were carried out on varying the values of the parameters (numerical results are not reported for space constraints). In all cases, we observed that the MCLU-ECBD and the MCLU-ABD techniques yield higher classification accuracies than the other AL techniques when small h values are considered. Thus, the values for the parameter m and λ, which should be defined by the user, are not critical for the accuracies of both the MCLU-ECBD and MCLU-ABD techniques.

Table 15.3 reports the computational time (in seconds) required for one trial by the MCLU-ECBD, the MCLU-ABD, and the MS-cSV techniques for different h values. In this case, the value of m for MCLU-ECBD and MCLU-ABD is fixed to $4h$. It can be noted that MCLU-ECBD and MCLU-ABD are fast both for small and high values of h. The computational time of MS-cSV is very high in the case of small h values, whereas it decreases by increasing the h value. In the table, some values of the computational time of MS-cSV are missing, because such a technique cannot work when the value of h is higher than the number of SVs (i.e., $h \leq |SVs|$).

The results obtained on the two data sets confirm that both the MCLU-ECBD and the MCLU-ABD are very effective in terms of both classification accuracy and computation complexity.

15.5.3 COMPARISON OF THE UNCERTAINTY CRITERION WITH THE COMBINATION OF UNCERTAINTY WITH DIVERSITY CRITERIA

Finally, we analyze the accuracy obtained by using only an uncertainty criterion and the combination of uncertainty with diversity criteria for different h values. Figure 15.6 shows the behavior of the average accuracy versus the number of training samples obtained by the MCLU (m is fixed equal to h for a fair comparison) and the MCLU-ECBD. In these experiments, $m = 4h$ with $h = 5$ and 100 is selected for Pavia data set, while $m = 2h$ with the same values of h is chosen for KSC data set. One can observe that, as expected, using only the uncertainty criterion results in poor accuracies when h is small, whereas the classification performances are significantly improved by using both uncertainty and diversity criteria. On the contrary, the choice of complex query functions is not justified when a large batch of samples is added to the training set at each iteration (i.e., similar results can be obtained with and without considering diversity). This mainly depends on the intrinsic higher probability of a large number of samples h to represent patterns in different positions of the feature space. Similar behaviors are observed with the other query functions.

15.6 DISCUSSION AND CONCLUSION

In this chapter, we reviewed the use of AL methods for the classification of RS images, presenting a survey of the literature and an experimental analysis of a set of promising AL techniques recently

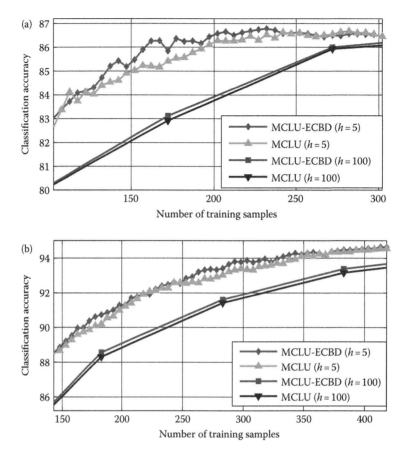

FIGURE 15.6 Overall classification accuracy versus the number of training samples for the uncertainty criterion and the combination of uncertainty and diversity criteria with different h values: (a) Pavia and (b) KSC data sets. The learning curves are reported starting from 92 samples for Pavia and 142 samples for KSC data sets to better highlight the differences.

proposed in RS. The results obtained on two different data sets confirm that the use of AL techniques for the classification of RS images reduces the computational time and the number of labeled samples used for training the supervised algorithm (which is associated to cost and time for defining the training set) and increases the classification accuracy with respect to traditional passive techniques.

In the experimental analysis, we compared different AL techniques defined on the basis of (1) uncertainty and (2) the combination of uncertainty and diversity. The experiments were carried out on the classification of both VHR multispectral and hyperspectral images. By this comparison, we observed that the MCLU-ECBD and MCLU-ABD methods resulted in higher accuracy with respect to other state-of-the-art methods for the same number of labeled samples. In addition, they can reach convergence with a smaller number of labeled samples than the other techniques. We underline that this is a very important advantage, because the main goal of AL is to perform an effective learning of a classifier with the minimum possible number of labeled samples. It was also observed that MCLU-ECBD and MCLU-ABD techniques are generally more effective than the other considered techniques also in terms of computational time (especially for small values of h). Thus, they are actually well suited for applications in which sample labeling is carried out with both ground survey and image photointerpretation. Moreover, we observed that (1) the MCLU technique is more effective in the selection of the most uncertain samples for multiclass problems than the

BLU technique; (2) the $c_{\text{diff}}(\mathbf{x})$ strategy is more precise than the $c_{\text{min}}(\mathbf{x})$ strategy to assess the confidence value in the MCLU technique; and (3) the use of both uncertainty and diversity criteria is necessary when h is greater than one but small, whereas high h values do not require the use of complex query functions.

The MCLU-ABD technique provides slightly lower or similar classification accuracies than the MCLU-ECBD method in most of the cases, with a similar computational time. It can be used for selecting a batch made up of any desired number of h samples. The MS-cSV technique provides quite good classification accuracies. However, in the case of small h values, the computational complexity of this technique is much higher than that of the other investigated techniques. This complexity decreases when h increases.

We assessed the compatibility of the considered AL techniques with the strategies to label unlabeled samples by image photointerpretation or ground data collection in order to provide some guidelines to the users under different conditions. As mentioned before, in the case of VHR images, in many applications, the labeling of unlabeled samples can be achieved by photointerpretation, which is compatible with several iterations of the AL process in which a small number h of samples are included in the training set at each step according to an interactive procedure of labeling carried out by an expert operator. On our VHR data set, we observed that batches of $h = 5$ or 10 samples can result in the best trade-off between accuracy and number of considered labeled samples. In the case of hyperspectral or medium/low-resolution multispectral data, expensive and time-consuming ground surveys are usually necessary for the labeling process. Under this last condition, only very few iterations of the AL process are realistic. Thus, it is reasonable to collect large batches (of, e.g., hundreds of samples) for each iteration. In this case, we observed that sophisticated query functions are not necessary, as with many samples often the uncertainty criterion alone is sufficient for obtaining good accuracies.

As a final remark, we point out that in real applications, some geographical areas may be not accessible for ground survey (or the process might be too expensive). Thus, the definition of the pool U should be carried out carefully, in order to avoid these areas and to take into account possible different costs of labeling samples in different geographical sites.

REFERENCES

1. B. M. Shahshahani and D. A. Landgrebe, The effect of unlabeled samples in reducing the small sample size problem and mitigating the Hughes phenomenon, *IEEE Transactions on Geoscience and Remote Sensing*, 32(5), 1087–1095, 1994.
2. L. Bruzzone, M. Chi, and M. Marconcini, A novel transductive SVM for the semisupervised classification of remote-sensing images, *IEEE Transactions on Geoscience and Remote Sensing*, 44(11), 3363–3373, 2006.
3. M. Chi and L. Bruzzone, Semi-supervised classification of hyperspectral images by SVMs optimized in the primal, *IEEE Transaction on Geoscience and Remote Sensing*, 45(6, Part 2), 1870–1880, 2007.
4. M. Marconcini, G. Camps-Valls, and L. Bruzzone, A composite semisupervised SVM for classification of hyperspectral images, *IEEE Geoscience and Remote Sensing Letters*, 6(2), pp. 234–238, 2009.
5. G. Camps-Valls, T. V. Bandos Marsheva, and D. Zhou, Semi-supervised graph-based hyperspectral image classification, *IEEE Transactions on Geoscience and Remote Sensing*, 45(10), 3044–3054, 2007.
6. L. Bruzzone and M. Marconcini, Domain adaptation problems: A DASVM classification technique and a circular validation strategy, *IEEE Transactions on Pattern Analysis and Machine Intelligence*, 32(5), 770–787, 2010.
7. M. Li and I. Sethi, Confidence-based active learning, *IEEE Transactions on Pattern Analysis and Machine Intelligence*, 28(8), 1251–1261, 2006.
8. D. D. Lewis and W. A. Gale, A sequential algorithm for training text classifiers, in W. B. Croft and C. J. van Rijsbergen (Eds.), *Proceedings of the 17th Annual International ACM SIGIR Conference on Research and Development in Information Retrieval*, London, pp. 3–12, 1994.
9. C. Campbell, N. Cristianini, and A. Smola, Query learning with large margin classifiers, *Proceedings of the 17th International Conference on Machine Learning (ICML'00)*, pp. 111–118, 2000.

10. G. Schohn and D. Cohn, Less is more: Active learning with support vector machines, *Proceedings of the 17th International Conference on Machine Learning (ICML'00)*, pp. 839–846, 2000.

11. S. Tong and D. Koller, Support vector machine active learning with applications to text classification, *Proceedings of the 17th International Conference on Machine Learning (ICML'00)*, pp. 999–1006, 2000.

12. T. Luo, K. Kramer, D. B. Goldgof, L. O. Hall, S. Samson, A. Remsen, and T. Hopkins, Active learning to recognize multiple types of plankton, *Journal of Machine Learning Research*, 6, 589–613, 2005.

13. P. Mitra, B. U. Shankar, and S. K. Pal, Segmentation of multispectral remote sensing images using active support vector machines, *Pattern Recognition Letter*, 25(9), 1067–1074, 2004.

14. K. Brinker, Incorporating diversity in active learning with support vector machines, *Proceedings of the International Conference on Machine Learning*, Washington DC, pp. 59–66, 2003.

15. Z. Xu, K. Yu, V. Tresp, X. Xu, and J. Wang, Representative sampling for text classification using support vector machines, *25th European Conference on Information Retrieval Research*, pp. 393–407, 2003.

16. H. T. Nguyen and A. Smeulders, Active learning using pre-clustering, *Proceedings of the 21th ICML*, Banff, AB, Canada, pp. 623–630, 2004.

17. B. Demir, C. Persello, and L. Bruzzone, Batch mode active learning methods for the interactive classification of remote sensing images, *IEEE Transactions on Geoscience and Remote Sensing*, 49(3), 1014–1031.

18. D. Cohn, Z. Ghahramani, and M. I. Jordan, Active learning with statistical models, *Journal of Artificial Intelligence Research*, 4, 129–145, 1996.

19. K. Fukumizu, Statistical active learning in multilayer perceptrons, *IEEE Transactions Neural Networks*, 11(1), 17–26, 2000.

20. N. Roy and A. McCallum, Toward optimal active learning through sampling estimation of error reduction, *Proceedings of the ICML*, Williamstown, MA, pp. 441–448, 2001

21. H. S. Seung, M. Opper, and H. Smopolinsky, Query by committee, *Proceedings of the 5th Annual ACM Workshop on Computational Learning Theory*, Pittsburgh, PA, pp. 287–294, 1992.

22. Y. Freund, H. S. Seung, E. Shamir, and N. Tishby, Selective sampling using the query by committee Algorithm, *Machine Learning*, 28, 133–168, 1997.

23. I. Dagan and S. P. Engelson, Committee-based sampling for training probabilistic classifiers, *Proceedings of the ICML*, San Francisco, CA, 1995, pp. 150–157.

24. N. Abe and H. Mamitsuka, Query learning strategies using boosting and bagging, *Proceedings of the ICML*, Madison, WI, pp. 1–9, 1998.

25. V. N. Vapnik, *The Nature of Statistical Learning Theory*, 2nd ed., New York: Springer, 2001.

26. F. Melgani and L. Bruzzone, Classification of hyperspectral remote sensing images with support vector machines, *IEEE Transactions on Geoscience and Remote Sensing*, 42(8), 1778–1790, 2004.

27. S. C. Hoi, R. Jin, J. Zhu, and M. R. Lyu, Batch mode active learning and its application to medical image classification, *Proceedings of the 23rd International Conference on Machine Learning (ICML'06)*, pp. 417–424, 2006.

28. S. C. Hoi, R. Jin, J. Zhu, and M. R. Lyu, Batch mode active learning with applications to text categorization and image retrieval, *IEEE Transactions on Knowledge and Data Engineering*, 21(9), 1233–1248, 2009.

29. A. Liu, G. Jun, and J. Ghosh, Active learning of hyperspectral data with spatially dependent label acquisition costs, *IEEE International Geoscience and Remote Sensing Symposium 2009, (IGARSS'09)*, pp. V-256–V-259, Cape Town, South Africa, 2009.

30. Y. Zhang, X. Liao, and L. Carin, Detection of buried targets via active selection of labeled data: Application to sensing subsurface UXO, *IEEE Transactions on Geoscience and Remote Sensing*, 42(7), 2535–2543, 2004.

31. Q. Liu, X. Liao, and L. Carin, Detection of unexploded ordnance via efficient semisupervised and active learning, *IEEE Transactions on Geoscience and Remote Sensing*, 46(9), 2558–2567, 2008.

32. S. Rajan, J. Ghosh, and M. M. Crawford, An active learning approach to hyperspectral data classification, *IEEE Transactions on Geoscience and Remote Sensing*, 46(4), 1231–1242, 2008.

33. D. Tuia, F. Ratle, F. Pacifici, M. Kanevski, and W. J. Emery, Active learning methods for remote sensing image classification, *IEEE Transactions on Geoscience and Remote Sensing*, 47(7), 2218–2232, 2009.

34. C. Burges, A tutorial on support vector machines for pattern recognition, *Data Mining and Knowledge Discovery*, 2(2), 121–167, 1998.

35. A. Vlachos, A stopping criterion for active learning, *Computer, Speech and Language*, 22(3), 295–312, 2008.

36. R. Zhang and A. I. Rudnicky, A Large scale clustering scheme for kernel k-means, *IEEE International Conference on Pattern Recognition*, 11–15 August 2002, Quebec, Canada, pp. 289–292.

37. B. Scholkopf, A. Smola, and K. R. Muller, Nonlinear component analysis as a kernel eigenvalue problem, *Neural Computation*, 10, 1299–1319, 1998.
38. M. Dalponte, L. Bruzzone, and D. Gianelle, Fusion of hyperspectral and LIDAR remote sensing data for the estimation of tree stem diameters, *IEEE International Geoscience and Remote Sensing Symposium 2009, (IGARSS'09)*, pp. II-1008–II-1011, Cape Town, South Africa, 2009.
39. L. Bruzzone and L. Carlin, A multilevel context-based system for classification of very high spatial resolution images, *IEEE Transactions on Geoscience and Remote Sensing*, 44(9), 2587–2600, 2006.
40. A. K. Jain and R. C. Dubes, *Algorithms for Clustering Data*. Upper Saddle River, NJ: Prentice-Hall, 1988.
41. J. Ham, Y. Chen, M. M. Crawford, and J. Ghosh, Investigation of the random forest framework for classification of hyperspectral data, *IEEE Transactions on Geoscience and Remote Sensing*, 43(3), 2005, 492–501.

16 Crater Detection Based on Marked Point Processes

Giulia Troglio, Jon Atli Benediktsson, Gabriele Moser, and Sebastiano Bruno Serpico

CONTENTS

16.1 INTRODUCTION

In the past few decades, a large number of images have been collected by different planetary missions and an increasing amount of images will be acquired by the current and the future missions. The collected data that are of different types (i.e., multitemporal, multisensor, and stereo images) need to be analyzed. Typically, an important task in such analysis is feature extraction, that is, the identification of spatial structures in the images.

For instance, in order to jointly exploit, integrate, or compare all these different data, image registration is a necessary task, which generally requires a prior accurate extraction of spatial features in the images to be registered. Moreover, feature extraction is an essential task for various other applications, such as identifying safe landing sites for rovers on the surface of the analyzed planet.

Feature extraction in planetary images can be manually performed by human experts but this is a very time-consuming endeavor. Therefore, there is the need of reliable automatic techniques for crater detection. In particular, the development of an automatic approach for the estimate of the position, the structure, and the dimension of each crater on the surface of the analyzed planet would be really useful. The identification of spatial features on planetary surfaces is a difficult task for various reasons. Indeed, the amount of data available is usually limited, planetary images generally present low contrast (being heavily affected by illumination, surface properties, and atmospheric state), and the features that are present in the images can be barely visible.

Different types of spatial structures of variable size and shape characterize planetary surfaces. Among the typical features, craters play a primary role. Detection of craters has been widely addressed and different approaches have recently been presented in the literature, based on the analysis of planetary topography data [30], satellite images in the visible spectrum, and the infrared spectrum [11].

Each of the previously published methodologies for automatic crater detection has its advantages and drawbacks, but so far, none looks robust enough to be applied as a stand-alone procedure with satisfactory final results (a review of the existing literature on crater detection from passive remote sensing images can be found in Section 16.1.1).

In order to overcome the typical problems of planetary images with limited contrast, poor illumination, and a lack of good features, in the present chapter, an unsupervised approach for the extraction of planetary craters (as proposed in Ref. [33]), based on a marked point process (MPP) [37], is reviewed here. The framework is stochastic and the goal is to minimize an energy on the state space of all possible configurations of objects, using a Markov chain Monte-Carlo (MCMC) algorithm and a simulated annealing (SA) scheme (see Section 16.2 for more details). More properly, MPPs are introduced to model the structure of the crater edges in the image. MMPs, which have been used in different areas of the terrestrial remote sensing [10], are here applied for the first time to the analysis of planetary images.

The chapter is organized as follows. The previous work about crater detection is revised in Section 16.1.1. Section 16.2 provides a theoretical background on MMPs. Section 16.3 describes the methodological issues involved by the approach reviewed here. Section 16.4 presents and discusses the experimental results obtained by applying the discussed technique to real planetary images. Finally, conclusions are drawn in Section 16.5.

16.1.1 Previous Work on Planetary Crater Detection

The techniques for crater detection can be divided into two different categories, that is, supervised and unsupervised. Supervised methods require the input of an expert and generally use supervised learning concepts to train the algorithm for feature extraction. These techniques contemplate a learning phase, in which a training set of images containing craters is labeled by human experts. Craters are then detected by applying the previously trained algorithm to new unlabeled sets of images. In Ref. [38], a continuously scalable detector, based on a supervised template matching technique, is applied. In Ref. [40], different supervised learning approaches, including ensemble methods, support vector machines (SVM), and continuously scalable template models, were employed to derive crater detectors from ground-truthed images. The SVM approach with normalized image patches provided the best detection and localization performance. In a different approach, Martins et al. [18] adopted a supervised boosting algorithm, originally developed by Viola and Jones [39] in the context of face detection, to identify craters on Mars.

Unsupervised methods do not make use of training information and are generally based on image analysis techniques. These approaches generally rely on the identification of circular or elliptical arrangements of edges along the crater boundary. A standard approach is based on the use of a Generalized Hough Transform (GHT) [35]. Examples include the works of Cross [9], Cheng et al. [7], Honda et al. [14], Leroy et al. [17], and Michael [20]. Instead, in Ref. [1], the identification of impact craters was achieved through the analysis of the probability volume created as a result of a template matching procedure, approximating the craters as objects of round shape. That unsupervised method enables the identification of round spatial features. Kim and Muller [15] presented a crater detection method based on texture analysis and ellipse fitting. That method was not robust when applied to optical images. Therefore, the authors needed to use also Digital Elevation Model (DEM) data and fuse them with the optical data.

In subsequent work [16], Kim et al. presented a combination of unsupervised and supervised techniques. In particular, edge detection, template matching, and supervised neural network-based schemes for the recognition of false positives were integrated, in order to automatically detect craters on Mars. In Ref. [36], Urbach and Stepinski presented a different approach, which combines unsupervised and supervised techniques, for crater detection in panchromatic planetary images. The method in Ref. [36] is based on using mathematical morphology for the detection of craters and on supervised techniques to distinguish between objects and false alarms.

16.2 MARKED POINT PROCESSES

MPPs represent a powerful tool for stochastic image modeling. Indeed, they enable to model the distribution of complex geometrical objects in a scene and have been exploited for different applications in image processing. MPPs have been successfully applied to address different problems in terrestrial remote sensing, including road network extraction [10] and building extraction in dense urban areas [6,21,26]. Moreover, in forestry applications, MPPs have been used to reproduce the spatial distribution of the stems [23].

An MPP is an abstract random variable whose realizations are configurations of objects, each object being described by a marked point. Similar to Markovian modeling, the "Maximum-A-Posteriori" (MAP) can be proved to be equivalent, under MPP assumption, to the minimization of a suitable energy function. An energy function, which takes into account the interactions between the geometric objects and the way they fit in the image, is minimized by using a Markov chain coupled with an SA scheme.

In Section 16.2.1, the theory behind MPP is described and in Section 16.2.2 the reversible jump Markov chain Monte-Carlo algorithm (RJMCMC) is detailed.

16.2.1 MARKED POINT PROCESS

Given a bounded subset P of $\mathbb{R}^2$, a point process X is a measurable mapping from a probability space $(\Omega, A, \mathbb{P})$ (where Ω is the certain event, A is the event space, and $\mathbb{P}$ is a probability measure [22]) to configurations of points on P, that is, a random variable whose realizations are random configurations x of points,

$$x = \{x_1,\ldots,x_n\}, \tag{16.1}$$

where x_i represents the position of the ith point in the image plane ($x_i \in P$, $i = 1,2,\ldots,n$).

These configurations belong to a suitable measure space defined over all subsets of P. An MPP is a point process defined by a density function with respect to the Poisson measure (see below) [27,32]. A configuration of an MPP consists of a set of marked points. A set of random parameters (called mark) is associated with each point. In image analysis, these parameters define some geometric property of an underlying object attached to that point, so that each realization of an MPP represents a model for the possible spatial distribution of several objects in the image plane. The marks are parameters that fully describe the related geometric object. For instance, a disk is fully described by one single parameter (i.e., the radius), whereas ellipses are described by three parameters (i.e., their major and minor axes and their orientation).

Hence, let X be an MPP defined in $S = P \times K$, a bounded set of $\mathbb{R}^d$, where P has the same meaning as above and K is the space of the marks describing the object geometry. The MPP X is still a measurable mapping from a probability space $(\Omega, A, \mathbb{P})$ to configurations of points in S.

The probability distribution of an MPP is uniformly continuous [29] with respect to the Poisson measure μ of intensity λ on S. We recall that, if S is a Polish space (i.e., a space homeomorphic to a complete metric space that has a countable dense subset) with a σ-finite measure λ on S, we call a random measure μ on S, a Poisson measure with intensity λ if it satisfies the following conditions: For every Borel subset B of S with $\lambda(B) < +\infty$, $\mu(B)$ is a random variable with Poisson distribution with parameter $\lambda(B)$, and if $B_1,\ldots,B_n$ are disjoint Borel sets, the variables $\mu(B_1),\ldots,\mu(B_n)$ are independent [3].

Operatively, it can be proven that X can be characterized by a density $f(\cdot)$ with respect to $\mu \cdot f(\cdot)$ is actually an unnormalized density and can be defined with respect to this dominating reference measure as

$$P_X(D) = \Pr\{X \in D\} = \frac{1}{Z}\int_D f\,\mathrm{d}\mu, \tag{16.2}$$

for any Borel subset D of S, where Z is a normalizing constant.

16.2.2 Reversible Jump Markov Chain Monte-Carlo

In Ref. [13], Green presented an algorithm to build a Markov chain, which ergodically converges to the probability distribution $P_X(\bullet)$ of the considered MPP. This algorithm can be summarized in the four steps that are listed below.

Given the probability space (Ω, A, Π), let π be a unnormalized measure on Ω. The RJMCMC consists of defining a proposition kernel $Q(x, S)$, where $x^{TM}\Omega$ and S belong to the σ-algebra A. The algorithm is based on the assumption that there exists a symmetrical measure ξ on $\Omega \times \Omega$ dominating the product measure $\pi(x)Q(x,S)$, that is, there exists a measurable function h (namely, the Radon–Nikodym derivative), such that

$$\int_{D \times S} h\,\mathrm{d}\xi = \int_D Q(x, S)\,\mathrm{d}\pi(x)\forall D, \quad S \in A. \tag{16.3}$$

The algorithm is as follows:

1. Initialize x.
2. Generate the candidate configuration x' at iteration k by choosing a proposition kernel $Q_m(x,.)$ with probability p_m.
3. Compute the Green's ratio $R = (h_m(x', x))/(h_m(x, x'))$, where h_m is the Radon–Nikodym derivative corresponding to the proposition kernel Q_m.
4. Accept the new state x' with probability $\alpha = \min(1, R)$ and go to step 2 until convergence.

When the RJMCMC is integrated within an iterative SA framework, the Green's ratio R is replaced by $R^{1/(T(k))}$, where $T(k)$ is the annealing schedule at the iteration k.

The efficiency of the algorithm depends highly on the variety of the proposition kernels Q. For the model considered in Ref. [33] and reviewed in this chapter, the proposition kernel is defined as a combination of kernels defining birth, death, and the different nonjumping moves (translation, rotation, scaling of each single object in the configuration). Birth and death are jumping perturbations, that is, they vary the number of objects in the configuration. If birth is chosen, a new marked point is randomly generated and added to the configuration, while if death is chosen a randomly selected point in the configuration is removed. Nonjumping moves are transformations that do not change the number of objects in the configuration. In particular, a marked point is randomly selected and is replaced by a "perturbed" version of it. Perturbations can be translation, rotation, and axis dilation.

16.3 METHODOLOGY

In this chapter, we review an MPP approach, as proposed in Ref. [33], for the detection of craters in planetary images. In the discussed method, a novel MPP model is defined to determine the statistical distribution of elliptical objects in the image. The boundaries of the regions of interest are considered as a realization of an MPP of ellipses. Hence, the optimum configuration of objects has to be estimated.

Section 16.3.1 provides an overview of the presented crater detection method. Section 16.3.2 describes the presented formulation of the MPP energy functional. Section 16.3.3 details the presented approach for the minimization of the energy functional.

16.3.1 OVERVIEW OF THE PRESENTED METHOD

Planetary images show the surface of a planet and its structures. The aim of this study is to automatically detect elliptical structures, such as craters, that are present on a considered planetary surface by using image analysis techniques. The extracted features can be used for the registration of multitemporal, multisensor, and stereo-images.

Different types of spatial features are present in the planetary images, but the most evident ones are generally craters, that is, objects of approximately elliptical shapes with shadows. Their extraction is a difficult task for several reasons. Indeed, planetary images are generally blurry, quite noisy, present lack of contrast and uneven illumination, and the represented objects are not well defined (examples of regions of a planetary surface are shown in Figure 16.1).

In order to address this problem, an MPP-based approach, aimed at detecting round and elliptical objects, is discussed here. In this method, the objects that are searched for are craters and a novel MPP model is defined to determine their statistical distribution in the image. The boundaries of the regions of interest are considered as a realization of an MPP of ellipses. Hence, the optimum configuration of objects has to be estimated.

The overall architecture of the MPP approach for crater detection reviewed here is shown in Figure 16.2. First, the noise is reduced by applying a smoothing filtering operation. Then, in order to produce an edge map I_g, showing the contours of the objects represented in the original image, the Canny edge detector [5] is applied. The result of this first step, I_g, is a binary image that shows the object boundaries.

Craters are objects of approximately elliptical shape, and their structure is complex and, due to their depth and uneven illumination, they exhibit shadows. Hence, their borders can be approximated with incomplete noncontinuous elliptical curves. Here, the 2D model, used to extract the features of interest, consists of an MPP of ellipses. The model, which describes each object, is represented in Figure 16.3. Using the notation of the previous section, each ellipse is represented by a 5-tuple (u, v, a, b, θ), taking values in the set space,

$$S = \overbrace{[0,M] \times [0,N]}^{P} \times \overbrace{[a_m,a_M] \times [b_m,b_M] \times [0,\pi]}^{K}, \tag{16.4}$$

(a)   (b)

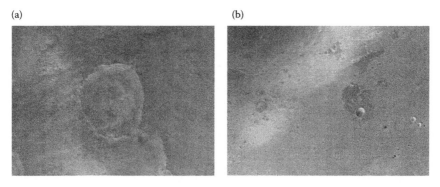

FIGURE 16.1 Portions of images representing a planetary surface. (a) Region 1 and (b) Region 2.

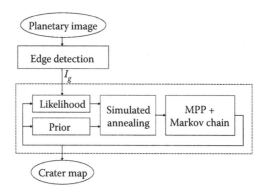

FIGURE 16.2 Flowchart of the presented approach. The original planetary image is first preprocessed, in order to smooth the noise and extract the image edges. An edge map of the original image, I_g, is obtained. Crater, which are objects of elliptical shape, are detected by using a marked point process (MPP). In particular, an MPP is integrated within a minimization scheme of an energy functional, based on a Markov chain Monte-Carlo and a simulated annealing algorithm. The result is a map of all the detected craters.

where $(u, v) \in [0, M] \times [0, N] = P$ are the coordinates of the ellipse center (M and N being the width and height of I_g and P being the bounded subset where the objects in X are located), a and b are the ellipse axes (ranging in $[a_m, a_M]$ and $[b_m, b_M]$, respectively), $\theta \in [0, \pi]$ is the ellipse orientation angle, and K denotes the set where the mark associated with each object can take values.

The probability distribution of this stochastic process is uniformly continuous [29] with respect to a suitable Poisson measure on S (see Section 16.2.1 for more details). Operatively, this means that it may be characterized by a density f with respect to this measure. Similarly, the posterior distribution of x conditioned to I_g can also be characterized by a density function f_p with respect to the same measure and a Gibbs formulation is proven to hold for f_p [21]. Hence, one may introduce an energy function U such that

$$f_p\left(x|I_g\right) = \frac{1}{Z}\exp\left\{-U\left(x|I_g\right)\right\}, \tag{16.5}$$

where Z is a normalizing constant. Hence, in order to minimize this posterior distribution, U will be minimized on the space of all configurations x in the feature extraction process.

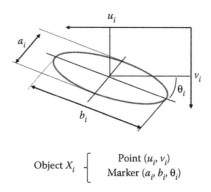

$$\text{Object } X_i \begin{cases} \text{Point } (u_i, v_i) \\ \text{Marker } (a_i, b_i, \theta_i) \end{cases}$$

FIGURE 16.3 Representation of an object i of our MPP model.

16.3.2 Energy Function

The energy function takes into account the interactions between the geometric objects $x_1, x_2, \ldots, x_n$ in the configuration x (the prior energy U_P), and the way they fit to the data (the likelihood energy U_L)

$$U\left(x|I_g\right) = U_P\left(x\right) + U_L\left(I_g|x\right). \tag{16.6}$$

The prior term characterizes the general aspect of the desired solution. According to the geometric properties of the configurations of craters, a basic rule is imposed on the prior term of our model. The prior energy, U_P, penalizes overlapping objects in x, which are very unlikely, by adding a repulsion between objects which intersect. The prior energy of our model is

$$U_P(x) = \frac{1}{n} \sum_{x_i * x_j} R(x_i, x_j), \tag{16.7}$$

where R is a repulsion coefficient which penalizes each pair of overlapping objects (denoted as $x_i * x_j$) in the configuration x. The repulsion coefficient R is calculated as follows:

$$R(x_i, x_j) = \frac{x_i \cap x_j}{x_i \cup x_j}, \tag{16.8}$$

where $x_i \cap x_j$ denotes the overlapping area between the two objects x_i and x_j in the configuration $(i, j = 1, 2, \ldots, n, i \neq j)$ and $x_i \cup x_j$ indicates the sum of the areas covered by the two objects x_i and x_j.
Then, the likelihood term U_L is defined as

$$U_L\left(I_g|x\right) = U_S\left(I_g|x\right) + U_D\left(I_g|x\right), \tag{16.9}$$

where U_S measures the similarity between the configuration and the data, whereas the data term U_D measures the distance between the objects in the configuration and the contours of the data. Different formulations for the likelihood energy, which have been presented in previous works on MPP [23,25], have proven to be unfeasible for planetary data. Hence, a new formulation for U_L, more appropriate for the analyzed planetary data, is proposed in Ref. [33] and presented here.
In particular, the similarity energy U_S between the data I_g and the current configuration x is defined as a correlation measure*

$$U_S\left(I_g|x\right) = \frac{\left|\left\{(u,v) : I_g(u,v) = 1 \text{ and } \Pi\left(u,v|x\right) = 1\right\}\right|}{\left|\left\{(u,v) : I_g(u,v) = 1\right\}\right|}, \tag{16.10}$$

where u and v are the spatial coordinates in the image plane; $\Pi(\cdot|x)$ is the projection of the configuration x such that $\Pi(u,v|x) = 1$ if (u,v) belongs to the boundary of at least one ellipse in the configuration x (i.e., if there exists $i \in \{1,2,\ldots,n\}$ such that (u, v) is on the boundary of x_i), and $\Pi(u, v|x) = 0$, otherwise. Consequently, U_S expressed as Equation 16.10 is equivalent to the definition of a correlation function between the binary images I_g and $\Pi(\cdot|x)$, representing the extracted and the modeled edges, respectively. According to the correlation definition, in the binary case, only nonzero pixels from both images contribute to the value of the correlation. This energy term, which is novel with respect to the MPP literature, resembles analogous correlation measures that have been used for

---

* Given a finite set A, we denote by $|A|$ the cardinality (i.e., the number of elements) of A.

registration purposes [19]. The correlation measure in Equation 16.10 is considered to be appropriate here because it enables to estimate the match between two binary images (I_g and Π) in a fast and accurate way.

Then, the data energy U_D is calculated at the object level: For each object x_i in the current configuration, x a weight parameter W_i, proportional to the distance from the closest detected edge pixel in the data I_g with respect to its dimension, is calculated, that is,

$$W_i = \frac{\inf\left\{\sqrt{(u - u')^2 + (v - v')^2} : I_g(u,v) = 1 \text{ and } \Pi\left(u',v'|x_i\right) = 1\right\}}{\max(a_i, b_i)}, \tag{16.11}$$

where $\Pi(\cdot|x_i)$ has a meaning similar to above and a_i and b_i are the two ellipse axes associated with the object x_i ($i = 1,2,\ldots,n$).

The resulting data energy will be

$$U_D\left(I_g|x\right) = \frac{1}{n}\sum_{i=1}^{n} W_i. \tag{16.12}$$

Then, objects with a low value of W will be favored in the configuration.

16.3.3 Energy Minimization

An MCMC algorithm [12], coupled with an SA (applied with a given annealing schedule $T(\cdot)$), is used in order to find the configuration x which minimizes U. We stress here that this minimization is carried out with respect to not only the locations and marks of the objects in the MPP realization but also the number of objects, that is, the presented method also automatically optimizes the choice of the number of detected craters. In particular, the MPP X, defined by f, is sampled by using a random jump MCMC algorithm (see Section 16.2.2 for more details). It allows building a Markov chain X_k ($k = 0,1,\ldots$), which jumps between the different dimensions of the space of all possible configurations and, in the ideal case, ergodically converges to the optimum distribution [28]. The final configuration of convergence does not depend on the initial state. The flowchart of the minimization scheme is shown in Figure 16.4.

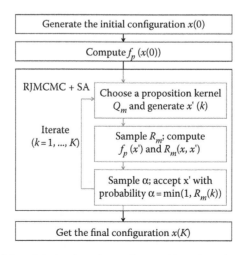

FIGURE 16.4 Flowchart of the minimization scheme discussed here, based on a Markov chain Monte-Carlo and a simulated annealing algorithm.

At each step, the transition of this chain depends on a set of "proposition kernels," which are random changes proposed to the current configuration. In order to find the configuration that maximizes the density $f_p(\cdot)$ on S, we sample within an SA scheme, which gives us the MAP estimator. SA is an iterative algorithm where at each iteration k, a perturbation is proposed to the current configuration at temperature $T(k)$, $k = 1, 2, \ldots, K$. This perturbation is accepted or rejected with a probability which ensures that the probability distribution of the Markov chain ergodically converges to $f_p(x)^{1/(T(k))}$. Here, the annealing schedule, $T(\cdot)$, is defined as

$$T(k) = T_1 \cdot \left(\frac{T_F}{T_I}\right)^{(k/K)}, \tag{16.13}$$

where T_I and T_F are the initial and the final temperatures, respectively, and K is the total number of allowed iterations. In practice, in order to cope with too long computational times, the decrease in the temperature is geometric (as usual in SA for Markov random fields) and does also not imply the ergodic convergence to a probability distribution localized at the minima of $U(x|I_g)$, in contrast, it follows the adaptive approach developed in Ref. [24].

The set of proposition kernels are birth and death, translation, dilation, and rotation [13], as described in Section 16.2.2. For each proposition kernel m, a Green ratio $R_m(x, x')$ is defined, which tunes the likelihood of replacing configuration x by configuration x' at each SA iteration (analytical details can be found in Section 16.2.2). More precisely, the birth and death kernel consists in proposing, with probability p_B, to uniformly add in S an object to the current configuration x or, with probability $p_D = 1 - p_B$, to remove a randomly chosen object of x. Green's ratios for birth and death (namely, R_B and R_D, respectively) are

$$R_B = \frac{p_D}{p_B} \frac{f_p(x'|I_g)}{f_p(x|I_g)} \frac{v(S)}{n(x) + 1}; \quad R_D = \frac{p_D}{p_B} \frac{f_p(x'|I_g)}{f_p(x|I_g)} \frac{n(x)}{v(S)} \tag{16.14}$$

where $n(x)$ is the number of objects in the current configuration x, x' is the candidate configuration, and $v(S)$ is the Poisson intensity measure, which represents the average expected number of objects and is an input tunable parameter of the method. For the selected nonjumping kernels (i.e., translation, dilation, and rotation), the suitable Green's ratio is given by the usual Metropolis–Hastings ratio

$$R(x, x') = \frac{f_p(x'|I_g)}{f_p(x|I_g)}. \tag{16.15}$$

16.4 EXPERIMENTAL RESULTS

Experiments were carried out using Mars data, collected during the 2001 Mars Odyssey mission, by the Thermal Emission Imaging System (THEMIS), an instrument on board the Mars Odyssey spacecraft. Such an instrument combines a five-band visual imaging system with a 10-band infrared imaging system [8]. Both visible (VIS) and infrared (IR) THEMIS images, with a resolution of 18 and 100 m/pixel, respectively, were used to test the approach reviewed here.

Few parameters of the presented method had to be assigned, concerning both the MPP state space S and the MCMC sampler. Let us recall that $S = P \times K$, where $P = [0, M] \times [0, N]$ corresponds to the size of the data (I_g). The resolution r varies for the two different types of images used; hence, the total area of interest is $M \times N \times r^2$ (m^2). The parameters of K (i.e., a_m, a_M, b_m, and b_M) depend on the size of the objects that need to be detected. In this study, the minimum size for both semiaxes was fixed to three pixels (i.e., $a_m = b_m = 3$) and the maximum size to 100 pixels (i.e., $a_M = b_M = 100$). The eccentricity e of each object, defined as

$$e = \sqrt{1 - \left(\frac{\min(a,b)}{\max(a,b)}\right)^2},$$ (16.16)

was constrained to $e \in [0, 0.6]$ (i.e., $\min(a, b) > 0.8 \cdot \max(a, b)$), being craters of bigger e unlikely.

For comparison purposes, a method for ellipse detection based on a GHT [35] has been implemented and tested on our data set. With this method, for every pair of pixels that was detected as edge points in the Canny gradient and exhibits opposite gradient directions, an accumulator, corresponding to the median point between them in the image plane, is incremented by a unit value. The maxima of the accumulator are taken as centers of ellipses. Then, the three parameters describing the ellipse centered in each detected maximum are computed and a 3D accumulator is used to estimate the two semiaxes and the direction angle of the ellipse from all the pairs of points that contribute to the accumulator in the considered center. The results obtained by the presented approach and by GHT were compared. This particular approach was chosen for comparison, being a standard technique for the detection of round and elliptical objects, commonly used for crater detection [14,17,20,34].

Reference data were generated by manually analyzing each image of the data set and identifying all the craters that are present. Only objects completely included within the images were considered (i.e., objects cut by the borders of the image were discarded). A quantitative assessment of the obtained results by the presented method was performed using these reference data. This was accomplished by comparing the obtained results with the labeled features in the correspondent reference map. The detection percentage D, the branching factor B, and the quality percentage Q were computed as follows:

$$D = \frac{100 \cdot TP}{TP + FN}; \quad B = \frac{FP}{TP}; \quad Q = \frac{100 \cdot TP}{TP + FP + FN}$$ (16.17)

where true positive (TP) is the number of detected features that correspond to labeled objects in the reference map, false positive (FP) is the number of features detected by the presented approach, which do not correspond to any object in the reference map, and false negative (FN) is the number of objects in the reference map that have not been detected by the presented approach. The global values of D, B, and Q obtained by the MPP approach and the standard GHT method used for comparison for both VIS and IR data are shown in Table 16.1. The global values of D for VIS data and IR data obtained by the presented approach were about 82% and 89%, respectively. These high values indicate a good detection rate (because of the high number of TP). B was about 0.22 for VIS and 0.13 for IR, which indicate a small amount of false detections with respect to the true detections

TABLE 16.1

Average Numerical Performance of Both the Presented Approach (MPP) and a Standard Method (GHT) as Measured by Detection Percentage (D), Branching Factor (B), and Quality Percentage (Q)

Data	Method	D	B	Q	Method	D	B	Q
VIS	GHT	73	0.24	62	MPP	82	0.22	71
IR	GHT	78	0.14	70	MPP	89	0.13	79
Average	GHT	75	0.20	65	MPP	85	0.18	74

in both cases, thanks to the small number of FP. The results obtained by applying the presented approach are more accurate when compared with the performance of the implemented standard technique based on the GHT. In particular, the average value of the detection rate D improved from 75% for the GHT to 85% for the MPP. This is explained by the increase in true detections with respect to the reference map. A similar improvement can be appreciated in terms of quality percentage Q. A relatively smaller improvement in the branching factor B is due to the fact that the number of FP was already small when applying GHT.

Moreover, the detection performance of the presented approach in terms of D, B, and Q also compares favorably with most of the results previously published for automatic crater detection methods [1,2,4,36]. Ideally, the performance of the presented approach should be compared with the results obtained by the previously published methods when applied to the same data. Unfortunately, the performance of each published approach has been assessed on different sites and distinct types of data (e.g., panchromatic images, topographic data). The most direct performance comparison can be made with the method proposed by Barata et al. [2]. That approach was tested on images acquired by the Mars Orbiter Camera (MOC). The method in Ref. [2] identified 546 craters, with TP = 171, FN = 93, and FP = 282. Hence, the resulting assessment factors were about $D = 65\%$, $B = 1.65$, and $Q = 31\%$. Bandeira et al. [1] presented an unsupervised approach for crater detection from MOC data based on template matching. The average performances of that approach were about $D = 86\%$ and $Q = 84\%$. However, they tested their algorithm on images having resolution of 200–300 m/ pixel. The high performances obtained in Ref. [1] may be attributed to the fact that large craters in the sites of analysis have a very regular shape and are relatively easy to identify by template matching. The performance of that approach for the detection of small and irregular craters is unknown. Bue and Stepinski [4] proposed a supervised approach for crater detection from topographic data. The average performances of that approach were about $D = 74\%$, $B = 0.29$, and $Q = 61\%$. The evaluation factors increased to $D = 92\%$, $B = 0.29$, and $Q = 73\%$ if degraded craters, which the method was not able to detect, were excluded. That approach is not fully comparable with the presented method being supervised. Urbach and Stepinski [36] proposed a supervised approach for crater detection from panchromatic images. The performance factors of their method were about $D = 68\%$, $B = 0.29$, and $Q = 57\%$, when detecting craters of diameter >200 m, and lower when taking into account also craters of smaller dimensions. However, a full comparison with our approach is again not possible. In general, the results obtained by the presented approach are comparable to, and in some cases better than results obtained by methods reported in the literature in terms of the assessment factors. Unfortunately, a full comparison is not possible because the methods are applied to different data.

Visual results of the feature extraction are shown for a visible image of our data set (Figure 16.5a). The image is first preprocessed in order to reduce the noise. In particular, Gaussian filtering and median filtering operations are applied in a cascade [31] in order to reduce the noise and preserve the edges at the same time. The Canny edge detector is applied to the smoothed image and the binary gradient I_g is shown in Figure 16.5b. The estimated optimum configuration of the MPP x^*, which identifies the feature contours, is shown in Figure 16.5c. The optimum configuration x^* is represented in red, transparently superimposed to the original image. By a visual inspection, it is possible to appreciate the accuracy of the detection, even when many false alarms are present in the binary image gradient I_g. Also, the reconstruction of the feature shape is very accurate.

Then, visual results obtained by applying the presented approach to another image of our data set (Figure 16.6a) are shown. In particular, the Canny gradient I_g is shown in Figure 16.6b and the estimated x^* is shown in Figure 16.6c, transparently superimposed to the original image. The contours of the represented crater appear noncontinuous in the binary image gradient I_g, due to the uneven quality of the image. Anyway, the feature is correctly detected and its shape is reconstructed.

In conclusion, the visual analysis of the detection results obtained by applying the method presented in this chapter (see Figures 16.5 and 16.6) confirms that the method is able to correctly identify the location and shape of the imaged craters, even though the input edge map detected only a

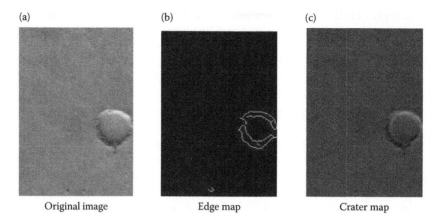

(a)   (b)   (c)

Original image   Edge map   Crater map

FIGURE 16.5 Experimental results obtained by applying the presented method to a visible image of our data set. (a) Original image, (b) Canny gradient, and (c) detected crater contours, transparently superimposed to the original image.

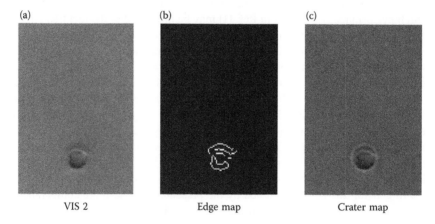

(a)   (b)   (c)

VIS 2   Edge map   Crater map

FIGURE 16.6 Experimental results obtained by applying the presented method to another image of our data set. (a) Original image, (b) Canny gradient, and (c) detected crater contours, transparently superimposed to the original image.

part of the crater borders, included many spurious contours unrelated with the craters, and was severely affected by the shadows in the crater area. More results are shown in Ref. [33].

16.5 CONCLUSIONS

In this chapter, the problem of crater detection in planetary images acquired by passive orbiting sensors has been addressed. An approach for automatic detection of craters that characterize planetary surfaces, proposed in Ref. [33], has been reviewed here. The identification is achieved by using a method based on a MPP, coupled with a Markov chain and an SA scheme.

Mars IR and VIS multiband images, captured by THEMIS during the Mars Odyssey Mission, were used. Before the algorithm could be used to estimate the MPP optimum configuration, the images had to go through a preprocessing stage, aimed at obtaining contour map of the analyzed image. The likelihood between the extracted map and the current configuration was measured and maximized, in order to identify the optimum configuration.

MPPs were developed in the context of computer vision and previously used in many different applications in terrestrial remote sensing. Here, they were applied to the analysis of planetary images and have proven to be effective. For such data, the features to be extracted are not as well contrasted nor defined as for Earth data. Nevertheless, their identification can be accurately achieved. The accuracy of the detection has been assessed by a comparison to a manually generated reference map. The method outperformed a standard method for crater detection based on a GHT, in terms of several indices based on TPs, FNs, and FPs. Moreover, the obtained results compared favorably to most previously proposed approaches, when performances reported in the literature were considered for the same indices. Finally, a visual inspection of the detection results confirmed that the proposed method was also able to correctly identify the location and shape of the detected craters.

The presented approach can be adopted as the first important step in several applications dealing with all the various data that are being collected during the current and future planetary missions. Among them selecting safe landing sites, identifying planetary resources, and preparing for subsequent planetary exploration by both humans and robots.

In future work, the method discussed may be applied to different types of planetary data and to the registration of multisensor and multitemporal images, by performing feature matching. Moreover, the approach could be used to extract other features of elliptical shape, such as volcanoes. Finally, features of other shapes, such as ridges or polygonal patterns among others, could be extracted, by adapting the model of the MPP.

ACKNOWLEDGMENT

This research was partially supported by the Research Fund of the University of Iceland. The support is gratefully acknowledged.

REFERENCES

1. L. Bandeira, J. Saraiva, and P. Pina. Impact crater recognition on mars based on a probability volume created by template matching. *IEEE Transactions on Geoscience and Remote Sensing*, 45:4008–4015, 2007.
2. T. Barata, E. I. Alves, J. Saraiva, and P. Pina. *Automatic Recognition of Impact Craters on the Surface of Mars*, Vol. 3212. Springer-Verlag, Berlin, Germany, 2004.
3. J. Bertoin. *Levi Processes*. Cambridge University Press, Cambridge, 1998.
4. B. D. Bue and T. F. Stepinski. Machine detection of Martian impact craters from digital topography data. *IEEE Transactions on Geoscience and Remote Sensing*, 45:265–274, 2007.
5. J. Canny. A computational approach to edge detection. *IEEE Transactions on Pattern Analysis and Machine Intelligence*, 10(6):679–698, 1986.
6. F. Cerdat, X. Descombes, and J. Zerubia. Urban scene rendering using object description. In *IEEE International Geoscience and Remote Sensing Symposium*, Toulouse, France, Vol. 1, pp. 62–64, 2003.
7. Y. Cheng, A. E. Johnson, L. H. Matthies, and C. F. Olson. Optical landmark detection for spacecraft navigation. In *13th Annual AAS/AIAA Space Flight Mechanics Meeting*, Ponce, Puerto Rico, February 2002.
8. P. Christensen, B. M. Jakosky, H. H. Kieffer, M. C. Malin, H. Y. Mcsween, K. Nealson, G. L. Mehalland et al. The thermal emission imaging system for the Mars 2001 odyssey mission. *Space Sciences Reviews*, 100:85–130, 2004.
9. A. M. Cross. Detection of circular geological features using the Hough transform. *International Journal of Remote Sensing*, 9:1519–1528, 1988.
10. X. Descombes and J. Zerubia. Marked point processes in image analysis. *IEEE Signal Processing Magazine*, 19(5):77–84, 2002.
11. A. Flores-Mendez. *Crater Marking and Classification Using Computer Vision*, Vol. 2905. Springer-Verlag, New York, 2003.
12. C. Geyer. Likelihood inference for spatial point processes. In *Stochastic Geometry: Likelihood and Computation*, eds. W. Kendall, O. Barndor-Nielsen, and M. N. M. van Lieshout, CRC Press/Chapman and Hall, London, Bocca Raton, FL, 79–140, 1999.
13. P. Green. Reversible jump Markov chain Monte-Carlo computation and Bayesian model determination. *Biometrika*, 82:711–732, 1995.

14. R. Honda, Y. Iijima, and O. Konishi. Mining of topographic feature from heterogeneous imagery and its application to lunar craters. In S. Arikawa and A. Shinohara, editors, *Progress in Discovery Science, Lecture Notes in Computer Science*, Vol. 1, pp. 27–44. Springer, Berlin, Heidelberg, 2002.

15. J. R. Kim and J.-P. Muller. Impact crater detection on optical images and DEM. In *ISPRS WG IV/9: Extraterrestrial Mapping Workshop*, Houston, TX, 2003.

16. J. R. Kim, J.-P. Muller, S. van Gasselt, J. G. Morley, and G. Neukum. Automated crater detection, a new tool for mars cartography and chronology. *Photogrammetric Engineering and Remote Sensing*, 71(10):1205–1217, 2005.

17. B. Leroy, G. Medioni, A. E. Johnson, and L. H. Matthies. Crater detection for autonomous landing on asteroids. *Image Vision Computation*, 19:787–792, 2001.

18. R. Martins, P. Pina, J. S. Marques, and M. Silveira. Crater detection by a boosting approach. *IEEE Geoscience and Remote Sensing Letters*, 6:127–131, 2009.

19. G. Matsopoulos, N. Mouravliansky, K. Delibasis, and K. Nikita. Automatic retinal image registration scheme using global optimization techniques. *IEEE Transactions on Information Technology in Biomedicine*, 3(1):47–60, 1999.

20. G. G. Michael. Coordinate registration by automated crater recognition. *Planetary and Space Science*, 51:563–568, 2003.

21. M. Ortner, X. Descombes, and J. Zerubia. A marked point process of rectangles and segments for automatic analysis of digital elevation models. *IEEE Transactions on Pattern Analysis and Machine Intelligence*, 30(1):105–119, 2008.

22. A. Papoulis. *Probability, Random Variables, and Stochastic Processes*, 3rd edition. McGraw-Hill International Editions, New York, 1991.

23. G. Perrin, X. Descombes, and J. Zerubia. Tree crown extraction using marked point processes. In *European Signal Processing Conference*, Vienna, Austria, 2127–2130, 2004.

24. G. Perrin, X. Descombes, and J. Zerubia. Adaptive simulated annealing for energy minimization problem in a marked point process application. In *Conference on Energy Minimization Methods in Computer Vision and Pattern Recognition*, Saint Augustine, FL, 3757:3–17, 2005.

25. G. Perrin, X. Descombes, and J. Zerubia. A marked point process model for tree crown extraction in plantations. In *IEEE International Conference on Image Processing*, Genova, Italy, pp. 661–664, 2005.

26. M. Quartulli and M. Datcu. Stochastic geometrical modeling for built-up area understanding from a single SAR intensity image with meter resolution. *IEEE Transactions on Geoscience and Remote Sensing*, 42(9):1996–2003, 2004.

27. B. D. Ripley. Modelling spatial patterns. *Royal Statistical Institute*, 39:172–212, 1997.

28. C. Robert and G. Casella. *Monte-Carlo Statistical Methods*. Springer-Verlag, New York, 1999.

29. W. Rudin. *Principles of Mathematical Analysis*. 2nd edition, McGraw-Hill, New York, 1976.

30. G. Salamuniccar and S. Loncaric. Method for crater detection from Martian digital topography data using gradient value/orientation, morphometry, vote analysis, slip tuning, and calibration. *IEEE Transactions on Geoscience and Remote Sensing*, 48:2317–2329, 2010.

31. L. G. Shapiro and G. C. Stockman. *Computer Vision*. Prentice-Hall, Englewood Cliffs, NJ, 2001.

32. D. Stoyan, W. Kendall, and J. Mecke. *Stochastic Geometry and Its Applications*. Wiley, New York, 1987.

33. G. Troglio, J. A. Benediktsson, G. Moser, and S. B. Serpico. Crater detection based on marked point processes. In *IEEE International Geoscience and Remote Sensing Symposium*, Honolulu, Hawaii, pp. 1378–1381, 2010.

34. G. Troglio, J. L. Moigne, J. A. Benediktsson, G. Moser, and S. B. Serpico. Automatic extraction of ellipsoidal features for planetary image registration. *IEEE Geoscience and Remote Sensing Letters*, 2011, doi: 10.1109/LGRS.2011.2161263.

35. S. Tsuji and F. Matsumoto. Detection of ellipses by a modified Hough transformation. *IEEE Transactions on Computers*, 27(9):777–781, 1978.

36. E. R. Urbach and T. F. Stepinski. Automatic detection of sub-km craters in high resolution planetary images. *Planetary Space Science*, 57:880–887, 2009.

37. M. van Lieshout. *Marked Point Processes and Their Applications*. London: Imperial College Press 2000.

38. T. Vinogradova, M. Burl, and E. Mjolsness. Training of a crater detection algorithm for mars crater imagery. In *IEEE Aerospace Conference Proceedings*, Big Sky, MN, Vol. 7, pp. 3201–3211, 2002.

39. P. Viola and M. Jones. Robust real-time face detection. *International Journal of Computer Vision*, 57:137–154, 2004.

40. P. G. Wetzler, B. Enke, W. J. Merline, C. R. Chapman, and M. C. Burl. Learning to detect small impact craters. In *IEEE Workshop on Applications of Computer Vision*, Breckenridge, Colorado, Vol. 1, pp. 178–184, 2005.

17 Probability Density Function Estimation for Classification of High-Resolution SAR Images

Vladimir A. Krylov, Gabriele Moser, Sebastiano Bruno Serpico, and Josiane Zerubia

CONTENTS

17.1 INTRODUCTION

In modern remote sensing, the use of synthetic aperture radar (SAR) represents an important source of information for Earth observation. Recent improvements have enabled modern satellite SAR missions, such as COSMO-SkyMed (CSK), TerraSAR-X, RADARSAT-2, to acquire high-resolution (HR) data (up to metric resolution) with a very short revisit time (e.g., 12 h for CSK). In addition, SAR is robust with respect to lack of illumination and atmospheric conditions. Together, these factors explain the rapidly growing interest in SAR imagery for various applications, such as flood/fire monitoring, urban mapping, and epidemiological surveillance. HR imagery allows to appreciate various ground materials resulting in highly mixed distributions. The resulting spatial heterogeneity is a critical problem in applications to image classification (estimation of class-conditional statistics) or filtering (estimation of local statistics, e.g., in moving-window approaches) [1]. Analysis and modeling of heterogenous HR SAR data pose a difficult statistical problem, which, to the best of our knowledge, has not been sufficiently addressed so far.

Accurate modeling of statistical information is a crucial problem in the context of SAR image processing and its applications. Specifically, an accurate probability density function (pdf) estimate can effectively improve the performance of SAR image denoising [1], classification [2], and target detection [3]. Over the years, a number of methods have been proposed for modeling SAR amplitude pdfs. Nonparametric methods, for example, Parzen window estimator [4], and support vector machines [5], do not assume any specific analytical model for the unknown pdf, thus providing a higher flexibility, although usually involving manual specification of internal architecture parameters [4]. Parametric methods postulate a given mathematical model for each pdf and formulate the pdf estimation problem as a parameter estimation problem. Empirical pdf models, including lognormal [1], Weibull [1], Fisher [2] and, recently, the generalized Gamma distribution (GΓD) [6], have been reported to accurately model amplitude SAR images with different heterogenous surfaces. Several theoretical models, such as Rayleigh [1], Nakagami [1], generalized Gaussian Rayleigh (GGR) [7], symmetric-α-stable generalized Rayleigh (SαSGR) [8], $\mathcal{K}$ [9] ($\mathcal{K}$-root for amplitudes), $\mathcal{G}$ [10], have been derived from specific physical hypotheses for SAR images with different properties. However, several parametric families turned out to be effective only for specific land cover typologies [1], making the choice of a single optimal SAR amplitude parametric pdf model a hard task, especially in the case of heterogenous imagery. To solve this problem, the dictionary-based stochastic expectation maximization (DSEM) approach [11] was designed and validated on medium-resolution SAR imagery. It addressed the problem by adopting a finite mixture model (FMM) [12] for the SAR amplitude pdf, that is, by postulating the unknown amplitude pdf to be a linear combination of parametric components, each corresponding to a specific statistical population.

In this chapter, we first address the general problem of modeling the statistics of single-channel SAR amplitude images and, specifically, HR SAR. Given the variety of approaches above, we extend and enhance the DSEM technique proposed in [11] for coarser-resolution SAR. We expect DSEM to be an appropriate tool for this modeling problem, since it is a flexible method, intrinsically modeling SAR statistics as resulting from mixing several populations, and it is not constrained to a specific choice of a given parametric model allowing us to benefit from many of them (dictionary approach). Thus, we extend the earlier DSEM approach to HR satellite SAR imagery and enhance it by introducing a novel procedure for estimating the number of mixture components, which enables us to appreciably reduce its computational complexity (as much as five times in some cases), resulting in an Enhanced DSEM (EDSEM) algorithm.

Building on the proposed method for single-channel SAR pdf estimation, we proceed to multi-channel joint pdf estimation and classification. Contemporary satellite SAR missions are capable of registering polarimetric SAR (PolSAR) imagery, which provides a more complete description of landcover scattering behavior than single-channel SAR data [1,13]. The potential for improved classification accuracy with data in several polarizations, compared to single-channel data, explains the special interest to PolSAR image classification. Furthermore, several current satellite SAR systems, for example, TerraSAR-X, CSK, RADARSAT-2, support, at least, dual-pol acquisition modes. In this chapter, we investigate the quad-polarization (quad-pol), dual-polarization (dual-pol), as well as single-polarization (single-pol) SAR imagery scenarios.

A wide variety of methods have been developed so far for the classification of PolSAR data [13]. We indicate some recent methods classified based on the employed methodological approach: maximum likelihood (ML) [14–18], neural networks [19,20], support vector machines [21], fuzzy methods [22,23], stochastic complexity [24], spectral graph partitioning [25], wavelet texture models [26], and other approaches [27,28]. In this chapter, we develop a classification method based on a ML approach. As such, this method explicitly specifies a pdf describing the statistics of PolSAR data. Previously, several models have been proposed for this purpose: the classical Wishart distribution [14,29], the K-distribution [15,30] for textured areas, the K-Wishart distribution [17] designed to improve the distinguishability of non-Gaussian regions, the G-distribution [16,31] for extremely heterogeneous areas, the Ali-Mikhail-Haq (AMH) copula-based model [32] combined with the Sinclair matrix representation, and the KummerU distribution [18] for Fisher distributed texture. These models were developed for the multilook complex-valued SAR image statistics. In this chapter, we study the problem of PolSAR image classification using only the amplitude data and not the complex-valued data. This is an important data typology since several image products provided by novel HR satellite SAR systems are geocoded ellipsoid-corrected amplitude (intensity) images, and also because several earlier coarser resolution sensors [e.g., European remote sensing satellite (ERS)] primarily used this modality.

The classification technique developed in this chapter combines the Markov random field (MRF) approach to Bayesian image classification with the finite mixture EDSEM amplitude pdf estimator. MRFs represent a general family of probabilistic image models that provide a convenient and consistent way to characterize context-dependent data [33]. The resulting DSEM-MRF technique is a simple and efficient tool for single-channel SAR classification. In order to support multichannel PolSAR data, copula theory is used for modeling the joint class-conditional distributions of the multiple channels, resulting in a Copula-DSEM-MRF (CoDSEM-MRF) approach. The employed joint distribution modeling tool, copulas [34], is a rapidly developing statistical tool that was designed for constructing joint distributions from marginals with a wide variety of allowable dependence structures. For every class, the choice of an optimal copula from a dictionary of copulas is performed by a dedicated criterion. The concept of copulas is relatively new in image processing, and has recently emerged in remote sensing methods [32,35,36]. We compare the results of the method proposed in this chapter with an earlier-developed copula-based classification technique [32] to demonstrate the higher adequacy of our model. The proposed CoDSEM-MRF HR PolSAR classification technique is based on three flexible statistical modeling concepts, that is, copulas, finite mixtures, and MRFs, and constitutes an efficient and robust approach with respect to possibly sophisticated statistics of the considered classes of interest.

The chapter is organized as follows. In Section 17.2, we introduce the EDSEM technique for the amplitude SAR pdf estimation. In Section 17.3, we proceed to multichannel pdf construction by first introducing the concept of copulas and then derive a novel method for PolSAR joint pdf estimation. In Section 17.4, we build on the result of the third section and introduce a supervised PolSAR classification approach. Section 17.5 presents experimental results on HR single-channel SAR and PolSAR data to demonstrate the performance of the developed pdf estimation and classification techniques along with comparisons with relevant benchmark approaches. Section 17.6 concludes the chapter.

17.2 SINGLE-CHANNEL SAR AMPLITUDE PROBABILITY DENSITY FUNCTION ESTIMATION BASED ON MIXTURE MODELS

In this section, we consider the single-channel SAR imagery amplitude pdf estimation problem. To take into account possible heterogenous scenarios, when several distinct land-cover typologies are present in the same SAR image, a FMM [12] for the distribution of gray levels is assumed. An amplitude SAR image is modeled as a set $I = \{r_1, \ldots, r_N\}$ of independent samples drawn from a mixture pdf with K components:

$$p_r(r) = \sum_{i=1}^{K} P_i p_i(r), \quad r \geq 0, \tag{17.1}$$

where $p_i(r)$ are parametric pdfs and $\{P_i\}$ are mixing proportions: $\Sigma_{i=1}^{K} P_i = 1$, with $0 \leq P_i \leq 1, i = 1, \ldots, K$. Each component $p_i(r)$ in Equation 17.1 is modeled by resorting to a finite dictionary $\mathcal{D} = \{f_1, \ldots, f_8\}$ (see Table 17.1) of SAR-specific distinct parametric pdf families $f_j(r|\theta_j)$, parameterized by $\theta_j \in A_j$, $j = 1, \ldots, 8$. Dealing with HR heterogenous SAR data, along with classical SAR pdf models, such as Lognormal, Weibull, Nakagami, $\mathcal{K}$-root, GGR, and SαSGR, we include the Fisher model, which was demonstrated to perform well for heterogenous HR SAR imagery [37], and GΓD, highly flexible empirical model, including Lognormal, Weibull, and Nakagami as special cases [6], into the considered dictionary $\mathcal{D}$.

As discussed in [11], considering the variety of estimation approaches for FMMs, an appropriate choice for this particular finite mixture estimation problem is the SEM scheme [12]. SEM was developed as a stochastic modification of the classical EM algorithm, involving stochastic sampling

TABLE 17.1
Pdfs and MoLC Equations for the Parametric pdf Families in $\mathcal{D}$

Family	Probability Density Function	MoLC Equations
f_1 Lognormal	$f_1(r\|m,\sigma) = \dfrac{1}{\sigma\sqrt{2\pi}}\exp\left[-\dfrac{(\ln r - m)^2}{2\sigma^2}\right], r > 0,$	$\kappa_1 = m,\ \kappa_2 = \sigma^2.$
f_2 Weibull	$f_2(r\|\eta,\mu) = \dfrac{\eta}{\mu^\eta}r^{\eta-1}\exp\left[-\left(\dfrac{r}{\mu}\right)^\eta\right], r \geq 0,$	$\kappa_1 = \ln\mu + \eta^{-1}\Psi(1),\ \kappa_2 = \eta^{-2}\Psi(1,1).$
f_3 Fisher	$f_3(r\|L,M,\mu) = \dfrac{\Gamma(L+M)}{\Gamma(L)\Gamma(M)}\dfrac{[ar]^L}{r[1+ar]^{L+M}}, r > 0,$ with $a = L/(M\mu),$	$\kappa_1 = \ln\mu + (\Psi(L) - \ln L) - (\Psi(M) - \ln M),$ $\kappa_j = \Psi(j-1, L) + (-1)^j\Psi(j-1,M),\ j = 2,3$
f_4 GTD	$f_4(r\|\nu,\kappa,\sigma) = \dfrac{\nu}{\sigma\Gamma(\kappa)}\left(\dfrac{r}{\sigma}\right)^{\kappa\nu-1}\exp\left[-\left(\dfrac{r}{\sigma}\right)^\nu\right], r \geq 0,$	$\kappa_1 = \Psi(\kappa)/\nu + \ln\sigma,\ \kappa_j = \Psi(j-1, \kappa)/\nu^j,\ j = 2,3.$
f_5 Nakagami	$f_5(r\|L,\lambda) = \dfrac{2}{\Gamma(L)}(\lambda L)^L r^{2L-1}\exp[-\lambda Lr^2], r \geq 0,$	$2\kappa_1 = \Psi(L) - \ln\lambda L,\ 4\kappa_2 = \Psi(1,L).$
f_6 $\mathcal{K}$-root	$f_6(r\|L,M,\mu) = \dfrac{4}{\Gamma(L)\Gamma(M)}[\lambda LM]^{(L+M)/2}r^{L+M-1}$ $\times K_{M-L}\left(2r[\lambda LM]^{1/2}\right), r \geq 0, L \geq M > 0,$	$2\kappa_1 = \Psi(L) + \Psi(M) - \ln LM\lambda.$ $2^j\kappa_j = \Psi(j-1, L) + \Psi(j-1,M),\ j = 2,3$
f_7 GGR	$f_7(r\|\lambda,\gamma) = \dfrac{\gamma^2 r}{\lambda^2\Gamma^2(\lambda)}\displaystyle\int_0^{\pi/2}\exp\left[-(\gamma r)^{1/A}s(\theta)\right]\,d\theta, r \geq 0.$ with $s(\theta) = \|\cos\theta\|^{1/A} + \|\sin\theta\|^{1/A},$	$\kappa_1 = \lambda\Psi(2\lambda) - \ln\gamma - \lambda G_1(\lambda)[G_0(\lambda)]^{-1},$ $\kappa_2 = \lambda^2\left[\Psi(1,2\lambda) + \dfrac{G_2(\lambda)}{G_0(\lambda)} - \left(\dfrac{G_1(\lambda)}{G_0(\lambda)}\right)^2\right].$
f_8 SαSGR	$f_8(r\|\alpha,\gamma) = r\displaystyle\int_0^{+\infty}\rho\exp[-\gamma\rho^\alpha]J_0(r\rho)\,d\rho, r \geq 0,$	$\alpha\kappa_1 = (\alpha - 1)\Psi(1) + \ln\gamma 2^\alpha,\ \kappa_2 = \alpha^{-2}\Psi(1,1)$

Note: Here $\Gamma(\cdot)$ is the Gamma Function [39], $K_\alpha(\cdot)$ the αth Order Modified Bessel Function of the Second King [39], $J_0(\cdot)$ is the Zero-th Order Bessel Function of the First Kind [39], $\Psi(\cdot)$ the Digamma Function [39], $\Psi(\nu,\cdot)$ the νth Order Polygamma Function [39] and $G_\nu(\cdot)$ are the Specific Integral Functions for GGR [7].

on every iteration, and demonstrating higher chances of avoiding local maxima of the likelihood function [12]. By definition, SEM is an iterative estimation procedure dealing with the problem of *data incompleteness*. In case of FMMs the complete data are represented by the set $\{(r_i, s_i), i = 1, \ldots, N\}$, where r_i are the observations (SAR amplitudes) and s_i-the missing labels: given an FMM with K components, $s_i \in \{\rho_1, \ldots, \rho_k\}$ denotes to which of the K components the ith observation belongs.

Instead of adopting the ML estimates, as the classical SEM scheme [12] suggests, in DSEM the Method of Log-Cumulants (MoLC) [2] for component parameter estimation is adopted, which has been demonstrated to be a feasible and effective estimation tool for all the pdfs in $\mathcal{D}$ [2,6,11]. MoLC has recently been proposed as a parametric pdf estimation technique suitable for distributions defined on $[0, +\infty)$, and has been widely applied in the context of SAR-specific parametric families for amplitude and intensity data modeling. MoLC adopts the Mellin transform by analogy to the Laplace transform in moment generating function [2]. Given a nonnegative random variable u, the second-kind characteristic function ϕ_u of u is defined as the Mellin transform [2] $\mathcal{M}$ of the pdf of u, that is:

$$\phi_u(s) = \mathcal{M}(p_u)(s) = \int_0^{+\infty} p_u(u)u^{s-1}\, du, \quad s \in \mathbb{C}.$$

The derivatives $\kappa_j = [\ln \phi_u]^{(j)} (1)$ are the jth order log-cumulants, where (j) denotes for the jth derivative, $j = 1, 2, \ldots$. If the Mellin transform converges for s lying in a neighborhood of 1, the following MoLC equations take place [2]:

$$\begin{cases} \kappa_1 = E\{\ln u\} \\ \\ \kappa_j = E\{(\ln u - \kappa_1)^j\} \end{cases}, \quad j = 2,3.$$

Analytically expressing κ_js, $j = 1, 2, 3$, as functions of unknown parameters and estimating these log-cumulants in terms of sample log-moments one derives a system of nonlinear equations. These equations have one solution for any observed values of log-cumulants for all the pdfs in $\mathcal{D}$ (see Table 17.1), except for, in some cases, GGR, $\mathcal{K}$-root, see [11], due to a complicated parameter estimation procedure [38].

For the purpose of K estimation we adopt a procedure similar to the one suggested in Ref. [12], that consists of initializing SEM with $K_0 = K_{max}$, and then allowing components to be eliminated from the mixture during the iterative process, once their priors P_i become too small, thus decreasing K. This strategy provides efficient K^* estimates consistent with DSEM estimates [11], allowing, however, to significantly reduce the computational complexity, especially for high values of K_{max}.

Thus, each iteration of EDSEM consists of the following steps:

- *E-step:* compute, for each graylevel z and ith component, the posterior probability estimates corresponding to the current pdf estimates, that is, $z = 0, \ldots, Z - 1$:

$$\tau_i^t(z) = \frac{P_i^t p_i^t(z)}{\sum_{j=1}^{K_t} P_j^t p_j^t(z)}, \quad i = 1,\ldots,K_t,$$

where $p_i^t(\cdot)$ is the ρ_i-conditional pdf estimate on the tth step.
- *S-step:* sample the label $s^t(z)$ of each graylevel z according to the current estimated posterior probability distribution $\{\tau_i^t(z) : i = 1,\ldots K_t\}, z = 0,\ldots,Z - 1$.

- *MoLC-step:* for the *i*th mixture component, compute the following histogram-based estimates of the mixture proportions and the first three log-cumulants:

$$P_i^{t+1} = \frac{\sum_{z \in Q_{it}} h(z)}{\sum_{z=0}^{Z-1} h(z)}, \quad \kappa_{1i}^t = \frac{\sum_{z \in Q_{it}} h(z) \ln z}{\sum_{z \in Q_{it}} h(z)}, \quad \kappa_{bi}^t = \frac{\sum_{z \in Q_{it}} h(z)(\ln z - \kappa_{1i}^t)^b}{\sum_{z \in Q_{it}} h(z)}, \quad i = 1, \ldots, K_t,$$

where $b = 2, 3$; $h(z)$ is the image histogram; $Q_{it} = \{z: s^t(z) = \rho_i\}$ is the set of gray levels assigned to the *i*th component; then, solve the corresponding MoLC equations (see Table 17.1) for each parametric family $f_j(\cdot|\theta_j)$ ($\theta_j \in A_j$) in the dictionary, thus computing the resulting MoLC estimate $\theta_{ij}^t, j = 1, \ldots, M$.

- *K-step:* $\forall i, i = 1, \ldots, K_t$: if $P_i^{t+1} < \gamma$, eliminate the *i*th component and correspondingly update $K_t + 1$. The choice of threshold γ does not appreciably affect EDSEM, provided it is small, for example, 0.005.
- *Model selection-step:* for each mixture component *i*, compute the log-likelihood of each estimated pdf $f_j(\cdot|\theta_{ij}^t)$ according to the data assigned to the *i*th component:

$$L_{ij}^t = \sum_{z \in Q_{it}} h(z) \ln f_j(z|\theta_{ij}^t), \quad i = 1, \ldots, K_{t+1},$$

and define $p_i^{t+1}(\cdot)$ as the estimated pdf $f_j(\cdot|\theta_{ij}^t)$ yielding the highest value of $L_{ij}^t, j = 1, \ldots, M$.

The sequence of estimates generated by SEM is a discrete-time random process $\{\Theta_t\}_{t=0}^{\infty}$ and converges to a unique stationary distribution, and the ML estimate of the mixture parameters is asymptotically equivalent to the mathematical expectation of this stationary distribution. This behavior has been proved under suitable assumptions [12], which do not hold strictly for all the pdfs in $\mathcal{D}$. This means that, in general, the convergence of the SEM estimation procedure is not guaranteed when dealing with finite mixtures drawn from the considered dictionary $\mathcal{D}$. This owes to the fact that in this chapter we suggest working with a complicated set of distributions, for most of which it is impossible to demonstrate the SEM convergence even if they were considered as the only types of distributions to be mixed. However, we recall that SEM, compared to the classical EM or other deterministic variants for FMM, was specifically designed to improve the exploratory properties of EM in the case of multimodal likelihood function [12]. In our experience, we have very rarely observed the lack of convergence, that is, if the sufficient number of SEM iterations was performed, the SEM estimates obtained after running the algorithm several times were consistently close even starting with the random initialization.

17.3 POLARIMETRIC SAR AMPLITUDE JOINT PROBABILITY DENSITY FUNCTION-ESTIMATION VIA COPULAS

In this section, we introduce a novel approach to PolSAR amplitude pdf-estimation based on the use of the EDSEM algorithm for single-channel SAR pdf estimation and the statistical concept of copulas for constructing the joint multichannel pdf function.

We recall that a *D*-dimensional copula is a multivariate joint distribution defined on $[0, 1]^D$ such that its every marginal distribution is uniform on $[0, 1]$. We will define any copula by its cumulative distribution function (cdf) C. Some basic facts and definitions from the copula theory are summarized in Appendix A. The copula approach to define a multivariate distribution is based on the idea that a simple transformation of marginal distributions can be made in such a way that each transformed marginal has a uniform distribution [34]. This approach develops the concept of multivariate cdf construction and, therefore, provides a more general and flexible method of joint pdf-estimation.

In this chapter the PolSAR pdf-estimation approach is discussed in the following two steps:

EDSEM step. The marginal pdfs of the polarization channels are separately estimated by applying the EDSEM approach. For the dth channel, $d = 1, \ldots, D$, the mixture EDSEM pdf estimator $p_d(x_d)$ and the corresponding cdf $F_d(x_d)$ are written in the following form:

$$p_d(x_d) = \sum_{i=1}^{K_d} P_{di} f_{di}(x_d), \quad F_d(x_d) = \sum_{i=1}^{K_d} P_{di} F_{di}(x_d), \tag{17.2}$$

where x_d represents the amplitudes of the given polarization channel d and K_d is the number of components in the dth mixture. F_{di} and p_{di} represent the ith mixture component in the cdf and pdf domains, respectively, and P_{di} are the related mixture proportions; p_{di} is automatically drawn by EDSEM from a dictionary of SAR-specific parametric families, $i = 1, \ldots, K_d$.

In the case of PolSAR pdf estimation, the calculation of cdfs is needed to merge the marginal EDSEM pdf estimates of polarization channels into joint pdfs via copulas (see Equation 17.3). Therefore, the list of pdfs in the dictionary $\mathcal{D}$ presented in Table 17.1 has been restricted, so that it includes the pdfs that either have an analytical closed-form expression (f_1, f_2), or a simple numerical approximation procedure (f_4, f_5) for the related cdf. For the rest of the pdf families there are no closed-form cdfs and numerical approximation is computationally intensive. Moreover, further experiments demonstrate that very accurate pdf estimates can be obtained with this reduced dictionary. Therefore, from now on we will consider the restricted dictionary $\bar{\mathcal{D}} = \{f_1, f_2, f_4, f_5\}$.

Now we address the reason why we suggest the use of EDSEM instead of adopting one of the single-family pdf models. First, EDSEM is an efficient and automatic tool for SAR amplitude pdf estimation, and it is capable of providing estimates of higher accuracy compared to single parametric pdf models [11,40]. Specifically, EDSEM was experimentally designed to accurately estimate the statistics of an HR SAR imagery [40]. Second, the underlying mixture assumption in EDSEM enables accurate characterization of the inhomogeneous classes of interest, that is, the classes that contain several different landcover subclasses. This is a very important property when dealing with HR imagery, since the corresponding image statistics are usually strongly mixed due to the high level of spatial detail appreciable at HR. However, even in the case of homogenous classes, EDSEM can be viewed as a tool for choosing the best single-pdf model from the set of pdfs in the dictionary.

Copula step. The goal of this step is to merge the marginal pdfs, which correspond to polarization channels estimated on the EDSEM-step, into a joint multivariate pdf describing the joint amplitude distribution of a PolSAR image. As recalled in Section 17.1, a number of multivariate pdf models have been proposed earlier to deal with this problem. In this chapter, we suggest a more flexible solution to this problem, which allows us to achieve higher pdf-approximation accuracy, thanks to a wider panel of dependency structures considered. We model the joint pdf $p(\mathbf{x})$ via an appropriate copula from the marginal distributions estimated in Equation 17.2. Thus, the joint pdf is constructed as

$$p(\mathbf{x}) = p_1(x_1) \ldots p_D(x_D) \frac{\partial^D C^*}{\partial x_1 \ldots \partial x_D} (F_1(x_1), \ldots, F_D(x_D)), \tag{17.3}$$

where C^* is coming from an appropriate family of copulas.

In this chapter, we consider several families of one-parametric copulas. By taking advantage of the connection between copulas and the Kendall's tau ranking coefficient τ, we obtain closed-form equations to estimate the copula parameter θ for each copula from the empirically observed value $\hat{\tau}$.

In order to fully exploit the modeling potential of copulas, we consider a dictionary $\mathcal{D}_C$ of several one-parametric copulas $C_c(\mathbf{x}|\theta)$. In the bivariate case (corresponding to dual-pol SAR imagery), we consider 10 copulas: 6 Archimedean (Clayton, Gumbel, Frank, AMH, A12, A14) [34], a copula with a quadratic section (Farlie-Gumbel-Morgenstern) [34], 2 elliptical (Gaussian and Student-t^* [41]), and a non-Archimedean copula with simultaneous presence of an absolutely continuous and a singular component (Marchal-Olkin) [34]. Here the names for A12 and A14 copulas originate from their positions in the list of Archimedean copulas in [34]. When $D \geq 3$ the same dictionary cannot be considered since the described parameter estimation strategy is not applicable to all the copula families. Therefore, in such case, we consider only the first three copula families in Table 17.2: Clayton, Gumbel, and Frank. Such a choice of copulas is capable of modeling a sufficiently wide variety of dependence structures, and covers most copula applications [42]. The considered dictionary of copulas is summarized in Table 17.2 (further information can be found in [34,41,42]).

The choice of a specific copula C^* from the dictionary (see Equation 17.3) is performed as follows. First, given an estimator $\hat{\tau}$ of τ, we need to decide whether a specific copula is appropriate for modeling the dependence with such a level of Kendall's tau rank correlation. Indeed, some copulas are specific to marginals with a low level of dependence, others deal with strongly dependent marginals, and still others are capable of modeling all levels of dependency. In other words, the effective list of copulas is limited to those which are capable of accurately modeling the specific empirically estimated value $\hat{\tau}$ (see Table 17.2). Thus, we first discard the copulas for which the current sample $\hat{\tau}$ is outside the related τ-relevance interval; for the remaining copulas, we derive an estimate $\hat{\theta}$ of the related parameter by the above-mentioned Kendall's tau method. Next, from the list of remaining copulas we choose the copula with the highest p-value provided by a Pearson chi-square test-of-fitness (PCS) [43]. In general, PCS tests the null hypothesis that the frequency distribution of certain events observed in a sample is consistent with a particular theoretical distribution. The χ^2 statistic is constructed as follows:

$$X^2 = \sum_{i=1}^{n} \frac{(O_i - E_i)^2}{E_i}, \tag{17.4}$$

where O_i and E_i are the observed and the hypothetical frequencies, respectively, and n is the number of outcomes. PCS is one of the statistical tests whose results follow a χ^2 distribution [43], that is, $X^2 \sim \chi^2_{n-r-1}$, where r is the number of reductions of degrees of freedom (typically, the number of parameters for parametric cdfs). The reference to χ^2 distribution allows p-values for the considered null hypothesis to be calculated. In our case, the null hypothesis in PCS is that the sample frequencies $(F_1(x_1), \ldots, F_D(x_D))$ are consistent with the theoretical frequencies (probabilities) predicted by the parametric copula C_c ($c = 1, \ldots, 10$). This is correct because if $\mathbf{x}$ is distributed with cdf $F(\cdot)$, and $\mathbf{x}_1, \ldots, \mathbf{x}_N$ are independent observations of $\mathbf{x}$, then $F(\mathbf{x}_i)$, $i = 1, \ldots, N$, are independent $[0, 1]$-uniformly distributed random variables [43].

17.4 POLARIMETRIC SAR IMAGE CLASSIFICATION

In this section, we propose a novel Bayesian supervised classification approach for amplitude PolSAR images. As a likelihood term, we propose the use of the joint pdf model developed in the previous sections based on the combined use of copulas and EDSEM for each thematic class. To introduce regularization and ensure robustness against speckle we adopt an MRF model in the form of the second-order Potts model. To optimize the resulting energy we use a Modified Metropolis Dynamics (MMD) approach. We will refer to the involved joint pdf model as CoDSEM, and to the related classification approach as CoDSEM-MRF. We notice that in the case of single-channel data there is no need to construct a joint pdf from the marginals; therefore, the classification can be obtained without copula modeling. We will refer to this simpler one-dimensional version as DSEM-MRF.

---

* Unlike all other considered copulas, the Student-t copula depends on two parameters: the linear correlation coefficient θ and the number of degrees of freedom ν. To avoid a cumbersome $\hat{\nu}$ ML estimation [41] we employ the following approach: we consider separately $\nu = 3k$, with k from 1 to 9 due to the fact that as ν grows large ($\nu \geq 30$) Student-t copula becomes indistinguishable from a Gaussian copula [41]. In other words, instead of a single two-parametric Student-t copula we consider hereunder nine one-parametric Student-t copulas with fixed values of ν.

TABLE 17.2

Considered Dictionary $\mathcal{D}_C$ of One-parametric Copulas: Clayton, Gumbel, Frank, AMH, A12, A14, Farlie-Gumbel-Morgenstern, Marchal-Olkin, Gaussian and Student-t, Each Defined by the Function $C_c(u|\theta)$, $u \in \mathbb{R}^D$, Along with Corresponding Dimensionalities, $\theta(\tau)$ Dependencies and τ-intervals. Here $\phi^{-1}(t)$ and $t_\nu^{-1}(t)$ Denote the Quantile Functions of a Standard Univariate Normal and a Standard Univariate t_ν Distributions Respectively

Copula	Dim	$C_c(u\|\theta)$	$\theta(\tau)$ Dependence	τ Interval
Clayton	D	$\left(u_1^{-\theta} + \cdots + u_D^{-\theta} - D + 1\right)^{-1/\theta}$	$\theta = \dfrac{2\tau}{1-\tau}$	$\tau \in (0, 1]$
Gumbel	D	$\exp\left(-\left((-\ln(u_1))^\theta + \cdots + (-\ln(u_D))^\theta\right)^{1/\theta}\right)$	$\theta = \dfrac{1}{1-\tau}$	$\tau \in [0, 1]$
Frank	D	$-\dfrac{1}{\theta}\ln\left(1 + \dfrac{(e^{-\theta u_1}-1)\cdots(e^{-\theta u_D}-1)}{(e^{-\theta}-1)^{D-1}}\right)$	$\tau = 1 - \dfrac{4}{\theta^2}\displaystyle\int_0^\theta \dfrac{t}{e^t-1}\,dt$	$\tau \in [-1, 0] \cup [0, 1]$
AMH	2	$\dfrac{u_1 u_2}{1 - \theta(1-u_1)(1-u_2)}$	$\tau = \dfrac{3\theta-2}{3\theta} - \dfrac{2}{3}\left(1-\dfrac{1}{\theta}\right)^2 \ln(1-\theta)$	$\tau \in [-0.1817, 0.3333]$
A12	2	$\left(1 + \left[(u_1^{-1}-1)^\theta + (u_2^{-1}-1)^\theta\right]^{1/\theta}\right)^{-1}$	$\theta = \dfrac{2}{3-3\tau}$	$\tau \in [0.3334, 1]$
A14	2	$\left(1 + \left[(u_1^{-1/\theta}-1)^\theta + (u_2^{-1/\theta}-1)^\theta\right]^{1/\theta}\right)^{-\theta}$	$\theta = \dfrac{1+\tau}{2-2\tau}$	$\tau \in [0.3334, 1]$
Farlie–Gumbel–Morgenstern	2	$u_1 u_2 (1 + \theta(1-u_1)(1-u_2))$	$\theta = \dfrac{9}{2}\tau$	$\tau \in [-0.2, 0.2222]$
Marchal–Olkin	2	$\min\left(u_1^{1-\theta}u_2, u_1 u_2^{1-\theta}\right)$	$\theta = \dfrac{2\tau}{\tau+1}$	$\tau \in [0,1]$
Gaussian	2	$\displaystyle\int_{-\infty}^{\phi^{-1}(u_1)}\int_{-\infty}^{\phi^{-1}(u_2)} \frac{1}{2\pi\sqrt{1-\theta^2}}\exp\left(\frac{2\theta xy - x^2 - y^2}{2(1-\theta^2)}\right) dx\,dy$	$\theta = \sin\left(\dfrac{\pi}{2}\tau\right)$	$\tau \in (-1, 1)$
Student-t	2	$\displaystyle\int_{-\infty}^{t_\nu^{-1}(u_1)}\int_{-\infty}^{t_\nu^{-1}(u_2)} \frac{1}{2\pi\sqrt{1-\theta^2}}\left(1 + \frac{x^2 - 2\theta xy + y^2}{\nu(1-\theta^2)}\right)^{-(\nu+2)/2} dx\,dy$	$\theta = \sin\left(\dfrac{\pi}{2}\tau\right)$	$\tau \in (-1, 1)$

Once the CoDSEM pdfs for each class $m = 1, \ldots, M$ are estimated on the training pixels, the following steps have to be performed in order to perform the classification.

MRF step: In order to take into consideration the contextual information disregarded by the pixelwise Copula-EDSEM technique, and to gain robustness against the inherent noise-like phenomenon of SAR known as speckle, we adopt a contextual approach based on an MRF model. Following the classical definitions of MRFs (see, e.g., [44,45]) on the two-dimensional lattice S of N observations $\mathbf{y} = \{y_1, \ldots, y_N\}$ and class labels $\mathbf{x} = \{x_1, \ldots, x_N\}$, $x_i \in \{1, \ldots, M\}$, we introduce an isotropic second-order neighborhood system C with cliques of size 2. The Hammersley–Clifford theorem [44] allows for the presentation of the joint probability distribution of an MRF as a Gibbs distribution:

$$P(\mathbf{x}) = Z^{-1} \exp(-U(\mathbf{x}|\beta)),$$

where $Z = \sum_z \exp(-U(\mathbf{z}|\beta))$ is a normalizing constant, β is a positive parameter and $U(\mathbf{x}|\beta)$ is the MRF energy function associated with class labels. More specifically $U(\mathbf{x}|\beta)$ takes the following form:

$$U(\mathbf{x}|\beta) = \sum_{\{s,s'\}\in C} [-\beta \delta_{x_s = x_{s'}}],$$

where δ is the Kronecker delta function:

$$\delta_{x_s = x_{s'}} = \begin{cases} 1, & \text{if } x_s = x_{s'} \\ 0, & \text{otherwise.} \end{cases}$$

Image classification poses a problem of recovering the unobserved data, that is, class labels. In the case of hidden MRFs, the unobserved data $\mathbf{x}$ are modeled by an MRF and the observed data $\mathbf{y}$, that is, SAR amplitudes, are assumed to be conditionally independent given $\mathbf{x}$, that is:

$$p(\mathbf{y}_i|\mathbf{y}_{s\backslash\{i\}}, x_i) = p(\mathbf{y}_i|x_i), \quad \forall i \in S,$$

where $\mathbf{y}_i$ is the D-vector of amplitudes $(y_1, \ldots, y_D)$ at pixel i, $\mathbf{y} = \{\mathbf{y}_1, \ldots, \mathbf{y}_N\}$, and $S\backslash\{i\}$ represents all the pixels of S except for i. Then involving Bayes formula, we get the following energy function for the full data $(\mathbf{y},\mathbf{x})$:

$$U(\mathbf{x}|\mathbf{y},\beta) = \sum_{i \in S} \left(-\sum_{m=1}^{M} \ln p(\mathbf{y}_i|x_i = m) - \beta \sum_{s:\{i,s\}\in C} \delta_{x_i = x_s} \right). \tag{17.5}$$

Here, the energy at each site i is constructed as a sum of two distinct contributions: probabilities of observing the amplitude vector $\mathbf{y}_i$ and the energy of label configuration in the neighborhood of i. In this formula, given the conditional pdfs $p(\cdot)$ are defined in Equation 17.3. The energy function U in Equation 17.5 has a single parameter β that must be estimated. At this stage, all of the parameters involved in defining the pdfs $p(\cdot)$ are estimated, and, therefore, the resulting energy function Equation 17.2 through (17.5) also has only one parameter. In order to estimate β, we employ a simulated annealing (SA) procedure [46,47] with a pseudo-likelihood function PL [44] of the following form:

$$\begin{cases} \ln \mathrm{PL}(\mathbf{x}|\beta) = \ln \left[\prod_{s \in S} P(x_s|\mathbf{x}_{S\backslash\{s\}}, \beta) \right] \\[2em] P(x_s|\mathbf{x}_{S\backslash\{s\}}, \beta) = \dfrac{\exp(-U(x_s|\mathbf{x}_{S\backslash\{s\}}, \beta))}{\sum\limits_{z_s \in X_S} \exp(-U(z_s|\mathbf{x}_{S\backslash\{s\}}, \beta))} \end{cases},$$

for $X_s = \{\omega_1, \ldots, \omega_M\}$.

The proposed SA procedure generates a sequence $\{\beta_t\}$ of estimates, and employs the normal proposal distribution $N(\beta_t, 1)$ [48] together with an exponentially decreasing cooling schedule $T_t = 0.95 \cdot T_{t-1}$. Considering the convergence in distribution of the related Metropolis algorithm, the final estimate $\beta*$ was set by averaging the estimates on the last n iterations. The estimation of β was performed on an ML preclassification of the image, which associated every pixel with the highest probability-density label according to the CoDSEM class models.

Energy minimization step: This step involves the search for a label configuration that minimizes the energy given by Equation 17.5, that permits to find the optimal pixel labeling as defined by the Maximum a posteriori (MAP) principle [45]. For this optimization problem the iterative deterministic MMD [49] algorithm is adopted. This is a compromise between the deterministic "iterated conditional modes" (ICM) algorithm [44], which is a fast local minimization technique, which is strongly dependent on the initial configuration, and the stochastic SA [46,47,50], which is computationally a very intensive global minimization approach. MMD is computationally feasible and provides reasonable results in real classification problems [49]. As well as SA, MMD requires a predefined cooling schedule and proceeds as follows:

1. Sample a random initial label configuration $\mathbf{x}^0$, define an initial temperature T^0 and parameters: proportion $\alpha \in (0, 1)$, number of labels visited per iteration $n_1 \in \mathbb{N}$, cooling schedule parameter $\tau \in (0, 1)$, and the termination criterion $\gamma \in \mathbb{R}^+$; initialize iteration counter $k = 0$ and temperature $T_0 = T^0$.
2. Set $i = 0$.
3. Using uniform distribution pick up a label configuration $\boldsymbol{\eta}$ which differs exactly in one element from $\mathbf{x}^k$.
4. Compute $\Delta = U(\boldsymbol{\eta}) - U(\mathbf{x}^k)$ and accept $\boldsymbol{\eta}$ according to the rule:

$$
\mathbf{x}^{k+1} = \begin{cases} \boldsymbol{\eta}, & \text{if } \Delta \leqslant 0 \\[2mm] \boldsymbol{\eta}, & \text{if } \Delta > 0 \quad \text{and} \quad \ln(\alpha) \leqslant -\frac{\Delta}{T_k}. \\[4mm] \mathbf{x}^k, & \text{otherwise} \end{cases}
$$

5. Calculate $\Delta U_i = \left| U(\mathbf{x}^{k+1}) - U(\mathbf{x}^k) \right|$.
6. If $i < n_1$, set $i = i + 1$ and go to Step 3.
7. Calculate $\Delta U = \sum_{i=1}^{n_1} \Delta U_i$.
8. If $| \Delta U / U(\mathbf{x}^{k+1})| > \gamma$ decrease the temperature $T_{k+1} = \tau T_k$, increase $k = k + 1$, and goto Step 2; stop otherwise.

During the early stages of this iterative procedure, the behavior of MMD is close to that of SA, and, on later stages, the MMD behaves as ICM. The proportion between the stages is controlled by the proportion parameter α. Therefore, MMD can be regarded as an ICM strategy with SA initialization and, as such, it provides better exploratory properties than ICM.

17.5 EXPERIMENTS

In this section, we present several experimental results with real space-borne SAR and PolSAR imagery that demonstrate the performance of the proposed single-channel SAR pdf estimation approach

EDSEM, multichannel/PolSAR joint pdf estimation approach CoDSEM and the derived supervised classification method CoDSEM-MRF. All the experiments are completed with comparisons to the relevant benchmark methods to further explore the pros and cons of the proposed approaches.

17.5.1 DATASETS FOR EXPERIMENTS

The experiments were performed on the following HR SAR datasets:

- Single-pol HH, single-look, 2.5 m ground resolution CSK®, Stripmap image acquired over Piemonte, Italy (©ASI). We present experiments on a 700×500 subimage CSK1 (HH pol, see Figure 17.1) and a 700×1000 subimage CSK2 (see Figure 17.6).
- Dual-pol HH/VV, 2.66-look, 6.5 m ground resolution TerraSAR-X Stripmap, geocorrected image acquired over Sanchagang, China (©Astrium GEO-Information Services). We present experiments with the following subimages: 700×500 subimage TSX1 (VV pol, see Figure 17.1), 700×500 subimage TSX2 (HH pol, see Figure 17.1), 1000×1200 subimage TSX3 (see Figure 17.2), and 750×750 subimage TSX4 (see Figure 17.5).
- Fine quad-pol HH/HV/VH/VV, single-look, 7.5 m resolution RADARSAT-2* image of Vancouver, Canada. The experiments are performed on a 500×500 subimage RS1 (HV pol, see Figure 17.1), a 1000×1200 subimage RS2 (Figure 17.7) and a 500×700 subimage RS3 (Figure 17.8).

17.5.2 SINGLE-CHANNEL PDF ESTIMATION EXPERIMENTS

In this section, we report the experiments of the EDSEM pdf estimation algorithm on several HR SAR datasets. We notice that EDSEM is also used as a structural component in the experiments of the next section.

The obtained EDSEM pdf estimates have been assessed quantitatively by computing Kolmogorov–Smirnov distances ($\mathcal{K}$) between estimates and normalized image histograms [43]. The estimates were also assessed by the Kolmogorov–Smirnov goodness-of-fit test [51] and the corresponding p-values are reported in Table 17.3. Here $1 - p$ can be interpreted as a confidence level at which the corresponding fit hypothesis will be rejected. The following EDSEM parameters were used: initial number of mixture components $K_0 = 6$, number of iterations $T = 200$. Both values were set manually high enough to give sufficient burn-in for SEM and their further increase did not appreciably affect the results. For the sake of comparison, for all test images we provide the

TABLE 17.3

EDSEM Results: K^* Estimate, the Estimated Mixture (f_is as in Table 17.1), Kolmogorov-Smirnov Distance $\mathcal{KS}$, p-value for the Kolmogorov–Smirnov Test, the Computation Time t (in seconds), Along with the Best-Fitting Model in $\mathcal{D}$ with $\mathcal{KS}$ and p-value

Image	K^*	EDSEM Estimate			t	Best Fit in $\mathcal{D}$		
		Mixture	$\mathcal{KS}$	p-value		Model	$\mathcal{KS}$	p-value
CSK1	4	f_5, f_4, f_1, f_5	0.007	1.0	192 s	f_5	0.020	0.9869
RS1	2	f_6, f_3	0.010	1.0	115 s	f_3	0.048	0.5341
TSX1	2	f_4, f_6	0.006	1.0	117 s	f_4	0.024	0.9252
TSX2	3	f_4, f_3, f_2	0.004	1.0	148 s	f_1	0.058	0.3552

* RADARSAT is an official mark of the Canadian Space Agency.

best fitting pdf (with the smallest $\mathcal{KS}$), with parameters estimated by MoLC, within the eight models in the dictionary $\mathcal{D}$ (see Table 17.3). For the considered images the accuracy improvement granted by EDSEM is significant ($\Delta\mathcal{KS}$ up to 0.05) and the employed goodness-of-fit test suggests the obtained estimates to be very accurate statistically. Visual analysis of the EDSEM estimate plots (see Figure 17.1) confirms an important improvement in the estimation accuracy. The computation times of the EDSEM reported in Table 17.3 were observed on an Intel Core 2 Duo 1.83 GHz, 1 Gb RAM, WinXP system.

We notice that, given a new model not contained in $\mathcal{D}$ and provided with corresponding MoLC-equations and solution, the process of adding such a model into the dictionary is straightforward. In the experiments of the next subsection we are employing a reduced dictionary $\bar{\mathcal{D}}$ of four pdfs and our experiments have demonstrated that this reduction still allows very accurate results to be obtained. Further still, the dictionary reduction, which was originally caused by the necessity to estimate cdfs, also enabled a significant computation acceleration (up to 50%).

17.5.3 Joint pdf Estimation and Classification Experiments

This section is devoted to experiments with PolSAR imagery. We have tested the CoDSEM method for multichannel joint pdf estimation and the CoDSEM-MRF model for supervised PolSAR image classification. The experiments with TerraSAR-X datasets were performed within an epidemiological study and involved the classification of humid regions into $M = 3$ classes: "water," "wet," and "dry soil." Same classification was performed on CSK datasets. RADARSAT-2 experiments investigated the urban area classification with $M = 3$ target classes: "water" for water surface and wet soil pixels, "urban," and "vegetation."

First, we introduce the experimental settings that were employed. In the EDSEM-step, given the relative homogeneity of the target classes on the experiment datasets, an average value of the initial number of components $K_0 = 3$ was selected, thus assuming mixtures with just a few components per class. In the Copula-step, the following settings in the PCS test were used. In the dual-pol case, the $[0, 1] \times [0, 1]$ square was divided into 25 equal squares S_i by four horizontal and four vertical lines parallel to the axes (i.e., $n = 25$). For every class $m = 1, ..., M$ and every cluster $S_i, i = 1, ..., n$:

$$O_{im} = \sum_{(y_1, y_2) \in \omega_m} I_{S_i}[F_1(y_1), F_2(y_2)],$$

where I is the indicator function and the summation is taken over all the training pixels (y_1, y_2) available for class m. For each copula $c = 1, ..., 10$ in $\mathcal{D}_C$, the value of E_{im} was calculated by integrating the related pdf over the square S_i, thus retrieving the probability that a pair of $[0, 1]$-uniform random variables, whose joint distribution is defined by the c-th copula, belong to $S_i (i = 1, ..., n)$. In the case of 4D quad-pol imagery, in order to pick the copula we have employed the strategy similar to the one employed for Kendall's tau estimation in Appendix A: we have used all 2D combinations of channels to calculate the values O_{im}. In the energy minimization step, the revisit scheme on Step 3 of MMD was implemented raster-wise and the following parameter values were used: $T^0 = 5.0$, $\alpha = 0.3$, $\tau = 0.97$, $\gamma = 10^{-4}$ and n_1 was set equal to the size of the image in pixels as in [49]. Compared to ICM, MMD generated significantly better results (above 5% of accuracy gain), starting at ML initialization (maximizing Equation 17.3 at each pixel). SA obtained slightly better results (roughly a 1% accuracy gain) compared to MMD. However, SA generated a drastic increase in computational complexity (around 200 iterations for MMD and over 1500 for SA).

For every dataset, the CoDSEM-MRF algorithm was trained on a small 250×250 subimage endowed with a manually annotated nonexhaustive ground truth (GT), which did not overlap with the test areas.

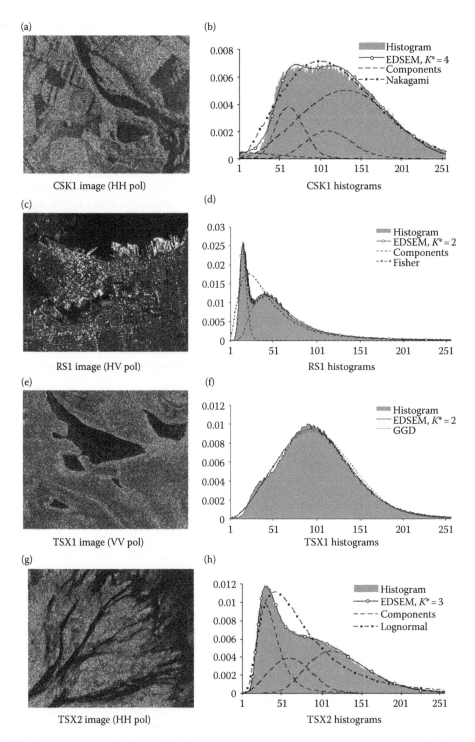

FIGURE 17.1 (a) CSK1 (©ASI), (c) RS1 (©MacDonald, Dettwiler and Associates Ltd.), (e) TSX1, and (g) TSX2 images (©Astrium) with corresponding plots of pdf estimates (b), (d), (f), (h). The plots contain: normalized image histogram, EDSEM pdf estimate with plots of K^* components in the estimated mixture, and the plot of the best-fitting pdf model from $\mathcal{D}$.

Table 17.4 reports the classification accuracies obtained on the test samples for each class ("class accuracy"), the resulting average accuracy (i.e., the arithmetic mean of the class accuracies) and overall accuracy (i.e., the percentage of correctly classified test samples, irrespective of their classes). The first test image TSX3 (see Figure 17.2a) covers an area of a river delta in Sanchagang, China. For this experiment, an exhaustive GT was manually created, and the classification results were referenced to it in order to calculate the accuracies and demonstrate the results visually. The automatically selected copulas were: Gumbel for the "water," and Frank for the "wet soil," and "dry soil" classes. The MRF parameter estimation provided $\beta^* = 1.408$ and the overall classification accuracy achieved by CoDSEM-MRF was 84.5%. The resulting CoDSEM-MRF classification map is shown in Figure 17.2c and referenced to the GT in Figure 17.2d.

In order to appreciate the classification gain from dual-pol data (HH/VV), compared to single-channel data, we performed an experiment of DSEM-MRF classification on the HH channel, which reported an overall accuracy of 80.6%. On average, the overall accuracy gain from adding the second polarization channel on the considered datasets was 3–7%.

For the sake of comparison, we also provide classification maps obtained by the benchmark 2D Nakagami-Gamma (2DNG) model [29] (Figure 17.2e) and K-nearest-neighbors (K-NN) method [45] (Figure 17.2f). In order to perform a fair comparison among contextual methods, the above models were combined with the contextual MRF approach. The same MRF parameter estimation and energy minimization procedures as in CoDSEM-MRF were employed in these benchmark experiments as well. For the 2DNG model, the equivalent number of looks was set to $L = 2.66$, and for the K-NN model, $K^* = 40$ was estimated by cross-validation [45]. The overall accuracies of these two benchmark approaches were 81.5% for 2DNG-MRF and 81.9% for K-NN-MRF on the dual-pol HH/VV TSX3 image. These accuracies were lower than the overall classification accuracy achieved by CoDSEM-MRF. K-NN-MRF reported higher accuracy for the "wet soil" class, however this was achieved at the expense of accuracy for the "dry soil."

To demonstrate the accuracy of the Copula-DSEM joint pdf model directly we present K-plots of the sample HH-VV dependence and the dependence estimated by copulas (see Figure 17.3) on the learning image for TSX3. A K-plot (from Kendall-plot) [52] is a rank-based graphical tool developed for visualization of dependence structure between two random variables. In our case, we build a

TABLE 17.411

Classification Accuracies on the Considered Test Images

Image	Method	Water	Wet Soil	Dry Soil	Average	Overall
TSX3	CoDSEM-MRF	90.02%	82.56%	84.80%	85.79%	**84.55%**
	2DNG-MRF	89.33%	82.12%	76.51%	82.78%	**81.49%**
	K-NN-MRF	87.15%	87.81%	71.45%	82.14%	**81.95%**
	DSEM-MRF on HH pol	90.00%	69.93%	91.28%	83.74%	**80.61%**
TSX4	CoDSEM-MRF, $\beta^* = 1.324$	92.48%	94.59%	85.16%	90.49%	**92.41%**
	CoDSEM-MRF, $\beta = 0.5$	91.52%	89.31%	77.48%	86.11%	**87.98%**
	2DNG-MRF	92.59%	89.33%	86.01%	89.31%	**89.74%**
	K-NN-MRF	90.21%	98.56%	78.91%	89.23%	**92.61%**
CSK2	DSEM-MRF	95.57%	92.22%	92.45%	93.41%	**93.14%**
	$\mathcal{K}$-root-MRF	89.64%	87.78%	96.83%	91.53%	**90.94%**
		Water	**Vegetation**	**Urban**		
RS2	CoDSEM-MRF	98.04%	90.33%	71.49%	86.61%	**88.69%**
	K-NN-MRF	98.85%	89.02%	62.89%	83.59%	**86.42%**
RS3	CoDSEM-MRF	97.30%	95.12%	86.12%	92.85%	**94.18%**
	K-NN-MRF	98.10%	96.06%	81.33%	91.83%	**93.51%**

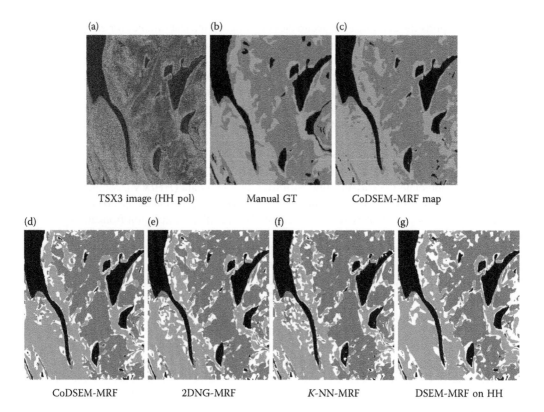

FIGURE 17.2 (See color insert.) (a) HH channel of dual-pol TSX3 image (©Astrium), (b) manually created GT, (c) CoDSEM-MRF classification map (water—black; wet soil—dark gray; and dry soil—light gray). Classification maps referenced to the GT: (d) CoDSEM-MRF, (e) 2DNG-MRF, (f) K-NN-MRF, and (g) DSEM-MRF on the HH pol image (correctly classified water—black; wet soil—dark gray; dry soil—light gray; misclassification of all types—white).

sample K-plot as well as K-plots corresponding to copula models. One can see good agreement between observations and the automatically selected models.

We now briefly compare the developed CoDSEM-MRF approach with the previously proposed copula-based method [32]. First, the approach [32] is based on the use of the AMH copula [34], which can model dependencies corresponding to Kendall's correlation coefficient $\tau \in [-0.1817, 0.3333]$.

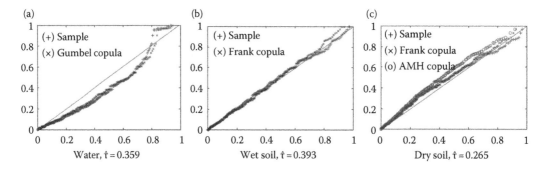

FIGURE 17.3 K-plots display the graphical goodness-of-fit for copula models on the learning stage for TSX3 image. The sample HH-VV dependences (marker +) for the (a) water, (b) wet soil and (c) dry soil classes are fitted by automatically selected copulas (marker ×). The K-plot for the AMH copula (marker ○) is also presented for the dry soil class. The diagonal (–) represents the independency scenario.

On the employed TerraSAR-X dataset we have observed empirical values $\hat{\tau}_{\text{water}} = 0.359$, $\hat{\tau}_{\text{wet}} = 0.393$, and $\hat{\tau}_{\text{dry}} = 0.265$. Therefore, the AMH copula can be used only for the dry soil class and its K-plot is presented in Figure 17.3c. However, even in this case, the goodness-of-fit provided by the automatically selected Frank copula is better. Thus, the use of dictionary-based copula selection approach is more accurate and constitutes a more flexible model with a wider range of applicability. Second, the use of the EDSEM finite mixture estimation approach for marginal pdf estimation provides higher accuracy estimates than singular pdf models, whose use was suggested in [32], especially for HR SAR images.

It is well known that the acquisition of the training sets required for supervised classification is a very costly procedure. Therefore, the classification models are designed to be as robust as possible with respect to small learning sets. To evaluate this characteristic of the proposed algorithm, we present an experimental study (see Figure 17.4) of the classification accuracy as a function of the size of the employed training set. We start with the learning image of 250×250 pixels, and then repeatedly reduce its size by a half at each iteration, that is by discarding 50% randomly selected pixels the learning set of each thematic class, until 1/32 of the initial learning image is kept. At each iteration we evaluate the class and overall accuracies of classification on the TSX3 image. To make the accuracy estimates more consistent, the whole process has been repeated three times and the averaged results are presented in Figure 17.4. We observe that the algorithm behaves robustly till 1/8 (90×90 pixels) of the initial learning image is kept. When very strong subsampling rates (≥ 16) are applied, the wet area classification rate drops significantly. This is due to the fact that the histogram of wet soil lies "between" those of water and dry soil. Thus, when the training set becomes unrepresentatively small, a lot of wet soil pixels are misclassified as dry soil. This result suggests the capability of the developed algorithm to perform fairly good and consistently on small training sets, that is, up to 100×100 training images for the considered dataset.

The most time-consuming stages of the algorithm are the MRF parameter estimation and energy minimization steps. Given the fairly low value $K_0 = 3$, the EDSEM method was very fast. The experiments were conducted on a Core 2 Duo 1.83 GHz, 1 Gb RAM, WinXP system. With the number of iterations equal to 200, the β-estimation on a roughly 1000×1000 image took around 80 s. The average of 200 iterations required for convergence with MMD (energy minimization step) took about 100 s. Thus, about 200 s were required for the complete classification process of a 1000×1000 dual-pol image with three classes.

The second test image TSX4 (see Figure 17.5) originates from the same TerraSAR-X dataset. Therefore, the same learning image has been employed, and, consequently, the same set of copulas were selected. The overall classification accuracy obtained by CoDSEM-MRF was 92.4% with $\beta^* = 1.324$. This accuracy is higher than that reported for the TSX3 image. However, this increase

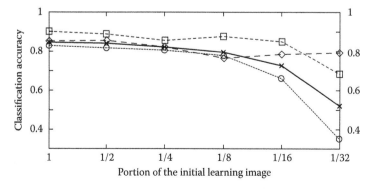

FIGURE 17.4   The sensitivity of CoDSEM-MRF classification accuracy to the size of the learning image. The class accuracies (water □, wet soil ○, dry soil ◊) and overall accuracies (solid line **x**) are reported for TSX3 image. The initial learning image is 250×250 pixels.

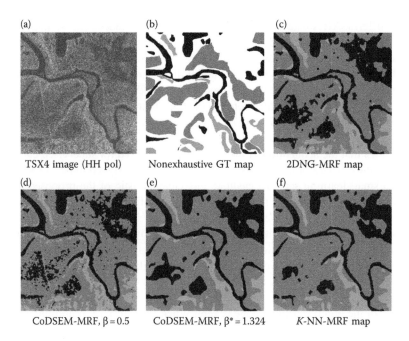

FIGURE 17.5 **(See color insert.)** (a) HH channel of the dual-pol TSX4 image (©Astrium), (b) nonexhaustive GT map (water—black; wet soil—dark gray; dry soil—light gray; and outside GT—white). Classification maps: (c) 2DNG-MRF, CoDSEM-MRF with (d) manually set $\beta = 0.5$ and (e) automatically estimated $\beta^* = 1.324$, (f) K-NN-MRF (same colors).

is mostly due to the use of a nonexhaustive test map in this experiment. This map did not contain many class transition regions where misclassified pixels may be visually spotted. Once again, we provide comparisons with the benchmark 2DNG-MRF (Figure 17.5c) and K-NN-MRF (Figure 17.5f) classification approaches. Consistently with our earlier observations, here CoDSEM-MRF outperformed appreciably 2DNG-MRF (89.7% of overall accuracy). K-NN-MRF with 92.6%, on the other hand, reported about the same level of overall accuracy as CoDSEM-MRF. Here we come to the same conclusion as with the TSX3 image: K-NN-MRF performs better on "wet soil" and far worse on "dry soil." We notice also that the average accuracy of K-NN-MRF is inferior to that of CoDSEM-MRF on both TSX3 and TSX4.

A slight oversmoothing effect can be noticed in the results obtained by CoDSEM-MRF on TSX4, see Figure 17.5e. Therefore, we also show the classification map obtained by manually setting a smaller value of the MRF parameter $\beta = 0.5$ (Figure 17.5d). One can see that the spatial details are more precise at the expense of a noisier segmentation. The classification accuracies achieved with $\beta = 0.5$ (see Table 17.4) are inferior to those obtained in the $\beta^* = 1.324$ case. This suggests that, at least for this dataset, stronger regularization is preferable.

We now move to the experiments with CSK imagery. Here the proposed classification approach was tested on a single-pol CSK2 image (see Figure 17.6a). A nonexhaustive GT was employed. The overall classification accuracy of DSEM-MRF (Figure 17.6b) was 93.1% with $\beta^* = 1.566$. We compare this approach with the $\mathcal{K}$-root-MRF contextual classification approach that is based on a $\mathcal{K}$-root model [9] for each class-conditional statistics. The $\mathcal{K}$ distribution is a well-known model for a possibly textured SAR multilook single-channel intensity and $\mathcal{K}$-root is the corresponding amplitude parametric pdf. The overall classification accuracy reported in this experiment (Figure 17.6c) was equal to 90.9%. The improved performance of DSEM-MRF is due to the more accurate pdf estimates generated by DSEM than by the $\mathcal{K}$-root model. Note here, that in case of single-channel SAR we do not use the K-NN-MRF classifier, as in case of one-dimensional discrete feature space (i.e., with amplitude values

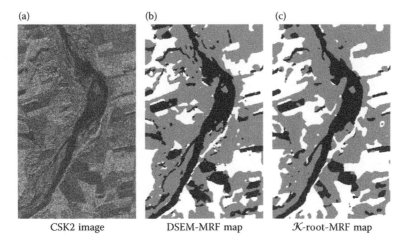

FIGURE 17.6   (a) HH polarization CSK2 image (©ASI), and classification maps: (b) DSEM-MRF and (c) $\mathcal{K}$-root-MRF (water—black, wet soil—dark gray, dry soil—white).

1, ..., 256) and target classes with close histograms the use of K-NN classifier becomes impossible. More specifically, if we consider the distance zero or one in feature space, we already have hundreds of neighbors of which we then need to somehow select K^* "voting" neighbors.

Finally, we consider the quad-pol RADARSAT-2 imagery (see Figures 17.7 and 17.8) classification into the following cases: "water," "vegetation," and "urban." We employ a nonexhaustive GT

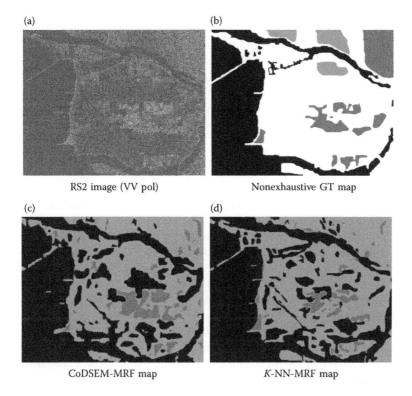

FIGURE 17.7   (**See color insert.**) (a) VV channel of the quad-pol RS2 image (RADARSAT-2 Data and Products ©MacDonald, Dettwiler and Associates Ltd., 2008—All Rights Reserved), (b) nonexhaustive GT map (water—black, urban—dark gray, vegetation—light gray, outside GT—white). Classification maps: (c) CoDSEM-MRF, (d) K-NN-MRF (same colors).

(a)   (b)   (c)   (d)

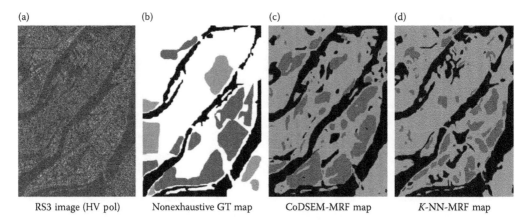

| RS3 image (HV pol) | Nonexhaustive GT map | CoDSEM-MRF map | *K*-NN-MRF map |

FIGURE 17.8   (a) HV channel of the quad-pol RS3 image (RADARSAT-2 Data and Products ©MacDonald, Dettwiler and Associates Ltd., 2008—All Rights Reserved), (b) nonexhaustive GT map (water—black, urban—dark gray, vegetation—light gray, outside GT—white). Classification maps: (c) CoDSEM-MRF, (d) *K*-NN-MRF (same colors).

map and compare the result with the *K*-NN-MRF classification, with $K^* = 35$ estimated by cross-validation [45] and $\beta = 1.419$. We remind again that only the three copula families from $\mathcal{D}_C$ were considered, and the following were automatically selected: Gumbel for "water," and Frank for "vegetation," and "urban" classes. We stress here that the comparison with the Nakagami–Gamma model is not feasible, because this model has not been defined for dimensionality higher than two [29]. In these experiments we have observed about the same level of classification accuracy for both CoDSEM-MRF and *K*-NN-MRF on "water" and "vegetation" classes, and an appreciable increase of accuracy for the "urban" class in case of CoDSEM-MRF classification. Therefore, the higher overall and average accuracies demonstrated by the CoDSEM-MRF method in these quad-pol RADARSAT-2 experiments suggest the developed algorithm to be a competitive and efficient supervised classification approach. Moreover, from the methodological point of view, CoDSEM might be preferable to *K*-NN as it provides an explicit description of a statistical model for the data, whereas the latter operates as a "black box."

17.6   CONCLUSIONS

In this chapter, we have considered the issues of amplitude PolSAR pdf estimation and supervised classification in the framework of this HR satellite imagery. We have proposed a general flexible pdf-estimation method for single-channel SAR images and specifically validated it with heterogeneous HR SAR data, which represent a very up-to-date and relevant case of SAR imagery. The developed model is based on the DSEM approach recently developed in [11] for medium resolution SAR. The developed EDSEM extended DSEM to the novel type of imagery (heterogenous HR), enhanced by a novel efficient procedure for estimating the number of mixture components, reported very accurate and computationally fast estimation results in experiments with HR SAR images acquired by the aforementioned modern satellite systems. We stress here that the problem of modeling the statistics of HR satellite SAR amplitude images has not been satisfactorily addressed so far, and the proposed EDSEM technique looks very promising for this type of imagery, since it is based on a finite mixture approach, and intrinsically takes into account heterogeneity, which is an inherent HR image property. The obtained results suggest EDSEM to be an attractive approach for various application problems, for example, it can be efficiently used for SAR image segmentation (to discriminate the resulting mixture components) [11] and to class-conditional pdf modeling in supervised HR SAR classification. The extension of the proposed method to intensity data (instead of amplitudes) is straightforward and can be performed by replacing the amplitude pdfs in the dictionary with the corresponding intensity pdfs.

We have introduced a novel model for the joint pdf of the amplitude PolSAR imagery. Based on flexible statistical tools, such as, copulas and EDSEM estimation technique, this approach proved to be accurate, easy to implement and fast from the parameter-estimation point of view. It enables a new level of flexibility in modeling the joint distribution from marginal single-channel distributions, which results in higher classification accuracies. The experiments with several datasets of HR SAR imagery have demonstrated the high applicability and accuracy of this model.

Based on the proposed joint pdf model we have developed a supervised classification algorithm for single-channel and PolSAR satellite imagery. Specifically, it combines the MRF approach to Bayesian image classification, finite mixture technique for pdf estimation, and copulas for multivariate pdf modeling. These three statistical concepts ensure high flexibility and applicability of the developed method to HR PolSAR image classification. Structurally, the proposed classification algorithm can be described as supervised and semiautomatic (a few EDSEM and MMD parameters have to be specified). The accuracy of the proposed algorithm was validated on classification into three target classes on several HR satellite SAR images: a single-pol CSK image, a dual-pol TerraSAR-X image, and a quad-pol RADARSAT-2 image. The experiments demonstrated a high level of accuracy on the experimental datasets and outperformed several parametric and nonparametric contextual benchmark algorithms.

Finally, as far as the future extensions of this work are concerned, a promising direction is to explore the use of more efficient optimization approaches. One of the best candidates to replace MMD is an appropriate graph cut-based approach [53]. Such approaches yield very good approximations in the MAP segmentation problem and are known to be fast. Another direction of development lies in the specialization of this model to urban area classification, which is an important and relevant application of SAR classification. To this end, the contextual MRF model would also need to incorporate geometrical information.

APPENDIX A: COPULA THEORY

A *D-dimensional copula* is a function $C: [0, 1]^D \rightarrow [0, 1]$, which satisfies:

1. $C(\mathbf{x}) = 0$ for any $\mathbf{x}$: $\exists i \in \{1, \ldots, D\}$ so that $x_i = 0$.
2. $C(\mathbf{x}) = x_d$ for any $\mathbf{x}$: $x_i = 1$, for all $i \neq d$.
3. The *D-increasing* condition: for any $0 \leq x_{i,1} \leq x_{i,2} \leq 1$, where $i = 1, \ldots, D$,

$$V_C \equiv \sum_{i_1=1}^{2} \cdots \sum_{i_D=1}^{2} (-1)^{(i_1+\cdots+i_D)} C(x_{1,i_1}, \ldots, x_{D,i_D}) \geqslant 0.$$

The important property of copulas is given by the *Sklar's theorem* [34], which states the existence of a copula C, that models the joint distribution H of arbitrary random variables $X_1, \ldots, X_D$ with cdfs $F_1, \ldots, F_D$:

$$H(\mathbf{x}) = C(F_1(x_1), \ldots, F_D(x_D)), \quad \forall \mathbf{x} \in \mathbb{R}^D. \tag{17.6}$$

Given absolutely continuous random variables with pdfs $f_1(x), \ldots, f_D(x)$ with corresponding cdfs $F_1(x), \ldots, F_D(x)$, the pdf of the joint pdf $h(\mathbf{x})$, $x \in \mathbb{R}^D$ corresponding to Equation 17.6 is given by

$$h(\mathbf{x}) = f_1(x_1) \ldots f_D(x_D) \frac{\partial^D C}{\partial x_1 \ldots \partial x_D} (F_1(x_1), \ldots, F_D(x_D)). \tag{17.7}$$

where $(\partial^D C / \partial x_1 \cdots \partial x_D)(F_1(x_1), \ldots, F_D(x_D))$ is the pdf corresponding to copula $C(\mathbf{x})$, provided this derivative exists in $[0, 1]^D$.

An important family of copulas are Archimedean copulas, which have a simple analytical form and yet provide a wide variety of modeled dependence structures. An *Archimedean* copula is a copula C, defined as

$$C(\mathbf{x}) = \phi^{-1}(\phi(x_1) + \cdots + \phi(x_D)), \tag{17.8}$$

where the *generator function* $\phi(u)$ is a function satisfying the following properties:

1. $\phi(u)$ is continuous on $[0, 1]$.
2. $\phi(u)$ is decreasing, $\phi(1) = 0$.
3. $\phi(u)$ is convex,

and $\phi^{-1}(u)$ denotes the inverse of $\phi(u)$.

A common way to perform copula parameter estimation is by using its connection with Kendall's tau, which is a ranking correlation coefficient [34]. *Kendall's tau* is a concordance–discordance measure between two independent realizations (X, Y) and $(\hat{X}, \hat{Y})$ from the same cdf $H(x, y)$ defined as

$$\tau = \text{Prob}\{(X - \hat{X})(Y - \hat{Y}) > 0\} - \text{Prob}\{(X - \hat{X})(Y - \hat{Y}) < 0\}.$$

Given realizations (x_l, y_l), $l = 1, \ldots, N$, the empirical estimator of Kendall's tau is given by

$$\hat{\tau} = \frac{4}{N(N - 1)} \sum_{i \neq j} I[x_j \leq x_j] I[y_i \leq y_j] - 1, \tag{17.9}$$

where $I[\cdot]$ denotes the indicator function.

In case of $D = 2$, that is, *bivariate* copulas, by integrating in the definition of τ over the distribution of $(\hat{X}, \hat{Y})$, we get the general relationship between Kendall's τ and the bivariate copula C associated with $H(x_1, x_2)$, expressed by the following Lebesgue–Stieltjes integral:

$$\tau + 1 = 4 \int_0^1 \int_0^1 C(u, v) \, dC(u, v), \tag{17.10}$$

In the specific case of bivariate Archimedean copulas, the relationship is expressed in terms of generator function $\phi(t)$:

$$\tau = 1 + 4 \int_0^1 \frac{\phi(t)}{\phi'(t)} \, dt. \tag{17.11}$$

In the general multivariate case $(D \geq 3)$, Equation 17.10 takes the same form, where $C(u, v)$ is a bivariate copula of the same type* and $\tau = \bar{\tau}$ is the average of $D(D - 1)/2$ consistent Kendall's τ

---

* Recall that by definition Equation 17.8 an Archimedean copula is defined solely by its generator function and, as such, can be defined for any dimensionality factor D.

estimates Equation 17.9 corresponding to all combinations of bivariate marginals (X_{d_1}, X_{d_2}), $0 \le d_1 < d_2 \le D$, see [54]. Equation 17.11 holds with the same modification $\tau = \bar{\tau}$.

The use of Kendall's tau-based parameter estimation, as compared to other copula parameter estimation strategies, is motivated by a relatively small number of samples N available in our experiments [54].

ACKNOWLEDGMENTS

This research was a collaborative effort between the Institut National de Recherche en Informatique et en Automatique (INRIA) Sophia Antipolis—Méditerranée Center, France, and the Department of Biophysical and Electronic Engineering (DIBE) of the University of Genoa, Italy. The work was carried out with the partial financial support of the INRIA, France, and the French Space Agency (CNES). This support is gratefully acknowledged. The authors would like to thank the Italian Space Agency for providing the CSK® image of Piemonte (CSK Product—©ASI—Agenzia Spaziale Italiana—2008. All Rights Reserved). The TerraSAR-X image of Sanchagang (©Astrium GEO-Information Services, 2008) was taken at http://www.infoterra.de/, and the RADARSAT-2 image of Vancouver (©MacDonald, Dettwiler, and Associates Ltd., 2008)—at http://www.radarsat2.info/.

REFERENCES

1. C. Oliver and S. Quegan, *Understanding Synthetic Aperture Radar Images* (2nd edition). NC, USA: SciTech, Raleigh, 2004.
2. C. Tison, J.-M. Nicolas, F. Tupin, and H. Maitre, A new statistical model for Markovian classification of urban areas in high-resolution SAR images, *IEEE Trans. Geosci. Remote Sens.*, 42(10), 2046–2057, 2004.
3. A. Banerjee, P. Burlina, and R. Chellappa, Adaptive target detection in foliage-penetrating SAR images using alpha-stable models, *IEEE Trans. Image Process.*, 8(12), 1823–1831, 1999.
4. R. O. Duda, P. E. Hart, and D. G. Stork, *Pattern Classification*. New York: Wiley Interscience, 2001.
5. P. Mantero, G. Moser, and S. B. Serpico, Partially supervised classification of remote sensing images using SVM-based probability density estimation, *IEEE Trans. Geosci. Remote Sens.*, 43(3), 559–570, 2005.
6. H.-C. Li, W. Hong, Y.-R. Wu, and P.-Z. Fan, An efficient and flexible statistical model based on generalized gamma distribution for amplitude SAR images, *IEEE Trans. Geosci. Remote Sens.*, 48(6), 2711–2722, 2010.
7. G. Moser, J. Zerubia, and S. B. Serpico, SAR amplitude probability density function estimation based on a generalized Gaussian model, *IEEE Trans. Image Process.*, 15(6), 1429–1442, 2006.
8. E. E. Kuruoglu and J. Zerubia, Modelling SAR images with a generalization of the Rayleigh distribution, *IEEE Trans. Image Process.*, 13(4), 527–533, 2004.
9. E. Jakeman and P. N. Pusey, A model for non-Rayleigh sea echo, *IEEE Trans. Antennas Propag.*, 24, 806–814, 1976.
10. A. C. Frery, H.-J. Muller, C. C. F. Yanasse, and S. Sant'Anna, A model for extremely heterogeneous clutter, *IEEE Trans. Geosci. Remote Sens.*, 35(3), 648–659, 1997.
11. G. Moser, S. Serpico, and J. Zerubia, Dictionary-based stochastic expectation maximization for SAR amplitude probability density function estimation, *IEEE Trans. Geosci. Remote Sens.*, 44(1), 188–199, 2006.
12. G. Celeux, D. Chauveau, and J. Diebolt, Stochastic versions of the EM algorithm: An experimental study in the mixture case, *J. Statist. Comput. Simul.*, 55(4), 287–314, 1996.
13. J.-S. Lee and E. Pottier, *Polarimetric Radar Imaging: From Basics to Applications*. New York: CRC Press, 2009.
14. J.-S. Lee, M. R. Grunes, and R. Kwok, Classification of multilook polarimetric SAR imagery based on complex Wishart distribution, *Int. J. Remote Sens.*, 15(11), 2299–2311, 1994.
15. J.-M. Beaulieu and R. Touzi, Segmentation of textured polarimetric SAR scenes by likelihood approximation, *IEEE Trans. Geosci. Remote Sens.*, 42(10), 2063–2072, 2004.
16. A. C. Frery, C. C. Freitas, and A. H. Correia, Classifying multifrequency fully polarimetric imagery with multiple sources of statistical evidence and contextual information, *IEEE Trans. Geosci. Remote Sens.*, 45(10), 3098–3109, 2007.
17. A. P. Doulgeris, S. N. Anfinsen, and T. Eltoft, Classification with a non-Gaussian model for PolSAR data, *IEEE Trans. Geosci. Remote Sens.*, 46(10), 2999–3009, 2008.

18. L. Bombrun, G. Vasile, M. Gay, and F. Totir, Hierarchical segmentation of polarimetric SAR images using heterogeneous clutter models, *IEEE Trans. Geosci. Remote Sens.*, 49(2), 726–737, 2011.

19. Y. Ito and S. Omatu, Polarimetric SAR data classification using competitive neural networks, *Int. J. Remote Sens.*, 19(14), 2665–2684, 1998.

20. K. S. Chen, W. P. Huang, D. H. Tsay, and F. Amar, Classification of multifrequency polarimetric SAR imagery using a dynamic learning neural network, *IEEE Trans. Geosci. Remote Sens.*, 34(3), 814–820, 1996.

21. C. Lardeux, P.-L. Frison, C. Tison, J.-C. Souyris, B. Stoll, B. Fruneau, and J.-P. Rudant, Support vector machine for multifrequency SAR polarimetric data classification, *IEEE Trans. Geosci. Remote Sens.*, 47(12), 4143–4152, 2009.

22. C. T. Chen, K. S. Chen, and J.-S. Lee, The use of fully polarimetric information for the fuzzy neural classification of SAR images, *IEEE Trans. Geosci. Remote Sens.*, 41(9), 2089–2100, 2003.

23. P. R. Kersten, J.-S. Lee, and T. L. Ainsworth, Unsupervised classification of polarimetric synthetic aperture radar images using fuzzy clustering and EM clustering, *IEEE Trans. Geosci. Remote Sens.*, 43(3), 519–527, 2005.

24. J. Morio, F. Goudail, X. Dupuis, P. Dubois-Fernandez, and P. Refregier, Polarimetric and interferometric SAR image partition into statistically homogeneous regions based on the minimization of the stochastic complexity, *IEEE Trans. Geosci. Remote Sens.*, 45(12), 3599–3609, 2007.

25. K. Ersahin, I. Cumming, and R. Ward, Segmentation and classification of polarimetric SAR data using spectral graph partitioning, *IEEE Trans. Geosci. Remote Sens.*, 48(1), 164–174, 2010.

26. G. D. De Grandi, J. S. Lee, and D. L. Schuler, Target detection and texture segmentation in polarimetric SAR images using a wavelet frame: Theoretical aspects, *IEEE Trans. Geosci. Remote Sens.*, 45(11), 3437–3453, 2007.

27. S. R. Cloude and E. Pottier, An entropy based classification scheme for land applications of polarimetric SAR, *IEEE Trans. Geosci. Remote Sens.*, 35(1), 68–78, 1997.

28. J.-S. Lee, M. R. Grunes, E. Pottier, and L. Ferro-Famil, Unsupervised terrain classification preserving polarimetric scattering characteristics, *IEEE Trans. Geosci. Remote Sens.*, 42(4), 722–731, 2004.

29. J.-S. Lee, K. W. Hoppel, S. A. Mango, and A. R. Miller, Intensity and phase statistics of multilook polarimetric and interferometric SAR imagery, *IEEE Trans. Geosci. Remote Sens.*, 32(5), 1017–1028, 1994.

30. S. H. Yueh, J. A. Kong, J. K. Jao, R. T. Shin, and L. M. Novak, K-distribution and polarimetric terrain radar clutter, *J. Electromagn. Waves Appl.*, 3, 747–768, 1989.

31. C. C. Freitas, A. C. Frery, and A. H. Correia, The polarimetric G distribution for SAR data analysis, *Environmetrics*, 16(1), 13–31, 2005.

32. G. Mercier and P.-L. Frison, Statistical characterization of the Sinclair matrix: Application to polarimetric image segmentation, in *Proceedings of IGARSS*, Cape Town, South Africa, pp. III-717–III-720, 2009.

33. Z. Kato, J. Zerubia, and M. Berthod, Unsupervised parallel image classification using Markovian models, *Pattern Recog.*, 32(4), 591–604, 1999.

34. R. B. Nelsen, *An Introduction to Copulas* (2nd edition). New York: Springer, 2006.

35. G. Mercier, S. Derrode, W. Pieczynski, J.-M. Nicolas, A. Joannic-Chardin, and J. Inglada, Copula-based stochastic kernels for abrupt change detection, in *Proceedings of IGARSS*, Denver, USA, pp. 204–207, 2006.

36. G. Mercier, G. Moser, and S. Serpico, Conditional copulas for change detection in heterogeneous remote sensing images, *IEEE Trans. Geosci. Remote Sens.*, 46(5), 1428–1441, 2008.

37. F. Galland, J.-M. Nicolas, H. Sportouche, M. Roche, F. Tupin, and P. Refregier, Unsupervised synthetic aperture radar image segmentation using Fisher distributions, *IEEE Trans. Geosci. Remote Sens.*, 47(8), 2966–2972, 2009.

38. K. Song, Globally convergent algorithms for estimating generalized gamma distributions in fast signal and image processing, *IEEE Trans. Image Process.*, 17(8), 1233–1250, 2008.

39. M. Abramowitz and I. Stegun (eds), *Handbook of Mathematical Functions*. New York: Dover, 1964.

40. V. Krylov, G. Moser, S. Serpico, and J. Zerubia, Dictionary-based probability density function estimation for high-resolution SAR data, in *Proceedings of SPIE*, 7246, San Jose, USA, p. 72460S, 2009.

41. G. Frahm, M. Junker, and A. Szimayer, Elliptical copulas: Applicability and limitations, *Stat. Probab. Lett.*, 63, 275–286, 2003.

42. D. Huard, G. Évin, and A.-C. Favre, Bayesian copula selection, *Comput. Stat. Data Anal.*, 51(2), 809–822, 2006.

43. E. Lehmann and J. Romano, *Testing Statistical Hypotheses* (3rd edition). New York: Springer, 2005.

44. J. Besag, On the statistical analysis of dirty pictures, *J. Royal Stat. Soc. B*, 48, 259–302, 1986.

45. C. M. Bishop, *Pattern Recognition and Machine Learning*. New York: Springer, 2006.

46. S. Geman and D. Geman, Stochastic relaxation, Gibbs distributions, and the Bayesian restoration of images, *IEEE Trans. Pattern Anal. Mach. Intell.*, 6, 721–741, 1984.
47. P. van Laarhoven and E. Aarts, *Simulated Annealing: Theory and Applications*. Dordrecht: Kluwer, 1992.
48. W. Hastings, Monte Carlo sampling method using Markov chains and their applications, *Biometrika*, 57, 97–109, 1970.
49. M. Berthod, Z. Kato, S. Yu, and J. Zerubia, Bayesian image classification using Markov random fields, *Image Vis. Comput.*, 14, 285–295, 1996.
50. W. Michiels, E. Aarts, and J. Korst, *Theoretical Aspects of Local Search*. Berlin: Springer, 2007.
51. M. Stephens, Test of fit for the logistic distribution based on the empirical distribution function, *Biometrika*, 66, 591–595, 1979.
52. C. Genest and J.-C. Boies, Detecting dependence with Kendall plots, *Am. Stat.*, 57(4), 275–284, 2003.
53. Y. Boykov, O. Veksler, and R. Zabih, Fast approximate energy minimization via graph cuts, *IEEE Trans. Pattern Anal. Mach. Intell.*, 23(11), 1222–1239, 2001.
54. I. Kojadinovic and J. Yan, Comparison of three semiparametric methods for estimating dependence parameters in copula models, *Insur. Math. Econ.*, 47, 52–63, 2010.

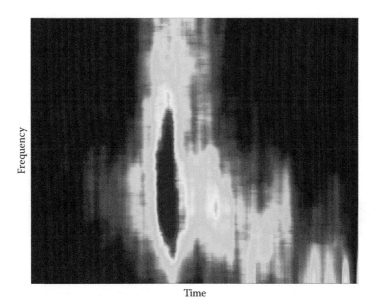

FIGURE 2.12 Time–frequency representation of earthquake accelerogram data.

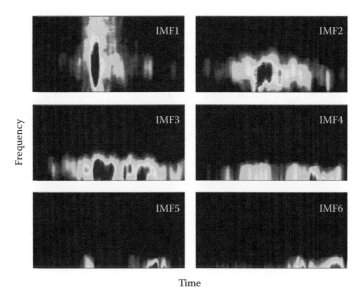

FIGURE 2.13 Time–frequency representations of first six IMFs produced by RCEMD operating on earthquake accelerogram data.

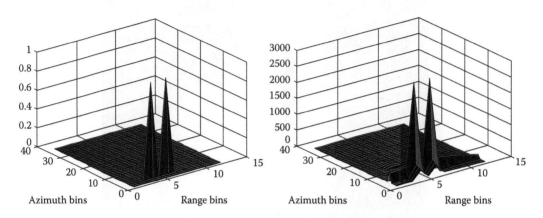

FIGURE 5.5 (a) Result of CS-based reconstruction. (b) Result of traditional reconstruction.

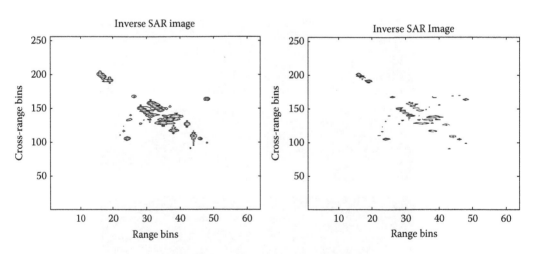

FIGURE 5.6 (a) Traditional reconstruction (30% samples). (b) CS-based reconstruction (30% samples).

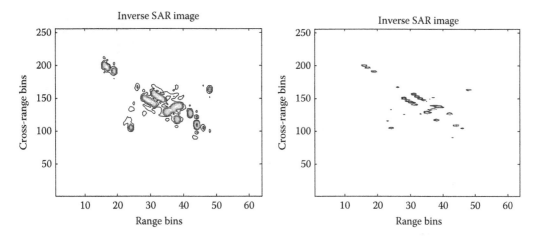

FIGURE 5.7 (a) Traditional reconstruction (12% samples). (b) CS-based reconstruction (12% samples).

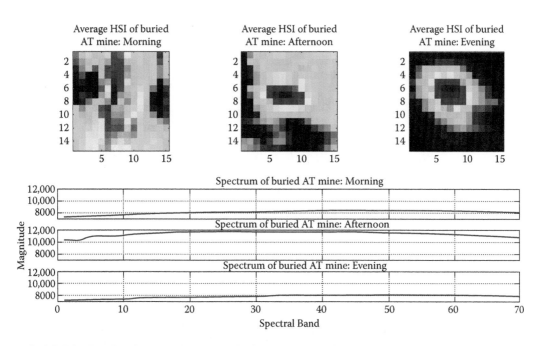

FIGURE 6.1 Top: Band-averaged HSI chips for buried antitank mines at three different locations at morning, afternoon, and evening times. Bottom: Spectrum of center pixel for each of the HSI chips shown in the top figure.

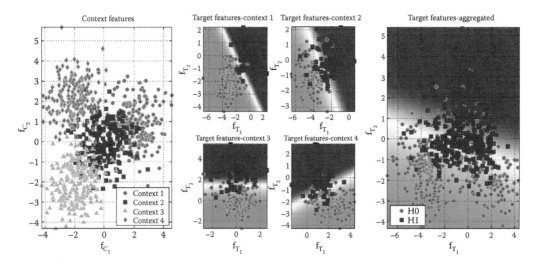

FIGURE 6.3 A synthetic data example of a context-dependent classification problem. Left: Two-dimensional (2D) context features $\mathbf{f_C}$, colored by context. Center: The context-specific linear classification problems in target features $\mathbf{f_T}$. Right: Linear classification boundaries obtained without incorporating contextual information.

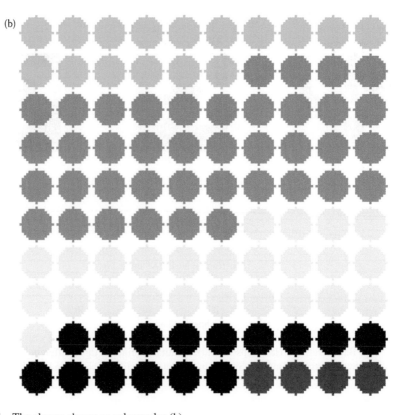

FIGURE 7.4 The classes shown as color codes (b).

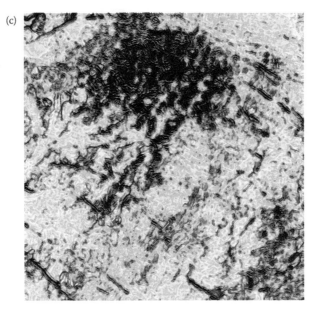

(c)

FIGURE 7.6 A SAR image area. (c) Image with the correlations D with the best-matching vectors of three from the five classes (1, 3, 5) on the R, G, and B channels, respectively.

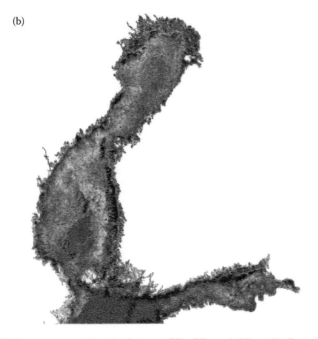

(b)

FIGURE 7.8 An RGB image representing the features GF_1, GF_2, and GF_3 on R, G, and B channels, respectively (b). GF_4 has a large correlation with GF_2.

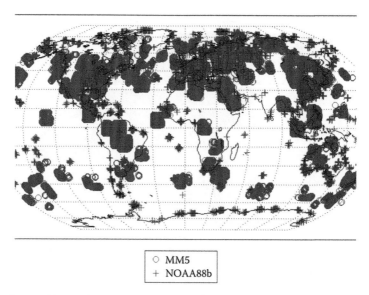

FIGURE 10.4 Geographical locations of the pixels contained in the Mesoscale Model, version 5 (MM5) and NOAA88b data sets. Global coverage including a variety of land and ocean pixels is achieved by each set.

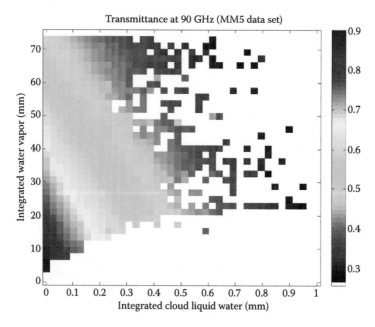

FIGURE 10.9 Atmospheric transmittance at nadir incidence at ~90 GHz as calculated from the MIT MM5 data set. Transmittance values are averaged over bins of integrated water vapor and integrated cloud liquid water content.

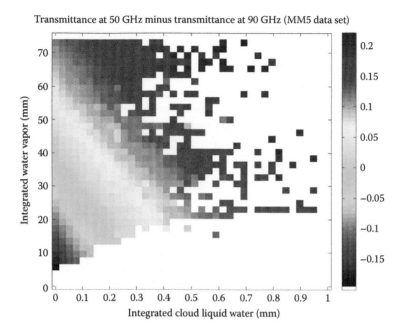

FIGURE 10.10 Atmospheric transmittance at nadir incidence at 50 GHz minus that at 90 GHz as calculated from the MIT MM5 data set. Transmittance values are averaged over bins of integrated water vapor and integrated cloud liquid water content.

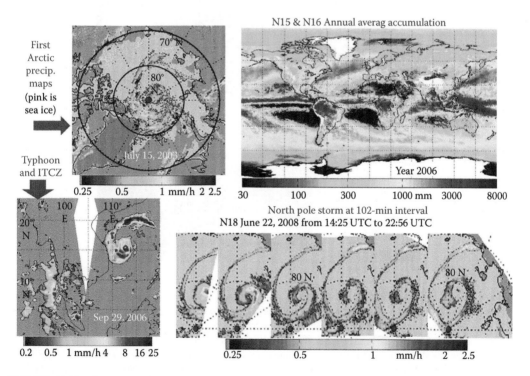

FIGURE 11.13 Examples of AMP-3 precipitation estimates. Light pink shows snow-covered land and sea ice. (Adapted from C. Surussavadee and D. H. Staelin, *IEEE Geosci. Remote Sens. Lett.*, 7(3), 440–444, 2010. With permission. Copyright 2010 IEEE.)

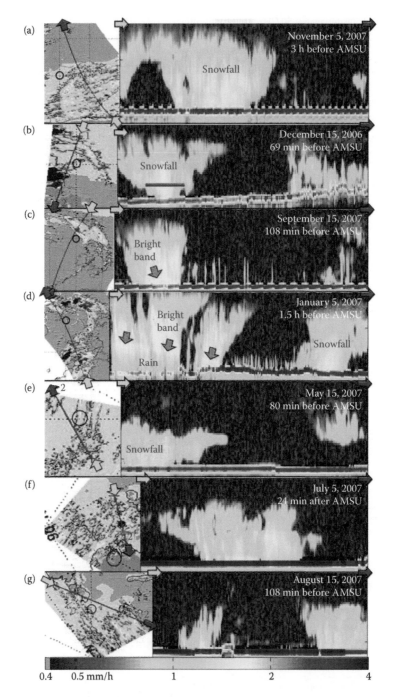

FIGURE 11.14 Comparisons of AMP-3 surface-precipitation retrievals and CloudSat echo strengths. The CloudSat orbit is red and bounded by red and green arrows; its peak illustrated echo altitude is ~12 km. Pink, dark pink, and black pixels indicate surface snow or ice, excessive surface elevation, and untrustworthy retrievals, respectively. Gray is land or water. The 200-km-diameter black circles are centered at (a) 60°S/20°E over Antarctic ice shelf, (b) 50°N/110°W over Canadian snow, (c) 60°N/10°W near a North Atlantic front, (d) 60°N/70°W over Labrador coastline, and over Arctic ice pack at (e) 82°N/180°E, (f) 82°N/30°W, and (g) 82°N/90°E. (From C. Surussavadee and D. H. Staelin, Satellite retrievals of arctic and equatorial rain and snowfall rates using millimeter wavelengths, *IEEE Trans. Geosci. Remote Sens.*, 47(11), 3697–3707, 2009. With permission. Copyright 2009 IEEE.)

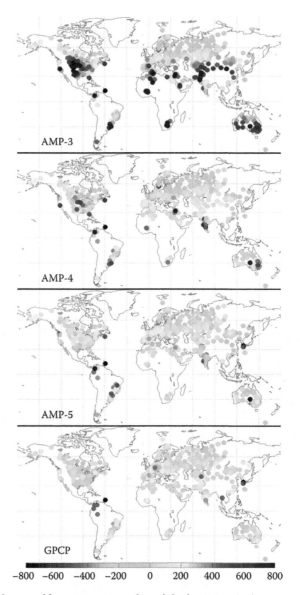

FIGURE 11.19 Global maps of 2-year mean annual precipitation errors (estimate–gauge) for AMP-3, AMP-4, AMP-5, and (after removal of gauge wind-loss corrections) GPCP for 509 wind-loss uncorrected gauge observations. (From C. Surussavadee and D. H. Staelin, Correcting microwave precipitation retrievals for near-surface evaporation, *IEEE Int. Geosci. Remote Sens. Symp. Proc.*, pp. 1312–1315, July 2010. With permission. Copyright 2010 IEEE.)

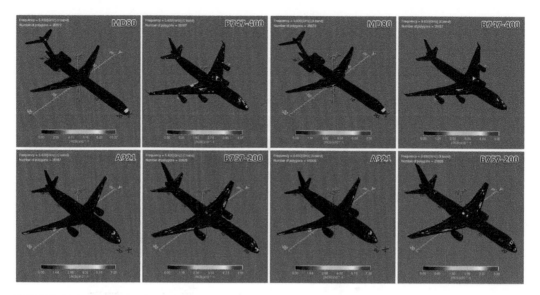

FIGURE 12.4 Computed RCS of commercial aircrafts for C-band HH polarization (left) and X-band VV polarization (right).

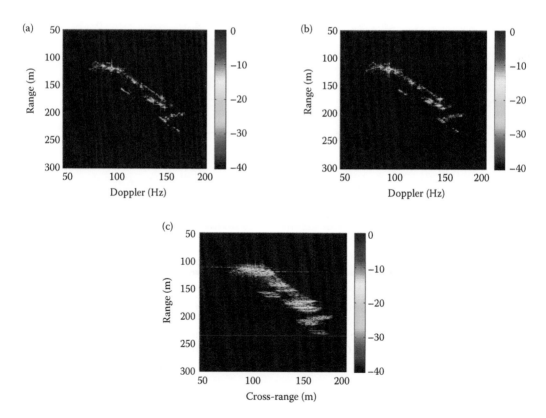

FIGURE 14.13 ISAR image of ship1 obtained by using the whole observation time and the IRD (a) and the IOK (b), and SAR image of ship 1 (c).

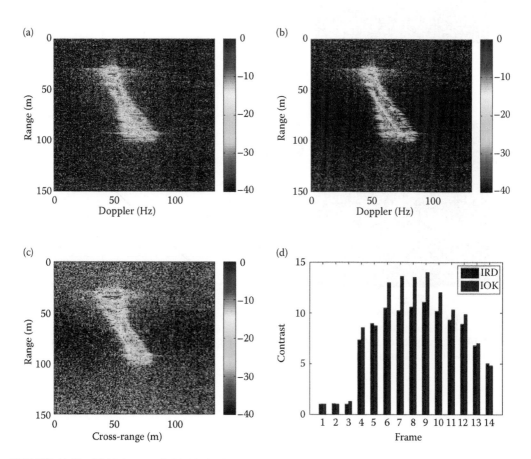

FIGURE 14.18 ISAR image of ship10 obtained by using the IRD (a), the IOK (b), SAR image of the target (c), and IC values relative to each frame (d).

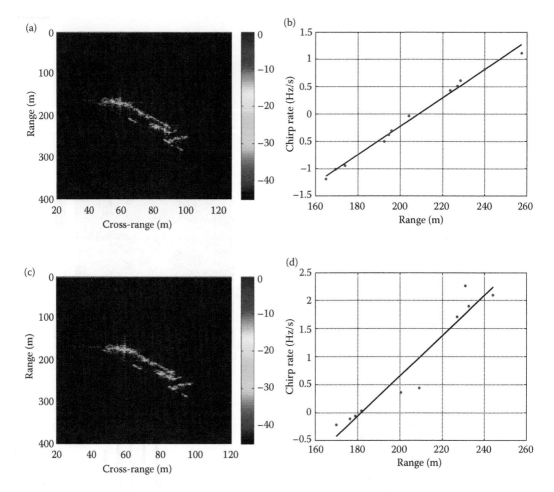

FIGURE 14.21 Fully scaled ISAR images obtained by applying the cross-range scaling technique and chirp rate estimates—IRD (a, b) and IOK (c, d).

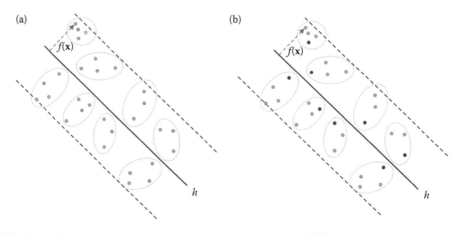

FIGURE 15.3 Comparison between the samples selected by (a) the CBD technique presented in Ref. [15] and (b) the ECBD technique.

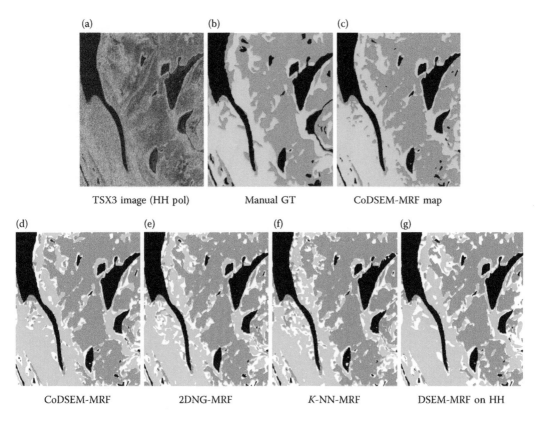

FIGURE 17.2 (a) HH channel of dual-pol TSX3 image (©Astrium), (b) manually created GT, (c) CoDSEM-MRF classification map (water ■, wet soil ■, and dry soil ▫). Classification maps referenced to the GT: (d) CoDSEM-MRF, (e) 2DNG-MRF, (f) *K*-NN-MRF, and (g) DSEM-MRF on the HH pol image (correctly classified water ■, wet soil ■, dry soil ▫, misclassification of all types ■).

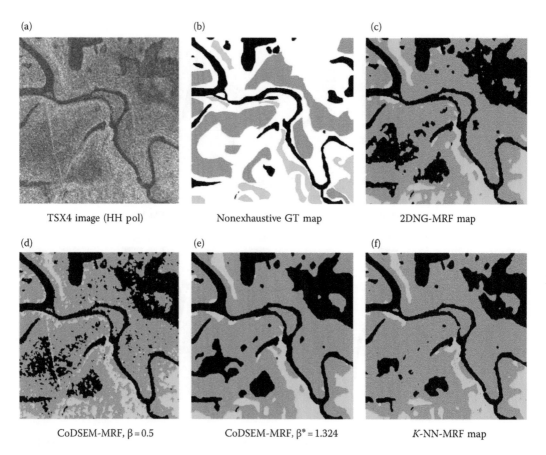

(a)

TSX4 image (HH pol)

(b)

Nonexhaustive GT map

(c)

2DNG-MRF map

(d)

CoDSEM-MRF, β = 0.5

(e)

CoDSEM-MRF, β* = 1.324

(f)

K-NN-MRF map

FIGURE 17.5   (a) HH channel of the dual-pol TSX4 image (©Astrium), (b) nonexhaustive GT map (water ■, wet soil ■, dry soil ■, and outside GT ■). Classification maps: (c) 2DNG-MRF, CoDSEM-MRF with (d) manually set β = 0.5 and (e) automatically estimated β* = 1.324, (f) K-NN-MRF (same colors).

(a)

(b)

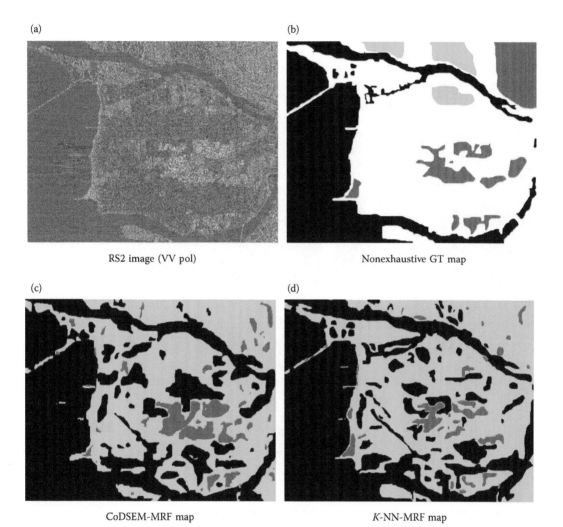

RS2 image (VV pol)

Nonexhaustive GT map

(c)

(d)

CoDSEM-MRF map

K-NN-MRF map

FIGURE 17.7 (a) VV channel of the quad-pol RS2 image (RADARSAT-2 Data and Products ©MacDonald, Dettwiler and Associates Ltd., 2008—All Rights Reserved), (b) nonexhaustive GT map (water ■, urban ■, vegetation ■, outside GT ■). Classification maps: (c) CoDSEM-MRF, (d) K-NN-MRF (same colors).

Landsat 5 TM (bands 4-3-2)   Envisat ASAR/ER S-2 (May-Jun.-Jul.)

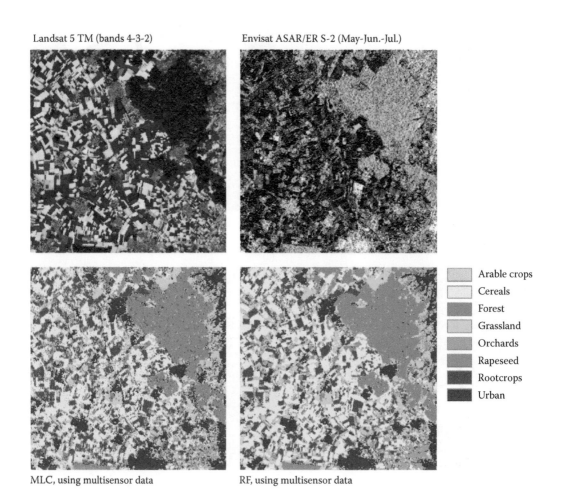

MLC, using multisensor data   RF, using multisensor data

Arable crops
Cereals
Forest
Grassland
Orchards
Rapeseed
Rootcrops
Urban

FIGURE 18.1   Data set and classification results using MLC and RF.

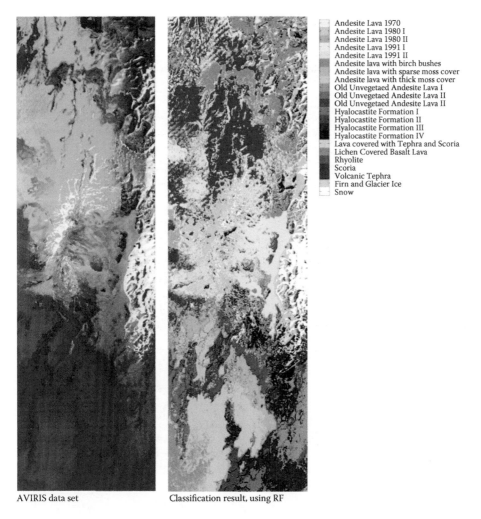

Andesite Lava 1970
Andesite Lava 1980 I
Andesite Lava 1980 II
Andesite Lava 1991 I
Andesite Lava 1991 II
Andesite lava with birch bushes
Andesite lava with sparse moss cover
Andesite lava with thick moss cover
Old Unvegetaed Andesite Lava I
Old Unvegetaed Andesite Lava II
Old Unvegetaed Andesite Lava II
Hyalocastite Formation I
Hyalocastite Formation II
Hyalocastite Formation III
Hyalocastite Formation IV
Lava covered with Tephra and Scoria
Lichen Covered Basalt Lava
Rhyolite
Scoria
Volcanic Tephra
Firn and Glacier Ice
Snow

AVIRIS data set   Classification result, using RF

FIGURE 18.2 AVIRIS data set and classification result using RF.

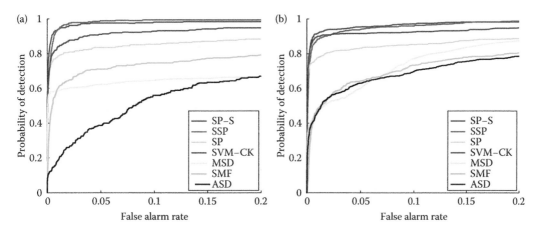

FIGURE 19.2 ROC curves using various detection and classification algorithms for (a) DR-II and (b) FR-I using local background dictionary (dual window approach), $N_t = 18$ and $N_b = 216$.

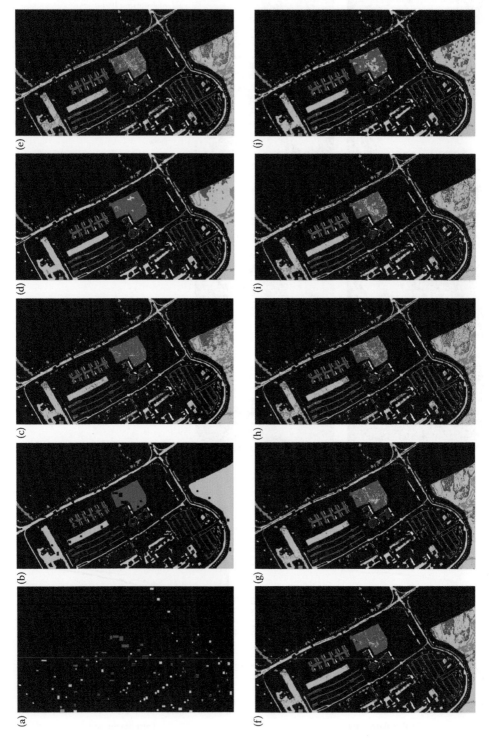

FIGURE 19.6 For the University of Pavia image: (a) training set and (b) test set. Classification maps obtained by (c) SVM, (d) SVM-CK, (e) SP, (f) SP-S, (g) SSP, (h) OMP, (i) OMP-S, and (j) SOMP.

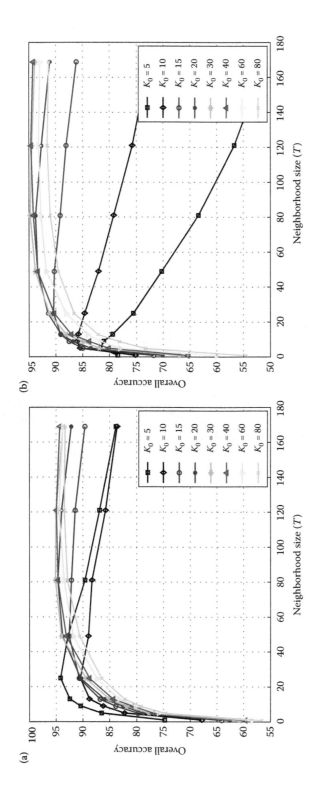

FIGURE 19.11 Effect of the sparsity level K_0 and size of neighborhood T for Indian Pines. (a) SOMP and (b) SSP.

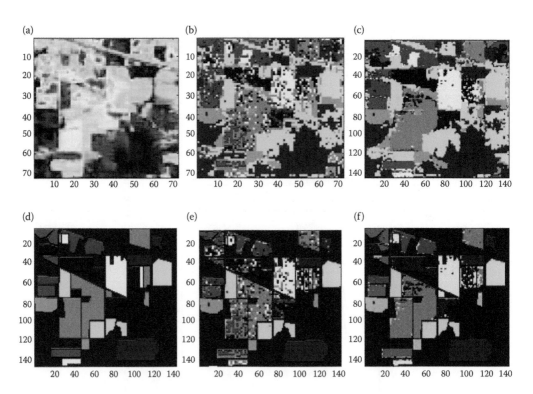

FIGURE 20.4 (a) AVIRIS Indian Pine original image, band 30. (b) Classification map obtained by using SVM. (c) Classification map obtained by using the proposed method. (d) Ground-truth map. (e), (f) Classification of the test data, in the case of SVM and the proposed method, respectively.

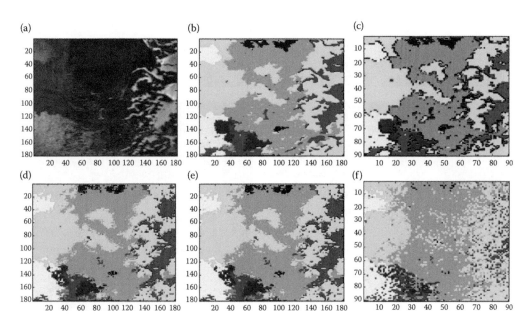

FIGURE 20.5 (a) Hekla data set, band 80. (b), (c) Ground truth in the high-resolution and low-resolution image, respectively (mixed pixels are shown in black). (d), (e) Classification maps obtained with the proposed method, before and after spatial regularization. (f) Classification map obtained with a classic SVM. (Adapted from A. Villa et al., 2011. *IEEE Journal of Selected Topics in Signal Processing*, 5:521–33, © 2011 IEEE.)

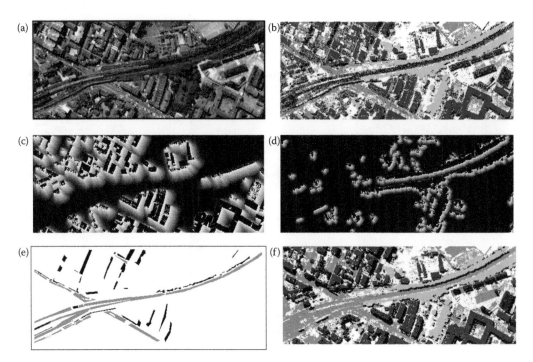

FIGURE 22.12 Classification of the Pavia image without and with using spatial contextual information. The classes in the classification and ground truth maps are asphalt (gray), shadow (black), tiles (red), and trees (green). The ground truth is produced by visual inspection. (a) True-color image, (b) classification map using decision rule (22.15) without spatial information, (c) directional landscape with respect to the detected tiles, (d) directional landscape with respect to the detected trees, (e) ground truth map, and (f) classifcation map using decision rule (22.17) with spatial information.

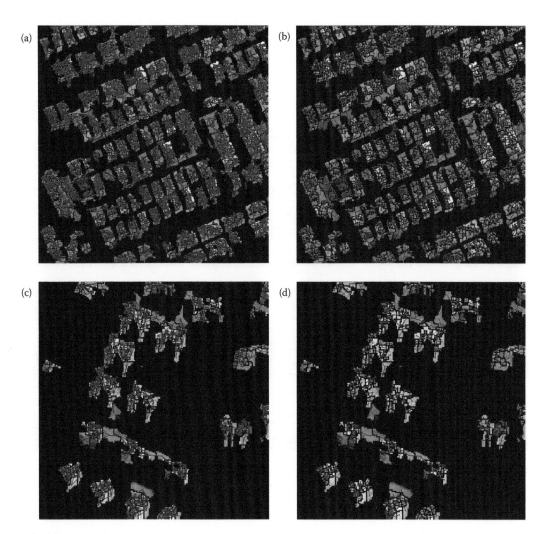

FIGURE 22.15 Examples of graph construction and minimum spanning tree-based clustering. The removed edges are colored in red. (a) Graph for Antalya 1, (b) clustering for Antalya 1, (c) graph for Antalya 2, and (d) clustering for Antalya 2.

FIGURE 22.16 Building detection results. The detected buildings are highlighted in red. (a) Results for Antalya 1, (b) results for Antalya 2, (c) results for Antalya 3, (d) results for Antalya 4, (e) results for Antalya 5, and (f) results for Antalya 6.

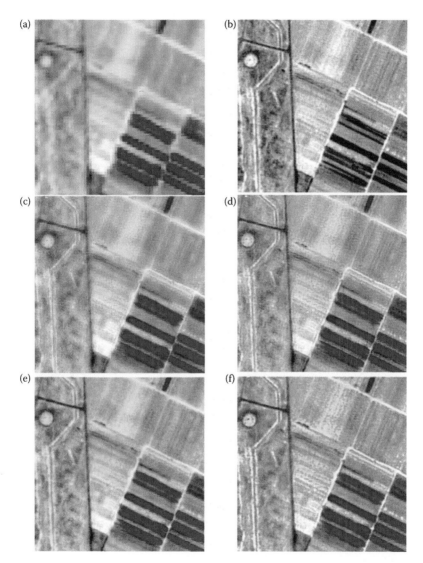

FIGURE 24.6 (a) Multispectral SPOT 5 image (original 10 m). (b) Panchromatic SPOT 5(2.5 m). (c) Fusion result with the decimated HT method ($N = 2$). (d) Fusion result with the undecimated HT ($N = 2$). (e) Fusion with eFIHS. (f) AWL false color composite scheme for displaying SPOT multispectral images is achieved with R = XS3 (NIR band), G = XS2 (red band), and B = XS1 (green band).

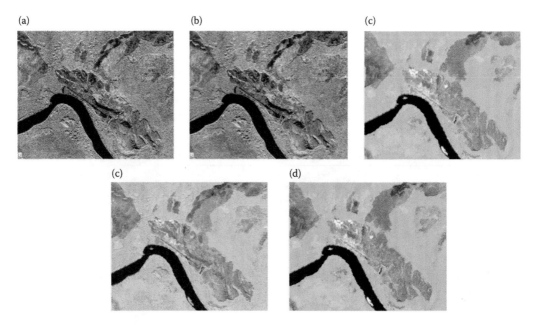

FIGURE 24.9 (a) Original SAR AeS-1 images with speckle. (b) SAR AeS-1 restored image. (c) Landsat 7 TM image. (d) uHT fused image. (e) GIM fused image.

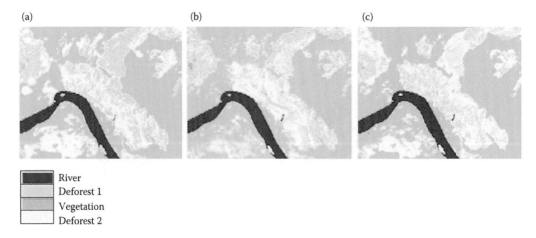

River
Deforest 1
Vegetation
Deforest 2

FIGURE 24.10 Four-class ISODATA classification applied to (a) original MS image, (b) uHT SAR-MS fused image, and (c) GIM fused image.

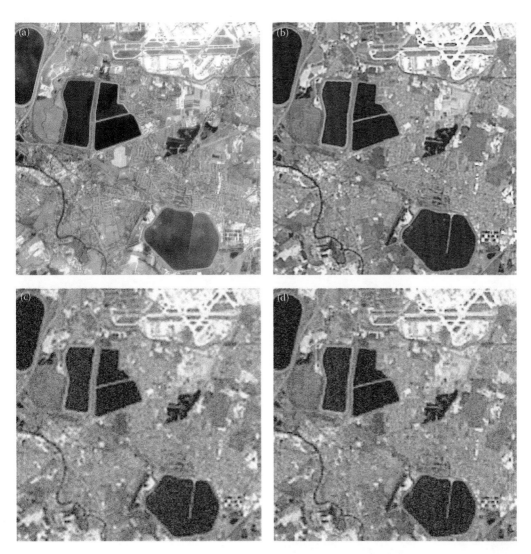

FIGURE 25.4 Comparison with state-of-the-art pansharpening techniques. (a) Pan image; (b) Reference image: original Landsat image; (c) low-resolution Landsat image with additive noise ($\sigma = 15$); (d)–(f) fused results of AWLP, GLP-CBD, and the proposed method; (g)–(i) zoomed results of (d)–(f). (Copyright [2009] IEEE.)

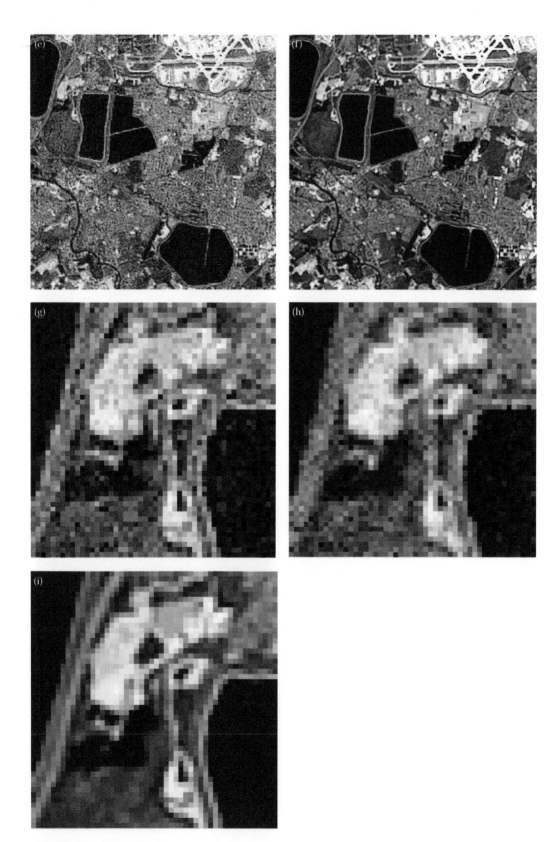

FIGURE 25.4 Continued.

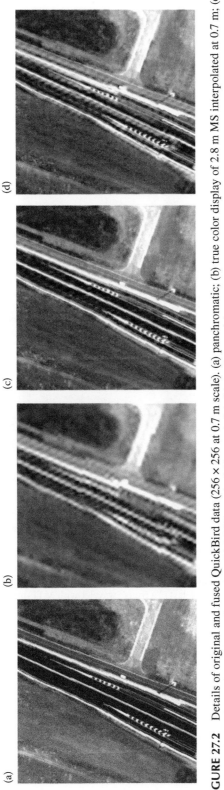

FIGURE 27.2 Details of original and fused QuickBird data (256 × 256 at 0.7 m scale). (a) panchromatic; (b) true color display of 2.8 m MS interpolated at 0.7 m; (c) enhanced Gram–Schmidt; (d) "à trous" wavelet with MTF-matched filter and global injection model.

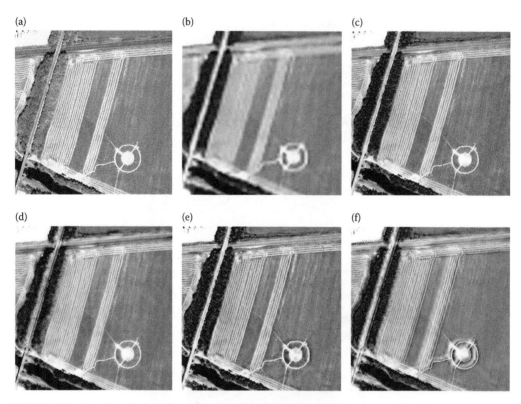

FIGURE 27.3 Details of original and fused IKONOS data (256 × 256 at 1 m scale). (a) panchromatic; (b) true color display of 4 m MS interpolated at 1 m; (c) enhanced Gram–Schmidt; (d) enhanced Gram–Schmidt with data misaligned by 4 pels at panchromatic scale (1 pel at MS scale); (e) "à trous" wavelet with MTF-matched filter and global injection model; (f) "à trous" wavelet with MTF-matched filter and global injection model with data misaligned by 4 pels at panchromatic scale.

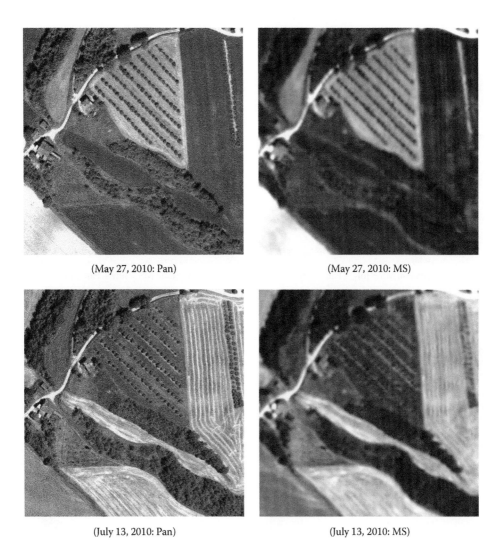

(May 27, 2010: Pan)   (May 27, 2010: MS)

(July 13, 2010: Pan)   (July 13, 2010: MS)

FIGURE 27.4 GeoEye-1 MS + Pan (512 × 512 at 0.5 m). Pan and MS images acquired both on May 27th 2010; Pan and MS images acquired both on July 13, 2010.

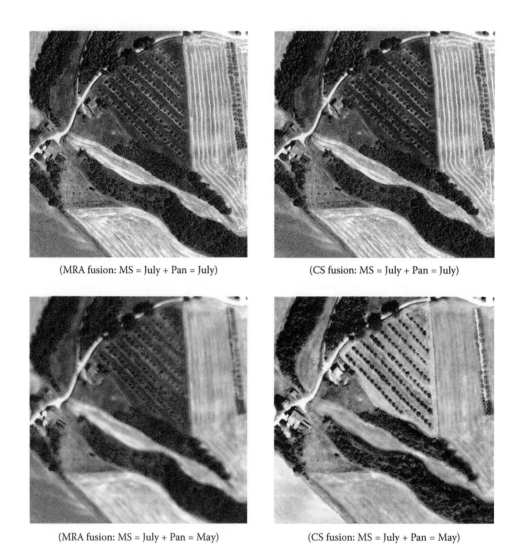

(MRA fusion: MS = July + Pan = July)   (CS fusion: MS = July + Pan = July)

(MRA fusion: MS = July + Pan = May)   (CS fusion: MS = July + Pan = May)

FIGURE 27.5   Details of pansharpened GeoEye-1 data (512 × 512 at 0.5 m scale).

18 Random Forest Classification of Remote Sensing Data

*Björn Waske, Jon Atli Benediktsson,
and Johannes R. Sveinsson*

CONTENTS

18.1 INTRODUCTION

Land cover classifications based on ensemble classifiers have been performed successfully during the recent years and seem particularly interesting for multisource and high-dimensional data sets. Contrary to standard classifiers, which are based on one single classifier to obtain a decision, ensemble methods to train several classifiers and combine their results through a voting process. Many ensemble classifiers have been proposed [1,2] and two strategies exist to construct an ensemble: (1) a combination of different classifier algorithms and (2) a combination of variants of the same classifier. Benediktsson and Kanellopoulos [3], for example, used a neural network and a statistical classifier for combining synthetic aperture radar (SAR) and multispectral data. The two data sets undergo a separate classification and the individual outputs are combined by decision fusion, which can be defined as a strategy for combining information from different data sources. Waske and Benediktsson [4] use a decision fusion strategy that is based on support vector machines (SVMs) for the classification of multitemporal SAR data and multispectral imagery.

However, most applications combine a set of independent variants of the same classifier, the so-called base classifier. Often a simple decision tree (DT) is used, because of the fast training time and simple handling. Nevertheless, the use of more sophisticated methods for ensembles was investigated [5,6]. Bagging [7], boosting [8], and random feature selection [9] are well-known methods for the generation of a classifier ensemble. Bootstrap aggregating (bagging) describes the generation of different training sets by a random selection of *n* samples from a training set of same size, that is, the selection is performed with replacement. Therefore, a training sample can be selected several times and perhaps other samples are not considered in a particularly training set. Afterward, the base classifier is trained on different training sets, resulting in various classifier outputs. The final classification map is generated by combining the individual outputs, often by a simple majority vote. In contrast to random sampling, boosting adaptively changes the distribution of the training samples during the training process by reweighting. After each individual classification, misclassified samples are assigned a higher weight than those classified correctly. The next base classifier within the ensemble is trained by the newly distributed samples. Boosting achieves promising classification accuracies and can reduce the variance and the bias of the classification. However, boosting usually does not perform well in terms of accuracies when the data are characterized by noise, such as SAR data. Waske and Braun [10], for example, classified multitemporal SAR data, using different classifiers. Although the results clearly show that a boosted DT increases the classification accuracy when compared with a single DT, other ensemble methods can outperform the boosting technique in terms of accuracy. Moreover, the computation time is relatively slow when compared with bagging, because boosting generates the different classifiers in a sequential procedure, while bagging can be performed simultaneously. Besides the modification of the training data, the manipulation of the input feature space, for example, by a random selection of features, is another concept for constructing a classifier ensemble. It has been shown that this approach can be superior to bagging and boosting [9,11] and is interesting for classifying hyperspectral and multitemporal SAR data [6,12]. Breiman's Random Forests (RFs) [13] use DT classifiers and the two latter methods, bagging and random feature selection. Each tree within the forest is trained on a subset of the original training samples. In addition, each split node of the DT is determined, using only a randomly selected feature subset of the input data. The size of the feature subset is user defined, and often set to the square root of the number of input features. Finally, a simple majority vote is used to create the classification result.

The method was used successfully for classification of multisource [14], multitemporal and multisensor [10,15], and hyperspectral imagery [16–18]. In Gislason et al. [14], for example, RF was applied to a multisource data set, containing a multispectral Landsat image and topographical data. RF outperforms a single DT and comparable to other ensembles methods in terms of accuracy. Waske and van der Linden [15] used RF, among others, to combine SAR and multispectral data from different segmentation scales. In this study, RF performs similar or can even outperform more sophisticated methods, such as SVMs in terms of accuracy. Although more complex approaches such as SVM may achieve higher accuracies on hyperspectral imagery when compared with RF, the approach is also attractive. RF outperforms methods such as a Gaussian maximum likelihood classifier (MLC) and spectral angle mapper (SAM) in terms of accuracy. Moreover, RF performs well, even with a limited number of training samples and is a relatively simple method that is easy to handle, because it mainly depends on only two user-defined values (i.e., the number of selected features and the number of DT). Overall, RF is an interesting and powerful classifier, particularly in the context of multisource and high-dimensional data sets. The chapter is organized as follows. First, RF concept is introduced. Then, two experimental studies are presented, using multisensor and hyperspectral remote sensing data. Finally, conclusions are given in Section 18.4.

18.2 THE RANDOM FOREST CLASSIFIER

18.2.1 BACKGROUND

A RF classifier is an ensemble that consists of a large number of diverse DT, all aiming on the separation of the same classes. The set of diverse DT is obtained by three factors: (1) each DT is trained using a random subset of the training samples (in the same way as bagging), (2) the split rule at each mode of a tree is determined, using only a randomly selected feature subset of size n, which is usually much smaller than the total number of features N, and (3) every DT is fully grown, that is, the DT is not pruned. Consequently, several different classification results are generated, each obtained by randomly chosen subset of the input data (i.e., training samples and input features). A simple majority vote is used to create the final classification result. The computational time of the individual DT classifier is simplified, by reducing the number of features at each split. Therefore, RFs are adequate to handle high-dimensional data sets. Moreover, the computation time is reduced when compared with bagging and boosting, because RF uses only subsets of the input data [14]. The user-defined value m should be ideally defined small enough to generate an independent set of DT, but large enough to address the classification problem. It is interesting to underline that increasing the correlation between individual DT by increasing m increases the error rate of the RF, whereas increasing classification accuracy of every DT by increasing m can *decrease* the error rate of the ensemble [19]. However, often the value is set approximately to the square root of the number of input features [14,15,18].

18.2.2 PARAMETERS DERIVED FROM RF

Besides classification, RF provides additional parameters, such as the out-of-bag error (OOB-error), the variable importance (VI), and the proximities. This information can be used for further analysis.

18.2.2.1 Out-of-Bag Error

The OOB-error provides an "independent" evaluation of the classification accuracy. While each DT is trained approximately on a randomly chosen 2/3 of the whole training sample set, 1/3 is left out for a specific training set. To estimate the OOB-error, these out-of-bag samples are classified by that particular tree. The final OOB-error is derived by the classification error of all OOB samples, averaged over the total number of DT in the forest. This error estimate has been shown to be unbiased in many tests [13,20].

18.2.2.2 Variable Importance

RF can provide an estimation of the VI, that is, the impact of a variable on the classification accuracy. This seems particularly interesting for multisource applications, to evaluate the impact of the different sources on the classification accuracy [14,19]. The values in the OOB samples of variable i are randomly permuted for each classifier and the biased accuracy is compared with the accuracy, achieved with the nonpermuted training data. An increased OOB-error indicates the importance of variable m. For a single DT, the OOB cases are classified. Afterward, the values of variable i are randomly permuted in the OOB cases. Now, subtract the number of correct votes achieved by the randomly permuted data from the number of correctly cast votes for the original unbiased data. An increased error indicates the importance of variable i. The average of this value over all DT in the forests is the raw importance score for the VI.

18.2.2.3 Proximities

After the RF is generated and used for classification, RF can derive a measure of proximity between two training samples (i.e., pixels), by counting how many individual DTs in the two pixels end up in the same terminal node. If sample k and l are in the same terminal node, their proximity is increased by one (this is normalized by dividing with the number of trees in the ensemble). Finally, the so-called proximity matrix is provided, which can be used to visualize high-dimensional data [7,21]. Moreover, because the proximities indicate the "distance" between samples, this measure can be used to detect outliers in the training data [14,17].

18.2.3 RF AND DECISION TREE IMPLEMENTATIONS

Different implementations of RF exist, using two types of tree-like classifier as base classifiers: (1) classification and regression trees (CART) classifier and (2) binary hierarchical classifier (BHC). Whereas a CART partitions the input data, BHC trees partition the labels (i.e., the output) [19]. Most applications perhaps use a CART classifier; however, the use of BHC trees for RF is possible [16,19]. In Ref. [19], the two implementations were used for the classification of hyperspectral and multi-source remote sensing data and both approaches performed very well in experiments. Although, RF-BHC can achieve higher overall accuracies when compared with an RF-CART, the latter approach is interesting and feasible. The RF-CART achieves accurate classification results as well. Moreover, the authors underlined that an RF-BHC can become instable, because the partitioning algorithm used for the generation of BHC trees can fail to converge in the context of "difficult" classification problems, and thus no BHC classifier can be realized [19]. Based on these facts one may argue that the RF-CART approach is more reliable. A detailed description of the general concept of DT is given in Ref. [22]. A brief summary of CART-like trees is given below.

A CART consists of a root node, which includes all training samples, split nodes, and the final leave nodes, indicating the different classes. After a tree is grown, the unknown sample is handed-on from one split node to the other, following the rules at the nodes. Each sample is assigned to the class that corresponds to the final leaf node in which it finally ends up. Although various methods for the definition of the split nodes have been introduced (e.g., Zambon et al., 2006), they are generally based on the measurement of impurity of the data within potential nodes. If all samples within a potential node belong to the same class, the node is pure and the measurement value is 0, whereas the value is large, if the samples are equally distributed over all classes. In the RF, the Gini index [23] is used. The Gini index finds the largest homogeneous group within the training data and discriminates it from the remaining training samples by measuring the impurity at a split node [24].

The Gini index is described as

$$\text{Gini}(t) = \sum_{i=1}^{L} p_{\omega i}(1 - p_{\omega i}) \tag{18.1}$$

with $p_{\omega i}$, as the probability or the relative frequency of class ω_i at node t, which is given by

$$p_{\omega i} = \frac{j_{\omega i}}{j} \tag{18.2}$$

with $j_{\omega i}$ being the number of samples belonging to class ω_i and j the total number of samples within the training set.

By using an impurity measurement as the Gini index, it is possible to define an adequate split rule, that is, resulting in the purest possible descendent nodes. Like other classifiers, a tree classifier can be easily overfitted and often pruning methods are applied after the tree is fully grown. These methods eliminate inefficient and weak branches, resulting in a more compact DT. However, to generate a set of diverse DT, RF is usually constructed with unpruned trees.

18.3 EXPERIMENTAL RESULTS

18.3.1 CLASSIFICATION OF MULTISENSOR IMAGERY

18.3.1.1 Data Set and Methods

The first experiment is aiming at the classification of agricultural crops using multitemporal SAR data. The almost flat study is located near Bonn, Germany. The region is dominantly used for agriculture

Landsat 5 TM (bands 4-3-2)   Envisat ASAR/ERS-2 (May-Jun.-Jul.)

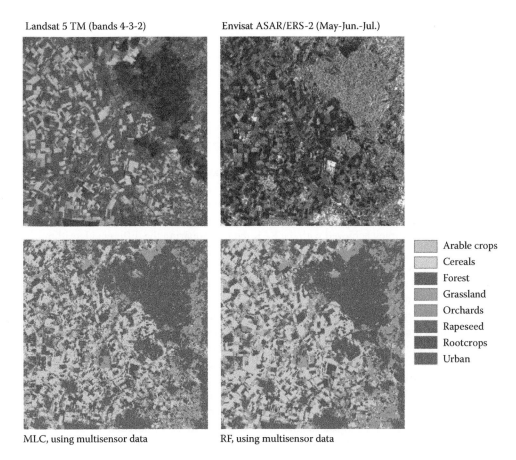

	Arable crops
	Cereals
	Forest
	Grassland
	Orchards
	Rapeseed
	Rootcrops
	Urban

MLC, using multisensor data   RF, using multisensor data

FIGURE 18.1   (**See color insert.**) Data set and classification results using MLC and RF.

and characterized by typical spatial patterns caused by differences in the phenology of planted crops. The data set includes Envisat ASAR and ERS-2 SAR images between the period of April and September 2005 (Figure 18.1). The SAR data was preprocessed and calibrated to backscatter intensity following common procedures. The imagery contains different polarizations and data from different tracks and swaths. Eight land cover classes were considered for the classification: *Arable crops*, *Cereals*, *Forest*, *Grassland*, *Orchard*, *Rapeseed*, *Rootcrops*, and *Urban*. The training data sets consist of 150 randomly selected samples per class, while an independent test set with 500 samples/class was generated. The results achieved by RF are compared to the classification outputs of a common Gaussian MLC and an SVM. The classification of the SVM was performed with a Gaussian kernel and the one-against-one (OAO) multiclass strategy, using imageSVM. ImageSVM is a freely available IDL/ENVI implementation (http://www.hu-geomatics/imageSVM) that uses the LIBSVM approach (http://www.csie.ntu.edu.tw/~cjlin/libsvm.) for the training of the SVM. The kernel parameters of C and γ are determined by a grid search, using a fivefold cross validation. For the RF classification, a freely available Fortran code was used (http://www.stat.berkeley.edu/~breiman/RandomForests/). The number of features was set to the square root of the number of input features and 500 iterations were performed.

18.3.1.2 Results and Discussion

The experiment was conducted on multisensor data sets consisting of SAR and multispectral imagery, using the three different classifier algorithms (MLC, SVM, and RF). All methods were applied three times on (1) the SAR, (2) the multispectral data, and (3) the multisensor data set (see Table 18.1). As expected, the experimental results clearly show the positive effect of multisensor classifications.

TABLE 18.1

Overall Accuracy (%) Using Different Classifiers and Data Sets

Method	SAR	Multispectral	SAR + Multispectral
MLC	57.1	66.0	72.1
SVM	63.2	70.0	74.8
RF	64.0	68.5	76.9

Irrespective of the classifier, the method is improved by the multisensor data set. Even when only a "weaker" classifier was in the case of MLC, the accuracy increased by using a multisensor data set. Comparing the multisensor-based results achieved by RF, the total accuracy increased by 8.4% compared to the classification results achieved on the multispectral Landsat image and 12.9% when compared with the accuracy achieved with the SAR data. Comparing the different algorithms, it can be assessed, that the standard MLC performed worst in terms of accuracies. The main reason for the weak performance of the MLC, particularly on the multitemporal SAR and multisensor data, might be the assumption of a Gaussian distribution, which may not be fulfilled in the context of these data sets. Whereas SVM and RF perform almost similar on the SAR and multispectral data in terms of classification accuracy, RF achieves the highest accuracy on the multisensory data set (76.9%).

These findings are confirmed by the class accuracies (Table 18.2). The producer's as well as the user's accuracies are clearly increased by using multisensor data. The classification accuracies achieved with single-source data sets for the land cover classes *arable crops* and *orchard* are relatively low, compared to the accuracies achieved for other classes. However, combining the two data sets by an adequate classifier, such as RF, significantly improves the producer and user accuracies. Even the accuracies of classes that are already classified very accurately by a single source (e.g., the class *forest* by multispectral data) can further improve the multisensor data set.

The visible evaluation of the classification maps confirms the good performance of RF (Figure 18.1). The maps from the MLC show the general structures of the classified area, but appear more noisy when compared with the map provided by RF. Sometimes, borders between individual agricultural parcels appear blurred and are more difficult to identify. This noise is significantly suppressed by the RF classification.

Overall, the experimental results underline the value of multisensor applications as well as the requirement for adequate classification algorithms. RFs demonstrate the full advantage of multiple

TABLE 18.2

Producer's and User's Accuracy (%) Using RF and Different Data Sets

	Data Set					
	SAR		Multispectral		SAR + Multispectral	
Class	Producer's Accuracy	User's Accuracy	Producer's Accuracy	User's Accuracy	Producer's Accuracy	User's Accuracy
Arable crops	51.8	50.9	61.0	61.5	69.2	65.5
Cereals	76.6	64.5	66.8	71.5	76.8	80.7
Forest	74.4	69.0	86.2	89.2	91.2	93.6
Grassland	71.4	72.4	61.0	53.3	75.8	72.3
Orchard	52.0	52.4	51.4	47.7	70.6	59.5
Rapeseed	63.8	63.2	72.2	86.4	76.6	88.9
Rootcrops	60.6	63.7	68.0	67.2	71.8	73.3
Urban	61.6	79.4	82.0	79.0	82.8	87.9

classifier ensembles and are probably one of the most powerful recent classifiers. The simple handling and fast training time, compared to SVM and ensemble techniques such as boosting (not presented in detail) are some issues making RF particularly interesting in the context of multisensory data analysis for operational monitoring systems.

18.3.2 CLASSIFICATION OF HYPERSPECTRAL IMAGERY

18.3.2.1 Data Set and Methods

The second experiment is aiming at the classification of lithological units, using airborne hyperspectral imagery. The study site lies around the volcano Hekla in South Iceland. Hekla is one of the most active volcanoes in the country and has erupted quite regularly every 10 years, in 1970, 1980–1981, 1991, and in 2000. The image was collected by airborne visible infrared imaging spectrometer (AVIRIS) on a cloud-free day, June 17, 1991 (Figure 18.2). The data set is 2048 × 614 pixels, with a

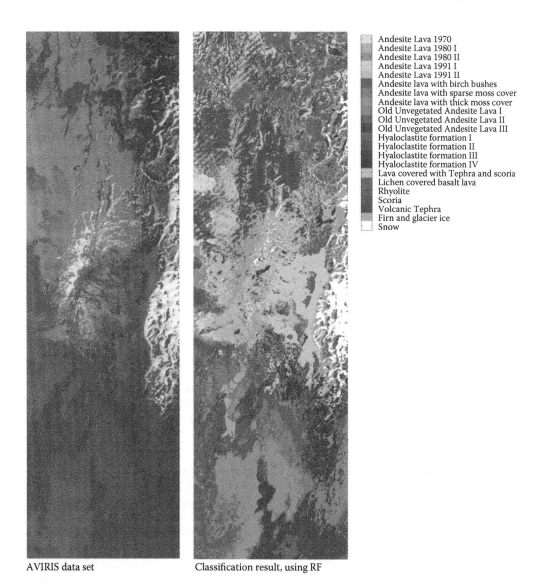

Andesite Lava 1970
Andesite Lava 1980 I
Andesite Lava 1980 II
Andesite Lava 1991 I
Andesite Lava 1991 II
Andesite lava with birch bushes
Andesite lava with sparse moss cover
Andesite lava with thick moss cover
Old Unvegetated Andesite Lava I
Old Unvegetated Andesite Lava II
Old Unvegetated Andesite Lava III
Hyaloclastite formation I
Hyaloclastite formation II
Hyaloclastite formation III
Hyaloclastite formation IV
Lava covered with Tephra and scoria
Lichen covered basalt lava
Rhyolite
Scoria
Volcanic Tephra
Firn and glacier ice
Snow

AVIRIS data set   Classification result, using RF

FIGURE 18.2 (**See color insert.**) AVIRIS data set and classification result using RF.

TABLE 18.3

Land Cover Classes and Number of Samples in the Test Set

Class Name	Test Set
Andesite Lava 1970	950
Andesite Lava 1980 I	424
Andesite Lava 1980 II	748
Andesite Lava 1991 I	1369
Andesite Lava 1991 II	205
Andesite lava with birch bushes	790
Andesite lava with sparse moss cover	511
Andesite lava with thick moss cover	2007
Old Unvegetated Andesite Lava I	1083
Old Unvegetated Andesite Lava II	799
Old Unvegetated Andesite Lava III	356
Hyaloclastite formation I	702
Hyaloclastite formation II	342
Hyaloclastite formation III	635
Hyaloclastite formation IV	341
Lava covered with Tephra and Scoria	350
Lichen covered basalt lava	511
Rhyolite	202
Scoria	275
Volcanic Tephra	1631
Firn and glacier ice	229
Snow	506

spatial resolution of 20 m. The sensor operates from the visible to the short-wave infrared region of the electromagnetic spectrum. However, due to a malfunction, spectrometer 4 was working incorrectly. In addition, some bands contained noise and thus were removed. Finally, 157 bands were left for the experiment. The classification aims at mapping 22 land cover classes, mainly different lithological units [18] (Table 18.3). The reference data set was generated by geological and vegetation maps, expert knowledge, and image interpretation. The data set was divided into independent training and test data. To investigate the impact of the number of training samples in the classification accuracy, two training data sets were generated by stratified random sampling, containing 50 and 200 samples/class. As in the first experiment, the results achieved by RF are compared to the classification outputs of an MLC and an SVM. In addition, an SAM was used, which is an example of a common technique used for classifying hyperspectral imagery [25]. The classification of the SVM and RF followed the procedure described above (see Section 18.3.11).

18.3.2.2 Results and Discussion

The accuracy assessment shows that RF and SVM perform very accurately and achieve high classification accuracies. Comparison of these results achieved by the MLC (67.7%) and SAM (69.7%) demonstrate that the two sophisticated methods significantly outperform the conventional methods in terms of accuracy (Table 18.4).

The class accuracies (not presented in detail) demonstrate the balanced performance between RF and SVM. Both methods perform weaker on the four land cover types *Old Unvegetated Andesite Lava III, Hyaloclastite formation I* and *III* in terms of classification accuracy, resulting in much lower accuracies, compared to the accuracies achieved for the other land cover classes. Although SVM outperforms the RF in terms of accuracy, it is worth underlining that RF achieves accurate

TABLE 18.4

Overall Accuracy and Kappa Coefficient Using Different Methods and Number of Training Samples

Method	Number of Training Samples	
	#50	#200
MLC	—	67.7
SAM	69.0	69.7
SVM	90.2	96.9
RF	88.1	92.7

results with overall accuracies of 88.1% and higher, even with a small number of training samples. Moreover, the relatively low computation time of RF seems particularly attractive when classifying high-dimensional data, such as hyperspectral imagery.

18.4 CONCLUSION

The use of RFs for classification of multisensory and hyperspectral remote sensing data has been discussed. Although more sophisticated classifiers (i.e., SVM) can outperform RF in terms of accuracy when classifying hyperspectral imagery, RF should be considered attractive for classification of both data types. RF is fast in training and classification and very simple to handle because it mainly depends on only two user-defined values. Moreover, RF provides additional information, such as VI and the OOB-error, which might be useful for remote-sensing applications [14,17]. Moreover, approaches such as RF and SVM are not exclusive and it is interesting to combine them [6,15]. For future work, the application of RF in context of semisupervised learning seems interesting [26]. Regarding near-real time applications, a further reduction of training and classification time could be useful, for example, by parallel implementation, which was recently discussed in the context of SVM and hyperspectral remote sensing [27].

ACKNOWLEDGMENTS

The authors wish to thank the European Space Agency for providing the remote sensing data for experiment 1 through a CAT 1 proposal (C1.3115). The images were acquired within the ENVILAND research project (FKZ 50EE0404), funded by the German Aerospace Center (DLR) and the Federal Ministry of Education and Research (BMBF). This work is partially funded by the Research Fund of the University of Iceland and The Icelandic Research Fund.

REFERENCES

1. R. Polikar, Ensemble based systems in decision making, *IEEE Circuits and Systems Magazine*, 6, 21–45, 2006.
2. J. A. Benediktsson, J. Chanussot, and M. Fauvel, Multiple classifier systems in remote sensing: From basics to recent developments. In: M. Haindl, J. Kittler, and F. Roli (Eds), *Multiple Classifier Systems. 7th International Workshop, MCS 2007*, Prague, Czech Republic, May 23–25, 2007, Proceedings. *Lecture Notes in Computer Science* 4472, Springer, 2007.
3. J. A. Benediktsson and I. Kanellopoulos, Classification of multisource and hyperspectral data based on decision fusion, *IEEE Transactions on Geoscience and Remote Sensing*, 37, 1367–1377, 1999.
4. B. Waske and J. A. Benediktsson, Fusion of support vector machines for classification of multisensor data, *IEEE Transactions on Geoscience and Remote Sensing*, 45(12), 3858–3866, 2007.
5. L. K. Hansen and P. Salamon, Neural network ensembles, *IEEE Transactions on Pattern Analysis and Machine Intelligence*, 12, 993–1001, 1990.

6. B. Waske, S. van der Linden, J. A. Benediktsson, A. Rabe, and P. Hostert, Sensitivity of support vector machines to random feature selection in classification of hyperspectral data, *IEEE Transactions on Geoscience and Remote Sensing*, 48(7), 2880–2889, 2010.

7. L. Breiman, Bagging predictors, *Machine Learning*, 24(2), 123–140, 1996.

8. Y. Freund and R. E. Schapire, Experiments with a new boosting algorithm, *Machine Learning: Proceedings of the Thirteenth International Conference*, Bari, Italy, pp. 148–156, 1996.

9. T. K. Ho, The random subspace method for constructing decision forests, *IEEE Transactions on Pattern Analysis and Machine Intelligence*, 20, 882–844, 1998.

10. B. Waske and M. Braun, Classifier ensembles for land cover mapping using multitemporal SAR imagery, *ISPRS Journal of Photogrammetry and Remote Sensing*, 64(5), 450–457, 2009.

11. R. Bryll, R. Gutierrez-Osuna, and F. Quek, Attribute bagging: Improving accuracy of classifier ensembles by using random feature subsets, *Pattern Recognition* 36, 1291–1302, 2003.

12. B. Waske, S. Schiefer, and M. Braun, Random feature selection for decision tree classification of multitemporal SAR data, *Proceedings of IGARSS'06 Symposium*, Denver, USA, 168–171, 2006.

13. L. Breiman, Random forests, *Machine Learning*, 40(1), 5–32, 2001.

14. P. O. Gislason, J. A. Benediktsson, and J. R. Sveinsson, Random forests for land cover classification, *Pattern Recognition Letters*, 27, 294–300, 2006.

15. B. Waske and S. van der Linden, Classifying multilevel imagery from SAR and optical sensors by decision fusion, *IEEE Transactions on Geoscience and Remote Sensing*, 46(5), 1457–1466, 2008.

16. J. Ham, Yangchi Chen, M. M. Crawford, and J. Ghosh, Investigation of the random forest framework for classification of hyperspectral data, *IEEE Transactions on Geoscience and Remote Sensing*, 43(3), 492–501, 2005.

17. S. R. Joelsson, J. A. Benediktsson, and J. R. Sveinsson, Random forest classifiers for hyperspectral data, *Proceedings of IGARSS'05 Symposium*, Seoul, Korea, July 25–29, 2005.

18. B. Waske, J. A. Benediktsson, K. Árnason, and J. R. Sveinsson, Mapping of hyperspectral AVIRIS data using machine-learning algorithms, *Canadian Journal of Remote Sensing*, 35, 106–116, 2009.

19. S. R. Joelsson, J. A. Benediktsson, and J. R. Sveinsson, Random forest classification of remote sensing data, In: C. H. Chen (Ed.), *Signal and Image Processing for Remote Sensing*, 1st ed, CRC Press, Boca Raton, FL, 61–78, 2007.

20. D. R. Cutler, T. C. Edwards, K. H. Beard, A. Cutler, and K. T. Hess, Random forests for classification in ecology, *Ecology*, 88(11), 2783–2792, 2007.

21. http://oz.berkeley.edu/users/breiman/RandomForests/cc_home.htm.

22. S. R. Safavian and D. A. Landgrebe. A survey of decision tree classifier methodology, *IEEE Transactions on Systems, Man, and Cybernetics*, 21, 660–674, 1991.

23. L. Breiman, J. Friedman, R. A. Olshen, and C. J. Stone, *Classification and Regression Trees*, CRC Press, Boca Raton, FL, 1984.

24. M. Zambon, R. Lawrence, A. Bunn, and S. Powell, Effect of alternative splitting rules on image processing using classification tree analysis. *Photogrammetric Engineering and Remote Sensing*, 72, 25–30, 2006.

25. F. A. Kruse, A. B. Lefkoff, J. W. Boardman, K. B. Heidebrecht, A. T. Shapiro, P. J. Barloon, and A. F. H. Goetz, The Spectral Image Processing System (SIPS). *Interactive Visualization and Analysis of Imaging Spectrometer Data. Remote Sensing of Environment*, 44(2–3), 145–163, 1993.

26. Z.-H. Zhou, When semi-supervised learning meets ensemble learning, In: J. A. Benediktsson, J. Kittler, and F. Roli (Eds.), *Multiple Classifier Systems, Proceedings 8th International Workshop, MCS 2009*, Reykjavik, Iceland, *Lecture Notes in Computer Science* 4472, Springer, 2009.

27. J. Muñoz, A. Plaza, J. A. Gualtieri, and G. Camps-Valls, Parallel implementation of SVM in Earth observation applications, In: F. Xhafa (Ed.), *Parallel Programming, Models and Applications in Grid and P2P Systems, Advances in Parallel Computing*, Vol. 17, pp. 292–213, 2009.

19 Sparse Representation for Target Detection and Classification in Hyperspectral Imagery

Yi Chen, Trac D. Tran, and Nasser M. Nasrabdi

CONTENTS

19.1 INTRODUCTION

Sparsity of signals has been an extremely powerful tool in many classical signal processing applications such as compression and denoising [1], as most natural signals can be compactly represented by only a few coefficients that carry the most important information in a certain basis or dictionary. Recently, applications of sparse data representation have been extended to the area of computer vision and pattern recognition [2] with the development of the compressed sensing (CS) framework [3,4] and sparse modeling of signals and images [5]. These applications are mainly based on the observation that despite the high dimensionality of natural signals, signals that belong to the same class usually lie in a low-dimensional subspace. Therefore, for every typical signal, there exists a sparse representation with respect to some proper basis which encodes the semantic information. The CS theories ensure that a sparse signal can be recovered from its incomplete but incoherent projections with a high probability. This enables the recovery of the sparse representation by decomposing the sample over a usually overcomplete dictionary generated by or learned from representative samples. Once the sparse representation vector is obtained, the semantic information can be directly extracted from the recovered vector. Applications of sparse representation in computer vision and pattern recognition can be found in various fields, including motion segmentation [6,7], image superresolution [8], image restoration [9,10], and discriminative tasks, including face recognition [11], iris recognition [12], tumor classification [13], and hyperspectral unmixing [14]. In these applications, the usage of sparsity as a prior often leads to the state-of-the-art performance.

In the hyperspectral case, the remote sensors capture digital images in hundreds of narrow spectral bands spanning the visible to infrared spectrum [15]. Pixels in hyperspectral imagery (HSI) are represented by vectors whose entries correspond to the spectral bands. Different materials usually reflect electromagnetic energy differently at specific wavelengths. This enables discrimination of materials based on the spectral characteristics. HSI has found many applications in various fields such as military [16–18], agriculture [19,20], and mineralogy [21]. Target detection and classification are two of the most important applications of HSI. A number of algorithms have also been proposed for target detection in HSI based on statistical hypothesis testing techniques [16]. Among these approaches, spectral matched filters [22,23], matched subspace detectors [24], and adaptive subspace detectors [25] have been widely used to detect targets of interests. Support vector machines [26,27] have been a powerful tool to solve supervised classification problems for high-dimensional data and have shown a good performance for hyperspectral classification [28,29]. Variations of the SVM-based algorithms have also been proposed to improve the classification accuracy. These variations include semisupervised learning which exploits both labeled and unlabeled samples [30], postprocessing of the individually labeled samples based on certain decision rules [31,32], and incorporating spatial information directly in the SVM kernels [33,34]. More recent HSI classification techniques can be found in Refs. [35–42].

In this chapter, we introduce a new technique for HSI target detection and classification that utilizes the sparsity of the input sample with respect to a given overcomplete training dictionary. This algorithm is based on a sparsity model where a test spectral pixel is approximately represented by a few training samples (atoms) among the entire training dictionary. The sparse vector representing the atoms and their associated weights for the test spectral pixel can be recovered by solving an optimization problem constrained by the sparsity level and reconstruction accuracy. The class of the test pixel can then be determined by the characteristics of the recovered sparse vector.

HSI usually has large homogeneous regions where the neighboring pixels within the regions consist of the same type of materials (same class) and has similar spectral characteristics. Previous works have shown that it is important to take into account the contextual information in HSI classification [31–34,43,44]. In this chapter, during the recovery of the sparse vector, in addition to the constraints on sparsity and reconstruction accuracy, we also exploit the spatial smoothness across neighboring HSI pixels. Two different approaches are presented to incorporate the contextual information directly in the sparse recovery problem. In the first approach, a local smoothing constraint is imposed to the optimization problem by forcing the *vector Laplacian* at the reconstructed pixel to be zero. Thus, the reconstructed pixel of interest is forced to have spectral characteristics similar to its four nearest neighbors. The proposed reconstruction problem with the explicit smoothing constraint can be reformulated as a standard sparsity-constrained optimization problem and then solved efficiently by available optimization tools. In the second approach, we exploit the interpixel correlation between neighboring HSI pixels by adopting a joint sparsity model [45,46], where pixels in a small neighborhood are assumed to be simultaneously represented by a few common training samples, but for each pixel these selected training samples are weighted with a different set of coefficients. In this way, the sparse vector representations of neighboring pixels are forced to have a common support corresponding to the common atoms in the given training dictionary. The support is recovered by simultaneously decomposing the neighboring pixels over the given training dictionary. In both approaches, the labels of the test samples are determined by the property of the recovered sparse vectors. Both proposed approaches enforce the smoothness constraint across neighboring pixels within the optimization process during the classification stage, rather than employing a postprocessing scheme to exploit the contextual information.

The remainder of this chapter is structured as follows. The proposed sparsity-based algorithm is introduced in Section 19.2. The details of the two approaches used to incorporate the spatial information are described in Section 19.3. The effectiveness of the proposed method is demonstrated in Section 19.4 by simulation results on several real hyperspectral images. Finally, Section 19.5 summarizes our work and makes some closing remarks.

19.2 SPARSE REPRESENTATION OF HYPERSPECTRAL PIXELS

In this section, we introduce a sparsity model where a test sample is represented by a sparse linear combination of training samples from a given dictionary. We then discuss the standard algorithms used to solve for the sparse representation as well as the procedure used for determining the label of the test pixel.

19.2.1 PIXEL-WISE SPARSITY MODEL

We first consider the case of target detection, which can be viewed as a binary classification problem where pixels are labeled as target (target present) or background (target absent) based on their spectral characteristics. Let x be a hyperspectral pixel observation, which is a B-dimensional vector whose entries correspond to responses to various spectral bands. If x is a background pixel, its spectrum approximately lies in a low-dimensional subspace spanned by the background training samples $\{a_i^b\}_{i=1,2,\dots,N_b}$. The pixel x can then be approximately represented as a linear combination of the training samples as follows:

$$x \approx \alpha_1^b a_1^b + \alpha_2^b a_2^b + \cdots + \alpha_{N_b}^b a_{N_b}^b = \underbrace{[a_1^b \quad \cdots \quad a_{N_b}^b]}_{A^b}\underbrace{[\alpha_1^b \quad \cdots \quad \alpha_{N_b}^b]^T}_{\alpha^b} = A^b \alpha^b, \qquad (19.1)$$

where N_b is the number of background training samples, A^b is the $B \times N_b$ background dictionary whose columns are the background training samples (also called atoms), and α is an unknown vector whose entries are the abundances of the corresponding atoms in A^b. In our model, α turns out to be a sparse vector (i.e., a vector with only few nonzero entries).

Similarly, a target pixel x approximately lies in the target subspace spanned by the target training samples $\{a_i^t\}_{i=1,2,\dots,N_t}$, which can also be sparsely represented by a linear combination of the training samples

$$x = \alpha_1^t a_1^t + \alpha_2^t a_2^t + \cdots + \alpha_{N_t}^t a_{N_t}^t = \underbrace{[a_1^t \quad \cdots \quad a_{N_t}^t]}_{A^t}\underbrace{[\alpha_1^t \quad \cdots \quad \alpha_{N_t}^t]^T}_{\alpha^t} = A^t \alpha^t, \qquad (19.2)$$

where N_t is the number of target training samples, A^t is the $B \times N_t$ target dictionary consisting of the target training pixels, and α^t is a sparse vector whose entries contain the abundances of the corresponding target atoms in A^t.

In our sparsity-based detection algorithm, an unknown test sample is modeled to lie in the union of the background and target subspaces. Therefore, by combining the two dictionaries A^b and A^t, a test sample x can be written as a sparse linear combination of all training pixels

$$x = A^b \alpha^b + A^t \alpha^t = \underbrace{[A^b \quad A^t]}_{A}\underbrace{\begin{bmatrix} \alpha^b \\ \alpha^t \end{bmatrix}}_{\alpha} = A\alpha, \qquad (19.3)$$

where $A = [A^b\ A^t]$ is a $B \times (N_b + N_t)$ matrix consisting of both background and target training samples, and $\alpha = [\alpha^b/\alpha^t]$ is a $(N_b + N_t)$-dimensional vector consisting of the two vectors α^b and α^t associated with the two dictionaries. The vector α is a concatenation of the two vectors associated with the background and target dictionaries and is also a sparse vector as follows. Since the background (e.g., trees, grass, road, soil) and target (e.g., metal, paint, glass) pixels usually consist of different materials, they have distinct spectral signatures and thus the spectrum of target and background pixels lie in different subspaces. For example, if x is a target pixel, then ideally it cannot be represented by the background training samples. In this case, α^b is a zero vector and α^t is a sparse vector. On the other hand, if x belongs to the background class, then α^b is sparse and α^t is a zero vector. Therefore, the test sample x can be sparsely represented by combined background and target dictionaries, and the locations of nonzero entries in the sparse vector α actually contains critical information about the class of the test sample x.

The sparse representation of x can also be written as a linear combination of only the K active dictionary atoms a_{λ_k} corresponding to the K nonzero entries α_{λ_k}, $k = 1,\ldots,K$:

$$x = \alpha_{\lambda_1} a_{\lambda_1} + \alpha_{\lambda_2} a_{\lambda_2} + \cdots + \alpha_{\lambda_M} a_{\lambda_K} = \underbrace{[a_{\lambda_1} \cdots a_{\lambda_K}]}_{A_{\Lambda_K}} \begin{bmatrix} \alpha_{\lambda_1} \\ \vdots \\ \alpha_{\lambda_K} \end{bmatrix}_{\alpha_{\Lambda_K}} = A_{\Lambda_K} \alpha_{\Lambda_K} \qquad (19.4)$$

where $K = \|\alpha\|_0$ denotes the ℓ_0-norm (or sparsity level) of α which is defined as the number of non-zero entries in α, the index set $\Lambda_K = \{\lambda_1, \lambda_2, \ldots, \lambda_K\}$ is the support of α, A_{Λ_K} is a $B \times K$ matrix whose columns are the K atoms $\{a_k\}_{k \in \Lambda_K}$ and α_{Λ_K} is a K-dimensional vector consisting of entries of α indexed by Λ_K.

The above sparsity model can be extended to the case of multiple classes. Suppose we have M distinct classes and the mth class has N_m training samples $\{a_j^m\}_{j=1,\ldots,N_m}$. Let x be a B-dimensional hyperspectral pixel observation belonging to the mth class. Its spectrum then approximately lies in a low-dimensional subspace spanned by the training samples $\{a_j^m\}_{j=1,\ldots,N_m}$ in the mth class and x can be compactly represented by a linear combination of these training samples

$$x \approx \alpha_1^m a_1^m + \cdots + \alpha_{N_m}^m a_{N_m}^m = \underbrace{[a_1^m \cdots a_{N_m}^m]}_{A^m} \underbrace{[\alpha_1^m \cdots \alpha_{N_m}^m]^T}_{\alpha^m} = A^m \alpha^m, \qquad (19.5)$$

where A^m is a $B \times N_m$ class subdictionary whose columns are the training samples in the mth class, and α^m is an unknown N_m-dimensional sparse vector whose entries are the weights of the corresponding atoms in A^m.

An unknown test sample is modeled to lie in the union of the M subspaces associated with the M classes. By combining the class subdictionaries $\{A^m\}_{m=1,\ldots,M}$, a test sample x is written as a sparse linear combination of all of the training pixels as

$$x = A^1 \alpha^1 + A^2 \alpha^2 + \cdots + A^M \alpha^M = \underbrace{[A^1 \cdots A^M]}_{A} \begin{bmatrix} \alpha^1 \\ \vdots \\ \alpha^M \end{bmatrix}_{\alpha} = A\alpha \qquad (19.6)$$

where A is a $B \times N$ structured dictionary consisting of training samples from all classes with $N = \sum_{m=1}^M N_m$ and α is an N-dimensional sparse vector formed by concatenating the sparse vectors $\{\alpha^m\}_{m=1,\ldots,M}$.

19.2.2 RECONSTRUCTION AND DECISION

In this section, we explain how to obtain α and how to classify a test sample from α. We first consider the reconstruction problem of finding the sparse vector α for a test sample x. Given the dictionary of training samples A, the representation α satisfying $A\alpha = x$ is obtained by solving the following optimization problem:

$$\hat{\alpha} = \arg\min \| \alpha \|_0 \text{ subject to } A\alpha = x. \tag{19.7}$$

To account for approximation errors in empirical data, the equality constraint in Equation 19.7 can be relaxed to an inequality one

$$\hat{\alpha} = \arg\min \| \alpha \|_0 \text{ subject to } \| A\alpha = x \|_2 \leq \sigma, \tag{19.8}$$

where σ is the error tolerance. The above problem can also be interpreted as minimizing the approximation error within a certain sparsity level

$$\hat{\alpha} = \arg\min \| A\alpha - x \|_2 \text{ subject to } \| \alpha \|_0 \leq K_0, \tag{19.9}$$

where K_0 is a given upper bound on the sparsity level [47]. The above problems are NP-hard, but they can be approximately solved by greedy pursuit algorithms such as orthogonal matching pursuit (OMP) [48] or subspace pursuit (SP) [49]. Both OMP and SP algorithms are used to locate the support of the sparse vector that approximately solves the problem in Equation 19.9, but the atoms are selected from the dictionary in different ways. The OMP algorithm augments the support set by one index at each iteration until K_0 atoms are selected or the approximation error is within a preset threshold. The SP algorithm maintains a set of K_0 indices. At each iteration, the index set is refined by adding K_0 new candidates to the current list and then discarding K_0 insignificant ones from the list of $2K_0$ candidates. With the backtracking mechanism, SP is able to find the K_0 most significant atoms.

The NP-hard problem in Equation 19.7 can also be relaxed to a linear programming problem, called basis pursuit (BP), by replacing $\|\cdot\|_0$ by $\|\cdot\|_1$ [50]:

$$\hat{\alpha} = \arg\min \| \alpha \|_1 \quad \text{subject to} \quad A\alpha = x. \tag{19.10}$$

Similarly, the problems in Equations 19.8 and 19.9 can also be relaxed to convex programming problems as

$$\hat{\alpha} = \arg\min \| \alpha \|_1 \quad \text{subject to} \quad \| A\alpha = x \|_2 \leq \sigma, \tag{19.11}$$

$$\hat{\alpha} = \arg\min \| A\alpha = x \|_2 \quad \text{subject to} \quad \| \alpha \|_1 \leq \tau, \tag{19.12}$$

respectively. The above problems can be solved efficiently by an interior point method [51] or gradient projection method [52,53] in polynomial time. In this chapter, we have employed the OMP and SP greedy algorithms to solve the sparsity-constrained optimization problem in Equation 19.9 and the SPG-L1 optimization toolbox [53] to solve the basis pursuit denoising (BPDN) problem in Equation 19.11.

The recovery process implicitly leads to a competition between the subspaces and therefore the recovered sparse representation is discriminative. Once the sparse vector $\hat{\gamma}$ is obtained, the class of x can be determined by comparing the residuals. In the case of target detection, the residuals with respect to the two subspaces are $r^b(x) = \| x - A^b\hat{\alpha}^b\|_2$ and $r^t(x) = \| x - A^t\hat{\alpha}^t\|_2$, where $\hat{\alpha}^b$ and $\hat{\alpha}^t$ represent the recovered sparse coefficients corresponding to the background and target dictionaries, respectively. The output of detector is then calculated by the difference of the two residuals:

$$D(x) = r^b(x) - r^t(x). \tag{19.13}$$

If $D(x) > \delta$ with δ being a prescribed threshold, the test sample x is determined as a target pixel; otherwise, x is labeled as background.

In the case of multiclass classification, define the mth residual (i.e., error between the test sample and the reconstruction from training samples in the mth class) to be

$$r^m(x) = \| x - A^m\hat{\alpha}^m\|_2, \quad m = 1, 2, ..., M, \tag{19.14}$$

where $\hat{\alpha}^m$ denotes the portion of the recovered sparse coefficients corresponding to the training samples in the mth class. The class of x is then determined as the one with the minimal residual

$$\text{Class}(x) = \arg \min_{m=1,...,M} r^m(x). \tag{19.15}$$

19.3 SPARSE REPRESENTATION WITH CONTEXTUAL INFORMATION

The above sparsity model is based on each pixel individually without considering the interpixel correlation. However, in hyperspectral images, neighboring pixels usually consist of similar materials and thus their spectral characteristics are highly correlated. It is necessary to incorporate the contextual information into the sparsity model. In this section, we show two different approaches to achieve this within the sparsity-constrained problem formulation. In the first approach, an explicit smoothing constraint is imposed on the problem formulation by forcing the *vector Laplacian* of the reconstructed image to become zero. In this approach, the reconstructed pixel of interest has similar spectral characteristics to its four nearest neighbors. The second approach is via a joint sparsity model where hyperspectral pixels in a small neighborhood around the test pixel are simultaneously represented by linear combinations of a few common training samples, which are weighted with a different set of coefficients for each pixel.

19.3.1 LAPLACIAN CONSTRAINT

Let I represent the hyperspectral image. Let x_1 be a pixel of interest in I, x_i, $i = 2, ..., 5$ be its four nearest neighbors in the spatial domain, as shown in Figure 19.1, and α_i be the sparse vector associated with x_i (i.e., $x_i = A\alpha_i$). Define *vector Laplacian* at the reconstructed point $\hat{x}_1$ to be the B-dimensional vector

$$\nabla^2(\hat{x}_1) = 4\hat{x}_1 - \hat{x}_2 - \hat{x}_3 - \hat{x}_4 - \hat{x}_5 = A(4\hat{\alpha}_1 - \hat{\alpha}_2 - \hat{\alpha}_3 - \hat{\alpha}_4 - \hat{\alpha}_5). \tag{19.16}$$

In order to incorporate the smoothness across the neighboring spectral pixels, we propose to force the vector Laplacian at $\hat{x}_1$ to become zero in addition to the constraints on sparsity level and reconstruction accuracy in Equation 19.7. In this way, the reconstructed test sample $\hat{x}_1$ is forced to

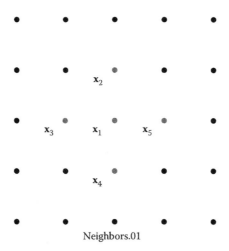

Neighbors.01

FIGURE 19.1 Four nearest neighbors of a pixel x_1.

have similar spectral characteristics to its four nearest neighbors. The new sparse recovery problem with the smoothing constraint is formulated as

$$
\begin{aligned}
&\text{minimize} \quad \sum_{i=1}^{5} \| \alpha_i \|_0 \\
&\text{subject to:} \quad A(4\alpha_1 - \alpha_2 - \alpha_3 - \alpha_4 - \alpha_5) = 0 \\
&\qquad\qquad\; x_i = A\alpha_i, \quad i = 1,\ldots,5.
\end{aligned}
\tag{19.17}
$$

In Equation 19.17, the first set of linear constraints forces the reconstructed image vector Laplacian to become zero such that the reconstructed neighboring pixels have similar spectral characteristics, and the second set minimizes reconstruction errors. The optimization problem in Equation 19.17 can be rewritten as

$$
\hat{\tilde{\alpha}} = \arg\min \| \tilde{\alpha} \|_0 \quad \text{subject to} \quad \tilde{A}\tilde{\alpha} = \tilde{x},
\tag{19.18}
$$

where

$$
\tilde{A} = \begin{bmatrix}
4\lambda A & -\lambda A & -\lambda A & -\lambda A & -\lambda A \\
A & & & & \\
& A & & & \\
& & A & & \\
& & & A & \\
& & & & A
\end{bmatrix}, \quad
\tilde{\alpha} = \begin{bmatrix} \alpha_1 \\ \vdots \\ \alpha_5 \end{bmatrix}, \quad
\tilde{x} = \begin{bmatrix} 0 \\ x_1 \\ \vdots \\ x_5 \end{bmatrix},
$$

$\lambda > 0$ is a weighting factor that controls the relative importance between the smoothing constraint and the reconstruction accuracy.

In practice, the equality constraints in Equation 19.17 (or equivalently in Equation 19.18) cannot be satisfied. Similar to the previous case in Section 19.2.2, the problem in Equation 19.18 can be relaxed to allow for smoothing and approximation errors in the form of Equation 19.8 or 19.9 as

$$\hat{\tilde{\alpha}} = \arg\min \| \tilde{\alpha} \|_0 \quad \text{subject to} \quad \| \tilde{A}\tilde{\alpha} - \tilde{x} \|_2 \le \sigma, \tag{19.19}$$

$$\text{or} \quad \hat{\tilde{a}} = \arg\min \| \tilde{A}\hat{\tilde{\alpha}} - \tilde{x} \|_2 \quad \text{subject to} \quad \|\tilde{\alpha}\|_0 \le K_0, \tag{19.20}$$

respectively. The above problems are standard sparse recovery problems and can be solved using the aforementioned optimization techniques.

Once the sparse vector in Equation 19.18 is obtained, the label of the test pixel can be determined based on the characteristics of the sparse coefficients as it was done in Section 19.2.2. For target detection, we calculate the total residuals obtained separately from the target and background dictionaries

$$r_b(x_1) = \sqrt{\sum_{i=1}^{5} \| x_i - A_b \hat{\alpha}_i^b \|_2^2} \quad \text{and} \quad r_t(x_1) = \sqrt{\sum_{i=1}^{5} \| x_i - A_t \hat{\alpha}_i^t \|_2^2}, \tag{19.21}$$

where $\hat{\alpha}_i^b$ and $\hat{\alpha}_i^t$ denote the recovered sparse coefficients for x_i associated with the background and target dictionaries, respectively. The output of the proposed sparsity-based detector for the center pixel x_1 is computed by the difference of residuals:

$$D(x_1) = r_b(x_1) - r_t(x_1). \tag{19.22}$$

If the output $D(x_1)$ is greater than a prescribed threshold δ, then the test sample x_1 is labeled as a target; otherwise, it is labeled as background.

In the case of M-class classification, the total error residuals between the original test samples and the approximations obtained from each of the M class subdictionaries are calculated as

$$r^m(x) = \sqrt{\sum_{i=1}^{5} \| x_i - A^m \hat{\alpha}_i^m \|_2^2}, \quad m = 1,2,\dots,M, \tag{19.23}$$

where x represents a concatenation of the five pixels $x_1, \dots, x_5$ in the 4-connected neighborhood centered at x_1, and $\hat{\alpha}_i^m$ denotes the portion of the recovered sparse vector for x_i associated with the mth-class subdictionary A^m. The label of the center pixel x_1 is then determined to be the class that yields the minimal total residuals

$$\text{Class}(x_1) = \arg\min_{m=1,\dots,M} r^m(x). \tag{19.24}$$

19.3.2 JOINT SPARSITY MODEL

An alternative way to exploit the spatial correlation across neighboring pixels is through a joint sparsity model [45,46]—assuming that the underlying sparse vectors associated with the neighboring pixels share a common sparsity pattern. That is, HSI pixels in a small spatial neighborhood are approximated by a sparse linear combination of a few common atoms from a given structured dictionary, but these atoms are weighted with a different set of coefficients for each pixel.

To illustrate the joint sparsity model, consider two neighboring hyperspectral pixels x_i and x_j consisting of similar materials. The sparse representation of x_i with respect to a given $B \times N$ structured dictionary A can be written as

$$x_i = A\alpha_i = \alpha_{i,\lambda_1} a_{\lambda_1} + \alpha_{i,\lambda_2} a_{\lambda_2} + \cdots + \alpha_{i,\lambda_K} a_{\lambda_K},$$

where the index set $\Lambda_K = \{\lambda_1, \lambda_2, \ldots, \lambda_K\}$ is the support of the sparse vector α_i. It is assumed that x_i and x_j consist of similar materials. Therefore, x_j can also be approximated by the same set of training samples $\{a_k\}_{k \in \Lambda_K}$, but with a different set of coefficients $\{\alpha_{j,k}\}_{k \in \Lambda_K}$

$$x_j = A\alpha_j = \alpha_{j,\lambda_1} a_{\lambda_1} + \alpha_{j,\lambda_2} a_{\lambda_2} + \cdots + \alpha_{j,\lambda_K} a_{\lambda_K}.$$

This can be extended to pixels in a small neighborhood $\mathcal{N}_\varepsilon$ consisting of T pixels. Let $X = [x_1 \ x_2 \ \ldots \ x_T]$ be a $B \times T$ matrix, where the columns $\{x_t\}_{t=1,\ldots,T} \in \mathcal{N}_\varepsilon$ are pixels in a spatial neighborhood in the hyperspectral image. Now, using the joint sparsity model, X can be represented by

$$\begin{aligned} X = [x_1 \quad x_2 \quad \cdots \quad x_T] &= [A\alpha_1 \quad A\alpha_2 \quad \cdots \quad A\alpha_T] \\ &= A\underbrace{[\alpha_1 \quad \alpha_2 \quad \cdots \quad \alpha_T]}_{S} = AS. \end{aligned} \quad (19.25)$$

The sparse vectors $\{\alpha_t\}_{t=1,\ldots,T}$ share the same support Λ_K and thus $S \in R^{N \times T}$ is a sparse matrix with only K nonzero rows. For convenience, we call the support Λ_K of α_t's also the support of the *row-sparse* matrix S.

Given the training dictionary A, the matrix S can be recovered by solving the following joint sparse recovery problem:

$$\hat{S} = \arg \min \| S \|_{\text{row},0} \quad \text{subject to} \quad AS = X, \quad (19.26)$$

where the notation $\|S\|_{\text{row},0}$ denotes the number of nonzero rows of S. The solution to the above problem $S = [\hat{\alpha}_1 \quad \hat{\alpha}_2 \quad \cdots \quad \hat{\alpha}_T]$ is an $N \times T$ sparse matrix with only few nonzero rows. For empirical data, the problem in (19.26) can also be rewritten to account for the approximation errors [54], as it was done in Equations 19.8 and 19.9:

$$\hat{S} = \arg \min \| S \|_{\text{row},0} \quad \text{subject to} \quad \| AS - X \|_F \leq \sigma, \quad (19.27)$$

$$\text{or } \hat{S} = \arg \min \| AS - X \|_F \quad \text{subject to} \quad \| S \|_{\text{row},0} \leq K_0, \quad (19.28)$$

where $\|\cdot\|_F$ denotes the Frobenius norm. Similar to the pixel-wise sparse recovery problems, the simultaneous sparse recovery problems Equations 19.26, 19.27, and 19.28 are NP-hard problems, which can be approximately solved by greedy algorithms [45,46], or relaxed to convex programming which can be solved in polynomial time [55,56].

In this chapter, two greedy pursuit algorithms are used to approximately solve the problem in Equation 19.28. The first one is a generalized OMP algorithm, called simultaneous orthogonal matching pursuit (SOMP) [45], which is summarized in Algorithm 1. In SOMP, the support of the solution is sequentially updated (i.e., the atoms in the dictionary A are sequentially selected). At each iteration, the atom that simultaneously yields the best approximation to all of the residual vectors is selected. Specifically, at the kth iteration, we calculate an $N \times T$ correlation matrix $C = A^T R_{k-1}$, where R_{k-1} is the residual between the data matrix X and its approximation. The (i,t)th entry in C is the correlation between the ith dictionary atom a_i and the residual vector for x_t at the current iteration k. In SOMP, we need to compute the ℓ_p-norm for some $p \geq 1$ for each of the N rows of C (step (1) in Algorithm 1). The row index corresponding to the largest ℓ_p-norm is then selected to augment the support set. In the literature, $p = 1$ is used in Ref. [45] and $p = 2$ is used in Ref. [46]. In Ref. [57], $p = 2$ and $p = \infty$ are also proposed for weak matching pursuit. The SOMP algorithm usually terminates when the residual is sufficiently small or the desired level of sparsity (controlled by the number of iterations) is achieved, corresponding to the problem in Equation 19.27 or 19.28, respectively. The implementation details of SOMP are summarized in Algorithm 1. It should be noted that the normalization of samples is not a requirement. Normalization is implicitly implemented by these greedy algorithms since the atoms are selected from the training dictionary by maximal correlation, regardless of the magnitude of the atoms.

The second simultaneous sparse recovery algorithm is our proposed simultaneous version of the SP algorithm (SSP), summarized in Algorithm 2. Similar to the pixel-wise SP algorithm [49], SSP also maintains a list of K_0 candidates. At each iteration, the K_0 atoms that yield the best simultaneous approximation to all of the T residual vectors are selected as the new candidates. Similar to SOMP, we also need to compute the ℓ_p-norm for some $p \geq 1$ for each of the N rows of the correlation matrix $C = A^T R_{k-1}$ (step (1) in Algorithm 2). The row indices corresponding to the rows with the K_0 largest ℓ_p-norm are then selected as the new candidates to augment the support set. A backtracking step is then implemented, where the K_0 significant atoms are selected in a similar fashion from the $2K_0$-candidate list (step (4) in Algorithm 2). The SSP algorithm terminates when the residual begins to increase, or a maximum number of iterations is reached.

After the sparse matrix $\hat{S}$ is recovered, the labels of the test samples can be determined based on the characteristics of the sparse coefficients as is done in Section 19.2.1. For target detection, we calculate and compare the total error residuals between the original test samples and the approximations obtained from the background and target subdictionaries. The output of the proposed sparsity-based detector is computed as in Equation 19.13 by the difference of the total residuals from all of the pixels in the neighborhood.

ALGORITHM 1: SIMULTANEOUS ORTHOGONAL MATCHING PURSUIT

Input: $B \times N$ dictionary $A = [a_1 \ \ a_2 \ \ \cdots \ \ a_N]$, $B \times T$ data matrix $X = [x_1 \ \ x_2 \ \ \cdots \ \ x_T]$, a stopping criterion {Make sure all columns in A and X have unit norm}
Initialization: residual $R_0 = X$, index set $\Lambda_0 = \emptyset$, iteration counter $k = 1$
while stopping criterion has not been met **do**

1. Find the index of the atom that best approximates all residuals: $\lambda_k = \arg \max\limits_{i=1,\dots,N} \| R_{k-1}^T a_i \|_p$, $p \geq 1$
2. Update the index set $\Lambda_k = \Lambda_{k-1} \bigcup \{\lambda_k\}$
3. Compute $P_k = (A_{\Lambda_k}^T A_{\Lambda_k})^{-1} A_{\Lambda_k}^T X \in \mathbb{R}^{k \times T}$, $A_{\Lambda_k} \in \mathbb{R}^{B}$: consists of the k atoms in A indexed in Λ_k
4. Determine the residual $R_k = X - A_{\Lambda_k} P_k$
5. $k \leftarrow k + 1$

end while
Output: Index set $\Lambda = \Lambda_{k-1}$, the sparse representation $\hat{S}$ whose nonzero rows indexed by Λ are the K rows of the matrix $(A_\Lambda^T A_\Lambda)^{-1} A_\Lambda^T X$

$$D(x) = \|X - A^b \hat{S}^b\|_F - \| X - A^t \hat{S}^t\|_F, \tag{19.29}$$

where $\hat{S}^b$ consists of the first N_b rows of the recovered matrix $\hat{S}$ corresponding to the background subdictionary A^b, and $\hat{S}^t$ consists of the remaining N_t rows in $\hat{S}$ corresponding to the target subdictionary A^t. If the output is greater than a prescribed threshold, then the test sample is labeled as a target; otherwise, it is labeled as background.

For the M-class classification, the error residuals between the original test samples and the approximations obtained from each class subdictionaries are calculated as follows:

$$r^m(X) = \| X - A^m \hat{S}^m \|_F, \quad m = 1, 2, \ldots, M, \tag{19.30}$$

where $\hat{S}^m$ consists of the N_m rows in $\hat{S}$ that are associated with the mth-class subdictionary A^m. The label of the center pixel x_1 is then determined by the minimal total residual as it was done previously in Section 19.2.2:

$$\text{Class}(x_1) = \arg \min_{m=1,\ldots,M} r^m(X). \tag{19.31}$$

ALGORITHM 2:   SIMULTANEOUS SUBSPACE PURSUIT

Input: $B \times N$ dictionary $A = [a_1 \quad a_2 \quad \cdots \quad a_N]$, $B \times T$ data matrix $X = [x_1 \quad x_2 \quad \cdots \quad x_T]$, sparsity level K_0, a stopping criterion {Make sure all columns in A and X have unit norm}

Initialization: index set $\Lambda_0 = \{K_0$ indices corresponding to the K_0 largest numbers in $\|X^T a_i\|_p, p \geq 1$, $i = 1, \ldots, N\}$, residual $R_0 = X - A_{\Lambda_0}(A_{\Lambda_0}^T A_{\Lambda_0})^{-1} A_{\Lambda_0}^T X$, iteration counter $k = 1$
while stopping criterion has not been met **do**

1. Find the indices of the K_0 atoms that best approximate all residuals: $\mathcal{I} = \{K_0$ indices corresponding to the K_0 largest numbers in $\| R_{k-1}^T a_i \|_p, p \geq 1, i = 1, \ldots, N$
2. Update the candidate index set $\tilde{\Lambda}_k = \Lambda_{k-1} \cup \mathcal{I}$
3. Compute $P_k = (A_{\tilde{\Lambda}_k}^T A_{\tilde{\Lambda}_k})^{-1} A_{\tilde{\Lambda}_k}^T X \in \mathbb{R}^{2K_0 \times T}$
4. Let p_k^i denote the ith row in P_k. Update the index set $\Lambda_k = \{K_0$ indices corresponding to the K_0 largest numbers in $\| (p_k^i)^T \|_p, p \geq 1, i = 1, \ldots, 2K_0$
5. Determine the residual

$$R_k = X - A_{\Lambda_k}(A_{\Lambda_k}^T A_{\Lambda_k})^{-1} A_{\Lambda_k}^T X$$

6. $k \leftarrow k + 1$

end while
Output: Index set $\Lambda = \Lambda_{k-1}$, the sparse representation $\hat{S}$ whose K_0 nonzero rows indexed by Λ are the rows of the matrix $(A_\Lambda^T A_\Lambda)^{-1} A_\Lambda^T X$.

19.4   EXPERIMENTAL RESULTS AND ANALYSIS

In this section, we demonstrate the effectiveness of the proposed sparsity-based techniques on target detection and classification of several hyperspectral images.

19.4.1   TARGET DETECTION

The proposed target detection algorithm as well as the classical techniques—spectral matched filters (SMF), matched subspace detectors (MSD), adaptive subspace detectors (ASD), and support vector machines (SVM) are applied to two HSI. The results are compared both visually and quantitatively by the receiver operating characteristics (ROC) curves. The ROC curve describes the probability of detection (PD) as a function of the probability of false alarms (PFA). To calculate the ROC curve, we pick thousands of thresholds between the minimum and maximum of the detector output. The class labels for all pixels in the test region are determined at each threshold. The PFA is calculated by the number of false alarms (background pixels determined as target) over the total number of pixels in the test region, and the PD is the ratio of the number of hits (target pixels determined as target) and the total number of true target pixels.

The two hyperspectral images used in the experiments, the desert radiance II data collection (DR-II) and forest radiance I data collection (FR-I), are from the hyperspectral digital imagery collection experiment (HYDICE) sensor [58]. The HYDICE sensor generates 210 bands across the whole spectral range from 0.4 to 2.5 μm which includes the visible and short-wave infrared bands. We use 150 of the 210 bands (23rd–101st, 109th–136th, and 152nd–194th) by removing the absorption and low-SNR bands. The DR-II image contains six military targets on the dirt road, while the FR-I image contains 14 targets along the tree line, as seen in Figures 19.3a and 19.4a, respectively. Every pixel on the targets is considered a target pixel when computing the ROC curves. We use a small target subdictionary constructed by $N_t = 18$ pixels on the leftmost target in the scene. For the background subdictionary, we use an adaptive local background dictionary. Specifically, the background subdictionary A^b is generated locally for each test pixel using a dual window centered at the pixel of interest. The inner window should be larger than the size of the targets in the scene and only pixels in the outer region will form the atoms in A^b. In this way, the subspace spanned by the background subdictionary becomes adaptive to the local statistics. For DR-II and FR-I, the outer and inner windows have size 21×21 and 15×15, respectively, so there are $N_b = 216$ background training samples in A^b.

The sparsity-based detection techniques are applied on these two HYDICE images. The SP algorithm [49] is used to solve the sparse recovery problems in Equations 19.9 and 19.20, and the results are denoted by SP and SP-S (SP with the Laplacian smoothing term), respectively. The SSP algorithm is used to solve the simultaneous sparse recovery problem in Equation 19.28 and the results are denoted by SSP. In all of the SP, SP-S, and SSP algorithms, the sparsity level is set to $K_0 = 10$. For SP-S, the weighting factor λ is fixed to 1. For the joint sparsity model, a 5×5 square neighborhood ($T = 25$ in the problem formulation Equation 19.25) is used. The ROC curves are shown in Figure 19.2 and the detector output for these three sparsity-based algorithms are shown in Figure 19.3b–d for DR-II and Figure 19.4b–d for FR-I. We see that the incorporation of the spatial correlation between neighboring pixels through the Laplacian smoothing constraint and joint sparsity model yields a significant improvement in detection performance compared to a direct application of the pixel-wise sparsity model (Equation 19.3). The results obtained by OMP-based recovery algorithms are similar to those obtained by the SP-based algorithms, and are omitted herein.

Under the same settings (i.e., same target and background training samples for all detectors), the classical statistical detectors SMF, MSD, and ASD as well as the binary classifier SVM are also applied to detect the targets of interest. The first three statistical detectors are pixel-wise detectors and their implementation details can be found in Ref. [59]. For SVM, we use a composite kernel (denoted SVM-CK) that combines the spectral and spatial information via a weighted summation, which has been shown to outperform the spectral-only SVM in HSI classification [33]. For each pixel, the spatial features (e.g., the mean and the standard deviation per spectral band) are explicitly extracted in the 5×5 neighborhood centered at that pixel. An SVM is then trained for each spectral-spatial pixel using atoms in A^b and A^t as belonging to two different classes with RBF kernels for both spectral and spatial features. The parameters associated with SVM and composite kernels are defined and explained in detail in Refs. [29,33]. For comparison, the detector output obtained from these detectors is also shown in Figure 19.3e–h and Figure 19.4e–h for DR-II and FR-I, respectively. The ROC curves for the two test HYDICE images using SMF, MSD, ASD, and SVM-CK are shown

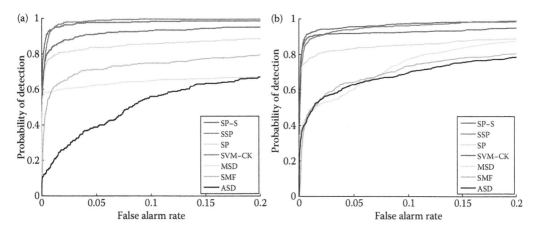

FIGURE 19.2 (**See color insert.**) ROC curves using various detection and classification algorithms for (a) DR-II and (b) FR-I using local background dictionary (dual window approach), $N_t = 18$ and $N_b = 216$.

in Figure 19.2. Overall from this figure, one can observe that for both images, the sparsity-based detectors incorporating the contextual information yield the best performance.

19.4.2 CLASSIFICATION

In this section, we examine the classification performance of the sparsity-based algorithms on three hyperspectral images. The classification results are then compared to those obtained by SVM and SVM-CK, which have shown good performances in hyperspectral classification [29,33,34].

The SP algorithm [49] and OMP algorithm [48] are used to approximately solve the sparsity-constrained problems in Equations 19.9 and 19.20, and the results are denoted as SP, OMP, SP-S (SP with smoothing), and OMP-S (OMP with smoothing). For comparison, the classification performance using the SPG-L1 package [53], which solves the linearized sparse recovery problem in (19.10), is also included. Algorithms 1 (SOMP) and 2 (SSP) are implemented to approximately solve the simultaneous sparse recovery problem in Equation 19.28. The SVM parameters (RBF-kernel parameter γ, regularization parameter C, and weight μ for composite kernels) are obtained by cross-validation. The one-against-one strategy is employed for M-class classification using SVM and SVM-CK.

19.4.2.1 AVIRIS Data Set: Indian Pines

The first hyperspectral image in our experiments is the commonly used airborne visible/infrared imaging spectrometer (AVIRIS) image Indian Pines [60]. The AVIRIS sensor generates 220 bands across the spectral range from 0.2 to 2.4 μm. In the experiments, the number of bands is reduced to 200 by removing 20 water absorption bands [28]. This image has spatial resolution of 20 m/pixel and spatial dimension 145 × 145. It contains 16 ground-truth classes, most of which are different types of crops (e.g., corns, soybeans, and wheats), as seen in Table 19.1. For each of the 16 classes, we randomly choose around 10% of the labeled samples for training and use the rest 90% for testing. The number of training and test samples for each class is shown in Table 19.1 and the training and test sets are visually shown in Figure 19.5a and b, respectively.

The classification accuracy for each class, the overall accuracy, average accuracy, and the κ coefficient measure [61] are shown in Table 19.2 using different classifiers on the test set. The overall accuracy is computed by the ratio between correctly classified test samples and the total number of test samples, and the average accuracy is the mean of the 16 class accuracies. The κ coefficient is computed by weighting the measured accuracies. It incorporates both the diagonal and off-diagonal entries of the confusion matrix and is a robust measure of the degree of agreement.

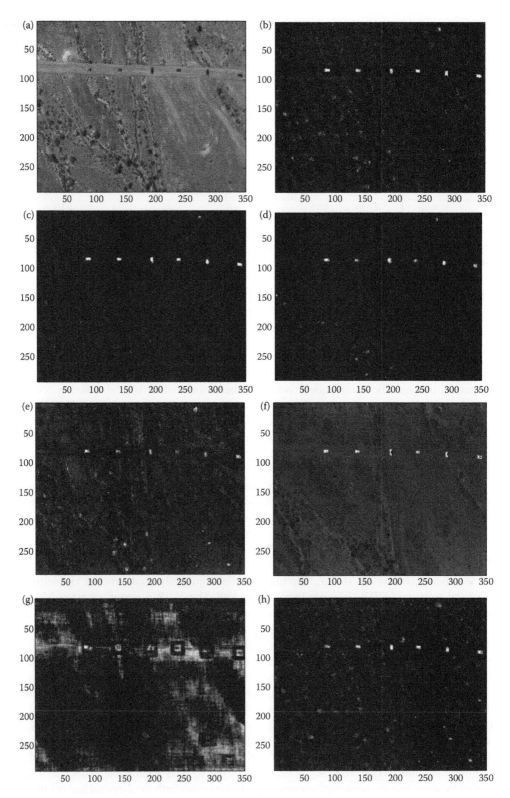

FIGURE 19.3 DR-II. (a) Averaged image over 150 bands. Detection results for DR-II using (b) SP, (c) SP-S, (d) SSP, (e) SMF, (f) MSD, (g) ASD, and (h) SVM-CK.

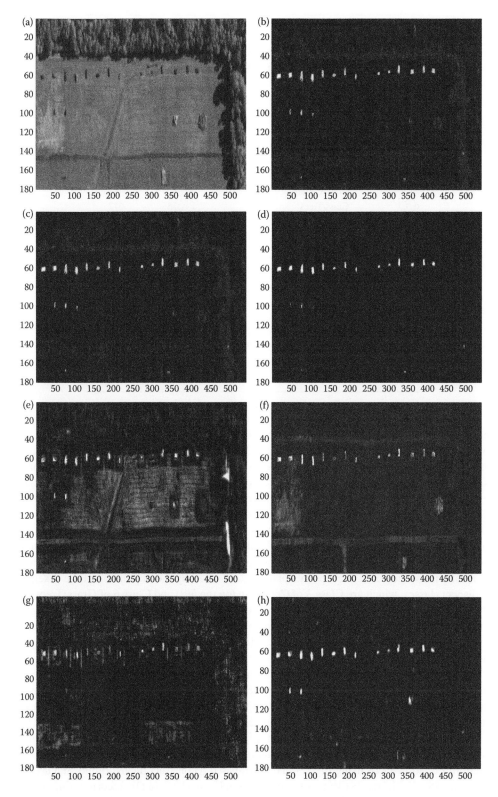

FIGURE 19.4 FR-I. (a) Averaged image over 150 bands. Detection results for FR-I using (b) SP, (c) SP-S, (d) SSP, (e) SMF, (f) MSD, (g) ASD, and (h) SVM-CK.

TABLE 19.1

The 16 Ground-Truth Classes in AVIRIS Indian Pines and the Training and Test Sets for Each Class

Class		Samples	
Number	Name	Train	Test
1	Alfalfa	6	48
2	Corn-notill	144	1290
3	Corn-min	84	750
4	Corn	24	210
5	Grass/Pasture	50	447
6	Grass/Trees	75	672
7	Grass/Pasture-mowed	3	23
8	Hay-windrowed	49	440
9	Oats	2	18
10	Soybeans-notill	97	871
11	Soybeans-min	247	2221
12	Soybean-clean	62	552
13	Wheat	22	190
14	Woods	130	1164
15	Building-Grass-Trees-Drives	38	342
16	Stone-steel towers	10	85
Total		1043	9323

The parameters for SVM and SVM-CK are obtained by fivefold cross-validation. For the sparsity-based algorithms, the sparsity level K_0 is chosen between $K_0 = 5$ and $K_0 = 30$. For SP-S and OMP-S, the weighting factor λ is fixed to 1. Since the Indian Pines image consists of large homogeneous regions, a large window of size 9×9 ($T = 81$) is used in SSP and SOMP. For pixel-wise algorithms, the ℓ_1-relaxation technique (Equation 19.10) yields comparable results to SP. In most cases, the proposed sparsity-based algorithm with spatial information outperforms the classical SVM and overall the SOMP algorithm provides the best performance. However, both SSP and SOMP fail to identify any samples belonging to the 9th class consisting of oats. This is partly due to the lack of training samples (20 ground-truth samples in total, and only two are used for training). Moreover, Oats pixels cover a very narrow region of size 10×2 located in the mid-left of the image. In SSP/SOMP, the 9×9 local window centered at each Oats pixel is dominated by the population of the pixels in the two adjacent classes, which are Class 3 of Corn-min (on the left) and Class 6 of Grass/Trees (on the right), where each of them occupies a large homogeneous region. By forcing the joint sparsity within this window size, the Oats-covered region is completely smoothed and misclassified as Corn-min and Grass/Trees. More results for this image using SVM and other supervised classifiers can be found in Ref. [34].

The classification maps on labeled pixels obtained from the various techniques are presented in Figure 19.5c–j. Figure 19.5c and d illustrates the results of the one-against-one multiclass SVM technique using a spectral-only kernel and a composite kernel combining both spectral and spatial information through a weighted summation, respectively. Figure 19.5e–j shows the visual results obtained by solving the original ℓ_0-norm minimization problem in Equation 19.9, the Laplacian-constrained sparse recovery problem in Equation 19.20, and the simultaneous sparse recovery problem in Equation 19.28 using the SP- and OMP-based reconstruction algorithms SP/OMP, SP-S/OMP-S, and SSP/SOMP, respectively. One can see from Figure 19.5 that by incorporating the spatial information, the sparsity-based algorithm leads to a much smoother classification map than the pixel-wise algorithms.

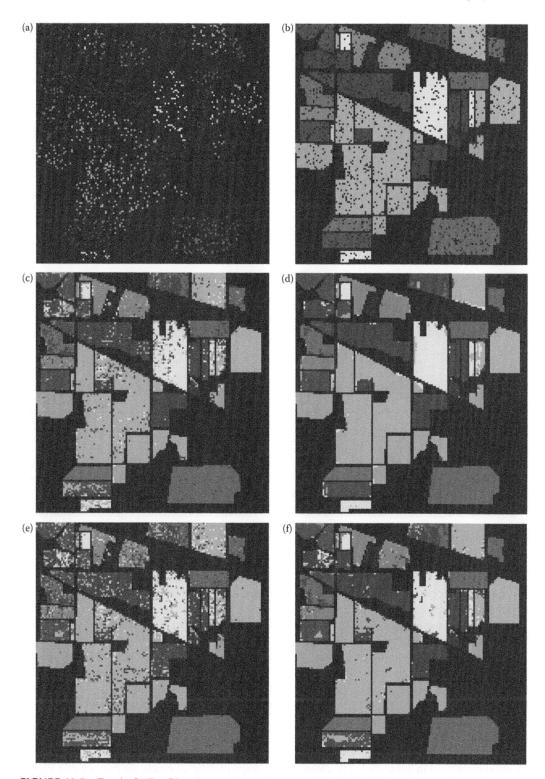

FIGURE 19.5   For the Indian Pines image: (a) training set and (b) test set. Classification maps obtained by (c) SVM, (d) SVM-CK, (e) SP, (f) SP-S, (g) SSP, (h) OMP, (i) OMP-S, and (j) SOMP.

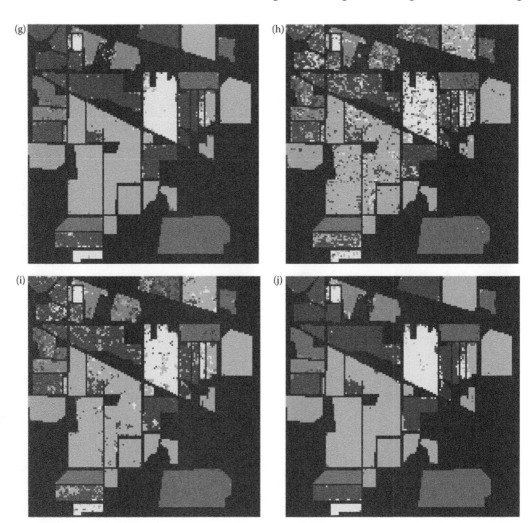

FIGURE 19.5 Continued.

19.4.2.2 ROSIS Urban Data Over Pavia, Italy

The next two hyperspectral images used in our experiments, University of Pavia and Center of Pavia, are urban images acquired by the reflective optics system imaging spectrometer (ROSIS). The ROSIS sensor generates 115 spectral bands ranging from 0.43 to 0.86 μm and has a spatial resolution of 1.3 m/pixel [34].

The University of Pavia image consists of 610 × 340 pixels, each having 103 bands with the 12 most noisy bands removed. There are nine ground-truth classes of interests, as shown in Table 19.3. For this image, we follow the same experiment settings for the training and test sets as used in [32], in which about 9% of all labeled data are used as training and the rest are used for testing. Details about the training and test sets are shown in Table 19.3 and displayed in Figure 19.6a and b, respectively.

The classification results using SVM, SVM-CK, SP, SP-S, SSP, OMP, OMP-S, SOMP, and SPG-L1 are summarized in Table 19.4. The classification maps on labeled pixels are presented in Figure 19.6c–j. For SVM and SVM-CK, we use a one-against-one strategy with RBF kernels, and the parameters are obtained by 10-fold cross-validation. The sparsity level for the SP- and OMP-based algorithms is chosen between $K_0 = 5$ and $K_0 = 30$ and the weighting factor λ is fixed to 1 for SP-S

TABLE 19.2

Classification Accuracy (%) for the Indian Pines Image on the Test Set

Class	SVM	SVM-CK	SP	SP-S	SSP	OMP	OMP-S	SOMP	ℓ_1
1	81.25	**95.83**	68.75	87.50	89.58	68.75	70.83	85.42	39.58
2	86.28	**96.67**	74.65	91.94	95.04	65.97	79.22	94.88	78.53
3	72.80	90.93	63.20	82.53	92.93	60.67	76.67	**94.93**	51.87
4	58.10	85.71	40.00	70.95	85.24	38.57	55.24	**91.43**	28.57
5	92.39	93.74	89.04	94.41	92.17	89.49	**95.30**	89.49	80.76
6	96.88	97.32	95.98	99.26	98.81	95.24	98.96	98.51	**99.40**
7	43.48	69.57	21.74	47.83	73.91	21.74	52.17	**91.30**	17.39
8	98.86	98.41	99.09	**99.77**	99.55	97.05	**99.77**	99.55	99.32
9	50.00	55.56	61.11	**94.44**	0	33.33	72.22	0	16.67
10	71.53	**93.80**	70.72	86.80	88.98	68.20	82.32	89.44	63.95
11	84.38	94.37	77.94	93.38	**97.34**	75.96	88.79	**97.34**	86.04
12	85.51	**93.66**	61.23	84.24	86.59	54.53	73.73	88.22	57.79
13	**100**	99.47	**100**	**100**	99.47	**100**	98.95	**100**	**100**
14	93.30	**99.14**	95.62	98.28	98.88	92.87	97.25	**99.14**	97.94
15	64.91	87.43	48.25	69.30	97.37	41.23	49.71	**99.12**	35.96
16	88.24	**100**	92.94	95.29	85.88	94.12	**100**	96.47	90.59
Overall	84.52	94.86	78.10	91.16	94.79	74.78	85.52	**95.28**	77.99
Average	79.24	**90.73**	72.52	87.25	86.36	68.61	80.70	88.45	65.27
κ	0.823	0.941	0.749	0.899	0.940	0.712	0.834	**0.946**	0.746

Note: The bold numbers correspond to the highest rate in each row.

and OMP-S. Since this image is obtained from an urban area with small buildings, it lacks the large spatial homogeneity that was present in the Indian Pines image. Thus, a smaller neighborhood (3×3 to 5×5 window) is used in the experiment for SOMP and SSP for joint sparse recovery. The proposed sparsity-based classification algorithms with spatial information achieve better or comparable performance in most cases than the SVM classifier. However, SVM-CK yields the best overall performance and a smoother visual effect.

TABLE 19.3

The Nine Classes in University of Pavia and the Training and Test Sets

	Class		Samples	
Number	Name		Train	Test
1	Asphalt		548	6304
2	Meadows		540	18,146
3	Gravel		392	1815
4	Trees		524	2912
5	Metal sheets		265	1113
6	Bare soil		532	4572
7	Bitumen		375	981
8	Bricks		514	3364
9	Shadows		231	795
Total			3921	40,002

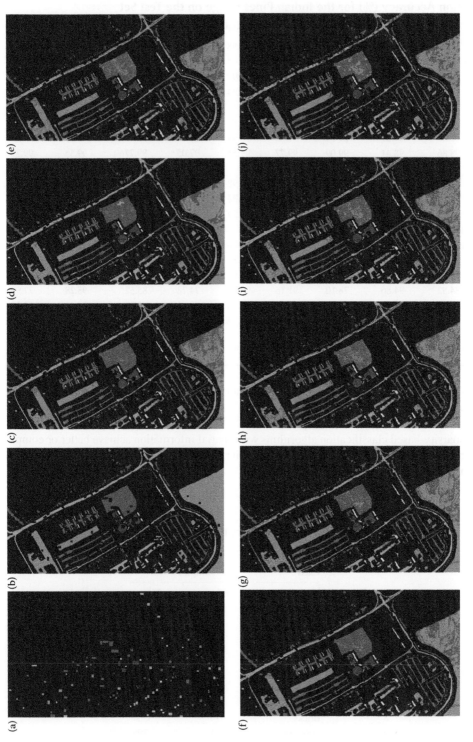

FIGURE 19.6 **(See color insert.)** For the University of Pavia image: (a) training set and (b) test set. Classification maps obtained by (c) SVM, (d) SVM-CK, (e) SP, (f) SP-S, (g) SSP, (h) OMP, (i) OMP-S, and (j) SOMP.

TABLE 19.4

Classification Accuracy (%) for University of Pavia on the Test Set Using Different Classifiers

Class	SVM	SVM-CK	SP	SP-S	SSP	OMP	OMP-S	SOMP	ℓ_1
1	**84.30**	79.85	69.70	83.79	69.59	68.07	72.53	59.33	80.65
2	67.01	**84.86**	67.69	72.35	72.31	67.07	73.12	78.15	64.74
3	68.43	81.87	67.11	71.85	74.10	65.45	75.87	**83.53**	73.22
4	97.80	96.36	97.46	**98.94**	95.33	97.32	98.18	96.91	98.35
5	99.37	99.37	99.82	**100**	99.73	99.73	**100**	99.46	99.91
6	92.45	**93.55**	75.79	92.63	86.72	73.29	77.69	77.41	92.54
7	89.91	90.21	88.79	91.44	90.32	87.26	94.90	**98.57**	86.95
8	92.42	92.81	84.96	**95.57**	90.46	81.90	87.60	89.09	81.54
9	97.23	95.35	93.58	98.24	90.94	94.72	97.48	91.95	**98.99**
Overall	79.15	**87.18**	74.45	82.09	78.39	73.27	78.48	79.00	76.87
Average	87.66	**90.47**	82.77	89.42	85.50	81.65	86.38	86.04	86.32
κ	0.737	**0.833**	0.675	0.772	0.724	0.661	0.724	0.728	0.709

Note: The bold numbers correspond to the highest rate in each row.

The third image, Center of Pavia, is the other urban image collected by the ROSIS sensor over the center of the Pavia city. This image consists of 1096×492 pixels, each having 102 spectral bands after 13 noisy bands are removed. The nine ground-truth classes and the number of training and test samples for each class are shown in Table 19.5 and illustrated in Figure 19.7a and b. For this image, about 5% of the labeled data are used as training samples.

The classification results are summarized in Table 19.6, and the classification maps are shown in Figure 19.7c–j. For SVM and SVM-CK, again, the one-against-one strategy is applied and the parameters are optimized by 10-fold cross-validation. For all sparsity-based algorithms, the sparsity level is chosen between $K_0 = 5$ and $K_0 = 25$. For SP-S and OMP-S, the weighting factor λ is set to 1. For SSP and SOMP, similar to the case of the University of Pavia, we use a small neighborhood of $T = 5$ for SSP and $T = 25$ for SOMP. In this case, all of the sparsity-based algorithms SP-S, SSP, OMP-S, and SOMP outperform SVM and SVM-CK, and SOMP yields the best overall performance.

TABLE 19.5

The Nine Ground-Truth Classes in Center of Pavia and the Training and Test Sets

Class		Samples	
Number	Name	Train	Test
1	Water	745	64,533
2	Trees	785	5722
3	Meadow	797	2094
4	Brick	485	1667
5	Soil	820	5729
6	Asphalt	678	6847
7	Bitumen	808	6479
8	Tile	223	2899
9	Shadow	195	1970
Total		5536	97,940

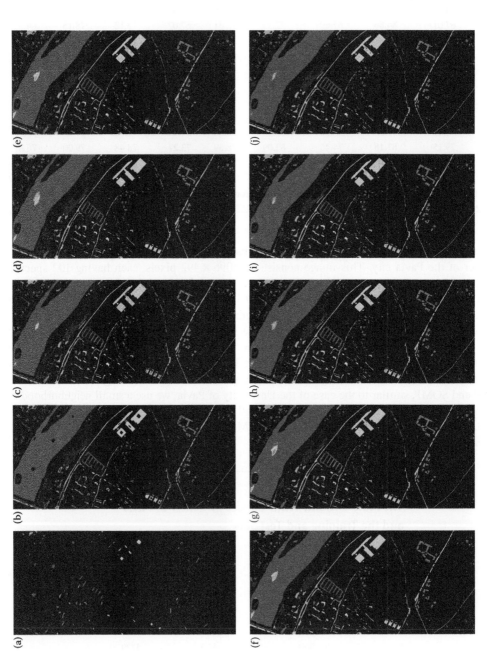

FIGURE 19.7 For Center of Pavia: (a) training set and (b) test set. Classification maps obtained by (c) SVM, (d) SVM-CK, (e) SP, (f) SP-S, (g) SSP, (h) OMP, (i) OMP-S, and (j) SOMP.

TABLE 19.6

Classification Accuracy (%) for Center of Pavia on the Test Set Using Different Classifiers

Class	SVM	SVM-CK	SP	SP-S	SSP	OMP	OMP-S	SOMP	ℓ_1
1	99.19	97.46	98.20	98.21	97.79	98.91	99.32	99.32	**99.35**
2	77.74	**93.08**	86.98	91.63	92.82	86.75	88.17	92.38	83.52
3	86.74	97.09	96.61	97.42	97.80	96.04	**97.99**	95.46	97.09
4	40.38	77.02	84.16	**94.24**	78.52	81.22	83.44	85.66	73.67
5	97.52	**98.39**	94.01	96.79	95.81	94.40	95.95	96.37	95.92
6	94.77	94.32	92.92	96.98	96.52	91.94	93.94	93.81	**97.01**
7	74.37	**97.50**	93.80	97.39	95.96	93.18	95.15	94.68	92.65
8	98.94	99.83	98.79	**99.86**	99.79	98.62	**99.86**	99.69	99.45
9	**100**	99.95	99.34	99.54	98.83	98.07	98.93	98.68	99.09
Overall	94.63	96.81	96.40	97.59	96.93	96.68	97.52	**97.66**	97.13
Average	85.52	94.96	93.87	**96.90**	94.87	93.24	94.74	95.01	93.08
κ	0.899	0.943	0.935	0.957	0.945	0.940	0.955	**0.958**	0.948

Note: The bold numbers correspond to the highest rate in each row.

19.4.3 Effects of Parameters in the Sparsity Models

For the sparsity-based algorithm with the Laplacian constraint, the scalar value λ in matrix $\tilde{A}$ used in Equation 19.20 controls the relative importance between smoothness and reconstruction accuracy, leading to different results. Now we demonstrate how λ affects the classification accuracy on the AVIRIS Indian Pines image. As previously done, 10% of the data are chosen as training samples and the remaining 90% are used for testing. The sparsity level K_0 is fixed to $K_0 = 10$ for SP-S and $K_0 = 5$ for OMP-S. The weighting factor λ varies from 10^{-2} to 10. The overall accuracy as a function of λ is illustrated in Figure 19.8. For both SP-S and OMP-S, with $\lambda \leq 1$, there is only a slight difference in the classification accuracy. As λ increases (i.e., more weight on smoothness and less weight on approximation accuracy), the classification performance degrades quickly.

Next, we illustrate the effect of neighborhood size T in the joint sparsity model (Equation 19.25) on the detection performance. In this experiment, the simultaneous sparse approximation problem in Equation 19.28 is solved by SOMP for different neighborhood at a fixed sparsity level $K_0 = 10$. Specifically, we use 1×1 neighborhood (equivalent to the pixel-wise sparsity model), 4-connected

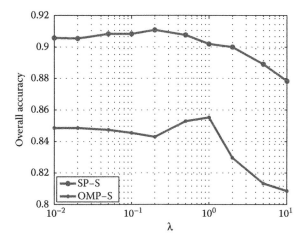

FIGURE 19.8   Effect of the weighting factor λ of the Laplacian constraint for Indian Pines.

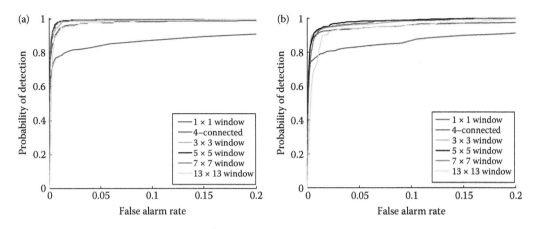

FIGURE 19.9 Effects of the neighborhood size T on detection performance for (a) DR-II and (b) FR-I.

neighborhood, 3×3, 5×5, 7×7, and 13×13 window neighborhood, corresponding to $T = 1, 5, 9,$ 25, 49, and 169, respectively. The ROC curves for the various types of neighborhood are shown in Figure 19.9. By incorporating the contextual interaction between neighboring pixels, the detector performance is significantly improved. As the neighborhood size increases, the performance tends to saturate.

Next, we demonstrate how the target detection results are affected by the sparsity level K_0 in the simultaneous sparse recovery process (Equation 19.28) on the two HYDICE images DR-II and FR-I. The sparsity level refers to the number of nonzero rows in S, which is also the number of common atoms selected from the dictionary by the greedy algorithms to simultaneously approximate all of the neighboring pixels. The neighborhood size T is fixed to 25 (i.e., a 5×5 window is used) in this experiment. The ROC curves for both images using SOMP with sparsity levels $K_0 = 1, 2, 5, 10,$ 15, and 50 are shown in Figure 19.10. For very small K_0, the sparsity-based technique reduces to simple template matching and leads to under-fitting. Generally the detection performance improves as the sparsity level K_0 increases to a certain level. However, if K_0 is too large, the solution becomes dense and involves both background and target atoms, and thus its discriminative power degrades.

Next, we examine the effect of the sparsity level K_0 and the neighborhood size T on the classification performance of the simultaneous sparse approximation algorithms SOMP and SSP on the

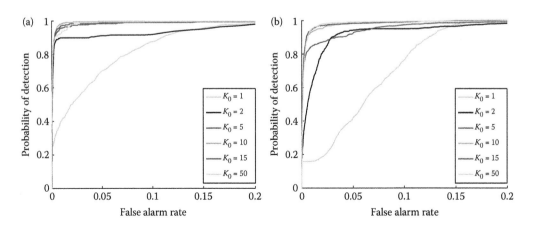

FIGURE 19.10 Effects of the sparsity level K_0 on detection performance for (a) DR-II and (b) FR-I.

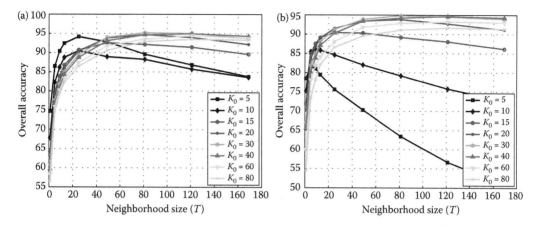

FIGURE 19.11 **(See color insert.)** Effect of the sparsity level K_0 and size of neighborhood T for Indian Pines. (a) SOMP and (b) SSP.

AVIRIS image of Indian Pines. In this experiment, we randomly choose 10% of the data in each class as training samples and use the remaining 90% as test samples. In each test, we apply the SOMP and SSP algorithms with different sparsity level K_0 and neighborhood size T to solve the problem in Equation 19.28. The sparsity level K_0 ranges from $K_0 = 5$ to $K_0 = 80$, and the neighborhood ranges from the 4-connected neighborhood ($T = 5$) to a 13×13 window ($T = 169$). The overall classification accuracy plots on the entire test set for Indian Pines are shown in Figure 19.11a and b for SOMP and SSP, respectively. The horizontal axis indicates the size of the neighborhood T and the vertical axis is the overall accuracy (%). For small sparsity level K_0, if the neighborhood size T is too large, then the neighboring pixels cannot be faithfully approximated by few training samples and the classification accuracy is significantly reduced. On the other hand, as K_0 increases toward the size of the dictionary, the solution converges to the pseudo-inverse solution, which is no longer sparse and may involve atoms in multiple subdictionaries, leading to a performance degradation. One can also see that for sufficiently large sparsity level K_0, in general the classification performance increases almost monotonically as T increases. However, large neighborhood may cause oversmoothing over neighboring classes, which would lead to a decrease in the overall classification accuracy.

19.5 CONCLUSIONS

In this chapter, we introduced a new technique for HSI target detection and classification based on sparse representation. In the proposed algorithm, an HSI pixel is assumed to be sparsely represented by a few atoms in a given training dictionary. The sparse representation of a test spectral sample is recovered by solving a sparsity-constrained optimization problem via greedy pursuit algorithms. To improve the classification performance, two different ways are proposed to incorporate the contextual information of HSI. One approach is to explicitly include a smoothing term through the vector Laplacian at the reconstructed pixel of interest in the optimization formulation, and the other is through a joint-sparsity model for neighboring pixels centered at the pixel of interest. Experimental results show that the sparsity-based algorithms incorporated with interpixel correlation outperform classical target detection and classification algorithms, especially for images with large homogeneous areas.

ACKNOWLEDGMENT

The authors would like to thank the University of Pavia and the HySenS project for kindly providing the ROSIS images of University of Pavia and Center of Pavia.

REFERENCES

1. S. Mallat, *A Wavelet Tour of Signal Processing: The Sparse Way*, Amsterdam, Boston: Academic Press, 2008.
2. J. Wright, Y. Ma, J. Mairal, G. Sapiro, T. Huang, and S. Yan, Sparse representation for computer vision and pattern recognition, *Proceedings of the IEEE*, 98(6), 1031–1044, 2010.
3. E. Candès, J. Romberg, and T. Tao, Robust uncertainty principles: Exact signal reconstruction from highly incomplete frequency information, *IEEE Transactions on Information Theory*, 52(2), 489–509, 2006.
4. D. L. Donoho, Compressed sensing, *IEEE Transactions on Information Theory*, 52(4), 1289–1306, 2006.
5. A. M. Bruckstein, D. L. Donoho, and M. Elad, From sparse solutions of systems of equations to sparse modeling of signals and images, *SIAM Review*, 51(1), 34–81, 2009.
6. S. Rao, R. Tron, R. Vidal, and Y. Ma, Motion segmentation via robust subspace separation in the presence of outlying, incomplete, or corrupted trajectories, in *Proceedings of the IEEE Conference on Computer Vision and Pattern Recognition*, June 2008, pp. 1–8.
7. E. Elhamifar and R. Vidal, Sparse subspace clustering, in *Proceedings of the IEEE Conference on Computer Vision and Pattern Recognition*, June 2009, pp. 2790–2797.
8. J. Yang, J. Wright, T. Huang, and Y. Ma, Image super-resolution as sparse representation of raw image patches, in *Proceedings of the IEEE Conference on Computer Vision and Pattern Recognition*, June 2008, pp. 1–8.
9. M. Elad and M. Aharon, Image denoising via sparse and redundant representations over learned dictionaries, *IEEE Transactions on Image Processing*, 15(12), 3736–3745, 2006.
10. J. Mairal, M. Elad, and G. Sapiro, Sparse representation for color image restoration, *IEEE Transactions on Pattern Analysis and Machine Intelligence*, 17(1), 53–69, 2008.
11. J. Wright, A. Y. Yang, A. Ganesh, S. Sastry, and Y. Ma, Robust face recognition via sparse representation, *IEEE Transactions on Pattern Analysis and Machine Intelligence*, 31(2), 210–227, 2009.
12. J. K. Pillai, V. M. Patel, and R. Chellappa, Sparsity inspired selection and recognition of iris images, in *Proceedings of the IEEE Third International Conference on Biometrics: Theory, Applications and Systems*, September 2009, pp. 1–6.
13. X. Hang and F.-X. Wu, Sparse representation for classification of tumors using gene expression data, *Journal of Biomedicine and Biotechnology*, 2009, doi: 10.1155/2009/403689.
14. Z. Guo, T. Wittman, and S. Osher, L1 unmixing and its application to hyperspectral image enhancement, in *Proceedings of the SPIE Conference on Algorithms and Technologies for Multispectral, Hyperspectral, and Ultraspectral Imagery XV*, April 2009, vol. 7334.
15. M. Borengasser, W. S. Hungate, and R. Watkins, *Hyperspectral Remote Sensing—Principles and Applications*, CRC Press, Boca Raton, FL, 2008.
16. D. Manolakis and G. Shaw, Detection algorithms for hyperspectral imaging applications, *IEEE Signal Processing Magazine*, 19(1), 29–43, 2002.
17. D. W. J. Stein, S. G. Beaven, L. E. Hoff, E. M. Winter, A. P. Schaum, and A. D. Stocker, Anomaly detection from hyperspectral imagery, *IEEE Signal Processing Magazine*, 19(1), 58–69, 2002.
18. M. T. Eismann, A. D. Stocker, and N. M. Nasrabadi, Automated hyperspectral cueing for civilian search and rescue, *Proceedings of the IEEE*, 97(6), 1031–1055, 2009.
19. N. K. Patel, C. Patnaik, S. Dutta, A. M. Shekh, and A. J. Dave, Study of crop growth parameters using airborne imaging spectrometer data, *International Journal of Remote Sensing*, 22(12), 2401–2411, 2001.
20. B. Datt, T. R. McVicar, T. G. Van Niel, D. L. B. Jupp, and J. S. Pearlman, Preprocessing EO-1 Hyperion hyperspectral data to support the application of agricultural indexes, *IEEE Transactions on Geoscience and Remote Sensing*, 41(6), 1246–1259, 2003.
21. B. Hörig, F. Kühn, F. Oschütz, and F. Lehmann, HyMap hyperspectral remote sensing to detect hydrocarbons, *International Journal of Remote Sensing*, 22(8), 1413–1422, 2001.
22. F. C. Robey, D. R. Fuhrmann, E. J. Kelly, and R. Nitzberg, A CFAR adaptive matched filter detector," *IEEE Transactions on Aerospace Electronic Systems*, 28(1), 208–216, 1992.
23. D. Manolakis, G. Shaw, and N. Keshava, Comparative analysis of hyperspectral adaptive matched filter detectors, in *Proceedings of the SPIE Conference on Algorithms for Multispectral, Hyperspectral, and Ultraspectral Imagery VI*, April 2000, vol. 4049, pp. 2–17.
24. L. L. Scharf and B. Friedlander, Matched subspace detectors, *IEEE Transactions on Signal Processing*, 42(8), 2146–2157, 1994.
25. S. Kraut, L. L. Scharf, and L. T. McWhorter, Adaptive subspace detectors, *IEEE Transactions on Signal Processing*, 49(1), 1–16, 2001.
26. B. E. Boser, I. M. Guyon, and V. N. Vapnik, A training algorithm for optimal margin classifiers, in *Proceedings of the Fifth Annual Workshop on Computational Learning Theory*, 1992, pp. 144–152.

27. V. N. Vapnik, *The Nature of Statistical Learning Theory*, New York: Springer, 1995.
28. J. A. Gualtieri and R. F. Cromp, Support vector machines for hyperspectral remote sensing classification, in *Proceedings of the SPIE*, 3584, 221–232, 1998.
29. F. Melgani and L. Bruzzone, Classification of hyperspectral remote sensing images with support vector machines, *IEEE Transactions on Geoscience and Remote Sensing*, 42(8), 1778–1790, 2004.
30. L. Bruzzone, M. Chi, and M. Marconcini, A novel transductive SVM for the semisupervised classification of remote sensing images, *IEEE Transactions on Geoscience and Remote Sensing*, 44(11), 3363–3373, 2006.
31. F. Bovolo, L. Bruzzone, and M. Marconcini, A novel context-sensitive SVM for classification of remote sensing images, in *Proceedings of the IEEE International Geoscience and Remote Sensing Symposium*, pp. 2498–2501, 2006.
32. Y. Tarabalka, J. A. Benediktsson, and J. Chanussot, Spectra-spatial classification of hyperspectral imagery based on partitional clustering techniques, *IEEE Transactions on Geoscience and Remote Sensing*, 47(8), 2973–2987, 2009.
33. G. Camps-Valls, L. Gomez-Chova, J. Muñoz-Marí, J. Vila-Francés, and J. Calpe-Maravilla, Composite kernels for hyperspectral image classification, *IEEE Geoscience and Remote Sensing Letters*, 3(1), 93–97, 2006.
34. A. Plaza, J. A. Benediktsson, J. W. Boardman, J. Brazile, L. Bruzzone, G. Camps-Valls, J. Chanussot, M. Fauvel, P. Gamba, A. Gualtieri, M. Marconcini, J. C. Tilton, and G. Trianni, Recent advances in techniques for hyperspectral image processing, *Remote Sensing of Environment*, 113(Supplement 1), S110–S122, 2009.
35. H.-Y. Huang and B.-C. Kuo, Double nearest proportion feature extraction for hyperspectral-image classification, *IEEE Transactions on Geoscience and Remote Sensing*, 48(11), 4034–4046, 2010.
36. H. R. Kalluri, S. Prasad, and L. M. Bruce, Decision-level fusion of spectral reflectance and derivative information for robust hyperspectral land cover classification, *IEEE Transactions on Geoscience and Remote Sensing*, 48(11), 4047–4058, 2010.
37. B. D. Bue, E. Merényi, and B. Csathó, Automated labeling of materials in hyperspectral imagery, *IEEE Transactions on Geoscience and Remote Sensing*, 48(11), 4059–4070, 2010.
38. B. Demir and S. Ertürk, Empirical mode decomposition of hyperspectral images for support vector machine classification, *IEEE Transactions on Geoscience and Remote Sensing*, 48(11), 4071–4084, 2010.
39. J. Li, J. M. Bioucas-Dias, and A. Plaza, Semisupervised hyperspectral image segmentation using multinomial logistic regression with active learning, *IEEE Transactions on Geoscience and Remote Sensing*, 48(11), 4085–4098, 2010.
40. L. Ma, M. M. Crawford, and J. Tian, Local manifold learning-based k-nearest-neighbor for hyperspectral image classification, *IEEE Transactions on Geoscience and Remote Sensing*, 48(11), 4099–4109, 2010.
41. W. Kim and M. M. Crawford, Adaptive classification for hyperspectral image data using manifold regularization kernel machines, *IEEE Transactions on Geoscience and Remote Sensing*, 48(11), 4110–4121, 2010.
42. Y. Tarabalka, J. A. Benediktsson, J. Chanussot, and J. C. Tilton, Multiple spectral-spatial classification approach for hyperspectral data, *IEEE Transactions on Geoscience and Remote Sensing*, 48(11), 4122–4132, 2010.
43. R. S. Rand and D. M. Keenan, Spatially smooth partitioning of hyperspectral imagery using spectral/spatial measures of disparity, *IEEE Transactions on Geoscience and Remote Sensing*, 41(6), 1479–1490, 2003.
44. Y. Tarabalka, M. Fauvel, J. Chanussot, and J. A. Benediktsson, SVM- and MRF-based method for accurate classification of hyperspectral images, *IEEE Geoscience and Remote Sensing Letters*, 7(4), 736–740, 2010.
45. J. A. Tropp, A. C. Gilbert, and M. J. Strauss, Algorithms for simultaneous sparse approximation. Part I: Greedy pursuit, *Signal Processing, special issue on Sparse Approximations in Signal and Image Processing*, 86, 572–588, 2006.
46. S. F. Cotter, B. D. Rao, K. Engan, and K. Kreutz-Delgado, Sparse solutions to linear inverse problems with multiple measurement vectors, *IEEE Transactions on Signal Processing*, 53(7), 2477–2488, 2005.
47. J. A. Tropp and S. J. Wright, Computational methods for sparse solution of linear inverse problems, *Proceedings of the IEEE*, 98(6), 948–958, 2010.
48. J. Tropp and A. Gilbert, Signal recovery from random measurements via orthogonal matching pursuit, *IEEE Transactions on Information Theory*, 53(12), 4655–4666, 2007.

49. W. Dai and O. Milenkovic, Subspace pursuit for compressive sensing signal reconstruction, *IEEE Transactions on Information Theory*, 55(5), 2230–2249, 2009.

50. S. S. Chen, D. L. Donoho, and M. A. Saunders, Atomic decomposition by basis pursuit, *SIAM Journal on Scientific Computing*, 20(1), 33–61, 1998.

51. S.J. Kim, K. Koh, M. Lustig, S. Boyd, and D. Gorinevsky, An interior-point method for large-scale l1-regularized least squares, *IEEE Journal on Special Topics in Signal Processing*, 1(4), 606–617, 2007.

52. M. A. T. Figueiredo, R. D. Nowak, and S. J. Wright, Gradient projection for sparse reconstruction: Application to compressed sensing and other inverse problems, *IEEE Journal of Selected Topics in Signal Processing*, 1(4), 586–597, 2007.

53. E. van den Berg and M. P. Friedlander, Probing the pareto frontier for basis pursuit solutions, *SIAM Journal on Scientific Computing*, 31(2), 890–912, 2008.

54. A. Rakotomamonjy, Surveying and comparing simultaneous sparse approximation (or group-lasso) algorithms, *Signal Processing*, 91(7), 1505–1526, 2011.

55. J. A. Tropp, A. C. Gilbert, and M. J. Strauss, Algorithms for simultaneous sparse approximation. Part II: Convex relaxation, *Signal Processing, special issue on Sparse Approximations in Signal and Image Processing*, 86, 589–602, 2006.

56. E. van den Berg and M. P. Friedlander, Theoretical and empirical results for recovery from multiple measurements, *IEEE Transactions on Information Theory*, 56(5), 2516–2527, 2010.

57. D. Leviatan and V. N. Temlyakov, Simultaneous approximation by greedy algorithms, *Advances in Computational Mathematics*, 25(1–3), 73–90, 2006.

58. R. W. Basedow, D. C. Carmer, and M. E. Anderson, HYDICE system: Implementation and performance, in *Proceedings of the SPIE Conference on Algorithms and Technologies for Multispectral, Hyperspectral, and Ultraspectral Imagery XV*, vol. 2480, pp. 258–267, April 1995.

59. H. Kwon and N. M. Nasrabadi, A comparative analysis of kernel subspace target detectors for hyperspectral imagery, *EURASIP Journal on Applied Signal Processing*, 2007(1), 193–193 2007.

60. AVIRIS NW Indiana's Indian Pines 1992 data set, http://cobweb.ecn.purdue.edu/~biehl/MultiSpec/documentation.html.

61. J. A. Richards and X. Jia, *Remote Sensing Digital Image Analysis: An Introduction*, Berlin: Springer, 4th edition, 2006.

20 Integration of Full and Mixed Pixel Techniques to Obtain Thematic Maps with a Refined Resolution

Alberto Villa, Jon Atli Benediktsson, Jocelyn Chanussot, and Christian Jutten

CONTENTS

20.1 INTRODUCTION

Remote sensing offers a very suitable possibility for obtaining geographically and temporally detailed data on land cover and its dynamic behavior. The main advantage of satellite and airborne systems is the possibility to obtain relatively cheap and rapid methods of acquiring up-to-date information over a large geographical area. Scientific advancements made possible the use of remote sensing in a variety of disciplines such as agriculture, forestry, hydrology, geology, cartography, and meteorology.

Hyperspectral imaging, in particular, is a continuously growing area of remote-sensing applications. The wide spectral range, varying from the ultraviolet to the near-infrared wavelengths, coupled with a very high spectral resolution, allows the detection and classification of surfaces and chemical elements in the observed image with improved discrimination capability.

One of the main limitations of remote sensing is handling the intrinsic scale of variation of land cover. A satellite image often contains objects smaller than the scale of sampling imposed by the satellite sensor (Atkinson 1997). The problem is particularly significant in the case of multispectral imagery, due to the trade-off which generally exists between spatial and spectral resolution. A common drawback of hyperspectral sensors is the relatively low spatial resolution, which may vary from few to tens of meters, especially in the case of high-altitude sensors or instruments covering wide areas. There are many factors (such as imperfect imaging optics, atmospheric scattering, secondary illumination effects, and sensor noise) that degrade the acquired image quality and make the development of new technology to improve the spatial resolution one of the most challenging tasks for sensor designers (Vane et al. 1993). The coarse spatial resolution can lead to inhomogeneity within a single pixel, which may contain features representing distinct classes. Also in the case of high spatial resolution, often provided by airborne sensors, the problem of mixed pixels (e.g., pixels containing more than a single land cover class), may occur for several reasons: (i) even if the dimension of the objects of interest are comparable to the pixel size, there will always be a considerable probability for an object to be shifted across pixel boundaries; (ii) due to the point spread function (PSF) shifting of the hyperspectral sensors, every pixel will have a (small) contribution from the materials lying in the neighboring positions; (iii) the presence of intimate mixtures of materials, corresponding to materials mixed at a microscopic level. Therefore, regardless of the spatial resolution of the acquired image, mixtures of classes will be likely observable in the scene.

In the case of structure detection and land cover classification, problems related to low spatial resolution and presence of mixed pixels may lead to inaccuracies in the analysis of the scene. The common assumption that every pixel of the image can be associated with a unique class label is no longer verified in this case, and mixed pixels cannot be correctly addressed by traditional classifiers (Plaza et al. 2004a). Sometimes, spatial structures can also become hard to detect, with a substantive loss of information.

In order to improve analysis performances of images affected by low spatial resolution, appropriate techniques are required. Several methods have been proposed to deal with the problem of mixed pixels and low spatial resolution of remote sensing images (Foody 1996, Nguyen et al. 2006, Keshava and Mustard 2002, Nachtegael et al. 2007, Robin et al. 2008). These techniques try to handle the problem considering different theoretical backgrounds, and can therefore be divided into three main groups. The first group includes methods performing data fusion: they consider a high spatial resolution image jointly with a low-resolution image covering the same area, in order to obtain a fused image with high spectral and spatial resolution (Eismann and Hardie 2005, Hardie et al. 2004, Akgun et al. 2005). The resulting image obtained from the fusion process can afterward be used as input for classification. This approach presents two main drawbacks: (i) the need for an accurate coregistration of the two images, and (ii) the need for ancillary data which are not always easy to obtain. Although some satellites carry both optical and panchromatic sensors to facilitate data fusion, such an implementation is still not available for hyperspectral sensors.

The second group of techniques is composed of superresolution techniques that try to improve the spatial resolution without the use of secondary sources of information providing high spatial resolution data. The first attempt in this direction was made by Tatem et al. (2001), who proposed an algorithm based on Hopfield neural network which does not need any secondary source of data to perform the superresolution mapping. This method shows good performances but has its major limitation in the high computational cost. An improvement of such technique in terms of computational complexity was proposed by Gu et al., which reduces the high computational burden through a fast learning-based algorithm to integrate the spatial and spectral information of the hyperspectral image, back propagation neural network, and some ground-truth information which is independent of the test data (Gu et al. 2008). The output provided by such methods is a series of fractional abundance maps at a finer spatial resolution, in a number equal to the land cover classes considered within the observed scene.

The third group of techniques includes methods which analyze the image assuming the possibility of quantifying the mixtures of pure materials at a subpixel level. A large number of these techniques have been proposed. Examples are represented by soft classification algorithms (Nachtegael et al. 2007), which, instead of providing a crisp classification map as output, compute, for each pixel of the image, a set of values (one per each considered class) expressing the degree of membership to the class. Another example is represented by linear spectral mixture analysis techniques (SMA) (Keshava and Mustard 2002, Plaza et al. 2010, Chang et al. 2011), assuming every pixel to be the weighted sum of some constituent spectra, also called *endmembers*. Similar techniques have been proposed for subpixel image labeling, based on the latest development of machine learning (Baraldi et al. 2001, Bovolo et al. 2010). On the one hand, soft classification and unmixing-based techniques can partially overcome the drawbacks of crisp classification methods when analyzing mixed pixels, providing more information about the mixtures of pure materials observed in the image. On the other hand, the final output is a classification map (or a series of maps) representing the membership degree (or the abundance) of a pixel with respect to a class. Therefore, these techniques are not useful to obtain an improvement of the resolution of a crisp output, since the additional information provided by the fuzziness nature of the classifier is lost.

Full pixel techniques and spectral unmixing methods can be seen as complementary approaches, which could be jointly used to improve the classification accuracy of thematic maps, especially in the case of images containing mixed pixels (Villa et al. 2010a,b). However, this possibility has only been touched on investigations so far, particularly in Plaza et al. (2004a) and Wang and Jia (2009). Plaza et al. proposed an unsupervised approach for the classification of mixed pixels in hyperspectral imagery, based on the generalization of the concept of extended morphological profiles to case of multispectral data. The method, when applied to images with a small number of land cover classes or in case of high spectral separability, showed interesting results compared to classical techniques. An extension of the support vector machine (SVM) classification technique to address the problem of mixed pixels was recently proposed by Wang and Jia (2009). This extension provides good results when applied to synthetic data, showing the interest of combining hard and soft classification methods.

The attempt of jointly using complementary techniques is the direction followed in this work. One of the first attempts to integrate full and mixed pixel techniques was proposed by Villa et al. (2011). The method, making use of SVMs and spectral unmixing concepts, is presented in order to obtain thematic maps at a finer spatial resolution. Finally, a spatial regularization to re-locate subpixels obtained in the previous step is discussed.

This chapter is organized as follows. The next section presents the problem of subpixel labeling, closely connected to the problem of spatial regularization. Section 20.3 describes in more detail the technique allowing obtaining classification maps at finer spatial resolution. Experimental validation of the method is presented in Section 20.4, while conclusions are drawn in Section 20.5.

20.2 SUBPIXEL LABELING

An improvement of the classification map spatial resolution, as aimed in this work, can be obtained in two steps. The first step corresponds to estimate the quantity of each land cover class at a subpixel level, in order to acquire a knowledge about the fractional abundances of land cover classes for each pixel (e.g., estimate that water covers 30% of a given pixel's surface and grass 70%), as estimated by classification methods, or spectral unmixing. In a second step, after resampling data at a finer spatial resolution (e.g., split each pixel into a previously selected number of subpixels), assign every subpixel to a class and identify a plausible location for each subpixel, through subpixel mapping.

Subpixel mapping can therefore be defined as the technique aiming at correctly locating, from a spatial point of view, the abundance fractions of a land cover class within a pixel. Subpixel mapping shows some similarity with the groups of techniques described in the introductory section to overcome the problem of low spatial resolution and mixed pixels; however, due to its intrinsic characteristics, it is different from all of them.

The idea of subpixel mapping was first introduced by Atkinson (1997). The method is based on the concept of spatial dependence, referring to the tendency that pattern belonging to the same land cover class will have a larger probability to be spatially close, rather than distant. This idea, which is intuitive when looking at the surrounding environment, was proven to be dependent both on the environment and on the nature of the observations. Since then, a large number of subpixel mapping methods were proposed in the literature.

Steinwendner et al. (1998) attempted to allocate field boundaries by introducing a knowledge-based analysis technique. The method was limited by the assumption of homogeneous fields with straight boundaries and was seen as a preprocessing step prior to automatic pixel-by-pixel land use classification. Their research aimed at making full use of the subpixel information.

This assumption of maximum class spatial correlation at the target resolution underpins several approaches to superresolution mapping. Atkinson (2005) examined a pixel-swapping optimization algorithm within a geostatistical framework. The algorithm predicted accurately when applied to relatively simple two class simulated and real images. Thornton et al. (2006) tested the algorithm with soft-classified imagery containing several classes. The pixel-swapping algorithm along with mathematical morphology, used to suppress error in the subpixel scale output, provided moderately accurate results. Atkinson (2004) developed the two-point histogram method, a geostatistical spatial simulated annealing (SA) algorithm, which could recreate any target spatial distribution. Foody et al. (2005) employed both two-point histogram and contour-based approach to map the waterline at subpixel scales. Muslim et al. (2006) examined both the two-point histogram method and the pixel-swapping method to accurately map the shoreline. These studies showed the considerable potential of superresolution mapping techniques. More recently, Ge et al. (2009) developed a computationally efficient method for subpixel mapping, still based on the concept of spatial correlation. An exhaustive comparison of subpixel mapping methods can be found in Mertens (2008).

The main difference between the work discussed here and the above-mentioned methods resides in the way fractional abundances of each endmember are computed. The great majority of subpixel mapping algorithms considers the abundance quantification as input data known in advance, in order to better evaluate the subpixel mapping performances. However, the computation of these abundances is not a straightforward task, and if it is not carefully addressed, it may lead to severe limitations in the proposed methods.

In our work, we integrate the concept of subpixel mapping into a supervised classification framework based on the use of SVMs. The abundances of each land cover class are computed within the whole image, and the subpixel mapping (based on the concept of SA) is applied only as final step to obtain the land cover map resolution improvement.

20.3 SUPERVISED SUPERRESOLUTION

The methodology discussed in this work aims at proposing a unifying framework to exploit the complementary characteristics of hard and soft classification, and to integrate in the algorithm a superresolution step to enhance the thematic map spatial resolution. The method discussed in this work for subpixel relocation is SA due to its simplicity and ease of use.

SA has already shown good performances in a number of optimization and real problems, and its wide range of parameters grants a high flexibility with respect to the analyzed problem. In multi-hyperspectral remote sensing, it has successfully been used for classification (Robin et al. 2008) and abundance estimation (Debba et al. 2006).

The proposed framework can be summarized in three steps:

1. Perform a raw classification by using a probabilistic classifier. The output corresponds to a preliminary classification map, where only pixels with a probability value higher than a chosen threshold are classified. The labeling of the other pixels will be addressed in the second step because they are considered as mixed or "anormal" pixel due to spectral variability.

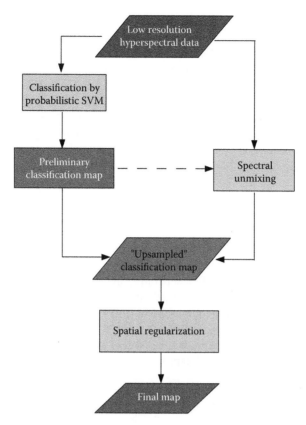

FIGURE 20.1 Flow chart of the presented method: Hyperspectral data at low spatial resolution are classified through SVM. The results will be the input for the spectral unmixing step used to model mixture of classes and obtain information about fractional abundances. A spatial regularization is finally applied to locate subpixels and obtain the final classification map.

2. In the second step, spectral unmixing is performed. The preliminary classification obtained after step 1 is used as the input. For each unlabeled pixel, possible "endmember candidates" are chosen among the samples classified in the first step, and fractional abundances of each land cover class is computed.
3. Each pixel is divided into a number of subpixels, assigned to a class according to the results of step 2. Subpixel location is optimized by using SA. The flow chart of the method presented in this chapter can be found in Figure 20.1.

20.3.1 CRISP CLASSIFICATION

The crisp classification of the hyperspectral image, by using a probabilistic algorithm, is addressed in the first step. The algorithm chosen in this work is the SVM (Vapnik 1998), in the probabilistic version proposed by Platt (2000). The reason of our choice is related to the characteristics of the SVM, which is an algorithm robust to the problems of high dimensionality and small training sets, having already proved to be very effective when used for hyperspectral image classification (more details on the SVM formulation and its application to hyperspectral image classification can be found in Vapnik (1998) and Melgani and Bruzzone (2004). However, it should be noted that any classifier could be used in the first step of the method discussed in this chapter, as long as it can provide a probabilistic output for each land cover class.

After the classification, only the maximum probabilities of a pixel belonging to the predominant class are considered. Pixels showing a high probability value are less likely to be mixed pixels, and can therefore be assigned to the considered class with low incertitude. On the contrary, pixels with a low probability value to belong to a class show a larger degree of incertitude, and their labeling should be addressed more carefully. In this work, we assume that such pixels could be either mixed pixels or pixels with high spectral variability, and their classification is performed in a second step.

The output of this first step is a preliminary classification map, where only labeled pixels are the ones that are more likely to belong to the corresponding class. This step has two aims: (i) classify pixels which can be considered as "pure" with a low degree of uncertainty and (ii) provide the input to the spectral unmixing step in order to have the possibility to exploit more information to classify the unlabeled pixels.

20.3.1.1 Probabilistic SVM

SVMs have been used in many works involving hyperspectral data, thanks to their good behavior with high-dimensional data. The main concept is to project the data into a higher-dimensional space through the kernel trick, and to fit a separating hyperplane to linearly separate the classes (Aronszajn 1950). The advantage of using a kernel function is the possibility to obtain the projection in an implicit way, while all the computations are done in the original space (Vapnik 1998).

Given a training set $S = (x^1, y_1), \dots, (x^l, y_l) \in \mathbb{R} \times \{-1, 1\}$, the SVM computes a decision function $f(x)$ such that sign $(f(x))$ can be used to predict the label of any test sample x. The decision function is found by solving the convex optimization problem:

$$\max g(\alpha) = \sum_{i=1}^{l} \alpha_i - \frac{1}{2} \sum_{i,j=1}^{l} \alpha_i \alpha_j y_i y_j k\left(x^i, x^j\right) \tag{20.1}$$

subject to

$$0 \leq \alpha_i \leq C \quad \text{and} \quad \sum_{i=1}^{l} \alpha_i y_i = 0$$

where α denotes the set of the Lagrange coefficients, $k(x^i, x^j) = k(x^i, x^j) + \delta_{ij}/C$, k the kernel function, C a constant that is used to penalize training errors, δ_{ij} a function such that $\delta = 1$ if $i = j, \delta = 0$ otherwise.

In Platt (2000), it was proposed to approximate the posterior class probabilities $P\langle y = 1 | x \rangle$ by using a sigmoid function:

$$P\langle y = 1 | x \rangle \approx P_{A,B}\left(f\right) = \frac{1}{1 + \exp\left(A\hat{f} + B\right)} \tag{20.2}$$

where $\hat{f}$ is an estimation of the decision function $f(x)$ computed by the SVM, and A and B are two parameters that need to be optimized. The best parameter setting (A^*, B^*) is determined by solving the following regularized maximum likelihood problem (with N_+ considered for the y_i's positive, and N_- for the negative ones):

$$\min_{z=A,B} F(z) = \sum_{i=1}^{l} \left(t_i \log(p_i) + (1 - t_i)(1 - \log(p_i))\right), \tag{20.3}$$

where

$$p_i = P_{A,B}(f_i), t_i = \begin{cases} \dfrac{(N_+ + 1)}{(N_+ + 2)} \text{ if } y_i = +1 \\ \dfrac{1}{(N_- + 2)} \text{ if } y_i = -1 \end{cases} \tag{20.4}$$

A detailed description of the method can be found in Platt (2000). In this work, we have used an improved implementation of the above algorithm presented by Lin et al. (2003), which is included in the LIBSVM library (Chang and Lin 2001).

In this first classification step, we consider two outputs:

1. A complete probability map, containing the probability estimates for each pixel to belong to the class with highest probability output.
2. A coarse classification map of the pixels considered as "not mixed," containing class labels for the samples with a probability belonging to the class higher than a chosen threshold.

The choice of the threshold to determine whether a pixel should be considered as "pure" in the first step rather than "mixed" is an important task. When labeling the pure pixels, we are interested in correctly classifying most of them, mainly because of two reasons: (i) pure pixels are more likely to be well classified with the SVM rather than with spectral unmixing algorithms, and (ii) the output of the first step will be considered later in the algorithm, and a large number of correctly classified pure pixel helps to provide suitable endmember candidates for the mixed pixels. For this same reason, the misclassification of pure pixels could lead to critical issues and cause a large error in the spectral unmixing step, and thus a trade-off is needed. Preliminary tests have been performed in order to estimate a suitable value of this parameter. The experiments have shown that in general a high threshold gives higher classification accuracy, since only the pixels which are reasonably sure to belong to a class are labeled. Therefore, in our experiments we have set the value to 80%. However, as will be shown later in this chapter, the choice of the threshold is not crucial for the performances of the method, since there is a wide range of values where the approach presented in this chapter performs better than the traditional SVM.

20.3.2 SPECTRAL UNMIXING

In the first step, a coarse thematic map is computed, where only the pixels estimated as "pure" with a low degree of uncertainty (thanks to the high probability to belong to a class) are labeled. The classification of the other pixels is addressed in the second step, through spectral unmixing.

Spectral mixture analysis (SMA) techniques have overcome some of the weaknesses of full pixel approaches by using statistical modeling and signal processing techniques (Keshava and Mustard 2002, Plaza et al. 2010).

Spectral unmixing algorithms can be divided in two groups: nonlinear techniques or linear techniques. Nonlinear mixed pixel analysis estimates multiple scattering effects that may arise when the different materials form intimate association at the microscopic level (Keshava and Mustard 2002). Although they can be useful for some types of analysis, when dealing with the classification of macroscopic land cover classes a linear mixing model can be used in the great majority of cases without significant loss of information (Keshava and Mustard 2002).

Two key tasks are performed by linear SMA: (i) to find an appropriate set of pure spectral constituents, also defined as "endmembers" in spectral unmixing terminology, and (ii) estimate the fractional abundances of each mixed pixel from its spectrum and the endmember spectra by using a linear mixture model. The spectrum of a mixed pixel is therefore represented as a linear combination of component spectra (endmembers).

We generally assume that the weight of each endmember spectrum (abundance) is proportional to the fraction of the pixel area covered by the endmember. If there are M spectral bands, the spectrum of the pixel and the spectra of the endmembers can be represented by M-dimensional vectors. Therefore, the general equation for LMM is described as a linear regression form

$$z = \sum_{i=1}^{L} a_i s_i + e = \mathbf{As} + e \tag{20.5}$$

where z is an $M \times 1$ column pixel vector which describes the spectrum of the mixed pixel, $s_i = [s_1, s_2, \ldots, s_L]$ is an $M \times L$ endmember matrix of material signature, where s are the M-dimensional spectra of the endmembers, a is an $L \times 1$ column vector and is composed of abundance coefficients a_i ($i = 1, 2, \ldots, L$), $\mathbf{e}$ is an M-dimensional error vector accounting for lack-fit and noise effects, and L is the number of the endmembers.

Due to physical reasons, Equation 20.6 has to respect the following constraints of nonnegativity (abundance fractions within a pixel cannot be negative) and sum to one (the sum of all the abundances fraction within a pixel must have 1 as a result):

$$a_i \geq 0 \tag{20.6}$$

$$\sum_{i=1}^{L} a_i = 1.$$

A large number of algorithms have been developed for automatic or semiautomatic extraction of spectral endmembers directly from the image data and to determine their fractional abundances within each pixel. A survey about several spectral unmixing techniques recently proposed in the literature can be found in Plaza et al. (2004b). In our experiments, we assume that a land cover ground truth is available in order to perform a supervised classification. Therefore, we do not need to estimate the pure constituent spectra composing the data, but simply the abundance of each land cover type within the pixels. A major issue related to the fraction abundance quantification is how to handle the spectral variability affecting hyperspectral data. As shown by Foody and Doan (2007), soft classification of hyperspectral images covering wide areas is negatively related to the intraclass spectral variability, and the assumption that a single endmember could extensively represent a class is in general not realistic. The choice of appropriate endmembers is very important in order to correctly estimate the fractional abundances. If the endmembers are not representative of the land cover classes, the estimates of the subpixel coverage can be highly biased and there will be a higher probability of misclassification errors.

In order to overcome this problem, we have proposed an adaptive approach to select the best endmember candidates for each pixel (Villa et al. 2010a). This approach is based on a trade-off between two main assumptions:

- The spatial correlation of the classes, that is, for each pixel, it is probable that the best endmember candidates lie in the spatial proximity of the considered pixel.
- The probabilistic output provided by the SVM, that is, if a candidate is not spatially close to the selected pixel, but the probabilistic value of the class to which it belongs is high, it is presumably a good candidate.

For each mixed pixel which has to be classified, we consider a set of 10 different spectra, which represent the endmember candidates. These candidates are chosen from the labeled samples of the training data and the results of the preliminary classification of step one, considered as a set of pure pixels correctly classified. If one of the land cover classes has a high probabilistic output (we consider a probabilistic output as high if its difference from the threshold chosen at step 1 is smaller than 5%), at least five spectra of this class are considered, chosen among the closest pixels labeled in the first step. Otherwise, all the 10 candidates are selected from the spectral signatures spatially closest to the considered pixel, after the coarse classification step.

Once the spectral signatures representative of each class are extracted from the image, the abundance fraction of the elements within each pixel should be determined. Several algorithms have been developed for the linear mixing model according to the required constraints of abundances fractions. The fully constrained least-squared unmixing algorithm is a widely adopted practical solution to avoid the appearance of theoretical problems, such as negative fractional abundances or abundances that sum up to more than one. Due to its computational efficiency, we have chosen this algorithm, which satisfies both abundance constraints and is optimal in terms of least squares error (Heinz and Chang 2001). After applying FCLS, we obtain the fractional abundances of each endmember. Due to the fact that in many cases several candidates belong the same class, by summing the fractional abundances of all the endmembers belonging to the same land cover class, we obtain the cover percentage of a class within a mixed pixel.

It should be noted that not all the pixels to be classified in this second step are mixed, but there are many "pure" pixels not labeled because of the low-probability output provided by the SVM. However, the method presented here allows the labeling of them as "pure" pixels also in this second step.

20.3.3 IMPROVING SPATIAL RESOLUTION

Spectral unmixing is useful to describe the scene at a subpixel level, but can only provide information about proportions of the endmembers within each pixel. Since the spatial location remains unknown, spectral unmixing does not perform any resolution enhancement. In this work, we present a subpixel mapping technique, which takes advantage of the information given by the spectral mixing analysis and uses it to enhance the spatial resolution of thematic maps. Our approach is as follows:

In a first step, each pixel is divided in a fixed number of subpixels, according to the desired resolution enhancement. Every subpixel is assigned to an endmember, in conformity with its fractional abundance within the pixel. For example, if we want to have a zoom factor of N, we have to divide each pixel into $N \times N$ subpixels. For each pixel, the number of subpixels n to assign to the class i is computed according to the equation

$$n_i = \text{round}\left(\frac{abd_i}{1/N^2}\right) \quad (20.7)$$

where abd_i is the fractional abundance of the class i within the considered pixel estimated with the FCLS and round(x) returns the value of the closest integer to x.

SA mapping function is then used, to create random permutation of these subpixels, in order to minimize a chosen cost function. Taking into consideration the spatial correlation tendency of land covers, we assume that each endmember within a pixel should be spatially close to the same endmembers in the surrounding pixels. Therefore, we consider the cost function C which needs to be minimized as the perimeter of the connected areas belonging to the same class:

$$C = \sum_{i=1}^{I} \sum_{j=1}^{N_i} P_j, \quad (20.8)$$

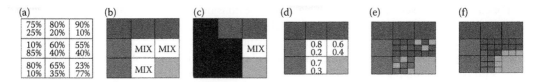

FIGURE 20.2 Basic steps of the presented method: (a) A probabilistic classification map is computed for each class. (b) The pixels with highest probability greater than a chosen threshold are considered as pure and classified (in the figure, we set the threshold to 70%). The other pixels are considered as mixed (MIX in the figure). (c) For each mixed pixel, a set of possible endmembers is selected, considering the results of the preliminary classification. The other pixels, pure or mixed, are just ignored. (d) Spectral unmixing provides information about the abundance fraction of a class within each pixel. (e) Pixels are split into n subpixels, according to the desired zoom factor, assigned to an endmember and randomly positioned within the pixel. The number of subpixels assigned to each class reflects the fractional value estimated in the previous step. (f) Simulated annealing performs random permutations of the subpixels position until *minimum* cost is reached. (Adapted from A. Villa et al. 2011. *IEEE Journal of Selected Topics in Signal Processing*, 5:521–33, © 2011 IEEE.)

where I is the number of the classes, N_i is the number of connected components of the class i, and P_j is the perimeter of the connected component j, computed according to the 8-connected border pixels model (Kulpa 1983).

SA is a well-established stochastic technique originally developed to model the natural process of crystallization (Metropolis et al. 1958). In analogy with thermodynamics, where a system is slowly cooled in order to reach its lowest energy state, the algorithm searches for a gradual solution in order to avoid suboptimal solution. More recently, SA has been proposed to solve global optimization problems (Kirkpatrick et al. 1983), and it has been used in various fields.

The basic idea of the method is that, in order to avoid to be trapped in local minima, uphill movements, that is, points corresponding to worse values of the objective function could, sometimes, be accepted for the following iteration. As it happens with a greedy search, all the changes decreasing the cost function are accepted. Changes increasing the cost function can also be accepted, but with a probability which is inversely proportional to the size of the degradation (small degradations are accepted with a higher probability). This probability also decreases as the search continues, or as the system cools down, allowing eventual convergence to the optimal solution.

An example of how SA spatial regularization works can be seen by looking at Figure 20.2, where Figure 20.2e represents the initial subpixel distribution and Figure 20.2f the optimal one. First, a mixed pixel is selected, according to the information provided by the spectral unmixing step. In a second step, a random permutation of the subpixels within the selected pixel is performed by SA. If this permutation leads to a decrease of the cost function (which, in our case, is the perimeter of connected components, i.e., components belonging to the same class), the change is accepted. Otherwise, as described above, the change will be probably rejected. The algorithm stops when *minimum* cost is obtained or when a previously fixed number of consecutively rejected changes are reached (in our case, we set the number of possible consecutive rejections to 100,000, since it is large enough to avoid suboptimal solutions and the MATLAB® code needs only a few seconds to perform this computation). Figure 20.3 shows a block diagram of the simulated annealing cost function considered in the proposed method.

20.4 EXPERIMENTS ON HYPERSPECTRAL REAL DATA

The experiments on real data were carried out considering two different data sets. The first data set is an airborne visible/infrared imaging spectrometer (AVIRIS) image taken over NW Indiana's Indian Pine test site in June 1992. This image has been widely used in the remote sensing community for both classification and spectral unmixing purposes, and thus represents an interesting benchmark for the method presented in this chapter. The experiment was carried out on the whole

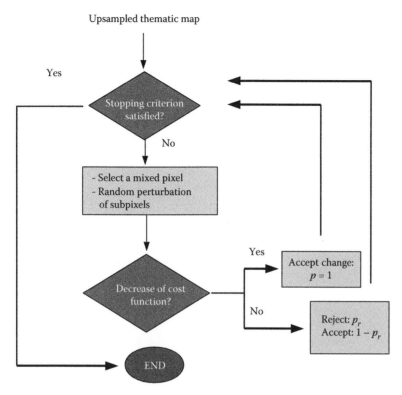

FIGURE 20.3 Simulated annealing block diagram. First, a mixed pixel is selected, and a random change of the subpixel positions is performed. If the cost function decreases, the change is accepted; otherwise, there will be a higher probability of rejection. The procedure continues until a stopping criterion is satisfied.

AVIRIS data set. Nine land cover classes were considered. The original image is composed of 145×145 pixels. The calibrated data are available online with detailed ground-truth information.

The second site is the region surrounding the central volcano Hekla in Iceland, one of the most active volcanoes in the country. Since 1970, Hekla has erupted quite regularly every 10 years, in 1970, 1980–1981, 1991, and in 2000. The volcano is located on the south-western margin of the Eastern volcanic zone in South Iceland. Hekla's products are mainly andesitic and basaltic lavas and tephra. AVIRIS data that were collected on a cloud-free day, June 17, 1991, were used for the classification. The AVIRIS sensor operates in the visible, near- and mid-infrared portions of the electromagnetic spectrum, its sensitivity range spanning wavelengths from 0.4 to 2.4 µm. As in the previous case, the sensor system has 224 data channels, utilizing four spectrometers, whereas each spectral band is ~10 nm in width. During the image acquisition, spectrometer 4 was not working properly.

This particular spectrometer operates in the wavelength range from 1.84 to 2.4 µm (64 bands). These 64 bands were deleted from the imagery along with the first channels for all the other spectrometers, and the remaining 157 data channels were left. A subset of 180×180 pixels has been used for this experiment. In order to address the issue of the random choice of the training samples, we have repeated the experiment with 10 different training sets for each data set.

Due to the difficulty to have a perfect knowledge of the fractional abundances of each land cover type, we decided to use the original ground-truth data only to compare the obtained results, and to decrease the spatial resolution of the image by applying an $n \times n$ low-pass filter, where n varies according to the considered data set. This way, we know exactly the quantity of each class within a pixel, and we can use the low-resolution image obtained after filtering as input for the presented method. The information about the classes and the training and the test sets can be found in Table 20.1.

TABLE 20.1

Information about Training–Test Samples for Each Class and Classification Accuracies of the Two Considered Data Sets

| Samples/ | AVIRIS Indian Pine | | | | AVIRIS Hekla | | | |
Approach	Train	Test	SVM	Proposed	Train	Test	SVM	Proposed
Class 1 (%)	40	1434	71.53	77.56	24	672	50.30	52.98
Class 2 (%)	40	834	63.81	87.32	126	3350	89.22	62.18
Class 3 (%)	40	497	51.23	76.99	523	11916	74.99	85.84
Class 4 (%)	40	747	61.45	84.55	220	4709	90.27	90.61
Class 5 (%)	40	489	70.63	90.12	279	6918	42.89	82.87
Class 6 (%)	40	968	52.18	61.83	103	2310	78.96	84.55
Class 7 (%)	40	2468	56.95	81.73	6	181	50.82	60.83
Class 8 (%)	40	614	60.02	83.00	51	1286	34.37	68.36
Class 9 (%)	40	1294	61.23	92.66	42	1058	53.68	84.31
AA (%)	–	–	61.45	82.78	–	–	62.83	74.72
OA (%)	–	–	90.02	73.17	–	–	69.19	81.71
K (%)	–	–	85.78	66.11	–	–	63.96	76.23
Mix pixels (%)	–	–	70.78	49.82	–	–	48.10	67.65
Spatial error (%)	–	–	–	2.56	–	–	–	2.92

Note: It has to be noticed that the training set is selected from the low-resolution image used as input of the method while the test set is selected from the high-resolution reference data used for comparison. "Mixed pixels" refers to the percentage of correctly classified mixed pixels (which are pixels considered as mixed in the low reference data but pure in the high-resolution image). Spatial error is the percentage error due to incorrect subpixel location.

The results were evaluated in terms of overall accuracy (OA), which is the percentage of high-resolution image pixels correctly classified, Kappa coefficient (K), which represents the percentage of correctly classified pixels without the amount given by the chance alone, single classes accuracy, and average class accuracy (AA). Further analysis were performed to estimate the percentage of mixed pixels correctly classified, and the percentage error due to the spatial regularization (i.e., the percentage of mixed pixels correctly retrieved after the unmixing step and not correctly located after applying the spatial regularization). When classifying the low-resolution image with a traditional classifier and comparing with the high-resolution ground truth, there will be a bias in the performances due to the impossibility for traditional classifiers to deal with mixed pixels. However, this is the issue the method presented in this work tries to handle.

20.4.1 AVIRIS INDIAN PINE DATA SET

The first experiment was carried out on the AVIRIS Indian Pine data set. Sixteen land cover classes composed the ground-truth data. Seven of them were discarded due to the small number of testing samples, and the nine remaining were considered in the experiment. The original image is composed of 145×145 pixels, and it was used as reference data. After applying a 2×2 low-pass filter, an image composed of 72×72 pixels was obtained. The land cover ground truth can be seen in Figure 20.4d. For training set, we have randomly selected, for each class, 40 samples which were considered as "pure" in the low-resolution image (that would correspond to about 10% of pixels of each class in the high-resolution image). To have the possibility to compare the results of the

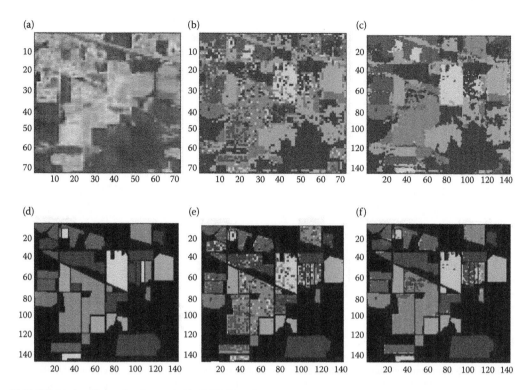

FIGURE 20.4 **(See color insert.)** (a) AVIRIS Indian Pine original image, band 30. (b) Classification map obtained by using SVM. (c) Classification map obtained by using the proposed method. (d) Ground-truth map. (e), (f) Classification of the test data, in the case of SVM and the proposed method, respectively.

proposed method with the available ground truth, we chose a zoom factor equal to 2. However, the high number of classes and their spectral similarity make this data set very challenging.

Figure 20.4b and c shows the classification maps obtained with a conventional SVM and the proposed method. The qualitative improvement obtained by the presented method can be clearly seen in the classification maps, resulting in a less noisy map and an improved detection of the borders of spatial structures (in this case, agricultural fields). To have a quantitative comparison of the results obtained with the two methods, the overall accuracy of pixels correctly classified has been compared.

The mean overall accuracy obtained in the five experiments with the SVM is 72.31%. The low value of accuracy provided by a traditional SVM is due to two main factors, which are the impossibility of a common hard classification technique to distinguish different land cover classes at a subpixel level, and the difficulty to handle the high spectral variability. The presented method obtained an average overall accuracy of 91.10%, showing the capability of the proposed approach to better deal with the aforementioned two main issues.

20.4.2 AVIRIS Hekla Data Set

For the second experiment, we consider a subset of the Hekla data set, located in the top-left corner of the scene. This subset is composed of 180×180 pixels, and it contains nine classes of interest. Also in this case a 2×2 low-pass filter was applied to the original image, leading to a low-resolution image of 90×90 pixels. Due to the insufficient availability of ground truth to quantify the results provided by the presented method, we have considered as ground truth the classification map obtained by a spectral-spatial method, proposed by Tarabalka et al. (2010), where the overall accuracy

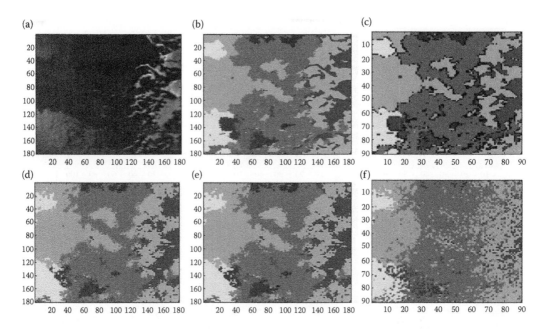

FIGURE 20.5 (**See color insert.**) (a) Hekla data set, band 80. (b), (c) Ground truth in the high-resolution and low-resolution image, respectively (mixed pixels are shown in black). (d), (e) Classification maps obtained with the proposed method, before and after spatial regularization. (f) Classification map obtained with a classic SVM. (Adapted from A. Villa et al., 2011. *IEEE Journal of Selected Topics in Signal Processing*, 5:521–33, © 2011 IEEE.)

computed on the reference test set was close to 100%. Thus, also if we have to keep in mind that the results are estimated by comparison with a classification map and not with a selected land cover ground-truth map, this classification map seems to be a reliable source of knowledge about the land class coverage of the area. The original image, the classification map obtained by Tarabalka et al. (2010), and the obtained classification maps are shown in Figure 20.5. 15% of the labeled pixels of the low-resolution data were randomly selected from each class and used to train the classifier, and the experiment repeated 10 times with different training sets.

The quantitative results in this experiment confirm those obtained in the previous ones: the proposed method provides not only a better classification map from a qualitative point of view, but also a large improvement of the overall accuracy of correctly classified pixels. Due to the irregular spatial structures in which the land cover classes are grouped, the spatial regularization method presented in this chapter was expected to be less effective than in the previous cases. The quantitative results agree with this supposition, also if the overall accuracy is penalized by only 2 percentage points. More advanced techniques could be investigated in our future works, in order to have an improvement of the classification accuracy of the mixed pixels.

20.4.3 DISCUSSION ABOUT THE CHOICE OF PARAMETERS AND COMPUTATIONAL BURDEN

In the proposed method, there are two parameters having an influence on the overall classification accuracy obtained (apart from the parameters of the SVM, which are automatically set during the training process through cross-validation): the threshold to distinguish between pure and mixed pixels and the number of "endmember candidates" to consider for the spectral unmixing, in the second step. How the classification accuracy changes by varying the value of the parameters can be seen in Figure 20.6. It can be noticed that the proposed method outperforms the traditional SVM in terms of accuracy over the whole range tested, being the choice of the parameters not crucial for the classification.

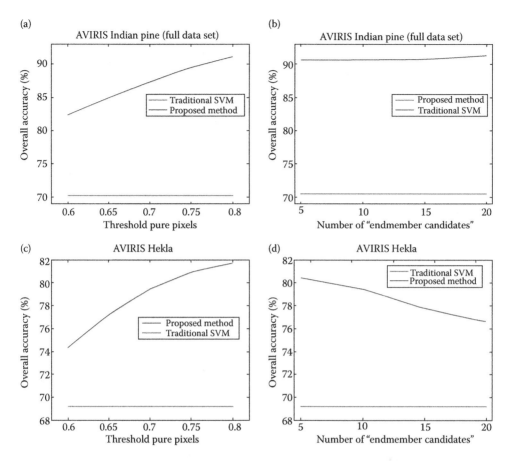

FIGURE 20.6 Variation of the overall classification accuracy versus the value of the parameter threshold to determine if a pixel can be considered as "pure" for (a), (b) AVIRIS Indian Pine (c), (d) AVIRIS Hekla data sets. (Adapted from A. Villa et al., 2011. *IEEE Journal of Selected Topics in Signal Processing*, 5:521–33, © 2011 IEEE).

As could be expected, a high value of the threshold to determine if a pixel can be considered as "pure" provides a higher accuracy, since only the most reliable pixels are labeled for the preliminary classification. When a low value of the threshold parameter is set, the preliminary classification map will be similar to the hard classification map obtained with a traditional SVM, thus decreasing the advantages obtained by integrating hard and soft classification techniques.

The number of "endmember candidates" considered in the spectral unmixing step points out the importance of the spatial information. The best results are in general obtained by considering a low number of candidates (which are spatially closest to the considered pixel). When setting a larger value of the parameter, endmember candidates spatially far from the analyzed pixel can be selected, introducing useless information and thus leading to a slight decrease in the classification accuracy.

The computational burden of the proposed method corresponds to the sum of the elapsed time of the three steps: classification with SVM, spectral unmixing, and spatial regularization. The training of the SVM, which quadratically depends on the size of the training set, is the most computationally expensive step of the proposed approach. The spectral unmixing step depends on the number of the mixed pixels to unmix, while the SA regularization depends on the number of mixed pixels and on the zoom factor desired. When requiring a larger zoom, the number of possible subpixel combinations grows exponentially, thus requiring a heavier computational burden to reach the optimal configuration. If the desired zoom factor is equal or higher than 4, the computational burden of the spatial regularization is expected to be highly penalizing for the proposed method. For the two experiments shown in this chapter, the computational burden of the presented approach was comparable to the SVM burden. The computational time of the proposed method was 23 and 43 min, respectively, while the elapsed time of SVM classification was 18 and 35 min.

20.5 CONCLUSIONS

Classification of hyperspectral images in the presence of mixed pixels was addressed in this work. A new method trying to improve the spatial resolution of the classification maps was discussed. The method makes use of both soft classification techniques and spectral unmixing algorithms, exploiting their characteristics to determine the fractional abundances of the classes at a subpixel scale. Once the fractional abundances have been determined, spatial regularization by SA is finally performed to spatially locate the land cover classes within each pixel. Experiments were carried out on two different AVIRIS data sets, showing the advantages of the method over traditional hard classification techniques when areas with mixtures of materials are located in the scene. The results are excellent both from a visually and quantitative point of view. Further research will be devoted to the creation of more realistic data sets to test the method, taking into account the point spread function of hyperspectral sensors to create low-resolution images.

REFERENCES

T. Akgun, Y. Altunbasak, and R. Mersereau, 2005. Super-resolution reconstruction of hyperspectral images, *IEEE Transactions on Image Processing*, 14:1860–75.

N. Aronszajn, 1950. Theory of reprodusing kernel, Harvard University, Division of Engineering Sciences, Technical Report.

P. M. Atkinson, 1997. Mapping sub-pixel boundaries from remotely sensed images, in *Innovations in GIS*, 4:166–80. New York: Taylor & Francis.

P.M. Akinson, 2004. Super-resolution land-cover classification using the two-point histogram, *GeoENV IV: Geostatistics for Environmental Applications* (X. Sánchez-Vila, J. Carrera, and J. Gómez-Hernández, editors), pp. 15–28, Dordrecht: Kluwer.

P.M. Atkinson, 2005. Sub-pixel target mapping from soft-classified, remotely sensed imagery, *Photogrammetric Engineering & Remote Sensing*, 71:839–46.

A. Baraldi, E. Binaghi, P. Blonda, P. Brivio, and A. Rampini, 2001. Comparison of the multilayer perceptron with neuro-fuzzy techniques in the estimation of cover class mixture in remotely sensed data, *IEEE Transactions on Geoscience and Remote Sensing*, 39:994–1005.

F. Bovolo, L. Bruzzone, and L. Carlin, 2010. A novel technique for subpixel image classification based on support vector machine, *IEEE Transactions on Image Processing*, 19:2983–99.

C. Chang and C. Lin, 2001. Libsvm: A library for support vector machines, Available at http://www.csie.ntu.edu.tw/ cjlin/libsvm.

C. Chang, C. Wu, and C. Tsai, 2011. Random n-finder (n-findr) endmember extraction algorithms for hyperspectral imagery, *IEEE Transactions on Image Processing*, 21:641–56.

P. Debba, E. Carranza, F. van der Meer, and A. Stein, 2006. Abundances estimation of spectrally similar minerals by using derivative spectra in simulated annealing, *IEEE Transactions on Geoscience and Remote Sensing*, 44:3649–58.

M. T. Eismann and R. C. Hardie, 2005. Hyperspectral resolution enhancement using high-resolution multispectral imagery with arbitrary response functions, *IEEE Transactions on Geoscience and Remote Sensing*, 43:455–65.

G.M. Foody, 1996. Approaches for the production and evaluation of fuzzy land cover classifications from remotely-sensed data, *International Journal of Remote Sensing*, 17:1317–40.

G.M. Foody, A.M. Muslim, and P.M. Atkinson, 2005. Super-resolution mapping of the waterline from remotely sensed data, *International Journal of Remote Sensing*, 26:5381–92.

G. Foody and H. Doan, 2007. Variability in soft classification prediction and its implications for sub-pixel scale change detection and super-resolution mapping, *Photogrammetry Engineering and Remote Sensing*, 73:923–33.

Y. Ge, S. Li, and V. C. Lakhan, 2009. Development and testing of a subpixel mapping algorithm, *IEEE Transactions on Geoscience and Remote Sensing*, 47:2155–64.

Y. Gu, Y. Zhang, and J. Zhang, 2008, Integration of spatial-spectral information for resolution enhancement in hyperspectral images, *IEEE Transactions on Geoscience and Remote Sensing*, 46:1347–58.

R. Hardie, M. Eismann, and G. Wilson, 2004. Map estimation for hyperspectral image resolution enhancement using an auxiliary sensor, *IEEE Transactions on Image Processing*, 9:1174–84.

D. Heinz and C. Chang, 2001. Fully constrained least squares linear spectral mixture analysis method for material quantification in hyperspectral imagery, *IEEE Transactions on Geoscience and Remote Sensing*, 39:529–45.

N. Keshava and J. F. Mustard, 2002. Spectral unmixing, *IEEE Signal Processing Magazine*, 19:44–57.

S. Kirkpatrick, C. Gelatt, and M. Vecchi, 1983. Optimization by simulated annealing, *Science*, 220:671–80.

Z. Kulpa, 1983. More about areas and perimeters of quantized objects, *Computer Vision, Graphics, and Image Processing*, 22:268–76.

H. Lin, C. Lin, and R. C. Weng, 2003. A note on Platt's probabilistic outputs for support vector machines, Department of Computer Science, National Taiwan University, Taipei, Taiwan, Technical Report.

F. Melgani and L. Bruzzone, 2004. Classification of hyperspectral remote sensing images with support vector machine, *IEEE Transactions on Geoscience and Remote Sensing*, 42:1778–90.

K. Mertens, 2008. Sub-pixel mapping: A comparison of techniques, PhD dissertation, Ghent University.

N. Metropolis, A. Rosenbluth, M. Rosenbluth, A. Teller, and E. Teller, 1958. Equations of state calculations by fast computing machines, *J. Chem. Phys.*, 21:1087–92.

A.M. Muslim, G.M. Foody, and P.M. Atkinson, 2006. Localized soft classification for super-resolution mapping of the shoreline, *International Journal of Remote Sensing*, 27:2271–85.

M. Nachtegael, D. Weken, E. van der Kerre, and W. Philips, Eds., 2007. *Soft Computing in Image Processing*. Berlin: Springer.

M. Nguyen, P. Atkinson, and H. Lewis, 2006. Superresolution mapping using a Hopfield Neural Network with fused images, *IEEE Transactions on Geoscience and Remote Sensing*, 44:736–49.

J. Platt, 2000. Probabilistic outputs for support vector machines and comparison to regularized likelihood methods, in *Advances in Large Margin Classifiers*, (A. J. Smola, P. L. Bartlett, B. Schölkopf, and D. Schuurmans editors) Cambridge, MA: MIT Press.

A. Plaza, P. Martinez, R. Perez, and J. Plaza, 2004a. A new approach to mixed pixel classification of hyperspectral imagery based on extended morphological profiles, *Pattern Recognition*, 37:1097–116.

A. Plaza, P. Martinez, R. Perez, and J. Plaza, 2004b. A quantitative and comparative analysis of endmember extraction algorithms from hyperspectral data, *IEEE Transactions on Geoscience and Remote Sensing*, 42:650–63.

A. Plaza, G. Martin, J. Plaza, M. Zortea, and S. Sanchez, 2010. Recent developments in spectral unmixing and endmember extraction, in *Optical Remote Sensing—Advances in Signal Processing and Exploitation Techniques* (S. Prasad, L. Bruce, and J. Chanussot editors) Berlin: Springer.

A. Robin, S. L. Hegarat-Mascle, and L. Moisan, 2008. Unsupervised subpixelic classification using coarse-resolution time series and structural information, *IEEE Transactions on Geoscience and Remote Sensing*, 46:1359–74.

J. Steinwendner, W. Schneider, and F. Suppan, 1998. Vector segmentation using spatial subpixel analysis for object extraction, *International Archives of Photogrammetry and Remote Sensing*, 32:265–71.

A. J. Tatem, H. G. Lewis, P. M. Atkinson, and M. S. Nixon, 2001. Superresolution target identification from remotely sensed images using a Hopfield Neural Network, *IEEE Transactions on Geoscience and Remote Sensing*, 39:781–96.

Y. Tarabalka, J. Chanussot, and J. A. Benediktsson, 2010. Segmentation and classification of hyperspectral images using minimum spanning forest grown from automatically selected markers, *IEEE Transactions on Systems, Man, and Cybernetics, Part B: Cybernetics*, 40:1267–79.

M.W. Thornton, P.M. Atkinson, and D.A. Holland, 2006. Superresolution mapping of rural land-cover features from fine spatial resolution satellite sensor imagery, *International Journal of Remote Sensing*, 27:473–91.

G. Vane, R. Green, T. Chrien, H. Enmark, E. Hansen, and W. Porter, 1993. The airborne visible/infrared imaging spectrometer (AVIRIS), *Remote Sensing of Environment*, 44:127–43.

V. Vapnik, 1998. *Statistical Learning Theory*. New York: Wiley.

A. Villa, J. Chanussot, J.A. Benediktsson, M. Ulfarsson, and C. Jutten, 2010. Super-resolution: An efficient method to improve spatial resolution of hyperspectral images, *IEEE IGARSS 2010*, Honolulu.

A. Villa, J. Chanussot, J.A. Benediktsson, and C. Jutten, 2010. Supervised super resolution to improve the resolution of hyperspectral images classification maps, *SPIE Remote Sensing Europe 2010*, Toulouse.

A. Villa, J. Chanussot, J.A. Benediktsson, and C. Jutten, 2011. Spectral unmixing for the classification of hyperspectral images at a finer spatial resolution, *IEEE Journal of Selected Topics in Signal Processing*, 5:521–33.

L. Wang and X. Jia, 2009. Integration of soft and hard classifications using extended support vector machines, *IEEE Geoscience and Remote Sensing Letters*, 6:543–47.

21 Signal Subspace Identification in Hyperspectral Imagery*

José M. P. Nascimento and José M. Bioucas-Dias

CONTENTS

21.1 INTRODUCTION

Terrestrial remote sensing imagery involves the acquisition of information from the Earth's surface without physical contact with the area under study. Among the remote sensing modalities, hyperspectral imaging has recently emerged as a powerful passive technology. This technology has been widely used in the fields of urban and regional planning, water resource management, environmental monitoring, food safety, counterfeit drugs detection, oil spill and other types of chemical contamination detection, biological hazards prevention, and target detection for military and security purposes [2–9].

Hyperspectral sensors sample the reflected solar radiation from the Earth surface in the portion of the spectrum extending from the visible region through the near-infrared and mid-infrared (wavelengths between 0.3 and 2.5 µm) in hundreds of narrow (of the order of 10 nm) contiguous bands [10]. This high spectral resolution can be used for object detection and for discriminating between different objects based on their spectral characteristics [6]. However, this huge spectral resolution yields large amounts of data to be processed. For example, the Airborne Visible/Infrared Imaging Spectrometer (AVIRIS) [11] collects a 512 (along track) × 614 (across track) × 224 (bands) × 12 (bits) data cube in 5 s, corresponding to about 140 MBs. Similar data collection ratios

---

* Work partially based on [1] copyright 2008 IEEE.

are achieved by other spectrometers [12]. Such huge data volumes put stringent requirements on communications, storage, and processing.

The problem of signal subspace identification of hyperspectral data represents a crucial first step in many hyperspectral processing algorithms such as target detection, change detection, classification, and unmixing. The identification of this subspace enables a correct dimensionality reduction (DR) yielding gains in data storage and retrieval and in computational time and complexity. Additionally, DR may also improve algorithms performance since it reduce data dimensionality without losses in the useful signal components. The computation of statistical estimates is a relevant example of the advantages of DR, since the number of samples required to obtain accurate estimates increases drastically with the dimensionality of the data (Hughes phenomenon) [13].

21.1.1 Brief Overview of DR Methods

Unsupervised DR has been approached in many ways. Band selection or band extraction, as the name suggests, exploits the high correlation existing between adjacent bands to select a few spectral components among those with higher *signal-to-noise ratio*, SNR [14–19]. Projection techniques seek for the best subspace to project data by minimizing an objective function. For example, principal component analysis (PCA) [20] computes the *Karhunen–Loéve* transform, which is the best data representation in the least-squares sense; singular-value decomposition (SVD) [21] provides the projection that best represents data in the maximum power sense; maximum noise fraction (MNF) [22] and noise-adjusted principal components (NAPC) [23] seek the projection that optimizes the ratio of noise power to signal power. NAPC is mathematically equivalent to MNF [23] and can be interpreted as a sequence of two principal component transforms: the first applies to the noise and the second applies to the transformed data set.

Topological methods are local approaches that infer the manifold, usually of low dimension, where data set live [24]. For example, curvilinear component analysis [25], curvilinear distance analysis [26], and manifold learning [27–30] are nonlinear projections based on the preservation of the local topology. Independent component analysis (ICA) [31,32], projection pursuit [33,34], wavelet decomposition [35,36], and averaged learning subspace method [37] have also been considered.

The identification of the signal subspace is a model order inference problem to which information theoretic criteria like the *minimum description length* (MDL) [38,39] or the *Akaike information criterion* (AIC) [40] come to mind. These criteria have in fact been used in hyperspectral applications [41] adopting the approach introduced by Wax and Kailath [42].

Harsanyi et al. [43] developed a Neyman–Pearson detection theory-based thresholding method (HFC) to determine the number of spectral endmembers in hyperspectral data, referred to in Ref. [41] as *virtual dimensionality* (VD). The HFC method is based on a detector built on the eigenvalues of the sample correlation and covariance matrices. A modified version, termed as noise-whitened HFC (NWHFC), includes a noise-whitening step [41]. These methods need to specify a false-alarm probability P_f to estimate the number of endmembers. As pointed out in Bajorski's work [44], VD has some undesirable properties such as being dependent on the average spectrum calculated over the whole image and being variant to rotations and translations, thus giving misleading results. *Second moment linear dimensionality* (SML) [44] is an alternative method to estimate the number of endmembers depending on a significance level α, and it is based on the detection of exceptionally large gaps in eigenvalues of the image covariance matrix.

Recently, a new method termed as *robust signal subspace estimation* (RSSE) [45,46], have been proposed in order to estimate the signal subspace in the presence of rare signal pixels. This method assumes that the signal subspace can be split into the abundant signal subspace and rare signal subspace, which are assumed to be independent. RSSE first uses the NWHFC methods to estimate the abundant signal subspace and then iteratively analyzes, at each step, the maximum residual energy (l_2^∞ norm) on the subspace that is orthogonal to the signal subspace. The estimation of the rare signal components is particularly relevant in the preprocessing step for small target detection applications.

Opposed to deterministic linear mixing model, *normal compositional model* [47,48] assume that the pixels of the hyperspectral image are linear combinations of random endmembers with known means. This model provides more flexibility regarding the observed pixels and the endmembers.

21.1.2 HYSIME METHOD

This chapter develops a minimum mean-squared error-based approach to infer the signal subspace in hyperspectral imagery. The method, termed as *hyperspectral signal identification by minimum error* (HySime), is eigen decomposition based, unsupervised, and fully-automatic (i.e., it does not depend on any tuning parameters). HySime starts by estimating the signal and the noise correlation matrices using multiple regression. A subset of eigenvectors of the signal correlation matrix is then used to represent the signal subspace. This subspace is inferred by minimizing the sum of the projection error power with the noise power, which are, respectively, decreasing and increasing functions of the subspace dimension. Therefore, if the subspace dimension is overestimated the noise power term is dominant, whereas if the subspace dimension is underestimated the projection error power term is the dominant. The overall scheme is computationally efficient, unsupervised, and fully automatic, yielding comparable or better results than the state-of-the-art algorithms.

The chapter is structured as follows. Section 21.2 formulates the signal subspace identification problem and reviews the SVD and MNF methods. Section 21.3 describes the fundamentals of the HySime method, including the noise estimation approach. Section 21.4 is devoted to denoising additive and multiplicative noises with HySime. Sections 21.5 and 21.6 evaluate the proposed algorithm using simulated and real data, respectively. Section 21.7 ends the chapter by presenting some concluding remarks.

21.2 PROBLEM FORMULATION

This section presents the signal model, formulates the signal subspace identification problem, and illustrates the limitations of DR classical methods.

21.2.1 SIGNAL MODEL DESCRIPTION

Spectral radiances read by a hyperspectral sensor from a given pixel can be represented as an L-dimensional vector, where L is the number of bands and each channel is assigned to one axis of $\mathbb{R}^L$. Thus, the observed spectral vectors, are given by

$$\mathbf{y} = \mathbf{x} + \mathbf{n}, \tag{21.1}$$

where $\mathbf{x}$ and $\mathbf{n}$ are L-dimensional vectors standing for signal and additive noise, respectively. Very often, the number of endmembers, p, present in a given scene is much smaller than the number of bands L. Therefore, hyperspectral vectors lie in an unknown p-dimensional subspace, that is,

$$\mathbf{x} = \mathbf{Ms}, \tag{21.2}$$

with $p < L$ and $\mathbf{M}$ being a full-rank $L \times p$ matrix. Under the linear mixing scenario (see, e.g., [2,6]), the columns of $\mathbf{M}$ are the endmember signatures and $\mathbf{s}$ is the *abundance fraction* vector. To be physically meaningful [49], abundance fractions are subject to nonnegativity and full additivity constraints, that is, abundance fractions are in the $p-1$ probability simplex $\{\mathbf{s} \in \mathbb{R}^p : s_j \geq 0, \sum_{j=1}^{p} s_j = 1\}$. Herein, we do not assume any special structure for the scattering mechanism; that is, our approach works both under the linear and nonlinear scenarios. Even in the nonlinear mixing scenario, it often happens that signal subspace dimension, although larger than the number of endmembers, is much smaller than the number of bands L. In these cases, it is still worthwhile, then, to estimate the signal subspace and represent the data on it. Note that this procedure does not preclude the application of future nonlinear projection techniques; on the contrary, it is an advantage, since the data are represented by vectors of smaller dimension, thus lightening the computational complexity of any posterior processing scheme.

21.2.2 EIGENANALYSIS OF DATA CORRELATION MATRIX

Let us assume, for a while, that noise $\mathbf{n}$ is zero-mean Gaussian i.i.d. (i.e., the components of $\mathbf{n}$ are independent and identically distributed) with variance σ_n^2 per band. Under these circumstances, the *maximum likelihood* (ML) estimate of the signal subspace is spanned by the p-dominant eigenvectors of the observed data correlation matrix of $\mathbf{y}$ [21, Ch. 6]; that is, $\langle \mathbf{M} \rangle = \langle [\mathbf{e}_1,...,\mathbf{e}_p] \rangle$ (the notation $\langle \mathbf{M} \rangle$ represents the subspace spanned by the columns of $\mathbf{M}$), where e_i for $i = 1, \ldots, p$, are the p-dominant eigenvectors of the observed data correlation matrix $\mathbf{R}_y$. The ML estimator assumes that the dimension of the subspace is known beforehand. However, this dimension is often *a priori* unknown. Nevertheless, a similar approach has been extensively used as a DR tool in hyperspectral image processing [6,10]. It consists in assuming that the noise is zero-mean i.i.d. Thus, the correlation matrix of the observed vectors may be written as $\mathbf{R}_y = \mathbf{E}(\boldsymbol{\Sigma} + \sigma_n^2 \mathbf{I}_L)\mathbf{E}^\mathrm{T}$, where $\mathbf{I}_L$ is the $L \times L$ identity matrix, $\mathbf{E}$ and $\boldsymbol{\Sigma}$ are the eigenvector and eigenvalue matrices of the signal correlation matrix $\mathbf{R}_x$, respectively. Assuming that $\mathbf{R}_x$ has just p positive eigenvalues and that they are ordered along the diagonal of $\boldsymbol{\Sigma}$ by decreasing magnitude, we have then $\langle \mathbf{M} \rangle = \langle [\mathbf{e}_1 \ldots \mathbf{e}_p] \rangle$; that is, the estimate of the signal subspace is the span of the eigenvectors of $\mathbf{R}_y$ whose respective eigenvalues are larger than σ_n^2 [21].

This is, basically, the idea behind SVD-based eigen-decomposition DR. Two limitations of this approach are as follows:

1. The noise present in most hyperspectral data sets is not i.i.d. and, thus, the signal subspace is no longer given by the span of the p eigenvectors corresponding to the largest eigenvalues nor by any other set of eigenvectors.
2. Even if the noise were i.i.d., the procedure described above to infer the subspace dimension would be prone to errors owing to random perturbations always present in the estimates of σ_n^2 $\mathbf{E}$, and $\boldsymbol{\Sigma}$.

We illustrate these limitations with an experiment built on a simulated hyperspectral image composed of 10^5 pixels and generated according to the linear mixing scattering mechanism. Each pixel is a mixture of live endmembers signatures ($p = 10$) selected from the United States geological survey (USGS) digital spectral library [50]. Abundance fractions are generated according to a Dirichlet distribution given by

$$D(s_1, s_2, \ldots, s_p \mid \theta_1, \theta_2, \ldots, \theta_p) = \frac{\Gamma\left(\sum_{j=1}^{p} \theta_j\right)}{\prod_{j=1}^{p} \Gamma(\theta_j)} \prod_{j=1}^{p} s_j^{\theta_j - 1}, \tag{21.3}$$

where $\Gamma(\cdot)$ denotes the Gamma function and $\{s_1, \ldots, s_p\}$ are in the $p - 1$ probability simplex. The mean value of the jth endmember fraction is $\mathbb{E}[s_j] = \theta_j / \sum_{l=1}^{p} \theta_l$ ($\mathbb{E}[\cdot]$ denotes the expectation operator) [51]. The Dirichlet density, besides enforcing positivity and full additivity constraints, displays a wide range of shapes, depending on its parameters. On the other hand, as noted in Ref. [52], the Dirichlet density is suited to model fractions.

Consider that the noise correlation matrix is $\mathbf{R}_n = \mathrm{diag}(\sigma_1^2,\ldots,\sigma_L^2)$ and that the diagonal elements follow a Gaussian shape centered at the band $L/2$, that is,

$$\sigma_i^2 = \sigma^2 \frac{e^{-(i-L/2)^2/(2\eta^2)}}{\sum_{j=1}^{L} e^{-(j-L/2)^2/(2\eta^2)}}, \tag{21.4}$$

for $i = 1, \ldots, L$. Parameter η plays the role of variance in the Gaussian shape ($\eta \to \infty$ corresponds to white noise; $\eta \to 0$ corresponds to *one-band* noise). Parameter σ^2 controls the total noise power. We set $\sigma^2 = 3.1 \times 10^{-4}$ leading to SNR = 30 dB, where

$$\mathrm{SNR} \equiv 10 \log_{10} \frac{\mathbb{E}[\mathbf{x}^T\mathbf{x}]}{\mathbb{E}[\mathbf{n}^T\mathbf{n}]}. \tag{21.5}$$

Figure 21.1 shows the shape of the noise variance σ_i^2 for $\eta \in \{40, \infty\}$.

To measure the dissimilarity between the signal subspace and the subspace inferred by SVD, we adopt the *chordal distance* [53,54], defined as

$$d = \frac{1}{\sqrt{2}} \|\mathbf{U}_p - \mathbf{U}_M\|_F, \tag{21.6}$$

where $\|\cdot\|_F$ denotes the Frobenius norm of a matrix, $\mathbf{U}_p = \mathbf{E}_p\mathbf{E}_p^T$ and $\mathbf{U}_M = \mathbf{E}_M\mathbf{E}_M^T$ are projection matrices onto the subspace of dimension p spanned, respectively, by the first p singular vectors of $\mathbf{R}_y$ and by the columns of $\mathbf{M}$. We note that the *chordal distance* is a measure of the projection error norm, that is, it is a measure of the errors $(\mathbf{U}_p - \mathbf{U}_M)\mathbf{x}$ for $\|\mathbf{x}\| = 1$ and $\mathbf{x} \in \mathbb{R}^L$. When this distance is zero, the two projections are equal.

Figure 21.2 (solid line) presents the chordal distance between the signal subspace and the subspace inferred by SVD, as a function of η. Note that the chordal distance increases as the noise correlation shape becomes less flat. The degradation of the signal subspace estimate, owing to the violation of the white noise assumption, is quite clear.

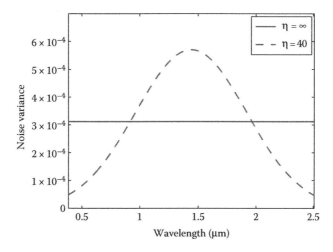

FIGURE 21.1 Noise variance in each band given by expression (Equation 19.4), nonwhite noise case ($\eta = 40$), and for white noise case ($\eta = \infty$).

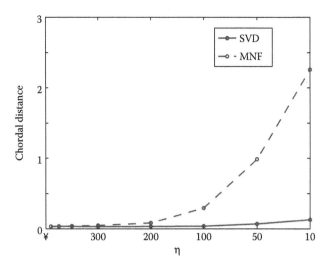

FIGURE 21.2 *Chordal distance* as a function of parameter η for SNR = 30 dB. Line P: subspace inferred by SVD; dashed line: subspace inferred by MNF.

In the example just presented, the subspace dimension was assumed known. However, this dimension is unknown in most real applications and must be inferred from data as already referred to. This is a model-order inference problem that, if based only on the eigenvalues of the observed data correlation matrix ($\mathbf{R}_y$), may lead to poor results. This aspect is illustrated in Figure 21.3, where we have plotted the eigenvalues of the signal correlation matrix $\mathbf{R}_x$ (large dots) and the eigenvalues of $\mathbf{R}_y$ (circles) computed in the experiment above with SNR = 15 dB and η = 40. From $\mathbf{R}_x$, eigenvalues (large dots) it is clear that the signal subspace dimension is $p = 10$. However, since $\mathbf{R}_x$ is not known in real applications, we resort to $\mathbf{R}_y$. An estimate of the signal subspace dimension based on the eigenvalues of $\mathbf{R}_y$ (circles) would lead, most probably, to $\hat{p} = 5$. We will see that HySime is able to infer the correct subspace dimension in similar scenarios.

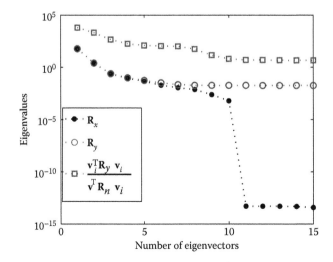

FIGURE 21.3 Eigen values of the signal correlation matrix $\mathbf{R}_x$ (large dots); eigen values of observed data correlation matrix $\mathbf{R}_y$ (circles); decreasing ratio of $(\mathbf{v}_i^T \mathbf{R}_y \mathbf{v}_i)/(\mathbf{v}_i^T \mathbf{R}_n \mathbf{v}_i)$ (squares) [$\mathbf{v}_i$, are the left-hand eigen vectors of $(\mathbf{R}_n \mathbf{R}_y^{-1})$]. Simulation parameters: SNR = 15 dB, η = 40, and $p = 10$.

21.2.3 Maximum Noise Fraction

MNF is a another popular subspace inference tool in remote sensing that takes into account the noise statistics. Nonetheless, it has limitations similar to SVD-based approaches, as illustrated below.

MNF finds orthogonal directions minimizing the noise fraction (or, equivalently, maximizing the SNR). Assuming that the noise correlation matrix $\mathbf{R}_n$ or an estimate is known, this procedure consists in finding orthogonal directions minimizing the ratio

$$\frac{\mathbf{v}_i^T \mathbf{R}_n \mathbf{v}_i}{\mathbf{v}_i^T \mathbf{R}_y \mathbf{v}_i}, \tag{21.7}$$

with respect to $\mathbf{v}_i$. This problem is known as the generalized Rayleigh quotient and its solution is given by the left-hand eigenvectors $\mathbf{v}_i$ of $\mathbf{R}_n \mathbf{R}_y^{-1}$, for $i = 1, \ldots, L$ [55].

For i.i.d. noise, we have $\mathbf{R}_n = \sigma_n^{-2} \mathbf{I}_L$ and $\mathbf{R}_y^{-1} = \mathbf{E}(\sigma + \sigma_n^2 \mathbf{I}_L)^{-1} \mathbf{E}^T$, and therefore MNF and SVD yield the same subspace estimate. However, if the noise is not i.i.d., the directions found by the MNF transform maximize the SNR but do not identify correctly the signal subspace. To illustrate this aspect, we apply the MNF transform to the data set generated in the previous section. The dashed line in Figure 21.2 represents the chordal distance between the signal subspace and the subspace inferred by the MNF transform for different values of parameter η and assuming $p = 10$. The chordal distance exhibits a pattern similar to that of the SVD-based approach being, however, larger for $\eta \neq \infty$ (white-noise case). If the space is unknown, it shall be inferred from the ratio $(\mathbf{v}_i^T \mathbf{R}_y \mathbf{v}_i)/(\mathbf{v}_i^T \mathbf{R}_n \mathbf{v}_i)$. Figure 21.3 (squares) plots this ratio by decreasing order for $p = 10$, $\eta = 40$, and SNR = 15 dB. The inference of the signal subspace dimension based on this plot may lead to misleading results, most probably, to $\hat{p} = 8$.

21.3 SIGNAL SUBSPACE ESTIMATION

This section introduces formally the HySime algorithm. The method starts by estimating the signal and the noise correlation matrices and then it selects the subset of eigenvectors that best represents the signal subspace in the minimum mean-squared error sense. The application of this criterion leads to the minimization of a two-term objective function. One term corresponds to the power of the signal projection error and is a decreasing function of the subspace dimension; the other term corresponds to the power of the noise projection and is an increasing function of subspace dimension.

21.3.1 Noise Estimation

Noise estimation is a classical problem in data analysis and particularly in remote sensing. Arguably, in hyperspectral imagery, the simplest noise estimation procedure is the *shift difference* method, also denoted as *nearest-neighbor difference* [22]. This approach assumes that noise samples taken from adjacent pixels are independent and have the same statistics, but the signal component is practically equal. To obtain meaningful noise estimates, the shift difference method shall be applied in homogeneous areas rather than on the entire image. This method has two weaknesses: first, it assumes that adjacent pixels have the same signal information, which is not valid in most hyperspectral data sets; second, to improve the noise estimation, selection of a homogeneous area must be carried out.

Herein, we follow a multiple regression theory [41,56]-based approach, which outperforms the shift difference method, used, for example, in the NAPC [23] algorithm. The high correlation between neighboring spectral bands is the main reason underlying the good performance of the multiple regression theory in hyperspectral applications.

Let $\mathbf{Y}$ denote an $L \times N$ matrix holding the N spectral observed vectors of size L. Define the matrix $\mathbf{Z} = \mathbf{Y}^{\mathrm{T}}$, the $N \times 1$ vector $\mathbf{z}_i; = [\mathbf{Z}]_{.,i}$, where $[\mathbf{Z}]_{.,i}$ stands for the ith column of $\mathbf{Z}$ (i.e., $\mathbf{z}_i$ contains the data read by the hyperspectral sensor at the ith band for all image pixels), and the $N \times (L - 1)$ matrix $\mathbf{Z}_{\partial_i} = [\mathbf{z}_1, \dots, \mathbf{z}_{i-1}, \mathbf{z}_{i+1}, \dots, \mathbf{z}_L]$.

Assume that $\mathbf{z}_i$ is explained by a linear combination of the remaining $L - 1$ bands. Formally, this consists in writing

$$\mathbf{z}_i = \mathbf{Z}_{\partial_i} \hat{\boldsymbol{\beta}}_i + \boldsymbol{\xi}_i, \tag{21.8}$$

where $\mathbf{Z}_{\partial_i}$ is the explanatory data matrix, $\boldsymbol{\beta}_i$ is the regression vector of size $(L - 1) \times 1$, and $\boldsymbol{\xi}_i$ is the modeling error vector of size $N \times 1$. For each $i \in \{1, \dots, L\}$, the least-squares estimator of the regression vector $\boldsymbol{\beta}_i$ is given by

$$\hat{\boldsymbol{\beta}}_i = (\mathbf{Z}_{\partial_i}^{\mathrm{T}} \mathbf{Z}_{\partial_i})^{-1} \mathbf{Z}_{\partial_i}^{\mathrm{T}} \mathbf{z}_i. \tag{21.9}$$

The noise is estimated by

$$\hat{\boldsymbol{\xi}}_i = \mathbf{z}_i - \mathbf{Z}_{\partial_i} \hat{\boldsymbol{\beta}}_i, \tag{21.10}$$

and the correlation matrix by $\hat{\mathbf{R}}_n = [\hat{\boldsymbol{\xi}}_1, \dots, \hat{\boldsymbol{\xi}}_N]^{\mathrm{T}} [\hat{\boldsymbol{\xi}}_1, \dots, \hat{\boldsymbol{\xi}}_N]/N$. Note that the determination of each noise vector $\hat{\boldsymbol{\xi}}_i$ implies the computation of the pseudo-inverse $\mathbf{Z}_{\partial_i}^{\#} = (\mathbf{Z}_{\partial_i}^{\mathrm{T}} \mathbf{Z}_{\partial_i})^{-1} \mathbf{Z}_{\partial_i}^{\mathrm{T}}$, of size $(L - 1) \times (L - 1)$, for each $i = 1, \dots, L$. This computational complexity can, however, be greatly reduced by taking advantage of the relation between $\mathbf{Z}_{\partial_i}^{\#}$ and $\mathbf{Z}$. Let the $L \times L$ symmetric and positive-definite matrices $\mathbf{R}$ and $\mathbf{R}^{-1}$ be partitioned into block matrices as follows:

$$\mathbf{R} = \begin{bmatrix} \mathbf{A} & \mathbf{b} \\ \mathbf{b}^{\mathrm{T}} & c \end{bmatrix}$$

$$\mathbf{R}^{-1} = \begin{bmatrix} \mathbf{A}' & \mathbf{b}' \\ \mathbf{b}'^{\mathrm{T}} & c' \end{bmatrix}$$

where $\mathbf{A}$ and $\mathbf{A}'$ are $(L - 1 \times L - 1)$ matrices, $\mathbf{b}$ and $\mathbf{b}'$ are $(L - 1 \times 1)$ vectors, and c and c' are scalars. Since $\mathbf{R}, \mathbf{R}^{-1}, \mathbf{A}, \mathbf{A}^{-1} c'$ are positive definite; thus,

$$\mathbf{A}\mathbf{A}' + \mathbf{b}\mathbf{b}'^{\mathrm{T}} = \mathbf{I}_{L-1} \tag{21.11}$$

$$\mathbf{A}\mathbf{b}' + \mathbf{b}c' = \mathbf{0}_{L-1}, \tag{21.12}$$

where $\mathbf{0}_L - 1$ is a $(L - 1 \times 1)$ vector with null entries. Replacing $\mathbf{A}^{-1}\mathbf{b} = -\mathbf{b}'/c'$, derived from Equation 21.12, into expression 21.11, we obtain

$$\mathbf{A}^{-1} = \mathbf{A}' - \mathbf{b}'\mathbf{b}'^{\mathrm{T}}/\mathbf{c}'. \tag{21.13}$$

Based on this relation, the inversion of the matrix $\mathbf{Z}_{\partial_i}^{\mathrm{T}} \mathbf{Z}_{\partial_i}$ for $i = 1, \dots, L$, can be obtained by removing the ith row and the ith column of the matrix $(\mathbf{Z}^{\mathrm{T}}\mathbf{Z})^{-1}$ and implementing the expression (Equation 21.13) with the necessary adjustments.

The pseudo-code for the noise estimation is shown in the Algorithm 1. Symbol $[\hat{\mathbf{R}}]_{\partial_i,\partial_i}$ denotes the matrix obtained from $\hat{\mathbf{R}}$ by deleting the ith row and the ith column, $[\hat{\mathbf{R}}]_{i,\partial_i}$ denotes the ith row of $[\hat{\mathbf{R}}]_{:,\partial_i}$ and $[\hat{\mathbf{R}}]_{\partial_i,i}$ denotes $[\hat{\mathbf{R}}]_{i,\partial_i}^{\mathrm{T}}$. Steps 3 and 4 compute matrix $\hat{\mathbf{R}} = \mathbf{Z}^{\mathrm{T}}\mathbf{Z}$ and its inverse, respectively. Steps 6 and 7 estimate, respectively, the regression vector $\hat{\beta}_i$ and the noise $\hat{\xi}_i$ for each $i = 1, \dots, L$.

The main advantage of Algorithm 1 is that the computation of $\hat{\mathbf{R}}$ and of $\mathbf{R}' = \hat{\mathbf{R}}^{-1}$ are out of the *for* loop. Thus, the computational complexity, that is., the number of floating point operations (flops), of Algorithm 1 is substantially lower than that of an algorithm implementing the multiple regression without using the relation (21.13). Note that the computation of the observed data correlation matrix and of its inversion demands, approximately, $2NL^2 + L^3$ flops, whereas the multiple regression algorithm without using the relation (21.13) has to compute L times the above matrices, thus demanding, approximately, $2NL^3 + L^4$ flops. The approximated expressions for the number of floating point operations used by Algorithm 1 is $4NL^2 + 6L^3$, whereas multiple regression method without relation (21.13) is $4NL^2 + 2NL^3 + L^4$. For $N \gg L$, Algorithm 1 demands, approximately, $L/2$ less flops, which is a significant figure, since $L/2$ takes, in many applications, values of the order of 100. For example, the computational complexity of Algorithm 1 when applied to a subimage (145×145 pixels and 224 bands) of the Indian Pine Test Site acquired by the AVIRIS instrument is 2.8 Gflops, whereas the algorithm implemented without using the relation (Equation 21.13) has a computational complexity of 253.1 Gflops.

ALGORITHM 1: NOISE ESTIMATION

1. INPUT $\mathbf{Y} \equiv [\mathbf{y}_1, \mathbf{y}_2, \dots, \mathbf{y}_N]$
2. $\mathbf{Z} = \mathbf{Y}^T$;
3. $\hat{\mathbf{R}} := (\mathbf{Z}^T\mathbf{Z})$;
4. $\mathbf{R}' := \hat{\mathbf{R}}^{-1}$;
5. **for** $i := 1$ to L **do**
6. $\quad \hat{\beta}_i := ([\mathbf{R}']_{\partial_i,\partial_i} - [\mathbf{R}']_{\partial_i,i}[\mathbf{R}']_{i,\partial_i} / [\mathbf{R}']_{i,i}[\hat{\mathbf{R}}]_{\partial_i,i}$
7. $\quad \hat{\xi}_i := \mathbf{z}_i - \mathbf{Z}_{\partial_i}\hat{\beta}_i$
8. **end for**
9. OUTPUT $\hat{\xi}$; {$\hat{\xi}$ is a $N \times L$ matrix with the estimated noise}

The next experiment illustrates Algorithm 1 working. The input data are a simulated hyperspectral image composed of 10^4 spectral vectors, each one following the linear mixing model (Equations 21.1 and 21.2). The abundance fractions are generated according to a Dirichlet distribution, the number of endmembers is set to $p = 10$, and their signatures are selected from the USGS digital spectral library. The noise is zero-mean independent with variances along the bands following the Gaussian shape (see expression 21.4), for $\eta = 40$, and SNR= 15 dB ($\sigma^2 = 9.1 \times 10^{-3}$). Figure 21.4 shows the noiseless spectral vector $\mathbf{x}$, the noisy version $\mathbf{y} = \mathbf{x} + \mathbf{n}$ and the estimated spectral vector $\hat{\mathbf{x}} = \mathbf{y} - \hat{\mathbf{n}}$. The improvement in the SNR (i.e., $\mathbb{E}[\|\mathbf{n}\|^2]/\mathbb{E}[\|\hat{\mathbf{x}}-\mathbf{x}\|^2]$ is about 13 dB, leading to a final $\text{SNR}_f = E[\|\mathbf{x}\|^2]/E[\|\hat{\mathbf{x}}-\mathbf{x}\|^2] = 28\,\text{dB}$. Figure 21.5 plots the noise covariance curve and the respective estimates, as a function of the wavelength, for the same experiment. Note that the noise covariance estimate is very accurate.

21.3.2 SIGNAL SUBSPACE ESTIMATION

This section presents the core structure of the HySime method. The first step, based on the noise estimation procedure introduced in the previous section, identifies a set of orthogonal directions of which an unknown subset spans the signal subspace. This subset is then determined by seeking the minimum mean-squared error between $\mathbf{x}$, the original signal, and a noisy projection of it obtained from the vector $\mathbf{y} = \mathbf{x} + \mathbf{n}$. In the following, we assume $\mathbf{n} \sim \mathcal{N}(0, \hat{\mathbf{R}}_n)$, that is, the noise is zero-mean Gaussian distributed with covariance matrix $\hat{\mathbf{R}}_n$.

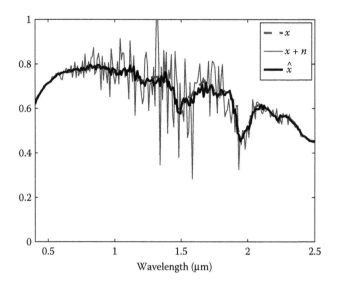

FIGURE 21.4 Illustration of the noise estimation algorithm for SNR = 15 dB, $\eta = 40$, and $p = 10$; dashed line: reflectance signal of a pixel; thin line: reflectance of the same pixel with noise (observed pixel); thick line: estimated reflectance of the same pixel.

Let the eigen-decomposition of the signal sample correlation matrix $\hat{\mathbf{R}}_x = [\hat{\mathbf{x}}_1, \ldots, \hat{\mathbf{x}}_N]$ $[\hat{\mathbf{x}}_1, \ldots, \hat{\mathbf{x}}_N]^\mathrm{T}/N$ be written as

$$\hat{\mathbf{R}}_x = \mathbf{E\Sigma E}^\mathrm{T}, \tag{21.14}$$

where $\mathbf{E} \equiv [\mathbf{e}_1, \ldots, \mathbf{e}_L]$ is a matrix with the eigenvectors of $\hat{\mathbf{R}}_x$. Given a permutation $\pi = \{i_1, \ldots, i_L\}$ of indices $i = 1, \ldots, L$, let us decompose the space $\mathbb{R}^L$ into two orthogonal subspaces: the k-dimensional subspace $\langle \mathbf{E}_k \rangle$ spanned by $\mathbf{E}_k \equiv [\mathbf{e}_{i_1}, \ldots, \mathbf{e}_{i_k}]$ and $\langle \mathbf{E}_k \rangle^\perp$, spanned by $\mathbf{E}_k^\perp \equiv [\mathbf{e}_{i_{k+1}}, \ldots, \mathbf{e}_{i_L}]$ that is, the orthogonal complement of subspace $\mathbf{E}_k$.

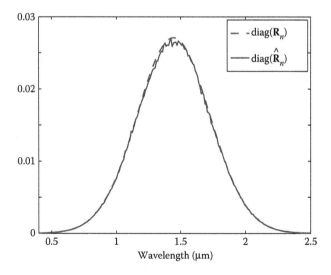

FIGURE 21.5 Illustration of the noise estimation algorithm for SNR = 15 dB, $\eta = 40$, and $p = 10$; solid line: diagonal of the estimated noise covariance matrix; dashed line: diagonal of the noise covariance matrix.

Let $\mathbf{U}_k = \mathbf{E}_k\mathbf{E}_k^{\mathrm{T}}$ be the projection matrix onto $\langle\mathbf{E}_k\rangle$ and $\hat{\mathbf{x}}_k \equiv \mathbf{U}_k\mathbf{y}$ be the projection of the observed spectral vector $\mathbf{y}$ onto the subspace $\langle\mathbf{E}_k\rangle$. The first- and the second-order moments of given $\mathbf{x}$ are

$$
\begin{aligned}
\mathbb{E}[\hat{\mathbf{x}}_k|\mathbf{x}] &= \mathbf{U}_k\mathbb{E}[\mathbf{y}|\mathbf{x}] \\
&= \mathbf{U}_k\mathbb{E}[\mathbf{x} + \mathbf{n}|\mathbf{x}] \\
&= \mathbf{U}_k\mathbf{x} \\
&\equiv \mathbf{x}_k,
\end{aligned}
\tag{21.15}
$$

$$
\begin{aligned}
\mathbb{E}[(\hat{\mathbf{x}}_k - \mathbf{x}_k)(\hat{\mathbf{x}}_k - \mathbf{x}_k)^{\mathrm{T}}|\mathbf{x}] &= \mathbb{E}[(\mathbf{U}_k\mathbf{y} - \mathbf{U}_k\mathbf{x})(\mathbf{U}_k\mathbf{y} - \mathbf{U}_k\mathbf{x}^{\mathrm{T}}|\mathbf{x}] \\
&= \mathbb{E}[(\mathbf{U}_k\mathbf{n}\mathbf{n}^{\mathrm{T}}\mathbf{U}_k^{\mathrm{T}}) \mid \mathbf{x}] \\
&= \mathbf{U}_k\hat{\mathbf{R}}_n\mathbf{U}_k^{\mathrm{T}}.
\end{aligned}
\tag{21.16}
$$

The mean-squared error between $\mathbf{x}$ and $\hat{\mathbf{x}}_k$ is

$$
\begin{aligned}
\mathrm{mse}(k|\mathbf{x}) &= \mathbb{E}[(\mathbf{x} - \hat{\mathbf{x}}_k)^{\mathrm{T}}(\mathbf{x} - \hat{\mathbf{x}}_k) \mid \mathbf{x}] \\
&= \mathbb{E}[\underbrace{(\mathbf{x} - \mathbf{x}_k - \mathbf{U}_k\mathbf{n})^{\mathrm{T}}}_{b_k}\underbrace{(\mathbf{x} - \mathbf{x}_k - \mathbf{U}_k\mathbf{n})}_{b_k} \mid \mathbf{x}] \\
&= \mathbf{b}_k^{\mathrm{T}}\mathbf{b}_k + \mathrm{tr}(\mathbf{U}_k\hat{\mathbf{R}}_n\mathbf{U}_k^{\mathrm{T}}).
\end{aligned}
\tag{21.17}
$$

Computing the mean of Equation 21.17 with respect to $\mathbf{x}$, noting that $\mathbf{b}_k = \mathbf{x} - \mathbf{x}_k = \mathbf{U}_k^{\perp}\mathbf{x}$, and using the properties $\mathbf{U} = \mathbf{U}^{\mathrm{T}}$, $\mathbf{U}^2 = \mathbf{U}$, and $\mathbf{U}^{\perp} = \mathbf{I} - \mathbf{U}$ of the projection matrices, we obtain

$$
\begin{aligned}
\mathrm{mse}(k) &= \mathbb{E}[(\mathbf{U}_k^{\perp}\mathbf{x})^{\mathrm{T}}(\mathbf{U}_k^{\perp}\mathbf{x})] + \mathrm{tr}(\mathbf{U}_k\hat{\mathbf{R}}_n\mathbf{U}_k^{\mathrm{T}}) \\
&= \mathrm{tr}(\mathbf{U}_k^{\perp}\mathbf{R}_x) + \mathrm{tr}(\mathbf{U}_k\hat{\mathbf{R}}_n) \\
&= \mathrm{tr}(\mathbf{U}_k^{\perp}\mathbf{R}_y) + 2\mathrm{tr}(\mathbf{U}_k\hat{\mathbf{R}}_n) + c,
\end{aligned}
\tag{21.18}
$$

where c is an irrelevant constant. The criterion we propose to estimate the signal subspace, let us call it X, is the minimization of $\mathrm{mse}(k)$ given by Equation 21.18, with respect to all the permutations $\pi = \{i_1, \ldots, i_L\}$ of size L and to k, with the correlation matrix $\mathbf{R}_y$ replaced with the sample correlation matrix $\hat{\mathbf{R}}_y = \mathbf{Y}\mathbf{Y}^{\mathrm{T}}/N$; that is,

$$
\hat{X} = \left\langle [\mathbf{e}_{\hat{i}_1}, \ldots, \mathbf{e}_{\hat{i}_k}] \right\rangle
\tag{21.19}
$$

$$
(\hat{k}, \hat{\pi}) = \arg\min_{k,\pi}\{\mathrm{tr}(\mathbf{U}_k^{\perp}\hat{\mathbf{R}}_y) + 2\mathrm{tr}(\mathbf{U}_k\hat{\mathbf{R}}_n)\},
\tag{21.20}
$$

where the dependence on the permutation π is through $\mathbf{U}_k = \mathbf{E}_k\mathbf{E}_k^{\mathrm{T}}$, since $\mathbf{E}_k \equiv [\mathbf{e}_{i_1}, \ldots, \mathbf{e}_{i_k}]$. For a given permutation π, each term of expression (21.20) has a clear meaning: the first term accounts for the projection error power and is a decreasing function of k; the second term accounts for the noise power and is an increasing function of k.

By exploiting, again, the fact that the $\mathbf{U}_k$ is a projection matrix and that $\mathrm{tr}(\mathbf{AB}) = \mathrm{tr}(\mathbf{BA})$, for $\mathbf{A}$, $\mathbf{B} \in \mathbb{R}^{L \times L}$, the minimization (Equation 21.20) can be rewritten as

$$
(\hat{k}, \hat{\pi}) = \arg\min_{k,\pi}\left\{c + \sum_{j=1}^{k}\underbrace{-p_{i_j} + 2\sigma_{i_j}^2}_{\delta_{ij}}\right\},
\tag{21.21}
$$

where c is an irrelevant constant and p_{i_j} and $\sigma_{i_j}^2$ are quadratic forms given by

$$p_{i_j} = \mathbf{e}_{i_j}^{\mathrm{T}} \hat{\mathbf{R}}_y \mathbf{e}_{i_j} \tag{21.22}$$

$$\sigma_{i_j}^2 = \mathbf{e}_{i_j}^{\mathrm{T}} \hat{\mathbf{R}}_n \mathbf{e}_{i_j}. \tag{21.23}$$

Based on the right-hand side of Equation 21.21, it follows that the corresponding minimization is achieved simply by including all the negative terms δ_i, for $i = 1, \ldots, L$, and only these, in the sum.

The pseudo-code for HySime is shown in Algorithm 2. HySime inputs are the spectral observed vectors and the noise estimates $\hat{\xi}$. Step 2 estimates the noise correlation matrix $\hat{\mathbf{R}}_n$. Step 3 estimates the signal correlation matrix $\hat{\mathbf{R}}_x$. Steps 4 and 5 calculate the eigenvectors of the signal correlation matrix and the terms δ_i based on the quadratic forms (Equations 21.22 and 21.23). Steps 6 and 7 implement the minimization (Equation 21.21). Finally, step 8 retrieves the signal subspace from $\hat{k}$ and $\hat{\pi}$.

ALGORITHM 2: HYSIME

1. INPUT $\mathbf{Y} = [\mathbf{y}_1, \mathbf{y}_2, \ldots, \mathbf{y}_N]$, and $\hat{\xi}$
2. $\hat{\mathbf{R}}_n := \frac{1}{N}\Sigma_i(\hat{\xi}_i\hat{\xi}_i^T)$; $\{\hat{\xi}_i$ is given by Equation 21.10$\}$
3. $\hat{\mathbf{R}}_x := \frac{1}{N}\Sigma_i((\mathbf{y}_i - \hat{\xi}_i)(\mathbf{y}_i - \hat{\xi}_i^T))$; $\{$estimate of $\hat{\mathbf{R}}_x\}$
4. $\mathbf{E} := [\mathbf{e}_1, \ldots, \mathbf{e}_L]$; $\{\mathbf{e}_i$ are the eigenvectors of $\hat{\mathbf{R}}_x \}$
5. $\delta = [\delta_1, \ldots, \delta_L]$; $\{\delta_i$ is given by Equation 21.21$\}$
6. $(\hat{\delta}, \hat{\pi}) = \text{sort}(\delta)$; $\{$sort δ_i by ascending order; save the permutation $\hat{\pi}\}$
7. $\hat{k} := $ number of terms $\hat{\delta}_i < 0$;
8. $\hat{X} = \langle[\mathbf{e}_{\hat{i}_1}, \ldots, \mathbf{e}_{\hat{i}_k}]\rangle$; $\{$signal subspace$\}$

21.4 DENOISING HYPERSPECTRAL IMAGES WITH HYSIME

There are several real scenarios where the Poisson sensor noise associated with the photon count process, also known as multiplicative noise, is not negligible. Techniques based on PCA [57], ICA [58], *Wavelets* [59], variable splitting and augmented Lagrangian framework [60], and variational approaches using total variation regularization methods [61] have been applied to denoise such data. Let us model a given observed spectral vector as

$$\mathbf{y} = \mathbf{x} + \mathbf{n} + \sqrt{\mathbf{x}}\mathbf{u}, \tag{21.24}$$

where $\mathbf{n}$ and $\mathbf{u}$ are L-dimensional vectors standing for additive and multiplicative noise, respectively.

In this section, we give evidence that Hysime can also be used to denoise both the additive and the multiplicative noises. Let us consider a scene containing 10 endmembers ($p = 10$). The spectral vectors are mixtures of this endmembers whose abundance fractions are Dirichlet distributed, as before. The noises $\mathbf{n}$ and $\mathbf{u}$ assumed to be zero-mean Gaussian distributed with variance $\sigma_n = 10^{-2}$ and $\sigma_u = 5 \times 10^{-2}$, respectively.

Figure 21.6 shows the noiseless spectral vector $\mathbf{x}$, the noisy version $\mathbf{y}$, and the estimated spectral vector $\hat{\mathbf{x}}$. This scenario corresponds to an initial $\text{SNR}_i = \mathbb{E}[\|\mathbf{x}\|^2]/\mathbb{E}[\|\mathbf{y} - \mathbf{x}\|^2] = 4$ dB and a final $\text{SNR}_f \equiv E[\|\mathbf{x}\|^2]/\mathbb{E}[\|\hat{\mathbf{x}} - \mathbf{x}\|^2] = 17$ dB. We conclude, therefore, both from the SNR improvements of 13 dB and from the visual inspection of Figure 21.6, that HySime is an effective tool to remove multiplicative noise as well.

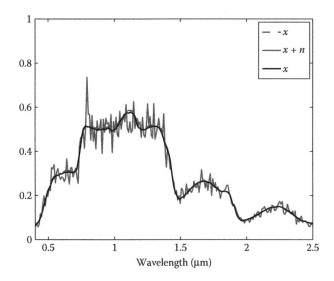

FIGURE 21.6 Illustration of image denoising performed by the Hysime method. Dashed line: reflectance signal of a pixel; thin line: reflectance of the same pixel with noise (observed pixel); thick line: estimated reflectance of the same pixel.

21.5 EVALUATION OF HYSIME WITH SIMULATED DATA

In this section, we apply the proposed HySime algorithm to simulated scenes and compare it with HFC and NWHFC eigen-based Neyman–Pearson detectors [41]. As concluded in [41], these algorithms are the state of the art in hyperspectral signal subspace identification, outperforming the information theoretical criteria approaches, namely the MDL [38,39] and the AIC [40].

The methods are evaluated with respect to the SNR (see Definition 21.5), the number of endmembers p, and the spectral noise shape (white and nonwhite). In the following experiments, the spectral signatures are selected from the USGS digital spectral library [50]. The abundance fractions are generated according to a Dirichlet distribution defined in expression 21.3.

Figure 21.7 shows the evolution of the mean-squared error for the HySime method as a function of the parameter k, for SNR = 30 dB, $\eta = 40$, and $p = 10$. The minimum of the mean-squared error

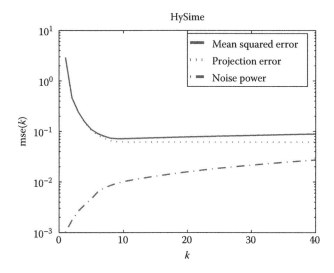

FIGURE 21.7 Mean squared error versus k, with $SNR = 30$ dB, $p = 10$, and $\eta = 20$, for the HySime method.

occurs at $k = 10$, which is exactly the number of endmembers present in the image. As expected, the projection error power and of noise power display decreasing and increasing behaviors, respectively, as a function of the subspace dimension k.

Table 21.1 presents the signal subspace order estimates yielded by HySime algorithm and the VD determined by the NWHFC and the HFC algorithms [41], as a function of the SNR, the number of endmembers, p, and the noise shape ($\eta = \infty$ and $\eta = 20$).

NWHFC algorithm is basically the HFC one [43] preceded by a noise-whitening step, based on the estimated noise correlation matrix. In implementing this step, we got poor results in very high SNRs and colored noise scenarios. This is basically because the noise estimation step in NWHFC needs to invert the noise correlation matrix, which gives inaccurate results when the noise power is small. For this reason, we have used both the true and estimated noise correlation matrices. The results based on the true correlation matrix are in brackets. We stress that, for the setting of this experiment, HySime method yields the same results, whether using the estimated or the true noise

TABLE 21.1

Signal Subspace Dimension $\hat{k}$, Based on 50 Monte Carlo Runs, as a Function of SNR, p, and η (Noise Shape)

SNR	Method	White Noise ($\eta = \infty$)			Gaussian-Shaped Noise ($\eta = 20$)		
		$p = 5$	$p = 10$	$p = 15$	$p = 5$	$p = 10$	$p = 15$
	HySime	**5**	**10**	**15**	**5**	**10**	**15**
	HFC ($P_f = 10^{-3}$)	5	7	11	—	—	—
	HFC ($P_f = 10^{-4}$)	5	7	8	—	—	—
50 dB	HFC ($P_f = 10^{-5}$)	4	6	8	—	—	—
	NWHFC ($P_f = 10^{-3}$)	5(5)	7(7)	10(11)	41(5)	61(10)	45(10)
	NWHFC ($P_f = 10^{-4}$)	5(5)	7(7)	8(8)	33(5)	54(10)	34(10)
	NWHFC ($P_f = 10^{-5}$)	4(4)	7(6)	8(8)	28(5)	41(9)	27(10)
	HySime	**5**	**10**	**15**	**5**	**10**	**15**
	HFC ($P_f = 10^{-3}$)	4	7	9	—	—	—
	HFC ($P_f = 10^{-4}$)	4	6	8	—	—	—
35 dB	HFC ($P_f = 10^{-5}$)	4	6	8	—	—	—
	NWHFC ($P_f = 10^{-3}$)	4(4)	7(7)	9(9)	10(5)	12(10)	10(10)
	NWHFC ($P_f = 10^{-4}$)	4(4)	7(6)	8(8)	9(5)	11(10)	8(10)
	NWHFC ($P_f = 10^{-5}$)	4(4)	6(6)	8(8)	7(5)	10(9)	8(10)
	HySime	**5**	**10**	**14**	**5**	**10**	**15**
	HFC ($P_f = 10^{-3}$)	5	6	8	—	—	—
	HFC ($P_f = 10^{-4}$)	5	6	7	—	—	—
25 dB	HFC ($P_f = 10^{-3}$)	4	5	7	—	—	—
	NWHFC ($P_f = 10^{-3}$)	5(5)	6(6)	9(8)	5(5)	11(10)	9(11)
	NWHFC ($P_f = 10^{-4}$)	5(5)	6(6)	7(7)	5(5)	11(10)	9(10)
	NWHFC ($P_f = 10^{-5}$)	4(4)	5(5)	7(7)	5(5)	11(9)	8(10)
	HySime	**5**	**8**	**12**	**5**	**8**	**12**
	HFC ($P_f = 10^{-3}$)	5	4	5	—	—	—
	HFC ($P_f = 10^{-4}$)	4	3	2	—	—	—
15 dB	HFC ($P_f = 10^{-3}$)	4	3	2	—	—	—
	NWHFC ($P_f = 10^{-3}$)	5(5)	5(4)	5(5)	5(5)	11(10)	10(10)
	NWHFC ($P_f = 10^{-4}$)	4(4)	3(3)	3(2)	5(5)	11(10)	8(10)
	NWHFC ($P_f = 10^{-5}$)	4(4)	3(3)	2(2)	5(5)	11(9)	8(10)

Note: Values within parentheses were computed based on the true noise statistics.

correlation matrices. The right side of Table 21.1 (noise-colored scenario, $\eta = 20$) does not present results for the HFC method because this method was designed only for white noise.

Another central issue of NWHFC and HFC algorithms is the false-alarm probability, P_f, they are parameterized with. This probability is used in a series of Neyman–Pearson tests, each one designed to detect a different orthogonal signal subspace direction. There is the need, therefore, to specify the false-alarm probability P_f of the tests. Based on the hints given in [41] and in our own results, we choose $P_f \in \{10^{-3}, 10^{-4}, 10^{-5}\}$.

The figures shown in Table 21.1, based on 50 Monte Carlo runs, display the following behavior:

1. HySime and NWHFC algorithms parameterized with $P_f = 10^{-3}$ show similar performances at low subspace dimension, say $p \le 5$, and white noise. This is also true for colored noise and NWHFC working with known noise covariance matrix. However, if the noise statistics are unknown, NWHFC performs much worse than HySime.
2. HySime performs better than NWHFC for high space dimensions, say $p > 5$.

We conclude, therefore, that the HySime algorithm yields systematically equal or better results than NWHFC and HFC algorithms. Another advantage of the HySime approach is that it does not depend on any tunable parameter.

21.6 EXPERIMENTS WITH REAL HYPERSPECTRAL DATA

In this section, the proposed method is applied to real hyperspectral data collected by the AVIRIS [11]. AVIRIS instrument covers the spectral region from 0.41 to 2.45 μm in 224 bands with a 10 nm bandwidth. Flying at an altitude of 20 km, it has an instantaneous field of view (IFOV) of 20 m and views a swath over 10 km wide. Two subsets are considered:

1. The Indian Pine Test Site in Northwestern Indiana.* The data set contains 145×145 pixels and 185 spectral bands (noisy and water absorption bands were removed). This observed region contains a mixture of agriculture and forestry. There is a major dual-lane highway (U.S. 52 and U.S. 231), a rail line crossing near the top, a major secondary road (Jackson Highway) near the middle, several other county roads, and houses (Figure 21.8a shows band 30 of the data set). A detailed ground truth map for this region is published in [62] where 16 materials were considered.

(a)    (b)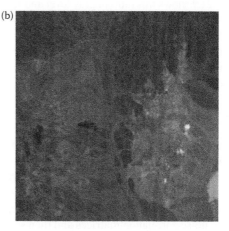

FIGURE 21.8 Band 30 of the Indiana Pines hyperspectral data set (a) and of Cuprite Nevada data set (b).

---

* Available at http://dynamo.ecn.purdue.edu/~biehl/MultiSpec/.

TABLE 21.2

Inferred Signal Subspace Dimension $\hat{k}$ for the Indian Pine and the Cuprite Data sets

Method	Indian Pine	Cuprite
HySime	**16**	**20**
HFC ($P_f = 10^{-3}$)	25	22
HFC ($P_f = 10^{-4}$)	22	19
HFC ($P_f = 10^{-5}$)	21	19
NWHFC ($P_f = 10^{-3}$)	18	23
NWHFC ($P_f = 10^{-4}$)	18	22
NWHFC ($P_f = 10^{-5}$)	18	20

2. The Cuprite mining area in southern Nevada.[*] The data set contains 350×350 pixels. The test site, located approximately 200 km northwest of Las Vegas, is a relatively undisturbed acid-sulfate hydrothermal system near highway 95 (Figure 21.8b shows band 29 of the data set). The geology and alteration were previously mapped in detail and a geologic summary and a mineral map can be found in [63]. In this data set, 18 materials were considered.

Table 21.2 shows the signal subspace dimension inferred by HySime, HFC, and NWHFC for the Indian Pine and the Cuprite datasets. The values $\{10^{-5}, 10^{-4}, \text{and } 10^{-3}\}$ were considered in the false-alarm probabilities, P_f, parameterizing the HFC and the NWHFC methods.

The value obtained by HySime coincides with the number of ground truth materials for the Indiana Pines data set, whereas the value obtained for the Cuprite data set is $k = 20$. According to the ground truth presented in Ref. [63], HySime overestimate the number of endmembers in the Cuprite data set. This difference is due to: (i) the presence of rare pixels not accounted for in Ref. [63] and (ii) spectral variability. Note that, methods NWHFC and HFC (for all P_f values considered) overestimate the number of endmembers in both data sets.

Although these results are an indicator of the HySime competitiveness, a better assessment of each algorithm performance is obtained by computing $\|\mathbf{E}_k^{\perp}\mathbf{x}\|^2 / \|\mathbf{x}\|^2$, that is, the relative power of the signal component orthogonal to the identified subspace and, therefore, not modeled by this subspace. Figure 21.9 shows this ratio in gray level and for each pixel; we have used $\hat{\mathbf{x}} = \mathbf{y} - \hat{\mathbf{n}}$, with $\hat{\mathbf{n}}$ given by expression (21.10), as an estimate of $\mathbf{x}$. HySime yielded the lowest values followed by HFC and NWHFC. The relative advantage of HySime over HFC and NWHFC is of the order of, 20 and 10^3, respectively. Furthermore, there is no perceptible structure in the HySime relative error, which is not the case at least with NWHFC, where the agriculture fields are clearly perceived, indicating the presence of signal where it should not be.

21.7 CONCLUSIONS

The huge volumes and rates of data generated by hyperspectral sensors demand expensive processors with very high performance and memory capacities. DR is, therefore, a relevant first step in the hyperspectral data processing chain. This chapter introduced the HySime method, which is an computationally efficient and automatic (in the sense that it dispenses with any tunable parameter) approach to infer the signal subspace in hyperspectral imagery. HySime first estimates the signal and the noise correlation matrices and then selects the subset of eigenvalues that best represents the

---

[*] Available at http://aviris.jpl.nasa.gov/html/aviris.freedata.html.

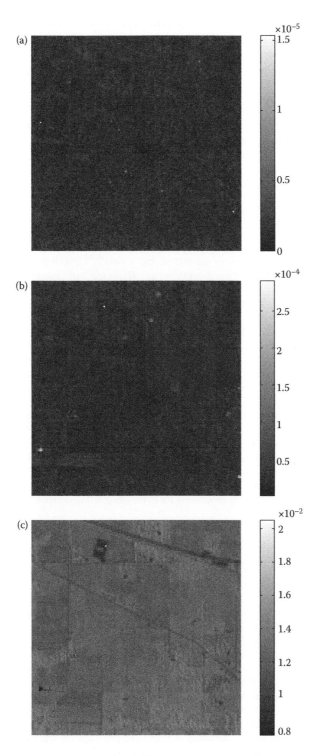

FIGURE 21.9 Relative power of the signal component orthogonal to identified subspace, $\langle \mathbf{E}_k \rangle$ for the Indian Pine data set: (a) HySime, (b) HFC, and (c) NWHFC.

signal subspace in the minimum mean-squared error sense. A set of experiments with simulated and real data led to the conclusion that the HySime is an effective and useful tool, yielding comparable or better results than the state-of-the-art algorithms.

ACKNOWLEDGMENTS

This work was supported by the Instituto de Telecomunicações and the Fundação para a Ciência e Tecnologia (IT/FCT) under project HoHus.

REFERENCES

1. J. M. Bioucas-Dias and J. M. P. Nascimento, Hyperspectral subspace identification, *IEEE Trans. Geosci. Remote Sens.*, 46(8), 2435–2445, 2008.
2. J. M. Bioucas-Dias and A. Plaza, Hyperspectral unmixing: Geometrical, statistical, and sparse regression-based approaches, In L. Bruzzone (ed.), 7830(1). SPIE, 2010, p. 78300A, doi:10.1117/12.870780.
3. M. Parente and A. Plaza, Survey of geometric and statistical unmixing algorithms for hyperspectral images, in *Second IEEE GRSS Workshop on Hyperspectral Image and Signal Processing-WHISPERS'2010*, 2010, pp. 1–4.
4. X. Briottet, Y. Boucher, A. Dimmeler, A. Malaplate, A. Cini, M. Diani, H. Bekman et al., *Military Applications of Hyperspectral Imagery*, in W. R. Watkins and D. Clement (eds), 6239(1). SPIE, 2006, p. 6239B, doi:10.1117/12.672030.
5. M. B. Lopes, J.-C. Wolff, J. M. Bioucas-Dias, and M. A. Figueiredo, Determination of the composition of Counterfeit Heptodin(TM) tablets by near infrared chemical imaging and classical least squares estimation, *Anal. Chim. Acta*, 641(1–2), 46–51, 2009.
6. N. Keshava and J. Mustard, Spectral unmixing, *IEEE Signal Process. Mag.*, 19(1), 44–57, 2002.
7. A. A. Gowen, C. P. O'Donnell, P. J. Cullen, G. Downey, and J. M. Frias. Hyperspectral imaging–an emerging process analytical tool for food quality and safety control. *Trends Food Sci. Technol.*, 18(12), 590–598, 2007.
8. B. Park, K. C. Lawrence, W. R. Windham, and D. P. Smith, Performance of hyperspectral imaging system for poultry surface fecal contaminant detection, *J. Food Eng.*, 75(3), 340–348, 2006.
9. C.-I. Chang (ed.), *Hyperspectral Data Exploitation: Theory and Applications*. Hoboken, New Jersey: John Wiley & Sons, Inc., 2007.
10. G. Shaw and D. Manolakis, Signal processing for hyperspectral image exploitation, *IEEE Signal Process. Mag.*, 19(1), 12–16, 2002.
11. R. O. Green, M. L. Eastwood, C. M. Sarture, T. G. Chrien, M. Aronsson, B. J. Chippendale, J. A. Faust et al., Imaging spectroscopy and the airborne visible/infrared imaging spectrometer (AVIRIS), *Remote Sens. Environ.*, 65(3), 227–248, 1998.
12. S. Ungar, J. Pearlman, J. Mendenhall, and D. Reuter, Overview of the Earth Observing One (EO-1) mission, *IEEE Trans. Geosci. Remote Sens.*, 41(6–1), 1149–1159, 2003.
13. D. Landgrebe, Hyperspectral image data analysis, *IEEE Signal Process. Mag.*, 19(1), 17–28, 2002.
14. C. Chang and S. Wang, Constrained band selection for hyperspectral imagery, *IEEE Trans. Geosci. Remote Sens.*, 44(6), 1575–1585, 2006.
15. S. D. Backer, P. Kempeneers, W. Debruyn, and P. Scheunders, A band selection technique for spectral classification, *IEEE Geosci. Remote Sens. Lett.*, 2(3), 319–323, 2005.
16. R. Huang and M. He, Band selection based on feature weighting for classification of hyperspectral data, *IEEE Geosci. Remote Sens. Lett.*, 2(2), 156–159, 2005.
17. S. S. Shen and E. M. Bassett, Information-theory-based band selection and utility evaluation for reflective spectral systems, in *Proceedings of the SPIE Conference on Algorithms and Technologies for Multispectral, Hyperspectral, and Ultraspectral Imagery VIII*, 4725, pp. 18–29, 2002.
18. C. Chang, Q. Du, T. Sun, and M. Althouse, A joint band prioritization and band-decorrelation approach to band selection for hyperspectral image classification, *IEEE Trans. Geosci. Remote Sens.*, 7(6), 2631–2641, 1999.
19. L. Sun and W. Gao, Method of selecting the best classification bands from hyperspectral images based on genetic Algorithm and rough set, in *Proceedings of the SPIE conference on Hyperspectral Remote Sensing and Application*, R. O. Green and Q. Tong (eds), 3502, pp. 179–184, 1998.

20. I. T. Jolliffe, *Principal Component Analysis*. New York: Spriger Verlag, 1986.
21. L. L. Scharf, *Statistical Signal Processing, Detection Estimation and Time Series Analysis*. New York: Addison-Wesley Pub. Comp., 1991.
22. A. Green, M. Berman, P. Switzer, and M. D. Craig, A transformation for ordering multispectral data in terms of image quality with implications for noise removal, *IEEE Trans. Geosci. Remote Sens.*, 26(1), 65–74, 1988.
23. J. B. Lee, S. Woodyatt, and M. Berman, Enhancement of high spectral resolution remote-sensing data by noise-adjusted principal components transform, *IEEE Trans. Geosci. Remote Sens.*, 28(3), 295–304, 1990.
24. J. Bruske and G. Sommer, Intrinsic dimensionality estimation with optimaly topologic preserving maps, *IEEE Trans. Pattern Anal. Mach. Intell.*, 20(5), 572–575, 1998.
25. P. Demartines and J. Herault, Curvilinear component analysis: A self-organizing neural network for non-linear mapping of data sets, *IEEE Trans. Neural Netw.*, 8(1), 148–154, 1997.
26. M. Lennon, G. Mercier, M. Mouchot, and L. Hubert-Moy, Curvilinear component analysis for nonlinear dimensionality reduction of hyperspectral images, in *Proceedings of the SPIE Symposium on Remote Sensing Conference on Image and Signal Proceedings for Remote Sensing VII*, 4541, pp. 157–169, 2001.
27. C. Bachmann, T. Ainsworth, and R. Fusina, Improved manifold coordinate representations of large-scale hyperspectral scenes, *IEEE Trans. Geosci. Remote Sens.*, 44(10), 2786–2803, 2006.
28. C. Bachmann, T. Ainsworth, and R. Fusina, Exploiting manifold geometry in hyperspectral imagery, *IEEE Trans. Geosci. Remote Sens.*, 43(3), 441–454, 2005.
29. C. Yangchi, M. Crawford, and J. Ghosh, Applying nonlinear manifold learning to hyperspectral data for land cover classification, in *Proceedings of the IEEE International Geoscience and Remote Sensing Symposium*, Vol. 6, pp. 4311–4314, 2005.
30. D. Gillis, J. Bowles, G. M. Lamela, W. J. Rhea, C. M. Bachmann, M. Montes, and T. Ainsworth, Manifold learning techniques for the analysis of hyperspectral Ocean data, in *Proceedings of the SPIE conference on Algorithms and Technologies for Multispectral, Hyperspectral, and Ultraspectral Imagery XI*, S. S. Shen and P. E. Lewis (eds), Vol. 5806, pp. 342–351, 2005.
31. J. Wang and C.-I. Chang, Independent component analysis-based dimensionality reduction with applications in hyperspectral image analysis, *IEEE Trans. Geosci. Remote Sens.*, 44(6), 1586–1600, 2006.
32. M. Lennon, M. Mouchot, G. Mercier, and L. Hubert-Moy, Independent component analysis as a tool for the dimensionality reduction and the representation of hyperspectral images, in *Proceedings of the IEEE International Geoscience and Remote Sensing Symposium*, 2001, pp. 2893–2895.
33. A. Ifarraguerri and C.-I. Chang, Unsupervised hyperspectral image analysis with projection pursuit, *IEEE Trans. Geosci. Remote Sens.*, 38(6), 127–143, 2000.
34. C. Bachmann and T. Donato, An information theoretic comparison of projection pursuit and principal component features for classification of Landsat TM imagery of central colorado, *Int. J. Remote Sens.*, 21(15), 2927–2935(9), 2000.
35. H. Othman and S.-E. Qian, Noise reduction of hyperspectral imagery using hybrid spatial–spectral derivative-domain wavelet shrinkage, *IEEE Trans. Geosci. Remote Sens.*, 44(2), 397–408, 2006.
36. S. Kaewpijit, J. L. Moigne, and T. El-Ghazawi, Automatic reduction of hyperspectral imagery using wavelet spectral analysis, *IEEE Trans. Geosci. Remote Sens.*, 41(4), 863–871, 2003.
37. H. Bagan, Y. Yasuoka, T. Endo, X. Wang, and Z. Feng, Classification of airborne hyperspectral data based on the average learning subspace method, *IEEE Geosci. Remote Sens. Lett.*, 5(3), 368–372, 2008.
38. G. Schwarz, Estimating the dimension of a model, *Ann. Stat.*, 6, 461–464, 1978.
39. J. Rissanen, Modeling by shortest data description, *Automatica*, 4, 465–471, 1978.
40. H. Akaike, A new look at the statistical model identification, *IEEE Trans. Autom. Control*, 19(6), 716–723, 1974.
41. C.-I. Chang and Q. Du, Estimation of number of spectrally distinct signal sources in hyperspectral imagery, *IEEE Trans. Geosci. Remote Sens.*, 42(3), 608–619, 2004.
42. M. Wax and T. Kailath, Detection of signals by information theoretic criteria, *IEEE Trans. Acoust. Speech Signal Process.*, 33(2), 387–392, 1985.
43. J. Harsanyi, W. Farrand, and C.-I. Chang, Determining the number and identity of spectral endmembers: An integrated approach using Neyman–Pearson eigen thresholding and iterative constrained RMS error minimization, in *Proceedings of the 9th Thematic Conference on Geologic Remote Sensing*, Pasadena, California, 1993.
44. P. Bajorski, Second moment linear dimensionality as an alternative to virtual dimensionality, *IEEE Trans. Geosci. Remote Sens.*, 49(2), 672–678, 2011.

45. N. A. M. Diani and G. Corsini, Hyperspectral signal subspace identification in the presence of rare signal components, *IEEE Trans. Geosci. Remote Sens.*, 48(4), 1940–1954, 2010.

46. N. A. M. Diani and G. Corsini, A new algorithm for robust estimation of the signal subspace in hyperspectral images in the presence of rare signal components, *IEEE Trans. Geosci. Remote Sens.*, 47(11), 3844–3856, 2009.

47. M. T. Eismann and D. W. J. Stein, *Stochastic Mixture Modeling*, ser. Hyperspectral data exploitation: Theory and applications. Hoboken, New Jersey: Jonh Wiley & sons, Inc., 2007, ch. 5.

48. O. Eches, N. Dobigeon, and J.-Y. Tourneret, Estimating the number of endmembers in hyperspectral images using the normal compositional model and a hierarchical Bayesian Algorithm, *IEEE J. Sel. Top. Signal Process.*, 4(3), 582–591, 2010.

49. D. Manolakis, C. Siracusa, and G. Shaw, Hyperspectral subpixel target detection using linear mixing model, *IEEE Trans. Geosci. Remote Sens.*, 39(7), 1392–1409, 2001.

50. R. N. Clark, G. A. Swayze, A. Gallagher, T. V. King, and W. M. Calvin, The U.S. Geological Survey Digital Spectral Library: Version 1:0.2 to 3.0 μm, U.S. Geological Survey, Open File Report 93–592, 1993.

51. A. Gelman, J. B. Carlin, H. S. Stern, and D. B. Rubin, *Bayesian Data Analysis*, 2nd ed. Boca Raton, Florida: CRC, 2004.

52. T. Minka, Estimating a Dirichlet distribution, MIT, Tech. Rep., 2000.

53. G. H. Golub and C. F. V. Loan, *Matrix Computations*, 3rd ed., ser. Mathematical Sciences. Baltimore and London: John Hopkins University Press, 1996.

54. G. W. Stewart and J.-G. Sun, *Matrix Perturbation Theory*. Michigan: Academic Press Inc., 1990.

55. R. O. Duda, P. E. Hart, and D. G. Stork, *Pattern Classification*, 2nd ed. New York: John Wiley & Sons, Inc., 2001.

56. R. Roger and J. Arnold, Reliably estimating the noise in AVIRIS hyperspectral imagers, *Int. J. Remote Sens.*, 17(10), 1951–1962, 1996.

57. G. Chen and S. Qian, Denoising of hyperspectral imagery using principal component analysis and wavelet shrinkage, *IEEE Trans. Geosci. Remote Sens.*, 49(3), 973–980, 2011.

58. M. Haritopoulos, H. Yin, and N. M. Allinson, Image denoising using self-organizing map-based nonlinear independent component analysis, *Neural Netw.*, 15, 1085–1098, 2002.

59. D. L. Donoho, De-noising by soft-thresholding, *IEEE Trans. Inform. Theory*, 41(3), 613627, 1995.

60. J. M. Bioucas-Dias and M. A. Figueiredo, Multiplicative noise removal using variable splitting and constrained optimization, *IEEE Trans. Image Process.*, 19(7), 1720–1730, 2010.

61. G. Aubert and J.-F. Aujol, A variational approach to removing multiplicative noise, *SIAM J. Appl. Math.*, 68(4), 925946, 2008.

62. D. Landgrebe, *Multispectral Data Analysis: A Signal Theory Perspective*, Purdue University, Tech. Rep., 1998.

63. G. Swayze, R. Clark, S. Sutley, and A. Gallagher, Ground-Truthing AVIRIS Mineral Mapping at Cuprite, Nevada, in *Summaries of the Third Annual JPL Airborne Geosciences Workshop*, Pasadena, California, 47–49, 1992.

22 Image Classification and Object Detection Using Spatial Contextual Constraints

Selim Aksoy, R. Gökberk Cinbiş, and H. Gökhan Akçay

CONTENTS

22.1 INTRODUCTION

Spatial information plays a fundamental role in the analysis and understanding of remotely sensed data sets. Common ways of incorporating spatial information into classification involve the use of textural, morphological, and object-based features. Features extracted using co-occurrence matrices, Gabor wavelets [1], morphological profiles [2], and Markov random fields [3] have been widely used in the literature to model spatial information in neighborhoods of pixels. However, problems such as scale selection and the detailed content of high-resolution imagery make the applicability of traditional fixed window-based methods difficult for such data sets.

Another powerful method for exploiting structural information is to perform region-based classification rather than classifying individual pixels. This is also referred to as object-oriented classification in the remote sensing literature. For example, Bruzzone and Carlin [4] performed classification using the spatial context of each pixel according to a hierarchical multilevel representation of the scene. In Ref. [5], we proposed an algorithm for selecting meaningful segments that maximize a measure consisting of spectral homogeneity and neighborhood connectivity in a hierarchy of segmentations, and described an algorithm for unsupervised grouping of candidate segments belonging to multiple hierarchical segmentations to find coherent sets of segments that correspond to actual objects. However, image segmentation is still an unsolved problem, and homogeneous regions obtained as a result of segmentation often correspond to very small details in high spatial resolution images obtained from the new-generation sensors.

Alternatively, contextual models that exploit spatial information can be used to resolve the ambiguities in the identification of structures having similar low-level spectral and textural properties. Contextual information has long been acknowledged for playing a very important role in both human and computer vision. Consequently, development of context models has become a challenging problem in both statistical and structural pattern recognition. A structural way of modeling context in images is through the quantification of spatial relationships. Typical relationships studied in the literature include geometric (based on size, position, shape, and orientation), topological (based on set relationships and neighborhood structure), semantic (based on similarity and causality), statistical (based on frequency and co-occurrence), and structural (based on spatial configuration and arrangement patterns) relationships [6]. The methods used for computing these relationships depend on the way how objects/regions are modeled. Widely used approaches include grid-based representations, centroids, and minimum bounding rectangles. However, even though centroids and minimum bounding rectangles can be useful when regions have circular or rectangular shapes, regions in natural scenes often do not follow these assumptions. Furthermore, fixed-sized grids are also not generally applicable as they cannot capture large number of structures with varying sizes and shapes.

When regions are represented as sets of points (pixels), spatial relationships can be modeled in terms of directional and distance information between pixel groups. In particular, adjacency of two regions can be measured as a fuzzy function of the distance between their closest points or using morphological dilations modeling connectivities [7]. Distance-based relationships can also be defined using fuzzy membership functions modeling symbolic classes such as near and far using the distance between boundary pixels. In previous work [8], we developed fuzzy models for pairwise topological spatial relationships such as bordering, invading, and surrounding based on overlaps between region boundaries, distance-based relationships such as near and far based on distances between region boundaries, and relative position-based relationships such as right, left, above and below using angles between region centroids. Then, we combined these pairwise relationships into higher-order relationship models using fuzzy logic, and illustrated their use in image retrieval [8]. We also developed a Bayesian framework that learns image classes based on automatic selection of distinguishing (e.g., frequently occurring, rarely occurring) relations between regions [9]. Finally, we built attributed relational graph structures to model scenes by representing regions by the graph nodes and their spatial relationships by the edges between such nodes [10], and used relational matching techniques to find similarities between graphs representing different scenes. We demonstrated the effectiveness of these approaches in scenarios that cannot be expressed by traditional approaches but where the proposed models can capture both feature and spatial characteristics of scenes and model them according to their high-level semantic content. Inglada and Michel also used attributed relational graphs where the relations are computed by using region connection calculus [11], and used these graphs for object matching.

This chapter presents an extension of our earlier work on modeling region spatial relationships [8] using relative position-based relationships: binary directional relationships [12] and the ternary between relationship [13]. Most of the existing methods for defining relative spatial positions rely on angle measurements between points of objects of interest where the angle corresponding to a pair of points is computed between the segment joining the points and a reference axis in the coordinate system [14]. For example, Miyajimaya and Ralescu [15] proposed to use a histogram that is constructed using the angles between all pairs of points from both objects where the mean or the maximum angle computed from this histogram can be used to represent the relative position of these objects. Matsakis and Wendling [16] introduced the histogram of forces as an alternative to the histogram of angles. This method computes the degree of satisfaction for a given angle using intersection of longitudinal sections of objects with lines having the desired direction. Wang et al. [17] proposed the F-templates that incorporate distance information with direction information. Bloch [18] proposed a morphological approach that is based on directional dilations where a fuzzy landscape for a reference object is created at a given angle and other objects are compared to this landscape to evaluate how well they match with the areas having high membership values.

Another relationship that is often used in daily life but has not been studied as thoroughly as the binary relationships is the between relationship (Ref. [19] provides an extensive review and a comparative study). Directional dilations are also useful for the between relationship. After obtaining an approximate relative angle between the reference regions, directional dilations are applied to both regions to extend them toward each other to generate the landscape. An angle histogram can be directly used to create the structuring element for dilation.

Intuitively, the influence of the shape of the object (e.g., concavities, extent) and the influence of the distance between objects are important points to be considered in the design of an algorithm for modeling spatial relationships. Mathematical morphology provides a strong basis for such a framework. Furthermore, the ambiguities and subjectiveness inherent in the definitions of the relationships make fuzzy representation a promising approach for modeling the imprecision in both the images and the results. In this chapter, we describe intuitive, flexible, and efficient methods for modeling pairwise directional spatial relationships and the ternary between relationship using fuzzy mathematical morphology. These methods define a fuzzy landscape where each point is assigned a value that quantifies its relative position with respect to the reference object(s) and the type of the relationship. Directional mathematical dilation with fuzzy structuring elements is used to compute this landscape. We provide flexible definitions of fuzzy structuring elements that are tunable along both radial and angular dimensions. Furthermore, for the pairwise directional relationships, the definitions for the fuzzy landscape are extended to support sensitivity to visibility to handle image areas that are fully or partially enclosed by a reference object but are not visible from image points along the direction of interest. Given a reference object and a direction of interest that specifies the spatial relationship, the degree of satisfaction of this relation by a target object can be computed by integrating the landscape corresponding to this relation over the support of the target region.

The definitions of the directional spatial relationships are also combined to generate a landscape in which the degree of each image area being located "between" the reference objects is quantified. Our definition also handles the cases where one object is significantly spatially extended relative to the other by taking spatial proximity into consideration. Similarly, the satisfaction of this ternary relation by a target object relative to two reference objects is computed by integrating the corresponding landscape.

We illustrate the use of the models described in this chapter as spatial contextual constraints for two image analysis tasks. First, we show how these spatial relationships can be incorporated into a Bayesian classification framework for land cover classification. The decision based on the maximum posterior probability rule produces limited accuracy for classes with similar appearance and spectral values when no spatial information is used. However, by constraining the classification of certain classes by using their spatial relationships to other classes as contextual information, we show that the classification accuracy can be significantly improved. Then, we show that detection of buildings with complex shapes and roof structures can be improved by using directional spatial relationships between candidate building regions and shadow regions along the sun azimuth angle.

The rest of the chapter is organized as follows. The fuzzy structuring elements and the morphological approach for quantifying the pairwise directional spatial relationships are described in Section 22.2. The approach for modeling the ternary between relationship and its computation for spatially extended objects is given in Section 22.3. Applications of these approaches to land cover classification and building detection are presented in Section 22.4. Finally, conclusions are discussed in Section 22.5.

22.2 DIRECTIONAL–SPATIAL RELATIONSHIPS

Directional relationships describe the spatial arrangement of two objects relative to each other. Although it is a common approach to use right (east), left (west), above (north), and below (south) as the directions, for generic modeling purposes it is more convenient and generalizable to use an angle-based definition of these relations where it is possible to calculate the degree of satisfaction of the relation for a given angle.

Given a reference object B and a direction specified by the angle α, our goal is to generate a landscape in which the degree of satisfaction of the directional relationship at each image area relative to the reference object is quantified. Then, given a second object, its relation to the reference object can be measured using this landscape. The landscape will be denoted by $\beta_\alpha (B)$ in the rest of the chapter. Definitions for crisp and fuzzy objects are available in the literature. However, only crisp objects are considered in this chapter.

22.2.1 Morphological Approach

The landscape $\beta_\alpha(B)$ around a reference object B along the direction specified by the angle α can be defined as a fuzzy set such that the membership value of an image point corresponds to the degree of satisfaction of the spatial relation under examination where points in areas that satisfy the directional relation with a high degree have high membership values. This relationship can be defined in terms of the angle between the vector from a point in the reference object to a point in the image and the unit vector along the direction α measured with respect to the horizontal axis. Bloch [18] suggested that the smallest such angle computed for a point in the image considering all points in the reference object corresponds to the visibility of the image point from the reference object in the direction α. Consequently, the value of the fuzzy landscape at an image point x can be computed as a function f: $[0, \pi] \rightarrow [0,1]$ of this angle as

$$\beta_\alpha(B)(x) = f\left(\min_{b\in B}\theta_\alpha(x,b)\right) \tag{22.1}$$

where b represents a point in B. $\theta_\alpha(x, b)$ is the angle between the vector $\overrightarrow{bx}$ and the unit vector $\vec{u}_\alpha = (\cos\alpha, \sin\alpha)^T$ along α, and can be computed as

$$\theta_\alpha(x,b) = \begin{cases} \arccos\left(\dfrac{\overrightarrow{bx}\cdot\vec{u}_\alpha}{\left\|\overrightarrow{bx}\right\|}\right) & \text{if } x \neq b, \\ 0 & \text{if } x = b. \end{cases} \tag{22.2}$$

Bloch [18] used a function that decreases linearly with θ as

$$f(\theta) = \max\left\{0, 1 - \frac{2\theta}{\pi}\right\} \tag{22.3}$$

for Equation 22.1. It can be shown this is equivalent to the morphological dilation of B,

$$\beta_\alpha(B)(x) = (B \oplus v_\alpha)(x) \cap B^c, \tag{22.4}$$

using the fuzzy structuring element

$$v_\alpha(x) = \max\left\{0, 1 - \frac{2}{\pi}\theta_\alpha(x,o)\right\} \tag{22.5}$$

where o is the origin (center) of the structuring element and B is removed from the result of dilation in Equation 22.4 (c represents complement). The fuzzy morphological dilation of the object B with the structuring element v is defined as

$$(B \oplus v)(x) = \max_y\{t[\varphi(y), v(x - y)]\} \tag{22.6}$$

where φ is the function representing object B, t is the t-norm operator for fuzzy intersection, and y is taken over all points in the image. An example synthetic image and fuzzy landscape examples using morphological dilation are given in Figure 22.1. In all figures in this chapter, white represents binary 1, black represents binary 0, and gray values represent the fuzziness in the range [0,1].

However, the linear function in Equation 22.3 and the corresponding structuring element in Equation 22.5 often lead to a landscape with a large spread and unintuitive transitions when the angle departs from α particularly at points that are farther away from the reference object. Thus, they may not give realistic results for many cases (see the examples in this section). Instead of using linearly decreasing membership values according to the angle, we developed a more intuitive and flexible structuring element using a nonlinear function with the shape of a Bézier curve

$$\nu_{\alpha,\lambda}(x) = g_\lambda\left(\frac{2}{\pi}\theta_\alpha(x,o)\right) \tag{22.7}$$

where $\lambda \in (0,1)$ determines the inflection point of the curve and increasing λ increases the spread around α (see the Appendix for the derivation). The nonlinear function enables different definitions of fuzziness for different cases. Fuzzy landscape examples using this structuring element definition are given in Figure 22.2.

The definition of the structuring element can be further extended to decrease the degree of a point's spatial relation to the reference object according to its distance to that object by introducing a new term

$$\nu_{\alpha,\lambda,\tau}(x) = g_\lambda\left(\frac{2}{\pi}\theta_\alpha(x,o)\right)\max\left\{0,1-\frac{\left\|\overrightarrow{ox}\right\|}{\tau}\right\} \tag{22.8}$$

where $\left\|\overrightarrow{ox}\right\|$ is the Euclidean distance of point x from the structuring element's center. In this definition, a point's spatial relation to the reference object decreases linearly with its distance to the object where τ corresponds to the distance where a point is no longer visible from the reference object. This definition also has a computational advantage because in the previous definitions (22.5) and (22.7) the structuring element must be at least twice as large as the landscape of interest in the image space whereas in definition (22.8) a structuring element with size of at most $2\tau \times 2\tau$ is sufficient. In fact, the resulting operations in Equation 22.8 have linear time complexity with respect to image size as opposed to the quadratic complexity in Equations 22.5 and 22.7, leading to dramatic improvements in the efficiency of the algorithm. Fuzzy landscape examples using this structuring element definition are given in Figure 22.3.

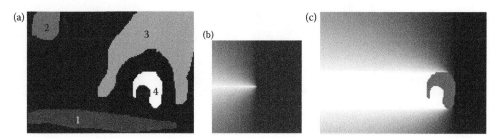

FIGURE 22.1 An example synthetic image and the directional landscape β_α for object labeled 4 using the structuring element ν_α defined in Equation 22.5 for $\alpha = \pi$. (a) Synthetic image, (b) ν_α, and (c) β_α.

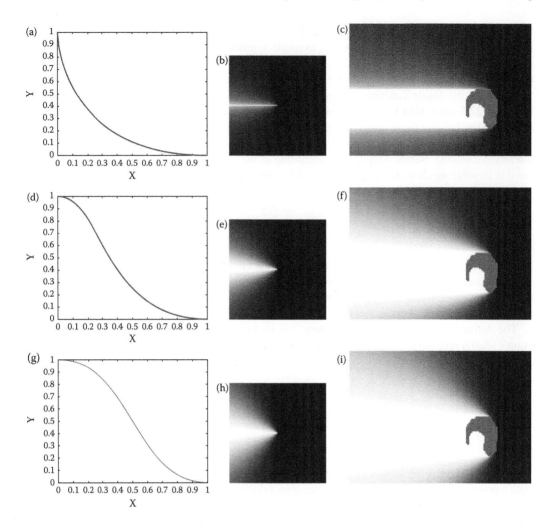

FIGURE 22.2 Structuring element $v_{\alpha,\lambda}$ defined in Equation 22.7 and directional landscape $\beta_{\alpha,\lambda}$ of object 4 (from Figure 22.1) for $\alpha = \pi$ and different values of λ. (a) g_λ for $\lambda = 0.001$, (b) $v_{\alpha,\lambda}$ for $\lambda = 0.001$, (c) $\beta_{\alpha,\lambda}$ for $\lambda = 0.001$, (d) g_λ for $\lambda = 0.3$, (e) $v_{\alpha,\lambda}$ for $\lambda = 0.3$, (f) $\beta_{\alpha,\lambda}$ for $\lambda = 0.3$, (g) g_λ for $\lambda = 0.5$, (h) $v_{\alpha,\lambda}$ for $\lambda = 0.5$, and (i) $\beta_{\alpha,\lambda}$ for $\lambda = 0.5$.

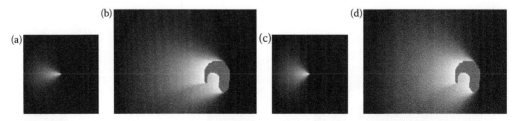

FIGURE 22.3 Structuring element $v_{\alpha,\lambda,\tau}$ defined in Equation 22.8 and directional landscape $\beta_{\alpha,\lambda,\tau}$ of object 4 (from Figure 22.1) for $\alpha = \pi$, $\tau = 100$, and different values of λ. (a) $v_{\alpha,\lambda,\tau}$ for $\lambda = 0.3$, (b) $\beta_{\alpha,\lambda,\tau}$ for $\lambda = 0.3$, (c) $v_{\alpha,\lambda,\tau}$ for $\lambda = 0.5$, and (d) $\beta_{\alpha,\lambda,\tau}$ for $\lambda = 0.5$.

22.2.2 VISIBILITY

In the directional dilation of Equation 22.4, the areas that are fully or partially enclosed by the reference object but are not visible from image points along the direction of interest may have high values as shown in Figures 22.3 and 22.4. To overcome this problem, we introduced the following definition:

$$\beta_{\alpha,\lambda,\lambda',\tau}(B)(x) = (B \oplus v_{\alpha,\lambda,\tau})(x) \cap (B \oplus v_{\alpha+\pi,\lambda'})(x)^c \qquad (22.9)$$

where the first dilation uses the structuring element defined in Equation 22.8 and the second dilation uses the structuring element defined in Equation 22.7. We compute fuzzy intersection using multiplication as the t-norm operator and compute fuzzy complement by subtracting the original values from 1. λ' can be set to a very small number to consider only the image points along $\alpha + \pi$. The proposed definition of visibility is illustrated in Figure 22.4.

Figure 22.5 shows additional examples of directional landscapes of objects using the definition in Ref. [18] (Equation 22.5 in this chapter) and our definition (22.9) on a synthetic image that was also used in Refs. [7,14,18,19]. Figure 22.5b and c illustrates the differences between the fuzzy directional landscapes obtained using the definition in Ref. [18] and our structuring element definitions. The latter is sensitive to the distance to the object according to the constant τ and the landscape's fuzziness is more centralized along the main direction of interest by the help of the constant λ. Figure 22.5d and e presents the importance of the support for visibility in our definition for directional

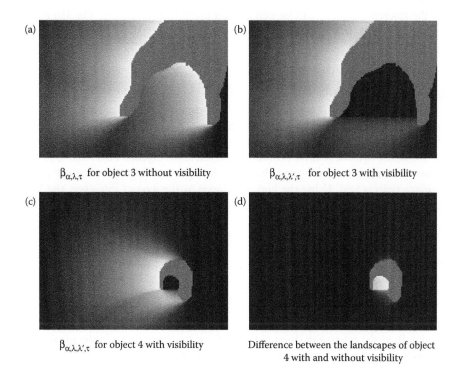

(a) $\beta_{\alpha,\lambda,\tau}$ for object 3 without visibility

(b) $\beta_{\alpha,\lambda,\lambda',\tau}$ for object 3 with visibility

(c) $\beta_{\alpha,\lambda,\lambda',\tau}$ for object 4 with visibility

(d) Difference between the landscapes of object 4 with and without visibility

FIGURE 22.4 Directional landscapes $\beta_{\alpha,\lambda,\tau}$ and $\beta_{\alpha,\lambda,\lambda',\tau}$ for objects 3 and 4 (from Figure 22.1) without and with the visibility extension, respectively, for $\alpha = \pi$, $\lambda = 0.3$, $\lambda' = 0.001$, and $\tau = 100$. (a) Uses the structuring element definition in Equation 22.8 without visibility, (b) and (c) use the definition in Equation 22.9 with visibility, and (d) illustrates the difference between the landscapes with and without visibility.

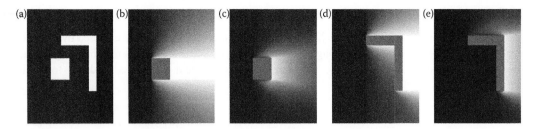

FIGURE 22.5 Directional landscape examples. (a) Synthetic image with two objects: square (A) and L-shaped (B). (b) $\beta_\alpha(A)$ for $\alpha = 0$ using the definition in [18]. (c) $\beta_{\alpha,\lambda,\lambda',\tau}(A)$ for $\alpha = 0$ using our definition. (d) $\beta_\alpha(B)$ for $\alpha = 0$ using the definition in Ref. [18], and (e) $\beta_{\alpha,\lambda,\lambda',\tau}(B)$ for $\alpha = 0$ using our definition. The constants are set as $\lambda = 0.3$, $\lambda' = 0.001$, and $\tau = 200$.

relationships. Although both landscapes for the direction "right" have similar distributions to the right of and above the reference object, the first one also has nonzero values on the left of the object, which contradicts the intuition.

22.3 BETWEEN RELATIONSHIP

Between relationship is a ternary relationship defined by two reference objects and a target object. Given two reference objects B and C, our goal is to generate a landscape in which the degree of each image area being located between the reference objects is quantified. Then, given a third object, its relation to the reference objects can be determined using this landscape. The landscape will be denoted by $\beta_\chi(B,C)$ in the rest of the chapter.

22.3.1 MORPHOLOGICAL APPROACH

Similar to the directional spatial relationships described in Section 22.2.1, the landscape $\beta_\chi(B,C)$ between two reference objects B and C can be defined as a fuzzy set such that image points with a high degree of the spatial relation have high membership values. This landscape can be computed as the intersection of the directional dilations of the reference objects along the directions $\alpha = \theta_\chi$ and $\alpha = \theta_\chi + \pi$ where θ_χ is the relative position of the reference objects. This relative position can be calculated using the maximum or mean value in the histogram of angles between all pairs of points of the reference objects [19]. Using the horizontal axis as the axis of reference, the histogram of angles for the objects B and C can be computed as

$$h_{B,C}(\theta) = \left| \left\{ (b,c) | b \in B, c \in C, \angle(\overrightarrow{bc}, \vec{u}_{\alpha=0}) = \theta \right\} \right| \tag{22.10}$$

and normalized as

$$H_{B,C}(\theta) = \frac{h_{B,C}(\theta)}{\max_{\theta'} h_{B,C}(\theta')}. \tag{22.11}$$

Then, using θ_χ as the relative position obtained from this histogram (as the maximum or mean value), the landscape between the reference objects B and C is computed as

$$\beta_\chi(B,C)(x) = \beta_{\alpha=\theta_\chi,\lambda,\lambda'}(B)(x) \cap \beta_{\alpha=\theta_\chi+\pi,\lambda,\lambda'}(C)(x) \tag{22.12}$$

where the directional landscape $\beta_{\alpha,\lambda,\lambda'}$ is computed as

$$\beta_{\alpha,\lambda,\lambda'}(B)(x) = (B \oplus v_{\alpha,\lambda})(x) \cap (B \oplus v_{\alpha+\pi,\lambda'})(x)^c \tag{22.13}$$

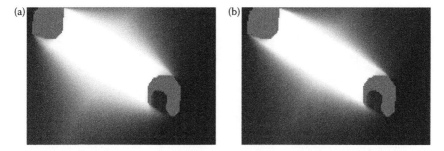

FIGURE 22.6 Between landscape $\beta_{\bar{\chi}}$ of objects 2 and 4 (from Figure 22.1) using the definition in Equation 22.12 with different values of λ and λ'. The relative angle for these objects is found as $\theta_{\bar{\chi}} = -30.04°$. (a) $\beta_{\bar{\chi}}$ for $\lambda = 0.3$ and $\lambda' = 0.001$, and (b) $\beta_{\bar{\chi}}$ for $\lambda = 0.15$ and $\lambda' = 0.001$.

using the structuring element definition in Equation 22.7. Since the landscape should include only the areas that are visible from both reference objects, the notion of visibility defined in Section 22.2.2 is used in the computation. Fuzzy landscape examples for the between relationship using this definition are given in Figure 22.6.

22.3.2 Myopic Vision

Although the histogram of angles generally provides a good approximation to the relative position of two objects, it fails in the cases where one object is significantly spatially extended relative to the other [19] (see Figure 22.7 for examples). We can solve this problem by taking into account only the

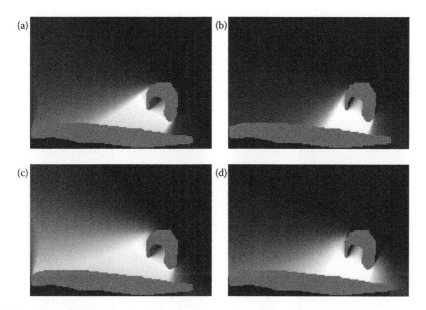

FIGURE 22.7 Between landscape $\beta_{\bar{\chi}}$ of objects 1 and 4 (from Figure 22.1) without and with myopic vision for different values of λ and λ' where object 1 is spatially extended relative to object 4. τ_{myopic} is taken as the half of the width of the image. The relative angles are 42.28° and 63.40° for the figures without and with myopic vision, respectively. For larger values of λ, error in landscape without myopic vision becomes more significant. (a) $\beta_{\bar{\chi}}$ for $\lambda = 0.15$ and $\lambda' = 0.001$ without myopic vision, (b) $\beta_{\bar{\chi}}$ for $\lambda = 0.15$ and $\lambda' = 0.001$ with myopic vision, (c) $\beta_{\bar{\chi}}$ for $\lambda = 0.5$ and $\lambda' = 0.001$ without myopic vision, and (d) $\beta_{\bar{\chi}}$ for $\lambda = 0.5$ and $\lambda' = 0.001$ with myopic vision.

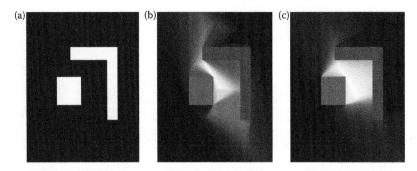

FIGURE 22.8 Between landscape examples. (a) Synthetic image with two objects: square (A) and L-shaped (B). (b) $\beta_{\bar{\chi}}(A,B)$ using the definition in Ref. [19], and (c) $\beta_{\bar{\chi}}(A,B)$ using our definition. The constants are set as $\lambda = 0.3$ and $\lambda' = 0.001$.

part of the spatially extended object close to the other object. (Bloch et al. [19] called this the "myopic vision" and suggested to use the distance map to find close parts of objects.)

Spatial proximity for handling spatially extended objects is incorporated into our morphological approach using a weighted histogram of angles where the contribution of the angle between each point pair in the histogram is weighted by the term $\max\{0, 1 - \|\vec{bc}\|/\tau_{\text{myopic}}\}$ (instead of a constant weight of 1 in Equation 22.10 as in Ref. [19]) where $\|\vec{bc}\|$ is the Euclidean distance between the points b and c, and τ_{myopic} is the threshold for the maximum distance between two points for allowing them to contribute to the histogram. The proposed definition of myopic vision is illustrated in Figure 22.7 using objects 1 and 4 where object 1 is spatially extended relative to object 4.

Figures 22.8 and 22.9 show additional examples of between landscapes of objects using the definition in Ref. [19] and our definition (22.12). The former calculates the landscape using dilation by a structuring element derived from the histogram of angles as defined in Equation 22.17 in Ref. [19]. These examples illustrate the differences in the two definitions when the objects have different spatial extents. For example, the landscape in Figure 22.8b, which is generated according to the definition in Ref. [19], is spatially too extended in the upper and lower parts of the image. It also includes nonsmooth transitions that are unintuitive. On the other hand, the landscape in Figure 22.8c, which is generated using Equation 22.12, is more compact and is fully covering the expected between area.

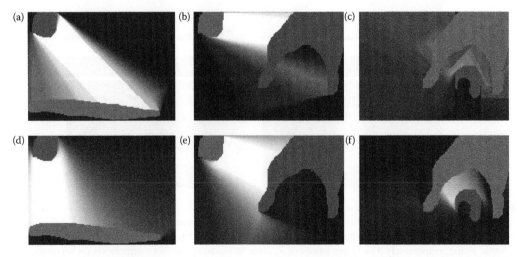

FIGURE 22.9 Between landscape examples for the synthetic image in Figure 22.1. (a) $\beta_{\bar{\chi}}(1, 2)$ using [19], (b) $\beta_{\bar{\chi}}(2, 3)$ using [19], (c) $\beta_{\bar{\chi}}(3, 4)$ using [19], (d) $\beta_{\bar{\chi}}(1, 2)$ using our definition, (e) $\beta_{\bar{\chi}}(2, 3)$ using our definition, and (f) $\beta_{\bar{\chi}}(3, 4)$ using our definition.

22.4   APPLICATIONS

We illustrate the use of the models described in this chapter as spatial contextual constraints for two image analysis tasks. The image classification task, described in Section 22.4.1, uses a Bayesian framework to incorporate contextual information in land cover classification to reduce the amount of commission among spectrally similar classes and improve the classification accuracy [12]. The object detection task, described in Section 22.4.2, illustrates the use of spatial constraints derived from shadow regions in improving the building detection accuracy [20]. As the *t*-norm operator, minimum is used in all definitions, except for visibility in directional relationships where multiplication is used as suggested in Section 22.2.2. After calculating the landscape β for a spatial relation as in Sections 22.2 or 22.3, the degree of satisfaction of this relation by a target object A is computed as

$$\mu(A) = \frac{1}{\text{area}(A)} \sum_{a \in A} \beta(a). \tag{22.14}$$

22.4.1   IMAGE CLASSIFICATION

The conventional method for automatically producing a land cover map in the remote sensing literature is to use a statistical classifier for supervised classification of pixels based on their spectral values. Even though these classifiers improve the processing time compared to manual digitization, their accuracy is limited with the discrimination ability of the spectral values of individual pixels as pixel-based classification does not take into account any spatial context.

A popular approach for incorporating spatial information into the classification process is to base the decisions on image regions using image segmentation techniques that automatically group neighboring pixels into contiguous regions based on similarity criteria on pixels' properties [20]. Even though image segmentation has been heavily studied, it is still an unsolved problem, especially for images with a very complex content with thousands of objects as in the examples that we study in this chapter.

Alternatively, spatial relationships can be used to model and quantify context in images. A commonly observed problem in pixel-level classification using spectral information is the confusion between the land covers with similar spectral values. In particular, a significant amount of confusion may occur between water pixels and shadow pixels and between asphalt pixels and shadow pixels that all appear dark in the image, as well as between snow and cloud classes that both have bright color values close to white. In this section, we illustrate the use of spatial relationships for improving the land cover classification accuracy.

Let **x** denote the feature vector of a pixel or an object at location x in a binary classification problem with two classes w_1 and w_2. As a widely used solution, the Bayesian classifier makes a decision using the posterior probabilities as

$$\text{Decide} \begin{cases} w_1 & \text{if } \dfrac{P(w_1 \mid \mathbf{x})}{P(w_2 \mid \mathbf{x})} > 1 \\ w_2 & \text{otherwise} \end{cases} \tag{22.15}$$

which is equivalent to

$$\text{Decide} \begin{cases} w_1 & \text{if } \dfrac{P(\mathbf{x} \mid w_1)}{P(\mathbf{x} \mid w_2)} > \dfrac{P(w_2)}{P(w_1)} \\ w_2 & \text{otherwise} \end{cases} \tag{22.16}$$

using the Bayes rule with the class-conditional and prior probabilities. The equal priors assumption $(P(w_1) = P(w_2))$ is often used when no additional information is available.

Assume that there is a third class w_3 that is related to w_2. The pixels/objects that are assigned to w_3 can be used as spatial constraints for improving the discrimination between w_1 and w_2. First, the directional landscape $\beta_\alpha(w_3)$ is computed for the whole scene by using w_3 as the reference. Then, the fuzzy landscape value in the range [0,1] at each image location is used as the spatial prior for w_2 at that location, that is, $P(w_2) = \beta_\alpha(w_3)(x)$ and $P(w_1) = 1 - P(w_2)$. The resulting contextual decision rule becomes

$$\text{Decide} \begin{cases} w_1 & \text{if } \dfrac{P(\mathbf{x} \mid w_1)}{P(\mathbf{x} \mid w_2)} > \dfrac{\beta_\alpha(w_3)(x)}{1 - \beta_\alpha(w_3)(x)} \\ w_2 & \text{otherwise} \end{cases} \tag{22.17}$$

using these spatial priors. The extension of Equation 22.17 for multiclass classification with multiple reference classes and multiple priors is straightforward.

This classification scheme is applied to two data sets. The first one consists of a LANDSAT scene covering the Washington State in the United States and the British Columbia in Canada. The multispectral image has six bands with 30 m spatial resolution and $7,680 \times 10,240$ pixels. We trained Bayesian classifiers for the water, shadow, and cloud classes as described in Ref. [20]. These classifiers produce the posterior probabilities $P(w_i|\mathbf{x})$, $i = 1,2,3$ for the water (w_1), shadow (w_2), and cloud (w_3) classes, respectively. A threshold is applied to the posterior probabilities to allow some of the pixels not being assigned to any class if none of the corresponding probabilities is high enough. Then, the maximum posterior probability rule is used for the final classification. Figure 22.10a shows a 2115×1070 pixel section of the scene. The resulting classification map is shown in Figure 22.10c. Binary classification between the water and shadow classes is also performed as in Equation 22.15. Table 22.1 shows the corresponding confusion matrix where Figure 22.10b is used as the ground truth (independent from the training set). The resulting 79.88% accuracy shows a significant amount of confusion between the water and shadow pixels when only pixel-based information is used.

To incorporate the spatial context into the decision process, we use the pixels classified as clouds (w_3) as reference objects, and compute the directional landscape $\beta_{\alpha,\lambda,\tau}(w_3)$ using the parameters $\alpha = 135°$, $\lambda = 0.3$, and $\tau = 1.6$ km. The α value of 135° measured from the horizontal axis in counterclockwise direction approximates the sun angle that can be obtained from the metadata of the image. The λ and τ values are determined empirically. The resulting landscape is shown in Figure 22.10d. The classification map and the confusion matrix resulting from the use of clouds as reference objects in the decision rule in Equation 22.17 are shown in Figure 22.10e and Table 22.1, respectively. Constraining the decision for classifying a pixel as shadow by requiring a high degree of directional spatial relationship with respect to clouds at a particular angle results in 96.48% accuracy, which corresponds to a net 16.60% improvement over the case where no spatial information is used. Figure 22.11 illustrates the results for the upper right portion of the scene in more detail.

The second data set consists of a 490×199 pixel section of a well-known hyperspectral image of Pavia, Italy obtained by the ROSIS sensor and has 102 spectral bands and 2.6 m spatial resolution. A pixel-based Bayesian classification was performed using spectral and textural features as described in Ref. [20]. The output of the Bayesian classifier is a probability value for each class at each pixel as in the LANDSAT case. Figure 22.12a and b shows the true-color image and the corresponding classification map, respectively. Table 22.2 shows the confusion matrix for the pixels where either the asphalt class (w_1) or the shadow class (w_2) has the highest probability. Figure 22.12e is used as the ground truth (independent from the training set). The 63.66% accuracy shows a significant amount of commission between these two classes when only pixel-based information is used.

The spatial context can be incorporated into the decision process by using the tiles (building roofs) and trees as additional information. The directional landscape is computed for the pixels

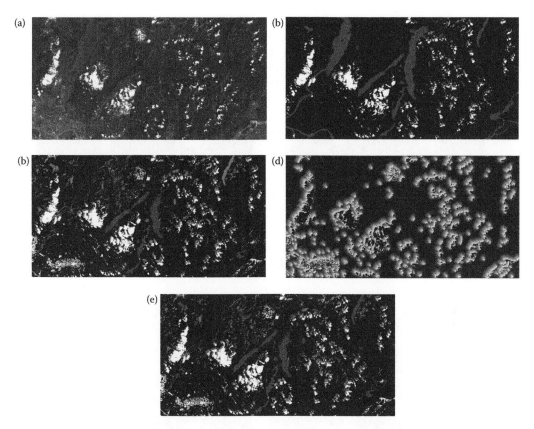

FIGURE 22.10   A 2115 × 1070 section of the LANDSAT scene and its classification without and with using spatial contextual information. The classes in the classification and ground truth maps are water (dark gray), shadow (light gray), and cloud (white). The ground truth is produced by careful visual inspection. (a) True-color image, (b) ground truth map, (c) classification map using decision rule (22.15) without spatial information, (d) directional landscape with respect to clouds, and (e) classifcation map using decision rule (22.17) with spatial information.

classified as tiles using the parameters $\alpha = -50°$, $\lambda = 0.3$, and $\tau = 25$. The α value measured from the horizontal axis in counter-clockwise direction is visually determined from the image to approximate the sun angle. The λ and τ values are determined empirically. Similarly, the directional landscape for the trees class is computed using the parameters $\alpha = -50°$, $\lambda = 0.3$, and $\tau = 10$. The two landscapes, shown in Figures 22.12c and d, are combined using the "max" operator (which is the

TABLE 22.1

Confusion Matrix for Water Versus Shadow Classification

| True | Assigned[a] | | True | Assigned[b] | |
	Water	Shadow		Water	Shadow
Water	51,502	20,533	**Water**	68,736	3299
Shadow	244	31,008	**Shadow**	337	30,915

[a] Using the decision rule (22.15) without spatial information. Overall accuracy is 79.88%.

[b] Using the decision rule (22.17) with spatial information. Overall accuracy is 96.48%.

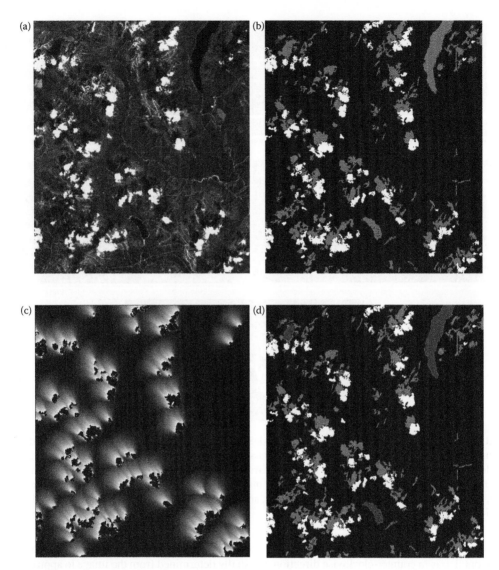

FIGURE 22.11 A 520×587 section in the upper right part of Figure 22.10a and its classification without and with using spatial contextual information. The classes in the classification map are water (dark gray), shadow (light gray), and cloud (white). (a) True-color image. (b) Classification map using decision rule (22.15) without spatial information. An accuracy of 61.32% is obtained where all water pixels are misclassified as shadow. (c) Directional landscape with respect to clouds. (d) Classification map using decision rule (22.17) with spatial information. Perfect detection (100%) is achieved.

equivalent of the Boolean "or" operator). Then, the contextual decision rule in Equation 22.17 is used to update the classification at each pixel by using tiles and trees as reference (w_3). Figure 22.12f and Table 22.2 show the classification results when spatial information is used. The updated contextual decision gives an 86.16% accuracy that corresponds to a net 22.50% improvement by classifying a pixel with shadow-like feature values as shadow only when it also has a high degree of directional spatial relationship with respect to buildings or trees at a particular angle. The significant improvement in accuracy for both the LANDSAT data set and the ROSIS data set confirms the importance of spatial information in classification and the effectiveness of the relationship models described in this chapter in modeling and quantifying this information.

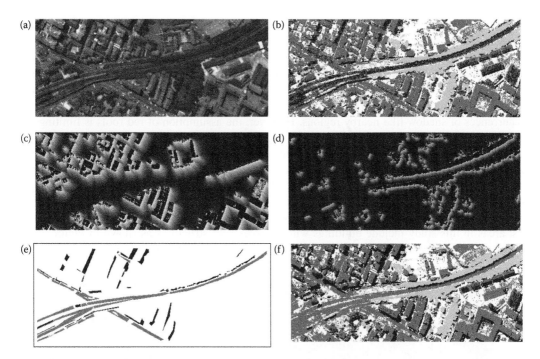

FIGURE 22.12 (**See color insert.**) Classification of the Pavia image without and with using spatial contextual information. The classes in the classification and ground truth maps are asphalt (gray), shadow (black), tiles (red), and trees (green). The ground truth is produced by visual inspection. (a) True-color image, (b) classification map using decision rule (22.15) without spatial information, (c) directional landscape with respect to the detected tiles, (d) directional landscape with respect to the detected trees, (e) ground truth map, and (f) classifcation map using decision rule (22.17) with spatial information.

TABLE 22.2
Confusion Matrices for Asphalt Versus Shadow Classification

	Assigned[a]				Assigned[b]	
True	**Asphalt**	**Shadow**	**True**		**Asphalt**	**Shadow**
Asphalt	3,302	2,000	**Asphalt**		5,054	248
Shadow	1,028	2,003	**Shadow**		905	2,126

[a] Using the decision rule (22.15) without spatial information. Overall accuracy is 63.66%.

[b] Using the decision rule (22.17) with spatial information. Overall accuracy is 86.16%.

22.4.2 Building Detection

Automatic detection of buildings in very high spatial resolution remotely sensed imagery has been an important problem because the detection results can be used in many applications such as change detection, urbanization monitoring, and digital map production. There is an extensive literature on building detection where both pixel level and object/region level processing have been used. However, most of the previous methods try to solve the problem for specific settings such as images having buildings with the same type of appearance and images where the buildings are isolated and have simple roof structures. With the increase in the spatial details in the images obtained from new generation sensors with meter and submeter spatial resolution, the buildings may have very complicated appearances and may have complex structures with very different spectral signatures.

Even though different buildings may appear in significantly different colors and shapes, a common property of such buildings can be the existence of shadows. The relationship between buildings and shadows has actually been exploited in earlier works [21,22]. More recently, Sirmacek and Unsalan [23] detected buildings with red roofs using color information and verified their existence with the occurrences of shadow-like nearby regions. However, the assumption of red roofs is limiting and there may be other sources of shadows in the image.

In this section, we describe a method for detection of buildings with complex shapes and roof structures in very high spatial resolution images by exploiting spectral, structural, and contextual information. The input to the method is a satellite image consisting of a panchromatic and multispectral data pair. First, the watershed segmentation algorithm is used to partition the panchromatic band into spectrally homogeneous regions. The results contain oversegmented regions because the test areas in this study include buildings with complex roof structures as shown in Figure 22.13. Then, among all regions, the ones that are likely to belong to shadows are selected using their

FIGURE 22.13   Examples from an Ikonos image of Antalya, Turkey and the corresponding watershed segmentation results. The segmentation boundaries are overlayed as white. (a) Antalya 1 image, (b) watershed segmentation of Antalya 1, (c) Antalya 2 image, and (d) watershed segmentation of Antalya 2.

spectral properties. This selection uses the normalized difference vegetation index (NDVI) that is computed using the pan-sharpened image where the regions whose average brightness values are lower than a brightness threshold and average NDVI values are lower than an NDVI threshold are denoted as shadow regions.

Next, candidate building regions are identified using the directional spatial relationships of all regions with respect to the detected shadow regions along the sun azimuth angle. Given the sun azimuth angle, we can find the directional landscapes of the shadow regions along this direction by using Equation 22.4. The resulting directional landscapes give high responses in areas close to the shadow regions along the sun azimuth angle. These areas correspond to the locations where the probability of the presence of buildings is high. Figure 22.14a and c shows the shadow regions and the corresponding landscapes. Consequently, the regions whose average satisfaction degrees are higher than a satisfaction threshold, average NDVI values are lower than the NDVI threshold, and sizes are lower than a size threshold are identified as candidate building regions. Figure 22.14b and d shows examples for candidate regions. As can be seen from the figures, most of the regions are correctly identified with a small number of misdetections and several false alarms.

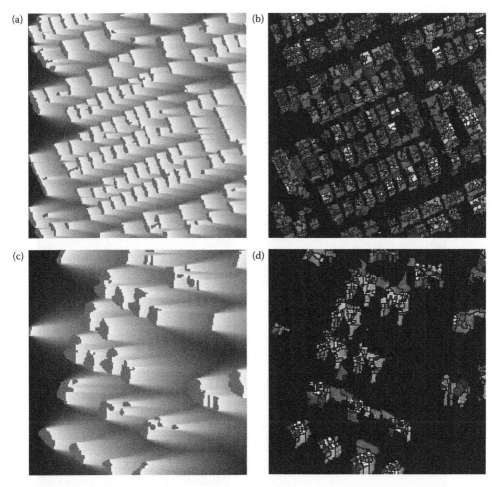

FIGURE 22.14 Examples of shadow regions, directional landscapes, and candidate building regions. (a) Shadows and spatial constraints in Antalya 1, (b) candidate building regions in Antalya 1, (c) shadows and spatial constraints in Antalya 2, and (d) candidate building regions in Antalya 2.

Finally, the building regions are selected by clustering the oversegmented regions that satisfy the spatial constraints using minimum spanning trees. An important observation is that regions forming a building are densely located whereas regions separating different buildings are found far from their neighbors. The distance between two regions is measured as the distance between their centroids. This seems to be a valid assumption because the regions are obtained from oversegmentation and mostly have compact shapes. Hence, we construct a graph where the graph nodes correspond to the candidate regions' centroids and the edges are created between two neighboring nodes with a weight corresponding to their spatial distance. What we expect is that the nodes representing parts of building regions will form dense subgraph components. These dense components are found by constructing the minimum spanning tree of the graph, and by eliminating some of the remaining edges that are longer than a length threshold. As a result, the nodes that are spatially close enough remain in the same cluster. Figure 22.15 shows examples for graph construction and clustering.

Six subscenes of 1 m spatial resolution Ikonos images of Antalya, Turkey are used to qualitatively evaluate the algorithm. Figure 22.16 shows example detection results. It can be seen that most of the building regions that cannot be obtained by traditional spectral segmentation methods that cannot incorporate structural and contextual information are correctly extracted.

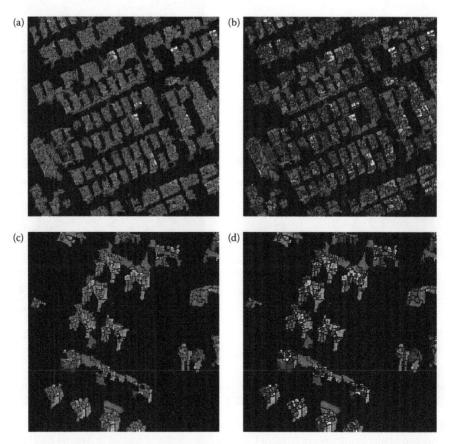

FIGURE 22.15   (**See color insert.**) Examples of graph construction and minimum spanning tree-based clustering. The removed edges are colored in red. (a) Graph for Antalya 1, (b) clustering for Antalya 1, (c) graph for Antalya 2, and (d) clustering for Antalya 2.

FIGURE 22.16 **(See color insert.)** Building detection results. The detected buildings are highlighted in red. (a) Results for Antalya 1, (b) results for Antalya 2, (c) results for Antalya 3, (d) results for Antalya 4, (e) results for Antalya 5, and (f) results for Antalya 6.

22.5 CONCLUSIONS

We presented new, intuitive, flexible, and efficient definitions for modeling pairwise directional spatial relationships and the ternary between relationship using fuzzy mathematical morphology techniques. Our contributions included flexible definitions for the fuzzy directional structuring elements that are tunable along both radial and angular dimensions, support for the notion of visibility for handling image areas that are partially enclosed by objects and are not visible from image points along the direction of interest, and handling of the cases where one object is significantly spatially extended relative to the other. Illustrations using synthetic data showed that our models produce more intuitive results than the state-of-the-art techniques. We also presented two applications with real data. First, we showed that incorporating the spatial relationships as contextual information in a Bayesian classification framework results in a significant improvement in land cover classification accuracy by reducing the amount of commission among spectrally similar classes in multispectral and hyperspectral data. Then, we showed that the use of spatial constraints derived from shadow regions improves building detection accuracy in very high spatial resolution imagery. The significant improvement in accuracy in these applications confirms the importance of spatial information in classification and the effectiveness of the relationship models described in this chapter in modeling and quantifying this information. Future work includes investigating ways of automating the selection of the parameters for different applications.

ACKNOWLEDGMENTS

This work was supported in part by the TUBITAK CAREER Grant 104E074 and European Commission Sixth Framework Programme Marie Curie International Reintegration Grant MIRG-CT-2005-017504.

APPENDIX: BÉZIER CURVES

Bézier curve is a parametric curve defined using a number of reference points. Four points a_0, a_1, a_2, a_3 on a plane define a cubic Bézier curve where the curve starts at a_0 going toward a_1 and arrives at a_3 coming from the direction of a_2. The parametric form of the curve is

$$b(t) = (1-t)^3 a_0 + 3t(1-t)^2 a_1 + 3t^2(1-t)a_2 + t^3 a_3 \tag{22.18}$$

where t is the parameter having values in $[0, 1]$.

To construct a one-dimensional function that has the shape of a Bézier curve and maps each $x \in [0, 1]$ to a $y \in [0, 1]$, we set the reference points $a = (x, y)^T$ as

$$a_0 = (0,1)^T, \ a_1 = (\lambda,1)^T, \ a_2 = (\lambda,0)^T, \ a_3 = (1,0)^T \tag{22.19}$$

where $\lambda \in (0, 1)$ so that the cubic curve has only one parameter. Then, Equation 22.18 reduces to

$$b_x(t) = 3t(1-t)^2 \lambda + 3t^2(1-t)\lambda + t^3 \tag{22.20}$$

$$b_y(t) = (1-t)^3 + 3t(1-t)^2 + 3t^2(1-t) \tag{22.21}$$

and for any $x \in [0, 1]$, $b_x(t)$ can be solved for t, and the corresponding $y \in [0, 1]$ can be computed using $b_y(t)$.

In this chapter, this function/mapping is denoted as $g_\lambda(x)$. The function has an inflection point at $x = \lambda$. Examples of $g_\lambda(x)$ for different λ values are shown in Figure 22.2.

REFERENCES

1. S. Bhagavathy and B. S. Manjunath. Modeling and detection of geospatial objects using texture motifs. *IEEE Transactions on Geoscience and Remote Sensing*, 44(12):3706–3715, December 2006.
2. M. Pesaresi and J. A. Benediktsson. A new approach for the morphological segmentation of high-resolution satellite imagery. *IEEE Transactions on Geoscience and Remote Sensing*, 39(2):309–320, February 2001.
3. F. Melgani and S. B. Serpico. A Markov random field approach to spatio-temporal contextual image classification. *IEEE Transactions on Geoscience and Remote Sensing*, 41(11):2478–2487, November 2003.
4. L. Bruzzone and L. Carlin. A multilevel context-based system for classification of very high spatial resolution images. *IEEE Transactions on Geoscience and Remote Sensing*, 44(9):2587–2600, September 2006.
5. H. G. Akcay and S. Aksoy. Automatic detection of geospatial objects using multiple hierarchical segmentations. *IEEE Transactions on Geoscience and Remote Sensing*, 46(7):2097–2111, July 2008.
6. S. Steiniger and R. Weibel. Relations among map objects in cartographic generalization. *Cartography and Geographic Information Science*, 34(3):175–197, 2007.
7. I. Bloch. Fuzzy spatial relationships for image processing and interpretation: A review. *Image and Vision Computing*, 23(2):89–110, February 2005.
8. S. Aksoy, C. Tusk, K. Koperski, and G. Marchisio. Scene modeling and image mining with a visual grammar. In C. H. Chen, editor, *Frontiers of Remote Sensing Information Processing*, pp.35–62. World Scientific, Singapore, 2003.
9. S. Aksoy, K. Koperski, C. Tusk, G. Marchisio, and J. C. Tilton. Learning Bayesian classifiers for scene classification with a visual grammar. *IEEE Transactions on Geoscience and Remote Sensing*, 43(3):581–589, March 2005.
10. S. Aksoy. Modeling of remote sensing image content using attributed relational graphs. In *Proceedings of IAPR International Workshop on Structural and Syntactic Pattern Recognition*, pp. 475–483, Hong Kong, August 17–19, 2006. *Lecture Notes in Computer Science*, Vol. 4109.

11. J. Inglada and J. Michel. Qualitative spatial reasoning for high-resolution remote sensing image analysis. *IEEE Transactions on Geoscience and Remote Sensing*, 47(2):599–612, February 2009.

12. S. Aksoy and R. G. Cinbis. Image mining using directional spatial constraints. *IEEE Geoscience and Remote Sensing Letters*, 7(1):33–37, January 2010.

13. R. G. Cinbis and S. Aksoy. Relative position-based spatial relationships using mathematical morphology. In *Proceedings of IEEE International Conference on Image Processing*, Vol. II, pp. 97–100, San Antonio, TX, September 16–19, 2007.

14. I. Bloch and A. Ralescu. Directional relative position between objects in image processing: A comparison between fuzzy approaches. *Pattern Recognition*, 36(7):1563–1582, July 2003.

15. K. Miyajima and A. Ralescu. Spatial organization in 2D segmented images: Representation and recognition of primitive spatial relations. *Fuzzy Sets and Systems*, 65(2–3):225–236, August 1994.

16. P. Matsakis and L. Wendling. A new way to represent the relative position between areal objects. *IEEE Transactions on Pattern Analysis and Machine Intelligence*, 21(7):634–643, July 1999.

17. X. Wang, J. Ni, and P. Matsakis. Fuzzy object localization based on directional (and distance) information. In *Proceedings of IEEE International Conference on Fuzzy Systems*, Vancouver, Canada, 2006.

18. I. Bloch. Fuzzy relative position between objects in image processing: A morphological approach. *IEEE Transactions on Pattern Analysis and Machine Intelligence*, 21(7):657–664, July 1999.

19. I. Bloch, O. Colliot, and R. M. Cesar. On the ternary spatial relation "between". *IEEE Transactions on Systems, Man, and Cybernetics, Part B: Cybernetics*, 36(2):312–327, April 2006.

20. S. Aksoy. Spatial techniques for image classification. In C. H. Chen, editor, *Signal and Image Processing for Remote Sensing*, pp. 491–513. Taylor & Francis Books, Boca Raton, Florida, 2006.

21. A. Huertas and R. Nevatia. Detecting buildings in aerial images. *Computer Vision, Graphics, and Image Processing*, 41(2):131–152, 1988.

22. R. B. Irvin and D. M. McKeown Jr. Methods for exploiting the relationship between buildings and their shadows in aerial imagery. *IEEE Transactions on Systems, Man, and Cybernetics*, 19(6):1564–1575, 1989.

23. B. Sirmacek and C. Unsalan. Building detection from aerial images using invariant color features and shadow information. In *Proceedings of International Symposium on Computer and Information Sciences*, Istanbul, Turkey, 2008.

23 Data Fusion for Remote-Sensing Applications

Anne H. S. Solberg

CONTENTS

23.1   INTRODUCTION

Earth observation is currently developing more rapidly than ever. During the last decade, the number of satellites has been growing steadily, and the coverage of the Earth in space, time, and the electromagnetic spectrum is increasing correspondingly fast.

The accuracy in classifying a scene can be increased by using images from several sensors operating at different wavelengths of the electromagnetic spectrum. The interaction between the electromagnetic radiation and the earth's surface is characterized by certain properties at different frequencies of electromagnetic energy. Sensors with different wavelengths provide complementary information about the surface. In addition to image data, prior information about the scene might be available in the form of map data from geographic information systems (GIS). The merging of multisource data can create a more consistent interpretation of the scene compared to an interpretation based on data from a single sensor.

This development opens up for a potential significant change in the approach of analysis of earth observation data. Traditionally, analysis of such data has been by means of analysis of a single satellite image. The emerging exceptionally good coverage in space, time, and the spectrum opens for analysis of time series of data, combining different sensor types, combining imagery of different scales, and better integration with ancillary data and models. Thus, data fusion to combine data from several sources is becoming increasingly more important in many remote-sensing applications.

This chapter provides a tutorial on data fusion for remote-sensing applications. The main focus is on methods for multisource image classification, but separate sections on multisensor image registration, multiscale classification, and multitemporal image classification are also included. The remainder of this chapter is organized in the following manner: in Section 23.2, the "multi" concept in remote sensing is presented. Multisensor data registration is treated in Section 23.3. Classification strategies for multisensor applications are discussed in Section 23.4. Multitemporal image classification is discussed in Section 23.5, while multiscale approaches are discussed in Section 23.6. Concluding remarks are given in Section 23.7.

23.2   "MULTI" CONCEPT IN REMOTE SENSING

The variety of different sensors already available or being planned creates a number of possibilities for data fusion to provide better capabilities for scene interpretation. This is referred to as the "multi" concept in remote sensing. The "multi" concept includes multitemporal, multispectral or multifrequency, multipolarization, multiscale, and multisensor image analysis. In addition to the concepts discussed here, imaging using multiple incidence angles can also provide additional information [1,2].

23.2.1   MULTISPECTRAL OR MULTIFREQUENCY ASPECT

The measured backscatter values for an area vary with the wavelength band. A land-use category will give different image signals depending on the frequency used, and by using different frequencies, a spectral signature that characterizes the land-use category can be found. A description of the scattering mechanisms for optical sensors can be found in Ref. [3], while Ref. [4] contains a thorough discussion of the backscattering mechanisms in the microwave region. Multispectral optical sensors have demonstrated this effect for a substantial number of applications for several decades; they are now followed by high-spatial-resolution multispectral sensors, such as Ikonos and Quickbird, and by hyperspectral sensors from satellite platforms (e.g., Hyperion).

23.2.2   MULTITEMPORAL ASPECT

The term multitemporal refers to the repeated imaging of an area over a period. By analyzing an area through time, it is possible to develop interpretation techniques based on an object's temporal variations and to discriminate different pattern classes accordingly. Multitemporal imagery allows the study of the variation of backscatter of different areas with time, weather conditions, and seasons. It also allows monitoring of processes that change over time.

The principal advantage of multitemporal analysis is the increased amount of information for the study area. The information provided for a single image is, for certain applications, not sufficient to properly distinguish between the desired pattern classes. This limitation can sometimes be resolved by examining the pattern of temporal changes in the spectral signature of an object. This is particularly important for vegetation applications. Multitemporal image analysis is discussed in more detail in Section 23.5.

23.2.3 Multipolarization Aspect

The multipolarization aspect is related to microwave-image data. The polarization of an electromagnetic wave refers to the orientation of the electric field during propagation. A review of the theory and features of polarization is given in Refs. [5,6].

23.2.4 Multisensor Aspect

With an increasing number of operational and experimental satellites, information about a phenomenon can be captured using different types of sensors.

Fusion of images from different sensors requires some additional preprocessing and poses certain difficulties that are not solved in traditional image classifiers. Each sensor has its own characteristics, and the image captured usually contains various artifacts that should be corrected or removed. The images also need to be geometrically corrected and coregistered. Because the multisensor images often are not acquired on the same data, the multitemporal nature of the data must also often be explained.

Figure 23.1 shows a simple visualization of two synthetic aperture radar (SAR) images from an oil spill in the Baltic sea, imaged by the ENVISAT ASAR sensor and the Radarsat SAR sensor.

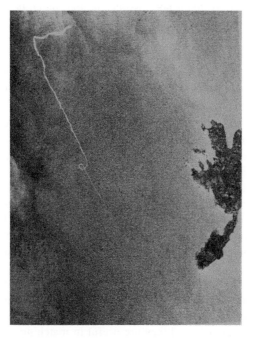

FIGURE 23.1 Example of multisensor visualization of an oil spill in the Baltic Sea created by combining an ENVISAT ASAR image with a Radarsat SAR image taken a few hours later.

The images were taken a few hours apart. During this time, the oil slick has drifted to some extent, and it has become more irregular in shape.

23.2.5 OTHER SOURCES OF SPATIAL DATA

The preceding sections have addressed spatial data in the form of digital images obtained from remote-sensing satellites. For most regions, additional information is available in the form of various kinds of maps, for example, topography, ground cover, elevation, and so on. Frequently, maps contain spatial information not obtainable from a single remotely sensed image. Such maps represent a valuable information resource in addition to the satellite images. To integrate map information with a remotely sensed image, the map must be available in digital form, for example, in a GIS system.

23.3 MULTISENSOR DATA REGISTRATION

A prerequisite for data fusion is that the data are coregistered, and geometrically and radiometrically corrected. Data coregistration can be simple if the data are georeferenced. In that case, the coregistration consists merely of resampling the images to a common map projection. However, an image-matching step is often necessary to obtain subpixel accuracy in matching. Complicating factors for multisensor data are the different appearances of the same object imaged by different sensors, and nonrigid changes in object position between multitemporal images.

The image resampling can be done at various stages of the image interpretation process. Resampling an image affects the spatial statistics of the neighboring pixel, which is of importance for many radar image feature extraction methods that might use speckle statistics or texture. When fusing a radar image with other data sources, a solution might be to transform the other data sources to the geometry of the radar image. When fusing a multitemporal radar image, an alternative might be to use images from the same image mode of the sensor, for example, only ascending scenes with a given incidence angle range. If this is not possible and the spatial information from the original geometry is important, the data can be fused and resampling done after classification by the sensor-specific classifiers.

An image-matching step may be necessary to achieve subpixel accuracy in the coregistration even if the data are georeferenced. A survey of image registration methods is given by Zitova and Flusser [7]. A full image registration process generally consists of four steps:

- *Feature extraction.* This is the step where regions, edges, and contours can be used to represent tie-points in the set of images to be matched are extracted. This is a crucial step, as the registration accuracy can be no better than what is achieved for the tie-points. Feature extraction can be grouped into area-based methods [8,9], feature-based methods [10–12], and hybrid approaches [7]. In area-based methods, the gray levels of the images are used directly for matching, often by statistical comparison of pixel values in small windows, and they are best suited for images from the same or highly similar sensors. Feature-based methods will be application dependent, as the type of features to use as tie-points needs to be tailored to the application. Features can be extracted either from the spatial domain (edges, lines, regions, intersections, and so on) or from the frequency domain (e.g., wavelet features). Spatial features can perform well for matching data from heterogeneous sensors, for example, optical and radar images. Hybrid approaches use both area-based and feature-based techniques by combining both a correlation-based matching with an edge-based approach, and they are useful in matching data from heterogeneous sensors.

- *Feature matching.* In this step, the correspondence between the tie-points or features in the sensed image and the reference image is found. Area-based methods for feature extraction use correlation, Fourier-transform methods, or optical flow [13–15]. Feature-based methods use the equivalence between correlation in the spatial domain and multiplication in the Fourier domain to perform matching in the Fourier domain [10,11]. Correlation-based methods are best suited for data from similar sensors. The optical flow approach involves estimation of the relative motion between two images and is a broad approach. It is commonly used in video analysis, but only a few studies have used it in remote-sensing applications [29,30].
- *Transformation selection* concerns the choice of mapping function and estimation of its parameters based on the established feature correspondence. The affine transform model is commonly used for remote-sensing applications, where the images normally are preprocessed for geometrical correction—a step that justifies the use of affine transforms.
- *Image resampling.* In this step, the image is transformed by means of the mapping function. Image values in no-integer coordinates are computed by the appropriate interpolation technique. Normally, either a nearest neighbor or a bilinear interpolation is used. Nearest-neighbor interpolation is applicable when no new pixel values should be introduced. Bilinear interpolation is often a good trade-off between accuracy and computational complexity compared to cubic or higher-order interpolation.

23.4 MULTISENSOR IMAGE CLASSIFICATION

The literature on data fusion in the computer vision and machine intelligence domains is substantial. For an extensive review of data fusion, we recommend the book by Abidi and Gonzalez [16]. Multisensor architectures, sensor management, and designing sensor setup are also thoroughly discussed in Ref. [17].

23.4.1 A GENERAL INTRODUCTION TO MULTISENSOR DATA FUSION FOR REMOTE-SENSING APPLICATIONS

Fusion can be performed at the *signal, pixel, feature,* or *decision* level of representation (see Figure 23.2). In signal-based fusion, signals from different sensors are combined to create a new signal with a better signal-to-noise ratio than the original signals [18]. Techniques for signal-level data fusion typically involve classic detection and estimation methods [19]. If the data are noncommensurate, they must be fused at a higher level.

Pixel-based fusion consists of merging information from different images on a pixel-by-pixel basis to improve the performance of image processing tasks such as segmentation [20]. Feature-based fusion consists of merging features extracted from different signals or images [21]. In feature-level fusion, features are extracted from multiple sensor observations, then combined into a concatenated feature vector, and classified using a standard classifier. Symbol-level or decision-level fusion consists of merging information at a higher level of abstraction. Based on the data from each single sensor, a preliminary classification is performed. Fusion then consists of combining the outputs from the preliminary classifications.

The main approaches to data fusion in the remote-sensing literature are statistical methods [22–25], Dempster–Shafer theory [26–28], and neural networks [22,29]. We will discuss each of these approaches in the following sections. The best level and methodology for a given remote-sensing application depends on several factors: the complexity of the classification problem, the available data set, and the goal of the analysis.

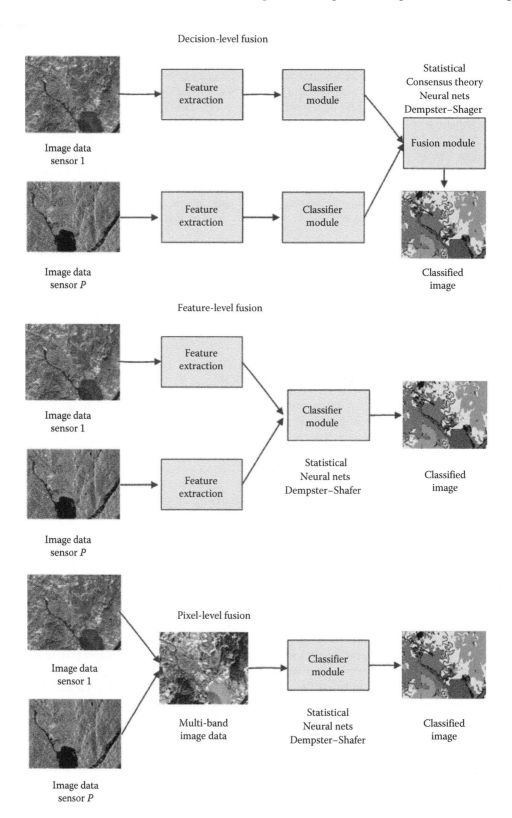

FIGURE 23.2 An illustration of data fusion at different levels.

23.4.2 Decision-Level Data Fusion for Remote-Sensing Applications

In the general multisensor fusion case, we have a set of images $X^1 \ldots X^P$ from P sensors. The class labels of the scene are denoted C. The Bayesian approach is to assign each pixel to the class that maximizes the posterior probabilities $P(C \mid X^1, \ldots, X^P)$

$$P(C|X^1,\ldots,X^P) = \frac{P(X^1,\ldots,X^P|C)P(C)}{P(X^1,\ldots,X^P)} \tag{23.1}$$

where $P(C)$ is the prior model for the class labels.

For decision-level fusion, the following conditional independence assumption is used:

$$P(X^1,\ldots,X^P|C) \equiv P(X^1|C)\cdots P(X^P|C)$$

This assumption means that the measurements from the different sensors are considered to be conditionally independent.

23.4.3 Combination Schemes for Combining Classifier Outputs

In the data fusion literature [30], various alternative methods have been proposed for combining the outputs from the sensor-specific classifiers by weighting the influence of each sensor. This is termed consensus theory. The weighting schemes can be linear, logarithmic, or of a more general form (see Figure 23.3).

The simplest choice, the linear opinion pool (LOP), is given by

$$\text{LOP}(X^1,\ldots,X^P) = \sum_{p=1}^{P} P(X^p|C)^{\lambda_p} \tag{23.2}$$

The logarithmic opinion pool (LOGP) is given by

$$\text{LOGP}(X^1,\ldots,X^P) = \prod_{p=1}^{P} P(X^p|C)^{\lambda_p} \tag{23.3}$$

which is equivalent to the Bayesian combination if the weights λ_p are equal. This weighting scheme contradicts the statistical formulation in which the sensor's uncertainty is supposed to be modeled by the variance of the probability density function.

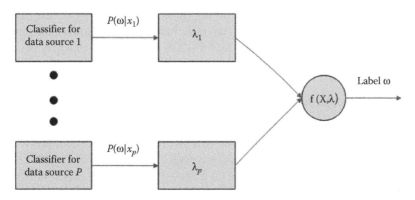

FIGURE 23.3 Schematic view of weighting the outputs from sensor-specific classifiers in decision-level fusion.

The weights are supposed to represent the sensor's reliability. The weights can be selected by heuristic methods based on their goodness [3] by weighting a sensor's influence by a factor proportional to its overall classification accuracy on the training data set. An alternative approach for a linear combination pool is to use a genetic algorithm [32].

An approach using a neural net to optimize the weights is presented in Ref. [30]. Yet another possibility is to choose the weights in such a way that they not only weigh the individual data sources but also the classes within the data sources [33].

Benediktsson et al. [30,31] use a multilayer perceptron (MLP) neural network to combine the class-conditional probability densities $P(X^p \mid C)$. This allows a more flexible, nonlinear combination scheme. They compare the classification accuracy using MLPs to LOPs and LOGPs, and find that the neural net combination performs best.

Benediktsson and Sveinsson [34] provide a comparison of different weighting schemes for an LOP and LOGP, genetic algorithm with and without pruning, parallel consensus neural nets, and conjugate gradient backpropagation (CGBP) nets on a single multisource data set. The best results were achieved by using a CGBP net to optimize the weights in an LOGP.

A study that contradicts the weighting of different sources is found in Ref. [35]. In this study, three different data sets (optical and radar) were merged using the LOGP, and the weights were varied between 0 and 1. Best results for all three data sets were found by using equal weights.

23.4.4 STATISTICAL MULTISOURCE CLASSIFICATION

Statistical methods for fusion of remotely sensed data can be divided into four categories: the augmented vector approach, stratification, probabilistic relaxation, and extended statistical fusion. In the augmented vector approach, data from different sources are concatenated as if they were measurements from a single sensor. This is the most common approach for many application-oriented applications of multisource classification, because no special software is needed. This is an example of pixel-level fusion.

Such a classifier is difficult to use when the data cannot be modeled with a common probability density function, or when the data set includes ancillary data (e.g., from a GIS system). The fused data vector is then classified using ordinary single-source classifiers [36]. Stratification has been used to incorporate ancillary GIS data in the classification process. The GIS data are stratified into categories and then a spectral model for each of these categories is used [37].

Richards et al. [38] extended the methods used for spatially contextual classification based on probabilistic relaxation to incorporate ancillary data. The methods based on extended statistical fusion [10,43] were derived by extending the concepts used for classification of single-sensor data. Each data source is considered independently and the classification results are fused using weighted linear combinations.

By using a statistical classifier one often assumes that the data have a multivariate Gaussian distribution. Recent developments in statistical classifiers based on regression theory include choices of nonlinear classifiers [11–13,18–20,26,28,33,38–56]. For a comparison of neural nets and regression-based nonlinear classifiers, see Ref. [57].

23.4.5 NEURAL NETS FOR MULTISOURCE CLASSIFICATION

Many multisensor studies have used neural nets because no specific assumptions about the underlying probability densities are needed [40,58]. A drawback of neural nets in this respect is that they act like a black box in that the user cannot control the usage of different data sources. It is also difficult to explicitly use a spatial model for neighboring pixels (but one can extend the input vector from measurements from a single pixel to measurements from neighboring pixels). Guan et al. [41] utilized contextual information by using a network of neural networks with which they built a quadratic regularizer. Another drawback is that specifying neural network architecture

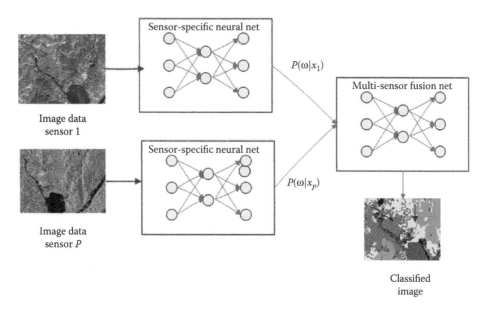

FIGURE 23.4 Network architecture for decision-level fusion using neural networks.

involves specifying a large number of parameters. A classification experiment should take care in choosing them and apply different configurations, making the complete training process very time consuming [52,58]. Hybrid approaches combining statistical methods and neural networks for data fusion have also been proposed [30]. Benediktsson et al. [30] apply a statistical model to each individual source and use neural nets to reach a consensus decision. Most applications involving a neural net use an MLP or radial basis function network, but other neural network architectures can be used [59–61].

Neural nets for data fusion can be applied both at the pixel, feature, and decision level. For pixel- and feature-level fusion a single neural net is used to classify the joint feature vector or pixel measurement vector. For decision-level fusion, a network combination like the one outlined in Figure 23.4 is often used [29]. An MLP neural net is first used to classify the images from each source separately. Then, the outputs from the sensor-specific nets are fused and weighted in a fusion network.

23.4.6 A Closer Look at Dempster–Shafer Evidence Theory for Data Fusion

Dempster–Shafer theory of evidence provides a representation of multisource data using two central concepts: plausibility and belief. Mathematical evidence theory was first introduced by Dempster in the 1960s, and later extended by Shafer [62].

A good introduction to Dempster–Shafer evidence theory for remote-sensing data fusion is given in Ref. [28].

Plausibility (Pls) and belief (Bel) are derived from a mass function m, which is defined on the [0,1] interval. The belief and plausibility functions for an element A are defined as

$$Bel(A) = \sum_{B \subseteq A} m(B) \tag{23.4}$$

$$Pls(A) = \sum_{B \cap A \neq \emptyset} m(B). \tag{23.5}$$

They are sometimes referred to as lower- and upper-probability functions. The belief value of hypothesis A can be interpreted as the minimum uncertainty value about A, and its plausibility as the maximum uncertainty [28].

Evidence from p different sources is combined by combining the mass functions $m_1 \ldots m_p$ by

$$\text{If } K \neq 1, \quad m(A) = \frac{\sum_{B_1 \cap \cdots \cap B_p = A}^{m(\emptyset)=0} \prod_{1 \leq i \leq p}}{1-K} m_i(B_i)$$

where $K = \sum_{B_1 \cap \cdots \cap B_p = 0} \prod_{1 < i < p} m_i(B_i)$ is interpreted as a measure of conflict between the different sources.

The decision rule used to combine the evidence from each sensor varies from different applications, either maximum of plausibility or maximum of belief (with variations). The performance of Dempster–Shafer theory for data fusion does, however, depend on the methods used to compute the mass functions. Lee et al. [20] assign nonzero mass function values only to the single classes, whereas Hegarat-Mascle et al. [28] propose two strategies for assigning mass function values to sets of classes according to the membership for a pixel for these classes.

The concepts of evidence theory belong to a different school than Bayesian multisensor models. Researchers coming from one school often have a tendency to dislike modeling used in the alternative theory. Not many neutral comparisons of these two approaches exist. The main advantage of this approach is its robustness in the method by which information from several heterogeneous sources is combined. A disadvantage is the underlying basic assumption that the evidence from different sources is independent. According to Ref. [43], Bayesian theory assumes that imprecision about uncertainty in the measurements is assumed to be zero and uncertainty about an event is only measured by the probability. The author disagrees with this by pointing out that in Bayesian modeling, uncertainty about the measurements can be modeled in the priors. Priors of this kind are not always used, however. Priors in a Bayesian model can also be used to model spatial context and temporal class development. It might be argued that the Dempster–Shafer theory can be more appropriate for a high number of heterogeneous sources. However, most papers on data fusion for remote sensing consider two or maximum three different sources.

23.4.7 CONTEXTUAL METHODS FOR DATA FUSION

Remote-sensing data have an inherent spatial nature. To account for this, contextual information can be incorporated in the interpretation process. Basically, the effect of context in an image-labeling problem is that when a pixel is considered in isolation, it may provide incomplete information about the desired characteristics. By considering the pixel in context with other measurements, more complete information might be derived.

Only a limited set of studies have involved spatially contextual multisource classification. Richards et al. [38] extended the methods used for spatial contextual classification based on probabilistic relaxation to incorporate ancillary data. Binaghi et al. [63] presented a knowledge-based framework for contextual classification based on fuzzy set theory. Wan and Fraser [61] used multiple self-organizing maps (MSOM) for contextual classification. Le Hégarat-Mascle et al. [28] combined the use of a Markov random field model with the Dempster–Shafer theory. Smits and Dellepiane [64] used a multichannel image segmentation method based on Markov random fields with adaptive neighborhoods. Markov random fields have also been used for data fusion in other application domains [65,66].

23.4.8 Using Markov Random Fields to Incorporate Ancillary Data

Schistad Solberg et al. [67,68] used a Markov random field model to include map data into the fusion. In this framework, the task is to estimate the class labels of the scene C given the image data X and the map data M (from a previous survey):

$$P(C|X,M) = P(X|C,M)P(C)$$

with respect to C.

The spatial context between neighboring pixels in the scene is modeled in $P(C)$ using the common Ising model. By using the equivalence between Markov random fields and Gibbs distribution

$$P(\cdot) = \frac{1}{Z}\exp - U(\cdot)$$

where U is called the energy function and Z is a constant; the task of maximizing $P(C|X,M)$ is equivalent to minimizing the sum

$$U = \sum_{i=1}^{P} U_{\text{data}(i)} + U_{\text{spatial, map}}$$

U_{spatial} is the common Ising model:

$$U_{\text{spatial}} = \beta_s \sum_{k \in N} I(c_i, c_k)$$

and

$$U_{\text{map}} = \beta_m \sum_{k \in M} t(c_i|m_k)$$

m_k is the class assigned to the pixel in the map, and $t(c_i|m_k)$ is the probability of a class transition from class m_k to class c_i. This kind of model can also be used for multitemporal classification [67].

23.4.9 A Summary of Data Fusion Architectures

Table 23.1 gives a schematic view on different fusion architectures applied to remote-sensing data. A summary of decision level fusion is given in Table 23.2.

23.5 MULTITEMPORAL IMAGE CLASSIFICATION

For most applications where multisource data are involved, it is not likely that all the images are acquired at the same time. When the temporal aspect is involved, the classification methodology must handle changes in pattern classes between the image acquisitions, and possibly also use different classes.

To find the best classification strategy for a multitemporal data set, it is useful to consider the goal of the analysis and the complexity of the multitemporal image data to be used. Multitemporal image classification can be applied for different purposes:

- *Monitor and identify specific changes.* If the goal is to monitor changes, multitemporal data are required either in the form of a combination of existing maps and new satellite imagery or as a set of satellite images. For identifying changes, different fusion levels can be considered. Numerous methods for change detection exist, ranging from pixel-level to

TABLE 23.1

A Summary of Data Fusion Architectures

	Pixel-Level Fusion
Advantages	Simple. No special classifier software needed
	Correlation between sources utilized
	Well suited for change detection
Limitations	Assumes that the data can be modeled using a common probability density function
	Source reliability cannot be modeled
	Feature-Level Fusion
Advantages	Simple. No special classifier software needed
	Sensor-specific features give advantage over pixel-based fusion
	Well suited for change detection
Limitations	Assumes that the data can be modeled using a common probability density function
	Source reliability cannot be modeled
	Decision-Level Fusion
Advantages	Suited for data with different probability densities
	Source-specific reliabilities can be modeled
	Prior information about the source combination can be modeled
Limitations	Special software often needed

TABLE 23.2

A Summary of Decision-Level Fusion Strategies

	Statistical Multisensor Classifiers
Advantages	Good control over the process
	Prior knowledge can be included if the model is adapted to the application
	Inclusion of ancillary data simple using a Markov random field approach
Limitations	Assumes a particular probability density function
	Dempster–Shafer Multisensor Classifiers
Advantages	Useful for representation of heterogeneous sources
	Inclusion of ancillary data simple
	Well suited to model a high number of sources
Limitations	Performance depends on selected mass functions
	Not many comparisons with other approaches
	Neural Net Multisensor Classifiers
Advantages	No assumption about probability densities needed
	Sensor-specific weights can easily be estimated
	Suited for heterogeneous sources
Limitations	The user has little control over the fusion process and how different sources are used
	Involves a large number of parameters and a risk of overfitting
	Hybrid Multisensor Classifiers
Advantages	Can combine the best of statistical and neural net or Dempster–Shafer approaches
Limitations	More complex to use

decision-level fusion. Examples of pixel-level change detection are classical unsupervised approaches like image math, image regression, and principal component analysis of a multitemporal vector of spectral measurements or derived feature vectors like normalized vegetation indexes. In this chapter, we will not discuss in detail these well-established unsupervised methods. Decision-level change detection includes postclassification comparisons, direct multidate classification, and more sophisticated classifiers.

- *Improved quality in discriminating between a set of classes.* Sometimes, parts of an area might be covered by clouds, and a multitemporal image set is needed to map all areas. For microwave images, the signature depends on temperature and soil moisture content, and several images might be necessary to obtain good coverage of all regions in an area as two classes can have different mechanisms affecting their signature. For this kind of application, a data fusion model that takes source reliability weighting into account should be considered. An example concerning vegetation classification in a series of SAR images is shown in Figure 23.5.

- *Discriminate between classes based on their temporal signature development.* By analyzing an area through time and studying how the spectral signature changes, it is possible to discriminate between classes that are not separable on a single image. Consider, for example, vegetation mapping. Based on a single image, we might be able to discriminate between deciduous and conifer trees, but not between different kinds of conifer or deciduous. By studying how the spectral signature varies during the growth season, we might also be able to discriminate between different vegetation species.

It is also relevant to consider the available data set. How many images can be included in the analysis? Most studies use bitemporal data sets, which are easy to obtain. Obtaining longer time series of images can sometimes be difficult due to sensor repeat cycles and weather limitations. In

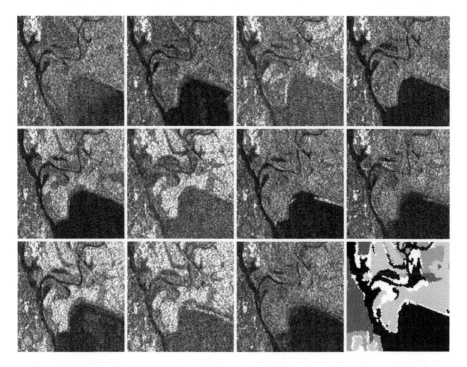

FIGURE 23.5 Multitemporal image from 13 different dates during August–December 1991 for agricultural sites in Norway. The ability to identify ploughing activity in a SAR image depends on the soil moisture content at the given date.

Northern Europe, cloud coverage is a serious limitation for many applications of temporal trajectory analysis. Obtaining long-time series tends to be easier for low- and medium-resolution images from satellites with frequent passes.

A principal decision in multitemporal image analysis is whether the images are to be combined on the pixel level or the decision level. Pixel-level fusion consists of combining the multitemporal images into a joint data set and performing the classification based on all data at the same time. In decision-level fusion, a classification is first performed for each time, and then the individual decisions are combined to reach a consensus decision. If no changes in the spectral signatures of the objects to be studied have occurred between the image acquisitions, then this is very similar to classifier combination [31].

23.5.1 MULTITEMPORAL CLASSIFIERS

In the following, we describe the main approaches for multitemporal classification. The methods utilize temporal correlation in different ways. *Temporal feature correlation* means that the correlation between the pixel measurements or feature vectors at different times is modeled. *Temporal class correlation* means that the correlation between the class labels of a given pixel at different times is modeled.

23.5.1.1 Direct Multidate Classification

In direct compound or stacked vector classification, the multitemporal data set is merged at the pixel level into one vector of measurements, followed by classification using a traditional classifier. This is a simple approach that utilizes temporal feature correlation. However, the approach might not be suited when some of the images are of lower quality due to noise. An example of this classification strategy is to use MSOM [69] as a classifier for compound bitemporal images.

23.5.1.2 Cascade Classifiers

Swain [70] presented the initial work on using cascade classifiers. In a cascade-classifier approach the temporal class correlation between multitemporal images is utilized in a recursive manner. To find a class label for a pixel at time t_2, the conditional probability for observing class ω given the images x_1 and x_2 is modeled as

$$P(\omega | x_1, x_2)$$

Classification was performed using a maximum likelihood classifier. In several papers by Bruzzone and coauthors [71,72] the use of cascade classifiers has been extended to unsupervised classification using multiple classifiers (combining both maximum likelihood classifiers and radial basis function neural nets).

23.5.1.3 Markov Chain and Markov Random Field Classifiers

Schistad Solberg et al. [67] describe a method for classification of multisource and multitemporal images where the temporal changes of classes are modeled using Markov chains with transition probabilities. This approach utilizes temporal class correlation. In the Markov random field model presented in Ref. [25], class transitions are modeled in terms of Markov chains of possible class changes and specific energy functions are used to combine temporal information with multisource measurements, and ancillary data. Bruzzone and Prieto [73] use a similar framework for unsupervised multitemporal classification.

23.5.1.4 Approaches Based on Characterizing the Temporal Signature

Several papers have studied changes in vegetation parameters (for a review see Ref. [74]). In Refs. [50,75] the temporal signatures of classes are modeled using Fourier series (using temporal feature correlation). Not many approaches have integrated phenological models for the expected development of vegetation parameters during the growth season. Aurdal et al. [76] model the phenological

TABLE 23.3
A Discussion of Multitemporal Classifiers

	Direct Multidate Classifiers
Advantages	Simple. Temporal feature correlation between image measurements utilized
Limitations	Is restricted to pixel-level fusion
	Not suited for data sets containing noisy images
	Cascade Classifiers
Advantages	Temporal correlation of class labels considered
	Information about special class transitions can be modeled
Limitations	Special software needed
	Markov Chain and MRF Classifiers
Advantages	Spatial and temporal correlation of class labels considered
	Information about special class transitions can be modeled
Limitations	Special software needed
	Temporal Signature Trajectory Approaches
Advantages	Can discriminate between classes not separable at a single point in time
	Can be used either at feature level or at decision level
	Decision-level approaches allow flexible modeling
Limitations	Feature-level approaches can be sensitive to noise
	A time series of images needed (can be difficult to get more than bitemporal)

evolution of mountain vegetation using hidden Markov models. The different vegetation classes can be in one of a predefined set of states related to their phenological development, and classifying a pixel consists of selecting the class that has the highest probability of producing a given series of observations. The performance of this model is compared to a compound maximum likelihood approach and found to give comparable results for a single scene, but more robust when testing and training on different images.

23.5.1.5 Other Decision-Level Approaches to Multitemporal Classification

Jeon and Landgrebe [46] developed a spatio-temporal classifier utilizing both the temporal and the spatial context of the image data. Khazenie and Crawford [47] proposed a method for contextual classification using both spatial and temporal correlation of data. In this approach, the feature vectors are modeled as resulting from a class-dependent process and a contaminating noise process, and the noise is correlated in both space and time. Middelkoop and Janssen [49] presented a knowledge-based classifier, which used land-cover data from preceding years. An approach to decision-level change detection using evidence theory is given in Ref. [43].

A summary of approaches for multitemporal image classifiers is given in Table 23.3.

23.6 MULTISCALE IMAGE CLASSIFICATION

Most of the approaches to multisensor image classification do not treat the multiscale aspect of the input data. The most common approach is to resample all the images to be fused to a common pixel resolution.

In other domains of science, much work on combining data sources at different resolutions exists, for example, in epidemiology [77], in the estimation of hydraulic conductivity for characterizing groundwater flow [78], and in the estimation of environmental components [44]. These approaches are mainly for situations where the aim is to estimate an underlying *continuous* variable.

The remote-sensing literature contains many examples of multiscale and multisensor data visualization. Many multispectral sensors, such as SPOT XS or Ikonos, provide a combination of multispectral band and a panchromatic band of a higher resolution. Several methods for visualizing such multiscale data sets have been proposed, and they are often based on overlaying a multispectral image on the panchromatic image using different colors. We will not describe such techniques in detail, but refer the reader to surveys like [51,55,79]. Van der Meer [80] studied the effect of multi-sensor image fusion in terms of information content for visual interpretation, and concluded that image fusion aiming at improving the visual content and interpretability was more successful for homogeneous data than for heteorogeneous data.

For classification problems, Puyou-Lascassies [54] and Zhukov et al. [81] considered unmixing of low-resolution data by using class label information obtained from classification of high-resolution data. The unmixing is performed through several sequential steps, but no formal model for the complete data set is derived. Price [53] proposed unmixing by relating the correlation between low-resolution data and high-resolution data resampled to low resolution, to correlation between high-resolution data and low-resolution data resampled to high resolution. The possibility of mixed pixels was not taken into account.

In Ref. [82], separate classifications were performed based on data from each resolution. The resulting resolution-dependent probabilities were averaged over the resolutions.

Multiresolution tree models are sometimes used for multiscale analysis (see, e.g., Ref. [48]). Such models yield a multiscale representation through a quad tree, in which each pixel at a given resolution is decomposed into four child pixels at higher resolution, which are correlated. This gives a model where the correlation between neighbor pixels depends on the pixel locations in an arbitrary (i.e., not problem-related) manner.

The multiscale model presented in Ref. [83] is based on the concept of a reference resolution and is developed in a Bayesian framework [84]. The reference resolution corresponds to the highest resolution present in the data set. For each pixel of the input image at the reference resolution it is assumed that there is an underlying discrete class. The observed pixel values are modeled conditionally on the classes. The properties of the class label image are described through an *a priori* model. Markov random fields have been selected for this purpose. Data at coarser resolutions are modeled as mixed pixels, that is, the observations are allowed to include contributions from several distinct classes. In this way it is possible to exploit spectrally richer images at lower resolutions to obtain more accurate classification results at the reference level, without smoothing the results as much as if we simply oversample the low-resolution data to the reference resolution prior to the analysis.

Methods that use a model for the relationship between the multiscale data might offer advantages compared to simple resampling both in terms of increased classification accuracy and being able to describe relationships between variables measured at different scales. This can provide tools to predict high-resolution properties from coarser resolution properties. Of particular concern in the establishment of statistical relationships is the quantification of what is lost in precision at various resolutions and the associated uncertainty.

The potential of using multiscale classifiers will also depend on the level of detail needed for the application, and might be related to the typical size of the structures one wants to identify in the images. Even simple resampling of the coarsest resolution to the finest resolution, followed by classification using a multisensor classifier, can help improve the classification result. The gain obtained by using a classifier that explicitly models the data at different scales depends not only on the set of classes used but also on the regions used to train and test the classifier. For scenes with a high level of detail, for example, in urban scenes, the performance gain might be large. However, it depends also on how the classifier performance is evaluated. If the regions used for testing the classifier are well inside homogeneous regions and not close to other classes, the difference in performance in

TABLE 23.4
A Discussion of Multiscale Classifiers

	Resampling Combined with Single-Scale Classifier
Advantages	Simple. Works well enough for homogeneous regions
Limitations	Can fail in identifying small or detailed structures
	Classifier with Explicit Multiscale Model
Advantages	Can give increased performance for small or detailed structures
Limitations	More complex software needed
	Not necessary for homogeneous regions

terms of overall classification accuracy might not be large, but visual inspection of the level of detail in the classified images can reveal the higher level of detail.

A summary of multiscale classification approaches is given in Table 23.4.

23.7 CONCLUDING REMARKS

A number of different approaches for data fusion in remote-sensing applications have been presented in the literature. A prerequisite for data fusion is that the data are coregistered and geometrically and radiometrically corrected.

In general, there is no consensus on which multisource or multitemporal classification approach works best. Different studies and comparisons report different results. There is still a need for a better understanding on which methods are most suited to different applications types, and also broader comparison studies. The best level and methodology for a given remote-sensing application depends on several factors: the complexity of the classification problem, the available data set, the number of sensors involved, and the goal of the analysis.

Some guidelines for selecting the methodology and architecture for a given fusion task are given below.

23.7.1 FUSION LEVEL

Decision-level fusion gives best control and allows weighting the influence of each sensor. Pixel-level fusion can be suited for simple analysis, for example, fast unsupervised change detection.

23.7.2 SELECTING A MULTISENSOR CLASSIFIER

If decision-level fusion is selected, three main approaches for fusion should be considered: the statistical approach, neural networks, or evidence theory. A hybrid approach can also be used to combine these approaches. If the sources are believed to provide data of different quality, weighting schemes for consensus combination of the sensor-specific classifiers should be considered.

23.7.3 SELECTING A MULTITEMPORAL CLASSIFIER

To find the best classification strategy for a multitemporal data set, the complexity of the class separation problem must be considered in light of the available data set. If the classes are difficult to separate, it might be necessary to use methods for characterizing the temporal trajectory of signatures. For pixel-level classification of multitemporal imagery, the direct multidate classification approach can be used. If specific knowledge about certain types of changes needs to be modeled, Markov chain and Markov random field approaches or cascade classifiers should be used.

23.7.4 Approaches for Multiscale Data

Multiscale images can either be resampled to a common resolution or a classifier with implicit modeling of the relationship between the different scales can be used. For classification problems involving small or detailed structures (e.g., urban areas) or heterogeneous sources, the latter is recommended.

ACKNOWLEDGMENT

The author would like to thank Line Eikvil for valuable input, in particular, regarding multisensor image registration.

REFERENCES

1. C. Elachi, J. Cimino, and M. Settle, Overview of the shuttle imaging radar-B preliminary scientific results, *Science*, 232, 1511–1516, 1986.
2. J. Cimino, A. Brandani, D. Casey, J. Rabassa, and S. D. Wall, Multiple incidence angle SIR-B experiment over Argentina: Mapping of forest units, *IEEE Trans. Geosci. Remote Sens.*, 24, 498–509, 1986.
3. G. Asrar, *Theory and Applications of Optical Remote Sensing*, Wiley, New York, 1989.
4. F. T. Ulaby, R. K. Moore, and A. K. Fung, *Microwave Remote Sensing, Active and Passive*, Vols. I–III, Artech House Inc., 1981, Norwood, MA, 1982, 1986.
5. F. T. Ulaby and C. Elachi, *Radar Polarimetry for Geoscience Applications*, Artec House Inc., Norwood, MA, 1990.
6. H. A. Zebker and J. J. Van Zyl, Imaging radar polarimetry: A review, *Proc. IEEE*, 79, 1583–1606, 1991.
7. B. Zitova and J. Flusser, Image registration methods: A survey, *Image Vis. Comput.*, 21, 977–1000, 2003.
8. P. Chalermwat and T. El-Chazawi, Multi-resolution image registration using genetics, in *Proceedings of the ICIP*, Kobe, Japan, 452–456, 1999.
9. H. M. Chen, M. K. Arora, and P. K. Varshney, Mutual information-based image registration for remote sensing data, *Int. J. Remote Sens.*, 24, 3701–3706, 2003.
10. X. Dai and S. Khorram, A feature-based image registration algorithm using improved chain-code representation combined with invariant moments, *IEEE Trans. Geosci. Remote Sens.*, 37, 17–38, 1999.
11. D. M. Mount, N. S. Netanyahu, and L. Le Moigne, Efficient algorithms for robust feature matching, *Pattern Recognit.*, 32, 17–38, 1999.
12. E. Rignot, R. Kwok, J. C. Curlander, J. Homer, and I. Longstaff, Automated multisensor registration: Requirements and techniques, *Photogramm. Eng. Remote Sens.*, 57, 1029–1038, 1991.
13. Z.-D. Lan, R. Mohr, and P. Remagnino, Robust matching by partial correlation, in *British Machine Vision Conference*, Edinburgh, UK, 651–660, 1996.
14. D. Fedorov, L. M. G. Fonseca, C. Kennedy, and B. S. Manjunath, Automatic registration and mosaicking system for remotely sensed imagery, in *Proceedings of the 9th International Symposium on Remote Sensing*, Crete, Greece, pp. 22–27, 2002.
15. L. Fonseca, G. Hewer, C. Kenney, and B. Manjunath, Registration and fusion of multispectral images using a new control point assessment method derived from optical flow ideas, in *Proceedings of the Algorithms for Multispectral and Hyperspectral Imagery V*, SPIE, Orlando, USA, pp. 104–111, 1999.
16. M. A. Abidi and R. C. Gonzalez, *Data Fusion in Robotics and Machine Intelligence*, Academic Press, Inc., New York, 1992.
17. N. Xiong and P. Svensson, Multi-sensor management for information fusion: Issues and approaches, *Inf. Fusion*, 3, 163–180, 2002.
18. J. M. Richardson and K. A. Marsh, Fusion of multisensor data, *Int. J. Robot. Res.* 7, 78–96, 1988.
19. D. L. Hall and J. Llinas, An introduction to multisensor data fusion, *Proc. IEEE*, 85(1), 6–23, 1997.
20. T. Lee, J. A. Richards, and P. H. Swain, Probabilistic and evidential approaches for multisource data analysis, *IEEE Trans. Geosci. Remote Sens.*, 25, 283–293, 1987.
21. N. Ayache and O. Faugeras, Building, registrating, and fusing noisy visual maps, *Int. J. Robot. Res.*, 7, 45–64, 1988.
22. J. A. Benediktsson and P. H. Swain, A method of statistical multisource classification with a mechanism to weight the influence of the data sources, in *IEEE Symposium on Geoscience and Remote Sensing (IGARSS)*, Vancouver, Canada, pp. 517–520, July 1989.

23. S. Wu, Analysis of data acquired by shuttle imaging radar SIR-A and Landsat Thematic Mapper over Baldwin county, Alabama, in *Proceedings of the Machine Processes Remotely Sensed Data Symposium*, West Lafayette, Indiana, pp. 173–182, June 1985.

24. A. H. Schistad Solberg, A. K. Jain, and T. Taxt, Multisource classification of remotely sensed data: Fusion of Landsat TM and SAR images, *IEEE Trans. Geosci. Remote Sens.*, 32, 768–778, 1994.

25. A. Schistad Solberg, Texture fusion and classification based on flexible discriminant analysis, in *International Conference on Pattern Recognition (ICPR)*, Vienna, Austria, pp. 596–600, August 1996.

26. H. Kim and P. H. Swain, A method for classification of multisource data using interval-valued probabilities and its application to HIRIS data, in *Proceedings of the Workshop Multisource Data Integration Remote Sensing*, NASA Conference Publication 3099, Maryland, pp. 75–82, June 1990.

27. J. Desachy, L. Roux, and E.-H. Zahzah, Numeric and symbolic data fusion: A soft computing approach to remote sensing image analysis, *Pattern Recognit. Lett.*, 17, 1361–1378, 1996.

28. S. L. Hégarat-Mascle, I. Bloch, and D. Vidal-Madjar, Application of Dempster–Shafer evidence theory to unsupervised classification in multisource remote sensing, *IEEE Trans. Geosci. Remote Sens.*, 35, 1018–1031, 1997.

29. S. B. Serpico and F. Roli, Classification of multisensor remote-sensing images by structured neural networks, *IEEE Trans. Geosci. Remote Sens.*, 33, 562–578, 1995.

30. J. A. Benediktsson, J. R. Sveinsson, and P. H. Swain, Hybrid consensys theoretic classification, *IEEE Trans. Geosci. Remote Sens.*, 35, 833–843, 1997.

31. J. A. Benediktsson and I. Kanellopoulos, Classification of multisource and hyperspectral data based on decision fusion, *IEEE Trans. Geosci. Remote Sens.*, 37, 1367–1377, 1999.

32. B. C. K. Tso and P. M. Mather, Classification of multisource remote sensing imagery using a genetic algorithm and Markov random fields, *IEEE Trans. Geosci. Remote Sens.*, 37, 1255–1260, 1999.

33. M. Petrakos, J. A. Benediktsson, and I. Kannelopoulos, The effect of classifier agreement on the accuracy of the combined classifier in decision level fusion, *IEEE Trans. Geosci. Remote Sens.*, 39, 2539–2546, 2001.

34. J. A. Benediktsson and J. Sveinsson, Multisource remote sensing data classification based on consensus and pruning, *IEEE Trans. Geosci. Remote Sens.*, 41, 932–936, 2003.

35. Solberg, G. Storvik, and R. Fjørtoft, A comparison of criteria for decision fusion and parameter estimation in statistical multisensor image classification, in *IEEE Symposium on Geoscience and Remote Sensing (IGARSS'02)*, Toronto, Canada, Vol. 1, pp. 72–74, 2002.

36. D. G. Leckie, Synergism of synthetic aperture radar and visible/infrared data for forest type discrimination, *Photogramm. Eng. Remote Sens.*, 56, 1237–1246, 1990.

37. S. E. Franklin, Ancillary data input to satellite remote sensing of complex terrain phenomena, *Comput. Geosci.*, 15, 799–808, 1989.

38. J. A. Richards, D. A. Landgrebe, and P. H. Swain, A means for utilizing ancillary information in multispectral classification, *Remote Sens. Environ.*, 12, 463–477, 1982.

39. J. Friedman, Multivariate adaptive regression splines (with discussion), *Ann. Stat.*, 19, 1–141, 1991.

40. P. Gong, R. Pu, and J. Chen, Mapping ecological land systems and classification uncertainties from digital elevation and forest-cover data using neural networks, *Photogramm. Eng. Remote Sens.*, 62, 1249–1260, 1996.

41. L. Guan, J. A. Anderson, and J. P. Sutton, A network of networks processing model for image regularization, *IEEE Trans. Neural Networks*, 8, 169–174, 1997.

42. T. Hastie, R. Tibshirani, and A. Buja, Flexible discriminant analysis by optimal scoring, *J. Am. Stat. Assoc.*, 89, 1255–1270, 1994.

43. S. Le Hégarat-Mascle and R. Seltz, Automatic change detection by evidential fusion of change indices, *Remote Sens. Environ.*, 91, 390–404, 2004.

44. D. Hirst, G. Storvik, and A. R. Syversveen, A hierarchical modelling approach to combining environmental data at different scales, *J. Royal Stat. Soc., Series C*, 52, 377–390, 2003.

45. J.-N. Hwang, D. Li, M. Maechelr, D. Martin, and J. Schimert, Projection pursuit learning networks for regression, *Eng. Appl. Artif. Intell.*, 5, 193–204, 1992.

46. B. Jeon and D. A. Landgrebe, Classification with spatio-temporal interpixel class dependency contexts, *IEEE Trans. Geosci. Remote Sens.*, 30, 663–672, 1992.

47. N. Khazenie and M. M. Crawford, Spatio-temporal autocorrelated model for contextual classification, *IEEE Trans. Geosci. Remote Sens.*, 28, 529–539, 1990.

48. M. R. Luettgen, W. Clem Karl, and A. S. Willsky, Efficient multiscale regularization with applications to the computation of optical flow, *IEEE Trans. Image Process.*, 3(1), 41–63, 1994.

49. J. Middelkoop and L. L. F. Janssen, Implementation of temporal relationships in knowledge based classification of satellite images, *Photogramm. Eng. Remote Sens.*, 57, 937–945, 1991.

50. L. Olsson and L. Eklundh, Fourier series for analysis of temporal sequences of satellite sensor imagery, *Int. J. Remote Sens.*, 15, 3735–3741, 1994.

51. G. Pajares and J. M. de la Cruz, A wavelet-based image fusion tutorial, *Pattern Recognit.*, 37, 1855–1871, 2004.

52. J. D. Paola and R. A. Schowengerdt, The effect of neural-network structure on a multispectral land-use/land-cover classification, *Photogramm. Eng. Remote Sens.*, 63, 535–544, 1997.

53. J. C. Price, Combining multispectral data of differing spatial resolution, *IEEE Trans. Geosci. Remote Sens.*, 37(3), 1199–1203, 1999.

54. P. Puyou-Lascassies, A. Podaire, and M. Gay, Extracting crop radiometric responses from simulated low and high spatial resolution satellite data using a linear mixing model, *Int. J. Remote Sens.*, 15(18), 3767–3784, 1994.

55. T. Ranchin, B. Aiazzi, L. Alparone, S. Baronti, and L. Wald, Image fusion—The ARSIS concept and some successful implementations, ISPRS, *J. Photogramm. Remote Sens.*, 58, 4–18, 2003.

56. B. D. Ripley, Flexible non-linear approaches to classification, in V. Cherkassky, J. H. Friedman, and H. Wechsler, (eds.), *From Statistics to Neural Networks. Theory and Pattern Recognition Applications*, NATO ASI series F: Computer and Systems Sciences, Springer-Verlag, Heidelberg, pp. 105–126, 1994.

57. A. H. Solberg, Flexible nonlinear contextual classification, *Pattern Recognit. Lett.*, 25, 1501–1508, 2004.

58. A. K. Skidmore, B. J. Turner, W. Brinkhof, and E. Knowles, Performance of a neural network: Mapping forests using GIS and remotely sensed data, *Photogramm. Eng. Remote Sens.*, 63, 501–514, 1997.

59. J. A. Benediktsson, J. R. Sveinsson, and O. K. Ersoy, Optimized combination of neural networks, in *IEEE International Symposium on Circuits and Systems (ISCAS'96)*, Atlanta, Georgia, pp. 535–538, May 1996.

60. G. A. Carpenter, M. N. Gjaja, S. Gopal, and C. E. Woodcock, ART neural networks for remote sensing: Vegetation classification from Landsat TM and terrain data, in *IEEE Symposium on Geoscience and Remote Sensing (IGARSS)*, Lincoln, Nebraska, pp. 529–531, May 1996.

61. W. Wan and D. Fraser, A self-organizing map model for spatial and temporal contextual classification, in *IEEE Symposium on Geoscience and Remote Sensing (IGARSS)*, Pasadena, California, pp. 1867–1869, August 1994.

62. G. Shafer, *A Mathematical Theory of Evidence*, Princeton University Press, Princeton, N.J., 1976.

63. E. Binaghi, P. Madella, M. G. Montesano, and A. Rampini, Fuzzy contextual classification of multisource remote sensing images, *IEEE Trans. Geosci. Remote Sens.*, 35, 326–340, 1997.

64. P. C. Smits and S. G. Dellepiane, Synthetic aperture radar image segmentation by a detail preserving Markov random field approach, *IEEE Trans. Geosci. Remote Sens.*, 35, 844–857, 1997.

65. P. B. Chou and C. M. Brown, Multimodal reconstruction and segmentation with Markov random fields and HCF optimization, in *Proceedings of the 1988 DARPA Image Understanding Workshop*, Cambridge, MA, pp. 214–221, 1988.

66. W. A. Wright, A Markov random field approach to data fusion and colour segmentation, *Image Vis. Comput.*, 7, 144–150, 1989.

67. A. H. Schistad Solberg, T. Taxt, and A. K. Jain, A Markov random field model for classification of multisource satellite imagery, *IEEE Trans. Geosci. Remote Sens.*, 34, 100–113, 1996.

68. A. H. Schistad Solberg, Contextual data fusion applied to forest map revision, *IEEE Trans. Geosci. Remote Sens.*, 37, 1234–1243, 1999.

69. W. Wan and D. Fraser, Multisource data fusion with multiple self-organizing maps, *IEEE Trans. Geosci. Remote Sens.*, 37, 1344–1349, 1999.

70. P. H. Swain, Bayesian classification in a time-varying environment, *IEEE Trans. Sys. Man Cyber.*, 8, 879–883, 1978.

71. L. Bruzzone and R. Cossu, A multiple-cascade-classifier system for a robust and partially unsupervised updating of land-cover maps, *IEEE Trans. Geosci. Remote Sens.*, 40, 1984–1996, 2002.

72. L. Bruzzone and D. F. Prieto, Unsupervised retraining of a maximum-likelihood classifier for the analysis of multitemporal remote-sensing images, *IEEE Trans. Geosci. Remote Sens.*, 39, 456–460, 2001.

73. L. Bruzzone and D. F. Prieto, An adaptive semiparametric and context-based approach to unsupervised change detection in multitemporal remote-sensing images, *IEEE Trans. Image Process*, 11, 452–466, 2002.

74. P. Coppin, K. Jonkheere, B. Nackaerts, and B. Muys, Digital change detection methods in ecosystem monitoring: A review, *Int. J. Remote Sens.*, 25, 1565–1596, 2004.

75. L. Andres, W. A. Salas, and D. Skole, Fourier analysis of multi-temporal AVHRR data applied to a land cover classification, *Int. J. Remote Sens.*, 15, 1115–1121, 1994.

76. L. Aurdal, R. B. Huseby, L. Eikvil, R. Solberg, D. Vikhamar, and A. Solberg, Use of hidden Markov models and phenology for multitemporal satellite image classification: Applications to mountain vegetation classification, in *MULTITEMP 2005*, 220–224, May 2005.
77. N. G. Besag, K. Ickstadt, and R. L. Wolpert, Spatial poisson regression for health and exposure data measured at disparate resolutions, *J. Am. Stat. Assoc.*, 452, 1076–1088, 2000.
78. M. M. Daniel and A. S. Willsky, A multiresolution methodology for signal-level fusion and data assimilation with applications to remote sensing, *Proc. IEEE*, 85(1), 164–180, 1997.
79. L. Wald, *Data Fusion: Definitions and Achitectures—Fusion of Images of Different Spatial Resolutions*, Ecole des Mines Press, Paris, 2002.
80. F. Van der Meer, What does multisensor image fusion add in terms of information content for visual interpretation? *Int. J. Remote Sens.*, 18, 445–452, 1997.
81. B. Zhukov, D. Oertel, F. Lanzl, and G. Reinhäckel, Unmixing-based multisensor multiresolution image fusion, *IEEE Trans. Geosci. Remote Sens.*, 37(3), 1212–1226, 1999.
82. M. M. Crawford, S. Kumar, M. R. Ricard, J. C. Gibeaut, and A. Neuenshwander, Fusion of airborne polarimetric and interferometric SAR for classification of coastal environments, *IEEE Trans. Geosci. Remote Sens.*, 37(3), 1306–1315, 1999.
83. G. Storvik, R. Fjørtoft, and A. Solberg, A Bayesian approach to classification in multiscale remote sensing data, *IEEE Trans. Geosci. Remote Sens.*, 43, 539–547, 2005.
84. J. Besag, Towards Bayesian image analysis, *J. Appl. Stat.*, 16(3), 395–407, 1989.

24 Image Fusion in Remote Sensing with the Steered Hermite Transform

Boris Escalante-Ramírez and Alejandra A. López-Caloca

CONTENTS

24.1 INTRODUCTION

Advances in sensor technology have produced a large variety of sensors capable of capturing different kinds of information from the Earth's observation satellites, with different characteristics and modalities, depending on their use, for example, multisensor, multitemporary, multiresolution, and multifrequency [1–3]. Sensors, however, present technological limitations that affect image acquisition characteristics, for instance, multispectral (MS) sensors may capture images with high spectral resolution, but with lower spatial resolutions than panchromatic (PAN) sensor. Climate conditions also pose limitations to sensor technology as is the case of cloudy conditions that limit the range of optical sensors. Radar sensors overcome this limitation, but they are seriously impaired by the presence of speckle. Owing to these facts, in recent years, image fusion has become one of the most important and useful tasks for the remote sensing community.

The goal of image fusion is to integrate information from multiple sources, in order to create new images containing more information. The process of image fusion should not introduce any artifact or inconsistency which may alter subsequent processes. It should be robust and tolerant to noise. As a result of a fusion process, two main characteristics are desirable: higher spatial resolution that can account for an adequate description of the shapes, features, and structures, and consistent spectral properties that allow the user to identify the different interacting objects.

There is a large variety of techniques described in the literature that tackle the problem of image fusion by incorporating high spatial resolution characteristics and keeping spectral properties from the same sensor. These approaches have evolved from simple linear combinations (intensity–hue–saturation) (IHS) [4,5] to methods based on principal component analysis (PCA) [6]. The main idea of fusion algorithms is to add spatial information to an image that is rich in spectral information, without modifying the latter. Methods like IHS and PCA improve the spatial quality, but they show spectral information distortion [7].

Recently, wavelet transform (WT)-based methods, with different approaches, have been widely used for image fusion. The discrete wavelet transform, either decimated [8–10] or nondecimated [11,12], has become very popular, multiresolution analysis being one of their most important properties. Decimated methods, implemented with dyadic structures based on the Mallat algorithm [13], present interesting properties, such as nonsymmetry and nonredundant descriptions. Their main disadvantage is the lack of shift invariance, meaning that small shifts of the input image produces very different wavelet coefficient contents. This limitation usually translates into artifacts introduced in the fused image. Shift invariance can be achieved with nondecimated methods, such as the "à trous" algorithm [14,15]. These methods imply redundant image descriptions that avoid decimation by inserting zeroes between the filter coefficients. Although they show better spatial reconstruction in comparison with decimated structures [7,16], they lose orientation selectivity. In this case, image decomposition consists of an approximated band and a single-detail image at each resolution level, resembling Laplacian or difference of Gaussian hierarchical structures. Among these methods, the additive wavelet method (AWL) developed by Nuñez et al. [12] has become a reference method. Later, González-Audícana et al. [17] found a way to incorporate the sensor's spectral properties to overcome the spectral distortion problem of traditional IHS fusion method, and developed the extended fast IHS method (eFIHS). They proved that this method performs as good as high computational cost methods based on nondecimated wavelet transforms.

Recently, the curvelet transform has been proposed as a tool for image fusion [18]. It consists of a multiresolution directional-oriented representation obtained from a nondecimated wavelet transform. Its results show the advantage of detecting and reconstructing oriented image patterns in image fusion products.

This chapter introduces the steered Hermite transform (HT) as an efficient representation model for image fusion.

Shift invariance and isotropic property (rotation) of the HT assures that no artifacts are introduced during the fusion process. The HT is also a good representation model for characteristic patterns such as edges, and lines, which may be extracted from the high-resolution PAN image and injected into the multispectral images, which results in images with a richer spatial content than the images obtained with processes like PCA and WT.

In particular, the rotation property of the HT [19,20] is relevant in the fusion process as it not only allows the detection of edges, but it also allows the estimation of other parameters, such as local orientation. Therefore, during the fusion process, patterns may be discriminated and selected according to their local energy and orientation. The use of the locally rotated HT provides a coefficient set with high-energy compaction, in such a way that few coefficients are needed to represent the more relevant image patterns needed for the fusion process.

One of the advantages of the HT over the WT is the use of a free subsampling parameter limited only by the support of the analysis window, that is, the only constraint for this parameter is that local analysis windows overlap with each other. This allows for the existence of decimated (subsampled) as well as the undecimated (no subsampling) HT decompositions. Both schemes are efficient for image fusion. Decimated structures are more efficiently computed, while undecimated structures provide shift invariance to the decomposition, a highly valuable property that produces no artifacts in image reconstruction problems. In this chapter, two fusion applications with the HT are shown, single sensor (MS-PAN) and multisensor (MS-SAR). In the case of MS and SAR image fusion, adaptive speckle reduction in SAR images can be readily achieved within the HT analysis–synthesis process of the fusion scheme. For the first case, we compare our results with the eFIHS [17] and AWL methods [12] and for the second one, with the multisensor image fusion algorithms based on the generalized intensity modulation proposed in Ref. [21]. This scheme combines three kinds of images (MS-PAN-SAR) with speckle reduction achieved previously to the fusion process. In contrast, we show a fusion scheme where both speckle reduction and fusion can be achieved together.

The layout of this chapter is as follows. Section 24.2 introduces the Hermite transform. Section 24.3 presents two fusion methodologies; in the first, we fuse multispectral and panchromatic images from the same satellite with different spatial resolutions and in the second, we fuse different sensor images, namely, multispectral and SAR. Section 24.4 presents experiments conducted on SPOT-5, SAR AeS-1, and Landsat 7 + ETM images, with their respective fusion results. We show how the proposed method can help improve spatial resolution and keep spectral properties of the original MS images. Quality assessing of the synthesized images was determined by spectral quality fusion [10], spatial quality [9], scatterplot red–NIR [22], SAM [21], and in the case of SAR-MS fusion, classification of the fused product was also used for evaluation purposes. Finally, Section 24.5 presents the conclusions of this chapter.

24.2 THE HERMITE TRANSFORM

24.2.1 THE HERMITE TRANSFORM AS AN IMAGE REPRESENTATION MODEL

With the development of the scale-space theory in the 1980s, it has become evident that an efficient description of the elements that conforms an image is obtained through a multiresolution decomposition. The scale-space theory proves that the Gaussian function is optimal for this task [23]. Most wavelet functions show irregular profiles, which makes them unsuitable for the representation of spatial phenomena in accordance to scale-space theory [24,25]. Several computational representation models include relevant properties of human vision, the Gabor transform being one of the more important [26–29]. More recently, however, several authors have proven the limitations of this model from both neurophysiological and mathematical points of view.

Stork and Wilson [30] reviewed neurophysiological measurements of others and analyzed psychophysical masking data and found that in many cases receptive-field functions other than Gabor functions fit better. They concluded that there are insufficient theoretical demonstrations and experimental data to favor Gabor functions over any of a number of other plausible receptive-field functions.

In contrast, Gaussian derivatives have been alternatively recognized as good models of the receptive field profiles of the human visual system [25,31–37].

Young made a comparison between both models and showed that the Gaussian model fits more accurately to the measurements of the signals at receptive fields of the human visual system, with the additional advantage of being orthogonal at the location of analysis [31–33].

Moreover, Koenderink and van Doorn [38,39] considered the problem of deriving linear operators from the scale-space representation considering that size invariance and the absence of spurious resolution are two requirements that characterize well-behaved spatial sampling in visual systems. They concluded that these operators must obey the time-independent Schrödinger equation, that is, a physical equation that governs the quantum mechanical oscillator. Thus, they provided a formal statement that Gaussian derivatives are *natural operators* to derive from scale-space.

The HT was originally developed as a mathematical model for explaining the receptive fields during early stages of human vision [40,41]. It is a special case of polynomial transforms whose basis functions are derivatives of Gaussian functions. The extension of this model to the multiresolution case was then formulated [42,43].

The HT uses overlapping Gaussian windows and projects images locally onto a basis of orthogonal polynomials.

First, windowing with a local function $\omega(x - p, y - q)$ at positions p, q that conform the sampling lattice S takes place. As argued before, the scale-space suggests using a Gaussian window, that is,

$$\omega(x, y) = \frac{1}{2\pi\sigma^2} \exp\left(-\frac{(x^2 + y^2)}{2\sigma^2}\right) \tag{24.1}$$

where σ is spread of the Gaussian window.

The Gaussian window is separable into Cartesian coordinates; it is isotropic, thus, it is rotationally invariant.

Through the replication of the window function over the sampling lattice, a periodic weighting function is defined as $W(x, y) = \sum_{(p,q) \in S} \omega(x - p, y - q)$. This weighting function must be different from zero for all coordinates (x, y). Next, local information at each analysis window is expanded in terms of a family of orthogonal polynomials $G_{m,n-m}(x, y)$ of the order m in x and $n - m$ in y. They are determined by the analysis window function, and satisfy the orthogonal condition:

$$\int_{-\infty}^{+\infty}\int_{-\infty}^{+\infty} \omega^2(x, y) G_{m,n-m}(x, y) G_{l,k-l}(x, y)\, dxdy = \delta_{nk}\delta_{ml} \tag{24.2}$$

for $n, k = 0, \ldots, \infty$; $m = 0, \ldots, n$ and $l = 0, \ldots, k$; where δ_{nk} denotes the Kronecker function.

In the case of a Gaussian window function, the associated orthogonal polynomials are the Hermite polynomials:

$$G_{n-m,m}(x, y) = \frac{1}{\sqrt{2^n (n - m)! m!}} H_{n-m}\left(\frac{x}{\sigma}\right) H_m\left(\frac{y}{\sigma}\right) \tag{24.3}$$

where $H_n(x)$ denotes the nth Hermite polynomial of degree n in x [44].

The polynomial coefficients $L_{m,n-m}(p, q)$ are calculated by convolution of the original image $L(x, y)$ with the filter function:

$$D_{m,n-m}(x, y) = G_{m,n-m}(-x, -y)\omega^2(-x, -y) \tag{24.4}$$

followed by subsampling at positions (p, q) of the sampling lattice S. For the case of the Hermite transform, it can be shown [40] that the filter functions $D_{m,n-m}(x, y)$ correspond to Gaussian derivatives of order m in x and $n - m$ in y, in agreement with the Gaussian derivative model of early vision.

The process of recovering the original image (synthesis) consists of interpolating the transform coefficients with the proper synthesis filters. This process is called an inverse polynomial transform and is defined by

$$\hat{L}(x, y) = \sum_{n=0}^{N}\sum_{m=0}^{n}\sum_{(p,q) \in S} L_{m,n-m}(p,q) P_{m,n-m}(x - p, y - q) \tag{24.5}$$

The synthesis filters $P_{m,n-m}(x,y)$ of order m and $n - m$ are defined by

$$P_{m,n-m}(x, y) = \frac{G_{m,n-m}(x, y)\omega(x, y)}{W(x, y)}$$

for $m = 0, \ldots, n$ and $n = 0, \ldots, \infty$.

In a discrete implementation, the Gaussian window function may be approximated by the binomial window function $\omega^2(x) = (1/2^N)C_N^x$ for $x = 0, \ldots, N$, where N is called the order of the binomial window and represents the function length, and $C_N^x = N!/(N-x)!x!$ for $x = 0, \ldots, M$. In this case, the orthogonal polynomials $G_{m,n-m}(x, y)$ associated with the binomial window are known as the Krawtchouck's polynomials:

$$K_n[x] = \frac{1}{\sqrt{C_N^n}} \sum_{k=0}^{n} (-1)^{n-k} C_{N-x}^{n-k} C_x^k$$

for $n = 0, 1, \ldots, N$.

For this discrete case, all previous relations hold, with some interesting modifications. First, support of the window function is finite (N); as a consequence, expansion with the Krawtchouck polynomials is also finite, and signal reconstruction from the expansion coefficients is perfect.

To define a polynomial transform, some parameters have to be chosen. First, we have to define the characteristics of the window function. As argued before, the Gaussian window is the best option from a perceptual point of view and from the scale-space theory, however, a discrete implementation may call for a discrete transform, and in this case the binomial window is a good choice. Another important parameter is the window spread. The choice may depend on the scale-space representation of the target objects in the image. Fine local changes are better detected with small windows, but on the contrary, representation of low-resolution objects need large windows. We have chosen binomial windows of orders $N = 2$; however, in order to overcome spatial resolution compromises, multiresolution representations are a good solution [42,43].

Last but not least, the subsampling factor, is a free parameter directly related to the subsampling positions (p, q) of the sampling lattice S. From a mathematical point of view, the only constraint for this parameter is that the weighting function $W(x, y)$ must be different from zero for all coordinates (x, y). In the case of the binomial window, for instance, this constraint translates into subsampling factors ranging from 1 to N. It can be noted that changing this subsampling factor will only produce different synthesis filters; analysis filters remain the same. A subsampling factor of 1 produces nondecimated image decompositions, which are known to yield shift invariance and are highly valuable for image reconstruction problems such as fusion. Larger subsampling factors produce decimated decompositions that in the case of pyramidal multiresolution schemes are computationally efficient. Figure 24.1a shows a Hermite transform decomposition. The original image is decomposed into a number of subimages which consist of a low-pass (approximation) image known as zero-order coefficient $(L_{0,0})$ and a series of high-pass coefficients containing detailed information. Figure 24.1a depicts coefficients of order zero $(L_{0,0})$, up to order three $(L_{2,1})$ and $(L_{1,2})$.

24.2.2 THE STEERED HERMITE TRANSFORM

The Hermite transform has the advantage that high-energy compaction can be obtained through adaptively steering the transform [19,20]. The term "steerable filters" describes a set of filters that are rotated copies of each other, and a copy of the filter in any orientation which is then constructed as a linear combination of a set of basis filters [45]. The resulting transform is self-inverting and translation- and rotation-invariant. Based on the steering property, the Hermite filters at each position in the image adapt to the local orientation content. This adaptability results in significant information compaction. The local rotation into the domain transform can be seen like a mapping of the

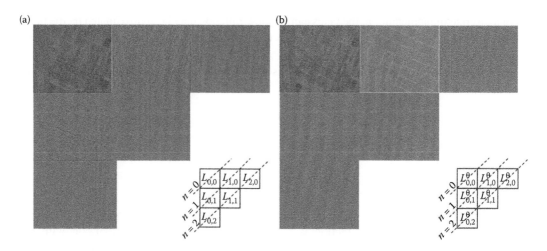

FIGURE 24.1 Decimated Hermite transform. (a) Original Hermite coefficients of image depicted in Figure 24.7a. Diagram shows the coefficient orders. Diagonals depict zero-order coefficients ($n = 0$), first-order coefficients ($n = 1$), and so on. Binomial window of order $N = 2$ and a subsampling factor of 2 were used. (b) Steered coefficients toward the local gradient angle. It can be noted that most coefficient energy is concentrated on the upper row of the steered coefficients and that coefficient $L^\theta_{1,0} = 0$.

expansion coefficients into a local coordinate system whose main axis corresponds to the direction of maximal signal energy. Rotation of the filter functions can be obtained by

$$D_{n-m,m}(x,y,\theta) = \sum_{k=0}^{n} \alpha_m(k,n,\theta) D_{k,n-k}(x,y)$$

where $\alpha_n(k,N,\theta) = s^k c^{-k} \Delta^n \{ C^{k-n}_{N-n} c^{2k-n} s^{N-2k+n} \}$ for $k, n = 0, \ldots, N$; $N \in \mathbb{N}$; and $c = \cos\theta$, $s = \sin\theta$, for $\theta \in [0, 2\pi)$.

Orientation θ of local maximum energy can be estimated by maximizing the coefficient energy measure at each window position. Furthermore, the steered Hermite transform offers a way to estimate one-dimensional (1D) energy,

$$E^{1D}_N(\theta) = \sum_{n=1}^{N} \left[L^\theta_{n,0} \right]^2$$

and 2D energy,

$$E^{2D}_N(\theta) = \sum_{n=1}^{N} \sum_{m=1}^{n} \left[L^\theta_{n-m,m} \right]^2.$$

By analyzing relations between these energies, it is possible to classify image patterns. Figure 24.2 shows a dimensional pattern classification obtained from a natural scene [42].

In practice, the local gradient angle, calculated from the expansion coefficients as $\theta = \arctan L_{0,1}/L_{1,0}$, where $L_{0,1}$ and $L_{1,0}$ are the first-order coefficients of the Hermite transform, can be an alternative estimator of the rotation angle θ. This choice would imply $L^\theta_{1,0} = 0$, as can be noted in Figure 24.1b.

(a)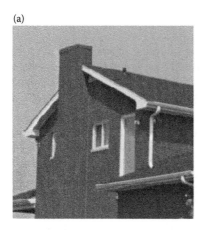

(b)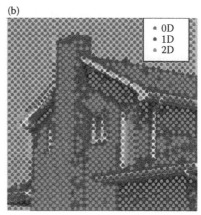

FIGURE 24.2 (a) Original scene. (b) Zero-dimensional (0D), one-dimensional (1D), and two-dimensional (2D) patterns found in scene (a) from dimensional energy analysis with the rotated Hermite Transform.

24.3 FUSION METHOD BASED ON THE HERMITE TRANSFORM

24.3.1 FUSION SCHEME WITH MULTISPECTRAL AND PANCHROMATIC IMAGES

Our objective in image fusion is to generate synthetic images that preserve the higher spatial resolution of the panchromatic (PAN) images while keeping the spectral characteristics of the original multispectral (MS) data.

In a way similar to other fusion techniques, our proposed fusion method requires that the multispectral images be resampled so that their pixel size will be the same as that of the panchromatic image.

The general framework for multispectral and panchromatic image fusion with the Hermite transform consists of several steps [46,47]:

I. Generate new panchromatic bands, whose histograms match each multispectral band's histogram. The purpose of doing this is that each pair of images has the same mean and standard deviation.

II. Perform Hermite transform decomposition over each of the two images, MS and PAN. The decompositions consist of a number of several subimages which represent a low-pass residue known as zero-order coefficients ($L_{0,0}^{MS}$ and $L_{0,0}^{PAN}$) and several high-pass bands containing detailed information coefficients: $L_{0,1}^{MS}$ and $L_{1,0}^{MS}$, and $L_{0,1}^{PAN}$ and $L_{1,0}^{PAN}$ are the first-order coefficients for the MS and PAN images; $L_{0,2}^{MS}$, $L_{1,1}^{MS}$ and $L_{2,0}^{MS}$, $L_{0,2}^{PAN}$, $L_{1,1}^{PAN}$ and $L_{2,0}^{PAN}$ are the second-order coefficients for the MS and PAN images, respectively, and so on, until the highest coefficient order N. For practical reasons, we implemented the discrete Hermite transform which, in fact, as argued before, uses a binomial window function of order N, and corresponding orthogonal polynomials known as Krawtchouck's polynomials. Theoretical and practical issues about this discrete polynomial transform have been well studied [42,43,48]. It is well known that the binomial function of order N approximates a Gaussian function with spread $\sigma = \sqrt{N/2}$. N also represents the maximum order of the transform coefficients since the binomial function has compact support. As explained in the previous section, the subsampling factor, this is, the distance between adjacent window functions is a free parameter. We found just perceivable, but significant, differences in fusion performance when setting this parameter to different values, and chose to fix this parameter to both its maximum allowed value, that is, N, and its minimum value, that is, one pixel. The former provides decimated structures with less computational complexity, while the latter produces nondecimated shift-invariance decompositions.

III. Locally rotate the HT coefficients toward the direction of maximum of energy. As previously explained, rotation angle can also be estimated by the local gradient angle. The steered Hermite transform has the advantage of energy compaction. Transform coefficients are selected with an energy compaction criterion from the steered Hermite transform; therefore, it is possible to reconstruct an image with few coefficients and still preserve details such as edges and textures. Hence, a set of rotated coefficients, that is, $L_{n,m}^{MS\theta}$ and $L_{n,m}^{PAN\theta}$ are obtained for the MS and PAN images, respectively.

IV. Select high-pass transform coefficients from each set (MS and PAN), according to a fusion rule based on the verification of consistency methods (Li et al. 1995) [49]. This approach considers the maximum absolute value within a 5×5 window over the image (area of activity). Increasing the size of the window may cause problems with lower salient patterns. The window variance is used as a measurement of the activity associated with the central pixel of the window. At each window position the maximum selection rule is used so that a significant value indicates the presence of a dominant pattern in the local area. A map of binary decision is then created and subjected to verification of consistency based on a majority filter in order to correct wrong selections. In practice, reconstruction can be achieved with a small number of coefficients, specifically those with high energy of compaction, that is, the upper-row coefficients as shown in Figure 24.1b, this is $L_{1,0}^{MS\theta}$, $L_{2,0}^{MS\theta}$, $L_{3,0}^{MS\theta}$ and $L_{1,0}^{PAN\theta}$, $L_{2,0}^{PAN\theta}$, $L_{3,0}^{PAN\theta}$ for the MS and PAN images, respectively. This is especially true for the case of one-dimensional patterns whose components are all concentrated in these coefficients. Texture patterns may indeed need all coefficients for perfect reconstruction; however, differences are little noticeable.

V. Add these new set of high-pass combined coefficients $L_{1,0}^{fus\theta}$, $L_{2,0}^{fus\theta}$, and $L_{3,0}^{fus\theta}$ to the zero-order coefficient obtained from the MS ($L_{0,0}^{MS}$) image, in order to create a new transform coefficient set that corresponds to the fused image.

VI. Finally, perform an inverse Hermite transform over the new rotated coefficient set.

Steps II through VI are repeated for each multispectral band.

Figure 24.3 shows a scheme of the proposed fusion method. This is executed band by band.

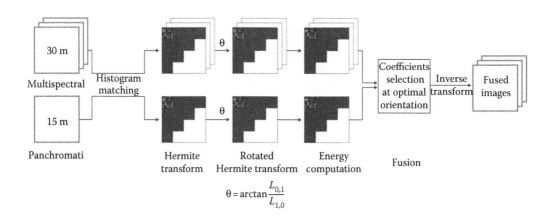

$$\theta = \arctan \frac{L_{0,1}}{L_{1,0}}$$

FIGURE 24.3 Hermite transform fusion method for multispectral and panchromatic images.

24.3.2 Fusion Scheme with Multispectral Images and SAR Image

The use of synthetic aperture radar (SAR) images as a complement to visible and multispectral images is becoming increasingly popular because of their capability of imaging even in the case of cloud-covered remote areas.

Unfortunately, the poor quality of SAR images makes it very difficult to perform direct information extraction tasks. Numerous filters have been proposed to remove speckle in SAR imagery; however, in most cases and even in the most elegant approaches, filtering algorithms have a tendency to smooth speckle as well as information. For numerous applications, low-level processing of SAR images remains a partially unsolved problem.

A wide variety of transform-based methods for speckle reduction have been proposed in the literature. Donoho and Johnstone [50,51] proposed to apply a threshold (T) to the wavelet detail coefficients. This nonlinear technique is fairly simple and implies the use of a binary decision map. Wavelet coefficient handling consists of keeping (or shrinking) and discarding (or killing) the coefficient. There are two alternatives for this scheme, hard thresholding and soft thresholding [50]. Let us assume that U is the coefficient value and D is the resulting coefficient value after thresholding. Hard thresholding is defined by $D(U,T) = U$ for all $|U| > T$, $D(U,T) = 0$ otherwise. This is known to account for a MIN-MAX binary decision solution. Soft thresholding is implemented by $D(U,T) = \text{sgn}(U)\max(0, |U| - T)$. In this case, coefficients are shrunk if their absolute value exceeds the threshold. The universal threshold proposed by Donoho and Johnstone [51] is defined by $T_{\text{universal}} = \sigma_n \sqrt{2\log(M)}$, where M is the sample size, and σ_n is the noise standard deviation. The universal threshold has been recognized as simple and efficient, especially when signal characteristics are unknown; however, it does not allow local adaptation. An alternative way to compute a suitable threshold has been proposed by Chang et al. [52], based on a minimum mean-square error criterion (MMSE). This method, called Bayes–Shrink, computes a signal-dependent threshold as $T = \sigma_n^2/\sigma_x^2$, where σ_n^2 y σ_x^2 are the local noise and signal variance, respectively. An adaptation of this method to speckle reduction with the undecimated wavelet transform was done by Argenti and Alparone [53].

We propose a method for thresholding the Hermite transform coefficients that locally adapts to the mean luminance value, thus compensating the multiplicative nature of speckle. Thresholding occurs only in homogeneous regions. According to the central limit theorem, the probability density function of transformed coefficients belonging to noisy homogeneous regions (i.e., in absence of image structures) approaches a Gaussian distribution. This means that the solution of MIN-MAX and MMSE detection criteria are equivalent. Our approach radically differs in the way noisy edges are treated. In this case, the rotated Hermite transform detects the orientation of edges, so that when edges are reconstructed, only those coefficients aligned with the corresponding edge orientation are included. All other coefficients are set to zero, thus eliminating the structure of speckle on edges, while preserving sharpness.

24.3.3 Noise Reduction with the Hermite Transform

The Hermite transform coefficients can be used to discriminate noise from relevant information such as edges and lines in SAR imagery [54,55]. A binary decision mask containing relevant image locations is built by properly thresholding (T) the first-order transform coefficient energy E_1: $E_1 = L_{0,1}^2 + L_{1,0}^2$ where $L_{0,1}$ and $L_{1,0}$ are the first-order coefficients of the Hermite transform.

As explained before, these coefficients are obtained by convolving the original image with the first-order derivatives of a Gaussian function, which are known to be quasi-optimal edge detectors; therefore, the first-order energy can be used to discriminate edges from noise by means of a threshold scheme.

The mask is then defined by

$$\text{Mask} = \begin{cases} 0 & \text{if } E_1 < T \\ 1 & \text{otherwise} \end{cases}$$

The optimal threshold is set considering two important characteristics of SAR images. First, one-look amplitude SAR images have a Rayleigh distribution and the signal-to-noise ratio (SNR) is ~1.9131. Second, in general, the SNR of multilook SAR images does not change over the whole image.

The threshold is calculated by

$$T = \frac{2\alpha}{\text{SNR}^2 N_{\text{look}}} \ln\left(\frac{1}{\text{Pr}}\right) L_{00}^2$$

- SNR is the signal-to-noise ratio, equal to 1.9131.
- N_{look} is the number of looks of the image.
- $\alpha = \left| R_L(x,y) * D_{1,0}(x,y) * D_{1,0}(-x,-y) \right|_{x=y=0}$, R_L is the normalized autocorrelation function of the input noise, and $D_{1,0}$ is the filter used to calculate the first-order coefficient.
- P_R is the probability (percentage) of noise left on the image and will be set by the user.
- L_{00} is the zero-order Hermite coefficient.

A careful analysis of this expression reveals that this threshold adapts to the local content of the image because of the dependence of σ on the local mean value μ_l, the latter being approximated by the Hermite coefficient L_{00}.

With the locations of relevant edges detected, the next step is to represent these locations as one-dimensional patterns. This can be achieved by steering the Hermite transform as described in the previous section so that the steering angle θ is determined by the local edge orientation. Next, only coefficients $L_{n,0}^{\theta}$ are preserved; all others are set to zero. This strategy is extremely effective for the restoration of noisy edges, since only oriented features are considered for edge reconstruction. An alternative interpretation of this strategy is that edge reconstruction is achieved by projecting edge transform coefficients over a one-dimensional space spanned toward the edge orientation. Two-dimensional structures, as is the case of noise located on edges, are thus eliminated.

In summary, the noise reduction strategy consists of classifying the image in either zero-dimensional patterns consisting of homogeneous noisy regions, or one-dimensional patterns containing noisy edges. The former are represented by the zero-order Hermite transform, that is, the local mean value, and the latter by oriented 1D Hermite coefficients.

When an inverse Hermite transform is performed over these selected coefficients, the resulting synthesized image consists of noise-free sharp edges and smoothed homogeneous regions. Therefore, the denoised image preserves sharpness and thus, image quality. Some speckle remains in the image since there is always a compromise between the degree of noise reduction and the preservation of low-contrast edges. The user controls the balance of this compromise by changing the percentage of noise left P_R on the image according to Equation 24.18. Figure 24.4 shows the algorithm for noise reduction.

24.3.4 Image Fusion

It is easy to figure out that local orientation analysis for the purpose of noise reduction can be combined with an image fusion scheme as the one described in Section 24.4.1. Figure 24.5 shows the complete methodology to reduce noise and fuse Landsat 7 TM with SAR images. After noise

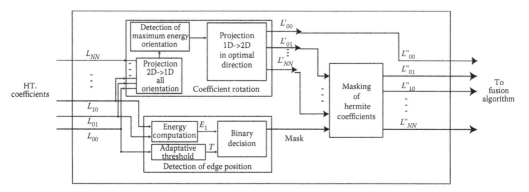

FIGURE 24.4 Noise reduction algorithm.

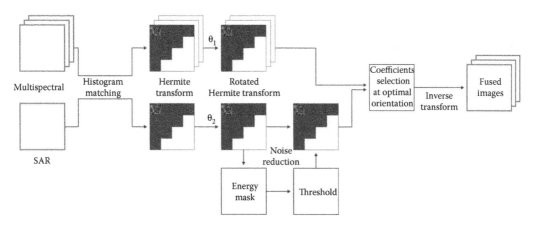

FIGURE 24.5 Noise reduction and fusion for multispectral and SAR images.

reduction is applied to the SAR image, histogram matching is applied on this image to adjust mean values with each MS band. With a similar method as the one previously described, a new coefficient set is generated, consisting of $L_{0,0}^{MS}$ and the detail coefficients selected either from the SAR or MS images. Finally, an inverse transform is performed to obtain the fused image.

24.4 EXPERIMENTAL RESULTS

As explained before, we use a discrete implementation of the Hermite transform based on a binomial filter of order N, and its corresponding orthogonal Krawtchouck's polynomials. We obtained best fusion performance with binomial window functions of orders $N = 2$.

24.4.1 MULTISPECTRAL AND PANCHROMATIC IMAGE FUSION

In this section, we show the results of image fusion with decimated and undecimated versions of the Hermite transform, and compare their performance with the well-known eFIHS [17] and AWL methods [12]; both have become widely accepted since they have overcome several of the limitations of traditional wavelet schemes. The proposed fusion scheme images have been tested on optical data, namely multispectral images from SPOT 5 (10 m spatial resolution and spectral ranges: B1 green 0.50–0.59 μm, B2 red 0.61–0.68 μm, B3 NIR 0.78–0.89 μm spectral range) and its panchromatic band (2.5 m resolution spatial, 0.48–0.71 μm spectral range). They were acquired on January 1, 2007.

Comparison of the proposed methods was based on spectral quality index [10] and spatial quality [9], Moreover, red versus NIR dispersion diagrams were also analyzed which are useful for bio-physical variable interpretation [22,56,57].

24.4.1.1   Visual Quality

Figure 24.6 shows an amplified (zoom) area of a fused image using decimated and undecimated HT schemes. $N = 2$ was used in both cases. It can be clearly seen that image reconstruction is better for the case of the undecimated HT. This is especially true for one-dimensional structures, such as edges and lines. This might be due to the shift invariance property of the undecimated HT. Figure 24.6 also shows a comparison of the HT results with the eFIHS and AWL methods. In this case too, the undecimated HT outperforms the eFIHS since it provides sharper performance, thus higher spatial resolution. This might be explained by the local orientation analysis performed within the

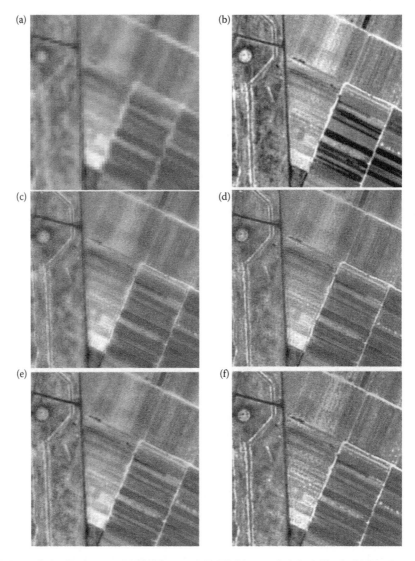

FIGURE 24.6   (**See color insert.**) (a) Multispectral SPOT 5 image (original 10 m). (b) Panchromatic SPOT 5(2.5 m). (c) Fusion result with the decimated HT method ($N = 2$). (d) Fusion result with the undecimated HT ($N = 2$). (e) Fusion with eFIHS. (f) AWL false color composite scheme for displaying SPOT multispectral images is achieved with R = XS3 (NIR band), G = XS2 (red band), and B = XS1 (green band).

HT methodology. The undecimated HT and the AWL methods show similar performance from a subjective point of view. Both present sharp results with similar spatial content.

24.4.1.2 Spectral Quality and Spatial Quality

Digital values were converted to spectral radiance units, measured in watts per square meter per steradian per micron (W/m^2 sr μm).

Spectral quality of a fused image can be assessed with respect to a reference image [2,10]. A new set of low-resolution images were created from the original multispectral and panchromatic images. Spatial resolution was reduced to about half the original value. A Gaussian filter was used for this purpose. The original multispectral image served as reference image B_k. The different fusion methods are then applied to the set of low-resolution images and the resulting images are then compared to reference image B_k.

Different objective quality metrics were calculated for all fused products with respect to the reference image B_k, namely the correlation coefficient, the difference mean value (bias), and the standard deviation of the difference image (sdd). Another quality metric for fused products, the ERGAS, was also calculated. ERGAS stands for *erreur relative globale adimensionnelle de synthese*, and is an estimator of the global spectral quality of fused products:

$$\text{ERGAS} = 100 \frac{h}{l} \sqrt{\frac{1}{N} \sum_{k=1}^{N} \left(\frac{\text{RMSE}(B_k)^2}{(M_k)^2} \right)}$$

(24.6)

where RMSE is the root mean square error, defined by

$$\text{RMSE}(B_k) = \sqrt{(\text{bias})^2 + (\text{sdd})^2},$$

where h is the resolution of the panchromatic image, l the resolution of multispectral images, S the number of spectral bands, B_k the reference spectral image, and M_k the mean value of the reference image B_k.

Ideally, both bias and the difference of standard deviations should tend to null, and the correlation coefficient should tend to one. The closer to zero an ERGAS value the better the image quality. Table 24.1 shows all these quality metrics for all fused methods, including the discrete Hermite transform. Results of Table 24.1 are presented for each multispectral band B1 to B3, in spectral radiance units. It can be noted that the best performance is achieved by the undecimated HT.

TABLE 24.1
Comparison of Spectral Quality Metrics for Different Fusion Methods

Correlation	AWL	eFIHS	HT	uHT
Spectral Quality				
B1	0.9790	0.9735	0.9791	0.9845
B2	0.9890	0.9855	0.9792	0.9846
B3	0.9918	0.9918	0.9634	0.9920
ERGAS	1.9801	2.2071	2.2191	2.0628
Spatial Quality				
B1	0.940	0.943	0.929	0.939
B2	0.934	0.932	0.923	0.945
B3	0.920	0.926	0.919	0.947

In order to estimate the spatial quality, an objective method proposed by Zhou et al. (1988) was used. A Laplacian filter is applied to all fusion products to be compared, as well as to the original panchromatic image. Then, correlation coefficients between each Laplacian-filtered fused image and original panchromatic image are calculated.

In this case too, undecimated HT outperforms decimated HT as well as eFIHS methods. Once again, AWL and undecimated HT perform very similarly with no significant differences in these parameters.

24.4.1.3 Red Versus NIR Dispersion Diagrams

The determination of spectral quality of the fused images is a complex task. Besides the spectral quality indexes presented before, we present next an analysis of parameters that describe soil line, and patterns of absorption/reflectance that represent relations between red and NIR. Due to the linear relation between red and NIR reflectance that describes the soil line, these parameters have been shown to be of major importance for the interpretation of remote sensed data [56]. Some studies refer to the importance of the soil line to extract relevant biophysical variables such as the leaf area index (LAI) [57].

In this work, we obtained the soil line parameters from the red–NIR spectral space as described in Ref. [22].

Soil line was obtained with the formula $\rho_{NIR} = \alpha/\rho_R + \beta$, where ρ_R and ρ_{NIR} are the reflectance in red and near-infrared bands, and α and β are the slope and intercept of the soil line.

Because many investigations have inferred vegetation measurements from near-infrared (NIR) and red data, we analyzed the integrity of information content before and after the fusion procedure. At first glance, little difference in the scatter plots is found between the different fusion methods with respect to the original. However, a deeper analysis shows relevant deviations of the eFIHS and the decimated HT methods. Soil line parameters were extracted from red and NIR data (Table 24.2) and plotted in Figure 24.7. Soil line plots show that undecimated HT and AWL remain very close to the original data, while the decimated HT slightly deviates from the original slope. The eFIHS slope remains close to the original; however, it deviates from its intercept, and its minimum and maximum reflectances also deviate considerably from the original data, meaning more substantial changes in spectra content with respect to the original multispectral data.

24.4.2 Multispectral and SAR Image Fusion

An interesting application of image fusion is to integrate information from different sensors, namely multispectral (MS) and synthetic aperture radar (SAR) images.

Despite the fact that MS and SAR images retrieve different kinds of information, it has been proved that the combination of both sources provides a means to better identify features in the scene. Recently, MS and SAR image data integration has been proposed by Alparone et al. [21].

TABLE 24.2

Soil Line Parameters Extracted from the Red–NIR Space of Different Fusion Methods

Fusion Method	α	β	Minimum Reflectances of Soil	Maximum Reflectances of Soil
Original	1.1758	0.0078	(0.0647,0.0839)	(0.1437,0.1768)
uHT	1.1591	0.0082	(0.0643,0.0818)	(0.1428,0.1727)
HT	1.1195	0.0049	(0.0639,0.0792)	(0.1419,0.1670)
eFIHS	1.1982	0.0036	(0.0581,0.0732)	(0.1604,0.1957)
AWL	1.168	0.0078	(0.06475,0.007811)	(0.1316,0.1682)

Note: α, slope; β, intercept.

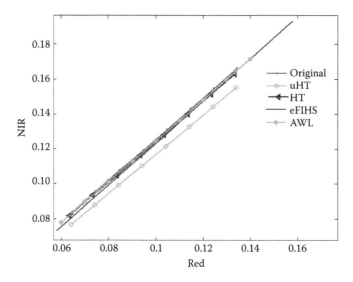

FIGURE 24.7   Soil line in red and NIR bands obtained from different fusion methods.

Here, a generalized intensity modulation (GIM) fusion scheme is proposed in combination with an IHS wavelet à trous transform. This method is capable of fusing MS, P, and SAR images within the same scheme. Detail information from an SAR image is injected in a multiplicative scheme, while P detail information is included within an additive model. This method assumes that noise on the SAR image has been previously reduced.

Our proposed fusion scheme integrates MS and SAR image information. As explained before, the Hermite transform is used with a double purpose, fuse image information and reduce speckle in the SAR image.

Landsat 7 TM 30 m spatial resolution and spectral ranges B1 (0.45–0.52 μm), B2 (0.52–0.60 μm), B3 (0.63–0.69 μm), B4 (0.76–0.90 μm), B5 (1.55–1.76 μm), B7 (2.08–2.35 μm), and SAR AeS-1 (5 m resolution spatial) images were used in this study. Landsat 7 ETM+ data was obtained on January 11, 2001, orbit 2647. AeS-1 data was acquired between October 1998 and January 1999.

We fused both sensor images with the Hermite transform and compared results with the GIM method [21]. GIM did not include a panchromatic image, only the Landsat and Radarsat images were fused as shown in Figure 24.8. Moreover, in order to make a fair comparison, speckle reduction with the Hermite transform was also applied to the SAR AeS-1 image before GIM fusion.

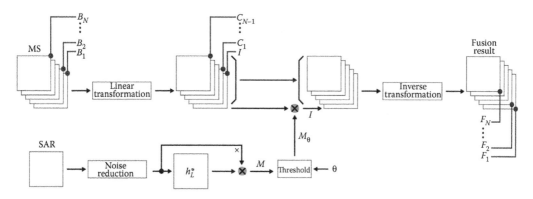

FIGURE 24.8   Modified GIM fusion algorithm.

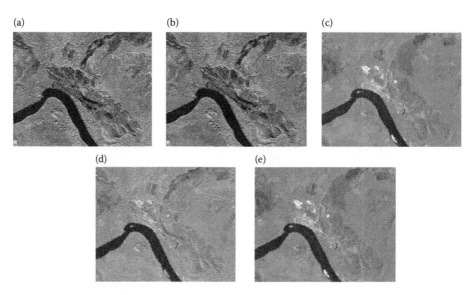

FIGURE 24.9 (See color insert.) (a) Original SAR AeS-1 images with speckle. (b) SAR AeS-1 restored image. (c) Landsat 7 TM image. (d) uHT fused image. (e) GIM fused image.

24.4.2.1 Visual Quality

Figure 24.9 illustrates the result of multispectral and SAR image fusion with the uHT and GIM methods. SAR images contain higher spatial resolution and more texture than MS images. In contrast, the former contain spectral information. Fusion results show that all these properties can be well incorporated into a single product. uHT fused images as well as GIM-fused images show significant spatial resolution improvement with respect to the original MS image, while no apparent spectral information is lost. The uHT, however, shows sharper results, better image structure reconstruction, and more natural texture content. This can be easily noted in the airport landing track and in the river contours.

24.4.2.2 Spectral Quality

The spectral angle mapper (SAM) is a measure of the spectral distortion and is defined as the absolute angle between the two vectors: $SAM(v, \hat{v}) = \arccos(\langle v, \hat{v}\rangle / \| v \|_2 \cdot \| \hat{v} \|_2)$, where $v = \{v_1, v_2, \ldots v_N\}$ is the original spectral vector and $\hat{v} = \{\hat{v}_1, \hat{v}_2, \ldots \hat{v}_N\}$ is the spectral vector obtained after fusion. Table 24.3 shows that the undecimated HT presents a 0.5° larger SAM than GIM.

24.4.2.3 Classification

In order to evaluate the algorithm performance in real applications, original and fused images were classified with the ISODATA algorithm in four different classes. Figure 24.10 shows the results of this experiment. Deforested and vegetation classified areas are difficult to evaluate, since a ground

TABLE 24.3

Spectral Distortion

Fusion Methods	uHT	GIM	Ideal
SAM	5.41°	4.89°	0°

Note: Average SAM between resampled original and fused image for both uHT and GIM methods.

(a)   (b)   (c)

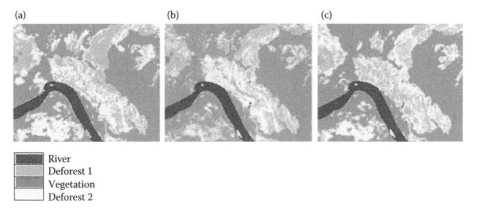

■ River
  Deforest 1
  Vegetation
□ Deforest 2

FIGURE 24.10 **(See color insert.)** Four-class ISODATA classification applied to (a) original MS image, (b) uHT SAR-MS fused image, and (c) GIM fused image.

truth is not available; however, the river and the airport landing track are easily identified in the fused images. In both cases, the uHT transform clearly separates classes more efficiently than GIM. It is important to note that the landing track high-resolution image primitives were not present in the original MS image but were injected from the SAR image.

24.5 CONCLUSIONS

In this chapter, we presented the Hermite transform as an efficient tool for image fusion in remote sensed data. The use of Gaussian derivatives as basis functions of the HT makes this transform especially suitable to represent relevant image structures such as edges. Moreover, the rotation property of the Hermite transform presented here is an important feature that allows detecting the orientation of relevant image structures. This translates into an energy compaction into few coefficients of the transform. Furthermore, the local orientation property of the HT is a key factor for the reconstruction of oriented patterns. We profit from this property in the proposed speckle reduction algorithm for SAR images. While noise in homogeneous regions is reduced by means of a local adaptive threshold scheme, noise present on edges is reduced by reconstructing them from transform coefficients that are oriented toward the proper edge direction. Another very important feature of the Hermite transform is the freedom to choose the subsampling factor used to compute the transform coefficients, the analysis window length being the only constraint. This implies that decimated and undecimated Hermite transform structures can be very easily generated by just changing a single parameter. This is especially relevant for the construction of shift invariant image analysis structures.

We presented two fusion methodologies, one for multispectral and panchromatic images, and the other for multispectral and SAR images. In the first case, fused products showed how to efficiently preserve the higher spatial resolution of the P image and the spectral content of the MS image. Evaluation was performed taking into account visual quality, spectral quality, and biophysical variable integrity. In all cases, the undecimated Hermite transform outperformed the eFIHS method and performed as well as AWL, both being two of the most referenced fusion methods in the recent literature. The proposed scheme for fusion between MS and SAR images also shows very good performance, since the higher resolution and relevant texture of the SAR image is incorporated into the MS image without losing spectral integrity. Noise reduction is a key factor in this case, since it is of extreme importance not to incorporate spurious information to the fused product. Comparison with the generalized intensity modulation fusion algorithm shows a better performance of the undecimated Hermite transform in terms of visual quality and spatial reconstruction. Analysis by means of the spectral angle mapper (SAM) shows that the Hermite transform method preserves the original spectral content with only a slightly larger spectral distortion in comparison with the GIM method. Moreover, an ISODATA classification experiment applied on the fused products confirms that the HT fusion method has the ability to better identify and separate classes in high-resolution image structures that were incorporated from the SAR image.

ACKNOWLEDGMENTS

This work was sponsored by UNAM PAPIIT grants IN113611 and IX100610, and the Centro de Investigación en Geografía y Geomática "Ing. Jorge L. Tamayo."

REFERENCES

1. C. Pohl and J. L. Van Genderen, Multisensor image fusion in remote sensing: Concepts, methods and applications, *International Journal of Remote Sensing*, 19(5), 823–854, 1998.
2. L. Wald, *Data Fusion Definitions and Architectures*, École des Mines de Paris, Paris, 2002.
3. G. Piella, A general framework for multiresolution image fusion from pixels to regions, *Information Fusion*, 4, 259–280, 2003.
4. W. J. Carper, T. M. Lillesand, and R. W. Kiefer, The use of intensity-hue-saturation transform for merging SPOT panchromatic and multi-spectral image data, *Photogrammetric Engineering and Remote Sensing*, 56(4), 459–467, 1990.
5. P. S. Chavez and J. A. Bowell, Comparison of the spectral information content of Landsat thematic mapper and SPOT for three different sites in the Phoenix, Arizona region, *Photogrammetric Engineering and Remote Sensing*, 54(12), 1699–1708, 1988.
6. P. S. Chavez and A. Y. Kwarteng, Extracting spectral contrast in Landsat thematic mapper image data using selective principal component analysis, *Photogrammetric Engineering and Remote Sensing*, 55(3), 339–348, 1989.
7. Z. Wang, D. Ziou, C. Armenakis, D. Li, and Q. Li. A comparative analysis of image fusion methods, *IEEE Transactions on Geoscience and Remote Sensing*, 43(6), 1391–1402, 2005.
8. D. A. Yocky, Image merging and data fusion by means of the two dimensional wavelet transform, *Journal of the Optical Society of America A*, 12(9), 1834–1841, 1995.
9. J. Zhou, D. L. Civco, and J. A. Silander, A wavelet transform method to merge Landsat TM and SPOT panchromatic data, *International Journal of Remote Sensing*, 19, 743–757, 1998.
10. T. Ranchin and L. Wald, Fusion of high spatial and spectral resolution images: The ARSIS concept and its implementation, *Photogrammetric Engineering and Remote Sensing*, 66, 49–61, 2000.
11. B. Aiazzi, L. Alparone, S. Baronti, and A. Garzelli, Context-driven fusion of high spatial and spectral resolution images based on oversampled multi-resolution analysis, *IEEE Transactions on Geoscience and Remote Sensing*, 40(10), 2300–2312, 2002.
12. J. Núñez, X. Otazu, O. Fors, A. Prades, and R. Arbiol, Multiresolution-based image fusion with additive wavelet decomposition, *IEEE Transactions on Geoscience and Remote Sensing*, 37(3), 1204–1211, 1999.
13. S. G. Mallat, A theory for multiresolution signal decomposition: The wavelets representation, *Transactions on Pattern Analysis and Machine Intelligence*, 11(7), 674–693, 1989.
14. M. Holschneider, R. Kronland-Martinet, J. Morlet, and P. Tchamitchian, A real-time algorithm for signal analysis with the help of wavelet transform, in *Wavelets, Time-Frequency Methods and Phase Space,* J. M. Combes, A. Grossmann, and P. Tchamitchian, editors. Springer-Verlag, Berlin, Germany, pp. 289–297, 1989.
15. M. J. Shensa, The discrete wavelet transform: Wedding the à trous and Mallat algorithm, *IEEE Transactions on Signal Processing*, 40, 2464–2482, 1992.
16. M. González-Audícana, J. L. Saleta, R. García Catalán, and R. García, Fusion of multispectral and panchromatic images using improved IHS and PCA mergers base don Wavelet decomposition, *IEEE Transactions on Geoscience and Remote Sensing*, 42(6), 1291–1299, 2004.
17. M. González-Audícana, X. Otazu, O. Fors, and J. A. Alvarez-Mozos, A low computational-cost method to fuse IKONOS images using the spectral response fuction of its sensors, *IEEE Transactions on Geoscience and Remote Sensing*, 44(6), 1683–1691, 2006.
18. F. Nencini, A. Garzelli, S. Baronti, and Luciano Alparone, Remote sensing image fusion using the curvelet transform, *Information Fusion*, 8, 143–156, 2007.

19. J. B. Martens, Local orientation analysis in images by means of the Hermite transform, *IEEE Transactions on Image Processing*, 6(8), 1103–1116, 1997.
20. A. M. Van Dijk and J. B. Martens, Image representation and compression with steered Hermite transform, *Signal Processing*, 56, 1–16, 1997.
21. Alparone L., Baronti S., Garzelli A., and Nencini F., Landsat ETM+ and SAR image fusion based on generalized intensity modulation, *IEEE Transactions on Geoscience and Remote Sensing*, 42(12), 2832–2839, 2004.
22. H. Fang and S. Liang, Retrieving leaf area index with a neural network method: Simulation and validation, *IEEE Transactions Geoscience and Remote Sensing*, 41(9), 2052–2062, 2003.
23. A. Witkin, Scale-space filtering: A new approach to multiscale description, *Image Understand*, 3, 79–95, 1984.
24. T. Lindeberg, *Scale-Scale Theory in Computer Vision*, Kluwer Academic, Boston, MA, 1994.
25. J. J. Koenderink and A. J. Van Doorn, Receptive field families, *Biological Cybernetics*, 63, 291–297, 1990.
26. S. Marcelja, Mathematical description of the responses of simple cortical cells, *Journal of the Optical Society of America.*, 70, 1297–1300, 1980.
27. J. G. Daugman, Two-dimensional spectral analysis of cortical receptive fields profiles, *Vision Research*, 20, 847–856, 1980.
28. M. Bastiaans, Gabor's signal expansion and degrees of freedom of a signal, *Optica Acta*, 29, 1223–1229, 1982.
29. M. Porat and M. Zeevi, The generalized gabor scheme of image representation in biological and machine vision, *IEEE Transactions* on *Pattern Analysis and Machine Intelligence*, 10(4), 454–467, 1988.
30. D. G. Stork and H. R. Wilson, Do Gabor functions provide appropriate descriptions of visual cortical receptive fields?, *Journal of the Optical Society of America A*, 7(8), 1362–1373, 1990.
31. R. A. Young, The Gaussian derivative theory of spatial vision: Analysis of cortical cell receptive field line-weighting profiles, Technical Report GMR-4920, General Motors Research Laboratories, Computer Science Department, Warren MI 48090, 1985.
32. R. A. Young, R. M. Lesperance, and W. W. Meyer, The Gaussian model for spatial-temporal vision: I. Cortical model, *Spatial Vision*, 14(3,4), 261–319, 2001.
33. R. A. Young, "Oh say, can you see?," The physiology of vision, in *Human Vision, Visual Processing, and Digital Display II*, 1453 in *Proc. SPIE*, 92–123, 1991.
34. A. C. den Brinker and J. A. J. Roufs, Evidence for a generalized Laguerre transform of temporal events by the visual system, *Biological Cybernetics*, 67(5), 395–402, 1992.
35. Y. Jian and A. Reeves, Visual pattern encoding with weighted Hermite polynomials *Spatial Vision*, 14(3–4), 391–412, 2004.
36. Y. Jian and A. Reeves, Bottom-up visual image processing probed with weighted Hermite polynomials, *Neural Networks*, 8(5), 669–691, 1995.
37. L. M. J. Florack, B. M. ter Haar Romeny, J. J. Koenderink, and M. A. Viergever, Scale and the differential structure of images, *Image and Vision Computing*, 10(6), 376–388, 1992.
38. J. J. Koenderink and A. J. V. Doorn, Representation of local geometry in the visual system, *Biological Cybernetics*, 55, 367–375, 1987.
39. J. J. Koenderink and A. J. V. Doorn, Generic neighborhood operators, *IEEE Transactions* on *Pattern Analysis and Machine Intelligence*, 14(6), 597–605, 1992.
40. J. B. Martens, The hermite transform—Theory, *IEEE Transactions on Acoustics, Speech and Signal Processing*, 38(9), 1607–1618, 1990.
41. J. B. Martens, The Hermite transform—Applications, *IEEE Transactions on Acoustics, Speech and Signal Processing*, 38(9), 1595–1606, 1990.
42. J. L. Silván-Cárdenas and B. Escalante-Ramírez, The multiscale Hermite transform for local orientation analysis, *IEEE Transactions on Image Processing*, 15(5), 1236–1253, 2006.
43. B. Escalante-Ramírez and J. L. Silván-Cárdenas, Advanced modeling of visual information processing: A multirresolution directional-oriented image transform based on Gaussian derivatives, *Signal Process: Image Communication*, 20(9–10), 801–812, 2005.
44. G. Szegö, *Orthogonal Polynomials*, American Mathematical Society, Colloquium Publications, 1959.
45. W. T. Freeman and E. H. Adelson, The design and use of steerable filters, *IEEE Transactions on Pattern Analysis and Machine Intelligence*, 13(9), 891–906, 1991.
46. B. Escalante-Ramírez and A. López-Caloca, Image fusion with the Hermite transform, in *Conf. ICIP 2003, International Conference on Image Processing* 14th, Barcelona, Spain.

47. B. Escalante-Ramírez and A. López-Caloca, The Hermite transform: An efficient tool for noise reduction and image fusion in remote-sensing, in *Signal and Image Processing for Remote Sensing*, C. H. Chen (ed.), Taylor & Francis, Boca Raton, FL, 2006.

48. M. Hashimoto and J. Sklansky, Multiple-order derivatives for detecting local image characteristics, *Computer Vision Graphics and Image Processing*, 39, 28–55, 1987.

49. H. Li, B. S. Manjunath, and S. K. Mitra, Multisensor image fusion using the wavelet transform, *Graphical Models and Image Processing*, 57(3), 235–245, 1995.

50. D. L. Donoho, Denoising by soft-thresholding, *IEEE Transactions on Information Theory*, 41, 613–627, 1995.

51. D. L. Donoho and I. M. Johnstone, Ideal spatial adaptation via wavelet shrinkage, *Biometrika*, 81, 425–455, 1994.

52. S. G. Chang, B. Yu, and M. Vetterli, Spatially adaptive wavelet thresholding with context modeling for image denoising, *IEEE Transactions on Image Processing*, 9(9), 1522–1531, 2000.

53. F. Argenti and L. Alparone, Speckle removal from SAR images in the undecimated wavelet domain, *IEEE Transactions on Geoscience and Remote Sensing*, 40(11), 2363–2373, 2002.

54. B. Escalante-Ramírez and J. Lira-Chávez, Performance-oriented analysis and evaluation of modern adaptive speckle reduction techniques in SAR images, *Visual Information Processing V*. SPIE-2753, 1996.

55. B. Escalante-Ramírez and J. B. Martens, Noise reduction in computerized tomography images by means of polynomial transforms, *Journal of Visual Communication and Image Representation*, 3(3), 272–285, 1992.

56. J. C. Price, Estimating leaf area index from satellite data, *IEEE Transactions on Geoscience and Remote Sensing*, 31(3), 727–734, 1993.

57. F. Baret, S. Jacquemoud and J. F. Hanocq, The soil concept in remote sensing, *Remote Sensing*, 7, 65–82, 1993.

25 Wavelet-Based Multi/ Hyperspectral Image Restoration and Fusion

Paul Scheunders, Arno Duijster, and Yifan Zhang

CONTENTS

25.1 INTRODUCTION

With the evolution of imaging technology, an increasing number of imaging modalities becomes available. In remote sensing, sensors are available that can generate multispectral or hyperspectral data, involving from a few to more than hundred bands. This increase in spectral accuracy is delivering more information, allowing a whole range of new and more precise applications. The detailed spectral information contained in hyperspectral images is helpful for interpretation, classification, and recognition purposes.

In optical sensors, usually a trade-off exists between signal to noise ratio (SNR), spatial and spectral resolutions. Also, remote sensing images are subject to degradation caused by atmospheric effects and physical limitations of the sensors. In order to recover the original scene, restoration of the images is required.

During recent years, several restoration techniques were developed for multi- and hyperspectral imagery. The most straightforward way of extending graylevel image processing techniques toward-multispectral images is to process each image band separately. However, in this way, the high redundancy between image bands is not accounted for. Since, there exists a high correlation between the bands, multiband processing is more advantageous.

Many restoration techniques rely on a model for the image probability density function (pdf). Such a model can be applied as prior information in a Bayesian framework for the purpose of image restoration. When models are applied for the image pdf, it may be advantageous to switch to another image representation. The wavelet domain has been a popular image representation because of some very interesting advantages. First, the wavelet domain provides a sparse representation, which makes it useful for separation of signal and noise for restoration purposes. Another major advantage of the wavelet transform is that it tends to spatially decorrelate the image information, which makes it dramatically more efficient for estimation of the model parameters. A model is then applied on the wavelet detail images. The most common model to be applied is the zero-mean Gaussian model. However, typically, the sparse wavelet detail images are better modeled by zero-mean heavy-tailed models. In the literature, models such as Laplacians, generalized Gaussians or Gaussian Scale Mixture (GSM) models were successfully applied.

When taking all previous consideration into account, multi- and hyperspectral images are best processed by a multiband procedure, requiring a multivariate model. When applied in the wavelet domain, a heavy-tailed multivariate pdf is best applied. In this chapter, we will investigate the use of such multivariate pdf's for modeling multi- and hyperspectral images. We will demonstrate their use in the application domains of image restoration and fusion.

In the next section, the wavelet-based framework is elaborated. In Section 25.3, it is applied to the problem of multi/hyperspectral image denoising. In Section 25.4, the problem of multi/hyperspectral image restoration is treated, and in Section 25.5, multi- and hyperspectral image fusion is elaborated.

25.2   THE WAVELET-BASED RESTORATION FRAMEWORK

25.2.1   Nondecimated Wavelet Transform

The wavelet transform reorganizes image content into a low-resolution approximation and a set of details of different orientations and different resolution scales. A fast algorithm for the discrete wavelet transform is an *iterative filter bank algorithm* of Mallat [24], where a pair of high-pass and low-pass filters followed by downsampling by two is iterated on the low-pass output. The outputs of the low-pass filter are the *scaling* coefficients and the outputs of the high-pass filter are the *wavelet* coefficients. At each decomposition level, the filter bank is applied sequentially to the rows and to the columns of the image. Low-pass filtering of both the rows and the columns yields the low-pass LL subband and other combinations of low-pass and high-pass filtering yield the wavelet subbands at different orientations: High-pass filtering of rows and low-pass filtering of columns (HL) yields horizontal edges and the opposite combination (LH) yields vertical edges, while high-pass filtering of both the rows and the columns (HH) yields diagonal edges. The jth decomposition level yields the coefficients at the *resolution scale* 2^j.

In image denoising, redundant wavelet transforms, like the nondecimated transform yield better results than the critically sampled one. In a *nondecimated* wavelet transform downsampling is excluded, and instead the filters are upsampled at each decomposition stage. In this chapter, we use a nondecimated wavelet transform implemented with the algorithm *à trous* [25]. The algorithm inserts $2^j - 1$ zeroes (i.e., "holes," French *trous*) between the filter coefficients at the resolution level j. The size of each wavelet subband equals the size of the input image.

For compactness, denote the spatial position vector $[m, n]$ by a single index l, and denote the scaling coefficients at the resolution level j by $a_l^{(j)}$ and the wavelet coefficients at the corresponding scale in three orientation subbands by $x_l^{(j,HL)}, x_l^{(j,LH)}$, and $x_l^{(j,HH)}$. Let $\mathbf{h}$ and $\mathbf{g}$ denote, respectively, the low-pass and the high-pass filters associated with the wavelet function ψ, and let $\bar{h}$ denote the conjugate complex of h, and $\mathbf{h}^j$ an up-sampled filter, where $2^j - 1$ zeroes are inserted between each two coefficients of $\mathbf{h}$. Denoting the discrete convolution by $*$, the nondecimated wavelet decomposition is formally given by

$$a_l^{(j+1)} = \overline{h^j h^j}_l * a_l^{(j)}$$

$$x_l^{(j+1,HL)} = \overline{g^j h^j}_l * a_l^{(j)}$$

$$x_l^{(j+1,LH)} = \overline{h^j g^j}_l * a_l^{(j)} \qquad (25.1)$$

$$x_l^{(j+1,HH)} = \overline{g^j g^j}_l * a_l^{(j)}.$$

Decomposing an image into J decomposition levels yields a wavelet image representation consisting of $3J + 1$ subbands: $[\mathbf{a}^{(J)}, \{\mathbf{x}^{(j,HL)}, \mathbf{x}^{(j,LH)}, \mathbf{x}^{(j,HH)}\}_{1 \le j \le J}]$. A reconstruction of the input image is obtained as

$$a_l^{(0)} = h^J h_l^J * a_l^{(J+1)} + \sum_{j=1}^{J} \left(g^j h_l^j * x_l^{(j+1,HL)} + h^j g_l^j * x_l^{(j+1,LH)} \right. \qquad (25.2)$$

$$\left. + g^j g_l^j * x_l^{(j+1,HH)} \right)$$

where the input image is approximated by $\mathbf{a}^{(0)}$ with a negligible error [25].

25.2.2 Wavelet Processing of Multi/Hyperspectral Images

In this work, all image bands of a multi/hyperspectral image are wavelet transformed separately. Let $x_l^{(j,o,b)}$ denote the noise-free wavelet coefficient at *spatial* position l, *resolution* level j, *orientation* subband o, and *image band* b. A vector processing approach groups the wavelet coefficients $x_l^{(j,o,b)}$ of all the B bands at a given spatial position, within a subband of a given orientation and resolution level into a B-dimensional vector,

$$\mathbf{x}_l^{(j,o)} = [x_l^{(j,o,1)}, \dots, x_l^{(j,o,B)}]^{\mathrm{T}}. \qquad (25.3)$$

Equivalent processing is typically applied to all the wavelet detail subbands, and hence we shall omit the indexes that denote the resolution level j and orientation o.

25.2.3 Imaging Model

The observed multi/hyperspectral image $\mathbf{X}$ with M pixels and B bands is denoted as $\mathbf{X} = [\mathbf{X}_1^{\mathrm{T}}, \mathbf{X}_2^{\mathrm{T}}, \dots, \mathbf{X}_M^{\mathrm{T}}]^{\mathrm{T}}$ where $\mathbf{X}_m = [X_{1,m}, X_{2,m}, \dots, X_{B,m}]^{T}$ expresses the spectral response at each spatial position m. $\mathbf{X}$ consists of an unknown signal $\mathbf{S}$, which is first degraded with a known impulse response, represented by a linear system $\mathbf{H}$, and then corrupted by some additive noise

$$\mathbf{X} = \mathbf{HS} + \mathbf{N}. \qquad (25.4)$$

Remark that the degradation is not limited to be purely spatial, which means that blurring in the spectral direction is allowed as well. Also, in general, the impulse response is not necessarily spatial or spectral invariant, but is allowed to vary from position to position. Usually, the operator is limited to be spatial invariant to reduce the complexity. We will do likewise. $\mathbf{N}$ is assumed to be zero-mean white Gaussian noise with covariance $\mathbf{C_N}$, with a pdf denoted by $p(\mathbf{N}) = \phi(0, \mathbf{C_N})$. The noise is assumed to be translation invariant, the covariance describes its spectral variation.

In the wavelet domain, one obtains,

$$\mathbf{x} = \mathcal{H}\mathbf{s} + \mathbf{n} \tag{25.5}$$

Here, $\mathcal{H}$ is obtained from the operator $\mathbf{H}$ in the image domain. With the orthogonal wavelet families [25] that we apply in this chapter, the wavelet transform $\mathbf{U}$ is a unitary transform, and thus $\mathcal{H} = \mathbf{U}\mathbf{H}\mathbf{U}^\mathsf{T}$.

The pdf of $\mathbf{n}$ is assumed to be a multivariate Gaussian of zero mean and covariance matrix $\mathbf{C_n}$: $p(\mathbf{n}) = \phi(\mathbf{n}; \mathbf{C_n})$. The noise covariance in each wavelet subband is in general a scaled version of the input image noise covariance, where the scaling factors depend on the wavelet filter coefficients (see, e.g., [21]). With the orthogonal wavelet families [25], which we apply, the noise covariance in all the wavelet subbands is equal to the input image noise covariance. In most cases, we will assume that the input image noise covariance is known. When conducting experiments with real noisy data, the noise covariance will be estimated separately.

25.2.4 ESTIMATION APPROACH AND OPTIMIZATION CRITERION

Various linear and nonlinear (adaptive) methods can be applied for data restoration. We focus on the *Bayesian* approach, where *a priori* knowledge about the distribution of $\mathbf{s}$ is assumed. In particular, we impose a multicomponent prior distribution (to be called hereafter *prior*) on the wavelet coefficients in a given subband and we differentiate between several approaches based on different specific priors.

As an optimization criterion, the *minimization of the mean squared error* can be adopted, where the Bayesian risk is a quadratic loss function. Estimation that uses this optimization criterion is referred to as least-squares estimation. The minimum mean-squared error (MMSE) estimate is the posterior conditional mean,

$$\mathbb{E}(\mathbf{s}|\mathbf{x}) = \int_{-\infty}^{\infty} \mathbf{s}p(\mathbf{s}|\mathbf{x})d\mathbf{x} = \frac{\int_{-\infty}^{\infty} \mathbf{s}p(\mathbf{x}|\mathbf{s})p(\mathbf{s})\,d\mathbf{s}}{\int_{-\infty}^{\infty} p(\mathbf{x}|\mathbf{s})p(\mathbf{s})\,d\mathbf{s}}. \tag{25.6}$$

In the case of denoising, for example, where the imaging model is given by

$$\mathbf{x} = \mathbf{s} + \mathbf{n} \tag{25.7}$$

the MMSE becomes,

$$\mathbb{E}(\mathbf{s}|\mathbf{x}) = \frac{\int_{-\infty}^{\infty} \mathbf{s}\phi(\mathbf{x} - \mathbf{s}; \mathbf{C_n})p(\mathbf{s})\,d\mathbf{s}}{\int_{-\infty}^{\infty} \phi(\mathbf{x} - \mathbf{s}; \mathbf{C_n})p(\mathbf{s})\,d\mathbf{s}}. \tag{25.8}$$

Assuming, for example, a Gaussian prior for the noise-free signal $p(\mathbf{s}) = \phi(\mathbf{s}; \mathbf{C_s})$, the above MMSE estimate becomes the Wiener filter

$$\hat{\mathbf{s}} = \mathbb{E}(\mathbf{s}|\mathbf{x}) = \mathbf{C_s}(\mathbf{C_s} + \mathbf{C_n})^{-1}\mathbf{x}. \tag{25.9}$$

Another estimator is the Maximum A Posteriori (MAP) estimator:

$$\hat{\mathbf{s}} = \arg\max_{s} p(\mathbf{s}|\mathbf{x}) \qquad (25.10)$$

which leads to the same result as the MMSE in the case of denoising using a Gaussian prior.

25.3 MULTI/HYPERSPECTRAL IMAGE DENOISING USING THE WAVELET-BASED FRAMEWORK

25.3.1 STATE-OF-THE-ART

Multi/hyperspectral image noise is usually treated as stochastic Gaussian distributed, where the noise in the different bands is not necessarily independent. Generally, in remote sensing multispectral noise removal is achieved by a transform, referred to as minimum noise fraction (MNF) [23]. MNF contains two principal component transformations. The first one diagonalizes the noise covariance, the second one decorrelates the noise-whitened data. MNF only uses spectral information for denoising. On the other hand, spatial smoothing of each spectral band separately is also a common practice in multispectral noise reduction. Recently, two-step approaches have been suggested, where spectral decorrelation of the noise is combined with band-wise denoising [30].

In this work, we aim at jointly denoising all bands of a multi/hyperspectral image, accounting for the interband correlations. We will focus on spatial wavelet denoising. The wavelet transform offers an efficient representation of spatial discontinuities within each spectral band [10,24,40]. It compresses the essential information of an image into a relatively few, large coefficients coinciding with the positions of image discontinuities. Such a representation naturally facilitates the construction of *spatially adaptive* denoising methods that can smooth noise without excessive blurring of image details. Typically, noise is reduced by *shrinking* the noisy wavelet coefficient magnitudes. Ideally, the coefficients that contain primarily noise (usually the smallest coefficients) are reduced to negligible values while the ones containing a "significant" noise-free component are reduced less [12]. Standard wavelet thresholding [12] treats the coefficients with magnitudes below a certain threshold as "nonsignificant" and sets these to zero; the remaining, "significant" coefficients are kept unmodified (hard-thresholding) or reduced in magnitude (soft-thresholding). Shrinkage estimators can also result from a *Bayesian* approach [1,9,26,32,33,37], which imposes a prior distribution on noise-free data. Common priors for noise-free data include (generalized) Laplacians [9,24,27], alpha-stable models [1], double stochastic (GSM) models [26,33], and mixtures of two distributions [41] where one distribution models the statistics of "significant" coefficients and the other one models the statistics of "insignificant" data.

Recently, several wavelet-based procedures for multicomponent images were proposed that account, to some extent, for the intercomponent correlations, applying Bayesian estimation, using different prior models [5,31,35]. In [11], a Bayesian framework was presented for wavelet-based denoising of multicomponent images to (1) fully account for the intercomponent covariances, and (2) use different prior models that optimally approximate the marginal densities of the wavelet coefficients. Within this framework, three different prior models were reviewed: the Bernouilli–Gaussian model, the GSM and the Generalized Gaussian models. In [34], the GSM model for denoising of multispectral images was elaborated. In the next section, we will briefly outline the followed procedure and demonstrate it using the GSM model.

25.3.2 GSM DENOISING

The standard MMSE denoising result (25.10) of the previous section is the result obtained using a multicomponent Gaussian prior model. It accounts for the multicomponent covariance, but it assumes that the marginal densities for the wavelet coefficients are Gaussian. It is well known that this assumption is not justified, and that these marginals are symmetric and zero mean, but heavier tailed than Gaussians. Different other priors were proposed to better approximate the marginal densities. In this section we present the GSM prior model and apply it within the MMSE estimation framework of the previous section.

The GSM prior [33] models the pdf $p(\mathbf{s})$ by a mixture of Gaussians

$$p(\mathbf{s}) = \int p(\mathbf{s}|z)p(z)\,dz \tag{25.11}$$

where $p(z)$ is the mixing density, and $p(\mathbf{s}|z)$ is a zero-mean Gaussian with covariance $\mathbf{C}_{\mathbf{s}|z}$. Under the GSM model, the additive noise model (25.7) becomes

$$\mathbf{x} = \mathbf{s} + \mathbf{n} = \sqrt{z}\mathbf{u} + \mathbf{n} \tag{25.12}$$

where both $\mathbf{u}$ and $\mathbf{n}$ are zero-mean Gaussians, with covariances given by $\mathbf{C}_{\mathbf{u}}$ and $\mathbf{C}_{\mathbf{n}}$ respectively. Then, $\mathbf{C}_{\mathbf{s}|z} = z\mathbf{C}_{\mathbf{u}}$ or, by taking expectations over z, with $\mathbb{E}(z) = 1$: $\mathbf{C}_{\mathbf{s}} = \mathbf{C}_{\mathbf{u}}$.

GSM densities are symmetric and zero-mean and heavier tailed than Gaussians. These are known to better model the shape of the wavelet coefficient marginals than Gaussians. In [33], GSMs were applied to model local spatial neighborhoods of wavelet coefficients in gray-level images. In this work, we apply the GSM to model multicomponent wavelet coefficients. In this way, the prior fully accounts for the interband covariances.

The Bayes least-squares estimate $E(\mathbf{s}|\mathbf{x})$ is given by

$$\mathbb{E}(\mathbf{s}|\mathbf{x}) = \int \mathbf{s}p(\mathbf{s}|\mathbf{x})\,d\mathbf{s}$$

$$= \int\int_0^\infty \mathbf{s}p(\mathbf{s}, z|\mathbf{x})\,dz\,d\mathbf{s}$$

$$= \int\int_0^\infty \mathbf{s}p(\mathbf{s}|\mathbf{x}, z)p(z \mid \mathbf{x})\,dz\,d\mathbf{s} \tag{25.13}$$

$$= \int_0^\infty p(z|\mathbf{x})\mathbb{E}(\mathbf{s}|\mathbf{x}, z)\,dz.$$

Since, using the GSM model $\mathbf{s}$, conditioned on z is Gaussian, the expected value within the integral is given by a Wiener estimate

$$\mathbb{E}(\mathbf{s}|\mathbf{x}, z) = z\mathbf{C}_{\mathbf{u}}(z\mathbf{C}_{\mathbf{u}} + \mathbf{C}_{\mathbf{n}})^{-1}\mathbf{x}. \tag{25.14}$$

The posterior distribution of z can be obtained, using Bayes' rule

$$p(z|\mathbf{x}) = \frac{p(\mathbf{x}|z)p(z)}{\displaystyle\int_0^\infty p(\mathbf{x}|\alpha)p(\alpha)\,d\alpha} \tag{25.15}$$

with $p(\mathbf{x}|z) = \phi(\mathbf{x}; z\mathbf{C}_{\mathbf{u}} + \mathbf{C}_{\mathbf{n}})$. In [33], the authors motivate the use of the so-called *Jeffrey's prior* [4] for the random multiplier z: $p(z) \propto 1/z$. We refer to [33] and [34] for further information about the practical implementation.

25.3.3 Experiments

In the experiment, we will demonstrate the proposed denoising technique on a multispectral image. For this, a seven-band Landsat TM image was taken over the Winnipeg area containing both rural as urban areas. From this image, the thermal band (band 6) was removed and six spectral bands were corrupted with additive Gaussian noise. The noise was chosen independently for each band: $C_n = \sigma_n^2 I$ with different standard deviations. The results are compared to single-component denoising, where each image band is denoised separately, and to multicomponent denoising with a multivariate Gaussian prior.

Table 25.1 and Figure 25.1 (part of which appeared in [34]) show the results of the denoising procedures for this image.

In both the table and the figure, one can observe that the multicomponent procedures outperform a single-component denoising, demonstrating that when accounting for the correlation between the image bands, the denoising performance is improved. The use of a heavy-tailed multivariate model is favorable compared to a multivariate Gaussian model.

25.4 MULTI/HYPERSPECTRAL IMAGE RESTORATION USING THE WAVELET-BASED FRAMEWORK

25.4.1 State-of-the-Art

Remote sensing images are subject to degradation caused by atmospheric effects and physical limitations of the sensors. The degradation is apparent as blurring affecting the spatial resolution and noise added on top. The goal of image restoration is to recover the original image. In case of multispectral images, each of the bands is degraded, in general with a different blurring and noise level. In fact, the blurring need not be spatially invariant, and can also be blurring in the spectral direction. The straightforward way to restore a multispectral image is to restore each band separately. However, this may destruct the spectral information that is contained in the multispectral image. Also, spectral blurring cannot be treated in this way. Moreover, since the different bands are in general highly correlated, a multispectral approach that exploits this correlation is favorable.

There exists a vast literature on grayscale image restoration. Usually, the image degradation is described by a linear space-invariant convolution (blurring) operator and additive Gaussian noise. Inversion of the blurring operator is generally done in the Fourier domain, requiring regularization (denoising) to avoid the singularities. Multispectral (multichannel) image restoration has been performed as well using similar strategies. A straightforward multiband restoration approach is to transform the multiband image to spectrally decorrelate the channels and restore the decorrelated images in a single-band fashion [20]. This will only work if no spectral blur is present. Multiband versions of linear methods such as Wiener filtering [17] and least-squares restoration [18] have been proposed. In [22], the authors present a generalization of frequency-domain single-band

TABLE 25.1
Peak signal to noise ratio (PSNR) (in dB) Values after Denoising of Multispectral Data Set

Initial	34.16	28.15	24.67	22.24
Single-component				
	37.40	34.03	32.38	31.27
Multi-component				
Gaussian prior	38.53	35.16	33.45	32.25
GSM prior	38.94	35.51	33.75	32.56

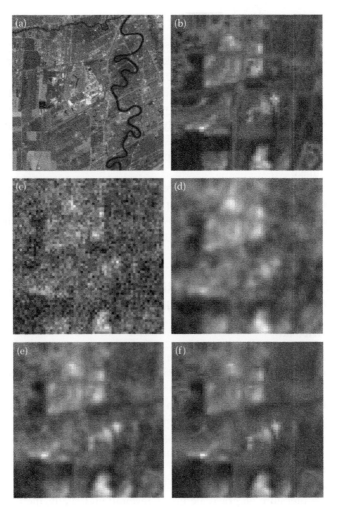

FIGURE 25.1   (a) One band of Landsat multispectral image; (b) detail of image; (c) detail image with simulated Gaussian noise ($\sigma = 15$); (d)–(f) detail of denoised images with the presented techniques ((d) single component, (e) multicomponent Gaussian prior, (f) multicomponent GSM prior). (Copyright [2007] IEEE.)

deconvolution techniques. In Ref. [36] a Bayesian MAP estimation have been proposed. Here, the Gibbs priors contain both spatial and spectral components.

Restoration is known to be an ill-posed inverse problem, since the deconvolution requires the inverse of the blurring operator, which may be nearly singular or may even not exist, resulting in magnification of noise. A disadvantage of the Fourier transform is that it does not efficiently represent image edges. So, only small amounts of regularization are allowed to avoid blurring of the edges in the image. In the wavelet domain, on the other hand, this problem is avoided since edges are represented by large coefficients which are better retained after regularization. This property makes the wavelet transform suited for image denoising, which has been demonstrated in a countless number of effective denoising schemes [12]. Therefore, for restoration, it is advantageous to separate the deconvolution and the denoising problems.

In the recent literature, this strategy has been applied, and several solutions for the deconvolution problem have been formulated. In Ref. [28], a technique, referred to as ForWaRD, applies a Wiener deconvolution followed by a wavelet shrinkage. Several iterative deconvolution methods were proposed based on the Expectation–Maximization (EM) algorithm [16], or generalizations of it [6] and the more recently developed Majorization–Minimization [15] and TwIST algorithms [7].

The EM algorithm is an iterative procedure developed to maximize the likelihood function corresponding to the observation model. Each iteration consists of two steps: an E-step solving the deconvolution and an M-step managing the noise. In Ref. [14], the technique of [16] was extended toward multispectral images.

In this section, we describe a multispectral image restoration technique by extending the recent restoration strategies based on the EM algorithm toward multispectral images. The procedure is constructed as follows:

1. The E-step is formulated by extending the spatial blurring operator towards an operator, allowing for spatial as well as spectral blurring.
2. In the M-step of the iterative procedure, the Bayesian wavelet-based denoising framework of the previous section is applied, including the GSM prior model for the pdf of the noise-free image.

25.4.2 EM RESTORATION

We extend the EM procedure of [16] towards multispectral images. The idea is to perform the deblurring and denoising in two separate steps. To do so, the imaging model $\mathbf{Y} = \mathbf{HS} + \mathbf{N}$ is decomposed as

$$\mathbf{Y} = \mathbf{HX} + \mathbf{N}_2 \tag{25.16}$$

$$\mathbf{X} = \mathbf{S} + \mathbf{N}_1 \tag{25.17}$$

where $\mathbf{HN}_1 + \mathbf{N}_2 = \mathbf{N}$ with $p(\mathbf{N}_1) = \phi\,(0, \mathbf{C_n})$ and $p(\mathbf{N}_2) = \phi\,(0, \mathbf{C_n} - \mathbf{HC_n}\mathbf{H}^\mathrm{T})$. The spatial invariance of $\mathbf{H}$ guarantees a semipositive definite covariance for $\mathbf{N}_2$. If $\mathbf{H}$ would be not translation invariant, a rescaling is required. In this way, the noise is decomposed into two independent parts.

In his paper [16], Figueiredo introduces a more general decomposition as $\alpha\mathbf{HN}_1 + \mathbf{N}_2 = \mathbf{N}$. The balance between both noise components is defined by α, with the restriction $0 \le \alpha \le 1$. In the experimental section of that paper, it is stated that $\alpha = 1$ leads to the best results. For a more detailed explanation, we refer to Ref. [16].

The specific choice of $p(\mathbf{N}_1)$ guarantees that $\mathbf{N}_1$ is Gaussian white noise, which makes (25.17) a simple denoising problem. In the first problem (25.16), this noise component $\mathbf{N}_1$ is embedded in $\mathbf{X}$, so that only $\mathbf{N}_2$ remains. We assume that $\mathbf{N}_2$ is small enough to be neglected. As a result, the original problem has been decoupled into a deblurring problem (25.16) and a denoising problem (25.17).

The EM algorithm is an iterative procedure developed to maximize the complete-data likelihood function. The goal is to find an estimation $\hat{\mathbf{S}}$ which maximizes the complete-data pdf

$$\hat{\mathbf{S}} = \arg\max_{\mathbf{S}} p(\mathbf{S}|\mathbf{Y},\mathbf{X}) \propto \arg\max_{\mathbf{S}} p(\mathbf{Y},\mathbf{X}|\mathbf{S})p(\mathbf{S}) \tag{25.18}$$

where $\mathbf{Y}$ denotes the observed data and $\mathbf{X}$ denotes missing data.

At each iteration k, the EM algorithm contains two steps. In the first step, the E-step, a likelihood, the so-called Q-function, is estimated as

$$Q\left(\mathbf{S},\hat{\mathbf{S}}^{(k-1)}\right) = \mathbb{E}\left[\log\,(p(\mathbf{Y},\mathbf{X} \mid \mathbf{S})p(\mathbf{S}))|\mathbf{Y},\hat{\mathbf{S}}^{(k-1)}\right] \tag{25.19}$$

which depends on the observed image $\mathbf{Y}$ and an estimate $\hat{\mathbf{S}}$ of the previous iteration. The second step, the M-step, maximizes this Q-function and calculates a new maximum likelihood estimate

$$\hat{\mathbf{S}}^{(k)} = \arg \max_{\mathbf{S}} \left[Q\left(\mathbf{S}, \hat{\mathbf{S}}^{(k-1)}\right) \right]. \qquad (25.20)$$

25.4.3   STEP 1: EXPECTATION STEP

Using Bayes' rule and the independency of $\mathbf{Y}$ on $\mathbf{S}$, when conditioned on $\mathbf{X}$, the conditioned pdf from Equation 25.19 can be written as

$$
\begin{aligned}
\log p(\mathbf{Y},\mathbf{X}|\mathbf{S}) &= \log[\, p(\mathbf{Y}|\mathbf{X},\mathbf{S})p(\mathbf{X}|\mathbf{S})] \\
&= \log[\, p(\mathbf{Y}|\mathbf{X})p(\mathbf{X}|\mathbf{S})] \\
&\propto -\frac{1}{2}(\mathbf{S} - \mathbf{X})^{\mathrm{T}}\mathbf{C_n}^{-1}(\mathbf{S} - \mathbf{X})
\end{aligned}
\qquad (25.21)
$$

omitting all terms not depending on $\mathbf{S}$.

To solve Equation 25.21, an estimate of $\mathbf{X}$ is required. The pdf $p(\mathbf{X}|\mathbf{Y},\hat{\mathbf{S}}^{(k-1)})$ is easily shown to be $\propto p(\mathbf{Y}|\mathbf{X})p(\mathbf{X}|\hat{\mathbf{S}}^{(k-1)})$. Following Equations 25.16 and 25.17: $p(\mathbf{Y}|\mathbf{X}) = \phi\ (\mathbf{Y} - \mathbf{HX},\ \mathbf{C_n} - \mathbf{HC_n}\mathbf{H}^{\mathrm{T}})$ and $p(\mathbf{X}|\hat{\mathbf{S}}^{(k-1)}) = \phi(\mathbf{X} - \hat{\mathbf{S}}^{(k-1)}, \mathbf{C_n})$. After some calculation, the MAP estimation of this product is obtained as

$$
\begin{aligned}
\hat{\mathbf{X}}^{(k)} &= \mathbb{E}\left[\mathbf{X}|\mathbf{Y},\hat{\mathbf{S}}^{(k-1)} \right] \\
&= \hat{\mathbf{S}}^{(k-1)} + \mathbf{H}^{\mathrm{T}}\left(\mathbf{Y} - \mathbf{H}\hat{\mathbf{S}}^{(k-1)} \right)
\end{aligned}
\qquad (25.22)
$$

which gives an estimate of the missing data $\mathbf{X}$. In fact, solving this equation is solving the deblurring problem of (25.16). As an initial value, $\hat{\mathbf{S}}^{(0)} = \mathbf{Y}$ can be used.

25.4.4   STEP 2: MAXIMIZATION STEP

A new estimate $\hat{\mathbf{S}}^{(k)}$ is then calculated as

$$
\begin{aligned}
\hat{\mathbf{S}}^{(k)} &= \arg \max_{\mathbf{S}} p(\mathbf{X}|\mathbf{S})p(\mathbf{S}) \\
&= \arg \max_{\mathbf{S}} \left[-\frac{1}{2}(\mathbf{S} - \hat{\mathbf{X}}^{(k)})^{\mathrm{T}}\mathbf{C_n}^{-1}(\mathbf{S} - \hat{\mathbf{X}}^{(k)}) - \log p(\mathbf{S}) \right]
\end{aligned}
\qquad (25.23)
$$

Solving this equation leads to solving the denoising problem of Equation 25.17. We will solve this denoising problem by applying the denoising framework of the previous section. For this, $\hat{\mathbf{X}}^{(k)}$ is wavelet transformed, and the MAP estimation is applied using a multivariate GSM prior for $p(\mathbf{s})$. The estimate gives $\hat{\mathbf{s}}^{(k)}$, which is inverse wavelet transformed to obtain $\hat{\mathbf{S}}^{(k)}$.

25.4.5   EXPERIMENTS

The first experiment is a validation of the proposed techniques. For this, a blurred and noisy image is simulated from a given multispectral Landsat image (with size $512 \times 512 \times 6$). First, spatial invariant blurring is applied. For each band, a Gaussian blurring kernel is applied with a standard deviation $\sigma_a = 0.8$ pixels sampled on a 5×5 window and Gaussian white noise ($\sigma_n^2 = 90$) is added. Since the operator $\mathbf{H}$ is band-specific in this case, the M-step is the only multicomponent step in the

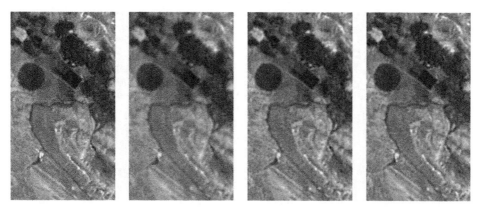

FIGURE 25.2 A part ($120 \times 210 \times 3$) of the restored multispectral Landsat image. From left to right, the original, the degraded image, and the restored images using the original (band by band) EM algorithm and the proposed method. (Copyright [2009] IEEE.)

proposed procedure. We compare the original EM algorithm (processed by band) with our multispectral restoration method.

In Figures 25.2 and 25.3 (appeared in Ref. [14]) the restored images are shown (as a color composite image of the bands 1–3). As can be seen, the restored image using the original EM algorithm contains noise. Using the proposed algorithm, the amount of noise is clearly reduced. The restored image is sharper for the proposed restoration method.

In the second experiment our algorithm is validated when there is also spectral blurring involved. Different image degradations are simulated on 80 bands of the AVIRIS image "Indian Pine." For each band a Gaussian blurring kernel is applied with a standard deviation σ_a varying from 0 to 1.6 pixels to simulate the spatial blurring. Furthermore, a Gaussian blurring kernel with a standard deviation σ_e between 0 and 6.8 bands is applied to simulate the spectral blurring. Finally, some Gaussian white noise ($\sigma_n^2 = 10$) is added. The E-step (25.23) is now fully three dimensional, because the blurring function $\mathbf{H}$ contains interband blurring.

To compare the results of the restoration the Spectral Angle Mapper (SAM) is used, which is a measure of the difference between two spectra.

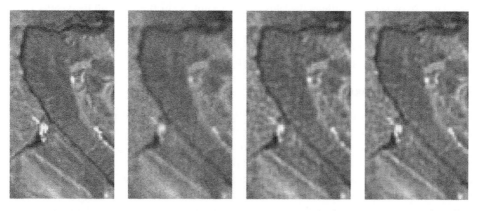

FIGURE 25.3 Detail from the restored multispectral Landsat image in Figure 25.2. From left to right, the original, the degraded image and the restored images using the original (band by band) EM algorithm and the proposed method. (Copyright [2009] IEEE.)

TABLE 25.2

The SAM in Degrees and PSNR in Decibels

		$\sigma_a = 0.0$		$\sigma_a = 0.8$		$\sigma_a = 1.2$		$\sigma_a = 1.6$	
		SAM	PSNR	SAM	PSNR	SAM	PSNR	SAM	PSNR
$\sigma_e = 0.0$	Degraded image	–	–	3.03	30.53	3.78	28.06	4.38	26.59
	Original EM	–	–	3.47	31.35	3.99	29.41	4.27	28.26
	Multispectral EM	–	–	1.97	35.86	3.08	31.01	3.88	28.79
$\sigma_e = 3.2$	Degraded image	5.75	27.58	6.01	26.26	6.40	25.27	6.74	24.52
	Original EM	5.59	27.90	4.96	28.47	5.14	27.60	5.30	26.90
	Multispectral EM	4.70	29.50	4.56	29.24	4.97	27.90	5.22	27.03
$\sigma_e = 4.4$	Degraded image	6.74	26.14	6.93	25.27	7.31	24.44	7.65	23.79
	Original EM	5.98	27.33	5.57	27.61	5.77	26.85	5.96	26.21
	Multispectral EM	5.50	28.04	5.28	28.11	5.65	27.06	5.89	26.32
$\sigma_e = 5.6$	Degraded image	7.59	25.05	7.67	24.33	7.98	23.69	8.26	23.18
	Original EM	6.71	26.30	6.39	26.26	6.57	25.70	6.74	25.21
	Multispectral EM	6.19	26.99	6.22	26.56	6.48	25.85	6.69	25.29
$\sigma_a = 6.8$	Degraded image	8.38	24.14	8.47	23.59	8.79	23.02	9.07	22.56
	Original EM	7.47	25.39	7.10	25.55	7.28	25.05	7.45	24.61
	Multispectral EM	7.03	25.91	6.94	25.78	7.20	25.17	7.41	24.68

Note: The values for the degraded image and the restored images using the original method and the proposed multispectral method.

The SAM is defined as

$$SAM(\mathbf{A},\mathbf{B}) = \arccos\frac{\mathbf{A}\cdot\mathbf{B}}{\|\mathbf{A}\|\,\|\mathbf{B}\|}. \tag{25.24}$$

In Table 25.2, the SAM and the PSNR are shown for this experiment. The differences between the two methods are most clearly visible when the blurring kernels are small. As can be seen, the multispectral restoration method always has a lower SAM than the original bandwise method. For larger blurring kernels the differences decrease, but the multispectral restoration method always is the best. The PSNR is a comparable measure: a decreasing SAM results in an increasing PSNR.

25.5 MULTI- AND HYPERSPECTRAL IMAGE FUSION USING THE WAVELET-BASED FRAMEWORK

25.5.1 STATE-OF-THE-ART

One way of dealing with the limitations in SNR, spatial and spectral resolutions of multi- and hyperspectral images has been image fusion, a well-studied field for more than 10 years. Most fusion techniques were developed for the specific purpose of enhancing multispectral images by using a panchromatic (Pan) image with higher spatial resolution (also referred to as Pansharpening). Principal component analysis (PCA) [38] and Intensity–Hue–Saturation (IHS) transform [8] based techniques are probably the most commonly used ones. The Pan image is applied to totally or partially substitute the 1st principal component or Intensity component of the coregistered and resampled multispectral image.

High-pass filtering and high-pass modulation techniques have been developed [38], in which spatial high-frequency information is extracted and injected into each band of the multispectral image. With the rise of multiresolution analysis, many researchers have proposed pansharpening techniques, using Gaussian and Laplacian pyramids as well as discrete decimated and undecimated wavelet transforms [2,29].

Improvement can be expected when a multispectral image of high spatial resolution is available to enhance a lower spatial resolution multispectral or hyperspectral image, mainly because it provides spatial high-resolution information combined with spectral information. In order to optimally make use of the multispectral image, both images need to be spectrally aligned, so that the spectrum scopes of the observed multispectral and hyperspectral images are identical. Because of this, the pansharpening techniques cannot be applied directly to the problem of multi- and hyperspectral image fusion. Usually, a spatial high-frequency component of the multispectral image is extracted (by PCA, IHS, etc.) first and then fused with the hyperspectral image, which may lead to spectral distortion.

Some researchers proposed statistical estimation techniques for multi- and hyperspectral image fusion. The method in Ref. [19] employed MAP estimation based on a spatially varying statistical model to enhance the hyperspectral image resolution. The framework developed was validated for pansharpening but allowed for any number of spectral bands in both the multi- and hyperspectral images. This statistical estimation method defined an observation model of the hyperspectral image, and a model between the high-resolution multi- and hyperspectral images. This leads to a complex framework with assumptions on the knowledge of the spectral and spatial response functions and the need to solve an incomplete deconvolution problem. Although additive noise was included in the imaging model, the technique worked in the image domain, so that the noise handling resorted to an image domain Wiener-type of filtering, which is known to be less effective than wavelet domain denoising. In Ref. [42], the technique was extended to work in the wavelet domain.

In this section, we employ the Bayesian framework of Ref. [19] to describe a simple but effective estimation approach. We perform the fusion in the wavelet domain, allowing for a scale- and subband-specific estimation. Besides this, we also include a noise model for the hyperspectral image, which is also much more effective in the wavelet domain, compared to the image domain. The enhancement of a hyperspectral image with the aid of a multispectral image of higher spatial resolution is treated, only assuming a joint normal distribution between the multi- and hyperspectral images. The deconvolution problem is avoided by interpolating the hyperspectral image *a priori*. We design a specific estimation strategy for the model parameters, based on an approximate knowledge on the resolution difference between the two images.

25.5.2 FUSION FRAMEWORK

The problem can be described as the fusion of a hyperspectral image ($\mathbf{X}$) with low spatial resolution and high spectral resolution, and a multispectral image ($\mathbf{Y}$) with high spatial resolution and low spectral resolution. Ideally, the fused result, $\mathbf{S}$ has the spatial resolution of $\mathbf{Y}$ and the spectral resolution of $\mathbf{X}$. In this work, we will assume that all images are equally spatially sampled at a grid of N pixels, sufficiently fine to reveal the spatial resolution of $\mathbf{Y}$.

The standard imaging model (25.4) between $\mathbf{S}$ and $\mathbf{X}$ is assumed, which becomes (25.5) in the wavelet domain. Also the multispectral image $\mathbf{Y}$ is wavelet transformed, leading to $\mathbf{y}$.

In a statistical framework, $\mathbf{s}$ can be estimated from the conditional probability $p(\mathbf{s}|\mathbf{x}, \mathbf{y})$, for example, as a MAP estimation: $\mathbf{s} = \arg\max_{\mathbf{s}} p(\mathbf{s}|\mathbf{x}, \mathbf{y})$.

Since $\mathbf{n}$ is independent of $\mathbf{y}$ and $\mathbf{s}$, it follows that $\mathbf{y}$ and $\mathbf{x}$, conditioned on $\mathbf{s}$ are independent. Applying Bayes rule, one obtains

$$\mathbf{s} = \arg\max_{\mathbf{s}} p(\mathbf{s}|\mathbf{x}, \mathbf{y})$$
$$= \arg\max_{\mathbf{s}} p(\mathbf{x}|\mathbf{s}) p(\mathbf{s}|\mathbf{y}) \tag{25.25}$$

The first pdf is given by the observation model Equation (25.5)

$$p(\mathbf{x}|\mathbf{s}) = \frac{1}{\sqrt{(2\pi)^{NP}|\mathbf{C}_n|}}\exp\left\{-\frac{1}{2}(\mathbf{x} - \mathcal{H}\mathbf{s})^T\mathbf{C}_n^{-1}(\mathbf{x} - \mathcal{H}\mathbf{s})\right\} \tag{25.26}$$

We will model the multispectral image $\mathbf{y}$ and the hyperspectral image $\mathbf{s}$ as jointly normally distributed. The pdf of $\mathbf{s}$ conditioned on $\mathbf{y}$ is then also normally distributed:

$$p(\mathbf{s}|\mathbf{y}) = \frac{1}{\sqrt{(2\pi)^{NP}|\mathbf{C}_{s|y}|}}\exp\left\{-\frac{1}{2}(\mathbf{s} - \mu_{s|y})^T\mathbf{C}_{s|y}^{-1}(\mathbf{s} - \mu_{s|y})\right\} \tag{25.27}$$

where $\mu_{s|y}$ and $\mathbf{C}_{s|y}$ denote the mean and covariance matrix of $\mathbf{s}$ conditioned on $\mathbf{y}$:

$$\begin{aligned}\mu_{s|y} &= \mathbb{E}(\mathbf{s}) + \mathbf{C}_{s,y}\mathbf{C}_{y,y}^{-1}[\mathbf{y} - \mathbb{E}(\mathbf{y})] \\ \mathbf{C}_{s|y} &= \mathbf{C}_{s,s} - \mathbf{C}_{s,y}\mathbf{C}_{y,y}^{-1}\mathbf{C}_{s,y}^T\end{aligned} \tag{25.28}$$

where $\mathbb{E}(\cdot)$ represents the expectation value and $\mathbf{C}_{.,.}$ the covariance matrix defined as

$$\mathbf{C}_{u,v} = \mathbb{E}([\mathbf{u} - \mathbb{E}(\mathbf{u})][\mathbf{v} - \mathbb{E}(\mathbf{v})]^T) \tag{25.29}$$

The observation model (25.26) was applied in the image domain in Ref. [19]. The wavelet transform represents the image information in a compact way. The detail images contain mostly zeroes with a few large coefficients representing the edges. This makes the distinction between real image information and noise more prominent, which is the major reason why denoising is superior in the wavelet domain.

The model (25.27) was applied in the image domain in Ref. [19]. The problem with this model is the complex covariance structure when spatial correlations are accounted for. In order to keep the calculations feasible, assumptions need to be made, for example, that the pixels are spatially independent, in which case the covariance matrices become diagonal (details will be explained in the next subsection). Such an assumption is rather accurate in the wavelet domain, since the wavelet transform spatially decorrelates the pixels [25]. In the image domain, however, no justification is available for such an assumption. This makes the modeling much more accurate in the wavelet domain.

Another advantage of the application of both models in the wavelet domain is that they are applied at each resolution and orientation level, with a separate estimation of the covariances for each level. This allows for a resolution- and orientation-specific adaption of the models to the image information, which is advantageous both for the denoising and for the fusion process.

Solving (25.25) using (25.26) and (25.27) leads to the following expression:

$$\hat{\mathbf{s}} = \mu_{s|y} + \mathbf{C}_{s|y}\mathcal{H}^T[\mathcal{H}\mathbf{C}_{s|y}\mathcal{H}^T + \mathbf{C}_n]^{-1}[\mathbf{x} - \mathcal{H}\mu_{s|y}] \tag{25.30}$$

When the applied models would not take the observation model of the hyperspectral image into account, the obtained estimation problem would reduce to: $\hat{\mathbf{s}} = \arg\max_s p(\mathbf{s}|\mathbf{y})$, leading to: $\hat{\mathbf{s}} = \mu_{s|y}$, which is a pure fusion problem. If, on the other hand, no spatial resolution difference were to be accounted for (i.e., $\mathbf{H} = \mathbf{I}$), the expression would reduce to

$$\hat{\mathbf{s}} = \mu_{s|y} + \mathbf{C}_{s|y}[\mathbf{C}_{s|y} + \mathbf{C}_{n}]^{-1}[\mathbf{x} - \mu_{s|y}]$$

$$= \mathbf{C}_{s|y}[\mathbf{C}_{s|y} + \mathbf{C}_{n}]^{-1}\mathbf{x} \tag{25.31}$$

$$+ \{1 - \mathbf{C}_{s|y}[\mathbf{C}_{s|y} + \mathbf{C}_{n}]^{-1}\}\mu_{s|y}$$

which represents a noise filtering, as a weighted average between a shrinkage of the noisy wavelet coefficients and a shrinkage of the fused result. When applying this estimation, a denoised hyperspectral image is obtained, without any resolution improvement.

25.5.3 Estimation Implementation and Parameter Calculation

As mentioned in the previous section, to make the calculation manageable, we assume that the wavelet coefficients of the fused hyperspectral image are conditionally spatially independent. In this case the model (25.27) is rewritten as

$$p(\mathbf{s}|\mathbf{y}) = \prod_{n=1}^{N} \frac{1}{\sqrt{(2\pi)^{NP}|\mathbf{C}_{s_n|y_n}|}}$$

$$\cdot \exp\left\{-\sum_{n=1}^{N} \frac{1}{2}(\mathbf{s}_n - \mu_{s_n|y_n})^{\mathrm{T}}\mathbf{C}_{s_n|y_n}^{-1}(\mathbf{s}_n - \mu_{s_n|y_n})\right\} \tag{25.32}$$

where $\mu_{s_n|y_n}$ and $\mathbf{C}_{s_n|y_n}$ are conditional expectation and covariance matrix for $\mathbf{s}_n$ given $\mathbf{y}_n$. Applying (25.28), the individual conditional expectation and covariance matrix can be estimated as

$$\mu_{s_n|y_n} = \mathbb{E}(\mathbf{s}_n) + \mathbf{C}_{s_n,y_n}\mathbf{C}_{y_n,y_n}^{-1}[\mathbf{y}_n - \mathbb{E}(\mathbf{y}_n)]$$

$$\mathbf{C}_{s_n|y_n} = \mathbf{C}_{s_n,s_n} - \mathbf{C}_{s_n,y_n}\mathbf{C}_{y_n,y_n}^{-1}\mathbf{C}_{s_n,y_n}^{\mathrm{T}} \tag{25.33}$$

The individual statistical parameters are related with the global ones by

$$\mu_{s|y} = [\mu_{s_1|y_1}^{\mathrm{T}}, \mu_{s_2|y_2}^{\mathrm{T}}, ..., \mu_{s_N|y_N}^{\mathrm{T}}]^{\mathrm{T}} \tag{25.34}$$

$$\mathbf{C}_{s|y} = \begin{pmatrix} \mathbf{C}_{s_1|y_1} & 0 & \cdots & 0 \\ 0 & \mathbf{C}_{s_2|y_2} & \cdots & 0 \\ \vdots & \vdots & \ddots & \vdots \\ 0 & 0 & \cdots & \mathbf{C}_{s_N|y_N} \end{pmatrix} = \bigoplus_{n=1}^{N} \mathbf{C}_{s_n|y_n} \tag{25.35}$$

As the global covariance matrix is diagonal, $\mathbf{s}_n$ for each spatial position can be estimated individually. In this chapter, a global estimation is implemented for $\mathbb{E}(\mathbf{u}_n)$ and $\mathbf{C}_{u_n,v_n}$, which assumes that they are constant over the complete detailed image. In particular, the expectation values $\mathbb{E}(\mathbf{s}_n)$ and $\mathbb{E}(\mathbf{y}_n)$ in Equation 25.33 are zero and the covariances are estimated globally over the entire wavelet detail image $\mathbf{C}_{u_n,v_n} = \mathbb{E}(\mathbf{u}_n\mathbf{v}_n^{\mathrm{T}}) = \frac{1}{N}\Sigma_{n'=1}^{N} \mathbf{u}_{n'}\mathbf{v}_{n'}^{\mathrm{T}}$.

To solve (25.30), $\mathcal{H}$ and thus $\mathbf{H}$ needs to be known. In practice, this knowledge is not always fully available. Even when sufficient knowledge of the response functions is available, the estimation of $\mathbf{s}$ requires an inverse operation, which is an incomplete deconvolution problem.

In this work, the knowledge of the blurring is assumed to be limited to an approximate knowledge on the resolution difference between the observed hyper- and multispectral image. In that case, we can replace the operation of $\mathbf{H}$ on the multispectral image by smoothing until a spatial resolution is obtained, similar to that of the observed hyperspectral image. Similar arguments hold for the cross-covariances related to Equation 25.30, where the smoothing holds for the left or right-hand argument in the covariance, depending on the position of $\mathcal{H}$ in the equation. Of course, when $\mathbf{s}$ appears as argument, the smoothing replaces $\mathbf{s}$ by $\mathbf{x}$. Considering that all the covariance matrices are preferably calculated on the same spatial resolution scale, the following expression is obtained:

$$\hat{\mathbf{s}}_n = \mu_{\mathbf{s}_n|\mathbf{y}_n} + \mathbf{C}_{\mathbf{x}_n|\tilde{\mathbf{y}}_n}[\mathbf{C}_{\mathbf{x}_n|\tilde{\mathbf{y}}_n} + \mathbf{C}_\mathbf{n}]^{-1}[\mathbf{x}_n - \tilde{\mu}_{\mathbf{s}_n|\mathbf{y}_n}] \tag{25.36}$$

with

$$
\begin{aligned}
\mu_{\mathbf{s}_n|\mathbf{y}_n} &= \mathbf{C}_{\mathbf{x}_n,\tilde{\mathbf{y}}_n}\mathbf{C}^{-1}_{\tilde{\mathbf{y}}_n,\tilde{\mathbf{y}}_n}\mathbf{y}_n \\
\tilde{\mu}_{\mathbf{s}_n|\mathbf{y}_n} &= \mathbf{C}_{\mathbf{x}_n,\tilde{\mathbf{y}}_n}\mathbf{C}^{-1}_{\tilde{\mathbf{y}}_n,\tilde{\mathbf{y}}_n}\tilde{\mathbf{y}}_n \\
\mathbf{C}_{\mathbf{x}_n|\tilde{\mathbf{y}}_n} &= \mathbf{C}_{\mathbf{x}_n,\mathbf{x}_n} - \mathbf{C}_{\mathbf{x}_n,\tilde{\mathbf{y}}_n}\mathbf{C}^{-1}_{\tilde{\mathbf{y}}_n,\tilde{\mathbf{y}}_n}\mathbf{C}^{\mathrm{T}}_{\mathbf{x}_n,\tilde{\mathbf{y}}_n}
\end{aligned}
\tag{25.37}
$$

where $\tilde{\mathbf{y}}$ is the spatially degraded version of $\mathbf{y}$ at the spatial resolution scale of $\mathbf{x}$. In practice, the smoothing of the multispectral image $\mathbf{Y}$ is performed in the image domain, after which it is wavelet transformed to obtain $\tilde{\mathbf{y}}$.

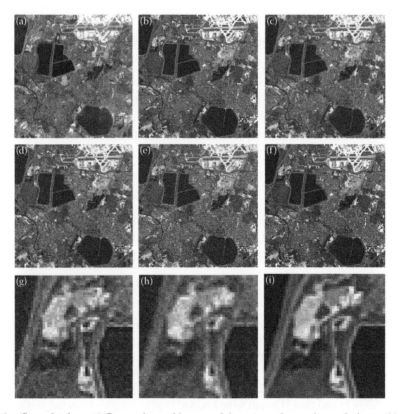

FIGURE 25.4 **(See color insert.)** Comparison with state-of-the-art pansharpening techniques. (a) Pan image; (b) Reference image: original Landsat image; (c) low-resolution Landsat image with additive noise ($\sigma = 15$); (d)–(f) fused results of AWLP, GLP-CBD, and the proposed method; (g)–(i) zoomed results of (d)–(f). (Copyright [2009] IEEE.)

In the experimental section, the proposed procedure will be demonstrated on the fusion of a multispectral image with a Pan image.

25.5.4 EXPERIMENTS

To demonstrate the proposed fusion approach, a set of color-composite Landsat images (three bands, 30 m resolution) and a SPOT Pan image (10 m resolution) covering an area near London are used as test data. To be able to use the original Landsat image as a reference, we degraded the Landsat image to 60 m and the SPOT image to 30 m resolution.

Most of the existing fusion methods are particularly designed for fusion of multispectral and Pan images (Pansharpening). In Ref. [3], eight Pansharpening algorithms were compared and it was concluded that two methods (AWLP and GLP-CBD) outperform all the others. AWLP is the $à$ trous wavelet transform based Pansharpening. The high-frequency details of the Pan image are injected into the multispectral image proportionally to reduce the spectral distortion of the fused image. Generalized Laplacian pyramid using context-based decision (GLP-CBD) employs local correlation coefficients to control the high-frequency detail injection.

To demonstrate the effectiveness and advantages of the proposed technique, we compare it to AWLP and GLP-CBD. Although both comparing methods have been demonstrated to produce good fusion results when no noise is present, we will compare their performance in a noisy case, to demonstrate the noise-resistance of the proposed fusion technique.

Figure 25.4 (appeared in Ref. [42]) shows the fusion results. From the images, we can conclude that the proposed technique produces good fusion results. The images are comparable to the state-of-the-art techniques but with these two fusion techniques, heavy spectral distortions become visible.

25.6 CONCLUSION

In this chapter, we presented a wavelet-based restoration framework for multi- and hyperspectral images, based on Bayesian principles, in which prior image information was introduced in the process in the shape of a statistical model of the image pdf. Heavy-tailed multivariate models that account for the correlation between the image bands were applied in the wavelet domain. Techniques were presented in the domains of multi- and hyperspectral image restoration and fusion. The techniques were demonstrated on remote sensing multi- and hyperspectral images.

The presented work is part of ongoing research on the use of wavelet-based Bayesian techniques with prior modeling for multi- and hyperspectral image processing and analysis applications. A multispectral image fusion method, including a deconvolution procedure, was developed using heavy-tailed multivariate prior models in the wavelet domain. A multispectral image segmentation procedure was developed, in which each processing step accounts for the correlation between the image bands, including a region-merging procedure, based on a hypothesis testing using a multivariate Gaussian model in the wavelet domain [14]. Finally, retrieval of multispectral images was studied, based on geodesic distance measures using a multivariate Generalized Gaussian model in the wavelet domain [39].

REFERENCES

1. A. Achim, P. Tsakalides, and A. Bezerianos. SAR image denoising via Bayesian wavelet shrinkage based on heavy-tailed modeling. *IEEE Trans. Geosci. Remote Sens.*, 41(8):1773–1784, 2003.
2. B. Aiazzi, L. Aparone, and A. Barducci. Estimating noise and information of multispectral imagery. *Opt. Eng.*, 41:(3):656–668, 2002.
3. L. Alparone, L. Wald, J. Chanussot, C. Thomas, P. Gamba, and L. Bruce. Comparison of pansharpening algorithms: Outcome of the 2006 grs-s data-fusion contest. *IEEE Trans. Geosci. Remote Sens.*, 45(10):3012–3021, 2006.

4. G. E. P. Bax and C. Tiao. *Bayesian Inference in Statistical Analysis*. Addison-Wesley, Reading, Mass, 1992.

5. A. Benazza-Benyahia and J.-C. Pesquet. Building robust wavelet estimators for multicomponent images using Steins' principle. *IEEE Trans. Image Process.*, 14(11):1814–1830, 2005.

6. J. M. Bioucas-Dias. Bayesian wavelet-based image deconvolution: A GEM algorithm exploiting a class of heavy-tailed priors. *IEEE Trans. Image Process.*, 15(4):937–951, 2006.

7. J. M. Bioucas-Dias and M. A. T. Figueiredo. A new TwIST: Two-step iterative shrinkage/thresholding algorithms for image restoration. *IEEE Trans. Image Process.*, 16(12):2992–3004, 2007.

8. W. J. Carper, T. M. Lillesand, and R. W. Kiefer. The use of intensity-hue-saturation transformations for merging spot panchromatic and multispectral image data. *Photogram. Eng. Remote Sens.*, 56:459–467, 1990.

9. S. G. Chang, B. Yu, and M. Vetterli. Adaptive wavelet thresholding for image denoising and compression. *IEEE Trans. Image Process.*, 9(9):1532–1546, 2000.

10. I. Daubechies. *Ten Lectures on Wavelets*. Society for Industrial and Applied Mathematics, Philadelphia, 1992.

11. S. de Backer, A. Pizurica, B. Huysmans, W. Philips, and P. Scheunders. Denoising of multispectral images using wavelet least-squares estimators. *Image Vis. Comput.*, 26(7):1038–1051, 2008.

12. D. L Donoho. De-noising by soft-thresholding. *IEEE Trans. Inform. Theory*, 41:613–627, 1995.

13. J. Driesen and P. Scheunders. A multicomponent image segmentation framework. *Lecture Notes Comput. Sci.*, 5259:589–600, 2008.

14. A. Duijster, P. Scheunders, and S. De Backer. Wavelet-based iterative multispectral image restoration. *IEEE Trans. Geosci. Remote Sens.*, 47(11):3892, 2009.

15. M. A. T. Figueiredo, J. M. Bioucas-Dias, and R. D. Nowak. Majorization–minimization algorithms for wavelet-based image restoration. *IEEE Trans. Image Process.*, 16(12):2980–2991, 2007.

16. M. A. T. Figueiredo and R. D. Nowak. An EM algorithm for wavelet-based image restoration. *IEEE Trans. Image Process.*, 12(8):906–916, 2003.

17. N. P. Galatsanos and R. T. Chin. Digital restoration of multichannel images. *IEEE Trans. Acoust., Speech Signal Process.*, 37(3):415–421, 1989.

18. N. P. Galatsanos, A. K. Katsaggelos, R. T. Chin, and A. D. Hillery. Least squares restoration of multichannel images. *IEEE Trans. Signal Process.*, 39(10):2222–2236, 1991.

19. R. C. Hardie, M. T. Eismann, and G. L. Wilson. Map estimation for hyperspectral image resolution enhancement using an auxiliary sensor. *IEEE TIP*, 13:(9):1174–1184, 2004.

20. B. R. Hunt and O. Kübler. Karhunen-Loeve multispectral image restoration, Part I: Theory. *IEEE Trans. Acoustics, Speech Signal Process.*, ASSP-32(3):592–600, 1984.

21. M. Jansen. *Noise Reduction by Wavelet Thresholding*. Springer-Verlag, New York, 2001.

22. A. K. Katsaggelos, K. T. Lay, and N. P. Galatsanos. A general framework for frequency domain multichannel signal processing. *IEEE Trans. Image Process.*, 2(3):417–420, 1993.

23. J. B. Lee, A. S. Woodyatt, and M. Berman. Enhancement of high spectral resolution remote-sensing data by noise-adjusted principal components transform. *IEEE Trans. Geosci. Remote Sens.*, 28(3):295–304, 1990.

24. S. Mallat. A theory for multiresolution signal decomposition: The wavelet representation. *IEEE Trans. Pattern Anal. Machine Intell.*, 11(7):674–693, 1989.

25. S. Mallat. *A Wavelet Tour of Signal Processing*. Academic Press, London, 1998.

26. M. K. Mihçak, I. Kozintsev, K. Ramchandran, and P. Moulin. Low-complexity image denoising based on statistical modeling of wavelet coefficients. *IEEE Signal Process. Lett.*, 6(12):300–303, 1999.

27. P. Moulin and J. Liu. Analysis of multiresolution image denoising schemes using generalized Gaussian and complexity priors. *IEEE Trans. Inform. Theory*, 45:909–919, 1999.

28. R. Neelamani, H. Choi, and R. Baraniuk. ForWaRD: Fourier-wavelet regularized deconvolution for ill-conditioned systems. *IEEE Trans. Image Process.*, 52(2):418–433, 2004.

29. J. Nunez, X. Otazu, O. Fors, A. Prades, V. Pala, and R. Arbiol. Image fusion with additive multiresolution wavelet decomposition; applications to spot + landsat images. *J. Opt. Soc. Am. A*, 16:467–474, 1999.

30. H. Othman and S.-E. Qian. Noise reduction of hyperspectral imagery using hybrid spatial–spectral derivative-domain wavelet shrinkage. *IEEE Trans. Geosci. Remote Sens.*, 44(2):397–408, 2006.

31. A. Pižurica and W. Philips. Estimating the probability of the presence of a signal of interest in multiresolution single- and multiband image denoising. *IEEE Trans. Image Process.*, 15(3):654–665, 2006.

32. A. Pižurica, W. Philips, I. Lemahieu, and M. Acheroy. A versatile wavelet domain noise filtration technique for medical imaging. *IEEE Trans. Med. Imaging*, 22(3):323–331, 2003.

33. J. Portilla, V. Strela, M. J. Wainwright, and E. P. Simoncelli. Image denoising using scale mixtures of Gaussians in the wavelet domain. *IEEE Trans. Image Process.*, 12(11):1338–1351, 2003.

34. P. Scheunders and S. De Backer. Wavelet denoising of multicomponent images, using Gaussian scale mixture models and a noise-free image as priors. *IEEE Trans. Image Process.*, 16(7):1865–1872, 2007.

35. P. Scheunders and J. Driesen. Least-squares interband denoising of color and multispectral images. In *Proceedings of the IEEE International Conference on Image Processing ICIP*, Singapore, Vol. 2, pp. 985–988, 2004.

36. R. R. Schultz and R. L. Stevenson. Stochastic modeling and estimation of multispectral image data. *IEEE Trans. Image Process.*, 4(8):1109–1119, 1995.

37 L. Şendur and I. W. Selesnick. Bivariate shrinkage functions for wavelet-based denoising exploiting interscale dependency. *IEEE Trans. Signal Process.*, 50(11):2744–2756, 2002.

38. V. K. Shettigara. A generalized component substitution technique for spatial enhancement of multispectral images using a higher resolution data set. *Photogram. Eng. Remote Sens.*, 58:561–567, 1992.

39. G. Verdoolaege, S. De Backer, and P. Scheunders. Multiscale colour texture retrieval using the geodesic distance between multivariate generalized Gaussian models. In *Proceedings of the IEEE, International Conference on Image Processing*, San Diego, CA, October 12–15, pp. 169–172, 2008.

40. M. Vetterli and J. Kovačević. *Wavelets and Subband Coding*. Prentice-Hall, Englewoods Cliffs, NJ, 1995.

41. B. Vidakovic. Nonlinear wavelet shrinkage with Bayes rules and Bayes factors. *J. Am. Stat. Assoc.*, 93:173–179, 1998.

42. Y. Zhang, S. De Backer, and P. Scheunders. Noise-resistant wavelet-based Bayesian fusion of multispectral and hyperspectral images. *IEEE Trans. Geosci. Remote Sens.*, 47(11):3834, 2009.

26 Land Cover Estimation with Satellite Image Using Neural Network

Yuta Tsuchida, Michifumi Yoshioka, Sigeru Omatu, and Toru Fujinaka

CONTENTS

26.1 INTRODUCTION

Remote sensing is the term used for observing the strength of electromagnetic radiation that is radiated or reflected from various objects on the ground level with a sensor installed in a space satellite or in an aircraft. The analysis of acquired data is an effective means to survey vast areas periodically. It is important to classify the surface of the earth into categories as water areas, forests, factories, and cities. In this chapter, we discuss an effective method for land map classification which uses Advanced Visible and Near Infrared Radiometer type 2 (AVNIR-2) and Neural Network. The sensors installed in the space satellite include optical and microwave sensors. AVNIR-2 , which is a passive-type optical sensor installed in the ALOS, a Japanese satellite, is used for the land map classification in this chapter.

AVNIR-2 has four band sensors, Band1: 0.24–0.50 μm (blue), Band2: 0.52–0.60 μm (green), Band3: 0.61–0.69 μm (red), and Band4: 0.76–0.86 μm, and the ground resolution, higher than those of conventional sensors, is 10 m.

As a classifier, Neural Network [1,2] is adopted, which is known as one of the most effective methods in pattern recognition. In this chapter, we propose to adjust the range of input signals to train the neural network.

The raw data ranges of the remote sensing image are integers such as 0–255. Therefore, it is necessary to consider whether this range is appropriate as input signals of neural network. When the signals of these ranges are inputted untouched, the neural network is difficult to converge. So, we propose the preprocessing for the input range.

Finally, we will show the land cover estimation by using an optimal range which is determined by our proposed method.

26.2 PROPOSED METHODS AND SIMULATION

As we described in the previous section, it is difficult for the neural network to converge by using the range of the raw data. This section shows the difference of the convergence of the network by changing the range of the input signals. Input signal of the sample data consists of two-dimensional vectors shown in Figure 26.1.

We used a total of three data sets in three types, where each type of data set consists of the data in two, three, or four categories, respectively. These categories are gathered artificially so that the boundaries draw easily. The output signals are set to the number of expected classes, and supervisor is 1 in a certain class or 0 in the others. For example, the output vector of target class indicates as $(1,0,...0)$. The number of hidden layers is set as 8, which is used for following training.

As a first step, the ranges of the input signal are set as 0 to n, where n is an integer. This range is similar to the range used for image pixel values. Figure 26.1 shows that the sample input signals and the lines indicate the boundary of the clustering for the learning result. The border lines are plotted as follows. The sample input data sets have discrete values and an infinite range. All the grid values as input to the neural network being trained are given. Then the one class which has maximum output value is selected when a set of coordinate data are inputted to the network. And the class value is set to the corresponding value at coordinate of the inputs. Finally the edge is extracted. And Figures 26.2 and 26.3 are other samples. As a result, the network is not converged using these input

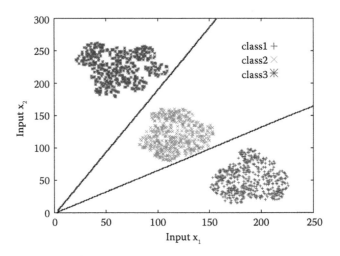

FIGURE 26.1 The result for sample 1.

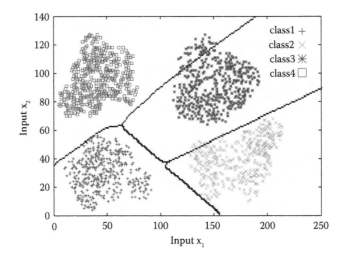

FIGURE 26.2 The result for sample 2.

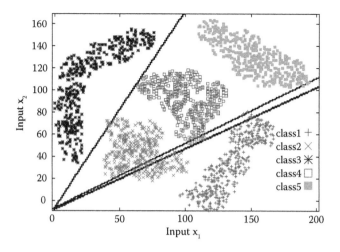

FIGURE 26.3   The result for sample 3.

signals of this range, because the domain of sigmoid function used in the neural network is between $-n$ and n, and in particular the gradient is larger near zero.

In the next step, the input range is moved to the original range to "$-n/2$ to $n/2$," which results in a zero-mean range. Figures 26.4 through 26.6 show the results where input range is changed using same samples as before. A more complicated sample can train the network by using this range compared with the previous step.

Although it is possible to classify larger number of categories using the revised input range, there are some samples which are erroneously categorized. Therefore, as the final step, these ranges are normalized to -1 to 1. These ranges are divided by the upper limit of these inputs. Figures 26.7 and 26.8 show the result with the normalized ranged using samples 2 and 3, respectively. As a result, the range -1 to 1 has the least error.

26.3   EXPERIMENT

In order to confirm the effectiveness of our proposed method, we have estimated the land cover using AVNIR-2, Advanced Visible and Near Infrared Radiometer type 2, with the proposed

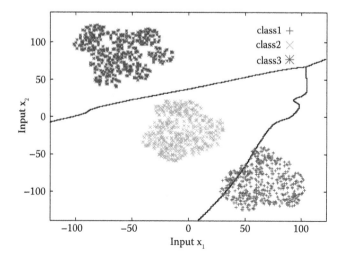

FIGURE 26.4   The result for sample 1 (range $[-n/2:n/2]$).

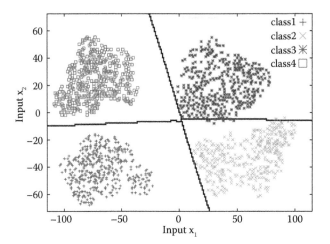

FIGURE 26.5 The result for sample 2 (range [−*n*/2:*n*/2]).

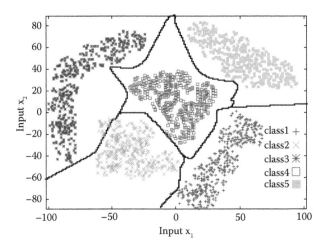

FIGURE 26.6 The result for sample 3 (range [−*n*/2:*n*/2]).

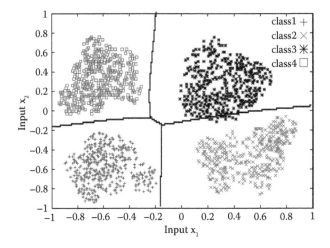

FIGURE 26.7 The result for sample 2 (range [−1:1]).

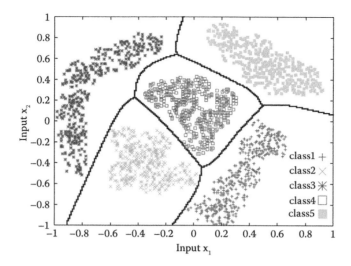

FIGURE 26.8 The result for sample 3 (range [−1:1]).

method [3]. This sensor is mounted to ALOS, Advanced Land Observing Satellite, a Japanese satellite. One of the characteristics of this sensor is that the spatial resolution is higher than conventional sensors in the same bands. The observation data were taken on March 31, 2007, around Chengdu city, in Sichen, China. We select the learning area based on the following conditions. The first condition is that there are few clouds. The second is that many kinds of areas to classify are contained. Thus, the 15 km square area around Xinji City in Sichen is picked up. The selected area is classified to seven categories which are Urban, Sand, Water, Mountain, Vegetation, Red soil, and Land along a river.

The input signals contain the data for four bands of frequency in every pixel. The signals are normalized. Preliminarily, we assigned one of seven classes for each pixel, and selected 100 pixels from each class. The training or test is executed for each pixel. The neural network which is used for experiments has three layers, and there are 4, 20, and 7 neurons in the input, hidden, and output layers, respectively. The supervisor signals are same as in the previous section. For example, the feature vector of the water area is $(1, 0, \ldots 0)$. A total of 70 patterns (10 patterns per class) are used for learning, and the remaining 630 are used for test.

Table 26.1 shows the test result in the classification accuracy (percent). The indexes in the first row and column indicate class numbers. The rows are supervised classes, and columns are the actual output. Approximately 87% classification accuracy is achieved in average. Figures 26.9 and 26.10 show the satellite true color image and the estimation result, respectively.

TABLE 26.1

The Result (Accuracy, %)

	1	2	3	4	5	6	7	
1	77	0	19	0	0	0	4	Urban
2	11	78	0	0	0	11	0	Sand
3	3	0	89	0	0	0	8	Water
4	13	0	0	84	0	0	2	Mountain
5	0	0	0	0	100	0	0	Vegetation
6	0	1	0	0	1	98	0	Red soil
7	13	0	0	0	0	0	87	Land along the river

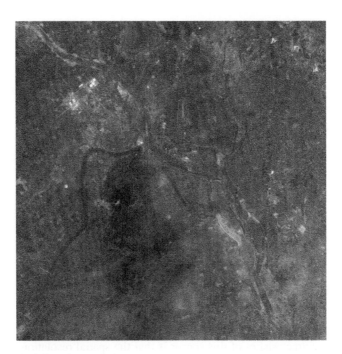

FIGURE 26.9   Satellite image (true color).

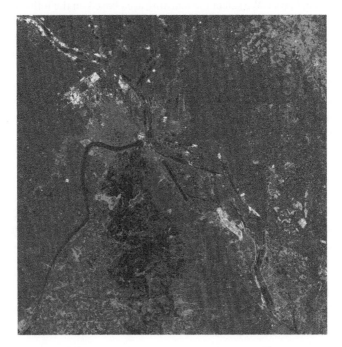

FIGURE 26.10   Result of estimation.

26.4   CONCLUSIONS

In this chapter, we have proposed the preprocessing method for the remote sensing data as the input signals of the neural network. When the proposed method is used, we have achieved the smooth convergence of neural networks with range of inputs. By using the nonoptimal ranges of input signals, it takes more computational costs for learning and the number of neurons in the hidden layer is increased. When many classes are needed to be trained, the input signals are normalized from −1 to 1. Finally the classification to seven categories is achieved with the accuracy of over 80%. As a future work, we need to improve the accuracy of classification in the area along a river.

REFERENCES

1. J. A. Anderson, A. Pellionisz, E. Rosenfeld, editors. *Neural Computing 2: Directions for Research*, The MIT Press, Cambridge, Massachusetts, 1990.
2. E. Sánchez-Sinencio, and C. Lau, editors. *Artificial Neural Networks: Paradigms, Applications, and Hardware Implementations*, IEEE Press, Piscataway, NJ, 1992.
3. Y. Tsuchida, S. Omatu and M. Yoshioka, The land cover estimation with ALOS satellite image using neural-network, *Proc. of the Fifteenth International Symposium on Artificial Life and Robotics* 2010, 428–431, 2010.

27 Twenty-Five Years of Pansharpening
A Critical Review and New Developments

Bruno Aiazzi, Luciano Alparone, Stefano Baronti,
Andrea Garzelli, and Massimo Selva

CONTENTS

27.1 INTRODUCTION

Pansharpening is a branch of data fusion, more specifically of image fusion, which is receiving an ever-increasing attention from the remote sensing community. New-generation spaceborne imaging sensors operating in a variety of ground scales and spectral bands provide huge volumes of data having complementary spatial and spectral resolutions. Constraints on the signal-to-noise ratio (SNR) impose that the spatial resolution must be lower, if the desired spectral resolution is larger. Conversely, the highest spatial resolution is obtained whenever no spectral diversity is required. The trade-off of spectral and spatial resolution makes it desirable to perform a spatial resolution enhancement of the lower-resolution multispectral (MS) data or, equivalently, to increase the spectral resolution of the data set having a higher ground resolution, but a lower spectral resolution; as a limit case, constituted by a unique panchromatic image (Pan) bearing no spectral information.

To pursue this goal, an extensive number of methods have been proposed in the literature over the last two decades. Most of them follow a general protocol, that can be summarized in the following two key points: (1) extract high-resolution geometrical information of the scene, not present in the MS image, from the Pan image; (2) incorporate such spatial details into the low-resolution MS bands, interpolated to the spatial scale of the Pan image, by properly modeling the relationships between the MS bands and the Pan image.

Progresses in pansharpening methods have been substantially motivated by advances in space-borne instruments. Since SPOT-1 (1986) the first system featuring two MS together with one Pan band, over a period of 25 years pansharpening methods have been progressively developed and adjusted to 3 MS bands + Pan (SPOT 2, 3, 4, 5), 6 MS bands + Pan (Landsat ETM+), 4 MS bands + Pan (IKONOS-2, QuickBird-2, GeoEye-1, GeoEye-2, Pleiades), up to 8 MS band + Pan (Worldview-2). All instruments launched during the last decade exhibit a ratio of scales between Pan and MS equal to four, instead of two, like in earlier missions, the presence of a narrow band in the blue (B) wavelengths and the bandwidth of Pan enclosing also part of the near infrared (NIR). While the change in scale ratios has not substantially influenced the development of methods, the presence of the B band, with possibility of true color display, and of a P, which embraces NIR, but not B, has created significant problems in earlier methods dramatically motivating the development of new alternatives.

Many image fusion methods have been proposed for combining MS and Pan data. Some methods, such as intensity–hue–saturation (IHS) [1], Brovey transform (BT) [2], and principal component analysis (PCA) [3] provide superior visual high-resolution multispectral images but ignore the requirement of high-quality synthesis of spectral information [4]. While these methods are useful for visual interpretation, high-quality synthesis of spectral information is very important for most remote-sensing applications based on spectral signatures, such as lithology and soil and vegetation analysis [5].

Over the last two decades, the existing image fusion methods have been classified into several groups. Schowengerdt [6] classified them into spectral domain techniques, spatial domain techniques, and scale space techniques. However, scale space techniques, for example, wavelets, are generally implemented by means of digital filters that are spatial domain techniques. Therefore, it is expected that methods such as HPF [7] and AWL [8], which differ by the type of digital filter, actually belong to the same class.

Ranchin and Wald [9,10] classified pansharpening methods into three groups: projection and substitution methods, relative spectral contribution methods, and those relevant to the ARSIS concept (Amélioration de la Résolution Spatiale par Injection de Structures), originally employing the decimated wavelet transform (DWT) [9]. It was found that many of the existing image fusion methods, such as HPF, GLP [11] and ATW methods, can be accommodated within the ARSIS concept. However, the first two classes, namely "projection and substitution," for example, IHS, and "relative spectral contribution," for example, BT are equivalent. In fact, Tu et al. [12] performed a mathematical development and found that IHS PCA, BT may not involve explicit calculation of the complete spectral transformation but only of the component that will be substituted, for example, intensity for both IHS and BT. Therefore, IHS and BT differ only in the way spatial details are weighted before their injection and not in the way they are extracted from the Pan image. Both IHS and BT fusion can be extrapolated to an arbitrary number of spectral bands.

The main objective of this chapter is to propose a comprehensive framework encompassing earlier classification attempts of pansharpening methods, which makes it possible to categorize, compare, and evaluate existing image fusion methods, as well as to develop new ones. The almost totality of methods can be accommodated into one of two classes. Such classes uniquely differ by the way spatial details are extracted from the Pan image. According to the new approach, all methods belonging to either of the two main classes possess complementary and predictable characteristics of spectral and spatial quality of fusion products, as well as typical behaviors in the presence of specific anomalies in the data, like aliasing of MS bands, mis-registration between MS and Pan and temporal changes between MS and Pan observations. A thorough experimental section with comparisons of several methods on very high-resolution (VHR) MS + Pan data highlight the assets of the new classification approach.

27.2 A CRITICAL REVIEW OF FUSION METHODS

According to the most recent studies carried out by the authors, the majority of image fusion methods can be divided into two main classes. Such classes uniquely differ in the way the spatial details are extracted from the Pan image.

- Techniques that employ linear space-invariant digital filtering of the Pan image to extract the spatial details, that is, the geometrical information, that will be added to the MS bands [13]; all methods employing multiresolution analysis (MRA) belong to this class.
- Techniques that yield the spatial details as pixel difference between the Pan image and a nonzero-mean component obtained from a spectral transformation of the MS bands, without any spatial filtering of the former. They are equivalent to substitution of such a component with the Pan image followed by reverse transformation to produce the sharpened MS bands [14], same as for plain CS-based methods.

Regardless of how spatial details have been obtained, their injection into the interpolated MS bands may be weighed by suitable gains, different for each band, possibly space varying, that is, a different gain at each pixel. Algorithms featuring context adaptive, that is, local, models generally perform better than schemes based on models fitting each band globally [15]. A pixel-varying injection model is capable of defining fusion algorithms based on modulation [16], for example, BT for the class of methods in Figure 27.1b and SFIM [4] for the methods outlined in Figure 27.1a.

The two classes of methods described above exhibit complementary spectral–spatial quality trade-off. Methods without spatial filtering, provide fused images with high geometrical quality of spatial details, but with possible spectral impairments. Methods employing spatial filtering are spectrally accurate in general, but may be unsatisfactory in terms of spatial enhancement. However, if the spectral combination of bands is optimized for spectral quality of pansharpened products and spatial filtering is optimized for spatial quality, the two categories of methods yield very similar results in terms of overall quality [15].

Figure 27.1 shows the flowcharts of the two approaches. In the former case, filtering is crucial: MTF filtering yields best results [17]. In the latter case, key point is the spectral transformation defined by the set of weights $\{w_k\}$ [14]. Detail-injection gains $\{g_k\}$ are used in both cases. The MTF lowpass filter in Figure 27.1b is used only to calculate the spectral weights $\{w_k\}$ and not to directly produce the spatial details, like in Figure 27.1a.

27.2.1 COMPONENT SUBSTITUTION

CS-based pansharpening is a typology of simple and fast techniques based on a spectral transformation of the original bands in a new vector space. Most widely used transformations are intensity–hue–saturation (IHS), principal components analysis (PCA) and Gram–Schmidt (GS) orthogonalization procedure [14,18]. IHS fusion technique, originally defined for three bands only, has been extended to an arbitrary number of spectral bands [19] and denoted as generalized IHS (GIHS). The rationale of CS fusion is that one of the transformed components (usually the first component or intensity, I) is substituted by the high-resolution panchromatic image, P, before the inverse transformation is applied. To ensure global preservation of radiometry, P is histogram-matched to I, in such a way that the histogram-matched sharpening P, once degraded to the spatial resolution of I, exhibits the same global mean and variance as I. However, since the histogram-matched P and I may not have the same local radiometry, spectral distortion, appearing as local color changes in a composition of three bands at a time, may occur in pansharpened products. To mitigate local spectral distortion, I may be taken as a linear combination of the MS bands with weighting coefficients adjusted to the extent of overlap between the spectral response of each MS channel and that of the P [19] (GIHSF in this chapter). In principle, if the lowpass approximation

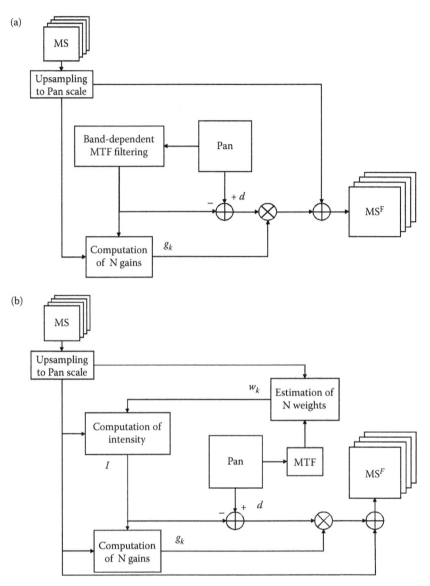

FIGURE 27.1 Flowchart of the two main pansharpening approaches: (a) based on filtering the Pan image, or more generally on multiresolution analysis; (b) based on a spectral combination of bands, without filtering the Pan image, or more generally on component/projection substitution.

of P, synthesized by combining the spectral channels, exactly matches the low-resolution version of P, spectral distortion does not occur [19,14].

It is noteworthy that the spectral transformation need not be explicitly calculated to perform fusion [14]. In principle, any nonsingular square matrix would define a possible transformation. However, fusion depends only on the first transformed component defined as a linear combination of the MS bands with weights $\{w_k\}$. Therefore, fusion methods labeled as *projection substitution* and *relative spectral contribution* [10] are equivalent. Both may be achieved by the flowchart in Figure 27.1b. In this perspective Table 27.1 reports the spectral weights $\{w_k\}$ and the injection gains $\{g_k\}$ for several of the CS schemes proposed so far.

Concerning GIHS, an exhaustive mathematical formulation is reported in Ref. [19]. Its description and implementation is immediate from the scheme of Figure 27.1b.

TABLE 27.1

Expressions for the Weights of the Scheme Reported in Figure 27.1b Relative to the Various CS-Based Methods

Method	w_1	w_2	w_3	w_4	g_1	g_2	g_3	g_4
IHS	1/3	1/3	1/3	0	1	1	1	0
GIHS	1/4	1/4	1/4	1/4	1	1	1	1
GIHSF	1/12	1/4	1/3	1/3	1	1	1	1
GIHSA	$\hat{w}_1$	$\hat{w}_2$	$\hat{w}_3$	$\hat{w}_4$	1	1	1	1
BT	1/4	1/4	1/4	1/4	$\dfrac{\tilde{\mathbf{B}}}{\mathbf{I}}$	$\dfrac{\tilde{\mathbf{G}}}{\mathbf{I}}$	$\dfrac{\tilde{\mathbf{R}}}{\mathbf{I}}$	$\dfrac{\tilde{\mathbf{NIR}}}{\mathbf{I}}$
PCA	x_{11}	x_{12}	x_{13}	x_{14}	x_{11}	x_{12}	x_{13}	x_{14}
GS	1/4	1/4	1/4	1/4	$\dfrac{cov(\mathbf{I},\tilde{\mathbf{B}})}{var(\mathbf{I})}$	$\dfrac{cov(\mathbf{I},\tilde{\mathbf{G}})}{var(\mathbf{I})}$	$\dfrac{cov(\mathbf{I},\tilde{\mathbf{R}})}{var(\mathbf{I})}$	$\dfrac{cov(\mathbf{I},\tilde{\mathbf{NIR}})}{var(\mathbf{I})}$
GSA	$\hat{w}_1$	$\hat{w}_2$	$\hat{w}_3$	$\hat{w}_4$	$\dfrac{cov(\mathbf{I},\tilde{\mathbf{B}})}{var(\mathbf{I})}$	$\dfrac{cov(\mathbf{I},\tilde{\mathbf{G}})}{var(\mathbf{I})}$	$\dfrac{cov(\mathbf{I},\tilde{\mathbf{R}})}{var(\mathbf{I})}$	$\dfrac{cov(\mathbf{I},\tilde{\mathbf{NIR}})}{var(\mathbf{I})}$

For GS, I is the average of the four MS bands. In any case, for all the GS-based methods the ith coefficient g_i is proportional to the covariance value between the synthesized intensity and the expanded ith MS band. Proof of this relationship is reported in Ref. [14].

GIHSA and GSA are the schemes reported in Ref. [14] whose weights $\{w_k\}$ are obtained as regression coefficients of the minimization of the difference between the P and I images according to the least mean square error (LMSE) criterion. The gains $\{g_k\}$ are global: all fixed and equal to 1 for GIHSA while computed same as for GS in GSA. The context-adaptive version of GSA (GSA-CA) exploits the same formula of GSA for gains; for GSA-CA $\{g_k\}$ are adaptive and computed on a local window set by users.

Concerning PCA, it can be easily shown that, if Pan is histogram-matched to the first principal component (PC1), $w_i = g_i$, both being the ith component of the first eigenvector x_1 constituting the first row or column of the unitary transformation matrix.

Also the fusion scheme based on BT [2] can be reported in the framework of CS schemes as shown in Ref. [20] with the weights indicated in Table 27.1.

27.2.2 Multiresolution Analysis

MRA-based techniques substantially split the spatial information of the MS bands and of the P image into a series of bandpass spatial frequency channels. The high-frequency channels are inserted into the corresponding channels of the interpolated MS bands. The sharpened MS bands are synthesized from their new sets of spatial frequency channels. The "à trous" wavelet (ATW) transform and the Laplacian pyramid are most widely used to perform the MRA [8,11]. In the former case, shown in Figure 27.1a, the zero-mean high-frequency spatial details are simply given as the difference between, P and its lowpass-filtered version, P_L. Recent studies [17] have demonstrated that if the lowpass filter is designed in such a way that its frequency response matches the modulation transfer function (MTF) of the spectral channel into which details will be injected, the spatial enhancement provided by MRA techniques becomes comparable to that of CS techniques. The gains $\{g_k\}$ are chosen, same as for CS schemes, according to the desired injection model. MRA-based techniques may be accommodated within the framework of ARSIS, originally employing the decimated discrete wavelet transform (DWT) [9], but later extended to other types of multiresolution analysis [13].

27.3   QUALITY ASSESSMENT OF FUSION PRODUCTS

Quality assessment of pansharpened MS images is not an easy task [7,21–23] since reference originals are generally not available. When spatially degraded MS images are processed for pansharpening, and thus reference MS images are available for objective comparisons, assessment of *fidelity* to the reference requires computation of a number of statistical indexes, the most widely used of which are reviewed in the following and utilized for experiments.

27.3.1   WALD'S PROTOCOL

A general paradigm usually accepted in the research community for quality assessment of fused images was first proposed by Wald et al. [21] and rediscussed in Ref. [13]. Such a paradigm is found on three properties the fused data have to cope with as much as possible.

The first property, known as *consistency*, requires that *any fused image Â, once degraded to its original resolution, should be as identical as possible to the original image A*. To achieve this, the fused image Â is spatially degraded to the same scale of A, thus obtaining an image Â*. Â* has to be very close to A. It is worthwhile that consistency measures spectral quality after spatial enhancement and is a condition necessary but not sufficient to state that a fused image possesses the necessary quality requirements, that is, both spectral and spatial quality.

The second property, known as *synthesis* states that *any image Â fused by means of a high-resolution (HR) image should be as identical as possible to the ideal image A$_I$ that the corresponding sensor, if existent, would observe at the resolution of the HR image*. Images are regarded here as scalar images, that is, one spectral band of an MS image. Similarity is measured by statistics of scalar pixels between fused image and ideal HR reference image. Besides scalar similarity indexes between individual bands of the MS image, the synthesis property is checked on the plurality of spectral bands constituting an MS image, in order to check the multispectral properties of the MS image, that is, of the whole set of fused bands: *the multispectral vector of images Â fused by means of a high resolution (HR) image should be as identical as possible to the multispectral vector of ideal images A$_I$ that the corresponding sensor, if existent, would observe at the spatial resolution of the HR image*. This second part of synthesis property is also known as third property.

Both the synthesis properties may not generally be directly verified, since A$_I$ is generally not available. Therefore, synthesis is usually checked at degraded spatial scales. Spatial degradation is achieved by means of proper lowpass filtering followed by decimation by a factor equal to the scale ratio of Pan to MS data sets. The multispectral image $A*$ and the panchromatic image P* are created from the original sets of images A and Pan. Pan is degraded to the resolution of the multispectral image and A to a lower resolution depending on the scale ratio for which fusion is assessed. The fusion method is applied to these two sets of images, resulting in a set of fused images at the resolution of the original MS image. The MS image serves now as reference and the second and third properties can be tested. It is noteworthy that fulfillment of synthesis properties is a condition both necessary and sufficient, provided that the similarity check performed at degraded spatial scale is consistent with the same check if it were hypothetically performed at the full scale; in other words, if the quality observed for the fused products is assumed to be close to the quality that would be observed for the fused products at the full scale. This point is crucial, especially for methods employing digital filters to analyze the Pan image. In fact, whenever simulations are carried out at degraded spatial scale, the lowpass filter of the fusion methods is cascaded with the lowpass filter used for decimation. Hence, fusion at full scale uses the former only; fusion at degraded scale uses the cascade of the former with the latter, that is, uses a different filter. This explains why methods providing acceptable spatial enhancement at degraded scale yield poor enhancement when they are used at full scale [22]. This issue has been largely discussed by the authors [17].

27.3.2 Quality Indexes

Quality indexes and/or distortion measurements have been defined in order to measure the similarity between images, either scalar or vector, as required, for example, by both consistency and synthesis properties of Wald's protocol. This review is limited to indexes that are established in the literature as providing results consistent with photo-analysis of pansharpened products. Under this perspective quality measures based on Shannon's entropy, or auto-information, though used sometimes, have never given evidence of being suitable for this task.

27.3.2.1 Indexes for Scalar Valued Images

Mean bias ($\Delta\mu$): given two scalar images A and B with means $\mu(A)$ and $\mu(B)$, approximated by averages $\bar{A}$ and $\bar{B}$, the mean bias is defined as

$$\Delta\mu \triangleq \mu(A) - \mu(B) \tag{27.1}$$

$\Delta\mu$ is a distortion; hence its ideal value is zero.

Root mean square error (RMSE): RMSE between A and B is defined as

$$\text{RMSE} \triangleq \sqrt{E[(A-B)^2]} \tag{27.2}$$

in which the expected value is approximated by a spatial average. RMSE is a distortion, whose ideal value is zero, if and only if $A = B$.

Cross-correlation, or correlation, coefficient (CC): CC between A and B is defined as

$$\text{CC} \triangleq \frac{\sigma_{A,B}}{\sigma_A \sigma_B} \tag{27.3}$$

where $\sigma_{A,B}$ is the covariance between A and B, given by $E[(A - \mu(A))(B - \mu(B))]$, and σ_A is the standard deviation of A given by $\sqrt{E[(A - \mu(A))^2]}$. In the same way $\sqrt{E[(B - \mu(B))^2]}$ represents the standard deviation of B. CC takes values in the range $[-1,1]$. CC = 1 means that A and B differ only by a global mean offset and gain factor. CC = -1 means that B is the negative of A (A and B still may differ by a gain and an offset). CC being a similarity index, its ideal value is one.

Universal image quality index (UIQI) [24] measures the similarity between two scalar images A and B and is defined as

$$Q = \frac{4\sigma_{A,B} \cdot \bar{A} \cdot \bar{B}}{(\sigma_A^2 + \sigma_B^2)[(\bar{A})^2 + (\bar{B})^2]} \tag{27.4}$$

in which $\sigma_{A,B}$ denotes the covariance between A and B, $\bar{A}$ and $\bar{B}$ are the means, and σ_A^2 and σ_B^2 the variances of A and B, respectively. Equation 27.4 may be equivalently rewritten as a product of three factors:

$$Q = \frac{\sigma_{A,B}}{\sigma_A \cdot \sigma_B} \cdot \frac{2\bar{A} \cdot \bar{B}}{[(\bar{A})^2 + (\bar{B})^2]} \cdot \frac{2\sigma_A \cdot \sigma_B}{(\sigma_A^2 + \sigma_B^2)}. \tag{27.5}$$

The first one is the correlation coefficient (CC) between A and B. The second one is always ≤ 1, from Cauchy–Schwartz inequality, and is sensitive to bias in the mean of B with respect to A. The third term is also ≤ 1 and accounts for changes in contrast between A and B. Apart from CC which ranges in $[-1, 1]$, being equal to 1 *iff* $A = B$, and equal to -1 *iff* $B = 2\bar{A} - A$, that is, B is the negative of A, all the other terms range in [0,1], if $\bar{A}$ and $\bar{B}$ are nonnegative. Hence, the dynamic range of Q is [−1,1] as well, and the ideal value $Q = 1$ is achieved *iff* $A = B$ for all pixels. To increase the discrimination capability of the three factors in (5), all statistics are calculated on suitable $N \times N$ image blocks and the resulting values of Q averaged over the whole image to yield a unique global score.

27.3.2.2 Indexes for Vector-Valued Images

Spectral angle mapper (SAM). Given two spectral vectors, v and $\hat{v}$, both having L components, in which $\mathbf{v} = \{v_1, v_2, \ldots, v_L\}$ is the original spectral pixel vector $v_l = A^{(l)}(m, n)$ while $\hat{\mathbf{v}} = \{\hat{v}_1, \hat{v}_2, \cdots, \hat{v}_L\}$ is the distorted vector obtained by applying fusion to the coarser resolution MS data, that is, $\hat{v}_l = \hat{A}^{(l)}(m, n)$, the spectral angle mapper (SAM) denotes the absolute value of the spectral angle between the two vectors:

$$\mathrm{SAM}(\mathbf{v}, \hat{\mathbf{v}}) \triangleq \arccos\left(\frac{< \mathbf{v}, \hat{\mathbf{v}} >}{\|\mathbf{v}\|_2 \cdot \|\hat{\mathbf{v}}\|_2} \right). \tag{27.6}$$

SAM(A,B) is defined according to (27.6) as E[SAM(a, b)] where a and b denote the generic pixel vector element of MS images A and B, respectively. SAM is usually expressed in degrees and is equal to zero when images A and B are spectrally identical, that is, all pixel vectors differ only by their moduli between A and B.

Relative dimensionless global error in synthesis (ERGAS). ERGAS was proposed by Ranchin et al. [13] as an error index that offers a global indication of the quality of a fused product, and is given by

$$\mathrm{ERGAS} \triangleq 100 \frac{d_h}{d_l} \sqrt{\frac{1}{L} \sum_{l=1}^{L} \left(\frac{RMSE(l)}{\mu(l)} \right)^2} \tag{27.7}$$

where d_h/d_l is the ratio between pixel sizes of Pan and MS, for example, 1/4 for Ikonos and QuickBird data, $\mu(l)$ is the mean (average) of the lth band, and L is the number of bands. Low values of ERGAS indicate similarity between multispectral data.

Q4 is a multispectral extension of UIQI suitable for images having four spectral bands, introduced by three of the authors for quality assessment of Pansharpened MS imagery [25]. For MS images with four spectral bands, let a, b, c, and d denote the radiance values of a given image pixel in the four bands, typically acquired in the B, G, R, and NIR wavelengths. Q4 is made up of different factors accounting for correlation, mean bias, and contrast variation of each spectral band, as well as of spectral angle. Since the modulus of the hypercomplex correlation coefficient (CC) measures the alignment of spectral vectors, its low value may detect when radiometric distortion is accompanied by spectral distortion. Thus, both radiometric and spectral distortions may be encapsulated in a unique parameter. Let

$$\begin{aligned} \mathbf{z}_A &= a_A + \mathbf{i}b_A + \mathbf{j}c_A + \mathbf{k}d_A \\ \mathbf{z}_B &= a_B + \mathbf{i}b_B + \mathbf{j}c_B + \mathbf{k}d_B \end{aligned} \tag{27.8}$$

denote the 4-bands reference MS image and the fusion product, respectively, both expressed as quaternions or hypercomplex numbers. The $Q4$ index is defined as

$$Q4 \triangleq \frac{4\left|\sigma_{z_A z_B}\right| \cdot \left|\overline{\mathbf{z}_A}\right| \cdot \left|\overline{\mathbf{z}_B}\right|}{\left(\sigma_{z_A}^2 + \sigma_{z_B}^2\right)\left(\left|\overline{\mathbf{z}_A}\right|^2 + \left|\overline{\mathbf{z}_B}\right|^2\right)}. \tag{27.9}$$

Equation 27.9 may be written as the product of three terms:

$$Q4 = \frac{\left|\sigma_{z_A z_B}\right|}{\sigma_{z_A} \cdot \sigma_{z_B}} \cdot \frac{2\sigma_{z_A} \cdot \sigma_{z_B}}{\sigma_{z_A}^2 + \sigma_{z_B}^2} \cdot \frac{2\left|\overline{\mathbf{z}_A}\right| \cdot \left|\overline{\mathbf{z}_B}\right|}{\left|\overline{\mathbf{z}_A}\right|^2 + \left|\overline{\mathbf{z}_B}\right|^2} \tag{27.10}$$

the first of which is the modulus of the hypercomplex CC between $\mathbf{z}_A$ and $\mathbf{z}_B$ and is sensitive both to loss of correlation and to spectral distortion between the two MS data sets. The second and third terms, respectively, measure contrast changes and mean bias on all bands simultaneously. Ensemble expectations are calculated as averages on $N \times N$ blocks. Hence, $Q4$ will depend on N as well. Eventually, $Q4$ is averaged over the whole image to yield the *global* score index. Alternatively, the minimum attained by $Q4$ over the whole image may represent a measure of *local* quality. $Q4$ assume values in the range [0,1] and is equal to 1 when A and B are equal.

27.3.3 QNR PROTOCOL

The Quality w/No Reference (QNR) protocol [26] calculates the quality of the pansharpened images without requiring a high-resolution reference MS image. QNR comprises two indexes, one pertaining to spectral and the other to spatial distortion. The two distortions may be combined together to yield a unique quality index. However, in many cases they are kept separate. Both spectral and spatial distortions are calculated through similarity measurements of couples of scalar images performed by means of UIQI.

The spectral distortion D_λ is calculated between the low-resolution MS images and the fused MS images. Hence, for determining the spectral distortion two sets of interband UIQI values are calculated separately at low and high resolutions. The differences of corresponding UIQI values at the two scales yields the spectral distortion introduced by the pansharpening process. Thus, spectral distortion can be represented mathematically as

$$D_\lambda = \sqrt[p]{\frac{1}{N(N-1)} \sum_{l=1}^{N} \sum_{r=1, r \neq l}^{N} \left| Q(\tilde{M}_l, \tilde{M}_r) - Q(\hat{M}_l, \hat{M}_r) \right|^p} \tag{27.11}$$

where, $\tilde{M}_l$ represents the low-resolution lth MS band, $\hat{M}$ the pansharpened MS band, $Q(A, B)$ represents UIQI between A and B and N is equal to the number of MS bands. The exponent p is an integer possibly chosen to emphasize large difference values by default p is set equal to 1.

The spatial distortion D_s is determined by calculating UIQI between each MS band and the Pan image degraded to the resolution of MS and again between fused MS and full resolution Pan. The difference between the two values yields the spatial distortion:

$$D_s = \sqrt[q]{\frac{1}{N} \sum_{l=1}^{N} \left| Q(M_l, P_L) - Q(\hat{M}_l, P) \right|^q} \tag{27.12}$$

in which P_L denotes the Pan image degraded to the resolution of MS and P the high-resolution Pan image. The exponent q is one by default.

The rationale of the QNR protocol is as follows:

1. The interrelationships (measured by UIQI) between couples of the low-resolution MS bands should not change with resolution, that is, once the MS image has been pansharpened.

2. The relationships between each MS band and a low-resolution version of the Pan image should be identical to those between each pansharpened MS band and the full-resolution Pan image.
3. Differences in similarity values computed at low- and high-resolution measure the distortion, either spectral (MS–MS) or spatial (MS–Pan).

27.4 EXPERIMENTAL RESULTS

A typical case study representative of what can be found in the literature is first reported concerning pansharpening and refers to an IKONOS observation. Afterwards a set of experiments has been carried out on VHR data, QuickBird, IKONOS, and Geo-Eye, to evidence the sensitiveness of fusion to aliasing, misregistration, and temporal misalignment. Two methods have been chosen as peculiar of the two classes described. The improved Gram–Schmidt spectral sharpening (GSA) [14] is a modified version of ENVI's GS [18] with improved spectral quality, thanks to a least-squares (LS) calculation of the spectral weights w_k defining the generalized intensity I. The second method is based on ATW [8,27] with MTF-matched Gaussian-like reduction filter [17]. The injection model, g_k, is equal to the covariance between interpolated kth MS band and either I or lowpass-filtered Pan (P_L), divided by the variance of either I or P_L [15].

27.4.1 PANSHARPENING RESULTS

Quantitative results are reported, on an IKONOS data set collected on the area of Toulouse, France, with spatial resolutions of 1 m for the Pan and 4 m for the MS bands. The Pan size is 2048×2048, while the MS size is four times lower. Table 27.2 reports quantitative scores for all the methods by considering average metrics as Q4 [25], SAM (Spectral Angle Mapper) [26], and ERGAS [9]. The quality index Q4 depends on mean, variance, and correlation coefficient (CC) variations, while ERGAS and SAM are sensitive to distortion figures, as RMSE and the spectral changes, respectively. The EXP entry concerns the plain (bicubic) resampling of the MS data set at the scale of the Pan [12]. Scores highlight the benefits of CA strategy on both GSA and ATW in all the metrics. The gains attained by SAM show that CA modeling can reduce spectral distortions, as well as a better recovery of local spatial features even if visual analysis could reveal some perturbation on texture and edge areas with the introduction of local spatial distortions.

It is to be noted that once optimized GSA and ATW obtain practically the same score. The performance of AWLP [27] is far lower than GSA and ATW methods on this image. This is mainly due to the SAM constraint of AWLP that operates by maintaining the same SAM of the expanded image (high in this case); conversely other methods well succeed in recovering the SAM of the fused images that significantly decrease with respect to the expanded image. Such recovery of SAM leads to further benefits that appear also on ERGAS and specially on $Q4$.

27.4.2 ALIASING

A point on which researchers have usually discussed is the tendency of some methods to introduce distortions and in particular aliasing [12]. Here we want to point out that pansharpening methods do

---

TABLE 27.2

Average Quality Scores of the Fused Images for IKONOS Data

Metric	EXP	GIHS	AWLP	GSA	GSA-CA	ATW	ATW-CA
$Q4$	0.630	0.855	0.797	0.921	0.926	0.922	0.928
SAM (°)	4.85	4.05	4.85	2.96	2.87	2.94	2.85
ERGAS	5.94	3.29	3.64	2.76	2.69	2.73	2.66

(a)   (b)   (c)   (d)

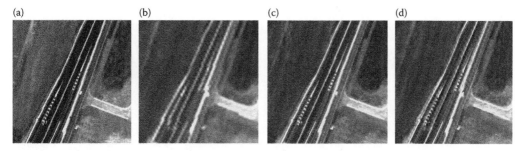

FIGURE 27.2 (**See color insert.**) Details of original and fused QuickBird data (256 × 256 at 0.7 m scale). (a) panchromatic; (b) true color display of 2.8 m MS interpolated at 0.7 m; (c) enhanced Gram–Schmidt; (d) "à trous" wavelet with MTF-matched filter and global injection model.

not produce aliasing when correctly applied but conversely have the capability to compensate the aliasing that is produced by the acquisition systems when sampling the analog input signal. A mathematical formulation for that can be found in Ref. [28]. We only report here the result and show an example that CS schemes have the capability to substantially recover aliasing artifacts.

QuickBird data have been chosen for this experiment, since the amount of aliasing is related to the value of the bell-shaped MTF at Nyquist frequency, which is greater for QuickBird than for IKONOS [29,30]. Figure 27.2a–d reports a representative experiment. The startup are aliasing-free Pan and aliased MS. Aliasing is due to insufficient sampling and is noticeable around sharp oblique contours of the scene as annoying jagged patterns. Such patterns appear in Figure 27.2b interpolated at the spatial scale of Pan. The outcomes of the two methods, GSA and ATW, are completely different: the former rejects aliasing almost totally; the latter leaves aliasing patterns almost unchanged. Spectral quality is always good, thanks to the good properties of all MRA methods in general and of GSA in particular: no significant change in color hues is perceivable in all fusion products. Under the perspective of aliasing rejection, a CS method optimized for spectral quality [14] is preferable to an MRA-based method, yet optimized for spatial quality [17].

27.4.3 MISREGISTRATION

As in the case of aliasing, CS schemes are more robust to misregistration between P and MS images than MRA schemes. In fact, CS schemes exhibit the capability to compensate misregistration and a mathematical demonstration of this property can be found in Ref. [28]. As an example an IKONOS data set has been chosen for simulations. For this image the amount of aliasing that could mask the effects of misregistration is negligible. The MS and Pan data sets have been misaligned along both x and y by 4 pixels at Pan scale, that is, one pixel at MS scale. The fusion methods compared are still GSA and ATW. Two distinct experiments are shown in Figure 27.3. For each experiment, Pan, interpolated MS, GSA fusion of overlapped MS and Pan, GSA fusion of misregistered MS and P, ATW fusion of overlapped data and ATW fusion of misregistered data are displayed. The visual results are stunning. While GSA and ATW behave quite similarly in the absence of misregistration, on misaligned data, the former produces an image with high resemblance to the ideal case, apart from colors, which are obviously shifted together with the MS original. The geometry of spatial details is preserved to a large extent notwithstanding the 4-pel shift. Conversely, ATW exhibits a better preservation of colors, but cannot tolerate the 4-pel shift. In Figure 27.3f the circular target is split into its lowpass and edge components.

In order to quantify the losses in quality, either spectral or spatial/geometric, of the two sample methods, GSA and ATW, in the presence of misregistration, separate measurements of SAM and of spatial distortion, D_S, according to the QNR protocol [26] have been performed between fused and nonfused MS data. The choice of distortion metrics stems from the requirement of carrying out

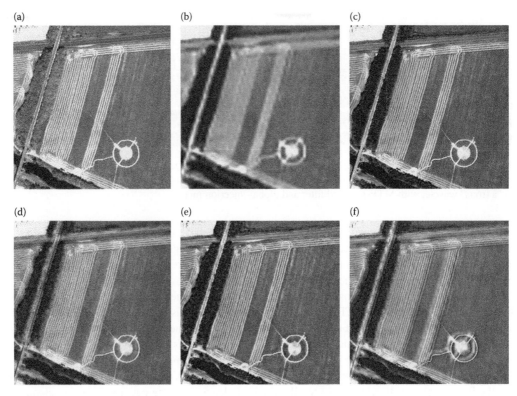

FIGURE 27.3 (**See color insert.**) Details of original and fused IKONOS data (256×256 at 1 m scale). (a) panchromatic; (b) true color display of 4 m MS interpolated at 1 m; (c) enhanced Gram–Schmidt; (d) enhanced Gram–Schmidt with data misaligned by 4 pels at panchromatic scale (1 pel at MS scale); (e) "à trous" wavelet with MTF-matched filter and global injection model; (f) "à trous" wavelet with MTF-matched filter and global injection model with data misaligned by 4 pels at panchromatic scale.

measurements at full spatial scale. Table 27.3 highlights that average SAM of the images obtained from misregistered data is simply achieved by adding the same constant offset to the corresponding distortions in the case of null shift. Conversely, ATW yields D_S 12% lower than GSA without misregistration; but when the data are shifted GSA attains a distortion that is almost 30% lower than ATW. Such distortion values match the higher degree of shift-tolerance of GSA, visually remarked in Figure 27.3.

TABLE 27.3

Geometrical (D_S) and Spectral (SAM) Distortions of Fusion Based on Filtering (ATW) and on Plain Projection (GSA)

	ATW	MIS ATW	GSA	MIS GSA
DS	0.0860	0.2869	0.0974	0.2017
SAM	0.4359	3.1315	0.4691	3.1562

Note: Test image IKONOS, 1024×1024. Interpolated MS and Pan exactly overlapped and misregistered (MIS) by 4 pels along x and y.

27.4.4 TEMPORAL MISALIGNMENT

The goal of this subsection is to investigate the effects of temporal misalignments between multi-spectral (MS) and panchromatic (Pan) observations when they are fused together to yield a pansharpened product. In fact, some satellite instruments do not produce a Pan image (e.g., ASTER) others have only the Pan image (EROS A and B).

Two images acquired in 2010 by GeoEye-1 on May 27th and July 13th, respectively, with different incidence angles on the area of Collazzone, in Central Italy, were available. The nominal spatial resolution for GeoEye-1 is 0.5 m for the Pan and 2 m for the MS images, respectively. The radiometric resolution is 11 bit. All the images have been orthonormalized by using a digital elevation model (DEM) available at 10 m resolution for all $x - y - z$ spatial coordinates. Some residual spatial misalignment is present after the orthonormalization procedure because of inadequate resolution of the DEM. However, the presence of misalignments does not influence this analysis.

Figure 27.4 shows a 512×512 detail of the whole scene that has been processed. True color display of MS and Pan is reported for the two dates in May and July. Many changes occurred between the two acquisitions. Such changes are of interest for the present analysis, since the main objective of this chapter is to investigate the behaviors of MRA- and CS-based fusion methods when temporal variations of the scene occur between the MS and Pan acquisitions. To this purpose, visual evaluations have been considered in this chapter. Fusion results are reported for two different modalities: (i) with both MS and Pan images of July; (ii) with MS of July and Pan of May. Two simple, yet optimized, methods developed by the authors have been selected as representative of MRA and CS fusion.

Figure 27.5 shows the details of the images. The upper part of Figure 27.5 refers to case (i). Visual analysis reveals that the quality of the fused products is very good and quite comparable.

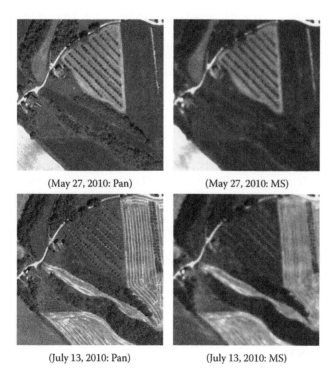

(May 27, 2010: Pan)   (May 27, 2010: MS)

(July 13, 2010: Pan)   (July 13, 2010: MS)

FIGURE 27.4 **(See color insert.)** GeoEye-1 MS + Pan (512×512 at 0.5 m). Pan and MS images acquired both on May 27th 2010; Pan and MS images acquired both on July 13, 2010.

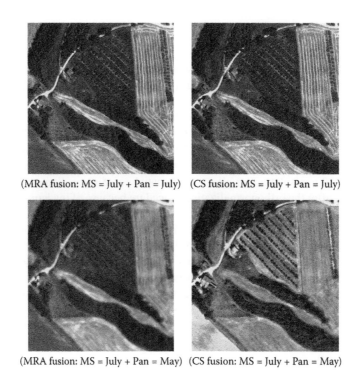

(MRA fusion: MS = July + Pan = July)   (CS fusion: MS = July + Pan = July)

(MRA fusion: MS = July + Pan = May)   (CS fusion: MS = July + Pan = May)

FIGURE 27.5   **(See color insert.)** Details of pansharpened GeoEye-1 data (512 × 512 at 0.5 m scale).

The middle part of Figure 27.5 refers to case (ii). MRA maintains most of the color feature of the original MS image and only some appears blurred due to temporal changes and due to misregistration. Conversely, CS exhibits a heavy color distortion even if spatial details are sharper than for MRA.

27.5   CONCLUDING REMARKS

Twenty-five years of pansharpening methods have been synthesized by the proposed categorization into two classes: methods using digital spatial filters to extract the geometrical details of the Pan image and methods not using spatial filters. The first class roughly corresponds to Wald's ARSIS that is based on MRA. The second class comprises all component/projection substitution methods (CS). However, according to the new classification, hybrid methods (CS + MRA) belong to the first class because they use digital spatial filter. Methods belonging to the first class may be improved by optimizing the digital filter; methods of the second class by optimizing the combination of bands yielding the intensity component.

The experimental section has demonstrated that the two classes exhibit complementary features and different behaviors in the presence of impairments of the data set, even though under standard conditions optimized versions of methods belonging to different classes yield almost identical results.

ACKNOWLEDGMENTS

The authors wish to thank Roberto Carlà and Leonardo Santurri of IFAC-CNR for kindly providing the GeoEye-1 dataset.

REFERENCES

1. Carper, W., Lillesand, T., Kiefer, R. The use of intensity–hue–saturation transformations for merging SPOT panchromatic and multispectral image data. *Photogrammetric Engineering and Remote Sensing* **56**, 1990, 459–467.

2. Gillespie, A.R., Kahle, A.B., Walker, R.E. Color enhancement of highly correlated images-II. Channel ratio and "chromaticity" transform techniques. *Remote Sensing of Environment* **22**, 1987, 343–365.

3. Shettigara, V.K. A generalized component substitution technique for spatial enhancement of multispectral images using a higher resolution data set. *Photogrammetric Engineering and Remote Sensing* **58**, 1992, 561–567.

4. Liu, J.G. Smoothing filter based intensity modulation: A spectral preserve image fusion technique for improving spatial details. *International Journal of Remote Sensing* **21**, 2000, 3461–3472.

5. Garguet-Duport, B., Girel, J., Chassery, J.M., Pautou, G. The use of multiresolution analysis and wavelet transform for merging SPOT panchromatic and multispectral image data. *Photogrammetric Engineering and Remote Sensing* **62**, 1996, 1057–1066.

6. Schowengerdt, R.A. *Remote Sensing: Models and Methods for Image Processing*. Academic Press, Orlando, FL, 1997.

7. Chavez, P.S., Stuart, J., Sides, C., Anderson, J.A. Comparison of three different methods to merge multi-resolution and multispectral data: Landsat TM and SPOT panchromatic. *Photogrammetric Engineering and Remote Sensing* **57**, 1991, 259–303.

8. Núnez, J., Otazu, X., Fors, O., Prades, A., Palà, V., Arbiol, R. Multiresolution-based image fusion with additive wavelet decomposition. *IEEE Transactions on Geoscience and Remote Sensing* **37**, 1999, 1204–1211.

9. Ranchin, T., Wald, L. Fusion of high spatial and spectral resolution images: The ARSIS concept and its implementation. *Photogrammetric Engineering and Remote Sensing* **66**, 2000, 49–61.

10. Thomas, C., Ranchin, T., Wald, L., Chanussot, J. Synthesis of multispectral images to high spatial resolution: A critical review of fusion methods based on remote sensing physics. *IEEE Transactions on Geoscience and Remote Sensing* **46**, 2008, 1301–1312.

11. Aiazzi, B., Alparone, L., Baronti, S., Garzelli, A. Context-driven fusion of high spatial and spectral resolution images based on oversampled multiresolution analysis. *IEEE Transactions on Geoscience and Remote Sensing* **40**, 2002, 2300–2312.

12. Tu, T.M., Su, S.C., Shyu, H.C., Haung, P.S. A new look at IHS-like image fusion methods. *Information Fusion* **2**, 2001, 177–186.

13. Ranchin, T., Aiazzi, B., Alparone, L., Baronti, S., Wald, L. Image fusion—The ARSIS concept and some successful implementation schemes. *ISPRS Journal of Photogrammetry and Remote Sensing* **58**, 2003, 4–18.

14. Aiazzi, B., Baronti, S., Selva, M. Improving component substitution Pansharpening through multivariate regression of MS + Pan data. *IEEE Transactions on Geoscience and Remote Sensing* **45**, 2007, 3230–3239.

15. Aiazzi, B., Baronti, S., Lotti, F., Selva, M. A comparison between global and context-adaptive pansharpening of multispectral images. *IEEE Geoscience and Remote Sensing Letters* **6**, 2009, 302–306.

16. Dou, W., Chen, Y., Li, X., Sui, D. A general framework for component substitution image fusion: An implementation using fast image fusion method. *Computers and Geoscience* **33**, 2007, 219–228.

17. Aiazzi, B., Alparone, L., Baronti, S., Garzelli, A., Selva, M. MTF-tailored multiscale fusion of high-resolution MS and Pan imagery. *Photogrammetric Engineering and Remote Sensing* **72**, 2006, 591–596.

18. Laben, C.A., Brower, B.V. Process for enhancing spatial resolution of multispectral imagery using pan-sharpening. US Patent Number 6,011,875, Eastman Kodak Company, 2000.

19. Tu, T.M., Huang, P.S., Hung, C.L., Chang, C.P. A fast intensity–hue–saturation fusion technique with spectral adjustment for IKONOS imagery. *IEEE Geoscience and Remote Sensing Letters* **1**, 2004, 309–312.

20. Wang, Z., Ziou, D., Armenakis, C., Li, D., Li, Q. A comparative analysis of image fusion methods. *IEEE Transactions on Geoscience and Remote Sensing* **43**, 2005, 1391–1402.

21. Wald, L., Ranchin, T., Mangolini, M. Fusion of satellite images of different spatial resolutions: Assessing the quality of resulting images. *Photogrammetric Engineering and Remote Sensing* **63**, 1997, 691–699.

22. Laporterie-Déjean, F., de Boissezon, H., Flouzat, G., Lefévre-Fonollosa, M.J. Thematic and statistical evaluations of five panchromatic/multispectral fusion methods on simulated PLEIADES-HR images. *Information Fusion* **6**, 2005, 193–212.

23. Alparone, L., Wald, L., Chanussot, J., Thomas, C., Gamba, P., Bruce, L.M. Comparison of pansharpening algorithms: Outcome of the 2006 GRS-S data fusion contest. *IEEE Transactions on Geoscience and Remote Sensing* **45**, 2007, 3012–3021.

24. Wang, Z., Bovik, A.C. A universal image quality index. *IEEE Signal Processing Letters* **9**, 2002, 81–84.

25. Alparone, L., Baronti, S., Garzelli, A., Nencini, F. A global quality measurement of pan-sharpened multispectral imagery. *IEEE Geoscience and Remote Sensing Letters* **1**, 2004, 313–317.

26. Alparone, L., Aiazzi, B., Baronti, S., Garzelli, A., Nencini, F., Selva, M. Multispectral and panchromatic data fusion assessment without reference. *Photogrammetric Engineering and Remote Sensing* **74**, 2008, 193–200.

27. Otazu, X., González-Audícana, M., Fors, O., Núñez, J. Introduction of sensor spectral response into image fusion methods. Application to wavelet-based methods. *IEEE Transactions on Geoscience and Remote Sensing* **43**, 2005, 2376–2385.

28. Baronti, S., Aiazzi, B., Selva, M., Garzelli, A., Alparone, L. A theoretical analysis of the effects of aliasing and misregistration on pansharpened imagery. *IEEE Journal of Selected Topics in Signal Processing* **5**, 2011, 446–453.

29. Rangaswamy, M.K. Quickbird II two-dimensional on-orbit Modulation Transfer Function analysis using convex mirror array. Master's thesis, South Dakota State University, 2003.

30. Cook, M.K., Peterson, B.A., Dial, G., Gibson, L., Gerlach, F.W., Hutchins, K.S., Kudola, R., Bowen, H.S. IKONOS technical performance assessment. In: *Proceedings of SPIE* **4381**, 2001, 94–108.

Index

Printed and bound by CPI Group (UK) Ltd, Croydon, CR0 4YY

01/11/2024

01782604-0016